Multiscale Biomechanics

Series Editor
Noël Challamel

Multiscale Biomechanics

Edited by

Jean-François Ganghoffer

ISTE PRESS

ELSEVIER

First published 2018 in Great Britain and the United States by ISTE Press Ltd and Elsevier Ltd

ISTE Press Ltd
27-37 St George's Road
London SW19 4EU
UK

www.iste.co.uk

Elsevier Ltd
The Boulevard, Langford Lane
Kidlington, Oxford, OX5 1GB
UK

www.elsevier.com

For information on all our publications visit our website at http://store.elsevier.com/

British Library Cataloguing-in-Publication Data
A CIP record for this book is available from the British Library
Library of Congress Cataloging in Publication Data
A catalog record for this book is available from the Library of Congress
ISBN 978-1-78548-208-3

Contents

Introduction

I.1. Biomechanics and its scientific challenges

The mechanics of soft or hard biological tissues is a multidisciplinary topic and rapidly expanding research area from different scientific areas, including biology, medicine, physics, materials science, mechanics, applied mathematics and computational physics. The word "Biomechanics" means mechanics applied to biology. Mechanics and living tissues and structures are often opposed: the first is rigid and repetitive, while the second is flexible and capable of improvisation. It thus seems paradoxical to want to explain mechanically biological phenomena and thus reduce the living to a simple human production, as the project of biology and mechanism wants. The mechanical model was nevertheless considered sufficient to describe biological structures by the French philosopher Descartes at the macroscopic level; at the microscopic level, biologists speak of "physico-chemical mechanisms". The analogy is reinforced by the etymology of the word "organ", which comes from "tool", and the organs are then natural tools. Even if this is only an analogy, living beings can thus be considered as technical objects; we are technically capable of treating living beings as machines, producing tissues or organs of substitution. In this mechanistic and deterministic perspective, it is the organ that creates the function, not the reverse, from a finalist point of view. Descartes in his time already compares the human body to a machine (automaton) to make medicine a science: the heart is assimilated to a pump, its two cavities to locks, veins to streams, nerves to nets, the joints to pulleys. The analogy is less structural than functional: it does not pretend to explain the nature of organs, but their functions, by identifying the causes of their movements.

The general purpose of mechanics for living organisms is to explore the mechanical properties of living organisms and to analyze the principles of engineering that make biological systems work. It deals with the existing relationships between structures and functions at all levels of life organization, from the molecular level (collagen, elastin) to tissues and organs. It can be divided into several sub-disciplines, each with its own field of investigation: biomechanics characterizes the spatio-temporal responses of biological materials – solid, fluid or viscoelastic – to internal and external constraints. The coinage *mechanobiology* has been invented to highlight the fact that a number of biological processes at different scales are sensitive to local mechanical fields. In addition to classical mechanics, biomechanics uses various disciplines and techniques such as rheology (study of the behavior of biological fluids such as blood) and resistance of materials to model stresses undergone by living tissues (e.g. cartilage of the joints, bones) and solid mechanics to analyze motricity and locomotion, from individual cells to whole organisms.

Bioengineering (contraction of Biomedical Engineering) is the science that applies engineering methods to solve biological problems. It integrates a set of disciplines for the advancement of medical and health care, such as the physical and chemical sciences, mathematics and engineering principles. It makes it possible to go beyond the empiricism of certain techniques (orthopedic surgery, vascular surgery) by improving them by a more global and constructed approach. Bioengineering innovates and advances fundamental concepts, creates knowledge of the molecular level at the systemic level and develops biomaterials, processes, computational approaches, implants and innovative devices. Bioengineering promotes the well-being of the population by preventing, diagnosing and treating diseases, rehabilitating patients and improving health. The living materials designed and produced by Bioengineering, also known as biomaterials, have as their therapeutic vocation the replacement of parts of the human body. The main challenge is to obtain an active biocompatibility, that is a good adaptation of the organism to the introduction of a new local stress (the implant) and an appropriate biological and/or mechanical response according the deficient functions to be restored.

In this exploration of living structures, whose specifications are demanding (variability of properties, heterogeneity, multiscale and multiphysical behaviors, structural and behavioral complexity must all ideally be considered), mechanics has already developed a wide range of themes and methods, amongst which we can cite: the strength of materials, hydraulics, elasticity, but also the mechanics of heterogeneous and anisotropic continuous media, soft or hard, the mechanics of multiphase deformable porous media, involving coupled phenomena including chemical reactions, transport and mass transfer, as well as electrical interactions due to the presence of charged ions and macromolecules.

Although the origin of Biomechanics can be traced back to Aristotle, as explained in the historical overview expounded in [FUN 93], the field of Biomechanics has expanded in a tremendous manner since the early 1980s and advances have been made on many fronts. The necessarily multidisciplinary approach and the multiscale aspects of living tissues make it difficult to apprehend all aspects of their mechanical response, starting from the cellular level up to the tissue level. Biology includes a lot of descriptive material and understanding the cellular mechanisms responsible for the tissue mechanical response at the macroscale as well as their time evolution due to growth and remodeling remains very challenging for non-biologists. On the other hand, mechanics is more quantitative in nature and requires some mathematical knowledge which is not easily understandable for researchers in medicine, biology or even materials science.

The classical framework of continuum mechanics which holds for continuous bodies with a constant mass has clearly to be enlarged to be able to treat and model biological phenomena under the umbrella of a field theory which treats all variables as continuous (fields) in space and time. Especially, the mechanical modeling of living tissues raises several challenges due to the following novel aspects to be considered in comparison to non-living materials and structures:

– biological tissues are prone to a biological evolution due to growth or remodeling phenomena, so that addition of mass due to transport of nutrients, modification of density and properties, shape and topology modifications of the tissue microstructures have ideally to be accounted for as new phenomena. This requires the reformulation of classical approaches (which were developed for inert materials) for the mechanics of continuous media, in order to take into account their adaptability to their environment. Moreover, one has to expand the thermodynamics of irreversible processes to account for a mass source and flux within the growing domain and to adopt the thermodynamics of open systems as a more natural framework;

– the microstructural evolution of biological tissues involves simultaneous genetic and epigenetic factors, including stress and strains, so that they adapt to mechanical signals and respond by changing their topology, shape, and mechanical properties. The functional adaptation of living tissues has raised a lot of work in the literature, and still remains challenging for researchers in mechanics;

– since the time evolution of biological tissues occurs at the cellular level, cellular phenomena ideally have to be incorporated into any modeling. Despite the many approaches to account for mechanotransduction, a clear view of the mechanisms though which cells feel signals from their environment and react to them is still not available;

– notwithstanding biological aspects, the mechanical response of living tissues involves mechanisms at different scales, so a multiscale integrated approach is needed so that the mechanical response at one scale is transferred to the next scale.

– since biological tissues experience interactions between transport phenomena, chemical reactions and generation of internal stresses, a multiphysical approach is further needed;

– the growth of living tissues involves the addition of new particles to the initial domain, which would require considering a change of the domain topology, which is not accounted for in the classical framework of continuum mechanics;

– from the point of view of measurements, the methods of investigation of living tissues require the probing of objects with heterogeneous characteristics, structurally complex and variable in terms of behavior. The characterization is done *in vitro* and *in vivo* and tries to take into account the variability and the uncertainty of the response. In addition to conventional mechanical tests (tensile compression, waves, ultrasound, etc.) and field measurements, new techniques have emerged, such as nanomechanics, microfluidics, acoustics, PIV and picoseconds. The following imaging methods are widely used: optical, NMR, MRI, SEM, ultrasound, confocal microscopy, X-ray microscopy, tomography, elastography and Raman spectroscopy.

I.2. Motivation of the book

Biomechanics of human tissues has been an expanding research field since the publication in 1981 of the book entitled *Biomechanics: Mechanical Properties of Living Tissues* by Yuan-Cheng Fung [FUN 81]. Since that date, many groups in the world have proposed biomechanical models of soft organs and hard tissues to study their physiology and complex mechanical behavior in the framework of nonlinear continuum mechanics, as reflected in e.g. [OGD 84, COW 00, TAB 04, HOL 00, HOL 06, GAS 06] and in the more recent monographs devoted to the physiology and mechanical response of all soft tissues of the human body [OHA 17] and references therein. The vivid research activity in Biomechanics is further reflected by the regular international conferences and courses organized on the topic.

The main motivation of this book is to present a comprehensive and self-contained perspective of Biomechanics from the standpoint of continuum mechanics and thermodynamics, by including the necessary information from physiology and biology. This book is intended to expose both the tools needed for an extension of continuum mechanics and thermodynamics to handle living tissues and to answer part of the aforementioned challenges.

It is hoped that students and researchers in both mechanical sciences and in other disciplines involved in Biomechanics will find an interest in a detailed exposition of the concepts and methods of continuum mechanics and thermodynamics like the one presented in this book. The involvement and application of these tools to model the mechanical response of biological tissues in stable and evolutive conditions constitutes a second objective of the book. The musculoskeletal system includes bone, cartilage, tendons, ligaments and muscles from the skeleton; all these entities except muscles are analyzed in this monograph from the point of view of their mechanical response, with a more focused view on bone since its microstructure, morphology and properties are evolving continuously during the bone remodeling cycle at the trabecular level; bone mechanics is thus very challenging for researchers in mechanics and biomechanics, since the continuously changing microstructure evolutions modify in turn the bone mechanical properties. Despite its small strains behavior as a hard tissue, bone shows large shape changes induced by external remodeling. All these features make bone a very good living tissue prototype for the applications of the methods of continuum mechanics and open systems mechanics to apprehend the mechanobiological couplings responsible for bone growth and remodeling across the scales.

Furthermore, it has become common that biomechanics deals with the complex phenomena witnessed by living tissues across multiple scales, from cellular up to tissue levels, trying to account for how quantities of interest transfer from one scale to the nest scale. It is an interesting challenge to propose approaches from multiscale continuum mechanics to model bone properties at the scale of the network of individual trabeculae, internal and external bone remodeling, exploring the extreme bone behaviors such as brittle and ductile bone fracture and collapse. The objective of predicting entire bone mechanics based on information available at tissue level and structural composition finds its motivation both from pure cognitive aspects but also from its clinical relevance in fracture predictions and optimization of bone implants in the context of a patient-specific approach. It is here challenging to formulate quantitative relationships between microscale variables characterizing the material and structural properties and the apparent mechanical response. Bone adaptation and remodeling is driven by a rather complex cellular machinery involving a cascade of cellular reactions, thereby highlighting the existing strong coupling between biological and mechanical phenomena. Similar multiscale aspects starting at the cellular level also prevail for other biological tissues studied in this book, including cartilage, tendons and the biological networks which define the generic constituents of biomembranes.

This book is unique in the sense that it provides the mechanical and thermodynamical foundations to enter the field of Biomechanics. There are very few books that truly integrate the mechanical foundations of Biomechanics and show students and researchers how to apply and integrate this knowledge to address the issues raised by living tissues in the context of Biomechanics. The book is intended for both students and researchers in mechanics and engineering interested in an overview of the concepts and methods used in Biomechanics and for researchers in Biomechanics, Bioengineering and more specialized sub-disciplines in the field of Biomechanics to have an overview of the applications of modern continuum mechanics to the description and modeling of living tissues.

I.3. Organization of the book

The book is divided into three main parts; the first part is devoted to an exposition of the necessary prerequisite in continuum mechanics and thermodynamics of growing solid bodies, also including a chapter presenting homogenization methods for network materials, that will be used to evaluate the mechanical response of trabecular bone and filament networks, those being the essential constituents of biological membranes.

Part 2 deals with multiscale bone mechanics, starting from static aspects of the mechanical response of bone from elasticity up to brittle and ductile collapse, and then treating the dynamical behavior associated with bone remodeling and cyclic fatigue. The investigation of the mechanical response of soft biological tissues including the intervertebral disk, soft networks as the building blocks of biological membranes and ligaments and tendons, is the object of part 3.

The standard background of tensor analysis and continuum mechanics is the starting point of the book in Part 1 (Chapters 1, 2 and 3), which is then further extended to account for growth in Chapter 5. Since the prediction of the effective mechanical properties of living tissues is a key aspect in biomechanics, we incorporate in Chapter 4 an exposition of a micromechanical approach to computing the effective properties of network materials, considering successively the small and large strains regimes. The so-called discrete homogenization which is the object of this chapter is particularly well suited for bone and biological networks, which both can be described by a repetitive network of beam elements.

The asymptotic homogenization technique is developed in Chapter 4 in the framework of geometrical nonlinearities to derive the large strains effective elastic response of network materials viewed as repetitive beam networks. A systematic methodology is established to compute the overall mechanical properties of these structures in the nonlinear regime, reflecting the influence of the geometrical and mechanical micro-parameters of the network structure on the overall response of the chosen equivalent continuum. Internal scale effects of the initially discrete structure are captured by consideration of a micropolar effective continuum model for which microtations enter the kinematics in addition to the displacement field. Applications to the large strain response of 3D hexagonal networks adopted as a prototype representative topology for trabecular bone exemplify the power of the proposed method.

We provide in Chapter 5 a mechanical and thermodynamical basis for the modeling of volumetric and surface growth of living tissues viewed as continua undergoing a change of mass, shape, and properties due to biological evolution. It is usually recognized that soft biological tissues grow by a change in their specific volume or density – this process deserving the name volumetric growth – whereas hard tissues like bone are prone to surface growth. The field equations of continuum mechanics are extended in order to account for the existing mass source and mass flux responsible for growth. The thermodynamics of coupled irreversible phenomena is then reviewed and evolution laws for a growth tensor in the context of volumetric growth are formulated, considering the interactions between transport of nutrients and the mechanical driving forces for growth. Surface growth phenomena are then introduced based on preliminary aspects of the differential geometry of surfaces and are further expanded in the framework of the thermodynamics of surfaces.

Chapter 6 deals with the construction of Cosserat models of trabecular bone in 2D and 3D situations from micromechanical approaches, in order to investigate microstructure-related scale effects on the macroscopic bone properties. Closed form expressions of the equivalent properties are obtained versus the geometrical and mechanical microparameters, accounting for the effects of bending and axial and transverse shear deformations; torsion is additionally considered for a 3D geometry. The classical and micropolar effective moduli and the internal flexural and torsional lengths are identified versus the micropolar material constants. The scaling laws of the effective moduli versus density are determined in situations of low and high effective densities to assess the impact of the transverse shear deformation.

Relying on the so-constructed micropolar effective continua, the static and dynamic effective behaviors of vertebral trabecular bone in the form of beam structures are analyzed in Chapter 7, in terms of the deflection, torsion and eigenfrequencies of deformations. The governing differential equations of static and dynamic bending and torsion of trabecular bone are derived using variational principles based on non-classical theory, and explicit solutions are derived, accounting for length-scale effects. The identified static and dynamical effective properties of bone are correlated to physiological factors, such as patient age and effective bone density, as well as pathologies leading to a modification of the bone internal architecture.

Considering next the extreme behavior of bone, the overall plastic yield and brittle failure behaviors of three-dimensional trabecular bone are investigated in Chapter 8 by a microstructural modeling approach based on the homogenization of the initially discrete architecture. The multiaxial yield behavior is analyzed to formulate a multiaxial plastic yield criterion. Furthermore, brittle bone fracture is modeled under triaxial stress states to construct the failure surfaces defined in the tension-tension quadrant. In plastic yielding, a micromechanical model based on a homogenization scheme is applied to a representative unit cell to determine the macroscopic plastic yield surfaces in stress space. This general framework is applied to evaluate the yield and failure properties of trabecular bone, which are of key interest in understanding and predicting the fracture of bones and bone implant systems. The effective strength of trabecular bone is evaluated in the two situations of fully brittle (fracture with no tissue ductility) and fully ductile (yield with no tissue fracture) failure of the trabecular tissue.

Models for internal and external bone remodeling are developed in Chapter 9 in the framework of the thermodynamics of irreversible processes. Internal remodeling is essentially accounted for by an evolution of the internal bone density versus the trace of the Eshelby stress, adopting here the ideas of configurational mechanics that microstructural evolutions occur in the reference configuration. The growing surface is endowed by a specific mechanical behavior elaborated from a surface potential, depending upon the elastic surface strain and the external normal. The surface remodeling velocity is related to the driving force for growth identified as the surface divergence of Eshelby stress. The developed formalism is applied to simulate bone internal and external remodeling, showing the influence of external mechanical stimuli on the evolution of the external shape of bone. Simulations of the evolution of the shape and density of bone structures are carried out successively at mesoscopic, microscopic and macroscopic levels, the former representing cellular level simulations of the trabecular unit cell. At the macroscopic level, a linear elastic constitutive law for trabecular bone is derived, relying on a discrete homogenization approach whereby the continuum parameters such as stiffness evolve with morphology as remodeling occurs. The developed numerical models are able to simulate combined internal and external bone remodeling at both trabecular and macroscopic scales.

Bone is a living tissue that continually undergoes damage under normal everyday mechanical loadings. It maintains its structural integrity by a continuous process of damage detection and repair, and by adaptation to changes in the mechanical environment. These adaptive changes are regulated by the so-called process of bone remodeling. Chapter 10 exposes an integrated remodeling-to-fracture model to simulate the bone remodeling process considering the anisotropy of trabecular bone. Models for density evolutions of continuum solid bodies are developed in the framework of the thermodynamics of irreversible processes, considering bone elasticity with damage properties, in which the damage stimuli is incorporated. The asymptotic homogenization technique is involved to derive the effective anisotropic continuum mechanical properties of the trabecular bone from an initially discrete structure representative of femur bone microstructure, consisting of a piling up of hexagonal planar cells. As a consequence, relationships are established between the apparent density of bone and its effective anisotropic properties, that is the elastic and shear moduli and Poisson's ratio. Damage is directly incorporated into the effective density, the evolution of which is computed versus time. This formalism is involved to simulate bone remodeling for 2D geometries in the context of damaged anisotropic bone elasticity, showing the influence of the external mechanical stimuli on the evolution of the density of bone.

Micromechanical models of intervertebral disks are examined in Chapter 11. Intervertebral disks are key elements of the spine; they allow spine motion like flexion extension, lateral bending and axial rotation, dissipating mechanical loadings during all these motions.

The asymptotic homogenisation technique is involved in Chapter 12 to derive the effective elastic response of biological membranes viewed as repetitive beam networks. Thereby, a systematic methodology is established, allowing the prediction of the overall mechanical properties of biological membranes in the nonlinear regime, reflecting the influence of the geometrical and mechanical micro-parameters of the network structure on the overall response of the equivalent continuum. The network irregularity is shown to have a small impact on the equivalent properties of the homogenized continua. Biomembrane networks are classified based on nodal connectivity, so that we analyze 3, 4 and 6-connectivity networks, which are representative of most biological networks. The individual filaments of the network are described as undulated beams prone to entropic elasticity, with tensile moduli determined from their persistence length. The effective micropolar continuum evaluated as a continuum substitute of the biological network has a kinematics reflecting the discrete network deformation modes, involving a nodal displacement and microrotation. The peptidoglycan network may exhibit a re-entrant hexagonal configuration due to thermal or pressure fluctuations, for which micropolar effects become important.

In Chapter 13, a brief description of the structure and main properties of ligaments and tendons is proposed, including their physiological function and environment. A non-exhaustive summary of the main *in vivo* and *in vitro* characterization methods used to determine their mechanical response is then exposed. A state of the art of modeling approaches applied to tendons and ligaments is then presented, including both phenomenological approaches and microstructural models. At last, a brief focus on ligament and tendon tissue engineering is proposed, in order to link the micromechanical models proposed in the literature with future applications concerning alternative reconstruction methods for these tissues. A conclusion concerning the barriers that are still to be broken down regarding tendon and ligament characterization or modeling will be proposed.

PART 1

Theoretical Basis: Continuum Mechanics, Homogenization Methods, Thermodynamics of Growing Solid Bodies

1

Tensor Calculus

1.1. A short historical vignette

The concepts of tensor analysis arose from the work of Carl Friedrich Gauss in differential geometry, and the formulation was much influenced by the theory of algebraic forms and invariants developed during the middle of the nineteenth century. The word "tensor" itself was introduced in 1846 by William Rowan Hamilton to describe something different from what is now meant by a tensor (Note 1) The contemporary usage was brought in by Voigt in 1898. Tensor calculus was developed around 1890 by Gregorio Ricci-Curbastro under the title absolute differential calculus, and originally presented by Ricci in 1892. It was made accessible to many mathematicians by the publication of Ricci and Tullio Levi-Civita's 1900 classic text *Méthodes de calcul différentiel absolu et leurs applications* (translated into *Methods of absolute differential calculus and their applications*).

In the 20th Century, the subject came to be known as tensor analysis, and was widely developed due to Einstein's theory of general relativity, around 1915. General relativity is indeed formulated completely in the language of tensors, a field Einstein had learned about with great difficulty from the geometer Marcel Grossmann. Levi-Civita then initiated a correspondence with Einstein to correct mistakes Einstein had made in his use of tensor analysis. Tensors were also found to be useful in other fields such as continuum mechanics. Some well-known examples of tensors in differential geometry are quadratic forms such as metric tensors, and the Riemann curvature tensor. In mathematics, the exterior algebra of Hermann Grassmann (middle of the nineteenth century), and the work of Élie Cartan on differential forms are more recent developments of the concept of tensor.

Chapter written by Jean-François GANGHOFFER.

Tensor fields are extensively used in field theories including electromagnetism, continuum mechanics and general relativity, as well as in numerous applications in physical sciences and engineering. A tensor is a generalization of a *scalar* (a pure number representing the value of some physical quantity) and a *vector* (a geometrical arrow in space), and a tensor field is a generalization of a *scalar field* or *vector field* that assigns, respectively, a scalar or vector to each point of space.

The main idea behind the concept of tensor is to delineate what is intrinsic (independent from a metric or a basis). Intrinsic means independent from any observer, thus from any basis or scalar product. Behind this stands the general notion of covariance: equations of physics have to be independent from any observer, thus should take the same mathematical form in all coordinate systems, which require the use of tensors. The strong form of covariance is invariance, which only hold for scalar valued quantities (they remain the same whatever the coordinate system), whereas vectors and tensors transform in a covariant manner under a change of coordinates (they follow the rule of matrix transformation you already know, but in an extended form).

In order to explain the need for the concept of tensor, while a vector is determined by 3 components, there are quantities requiring more than 3 components to be completely characterized. To be specific, consider a unitary cubic volume element submitted to forces (weight, contact forces) so that internal forces are generated (Figure 1.1).

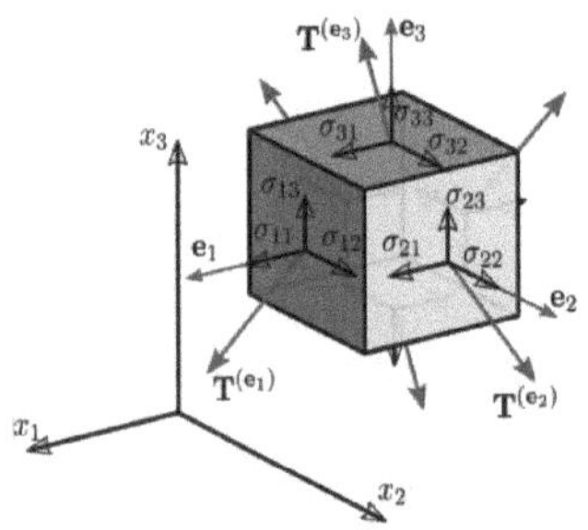

Figure 1.1. *Cubic material element subjected to contact forces. For a color version of this figure, see www.iste.co.uk/ganghoffer/biomechanics.zip*

Let $(S_i)_{i=1..3}$ be the face of the cube orthogonal to the basis vector $\mathbf{e}_i$ of the same index; the face S_i is submitted to a force (vector) $\mathbf{F}_i$ with *a priori* an arbitrary orientation: it can be accordingly decomposed along the three basis vectors

$$\mathbf{F}_i = \sigma_{i1}\mathbf{e}_1 + \sigma_{i2}\mathbf{e}_2 + \sigma_{i3}\mathbf{e}_3$$

This vector represents a contact force exerted by the neighboring cubes: the nine scalar quantities $(\sigma_{ij})_{i,j=1..3}$ are the components of the stress tensor; they completely specify the internal forces acting at each point. Since they are *a priori* independent, it means that the concept of vector is here clearly inadequate.

Areas of concern in mathematics: multilinear algebra, differential geometry. A vector is a fairly general object, it belongs to a vector space; by extension, one shall introduce the notion of tensor field. Representation of multilinear applications: in this sense, a tensor is intrinsic!

1.1.1. *Difference between matrix and tensor*

A matrix is a mere table of numbers, whereas a tensor is an abstract object whose components are changing when the basis is changing, thus they shall satisfy some consistency.

In physics and engineering, tensors are used to describe various physical quantities such as electrical and magnetic fields, many physical properties (direction dependent) such as permittivity, conductivity, and general relativity (space-time as a 4D curved manifold).

In differential geometry, tensors are used to define on a differentiable manifold the geometrical notions of distance, angle, volume. This is done in metric spaces via the choice of a metric, elaborated as a scalar product of vectors in the tangent space at each point.

1.2. Vector spaces

DEFINITION.– Examples of free vectors: $\mathbf{x}, \mathbf{y}, ... \in \mathbb{R}^3$.

Operations of vector addition: commutative $(\mathbf{x}+\mathbf{y}=\mathbf{y}+\mathbf{x})$, associative $(\mathbf{x}+(\mathbf{y}+\mathbf{z})=(\mathbf{x}+\mathbf{y})+\mathbf{z})$, existence of nil vector $(\mathbf{x}+\mathbf{0}=\mathbf{x})$ and opposite vector $(\mathbf{x}+(-\mathbf{x})=\mathbf{0})$.

Multiplication by a scalar $\alpha \in \mathrm{R}$: $1\mathbf{x}=\mathbf{x}$, associative $\alpha(\beta\mathbf{x})=(\alpha\beta)\mathbf{x}$. Distributive for addition of scalars: $(\alpha+\beta)\mathbf{x}=\alpha\mathbf{x}+\beta\mathbf{x}$; distributive for vector addition: $\alpha(\mathbf{x}+\mathbf{y})=\alpha\mathbf{x}+\alpha\mathbf{y}$.

Any vector can be expressed using the basis vectors; the number n of those vectors dictates the dimension of the vector space E, denoted E_n.

NOTE (Generalization).– Any set E of elements $\mathbf{x}, \mathbf{y}, ...$ endowed with two laws 'addition' and 'external multiplication' on a commutative field K having these properties has the status of a vector space; $\mathbf{x}, \mathbf{y}, ...$ are called vectors (they may be functions or more general objects).

For isomorphisms of vector spaces let us consider circular uniform motion as an example (Figure 1.2).

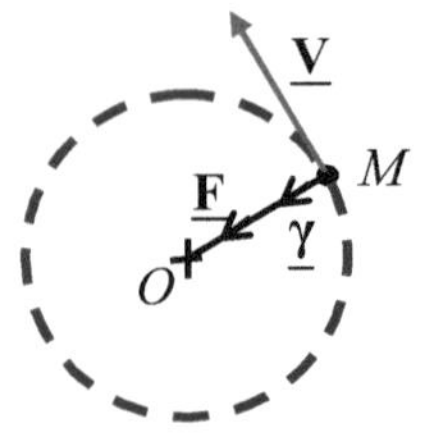

Figure 1.2. *Vectors representing different physical quantities at the same point. For a color version of this figure, see www.iste.co.uk/ganghoffer/biomechanics.zip*

Figure 1.2 pictures different vectors, **OM** distance, **V** velocity, **F** force, $\boldsymbol{\gamma}$ acceleration. These define quantities belonging to vector spaces of a different physical nature (dimensions are different), which are nevertheless represented in the same geometrical space. It is sufficient for this purpose to fix scales allowing interpretation of the norms: scale of distances, velocities, forces, accelerations. This is possible since all those vectors have the same dimension and obey the same calculus rules: there is an isomorphism between the vector spaces they belong to, which allows each of them to be represented in either space. This can be done e.g. in R^3.

NOTE (Einstein notation).– For a $n \times n$ matrix, $\tilde{A} = \left[a^i_{.j} \right]$, $i,j=1,...,n$, with i row index and j column index.

EXAMPLE.– The product of two matrices is constructed as $\tilde{P} = \tilde{A}.\tilde{B} \Rightarrow p^i_{.j} = \sum_{k=1}^{n} a^i_{.k} b^k_{.j}$, $i,j,k \in 1,...,n$. Since there is summation for each occurrence of the index in a lower (subscript) and upper position (superscript) in any monomial, the summation symbol can be removed: the summation integer k is mute, so that it can be replaced by any letter (similar to integration variable):

$$p^{i}_{.j} = a^{i}_{.k} b^{k}_{.j} = a^{i}_{.l} b^{l}_{.j} = ..., i, j, k, l \in 1...n$$

EXAMPLE.– Let $\tilde{Q} = \tilde{C}.\tilde{D}$, and we want to represent the product $\tilde{R} = \tilde{P}.\tilde{Q}$; we shall then write

$$r^{i}_{.j} = a^{i}_{.k} b^{k}_{.l} c^{l}_{.m} d^{m}_{.j}$$

so that each repeated mute index is not used twice. In the geometrical space R^3, all indices vary from 1 to 3.

1.3. Covariant and contravariant tensors

For a given basis $\mathbf{e}_1, \mathbf{e}_2, ..., \mathbf{e}_n$ of E_n, denoted collectively $\{\mathbf{e}_i\}$, any vector $\mathbf{V}$ can be represented as $\mathbf{V} = V^i \mathbf{e}_i$; quantities V^i are the components of $\mathbf{V}$ in this basis. For another basis $\mathbf{E}_1, \mathbf{E}_2, ..., \mathbf{E}_n$, denoted $\{\mathbf{E}_I\}$, each vector of the new basis is a linear combination of the old basis vectors: $\mathbf{E}_I = a^{i}_{.I} \mathbf{e}_i$, so that the matrix representing the change of basis is $\tilde{A} = \left[a^{i}_{.I}\right]$. This can be represented in matrix form as

$$[\mathbf{E}_1 \quad \mathbf{E}_2 \quad \mathbf{E}_3] = [\mathbf{e}_1 \quad \mathbf{e}_2 \quad \mathbf{e}_3] \begin{bmatrix} a^{1}_{.1} & a^{1}_{.2} & a^{1}_{.3} \\ a^{2}_{.1} & a^{2}_{.2} & a^{2}_{.3} \\ a^{3}_{.1} & a^{3}_{.2} & a^{3}_{.3} \end{bmatrix}$$

or symbolically $\mathbf{E}_I = \mathbf{e}_i a^{i}_{.I}$, denoting a general positioning of indices for the summation to make a correspondence with matrix product.

REMARK.– In continuum mechanics, one very often considers change of bases given by an orthogonal transformation, so that

$$\mathbf{E}_I = \cos(\mathbf{e}_j, \mathbf{E}_I)\mathbf{e}_j \equiv q_{ji}\mathbf{e}_j \leftrightarrow \mathbf{E}_I = Q.\mathbf{e}_j$$

In which the scalars $q_{ji} := \cos(\mathbf{e}_j, \mathbf{E}_I)$ define the components of an orthogonal transformation $Q = [q_{ij}]$. The ith column of Q is formed by the components of $\mathbf{E}_I$ in the initial basis.

For a vector $\mathbf{V}$, we have the correspondence $\mathbf{V} = V^i\mathbf{e}_i = V^I\mathbf{E}_I$, with

$$\mathbf{E}_I = \mathbf{e}_i a^i_{.I}, \text{ hence } \mathbf{V} = V^I a^i_{.I}\mathbf{e}_i \text{ Let } v^i = a^i_{.I}V^I \rightarrow \tilde{v} = \tilde{A}.\tilde{V}$$

and define

$$\tilde{V} = \tilde{B}.\tilde{v} \rightarrow \tilde{B} = \tilde{A}^{-1}, \text{ and } V^I = b^I_{.i}v^i.$$

Vectors transform in the following manner under such a change of basis:

$$\mathbf{x} = x_j\mathbf{e}_j = X_J\mathbf{E}_J = X_J q_{ji}\mathbf{e}_j = x_j q_{ji}\mathbf{E}_I \Rightarrow x_j = q_{ji}X_j \rightarrow \mathbf{x} = Q.\mathbf{X}$$

SUMMARY.– Basis vectors and components of vectors transform in an opposite manner, so that *covariance* denotes a transformation rule similar to the basis vectors (subscript), and *contravariance* means a transformation rule opposite to basis vectors (superscripts).

1.4. Linear forms and duality

Let us start with an example in R^3. A form F provides a correspondence between any vector $\mathbf{V} \in R^3$ and a scalar $\in \mathbb{R}$, called the transformed of $\mathbf{V}$: duality product, defined an intrinsic scalar (independent from any basis).

EXAMPLE.– a) $\langle F, \mathbf{V}\rangle = |\mathbf{V}|$, b) $\langle F, \mathbf{V}\rangle = \mathbf{U}.\mathbf{V}, \ \forall \mathbf{V} \in R^3, \mathbf{U}$ is fixed.

A form is defined in an intrinsic manner, so that the associated scalar $\langle F, \mathbf{V}\rangle = \alpha$ is independent from any basis; a form is linear iff it satisfies:

$$\forall \mathbf{V} \in R^3, \ \langle F, \mathbf{V} \rangle \in R$$
$$\langle F, \lambda \mathbf{V} + \mathbf{V}' \rangle = \lambda \langle F, \mathbf{V} \rangle + \langle F, \mathbf{V}' \rangle$$

– Previous examples: a) is not linear, b) is a linear form.

– Coefficients of a linear form: by linearity, $\langle F, \mathbf{V} \rangle = \langle F, v^i \mathbf{e}_i \rangle$ in basis $\{\mathbf{e}_i\}$, then

$$\langle F, v^i \mathbf{e}_i \rangle = v^i \langle F, \mathbf{e}_i \rangle = v^i f_i$$

with $f_i := \langle F, \mathbf{e}_i \rangle$ the coefficients of **F** in the basis $\{\mathbf{e}_i\}$.

Under a change of basis, it holds that

$\langle F, \mathbf{V} \rangle = f_i v^i = F_I V^I$, with $v^i = a^i_{.I} V^I$, thus $F_I = a^i_{.I} f_i$ and vice versa $f_i = b^I_{.i} F_I$.

The coefficients of F define a covariant series.

NOTE (Dual space).– The set of all linear forms on E_n constitutes a vector space:

$\forall \mathbf{V} \in E_n, \forall \lambda \in K$, we have the following properties:

a) Addition: $S = F + G \Leftrightarrow \langle S, \mathbf{V} \rangle = \langle F, \mathbf{V} \rangle + \langle G, \mathbf{V} \rangle$

b) External product: $P = \lambda F \Leftrightarrow \langle P, \mathbf{V} \rangle = \lambda \langle F, \mathbf{V} \rangle$

c) Equality of forms: $F = G \Leftrightarrow \langle F, \mathbf{V} \rangle = \langle G, \mathbf{V} \rangle$

d) Nill form: $F = 0 \Leftrightarrow \langle F, \mathbf{V} \rangle = 0$

This vector space, of dimension n, is called the dual space of E_n, denoted E_n^*. In all previous examples, $F, G, S, P, O \in E_n^*$. The set of linear forms is the set of linear applications $E_n^* \simeq L(E_n; \mathbb{R})$. Search of convenient bases in dual space: once a choice of the basis $\{\mathbf{e}_i\}$ of E_n has been made, one may define a linear form by its action on the basis vectors; we introduce a first form $\mathbf{e}^{*1}$ such that:

$$\langle \mathbf{e}^{*1}, \mathbf{e}_1 \rangle = 1; \ \langle \mathbf{e}^{*1}, \mathbf{e}_j \rangle = 0 \text{ for } j \neq 1$$

Similarly, let us introduce

$$\left\langle \mathbf{e}^{*2},\mathbf{e}_2\right\rangle = 1;\ \left\langle \mathbf{e}^{*2},\mathbf{e}_j\right\rangle = 0 \text{ for } j \neq 2, \ldots, \left\langle \mathbf{e}^{*n},\mathbf{e}_n\right\rangle = 1;\ \left\langle \mathbf{e}^{*n},\mathbf{e}_j\right\rangle = 0 \text{ for } j \neq n$$

Notation: $\left\langle \mathbf{e}^{*i},\mathbf{e}_j\right\rangle = \delta^i_{.j} = \begin{vmatrix} 1 \text{ if } i = j \\ 0 \text{ if } i \neq j \end{vmatrix}$ denotes the Kronecker symbol, such that

$$\delta^i_{.j} = \tilde{1} = \begin{bmatrix} 1 & 0 & & \\ & 1 & & \\ & & . & \\ 0 & & & . \\ & & & & 1 \end{bmatrix}$$, the identity matrix.

The dual basis vector $\mathbf{e}^{*i}$ is the projection vector on the vector space generated by vector $\mathbf{e}_i$ parallel to the other directions $\mathbf{e}_j$.

To show that the $\{\mathbf{e}^{*i}\}$ define a basis, one has to show their linear independency:

$$\alpha_i \mathbf{e}^{*i} = 0 \Rightarrow \alpha_i = 0, \forall i:$$

$$\alpha_i \left\langle \mathbf{e}^{*i}, \mathbf{V}\right\rangle = 0,\quad \forall \mathbf{V} = v^i \mathbf{e}_i \rightarrow \alpha_i v^j \left\langle \mathbf{e}^{*i}, \mathbf{e}j\right\rangle = \alpha_i v^j \delta^i_{.j} = \alpha_i v^i = 0, \forall v^i$$
$$\Rightarrow \alpha_i = 0, \forall i$$

By extension, the linear forms of E^*_n are called the vectors of E^*_n.

Change of basis in dual space: for a new basis $\{E_I\}$ in E_n, the new dual basis $\{E^{*I}\}$ is s.t.

$$\left\langle \mathbf{E}^{*I}, \mathbf{E}_J\right\rangle = \delta^I_{.J}$$

For any vector $\mathbf{V} \in E_n$, it holds that:

$$\left\langle \mathbf{e}^{*i}, v^j \mathbf{e}_j \right\rangle = v^j \delta^i_{.j} = v^i \;;\; \left\langle \mathbf{E}^{*I}, V^j \mathbf{E}_j \right\rangle = V^j \delta^I_{.J} = V^I$$

Since $V^I = b^I_{.i} v^i$, we have $\left\langle \mathbf{E}^{*I}, \mathbf{V} \right\rangle = b^I_{.i} \left\langle \mathbf{e}^{*i}, \mathbf{V} \right\rangle$, and has here a contravariant behavior. The components of vectors of E_n^* are then evaluated: by linearity, it holds that $\left\langle F, \mathbf{V} \right\rangle = f_i v^i$, with

$$f_i := \left\langle F, \mathbf{e}_i \right\rangle, \quad \text{since} \quad v^i = \left\langle \mathbf{e}^{*i}, \mathbf{V} \right\rangle \rightarrow \left\langle F, \mathbf{V} \right\rangle = f_i \left\langle \mathbf{e}^{*i}, \mathbf{V} \right\rangle, \forall \mathbf{V}, \quad \text{thus}$$
$$F = f_i \mathbf{e}^{*i}$$

The coefficients appear as the components of F in basis $\left\{ \mathbf{e}^{*i} \right\}$; they form a covariant suite.

We have thus introduced so far two different types of basis vectors: the covariant vectors $\left\{ \mathbf{e}_i \right\}$ which in fact span the tangent space (note that the index is a subscript), and the dual vectors $\left\{ \mathbf{e}^{*i} \right\}$ spanning the cotangent space (the index there is a superscript), and coined covectors. The duality between covectors and vectors, equalities $\left\langle \mathbf{e}^{*i}, \mathbf{e}_j \right\rangle = \delta^i_{.j}$, will later on (when a metric will be given) have the sense of a scalar product between $\mathbf{e}^{*i}$ and $\mathbf{e}_j$, the first written as a row vector, and the second one as a column vector.

NOTE (Bidual).– Let new linear forms be denoted G, associating scalars $\alpha^i \in K$ (commutative field) with 'vectors' $F \in E_n^*$:

$$\left\langle G, F \right\rangle = \left\langle G, f_i \mathbf{e}^{*i} \right\rangle = f_i \left\langle G, \mathbf{e}^{*i} \right\rangle = f_i g^i$$
$$\left\langle G, F \right\rangle = \left\langle G, F_I \mathbf{E}^{*I} \right\rangle = F_I G^I, \forall F \in E_n^*$$

It holds that $F_I = a^i_{.I} f_i \rightarrow g^i = a^i_{.I} G^I$ and $G^I = b^I_{.i} g^i$. Vectors G of E_n^{**} are such that their components form a contravariant suite; there is an isomorphism between E_n and E_n^{**}, so one can identify both spaces (Figure 1.3).

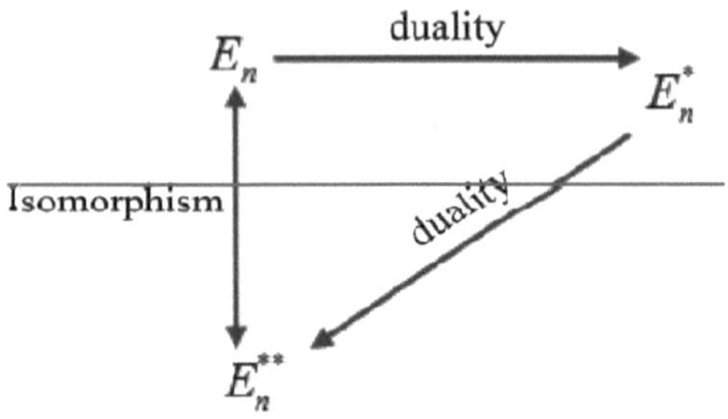

Figure 1.3. *Duality and isomorphism between a vector space, its dual and bidual*

1.5. Tensor algebra

1.5.1. *Notion of tensor product: tensor product of two vector spaces*

Consider two vectors

$$\left.\begin{array}{l}\mathbf{x}\in E_n \ (\dim n)\\ \mathbf{y}\in E_p \ (\dim p)\end{array}\right\}; \quad (\mathbf{x},\mathbf{y})\rightarrow \text{correspondance} \quad (\mathbf{x}\otimes\mathbf{y})\in E_n\otimes E_p .$$

The tensor product is in general non-commutative, so that $\mathbf{x}\otimes\mathbf{y}\neq\mathbf{y}\otimes\mathbf{x}$ *; this can be easily checked by looking at components.*

The following three necessary and sufficient conditions shall be satisfied:

a) Distributivity of $\otimes$ w.r. to addition and of addition w.r. to $\otimes$:

$$\mathbf{x}\otimes(\mathbf{y}_1+\mathbf{y}_2)=\mathbf{x}\otimes\mathbf{y}_1+\mathbf{x}\otimes\mathbf{y}_2$$
$$(\mathbf{x}_1+\mathbf{x}_2)\otimes\mathbf{y}=\mathbf{x}_1\otimes\mathbf{y}+\mathbf{x}_2\otimes\mathbf{y}$$

b) Associativity relative to exterior product: $(\alpha\mathbf{x})\otimes\mathbf{y}=\mathbf{x}\otimes(\alpha\mathbf{y})=\alpha(\mathbf{x}\otimes\mathbf{y})$

c) Existence of a basis: let $\{\mathbf{e}_i\}$ be a basis of E_n, and $\{\mathbf{f}_\alpha\}$ a basis of F_p, with $i=1\ldots n,\ j=1\ldots p$, then $\{\mathbf{e}_i\otimes\mathbf{f}_\alpha\}$ forms a basis of $E_n\otimes F_p$ (with np elements).

DEFINITION.– $\mathbf{x} \otimes \mathbf{y}$ is the tensor product of vectors $\mathbf{x}, \mathbf{y}$, and $E_n \otimes F_p$ the tensor product of the vector spaces E_n and F_p.

Analytical expression of the tensor product $\mathbf{x} \otimes \mathbf{y}$:

$$\mathbf{x} = x^i \mathbf{e}_i,\ \mathbf{y} = y^\alpha \mathbf{f}_\alpha \rightarrow \mathbf{x} \otimes \mathbf{y} = x^i y^\alpha \mathbf{e}_i \otimes \mathbf{f}_\alpha \equiv x^i y^\alpha \boldsymbol{\varepsilon}_{i\alpha}$$

introducing the basis $\boldsymbol{\varepsilon}_{i\alpha}$ of $E_n \otimes F_p$. More generally, any element $T \in E_n \otimes F_p$ can be written

$$T = t^{i\alpha} \boldsymbol{\varepsilon}_{i\alpha}$$

NOTE (Tensors).– Previous notions obviously extend to more than two vector spaces: one can take the tensor product of $\mathbf{x} \otimes \mathbf{y} \in E_n \otimes F_p$ by $\mathbf{z} \in G_q$ with the property:

$$(\mathbf{x} \otimes \mathbf{y}) \otimes \mathbf{z} = \mathbf{x} \otimes (\mathbf{y} \otimes \mathbf{z}) = \mathbf{x} \otimes \mathbf{y} \otimes \mathbf{z} \in E_n \otimes F_p \otimes G_q$$

We call a tensor built on the vector spaces $E_n, F_p, G_q, \ldots$ any element of the vector space $E_n \otimes F_p \otimes G_q \otimes \ldots$

In order to distinguish the tensor product from the Cartesian product, let us consider two vectors $\mathbf{V} = V_i \mathbf{e}_i$, $\mathbf{V}' = V'_i \mathbf{e}_i$ in E_2, and their tensor product $\mathbf{V} \otimes \mathbf{V}' = V_i V'_j \mathbf{e}_i \otimes \mathbf{e}_j' \equiv t_{ij} \mathbf{e}_i \otimes \mathbf{e}_j'$. The components of second order tensor $\mathbf{t}$ are not independent, since we have

$$\frac{t_{11}}{t_{12}} = \frac{t_{21}}{t_{22}} = \frac{V_1'}{V_2'} \Rightarrow t_{22} = \frac{t_{12} t_{21}}{t_{11}} \quad \text{if } t_{11} \neq 0$$

Although components (t_{ij}) are dependent, the tensor products $(\mathbf{e}_i \otimes \mathbf{e}_j)$ are independent thus the vector space $E_2 \otimes E_2$ is richer than the Cartesian product $E_2 \times E_2$: $E_2 \otimes E_2$ is the set of all linear combinations of tensor products $\mathbf{e}_1 \otimes \mathbf{e}_1, \mathbf{e}_1 \otimes \mathbf{e}_2, \mathbf{e}_2 \otimes \mathbf{e}_1, \mathbf{e}_2 \otimes \mathbf{e}_2$, thus it gives the cartesian product $E_2 \times E_2$ the structure of a vector space.

1.5.2. *Affine tensors and affine tensor algebra*

– Affine tensors attached to a vector space E_n : the qth tensor power of E_n, denoted $E_n^{(q)}$, is the tensor product $E_n \otimes E_n \otimes .... \otimes E_n$ of q identical spaces, with basis

$$\varepsilon_{i_1 i_2 ... i_n} = \mathbf{e}_{i_1} \otimes \mathbf{e}_{i_2} \otimes ... \otimes \mathbf{e}_{iq} \quad \text{(dimension } n^q\text{)}$$

Henceforth,

$$\text{if } x_{(j)} \in E_n, \text{ for j=1,2,..,q, } \mathbf{x}_{(1)} \otimes \mathbf{x}_{(2)} \otimes ... \otimes \mathbf{x}_{(n)} = x_{(1)}^{i_1} x_{(2)}^{i_2} ... x_{(q)}^{i_q} \boldsymbol{\varepsilon}_{i_1 i_2 ... i_n}$$

We can also use the dual space E_n^* and build a vector space of dimension n^q :

$$E_n^{(r_1)} \otimes E_n^{*(s_1)} \otimes ... \otimes E_n^{*(s_p)} \otimes E_n^{(r_m)}, \quad q = (r_1 + ... + r_m) + (s_1 + ... + s_p)$$

Any element of this space is an affine tensor of order q, $r_1 + ... + r_m$ - contravariant and $s_1 + ... + s_p$ - covariant.

One calls affine tensor attached to E_n any element of the vector spaces obtained by taking the tensor product of vector spaces identical to E_n or to its dual E_n^*.

– Components of an affine tensor: let for instance $T \in E_n^{(2)} \otimes E_n^{*(2)}$ be an affine tensor of order 4 with components $t^{ij}_{..kl}$, $i,j,k,l \in 1,...,n$, with $\boldsymbol{\varepsilon}_{ij}^{..kl} = \mathbf{e}_i \otimes \mathbf{e}j \otimes \mathbf{e}^{*k} \otimes \mathbf{e}^{*l}$. According to its components, this tensor is twice contravariant and twice covariant.

REMARK.– Since the position of indices is important, one can use points to denote missing indices on a given row, so that the position of each index (subscript or superscript is clear).

NOTE (Change of basis).– Let old bases $\{\mathbf{e}_i\},\{\mathbf{e}^{*i}\}:\langle \mathbf{e}^{*i},\mathbf{e}_i \rangle = \delta^i_{.j}$, and new bases

$\{\mathbf{E}_I\},\{\mathbf{E}^{*I}\}:\langle \mathbf{E}^{*I},\mathbf{E}_I \rangle = \delta^I_{.J}$ and let

$$\mathbf{T} = \mathbf{x} \otimes \mathbf{y} \otimes \mathbf{u}^* \otimes \mathbf{v}^* \in E_n^{(2)} \otimes E_n^{*(2)}, \text{ with } \mathbf{x}, \mathbf{y} \in E_n, \ \mathbf{u}^*, \mathbf{v}^* \in E_n^*.$$

Components: $t^{ij}_{..kl} = x^i y^j u_k v_l$ in basis $\{\boldsymbol{\varepsilon}_{ij}^{..kl}\}$, and $T^{MN}_{..PQ} = X^M Y^N U_P V_Q$ in basis $\{E_{MN}^{..PQ}\}$, with the transformation rules $X^M = b^M_{.i} x^i, \ldots$, hence giving

$T^{MN}_{..PQ} = b^M_{.i} b^N_{.j} a^k_{.P} a^l_{.Q} t^{ij}_{..kl}$ new components versus old ones, or
$t^{ij}_{..kl} = a^i_{.M} a^j_{.N} b^P_{.k} b^Q_{.p} T^{MN}_{..PQ}$ old components versus new ones.

The principle is general $\forall \mathbf{T}$ in an arbitrary affine space. Reciprocally: if n^q components transform in this way in a change of basis, then they are the components of a certain tensor of order q.

1.5.2.1. *Affine tensor algebra*

– Tensor addition: let e.g. two 4th order tensors $T, U \in E_n^{(2)} \otimes E_n^{*(2)}$, then $S \in E_n^{(2)} \otimes E_n^{*(2)}$ is their sum when $s^{ij}_{..kl} = t^{ij}_{..kl} + u^{ij}_{..kl}$ are the components of S.

One denotes the sum of two tensors $S = T + U$, and the generalization is immediate (S, T, U have same variance and order);

– Tensor multiplication: let e.g. $U \in E_n^{(2)} \otimes E_n^{*(2)}$, $V \in E_n^{*(2)}$, then the scalar quantities $p^{ij}_{..klmn} = u^{ij}_{..kl} v_{mn}$ are the components of $P = U \otimes V \in E_n^{(2)} \otimes E_n^{*(4)}$, order 6.

The generalization is immediate (note the addition of orders and variances).

REMARK.– External multiplication by a scalar is a specific case of tensor multiplication, provided one considers the scalar (invariant) as a zero order tensor.

NOTE (Tensoriality criterion).– A system of n^5 quantities t_{ijklm} may be considered as the components of a 5-covariant tensor iff for any three vectors $\mathbf{x}, \mathbf{y}, \mathbf{z} \notin E_n$, the quantities $t_{ijklm} x^i y^j z^k$ are the components of a (5-3)=2-covariant tensor.

Indeed, letting $T_{IJ} = a^i_{.I} a^j_{.J} t_{ij}$, it follows

$$T_{IJ}X^IY^J = a^i_{.I}b^I_{.k}a^j_{.J}b^J_{.l}t_{ij}x^ky^l$$

$$\tilde{A}.\tilde{B} = \tilde{1} \rightarrow T_{IJ}X^IY^J = \delta^i_{.k}\delta^j_{.l}t_{ij}x^ky^l \text{ , thus } T_{IJ}X^IY^J = t_{ij}x^iy^j$$

Conversely, admitting this last equality, it follows that the components t_{ij}, T_{IJ} are necessarily linked by the initial relation.

NOTE (Contraction of Indices).– Let us show that the quantity $t^{.i}_{i}$ obtained from the components $t^{.j}_{i}$ of a mixed second order tensor is invariant, since

$$T^{.I}_{I} = a^i_{.I}a^I_{.k}t^{.k}_{j} = \delta^{.j}_{k}t^{.k}_{j} = t^{.j}_{j}$$

Similarly, the components $t^{.i}_{i.klm}$ obtained by contraction from $t^{.j}_{i.klm}$ are the components of a 3-covariant tensor, since the quantities

$$t^{.j}_{i.klm}x^ky^lz^m$$

define a mixed second order tensor, thus $t^{.i}_{i.klm}x^ky^lz^m$ is invariant, hence $t^{.j}_{i.klm}$ is 3-covariant.

DEFINITION.– The contraction is the operation consisting, after choosing two indices – one covariant and the other one contravariant – to equate them and then sum this repeated index. The contraction of a tensor of order q generates a tensor of order q-2.

EXAMPLE.– $T' = t^{.i}_{i.kl}\mathbf{e}^{*k} \otimes \mathbf{e}^{*l}$ is generated by the contraction on I and j from $T = t^{.j}_{i.kl}\mathbf{e}^{*i} \otimes \mathbf{e}_j \otimes \mathbf{e}^{*k} \otimes \mathbf{e}^{*l}$.

1.5.3. *Contracted multiplication and general tensoriality criterion*

The combination of tensor product and contraction of indices bears the name of contracted product.

EXAMPLE.– $p^{....mnr}_{ijkl} = t_{ijkl}u^{mnr}$ are the components of the tensor product of T and U, with respective components t_{ijkl} and u^{mnr}. Next, $p^{....klr}_{ijkl} = t_{ijkl}u^{klr}$ are the components of their contracted product, obtained by contracting k and m, and l and n. One can also deduce a *tensoriality criterion*:

The set of quantities t^{ijkl} are the components of a 4-contravariant tensor iff for any 2-covariant tensor S_{kl}, the system of quantities $t^{ijkl}S_{kl}$ is the system of components of a (4-2)=2-contravariant tensor.

NOTE (Difference between matrices and tensors).– A matrix is a representation of tensors, but it has no intrinsic feature the tensor has, so for instance, a matrix associated with a change of basis depends on the choice of the bases!

1.6. Euclidean tensors

DEFINITION.– A vector space E_n is Euclidean if one can define a scalar product

$$\mathbf{x}.\mathbf{y} = \alpha, \quad \forall \mathbf{x}, \mathbf{y} \in E_n, \ \forall \alpha \in K$$, here denoted by a dot.

– Scalar product: let $\mathbf{x}.\mathbf{y} \in K$, an abelian field (usually $\mathbb{R}$) of E_n, with the properties:

a) Commutativity: $\mathbf{x}.\mathbf{y} = \mathbf{y}.\mathbf{x}, \quad \forall \mathbf{x}, \mathbf{y} \in E_n$

b) Distributivity of internal product w.r. to addition: $\mathbf{x}.(\mathbf{y}+\mathbf{z}) = \mathbf{x}.\mathbf{y} + \mathbf{x}.\mathbf{z}$;

c) Associativity w.r. to external product: $(\alpha\mathbf{x}).\mathbf{y} = \mathbf{x}.(\alpha\mathbf{y}) = \alpha\mathbf{x}.\mathbf{y}$;

d) $\mathbf{x}.\mathbf{y} = 0 \Rightarrow \mathbf{y} = 0, \forall \mathbf{x} \neq \mathbf{0}$.

– Scalar product as symmetrical bilinear form: by analogy with the notion of linear form,

$$\mathbf{x} \in E_n;\ \forall \mathbf{y} \in E_n \rightarrow \alpha = \mathbf{x}.\mathbf{y} \in K$$
$$\mathbf{y} \in E_n;\ \forall \mathbf{x} \in E_n \rightarrow \alpha = \mathbf{y}.\mathbf{x} \in K$$

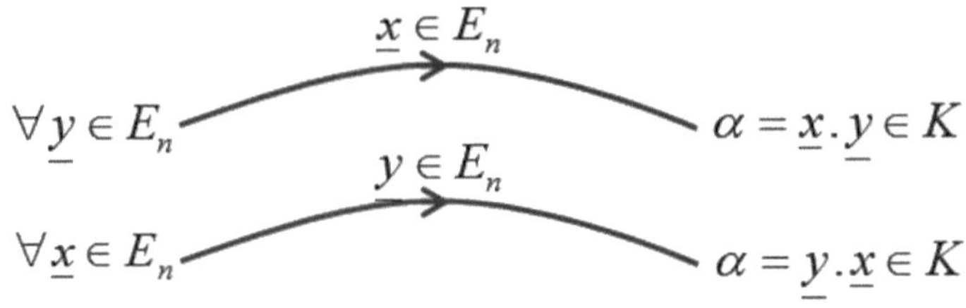

Figure 1.4. *Bilinear form defined from a linear form based on scalar product*

Thus each vector can play the role of linear form w.r. to the other one, hence this in fact defines a bilinear symmetrical form (symmetrical by commutativity), Figure 1.4.

1.6.1. *Fundamental metric tensor in E_n*

Let $\mathbf{x}, \mathbf{y} \in \mathrm{E_n}$ and a basis $\{\mathbf{e}_i\}$; by linearity, it follows

$$\mathbf{x}.\mathbf{y} = x^i \mathbf{e}_i . y^j \mathbf{e}_j = x^i y^j \mathbf{e}_i . \mathbf{e}_j$$

Let set $g_{ij} := \mathbf{e}_i . \mathbf{e}_j \rightarrow \mathbf{x}.\mathbf{y} = \alpha = g_{ij} \mathbf{e}_i . \mathbf{e}_j$. This last relation is sufficient to prove that the g_{ij} are the components of a 2-covariant tensor called the fundamental metric tensor in $\mathrm{E_n}$:

$$G = g_{ij} \mathbf{e}^{*i} \otimes \mathbf{e}^{*j}$$

Commutativity gives

$$g_{ij} = g_{ji} \rightarrow G = G^T$$

hence the metric tensor is symmetrical.

The metric tensor defines the length of an infinitesimal line element between two neighboring points

$$ds^2 = g_{ij} dx^i dx^j$$

Orthogonality and norm: one says that $\mathbf{x},\mathbf{y}\in E_n$ are orthogonal when $\mathbf{x}.\mathbf{y}=0$. The norm of vector $\mathbf{x}$ is scalar $N\mathbf{x} := \mathbf{x}.\mathbf{x} = g_{ij}x_i x_j$. When $N\mathbf{x}=1$, $\mathbf{x}$ is a normed vector or is said unitary.

In mechanics, the scalar product generally is such that the metric is the identity tensor when the canonical basis is used; but, one very often uses a basis attached to a system of coordinates (curvilinear coordinates, such as cylindrical or spherical coordinates), and such a basis is not orthonormal. In addition, continuum mechanics involves different spaces, amongst them the space of functions with finite energy when the sole displacement is considered, the space of functions with derivative of finite energy accounting for strain dependency of the energy.

NOTE (Proper Euclidean vector space).– A Euclidean space E_n is a proper Euclidean when $N\mathbf{x}>0, \forall\mathbf{x}\neq 0$. The module of $\mathbf{x}$ is the square root of the norm,

$$\mathrm{mod}\,\mathbf{x} := \sqrt{N\mathbf{x}}$$

In a proper Euclidean space, the Cauchy–Schwarz inequality is satisfied:

$$\mathrm{mod}(\mathbf{x}.\mathbf{y}) \leq \mathrm{mod}(\mathbf{x})\,\mathrm{mod}(\mathbf{y})$$

and as a corollary:

$$\mathrm{mod}(\mathbf{x}+\mathbf{y}) \leq \mathrm{mod}(\mathbf{x}) + \mathrm{mod}(\mathbf{y})$$

Recall that $\mathrm{mod}\,\alpha = |\alpha|, \ \forall\alpha\in K$. We can next define the angle between two vectors:

$$\cos\varphi := \frac{\mathbf{x}.\mathbf{y}}{\mathrm{mod}\,\mathbf{x}\,\mathrm{mod}\,\mathbf{y}}.$$

We also show that any proper Euclidean space admits orthonormal bases

$$\mathbf{e}_i.\mathbf{e}_j = \delta_{ij} \ (g_{ij} = \delta_{ij}).$$

1.6.2. *Covariant and contravariant components of a vector*

The covariant components of $\mathbf{x}$, denoted x_i, are defined (in a non-orthonormal basis) by

$$x_i := \mathbf{x}.\mathbf{e}_i \rightarrow x_i = x^j \mathbf{e}_j.\mathbf{e}_i = g_{ij} x^j$$

The previous relation can be inverted iff $\det(g_{ij}) \neq 0$: indeed let $x^j = g^{jk} x_k$, and insert this equality into x_i, thus

$$x_i = g_{ij} g^{jk} x_k \rightarrow g_{ij} g^{jk} = \delta_i^{.k}$$

which is invariant, which is sufficient to prove that the quantities g^{jk} are the components of a 2-contravariant symmetrical tensor:

$$G^* = g^{ij} \mathbf{e}_i \otimes \mathbf{e}_j$$

The metric tensor plays the role of a 'lift of indices', allowing the mutual transformation of covariant to contravariant components of tensors.

REMARK.– Property d) $\mathbf{x}.\mathbf{y} = 0 \Rightarrow \mathbf{y} = 0, \forall \mathbf{x} \neq \mathbf{0}$ expresses now as $g_{ij} x^i y^j = 0, \forall \mathbf{x}$, thus $g_{ij} y^j = 0, \forall j$, thus the system shall only admit the nil solution $y^j = 0, \forall j$. This proves that $\det(g_{ij}) \neq 0$.

1.6.3. *Linear form associated to a vector*

From the previous paragraph, the scalar product may be written as

$$\mathbf{x}.\mathbf{y} = g_{ij} x^i y^j = x_i y^i = x^i y_i = g^{ij} x_i y_j .$$

The linear form $\mathbf{x}^*$ can be associated with $\mathbf{x}$ in the following manner (Figure 1.5):

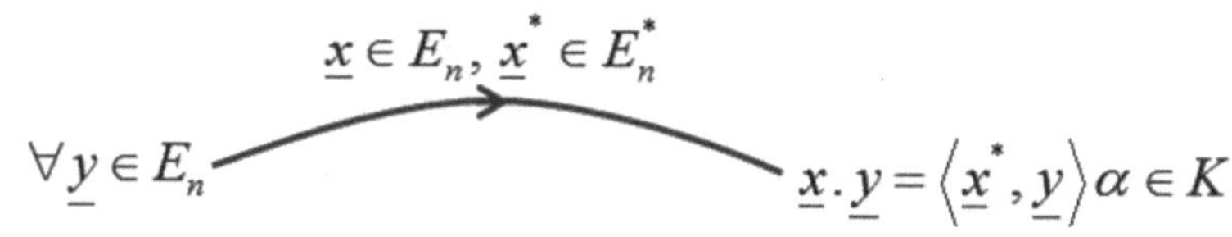

Figure 1.5. *Linear form associated with a vector*

Noting temporarily $\overline{x}_i$ the components of $\mathbf{x}^*$, it holds that

$$\mathbf{x}^* = \overline{x}_i \mathbf{e}^{*i}; \ \langle \mathbf{x}^*, \mathbf{y} \rangle = \overline{x}_i y^i = \mathbf{x}.\mathbf{y} = x_i y^i \Rightarrow \overline{x}_i = x_i$$

Hence, there is an identity between the components of $\mathbf{x}^*$ and the covariant components of $\mathbf{X}$: $\mathbf{x}^* = x_i \mathbf{e}^{*i}$, with $x_j = \mathbf{x}.\mathbf{e}_j$.

REMARK.– One can also define a scalar product in E_n^* , as $\mathbf{x}^i.\mathbf{y}^* = g^{ij} x_i y_j$, satisfying the property

$$\mathbf{x}^*.\mathbf{y}^* = \mathbf{x}.\mathbf{y}, \ \text{i.e.} \langle \mathbf{x}^*, \mathbf{y} \rangle = \langle \mathbf{y}^*, \mathbf{x} \rangle = \mathbf{x}.\mathbf{y} = \mathbf{x}^*.\mathbf{y}^* = x^i y_i = x_i y^i.$$

Identification of E_n and E_n^* : summarizing, it holds that

$$\boxed{\begin{array}{l} \mathbf{x} = x^i \mathbf{e}_i \quad \text{and} \quad \mathbf{x}^* = x_i \mathbf{e}^{*i} \\ x^i = g^{ij} x_j \quad \text{and} \quad x_i = g_{ij} x^j \end{array}}$$

There is no need to attribute different physical significance to $\mathbf{X}$ and $\mathbf{X}^*$, which can thus be identified, as well as the corresponding spaces E_n and E_n^* : this result bears the name *Riesz representation theorem.*

1.6.4. *Riesz representation theorem*

This theorem associates a unique vector representant with a linear form, so that

$$\mathbf{l} = l_i \mathbf{e}^i \rightarrow \exists! \ \tilde{\mathbf{l}} = l^i \mathbf{e}_i, \ \text{ with } l^i = g^{ij} l_j$$

The contravariant vector $\tilde{\mathbf{l}}$ associated with the linear form $\mathbf{l}$ is obtained by contraction with the metric tensor; it allows to express the action of the linear form on a vector as an internal product of $\tilde{\mathbf{l}}$ with the vector:

$$\mathbf{l}(\mathbf{v}) = \langle \tilde{\mathbf{l}}., \mathbf{v} \rangle \equiv \tilde{\mathbf{l}}.\mathbf{v}$$

in which the inner product (denoted by the dot) involves the metric. Due to this identification of a vector representative of a linear form, one may identify a vector space and its dual, and thus also the elements of these vector spaces:

$$\boxed{\mathbf{x} = \mathbf{x}^* = x^i \mathbf{e}_i = x_i \mathbf{e}^{*i}, \qquad E_n \equiv E_n^*}$$

The duality product identifies with the scalar product, $\langle \mathbf{x}^*, \mathbf{y} \rangle = \mathbf{x}^*.\mathbf{y}$, thus

$$\mathbf{x}.\mathbf{y} = \mathbf{x}^*.\mathbf{y}^* = \mathbf{x}^*.\mathbf{y} = \mathbf{x}.\mathbf{y}^*.$$

It also holds that $\langle \mathbf{e}^{*i}, \mathbf{e}_j \rangle = \mathbf{e}^{*i}.\mathbf{e}_j = \delta^i_{.j}$, allowing a graphical representation of $\mathbf{x} = \mathbf{x}^*$ (Figure 1.6).

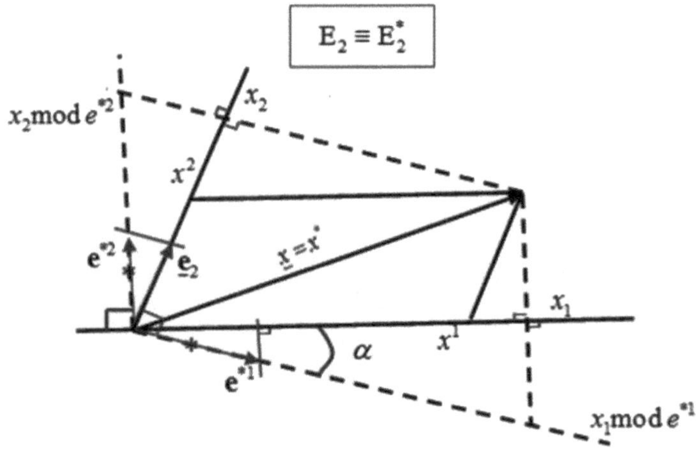

Figure 1.6. *Covariant and contravariant components of a vector*

The figure is done when the basis $\mathbf{e}_1, \mathbf{e}_2$ consists of normed vectors. By identification, one can also note that $\mathbf{e}^{*1} = \mathbf{e}^1, \ \mathbf{e}^{*2} = \mathbf{e}^2$, and the dual basis is thus not normed!

REMARK.– If $\mathbf{e}_2 \perp \mathbf{e}_1$, then the dual basis is identical to the initial basis, $\{\mathbf{e}_i\} = \{\mathbf{e}^{*i}\}$. We shall then see later on that the notion of variance is no longer necessary.

REMARK.– The dual basis is used by physicist to visualize Bravais lattices of crystalline structures in the so-called reciprocal space.

Note that the notion of covariant or a contravariant component is immaterial: the same tensor can be decomposed into either type of component. It is nevertheless admitted that strain components should be covariant whereas stress components are essentially contravariant.

As indicated in Figure 1.7, the dual basis vectors can be constructed as

$$\mathbf{g}^1 = \frac{\mathbf{g}^2 \times \mathbf{g}^3}{V}, \quad \mathbf{g}^2 = \frac{\mathbf{g}^3 \times \mathbf{g}^1}{V}, \quad \mathbf{g}^3 = \frac{\mathbf{g}^1 \times \mathbf{g}^2}{V}, \quad V = \mathbf{g}_1 . (\mathbf{g}_2 \times \mathbf{g}_3)$$

where V is the volume spanned by the three basis vectors.

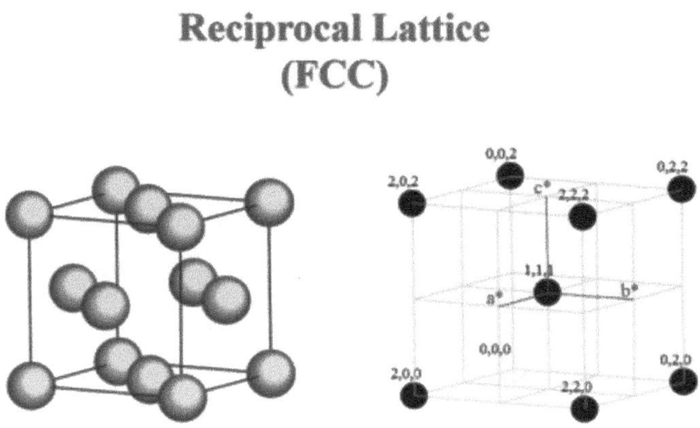

Importance of Reciprocal Space

- When a diffraction event occurs, the diffracted waves/pattern will "match" the reciprocal lattice.

► Radiation 'scatters' in inverse proportion to the spacings between diffraction centers (i.e., planes in crystals).

- We make the use of this fact in x-ray diffraction and transmission electron microscopy.

- For an arbitrary lattice [$a \neq b \neq c$ and $\alpha \neq \beta \neq \gamma (\neq 90°)$] in real space.

$$a^* = \frac{(b \times c)}{a \cdot (b \times c)} = \frac{(b \times c)}{V}$$

$$b^* = \frac{(c \times a)}{b \cdot (c \times a)} = \frac{(c \times a)}{V}$$

$$c^* = \frac{(a \times b)}{c \cdot (a \times b)} = \frac{(a \times b)}{V}$$

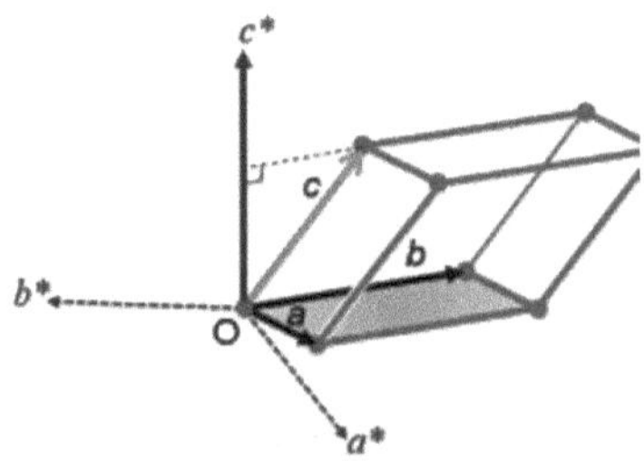

$V \equiv$ unit cell volume $= a \cdot (b \times c) = b \cdot (c \times a) = c \cdot (a \times b)$

$a^* \perp bc$ plane in real space; it is the reciprocal lattice vector for a

$b^* \perp ac$ plane in real space; it is the reciprocal lattice vector for b

$c^* \perp ab$ plane in real space; it is the reciprocal lattice vector for c

Figure 1.7. *Reciprocal space vectors. For a color version of this figure, see www.iste.co.uk/ganghoffer/biomechanics.zip*

NOTE (Components of a Euclidean tensor).– Let e.g. $\mathbf{T} = \mathbf{x} \otimes \mathbf{y} \otimes \mathbf{z}$ be a third order tensor. Since one can replace $\mathbf{x}$ by $\mathbf{x}^*$, $\mathbf{y}$ by $\mathbf{y}^*$, and $\mathbf{z}$ by $\mathbf{z}^*$, it also holds that

$$\mathbf{T} = \mathbf{x}^* \otimes \mathbf{y} \otimes \mathbf{z} = \mathbf{x} \otimes \mathbf{y}^* \otimes \mathbf{z} = \ldots = \mathbf{x}^* \otimes \mathbf{y}^* \otimes \mathbf{z}^*$$

Let us search for the existing relations between the different types of components of $\mathbf{T}$:

$$t^{ijk} = x^i y^j z^k \quad \text{for } \mathbf{x} \otimes \mathbf{y} \otimes \mathbf{z}, \quad t_i^{\,jk} = x_i y^j z^k \quad \text{for } \mathbf{x}^* \otimes \mathbf{y} \otimes \mathbf{z}, \ \ldots$$

$$t_{ijk} = x_i y_j z_k \ \text{for } \mathbf{x}^* \otimes \mathbf{y}^* \otimes \mathbf{z}^*$$

We use the property: $x^i = g^{il} x_l$ and $x_i = g_{il} x^l$, resulting in $t^{ijk} = g^{il} x_l y^j z^k = g^{il} t_l^{\cdot jk}$ and reciprocally, $t_i^{\cdot jk} = g_{il} x^l x^j x^k$.

THEOREM.– The different types of components of an Euclidean tensor, contravariant, covariant or mixed, are expressible from each other by multiplication by g_{ij} or g^{ij} and summation; this operation is repeated as many times as necessary.

EXAMPLE.– Lowering of indices, $u_{ijkl} = g_{im} g_{jn} g_{ko} g_{lp} u^{mnop}$. Raising indices: $v^{abcd} = g^{al} g^{bf} g^{cg} g^{dh} v_{efgh}$.

– Case of the fundamental metric tensor: the mixed components are given by $g^{\cdot j}_{\cdot k} = g^{ij} g_{ik} = \delta^{\cdot i}_{\cdot k}$, and by symmetry, $g^{\cdot j}_{\cdot k} = g^{\cdot k}_{\cdot j} = g_k^{\cdot j} = g_j^{\cdot k}$.

THEOREM.– The quantities g_{ij} and g^{ij} are respectively the covariant and contravariant components of the symmetrical fundamental metric tensor G of the Euclidean space. The quantities $g^{\cdot j}_{\cdot k} = \delta^{\cdot i}_{\cdot k}$ are the mixed components. This tensor is thus denoted simply $\mathbf{I}$n, employing more specifically the 2-covaruiant or 2-contravariant representations G or G^* respectively.

– Symmetry and antisymmetry: the symmetry of a second order tensor, exchange of its components (row and columns of the matrix representation), $t_{ij} = t_{ji}$, is an intrinsic property, not depending upon the basis or upon the components. If for instance $t_{ij} = g_{ik} g_{jl} t^{kl} = t_{ji} = g_{jl} g_{ik} t^{lk}$, then $t^{kl} = t^{lk}, t^{\cdot k}_{\cdot l} = t^{\cdot l}_{\cdot k}, \ldots$

One could also easily show that if $t_{ij} = t_{ji}$ in basis $\{\mathbf{e}^{*k}\}$, then $T_{IJ} = T_{JI}$ in the new basis $\{\mathbf{E}^{*K}\}$.

The same also holds for antisymmetry, that is $t_{ij} = -t_{ji}$. These considerations extend to any set of two indices taken amongst the indices of an Euclidean tensor of order $q > 2$.

1.6.5. *Algebra of Euclidean tensors*

The possibility to freely change the variance of components of a Euclidean tensor allows the removal of the restrictions of affine tensor algebra:

Addition of two tensors: they should have the same order.

Multiplication, contraction of two indices and contracted products: no restrictions.

EXAMPLE.– Contraction of t^{ijkl} on the second and third indices j,k leads to different possible operations, including:

– Lowering of indices: $t^{i.kl}_{.m} = g_{mj} t^{ijkl}$,

– Contraction: $t^{i.ml}_{.m} = g_{mj} t^{ijml}$,

– Symmetry: $t^{i.ml}_{.m} = t^{im.l}_{m}$.

We more frequently use the following operations defined in the frame of affine tensor algebra:

– *Simply contracted product*: $T = U \odot V \left(\Leftrightarrow \text{e.g.}: t^{ij}_{.m} = u^{ijk} v_{km}\right)$, contraction of the indices located on both sides of the contraction $\odot$. A simple point is used most often for this inner product.

– *Double contracted product*: $T = U : V \left(\Leftrightarrow \text{e.g.}: t^{i} = u^{ilk} v_{kl}\right)$. This operation can be repeated for the two following indices. It is noted $\mathbf{T} = \mathbf{U} : \mathbf{V}$ for Euclidean tensors.

REMARK.– In each pair of indices being summed, one is covariant and the second one is contravariant. One can never sum indices in the same position, unless Cartesian coordinates are used, in which case the variance does not matter!

A covariant vector is an element of the dual E^* (definition from differential geometry). The definition from change of basis says that components of a vector are covariant since they transform like the basis vector, $[1]_{new} = [1]_{old}.P$. Components of a vector transform in an opposite manner, and are thus denoted as contravariant, $[v]_{new} = P^{-1}.[v]_{old}$.

EXAMPLE.– The gradient of a function is a covector, but can be represented as a vector (contravariant) thanks to Riesz representation theorem.

– *Case of orthonormal bases*: if we only consider orthonormal bases $\{\mathbf{e}_i\}$ or $\{\mathbf{E}_i\}$, the transition matrix is a rotation matrix: $\tilde{A} = \tilde{B}^{-1} = \tilde{B}^T$, so that $a^i_{.I} = b^I_{.i}$.

Recalling previous results, we can write

$$\mathbf{E}_{I.} = a^i_{.I}\mathbf{e}_i \;;\; \mathbf{e}_i = b^I_{.i}\mathbf{E}_I$$

$$V^I = b^I_{.i}a^i \;; v^i = a^i_{.I}V^I$$

We notice in the case of orthonormal bases that no more difference exists between covariance and contravariance (transformation rules are the same): we then use, by convention, lower indices.

New = f(old)	Old = f(new)
$\mathbf{E}_{I.} = a^i_{.I}\mathbf{e}_i$ $V^I = b^I_{.i}a^i$	$\mathbf{e}_i = b^I_{.i}\mathbf{E}_I$ $v^i = a^i_{.I}V^I$
with $a_{iI} = b_{Ii}$	

1.7. Algebraic operations on tensors

Subsequent developments are done assuming Cartesian tensors.

1.7.1. *Useful tensors (in calculus)*

Unity tensor (Kronecker):

$$\mathbf{I}:\ \mathbf{x} \in R^n \mapsto \mathbf{x} \in R^n, \quad \mathbf{I} = \delta_{ij}\mathbf{e}_i \otimes \mathbf{e}_j$$

– Orientation tensor (fundamental alternated tensor or permutation tensor): For $\mathbf{V} \in \mathrm{R}^3$, let $\Pi_V : \mathbf{X} \mapsto \mathbf{X} \wedge \mathbf{V}$ second order tensor **t**, thus $\mathbf{X} \wedge \mathbf{V} = \mathbf{t}.\mathbf{X}$.

Let then $\boldsymbol{\varepsilon} : \mathbf{V} \mapsto \Pi_V$, third order tensor, thus $\Pi_V = \boldsymbol{\varepsilon}.\mathbf{V} \rightarrow t_{ij} = \varepsilon_{ijk} V_k$. It follows

$$\mathbf{X} \wedge \mathbf{V} = \begin{pmatrix} 0 & V_3 & -V_2 \\ -V_3 & 0 & V_1 \\ V_2 & -V_1 & 0 \end{pmatrix} \begin{pmatrix} X_1 \\ X_2 \\ X_3 \end{pmatrix} \Rightarrow \mathbf{t} = \begin{pmatrix} 0 & V_3 & -V_2 \\ -V_3 & 0 & V_1 \\ V_2 & -V_1 & 0 \end{pmatrix}.$$

The relation $t_{ij} = \varepsilon_{ijk} V_k$ then gives

$$\varepsilon_{ijk} = \begin{cases} 1 \text{ if } (i,j,k) \text{ even permutation of the list } (1,2,3) \\ -1 \text{ if } (i,j,k) \text{ odd permutation of the list } (1,2,3) \\ 0 \text{ if at least two of the indices out of 3 are equal} \end{cases}$$

This gives the list:

$$\left| \begin{array}{l} \varepsilon_{123} = \varepsilon_{231} = \varepsilon_{312} = 1 \\ \varepsilon_{132} = \varepsilon_{213} = \varepsilon_{321} = -1 \\ \text{other } \varepsilon_{ijk} \equiv 0 \end{array} \right.$$

A graphical representation of the permutation tensor in 3D is as follows (Figure 1.8).

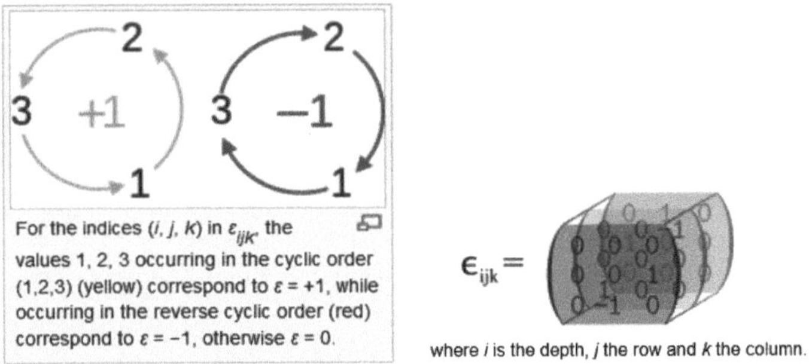

Figure 1.8. *Graphical representation of the permutation tensor in 3D. For a color version of this figure, see www.iste.co.uk/ganghoffer/biomechanics.zip*

It allows performing operations in a tensor manner (independent of any basis).

REMARK.– There also exist permutation tensors in 2D and 4D.

The two dimensional Levi-Civita symbol is defined by

$$\varepsilon_{ij} = \begin{vmatrix} 1 \text{ if i=1, j=2} \\ -1 \text{ if i=2, j=1} \\ 0 \text{ if i=j} \end{vmatrix}$$

The values can be arranged into a 2*2 antisymmetric matrix,

$$\varepsilon_{ij} = \begin{pmatrix} 0 & 1 \\ -1 & 0 \end{pmatrix}.$$

– Example in 3D:

$$\det \mathbf{A} \to \varepsilon_{lmn} \det \mathbf{A} = \varepsilon_{ijk} a_{il} a_{jm} a_{kn} \Rightarrow \left(\mathbf{A}^{-1}\right)_{ij} = \frac{1}{2\det \mathbf{A}} \varepsilon_{jlm} \varepsilon_{ipq} a_{lp} a_{mq}.$$

EXAMPLE.– $\mathbf{T} = \mathbf{T}^{T} \Leftrightarrow \varepsilon_{ijk} T_{ij} = 0$.

Useful formulae (exercise) $\varepsilon - \delta$ identity: $\varepsilon_{ijk}\varepsilon_{imn} = \delta_{jm}\delta_{kn} - \delta_{jn}\delta_{km}$, thus $\varepsilon_{ijk}\varepsilon_{ijn} = 2\delta_{kn}$, and finally by full contraction $\varepsilon_{ijk}\varepsilon_{ijk} = 6$.

1.7.2. *Vector product, alternated product, triple scalar and vector products*

– Euclidean framework: $\mathbf{u} = u_i \mathbf{e}_i \to \mathbf{u} = (\mathbf{u}.\mathbf{e}_i)\mathbf{e}_i$

– Vector product: $\mathbf{e}_j \times \mathbf{e}_k = \varepsilon_{ijk} \mathbf{e}_i$, thus

$$\mathbf{e}_m.(\mathbf{e}_i \times \mathbf{e}_k) = \varepsilon_{ijk}\delta_{im} = \varepsilon_{mjk} \to \varepsilon_{ijk} = \mathbf{e}_i.(\mathbf{e}_j \times \mathbf{e}_k).$$

– Distributivity: $(\alpha\mathbf{u} \times \beta\mathbf{v}) = \alpha\beta(\mathbf{u} \times \mathbf{v}),\ \forall \alpha, \beta \in R,\ \forall \mathbf{u}, \mathbf{v} \in V$, hence

$$\mathbf{w} = \mathbf{u} \times \mathbf{v} = (u_j \mathbf{e}_j) \times (v_k \mathbf{e}_k) = \varepsilon_{ijk} u_j v_k \mathbf{e}_i$$

$$\mathbf{u} \times \mathbf{v} = -\mathbf{v} \times \mathbf{u} \Rightarrow \mathbf{u} \times \mathbf{u} = \mathbf{0}$$

– Triple scalar product: $[\mathbf{u},\mathbf{v},\mathbf{w}] = \mathbf{u}.(\mathbf{v}\times\mathbf{w}) = \varepsilon_{ijk}u_i v_j w_k$.

– Linearity: $[\alpha\mathbf{u}+\beta\mathbf{v},\mathbf{x},\mathbf{y}] = \alpha[\mathbf{u},\mathbf{x},\mathbf{y}]+\beta[\mathbf{v},\mathbf{x},\mathbf{y}], \quad \forall\mathbf{u},\mathbf{v},\mathbf{x},\mathbf{y}\in V$

– Determinant form: $[\mathbf{u},\mathbf{v},\mathbf{w}] = \det\begin{bmatrix} u_1 & u_2 & u_3 \\ v_1 & v_2 & v_3 \\ w_1 & w_2 & w_3 \end{bmatrix}$, thus

$$\varepsilon_{ijk} = [\mathbf{e}_i,\mathbf{e}_j,\mathbf{e}_k] = \det\begin{bmatrix} \mathbf{e}_i.\mathbf{e}_1 & \mathbf{e}_i.\mathbf{e}_2 & \mathbf{e}_i.\mathbf{e}_3 \\ \mathbf{e}_j.\mathbf{e}_1 & \mathbf{e}_j.\mathbf{e}_2 & \mathbf{e}_j.\mathbf{e}_3 \\ \mathbf{e}_k.\mathbf{e}_1 & \mathbf{e}_k.\mathbf{e}_2 & \mathbf{e}_k.\mathbf{e}_3 \end{bmatrix} = \det\begin{bmatrix} \delta_{i1} & \delta_{i2} & \delta_{i3} \\ \delta_{j1} & \delta_{j2} & \delta_{j3} \\ \delta_{k1} & \delta_{k2} & \delta_{k3} \end{bmatrix}$$

EXAMPLE.– Prove the equality $\varepsilon_{ijk}\varepsilon_{pqr} = \det\begin{bmatrix} \delta_{ip} & \delta_{iq} & \delta_{ir} \\ \delta_{jp} & \delta_{jq} & \delta_{jr} \\ \delta_{kp} & \delta_{kq} & \delta_{kr} \end{bmatrix}$, using the identities $\det\mathbf{T} = \det(\mathbf{T}^T)$, $\det(\mathbf{R}\mathbf{S}) = \det(\mathbf{R})\det(\mathbf{S})$, and then the $\varepsilon-\delta$ identity.

EXAMPLE.– Show that $(\mathbf{u}\times\mathbf{v}).(\mathbf{u}\times\mathbf{v}) = (\mathbf{u}.\mathbf{u})(\mathbf{v}.\mathbf{v})-(\mathbf{u}.\mathbf{v})^2$ using index notation.

PROOF.– $\forall\mathbf{u},\mathbf{v}, \quad \mathbf{u}\times\mathbf{v} = \mathbf{0} \Leftrightarrow \mathbf{u},\mathbf{v}$ are linearly independent.

THEOREM.– $[\mathbf{u},\mathbf{v},\mathbf{w}] = 0$ iff $\mathbf{u},\mathbf{v},\mathbf{w}$ are linearly independent.

– Triple vector product: this is defined as

$$\mathbf{u}\times(\mathbf{v}\times\mathbf{w}) = \varepsilon_{ijk}u_j(\mathbf{v}\times\mathbf{w})_k\,\mathbf{e}_i = \varepsilon_{ijk}\varepsilon_{kmn}u_j v_m w_n \mathbf{e}_i = (\delta_{im}\delta_{jn}-\delta_{in}\delta_{jm})u_j v_m w_n \mathbf{e}_i =$$
$$= (u_m w_m v_i - u_m v_m w_i)\mathbf{e}_i = (\mathbf{u}.\mathbf{w})\mathbf{v}-(\mathbf{u}.\mathbf{v})\mathbf{w}$$

EXERCISE.– Show the identity $(\mathbf{u}\times\mathbf{v})\times\mathbf{w} = (\mathbf{u}.\mathbf{w})\mathbf{v}-(\mathbf{v}.\mathbf{w})\mathbf{u}$.

1.7.3. *Tensor product and higher order tensors*

Dyadic product method to construct tensors from vectors:

$$\mathbf{a}\otimes\mathbf{b}\rightarrow(\mathbf{a}\otimes\mathbf{b}).\mathbf{c}=(\mathbf{b}.\mathbf{c})\mathbf{a},\forall\mathbf{c}\in E_n$$

Here, the constructed second order tensor is the sum of the 9 dyads $a_i b_j \mathbf{e}_i\otimes\mathbf{e}_j$, a linear operation.

Any second order tensor can be decomposed into the tensor basis $\{\mathbf{e}_i\otimes\mathbf{e}_j\}$:

$$\mathbf{T}=T_{ij}\mathbf{e}_i\otimes\mathbf{e}_j.$$

Note that this implies in particular the decomposition of the identity tensor

$$\mathbf{I}=\mathbf{e}_i\otimes\mathbf{e}_i$$

Hence

$$\mathbf{T}.\mathbf{u}=(\mathbf{T}.\mathbf{u})_i\,\mathbf{e}_i=\left[\mathbf{e}_i.(\mathbf{T}.\mathbf{u})\right]\mathbf{e}_i=\left\{\mathbf{e}_i.\left[\mathbf{T}\left(u_j\mathbf{e}_j\right)\right]\right\}\mathbf{e}_i=\left\{u_j\left[\mathbf{e}_i.\mathbf{T}\left(\mathbf{e}_j\right)\right]\right\}\mathbf{e}_i=\left\{\left(\mathbf{u}.\mathbf{e}_j\right)\left[\mathbf{e}_i.\mathbf{T}\left(\mathbf{e}_j\right)\right]\right\}\mathbf{e}_i=$$
$$=\left[\mathbf{e}_i.\mathbf{T}\left(\mathbf{e}_j\right)\right]\left(\mathbf{u}.\mathbf{e}_j\right)\mathbf{e}_i=\left\{\left[\mathbf{e}_i.\mathbf{T}\left(\mathbf{e}_j\right)\right]\mathbf{e}_i\otimes\mathbf{e}_j\right\}.\mathbf{u}\equiv\left(T_{ij}\mathbf{e}_i\otimes\mathbf{e}_j\right).\mathbf{u}$$

T is a linear mapping, $T\in L\left(E_n,E_n\right)$.

NOTE (Generalization).– A third order tensor $\underline{\underline{\underline{\mathbf{B}}}}$ such that $\underline{\underline{\underline{\mathbf{B}}}}.v=\underline{\underline{\mathbf{T}}}$ is a linear application $\mathbf{B}:E_n\rightarrow E_n\otimes E_n$

– Construction of higher order tensors by successive dyadic products:

$$(\mathbf{a}\otimes\mathbf{b}\otimes\mathbf{c})=(\mathbf{a}\otimes\mathbf{b})(\mathbf{c}.\mathbf{d}),\quad\forall\mathbf{d}\in E_n.$$

Any third order tensor $\underline{\underline{\underline{\mathbf{B}}}}$ is the sum of 27 (3^3) dyads:

$$\underline{\underline{\underline{\mathbf{B}}}}=B_{ijk}\mathbf{e}_i\otimes\mathbf{e}_j\otimes\mathbf{e}_k,\text{ with }B_{ijk}=\mathbf{e}_i.\left(\underline{\underline{\underline{\mathbf{B}}}}.\mathbf{e}_k\right).\mathbf{e}_j$$

EXERCISE.– Show that $\underline{\underline{\mathbf{B}}}.\mathbf{u} = \mathbf{e}_i.\left[\left(\underline{\underline{\mathbf{B}}}.\mathbf{e}_k\right).\mathbf{e}_j\right]\left[\left(\mathbf{e}_i \otimes \mathbf{e}_j \otimes \mathbf{e}_k\right).\mathbf{u}\right]$, $\forall \mathbf{u}$.

– Tensor of order n : linear mapping, transforms a vector into a tensor of order $n-1$; can be written $T = T_{i_1 i_2 \ldots i_n} \mathbf{e}_{i_1} \otimes \mathbf{e}_{i_2} \otimes \ldots \otimes \mathbf{e}_{i_n}$, sum of 3^n dyads, belongs to the vector space of tensors of order n, with components given by

$$T_{i_1 i_2 \ldots i_n} = \mathbf{e}_{i_1}\left[\left(\left(\mathbf{T}.\mathbf{e}_{i_n}\right).\mathbf{e}_{i_{n-1}}\right)\ldots\mathbf{e}_{i_2}\right]$$

EXAMPLE.– In continuum mechanics, the 4th order tensor of elastic moduli $\underset{\approx}{\mathbf{C}}$, such that $\underset{\sim}{\mathbf{S}} = \underset{\approx}{\mathbf{C}}.\underset{\sim}{\mathbf{E}}$, relates the stress tensor $\underset{\sim}{\mathbf{S}}$ to the strain tensor $\underset{\sim}{\mathbf{E}}$. We use here as a compact notation the single tilda to indicate second order tensors and the double tilda to denote fourth order tensors.

Most tensors in continuum mechanics are of second order (stress and strain tensors). Higher order gradient theories (theories involving e.g. second order or even higher order gradients of the displacement) require tensors of order greater than four.

Properties:

– Distributivity: $\left(\underset{\approx}{\mathbf{C}} + \beta\underset{\approx}{\mathbf{D}}\right) : \underset{\approx}{\mathbf{A}} = \underset{\approx}{\mathbf{C}} : \underset{\approx}{\mathbf{A}} + \beta\underset{\approx}{\mathbf{D}} : \underset{\approx}{\mathbf{A}}, \ \forall \underset{\approx}{\mathbf{A}}, \ \ \forall \beta \in \mathbb{R}$.

– Contraction: $\left(\underset{\approx}{\mathbf{C}} : \underset{\approx}{\mathbf{D}}\right) : \underset{\approx}{\mathbf{A}} = \left(\underset{\approx}{\mathbf{C}}\right) : \left(\underset{\approx}{\mathbf{D}} : \underset{\approx}{\mathbf{A}}\right), \ \forall \underset{\approx}{\mathbf{A}}$.

The trace operator is a scalar invariant allows us to properly define the double contraction of two second order tensors (the order of tensors shall be identified from the context, tensors not underlined will be implicitly second order):

$$\mathbf{A} : \mathbf{B} = \mathrm{tr}\left(\mathbf{A}.\mathbf{B}^T\right) = \mathbf{B} : \mathbf{A} = \mathrm{tr}\left(\mathbf{B}.\mathbf{A}^T\right) = \mathrm{tr}\left\{\left(\mathbf{A}.\mathbf{B}^T\right)^T\right\}$$

EXAMPLE.– $(\mathbf{a} \otimes \mathbf{b}) : (\mathbf{c} \otimes \mathbf{d}) = \mathrm{Tr}\left((\mathbf{a} \otimes \mathbf{b}).(\mathbf{c} \otimes \mathbf{d})^T\right) =$

$$\mathrm{Tr}\left((\mathbf{a} \otimes \mathbf{b}).(\mathbf{d} \otimes \mathbf{c})\right) = (\mathbf{b}.\mathbf{d})\,\mathrm{Tr}(\mathbf{a} \otimes \mathbf{c}) = (\mathbf{b}.\mathbf{d})(\mathbf{a}.\mathbf{c})$$

Components of a 4th order tensor: let define the elementary second order tensors

$$\mathbf{T}:\mathbf{E}_{ij}=\mathrm{tr}\left(\mathbf{T}.\mathbf{e}_i\otimes\mathbf{e}_j\right)=\mathrm{T}_{kj}\mathrm{tr}\left(\mathbf{e}_j\otimes\mathbf{e}_i\right)=\mathrm{T}_{kj}\delta_{ki}=\mathrm{T}_{ij}$$

Take next $\underline{\underline{\mathbf{T}}}=\underline{\underline{\mathbf{E}}}_{kl}$, thus $\underline{\underline{\mathbf{E}}}_{ij}:\underline{\underline{\mathbf{E}}}_{kl}=\delta_{ik}\delta_{jl}$, thus $\underset{\approx}{\mathbf{C}}:\underline{\underline{\mathbf{E}}}_{kl}=\mathrm{C}_{ijkl}\underline{\underline{\mathbf{E}}}_{ij}$ by extension.

REMARK.– We have first formed $\underline{\underline{\mathbf{E}}}_{ij}\otimes\underline{\underline{\mathbf{E}}}_{kl}=\mathbf{e}_i\otimes\mathbf{e}_j\otimes\mathbf{e}_k\otimes\mathbf{e}_l$, and then contract

$$\mathrm{tr}\left(\underline{\underline{\mathbf{E}}}_{ij}.\left(\underline{\underline{\mathbf{E}}}_{kl}\right)^{\mathrm{T}}\right)=\mathrm{tr}\left(\underline{\underline{\mathbf{E}}}_{ij}.\mathbf{e}_l\otimes\mathbf{e}_k\right)=\mathrm{tr}\left(\left(\mathbf{e}_i\otimes\mathbf{e}_j\right).\left(\mathbf{e}_l\otimes\mathbf{e}_k\right)\right)\equiv\mathrm{tr}\left(\delta_{jl}\delta_{ik}\right).$$

This is analogous to the similar relation for second order tensors, $\underline{\underline{\mathbf{T}}}.\mathbf{e}_j=\mathrm{T}_{ij}\mathbf{e}_i$.

– Equality of 4th order tensors:

$$\underset{\approx}{\mathbf{C}}.\underset{\sim}{\mathbf{B}}=\underset{\approx}{\mathbf{D}}.\underset{\sim}{\mathbf{B}},\ \ \forall\mathbf{B}\text{, or }\left(\underset{\sim}{\mathbf{A}},\underset{\approx}{\mathbf{C}}.\underset{\sim}{\mathbf{B}}\right)=\left(\underset{\sim}{\mathbf{A}},\underset{\approx}{\mathbf{D}}.\underset{\sim}{\mathbf{B}}\right),\ \ \forall\underset{\sim}{\mathbf{A}},\underset{\sim}{\mathbf{B}}.$$

For the choice $\underline{\underline{\mathbf{A}}}=\underline{\underline{\mathbf{E}}}_{ij}$, $\underline{\underline{\mathbf{B}}}=\underline{\underline{\mathbf{E}}}_{kl}\rightarrow\mathrm{C}_{ijkl}=\mathrm{D}_{ijkl}$, thus tensors have the same components (intuitive idea of equality of tensors).

Let next express the relation between the components of tensors $\underset{\sim}{\mathbf{A}},\underset{\sim}{\mathbf{B}},\underset{\approx}{\mathbf{C}}$: write $\underset{\sim}{\mathbf{B}}=\underset{\approx}{\mathbf{C}}:\underset{\sim}{\mathbf{A}}\rightarrow\mathrm{B}_{ij}\underline{\underline{\mathbf{E}}}_{ij}=\underset{\approx}{\mathbf{C}}:\left(\mathrm{A}_{kl}\underline{\underline{\mathbf{E}}}_{kl}\right)=\mathrm{A}_{kl}\underset{\approx}{\mathbf{C}}:\underline{\underline{\mathbf{E}}}_{kl}=\mathrm{C}_{ijkl}\mathrm{A}_{kl}\underline{\underline{\mathbf{E}}}_{ij}$

thus $\mathrm{B}_{ij}=\mathrm{C}_{ijkl}\mathrm{A}_{kl}$.

Similarly, it holds that $\left(\underset{\approx}{\mathbf{C}}:\underset{\approx}{\mathbf{D}}\right)_{ijkl}=\left(\underset{\approx}{\mathbf{C}}\right)_{ijmn}\left(\underset{\approx}{\mathbf{D}}\right)_{mnkl}$, by contracting the two last repeated indices of $\underset{\approx}{\mathbf{C}}$ with the two first of $\underset{\approx}{\mathbf{D}}$.

1.7.4. *Useful fourth order tensors*

Identity tensor of order four, $\mathbb{I}_4$, such that $\mathbb{I}_4:\mathbf{A}=\mathbf{A}$, transforming any second order tensor **A** into itself. Note that double contraction is used here, while single contraction was used to define the identity second order tensor transforming any vector into itself.

– Transpose: $\mathbb{T} : \mathbf{A} = \mathbf{A}^{\mathrm{T}}$, with $\mathbb{T} : \mathbb{T} = \mathbb{I}_4$.

– Components: $\mathbb{T}_{ijmn} = \delta_{in}\delta_{jm}$.

– Symmetrization tensor: $\mathbb{S} : \mathbf{A} = \frac{1}{2}\left(\mathbf{A} + \mathbf{A}^{\mathrm{T}}\right)$.

– Antisymmetrization tensor: $\mathbb{W} : \mathbf{A} = \frac{1}{2}\left(\mathbf{A} - \mathbf{A}^{\mathrm{T}}\right) \rightarrow \mathbb{T} + \mathbb{W} = \mathbb{I}_4$.

A 4th order tensor $\mathbb{C}$ is invertible iff there exists an inverse tensor $\mathbb{C}^{-1}$ s.t.

$$\mathbb{C}^{-1}.\mathbb{C} = \mathbb{I}_4 = \mathbb{C}.\mathbb{C}^{-1}$$

– Inverse of a product: $\left(\mathbb{U} : \mathbb{V}\right)^{-1} = \mathbb{V}^{-1} : \mathbb{U}^{-1}$.

EXERCISE.– Show the following identities $\mathbb{S} : \mathbb{T} = \mathbb{T} : \mathbb{S} = \frac{1}{2}\left(\mathbb{T} + \mathbb{I}_4\right)$;

$$\mathbb{W} : \mathbb{T} = \mathbb{T} : \mathbb{W} = \frac{1}{2}\left(\mathbb{T} - \mathbb{I}_4\right);$$

$$\mathbb{S} : \mathbb{S} = \mathbb{S};\ \mathbb{W} : \mathbb{W} = \mathbb{W};\ \mathbb{S} : \mathbb{W} = 0 = \mathbb{W} : \mathbb{S}.$$

The components of those tensors are:

$$\left(\mathbb{I}_4\right)_{ijkl} = \delta_{ik}\delta_{jl};\ \mathbb{T}_{ijkl} = \delta_{il}\delta_{jk};\ \mathbb{S}_{ijkl} = \frac{1}{2}\left(\delta_{ik}\delta_{jl} + \delta_{il}\delta_{jk}\right);\ \mathbb{W}_{ijkl} = \frac{1}{2}\left(\delta_{ik}\delta_{jl} - \delta_{il}\delta_{jk}\right)$$

The transpose of a 4th order tensor is defined as: $\left(\mathbb{C}^{\mathrm{T}}\right)_{ijkl} = \mathbb{C}_{klij}$

such that $\left(\mathbf{A}, \mathbb{C}^{\mathrm{T}} : \mathbf{B}\right) = \left(\mathbf{B}, \mathbb{C}^{\mathrm{T}} : \mathbf{A}\right),\ \forall \mathbf{A}, \mathbf{B}$

PROOF.– $\left(\mathbb{C}^{T}\right)^{T}=\mathbb{C}$; $(\mathbb{C}:\mathbb{D})^{T}=\mathbb{D}^{T}:\mathbb{C}^{T}$; $\left(\mathbb{C}^{T}\right)^{-1}=\left(\mathbb{C}^{-1}\right)^{T}=\mathbb{C}^{-T}$

Dyadic product of second order tensors: $\mathbb{C}=\mathbf{A}\otimes\mathbf{B}$, s.t. $\mathbb{C}.\mathbf{E}_{kl}=\mathbb{C}_{ijkl}\mathbf{E}_{ij}$

thus the components of $\mathbb{C}$ are

$$(\mathbf{A}\otimes\mathbf{B})_{ijkl}=\mathbf{E}_{ij}:(\mathbf{A}\otimes\mathbf{B}):\mathbf{E}_{kl}=\left(\mathbf{A}:\mathbf{E}_{ij}\right)\left(\mathbf{B}:\mathbf{E}_{kl}\right)=A_{ij}B_{kl}$$

$\mathbb{C}$ maps a second order tensor into another second order tensor.

– Transposition of a dyadic product:

$(\mathbf{A}\otimes\mathbf{B})^{T}=\mathbf{B}\otimes\mathbf{A}$, such that $\forall\mathbf{C},\mathbf{D}$, $\left(\mathbf{C},(\mathbf{A}\otimes\mathbf{B})^{T}.\mathbf{D}\right)=\left(\mathbf{D},(\mathbf{A}\otimes\mathbf{B}).\mathbf{C}\right)=$
$(\mathbf{B}:\mathbf{C})(\mathbf{A}:\mathbf{D})=\left(\mathbf{C},(\mathbf{B}\otimes\mathbf{A}).\mathbf{D}\right)$

For $\mathbf{C}=\mathbf{u}\otimes\mathbf{v}\rightarrow(\mathbf{A}\otimes\mathbf{B}).(\mathbf{u}\otimes\mathbf{v})=(\mathbf{u}.\mathbf{B}.\mathbf{v}).\mathbf{A}$

– Expression of a 4th order tensor in a basis: $\mathbb{C}=C_{ijkl}\mathbf{E}_{ij}\otimes\mathbf{E}_{kl}$, due to

$\mathbb{C}:\mathbf{A}=(\mathbb{C}:\mathbf{A})_{ij}\mathbf{E}_{ij}=\left\{(\mathbb{C}:\mathbf{A})_{ij}\mathbf{E}_{ij}\right\}\mathbf{E}_{ij}=A_{kl}\left[\mathbf{E}_{ij}:(\mathbb{C}:\mathbf{E}_{kl})\right]\mathbf{E}_{ij}=$
$\mathbb{C}_{ijkl}\left(\mathbf{A}:\mathbf{E}_{kl}\right)\mathbf{E}_{ij}=\mathbb{C}_{ijkl}\left(\mathbf{E}_{ij}\otimes\mathbf{E}_{kl}\right)\mathbf{A},\ \forall\mathbf{A}$

Note that $\left\{\mathbf{E}_{ij}\otimes\mathbf{E}_{kl}\right\}$ is a basis of the set of 4th order tensors.

– Square tensor product: $\left(\mathbf{A}\overline{\otimes}\mathbf{B}\right):\mathbf{C}=\mathbf{A}.\mathbf{C}.\mathbf{B}^{T}$, $\forall\mathbf{C}$, with components
$\left(\mathbf{A}\overline{\otimes}\mathbf{B}\right)_{ijkl}=\mathbf{E}_{ij}:\left(\mathbf{A}\overline{\otimes}\mathbf{B}\right):\mathbf{E}_{kl}=\mathbf{E}_{ij}:\left(\mathbf{A}.\mathbf{E}_{kl}.\mathbf{B}^{T}\right)=\left(\mathbf{e}_{i}\otimes\mathbf{e}_{j}\right):\left[(\mathbf{A}.\mathbf{e}_{k})\otimes(\mathbf{B}.\mathbf{e}_{l})\right]=$
$=\left(\mathbf{e}_{i}.\mathbf{A}.\mathbf{e}_{k}\right)\left(\mathbf{e}_{j}.\mathbf{B}.\mathbf{e}_{l}\right)=A_{ik}B_{jl}$

EXAMPLE.– $\left(\mathbf{I}\overline{\otimes}\mathbf{I}\right):\mathbf{A}=\mathbf{I}.\mathbf{A}.\mathbf{I}^{T}=\mathbf{A},\quad\forall\mathbf{A}\Rightarrow I_{4}=\mathbf{I}\overline{\otimes}\mathbf{I}$

– Transpose of the square product: $\left(\mathbf{A}\overline{\otimes}\mathbf{B}\right)^{\mathrm{T}} = \mathbf{A}^{\mathrm{T}}\overline{\otimes}\mathbf{B}^{\mathrm{T}}$, according to

$$\left(\mathbf{C},\left(\mathbf{A}\overline{\otimes}\mathbf{B}\right)^{\mathrm{T}}:\mathbf{D}\right)=\left(\mathbf{D},\left(\mathbf{A}\overline{\otimes}\mathbf{B}\right):\mathbf{C}\right)=(\mathbf{B}:\mathbf{C})(\mathbf{A}:\mathbf{D})=\left(\mathbf{C},\left(\mathbf{B}\overline{\otimes}\mathbf{A}\right):\mathbf{D}\right)$$

PROOF (Exercises).– $\left(\mathbf{H}\overline{\otimes}\mathbf{I}\right):\mathbf{A}=\mathbf{H.A}$; $\left(\mathbf{I}\overline{\otimes}\mathbf{H}^{\mathrm{T}}\right):\mathbf{A}=\mathbf{A.H}$

– Composition: $\left(\mathbf{A}\overline{\otimes}\mathbf{B}\right).\left(\mathbf{C}\overline{\otimes}\mathbf{D}\right)=(\mathbf{A.C})(\mathbf{B.D})$

– Inversion: $\left(\mathbf{A}\overline{\otimes}\mathbf{B}\right)^{-1}=\mathbf{A}^{-1}\otimes\mathbf{B}^{-1}$

– Product with a vector: $\left(\mathbf{A}\overline{\otimes}\mathbf{B}\right):(\mathbf{u}\otimes\mathbf{v})=(\mathbf{A.u})(\mathbf{B.v})$

$$(\mathbf{A}\otimes\mathbf{B}):\left(\mathbf{C}\overline{\otimes}\mathbf{D}\right)=\mathbf{A}\otimes\left(\mathbf{C}^{\mathrm{T}}.\mathbf{B.D}\right);\quad\left(\mathbf{A}\overline{\otimes}\mathbf{B}\right):(\mathbf{C}\otimes\mathbf{D})=\left(\mathbf{A.C.B}^{\mathrm{T}}\right)\otimes\mathbf{D}$$

– Transposition: $\mathbb{T}:\left(\mathbf{A}\overline{\otimes}\mathbf{B}\right)=\left(\mathbf{B}\overline{\otimes}\mathbf{A}\right):\mathbb{T}$ commutative contraction.

– Symmetrization: $\mathbb{S}:\left(\mathbf{A}\overline{\otimes}\mathbf{B}+\mathbf{B}\overline{\otimes}\mathbf{A}\right)=\left(\mathbf{A}\overline{\otimes}\mathbf{B}+\mathbf{B}\overline{\otimes}\mathbf{A}\right)$

– Antisymmetrization defined similarly, but with negative sign.

Those tensors allow extracting the symmetrical and antisymmetrical parts of any second order tensor:

$$\mathbf{T}=\frac{1}{2}\left(\mathbf{T}+\mathbf{T}^{\mathrm{T}}\right)+\frac{1}{2}\left(\mathbf{T}-\mathbf{T}^{\mathrm{T}}\right)\equiv\mathbb{S}:\mathbf{T}+\mathbb{W}:\mathbf{T}$$

Summary of the various dyadic products in component forms:

$$(\mathbf{A}\otimes\mathbf{B})_{\mathrm{ijkl}}=\mathrm{A_{ij}B_{kl}},\quad\left(\mathbf{A}\overline{\otimes}\mathbf{B}\right)_{\mathrm{ijkl}}=\mathrm{A_{ik}B_{jl}}$$

$$\left\{\left(\mathbf{A}\overline{\otimes}\mathbf{B}\right):\mathrm{T}\right\}_{\mathrm{ijkl}}=\mathrm{A_{il}B_{jk}},\quad\left\{\mathrm{T}:\left(\mathbf{A}\overline{\otimes}\mathbf{B}\right)\right\}_{\mathrm{ijkl}}=\mathrm{A_{jk}B_{il}}$$

– Tensor symmetries: A 4th order tensor $\mathbb{C}$ is said to have a *minor symmetry* iff $\mathbb{C}=\mathbb{T}:\mathbb{C}\leftrightarrow\mathrm{C_{ijkl}}=\mathbb{T}_{\mathrm{ijmn}}\mathrm{C_{mnkl}}=\delta_{\mathrm{im}}\delta_{\mathrm{jm}}\mathrm{C_{mnkl}}=\mathrm{C_{jikl}}$ (permutation of the two first indices).

– A 4th order tensor $\mathbb{C}$ has a *minor second symmetry* iff $\mathbb{C}=\mathbb{C}:\mathbb{T} \leftrightarrow C_{ijkl}=C_{ijlk}$ (permutation of the two last indices).

– A 4th order tensor $\mathbb{C}$ has a *major symmetry* iff $\mathbb{C}=\mathbb{C}^T \leftrightarrow C_{ijkl}=C_{klij}$ (permutation of the two pairs of indices).

Symmetrical and antisymmetrical parts of a 4th order tensor: they are respectively

$$\left(\mathbb{C}+\mathbb{C}^T\right)/2 \text{ and } \left(\mathbb{C}-\mathbb{C}^T\right)/2$$

EXAMPLE.– $\mathbb{I}_4, \mathbb{T}, \mathbb{S}, \mathbb{W}$ are symmetrical tensors.

1.7.5. *Principal invariants of a second order tensor*

THEOREM.– Let $(\mathbf{u},\mathbf{v},\mathbf{w})$ and $(\mathbf{a},\mathbf{b},\mathbf{c})$ be two sets of linearly independent vectors, and $\mathbf{T}$ a second order tensor. It then holds:

1) $$\frac{[\mathbf{T.u},\mathbf{v},\mathbf{w}]+[\mathbf{u},\mathbf{T.v},\mathbf{w}]+[\mathbf{u},\mathbf{v},\mathbf{T.w}]}{[\mathbf{u},\mathbf{v},\mathbf{w}]}=\frac{[\mathbf{T.a},\mathbf{b},\mathbf{c}]+[\mathbf{a},\mathbf{T.b},\mathbf{c}]+[\mathbf{a},\mathbf{b},\mathbf{T.c}]}{[\mathbf{a},\mathbf{b},\mathbf{c}]}$$

COROLLARY.– $[\mathbf{T.u},\mathbf{v},\mathbf{w}]+[\mathbf{u},\mathbf{T.v},\mathbf{w}]+[\mathbf{u},\mathbf{v},\mathbf{T.w}]\equiv I_1[\mathbf{u},\mathbf{v},\mathbf{w}]$, with I_1 some invariant scalar function of $\mathbf{T}$.

2) The second scalar invariant can be defined from the following identity

$$\frac{[\mathbf{T.u},\mathbf{T.v},\mathbf{w}]+[\mathbf{u},\mathbf{T.v},\mathbf{T.w}]+[\mathbf{T.u},\mathbf{v},\mathbf{T.w}]}{[\mathbf{u},\mathbf{v},\mathbf{w}]}=\frac{[\mathbf{T.a},\mathbf{T.b},\mathbf{c}]+[\mathbf{a},\mathbf{T.b},\mathbf{T.c}]+[\mathbf{T.a},\mathbf{b},\mathbf{T.c}]}{[\mathbf{a},\mathbf{b},\mathbf{c}]}$$
$$\Rightarrow [\mathbf{T.u},\mathbf{T.v},\mathbf{w}]+[\mathbf{u},\mathbf{T.v},\mathbf{T.w}]+[\mathbf{T.u},\mathbf{v},\mathbf{T.w}]=I_2[\mathbf{u},\mathbf{v},\mathbf{w}]$$

with I_2 some invariant scalar functions of $\mathbf{T}$.

3) $$\frac{[\mathbf{T.u},\mathbf{T.v},\mathbf{T.w}]}{[\mathbf{u},\mathbf{v},\mathbf{w}]}=\frac{[\mathbf{T.a},\mathbf{T.b},\mathbf{T.c}]}{[\mathbf{a},\mathbf{b},\mathbf{c}]}\Rightarrow[\mathbf{T.u},\mathbf{T.v},\mathbf{T.w}]=I_3[\mathbf{u},\mathbf{v},\mathbf{w}]$$

with I_3 some invariant scalar functions of $\mathbf{T}$.

DEFINITION.– The scalar valued quantities I_1, I_2, I_3 are the principal invariants of tensor $\mathbf{T}$, and any other invariant built on it will depend on these three invariants.

Relation to trace and determinant: $I_1 = \text{tr}(\mathbf{T}),\ I_2 = \text{tr}(\text{cof}\mathbf{T}),\ I_3 = \det \mathbf{T}$

These are useful for writing constitutive laws.

Properties of tensor invariants:

– I_1 is a linear operator (relative to $\mathbf{T}$).

– $\text{tr}(\mathbf{T}^T) = \text{tr}(\mathbf{T}) = \text{tr}(T_{ij}\mathbf{e}_i \otimes \mathbf{e}_j) = T_{ij}\text{tr}(\mathbf{e}_i \otimes \mathbf{e}_j) = T_{ij}\mathbf{e}_i.\mathbf{e}_j = T_{ii}$

– $\text{tr}(\mathbf{a} \otimes \mathbf{b}) = \mathbf{a}.\mathbf{b}$

– $\text{tr}(\mathbf{R}.\mathbf{S}) = \text{tr}(\mathbf{S}.\mathbf{R})$

PROOF.– Let $(\mathbf{u}, \mathbf{v}, \mathbf{w})$ be an orthonormal basis, then

$$\text{tr}(\mathbf{R}.\mathbf{S}) = (\mathbf{R}.\mathbf{S}.\mathbf{u}, \mathbf{u}) + (\mathbf{R}.\mathbf{S}.\mathbf{v}, \mathbf{v}) + (\mathbf{R}.\mathbf{S}.\mathbf{w}, \mathbf{w}) = (\mathbf{S}.\mathbf{u}, \mathbf{R}^T.\mathbf{u}) + (\mathbf{S}.\mathbf{v}, \mathbf{R}^T.\mathbf{v}) + (\mathbf{S}.\mathbf{w}, \mathbf{R}^T.\mathbf{w})$$

'with $\mathbf{R}^T.\mathbf{u} = (\mathbf{R}^T.\mathbf{u}, \mathbf{u})\mathbf{u} + (\mathbf{R}^T.\mathbf{u}, \mathbf{v})\mathbf{v} + (\mathbf{R}^T.\mathbf{u}, \mathbf{w})\mathbf{w}$, similar expressions for $\mathbf{R}^T.\mathbf{v}, \mathbf{R}^T.\mathbf{w}$, and the proofs follows by substituting those expressions into $\text{tr}(\mathbf{R}.\mathbf{S})$.

– Norm of a second order tensor: $\|\mathbf{T}\| = (\mathbf{T} : \mathbf{T})^{1/2}$.

It follows the Cauchy-Schwarz inequality: $\|\mathbf{R}.\mathbf{S}\| \le \|\mathbf{R}\|\|\mathbf{S}\|$.

$$\mathbf{R} : (\mathbf{S}.\mathbf{T}) = (\mathbf{S}^T.\mathbf{R}) : \mathbf{T} = (\mathbf{R}.\mathbf{T}^T) : \mathbf{S} = (\mathbf{T}.\mathbf{R}^T) : \mathbf{S}^T$$

allows to move tensors through the double contraction sign.

EXERCISE.– Prove those equalities.

$$\det(\alpha\mathbf{T}) = \alpha^3 \det\mathbf{T}\,;\ \det(\mathbf{T}) = \det\left(\mathbf{T}^{\mathrm{T}}\right)$$

$$\det(\mathbf{R.S})[\mathbf{u},\mathbf{v},\mathbf{w}] = \det(\mathbf{R})\det(\mathbf{S})[\mathbf{u},\mathbf{v},\mathbf{w}], \quad \forall\mathbf{u},\mathbf{v},\mathbf{w}$$

(from previous equality), thus $\det(\mathbf{R.S}) = \det(\mathbf{R})\det(\mathbf{S})$

Let $\mathbf{T}$ a given second order tensor; then there exists a vector $\mathbf{n} \neq \mathbf{0}$ such that

$$\mathbf{T.n} = \mathbf{0} \Leftrightarrow \det\mathbf{T} = 0$$

(this means the kernel of T viewed as a linear mapping is not reduced to the nil vector).

1.7.6. *Expression of the inverse of a second order tensor from the cofactor*

$$\mathrm{cof}\mathbf{T}.(\mathbf{u}\times\mathbf{v}) = \mathbf{T.u}\times\mathbf{Tv}, \quad \forall\mathbf{u},\mathbf{v}.$$

This identity then leads to

$$\begin{aligned}
&\mathrm{I}_2[\mathbf{u},\mathbf{v},\mathbf{w}] = [\mathbf{T.u},\mathbf{T.v},\mathbf{w}] + [\mathbf{v},\mathbf{T.w},\mathbf{T.u}] + [\mathbf{w},\mathbf{T.u},\mathbf{T.v}] = \\
&\left[(\mathrm{cof}\mathbf{T})^{\mathrm{T}}.\mathbf{u},\mathbf{v},\mathbf{w}\right] + \left[(\mathrm{cof}\mathbf{T})^{\mathrm{T}}.\mathbf{v},\mathbf{w},\mathbf{T.u}\right] + \left[\mathbf{w},\mathbf{u},(\mathrm{cof}\mathbf{T})^{\mathrm{T}}.\mathbf{v}\right] = \\
&= \left[(\mathrm{cof}\mathbf{T})^{\mathrm{T}}.\mathbf{u},\mathbf{v},\mathbf{w}\right] + \left[\mathbf{u},(\mathrm{cof}\mathbf{T})^{\mathrm{T}}.\mathbf{v},\mathbf{w}\right] + \left[\mathbf{u},\mathbf{v},(\mathrm{cof}\mathbf{T})^{\mathrm{T}}.\mathbf{w}\right] = \mathrm{tr}\left\{(\mathrm{cof}\mathbf{T})^{\mathrm{T}}\right\}[\mathbf{u},\mathbf{v},\mathbf{w}], \quad \forall\mathbf{u},\mathbf{v},\mathbf{w}
\end{aligned}$$

Thus, $\mathrm{I}_2 = \mathrm{tr}\left\{(\mathrm{cof}\mathbf{T})^{\mathrm{T}}\right\}$

We have here used the equalities:

$$\mathbf{u}.(\mathbf{T.v}\times\mathbf{T.w}) = \mathbf{u}.\mathrm{cof}\,\mathbf{T}.(\mathbf{v}\times\mathbf{w}) = (\mathrm{cof}\,\mathbf{T})^{\mathrm{T}}.\mathbf{u}.(\mathbf{v}\times\mathbf{w}) = \left[(\mathrm{cof}\,\mathbf{T})^{\mathrm{T}}.\mathbf{u},\mathbf{v},\mathbf{w}\right]$$

COROLLARY.–

$$(\mathbf{T.u},(\mathrm{cof}\mathbf{T}).(\mathbf{u}\times\mathbf{w})) = \det\mathbf{T}\,(\mathbf{I.u},\mathbf{v}\times\mathbf{w}) \equiv \det\mathbf{T}[\mathbf{u},\mathbf{v},\mathbf{w}] \Rightarrow \left((\mathrm{cof}\mathbf{T})^{\mathrm{T}}.\mathbf{Tu},\mathbf{v}\times\mathbf{w}\right) = ((\det\mathbf{T})\mathbf{I.u},\mathbf{v}\times\mathbf{w})$$

since any vector can be written $\mathbf{a}=\mathbf{u}\times\mathbf{w}$; it follows the identity

$$\boxed{(\mathrm{cof}\mathbf{T})^{\mathrm{T}}.\mathbf{T}=(\det\mathbf{T})\mathbf{I}}$$

– Components of the cofactor: $(\mathrm{cof}\mathbf{T})_{ij}=\frac{1}{2}\varepsilon_{imn}\varepsilon_{jpq}T_{mp}T_{nq}$

– Properties: $\mathrm{cof}(\mathbf{R}.\mathbf{S})=\mathrm{cof}(\mathbf{R})\mathrm{cof}(\mathbf{S})$; $\mathbf{T}^{-1}=\frac{1}{\det\mathbf{T}}(\mathrm{cof}\mathbf{T})^{\mathrm{T}}$ if $\det\mathbf{T}\neq 0$

– Expression of the second invariant:

$$(\mathrm{tr}\mathbf{A})_{,\mathbf{A}}=\mathbf{I},\quad (\mathrm{tr}\mathbf{A})^2{}_{,\mathbf{A}}=2(\mathrm{tr}\mathbf{A})\mathbf{I},\quad (\mathrm{tr}\mathbf{A}^2)_{,\mathbf{A}}=(A^{r}_{.s}A^{s}_{.r})_{,\mathbf{A}}=2\mathbf{A}^{\mathrm{T}},\quad (\mathrm{tr}\mathbf{A}^3)_{,\mathbf{A}}=3(\mathbf{A}^2)^{\mathrm{T}}$$

– Inverse of a product: $(\mathbf{R}.\mathbf{S})^{-1}=\mathbf{S}^{-1}.\mathbf{R}^{-1}$

– Inverse of the transpose: $(\mathbf{T}^{\mathrm{T}})^{-1}=(\mathbf{T}^{-1})^{\mathrm{T}}$

– Woodbury formula (exercise):

$$(\mathbf{T}+\mathbf{U}.\mathbf{C}.\mathbf{V})^{-1}=\mathbf{T}^{-1}-\mathbf{T}^{-1}.\mathbf{U}.(\mathbf{C}^{-1}+\mathbf{V}.\mathbf{T}^{-1}.\mathbf{U}).\mathbf{V}.\mathbf{T}^{-1}.$$

In the specific case $\mathbf{C}=\mathbf{I}$, one obtains the Sherman-Morrison formula

$$(\mathbf{T}+\mathbf{U}\otimes\mathbf{V})^{-1}=\mathbf{T}^{-1}-\frac{\mathbf{T}^{-1}.(\mathbf{U}\otimes\mathbf{V}).\mathbf{T}^{-1}}{1+\mathbf{V}.\mathbf{T}^{-1}.\mathbf{U}}$$

1.7.7. *Spectral analysis: eigenvectors and eigenvalues of second order tensors*

Let $\mathbf{T}$ be a second order tensor; $\mathbf{n}$ is an eigenvector of $\mathbf{T}$ if $\exists\lambda\in\mathbb{R}\,/\,\mathbf{T}.\mathbf{n}=\lambda\mathbf{n}$

Necessary condition: $\det(\mathbf{T}-\lambda\mathbf{I})=0$, coined the characteristic equation (scalar valued), with roots $\lambda_1,\lambda_2,\lambda_3$ (3 complex valued roots a priori: 3 real roots exist under certain conditions on $\mathbf{T}$); it follows

$$[\mathbf{T}.\mathbf{u}-\lambda\mathbf{u},\mathbf{T}.\mathbf{v}-\lambda\mathbf{v},\mathbf{T}.\mathbf{w}-\lambda\mathbf{w}]=0$$

(from the definition of the third invariant)

Thus λ is a root of the third order polynomial equation

$$\lambda^3 - I_1\lambda^2 + I_2\lambda - I_3 = 0$$

The invariants are expressible versus the eigenvalues as

$$I_1 = \mathrm{tr}\mathbf{T} = \lambda_1 + \lambda_2 + \lambda_3$$
$$I_2 = \mathrm{tr}\left(\mathrm{cof}\mathbf{T}\right) = \lambda_1\lambda_2 + \lambda_2\lambda_3 + \lambda_3\lambda_1$$
$$I_3 = \det \mathbf{T} = \lambda_1\lambda_2\lambda_3$$

THEOREM (Cayley-Hamilton theorem).– The tensor **T** is the solution of a tensor equation of the same type as the characteristic equation.

$$\mathbf{T}^3 - I_1\mathbf{T}^2 + I_2\mathbf{T} - I_3\mathbf{I} = \mathbf{0}, \quad \forall\mathbf{T}$$

PROOF.– It results from the equality $\left(\mathrm{cof}\mathbf{T}\right)^{\mathrm{T}} = I_2\mathbf{I} - \left(\mathrm{tr}\mathbf{T}\right)\mathbf{T} + \mathbf{T}^2$ scaled by **T**, and the equality $\left(\mathrm{cof}\mathbf{T}\right)^{\mathrm{T}}.\mathbf{T} = I_3\mathbf{I}$.

EXAMPLE.– Taking the trace of Cayley-Hamilton identity and considering the definition of I_2 leads to

$$I_3 = \frac{1}{6}\left[I_1^{\,3} - 3I_1\mathrm{tr}\left(\mathbf{T}^2\right) + 2\mathrm{tr}\left(\mathbf{T}^3\right)\right]$$

Expressing next $\mathrm{tr}\left(\mathbf{T}^2\right)$ and $\mathrm{tr}\left(\mathbf{T}^3\right)$ versus I_1, I_2 by the Cayley-Hamilton theorem leads to a relation between I_3 and the two first invariants I_1, I_2.

NOTE (Important case of symmetrical real valued tensors).– $\mathbf{T} = \mathbf{T}^{\mathrm{T}} \Rightarrow \mathbf{T}$ has n real eigenvalues in $\mathbb{R}^n$. Two distinct eigenvalues have two mutually orthogonal principal directions.

PROOF.– Denoting the complex conjugate with an overhead bar, one has

$$T_{ij}v_j = \lambda v_i \Leftrightarrow T_{ij}\overline{v}_j = \overline{\lambda v_i} \Leftrightarrow \left| \begin{array}{l} T_{ij}v_j\overline{v}_i = \lambda v_i\overline{v}_i = \lambda\|\mathbf{v}\|^2 \\ T_{ij}\overline{v}_j v_i = \overline{\lambda}\overline{v}_i v_i = \overline{\lambda}\|\mathbf{v}\|^2 \end{array} \right. \rightarrow (\lambda - \overline{\lambda})\|\mathbf{v}\|^2 = 0 \Rightarrow \lambda = \overline{\lambda} \in \mathbb{R}$$

Let $\lambda_1 \neq \lambda_2$ and $\mathbf{V}^1, \mathbf{V}^2$ two eigenvectors associated to the eigenvalues λ_1, λ_2:

$$\left. \begin{array}{l} T_{ij}V_j^1 = \lambda_1 V_i^1 \\ T_{ij}V_j^2 = \lambda_2 V_i^2 \end{array} \right| \Rightarrow T_{ij}V_j^1V_i^2 = \lambda_1 V_i^1V_i^2 \text{ and } T_{ij}V_j^2V_i^1 = \lambda_2 V_i^2V_i^1 \Rightarrow (\lambda_1 - \lambda_2)\mathbf{V}^1.\mathbf{V}^2 = 0$$

Since $\mathbf{T} = \mathbf{T}^T$, hence $\mathbf{V}^1 \perp \mathbf{V}^2$ due to $\lambda_1 \neq \lambda_2$.

DEFINITION.– The multiplicity of each eigenvalues is the number of linearly independent eigenvectors. So it is the dimension of the eigenspace (but *a priori* different from the order the root of the characteristic polynomial).

PROOF.– The multiplicity of each eigenvalues is the order of the corresponding root of the characteristic equation.

COROLLARY.– Any symmetrical tensor has at least one principal basis in which it takes a diagonal form. The basis vectors define the principal directions, and the eigenvalues then enter as the diagonal elements of the tensor expressed in the principal directions.

– Classification on $\mathbb{R}^3$:

i) $\lambda_1 \neq \lambda_2 \neq \lambda_3$: there are three orthogonal principal directions.

ii) $\lambda_1 = \lambda_2 \neq \lambda_3$: λ_1 is a double root, thus its eigenspace has dimension 2; $\mathbf{T}$ is a cylindrical tensor.

iii) $\lambda_1 = \lambda_2 = \lambda_3$: any vector is an eigenvector; $\mathbf{T}$ is a spherical tensor.

NOTE (Possibilities of decomposition of second order tensors).– The symmetrical and antisymmetrical parts of a second order tensor are elaborated as:

$$\mathbf{T} = \frac{1}{2}(\mathbf{T} + \mathbf{T}^T) + \frac{1}{2}(\mathbf{T} - \mathbf{T}^T) \equiv \mathbb{S} : \mathbf{T} + \mathbb{W} : \mathbf{T}$$

EXAMPLE.– In continuum mechanics, strain tensors tend to be symmetrical. Antisymmetry is usually associated to rotation (spin) of material elements.

Spherical and deviatoric parts: $\mathbf{T}=\frac{1}{3}(\text{tr}\mathbf{T})\mathbf{I}+\mathbf{T}^D$. The second order tensor $\mathbf{T}^D$ is its deviatoric part, it is traceless.

EXAMPLE.– In the kinematic description of continuum mechanics, one makes a distinction between volume changes (isotropic, thus represented by a spherical tensor) and shape distortions (at constant volume), described by deviatoric tensors.

EXAMPLE.– The pressure is defined as the trace of the stress tensor.

Antisymmetrical tensors: let $\mathbf{W}\in \text{Skw}$, the set of skew symmetrical tensors, and $\mathbf{u},\mathbf{v},\mathbf{w}\in \text{E}_n$. It follows

$$(\mathbf{u},\mathbf{W}.\mathbf{v})=(\mathbf{W}^T.\mathbf{u},\mathbf{v})=-(\mathbf{W}.\mathbf{u},\mathbf{v})=-(\mathbf{v},\mathbf{W}.\mathbf{u})$$

Letting in particular $\mathbf{v}=\mathbf{u}\rightarrow(\mathbf{u},\mathbf{W}.\mathbf{u})=0\Rightarrow \mathbf{W}.\mathbf{u}\perp\mathbf{u},\quad\forall\mathbf{u}$

Selecting $\mathbf{u}=\mathbf{e}_1,\mathbf{e}_2,\mathbf{e}_3$ one of the three basis vectors, results in $\mathbf{W}$ having three independent components, thus there exists a unique vector $\mathbf{w}$ such that

$$\mathbf{W}.\mathbf{u}=\mathbf{w}\times\mathbf{u},\quad\forall\mathbf{u}\,.$$

This vector is called the axial or dual vector; the antisymmetrical tensor and its dual are related by the equalities

$$W_{ij}=-\varepsilon_{ijk}w_k\Leftrightarrow w_i=-\tfrac{1}{2}\varepsilon_{ijk}W_{jk}$$

If $\mathbf{w}=\begin{pmatrix}w_1 & w_2 & w_3\end{pmatrix}$, then $\mathbf{W}=\begin{pmatrix}0 & -w_3 & w_2\\ w_3 & 0 & -w_1\\ -w_2 & w_1 & 0\end{pmatrix}$.

PROOF.– It holds the relation

$$\mathbf{W}:\mathbf{W}=-\text{tr}(\mathbf{W}^2)=2\text{ww}=2I_2.\ \text{tr}(\mathbf{W})=\det(\mathbf{W})=0\,.$$

1.7.8. *Orthogonal tensors*

Such tensors satisfy the property $Q^T = Q^{-1} \Leftrightarrow Q^T.Q = Q.Q^T = \mathbf{I}$. As a corollary, it holds: $\det(Q) = \pm 1$.

The set of orthogonal tensors is denoted O_3; the set of proper orthogonal transformations (with determinant equal to +1) is the special orthogonal group (it does not include reflections), denoted SO_3. It holds that $O_3 = \{\pm R \,/\, R \in SO_3\}$.

THEOREM.– Q is orthogonal iff $(Q.\mathbf{u}, Q.\mathbf{v}) = (\mathbf{u}, \mathbf{v}),\ \forall \mathbf{u}, \mathbf{v}$, so Q preserves the scalar product between two vectors.

COROLLARY.– $\|Q.\mathbf{u}\| = \|\mathbf{u}\|$, thus Q preserves the norm of any vector, and $\|Q.\mathbf{u} - Q.\mathbf{v}\| = \|\mathbf{u} - \mathbf{v}\|$.Q corresponds to a *rigid body motion* (it does not introduce deformations), and it preserves angles and distances between material elements.

– Rotation axis: if $R \neq I,\ R \in SO_3$, it is defined as the set $\{\mathbf{e} \in \mathbb{R}^3 \,/\, R.\mathbf{e} = \mathbf{e}\} \subset \mathbb{R}^3$. Such a vector $\mathbf{e}$ exists since $\det(R - \mathbf{I}) = 0$ (exercise). One can show that the axis (a set) has dimension one.

– Representation theorem for rotations: let $\mathbf{W} \in \mathrm{Skw}$, and $\{\mathbf{e}, \mathbf{q}, \mathbf{r}\}$ denote three orthonormal vectors; any tensor $R \in SO_3$ can be written

$$R = \mathbf{I} + \sin\alpha \mathbf{W} + (1 - \cos\alpha)\mathbf{W}^2 \text{ , or}$$

$$R = \mathbf{e} \times \mathbf{e} + \cos\alpha(\mathbf{I} - \mathbf{e} \times \mathbf{e}) + \sin\alpha(\mathbf{r} \times \mathbf{q} - \mathbf{q} \times \mathbf{r})$$

This leads to Rodrigues' formula giving the finite rotation of angle α and rotation axis $\mathbf{W}$ (Figure 1.9).

$$R(\mathbf{w}, \alpha) = \mathbf{I} + \frac{1}{\|\mathbf{w}\|}\sin\alpha \mathbf{W} + \frac{1}{\|\mathbf{w}\|^2}(1 - \cos\alpha)\mathbf{W}^2$$

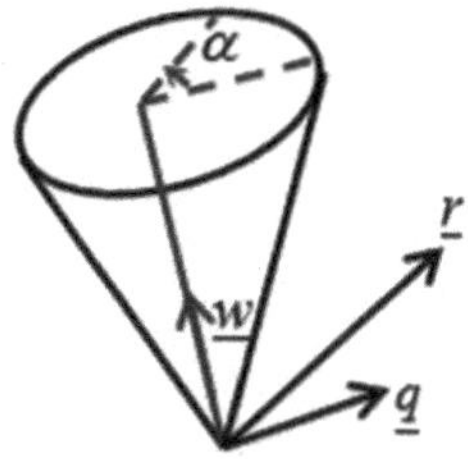

Figure 1.9. *Axis and angle of a 3D rotation*

The axial vector $\mathbf{w}$ is in the null space of $\mathbf{W} \in \mathrm{Skw}$: $\mathbf{W.w} = 0$, and α is the rotation angle in a plane orthogonal to $\mathbf{w}$, so that $\alpha = \|\mathbf{w}\|$.

We can show that $\mathrm{R}(\mathrm{t}) = \mathrm{e}^{\mathbf{W}(\mathrm{t})} = \mathbf{I} + \mathbf{W}\mathrm{t} + \frac{1}{2!}\mathbf{W}^2\mathrm{t}^2 + \ldots$

Then let form the orthonormal basis $\left(\frac{\mathbf{w}}{\|\mathbf{w}\|}, \mathbf{q}, \mathbf{r}\right)$; then the successive powers of $\mathbf{W}$ are

$$\mathbf{W} = \|\mathbf{w}\|(\mathbf{r} \times \mathbf{q} - \mathbf{q} \times \mathbf{r}) \Rightarrow \mathbf{W}^2 = \mathbf{w} \times \mathbf{w} = \|\mathbf{w}\|^2 \mathbf{I}$$

$$\mathbf{W}^3 = -\|\mathbf{w}\|^2 \mathbf{W}(\mathrm{t}) \text{ by the Cayley-Hamilton theorem,}$$

$$\mathbf{W}^4 = -\|\mathbf{w}\|^2 \mathbf{W}^2(\mathrm{t}),\ \mathbf{W}^5 = -\|\mathbf{w}\|^4 \mathbf{W}(\mathrm{t}),$$

thus proving Rodrigues' formula by a Taylor series development of the cosine and sine functions.

1.7.9. *Polar decomposition theorem for positive definite tensors*

Let $\mathbf{S}$ be a second order positive definite tensor, i.e. such that $(\mathbf{u}, \mathbf{S.u}) \geq 0,\ \forall \mathbf{u}$, and $(\mathbf{u}, \mathbf{S.u}) = 0 \Leftrightarrow \mathbf{u} = \mathbf{0}$. It turns out that only the symmetrical part of $\mathbf{S}$ plays a role.

THEOREM.– $\mathbf{S}$ is definite positive iff its principal values are positive, so iff the principal invariants of $\mathbf{S}$ are positive.

THEOREM.– If $\mathbf{S}$ is positive definite, there exists a unique tensor $\mathbf{U}$ such that $\mathbf{U}^2 = \mathbf{S} \to \mathbf{U} = \sqrt{\mathbf{S}}$.

Polar decomposition theorem: let $\mathbf{F}$ F be invertible (second order mapping), then the following decomposition holds, $\mathbf{F} = \mathrm{R}.\mathbf{U} = \mathbf{V}.\mathrm{R}$, with

$$\mathbf{U} = \sqrt{\mathbf{F}^{\mathrm{T}}.\mathbf{F}}, \quad \mathbf{V} = \sqrt{\mathbf{F}.\mathbf{F}^{\mathrm{T}}} \text{ , and R orthogonal tensor.}$$

This decomposition is unique.

In continuum mechanics, R rotates the principal directions, while $\mathbf{U}, \mathbf{V}$ represent pure deformations (generating no rotations of the principal directions), and $\mathbf{F} = \nabla_{\mathrm{X}}\mathbf{x}$ is the transformation gradient.

Let then $\left(\lambda_i, \mathbf{e}_i^*\right)$ be the spectral elements of $\mathbf{U}$ (eigenvalues and eigendirections); then $\mathbf{V}\left(\mathrm{R}.\mathbf{e}_i^*\right) = \mathrm{R}.\mathbf{U}.\mathbf{e}_i^* = \lambda_i\left(\mathrm{R}.\mathbf{e}_i^*\right) \equiv \lambda_i \mathbf{f}_i^*$, hence the spectral decomposition of the transformation gradient

$$\mathbf{F} = \lambda_i \mathbf{f}_i^* \otimes \mathbf{e}_i^*, \quad \mathbf{R} = \mathrm{R}\mathbf{I} = \mathbf{f}_i^* \otimes \mathbf{e}_i^*.$$

NOTE (Isotropic tensors).–

1) For a second order tensor $\mathbf{T} = \mathrm{T}_{ij}\mathbf{e}_i \otimes \mathbf{e}_j = \overline{\mathrm{T}}_{ij}\overline{\mathbf{e}}_i \otimes \overline{\mathbf{e}}_j$ in two different orthonormal bases $\{\mathbf{e}_i\}$ and $\{\overline{\mathbf{e}}_i\}$, related by $\overline{\mathbf{e}}_i = \mathrm{Q}.\mathbf{e}_i$, and so that the components of $\mathbf{T}$ are identical in both basis, this being a definition of isotropy, so that $\mathrm{T}_{ij} = \overline{\mathrm{T}}_{ij}$. It then follows $\mathrm{Q}.\mathbf{T}.\mathrm{Q}^{\mathrm{T}} = \mathbf{T}, \ \forall \mathrm{Q} \in \mathrm{SO}_3$, or equivalently $\mathrm{Q}.\mathbf{T} = \mathbf{T}.\mathrm{Q}$. The meaning is that the components of $\mathbf{T}$ are the same in any orthonormal basis: this defines an *isotropic tensor*.

THEOREM.– $\mathbf{T}$ is an isotropic tensor iff $\mathbf{T} = \lambda\mathbf{I}$, for some $\lambda \in \mathbb{R}$.

PROOF.– If $\mathbf{T} = \lambda\mathbf{I}$, trivial. Reciprocally, let $\mathbf{v} \in \mathbb{R}^3; \mathrm{R} \in \mathrm{SO}_3$ with axis $\mathbf{V}$; thus $\mathrm{R}.\mathbf{v} = \mathbf{v} \rightarrow \mathrm{R}.\mathbf{T}.\mathbf{v} = \mathbf{T}.\mathrm{R}.\mathbf{v}$, so that $\mathbf{T}.\mathbf{v}$ belongs to the axis of R. But, since $\dim\{\mathrm{axis}(\mathrm{R})\} = 1$, it follows the existence of $\lambda \in \mathbb{R} /\ \mathrm{T}.\mathbf{v} = \lambda\mathbf{v} = \lambda\mathbf{I}.\mathbf{v}$, thus $\mathbf{T} = \lambda\mathbf{I}$.

2) Let $\mathbb{C}$ be a 4th order tensor; $\mathbb{C}$ being isotropic means that it has the same components in any orthonormal basis. $\mathbb{C}$ is isotropic when $\mathbb{C} = \mathbb{Q}.\mathbb{C}.\mathbb{Q}^{\mathrm{T}}$, with the fourth order rotation $\mathbb{Q}$ elaborated as $\mathbb{Q} = \mathrm{Q}\overline{\otimes}\mathrm{Q},\ \mathrm{Q} \in \mathrm{SO}_3$. It holds that $\mathbb{Q}^{\mathrm{T}} : \mathbb{Q} = \mathbb{I}_4 = \mathbb{Q} : \mathbb{Q}^{\mathrm{T}}$.

$$(\mathbb{Q} : \mathbf{A}, \mathbb{Q} : \mathbf{B}) = (\mathbf{A}, \mathbf{B}), \quad \forall \mathbf{A}, \mathbf{B}.$$

THEOREM.– $\mathbb{C}$ is isotropic iff $\mathbb{C} = \lambda\mathbf{I} \otimes \mathbf{I} + 2\mu\mathbb{S} + 2\gamma\mathbb{W}$, with $\lambda, \mu, \gamma \in \mathbb{R}$, so that the components of $\mathbb{C}$ are a linear combination of the terms $\delta_{ij}\delta_{kl}$, $\delta_{ik}\delta_{jl}$, $\delta_{il}\delta_{jk}$.

In case $\mathbb{C}$ has the first or second minor symmetry, that is when $\mathbb{C} = \mathbb{S} : \mathbb{C}$, or $\mathbb{C} = \mathbb{C} : \mathbb{S}$, then one can show that $\mathbb{C} = \lambda\mathbf{I} \otimes \mathbf{I} + 2\mu\mathbb{S}$. This shall correspond to Hooke's law in isotropic elasticity, with $\mathbb{C}$ the tensor of elastic moduli, and λ, μ Lamé constants.

Isotropic tensors are important in the formulation of constitutive laws.

1.8. Differential calculus on tensors: tensor analysis

1.8.1. *Definition: Frechet and Gateaux derivatives*

Let $g(t)$ a scalar, vector or tensor valued function of scalar argument $t \in \mathbb{R}$, typically time; the notion of derivative is elaborated as follows

$$\mathrm{D}g(t) = \frac{d}{dt}g(t) = \lim_{\varepsilon \to 0}\frac{1}{\varepsilon}[g(t+\varepsilon) - g(t)]$$

This entails the expansion $g(t+\varepsilon)=g(t)+\varepsilon Dg(t)+o(\varepsilon)$, with $o(\varepsilon)/\lim\limits_{\varepsilon\to 0}\frac{o(\varepsilon)}{\varepsilon}=0$

It then appears, based on a previous expansion, that the derivative of a tensor is a tensor.

– Extension: let $g:\Omega\subset X\to Y$ be a mapping between the two topological spaces X,Y . Then g is differentiable at point $\mathbf{x}\in\Omega$ iff there exists a mapping

$$Dg(\mathbf{x})\in L(X,Y)/\quad g(\mathbf{x}+\mathbf{u})=g(\mathbf{x})+Dg(\mathbf{x})[\mathbf{u}]+o(\mathbf{u})$$

in which we have introduced the directional derivative or *Gateaux derivative* $Dg(\mathbf{x})[\mathbf{u}]$, acting on any vector $\mathbf{u}\in X$ and vector or tensor valued; note that the quantity $Dg(\mathbf{x})\in L(X,Y)$, the space of linear functions between X and Y, is the *Frechet derivative* at point $\mathbf{X}$; it is a stronger notion compared to the Gateaux derivative, since it does not involve any direction in space (it means that the existence of Frechet derivative warrants that of the Gateau derivative).

This notion is important for the purpose of writing the constitutive law.

EXAMPLE.– Let $\phi(\mathbf{v})=\mathbf{v}.\mathbf{v}\to\phi(\mathbf{v}+\mathbf{u})=\phi(\mathbf{v})+2\mathbf{v}.\mathbf{u}+o(\mathbf{u})$, thus $D\phi(\mathbf{v})[\mathbf{u}]=2\mathbf{v}.\mathbf{u}$

Here and in the sequel, the dot denotes an inner product based on a metric.

EXAMPLE.– Let G a linear mapping sending a second order tensor to its square, $G(\mathbf{T})=\mathbf{T}^2$

We have

$$G(\mathbf{T}+\mathbf{U})=\mathbf{T}^2+\mathbf{U}.\mathbf{T}+\mathbf{T}.\mathbf{U}+\mathbf{U}^2=G(\mathbf{T})+\mathbf{U}.\mathbf{T}+\mathbf{T}.\mathbf{U}+o(\mathbf{U})$$
$$\Rightarrow DG(\mathbf{T})[\mathbf{U}]=\mathbf{U}.\mathbf{T}+\mathbf{T}.\mathbf{U}$$

EXAMPLE.– $\phi(\mathbf{T})=\mathrm{tr}\mathbf{T}$; $\phi(\mathbf{T}+\mathbf{U})=\phi(\mathbf{T})+\mathrm{tr}\mathbf{U}\Rightarrow D(\mathrm{tr}\mathbf{T})[\mathbf{U}]=\mathrm{tr}\mathbf{U}=\mathbf{I}:\mathbf{U}$

EXAMPLE.– $\phi(\mathbf{T})=\det\mathbf{T}$; $\det(\mathbf{T}+\mathbf{U})=\det(\mathbf{T})+\mathrm{cof}\mathbf{T}:\mathbf{U}+\mathbf{T}:\mathrm{cof}\mathbf{U}+\det\mathbf{U}=\det(\mathbf{T})+\mathrm{cof}\mathbf{T}:\mathbf{U}+o(\mathbf{U})$

Hence, $D(\det\mathbf{T})[\mathbf{U}]=\mathrm{cof}\mathbf{T}:\mathbf{U}=(\det\mathbf{T})\mathbf{T}^{-T}:\mathbf{U}$

1.8.2. *Gradient, divergence, curl*

Let $\phi : \Omega \subset X \to \mathbb{R}$ a mapping, with a Hilbert space (generalization of Euclidean space, with existence of a norm defined from a scalar product). The gradient is a familiar operator largely used in engineering disciplines: it is used to define the strains in mechanics and the driving forces in thermal engineering and in mineral chemistry (gradient of the chemical potential). It is introduced through the more general notion of differential.

The derivative or differential $D\phi(\mathbf{x})$ is a linear mapping from X to $\mathbb{R}$; $D\phi(\mathbf{x}) \in X^* = L(X, \mathbb{R})$, thus there exists by Riesz representation theorem of linear forms a vector called the *gradient* of $\phi(\mathbf{x})$, denoted $\nabla\phi(\mathbf{x}) \in X \backslash (\nabla\phi(\mathbf{x}), \mathbf{u}) = D\phi(\mathbf{x})[\mathbf{u}]$; here, $(.,.)$ is an internal product in X.

Note that $\nabla\phi(\mathbf{x}) \in X$, but $D\phi(\mathbf{x}) \in X^*$, the dual of X.

EXAMPLE.– Let $X = \mathbb{R}^3$; $(\mathbf{u}, \mathbf{v}) := \mathbf{u}.\mathbf{v}$ the standard inner product of two vectors; we get the expansion of the scalar-valued function

$$\phi(\mathbf{x} + \mathbf{u}) = \phi(\mathbf{x}) + \nabla\phi(\mathbf{x})[\mathbf{u}] + o(\mathbf{u})$$

The components of the gradient of ϕ are obtained selecting $\mathbf{u} = \mathbf{e}_i$:

$$(\nabla\phi)_i = D\phi(\mathbf{x})[\mathbf{e}_i] = \lim_{\varepsilon \to 0}\left[\phi(\mathbf{x} + \varepsilon\mathbf{e}_i)\phi(\mathbf{x})\right] = \frac{\partial\phi}{\partial x_i}$$

which is a notation. This leads by linearity to

$$D\phi(\mathbf{x})[\mathbf{u}] = \nabla\phi.\mathbf{u} = \frac{\partial\phi}{\partial x_i} u_i$$

EXAMPLE.– Let X the space of second order tensors, with the inner product

$$(\mathbf{A}, \mathbf{B}) := \mathbf{A} : \mathbf{B} = \mathrm{tr}\left(\mathbf{A}.\mathbf{B}^T\right).$$

Thus, the gradient of any function with tensor argument writes

$$D\phi(\mathbf{T})[\mathbf{U}] = \frac{\partial \phi}{\partial \mathrm{T}} : \mathbf{U} = \frac{\partial \phi}{\partial \mathrm{T}_{ij}} : \mathrm{U}_{ij}$$

If $\mathbf{V} = \mathbf{V}(\mathbf{T})$ is a tensor-valued function, then $D\mathbf{V}(\mathbf{T})[\mathbf{U}] = \frac{\partial \mathbf{V}}{\partial \mathbf{T}} : \mathbf{U}$

The Gradient of a vector field at point $\mathbf{X}$ is then intrinsically defined from the differential as

$$\nabla \mathbf{v}(\mathbf{x}).\mathbf{u} := \mathrm{grad}\mathbf{v}(\mathbf{x}).\mathbf{u} = D\mathbf{v}(\mathbf{x}).[\mathbf{u}], \quad \forall \mathbf{u} \to D\mathbf{v}(\mathbf{x}) = \mathrm{grad}\mathbf{v}(\mathbf{x}).d\mathbf{x}, \quad \text{for } \mathbf{u} = d\mathrm{x}$$

Previous equality can further be developed as

$$\mathrm{grad}\mathbf{v}(\mathbf{x}).d\mathbf{x} = \mathbf{v}_{,i} dx^i = \left(v^j \mathbf{g}_j\right)_{,i} dx^i$$

We note that the partial derivative therein $\left(v^j \mathbf{g}_j\right)_{,i} = v^j{}_{,i} \mathbf{g}_j + v^j \mathbf{g}_{j,i}$ involves the partial derivative of the vector components and the partial derivative of the basis vectors, term $\mathbf{g}_{j,i}$, which will be treated in the next subsection.

NOTE (Case of Cartesian coordinates).– For the specific choice

$$\mathbf{u} = \mathbf{e}_j \to (\nabla \mathbf{v})_{ij} = \frac{\partial v_i}{\partial x_j} \Rightarrow \nabla \mathbf{v} = \frac{\partial v_i}{\partial x_j} \mathbf{e}_i \otimes \mathbf{e}_j$$

The divergence of a vector field is then defined as the contraction of nabla operator with the vector

$$\nabla.\mathbf{v} = \mathrm{tr}(\nabla \mathbf{v})$$

EXAMPLE.– For incompressible media (the case of many polymers in solid mechanics, the case of water in fluid mechanics), one writes a corresponding kinematic condition $\nabla.\mathbf{v} = 0$; this means physically that no net flux will occur through any volume centered around point $\mathbf{X}$.

In cartesian coordinates, it holds that

$$\frac{\partial v_i}{\partial x_j} \mathrm{tr}\left(\mathbf{e}_i \otimes \mathbf{e}_j\right) = \frac{\partial v_i}{\partial x_j} \mathbf{e}_i.\mathbf{e}_j = \frac{\partial v_i}{\partial x_j} \delta_{ij} = \frac{\partial v_i}{\partial x_i} = v_{i,i}$$

This can be extended to second order tensors: let $\mathbf{T}(\mathbf{x})$ be a second order tensor; its gradient is similarly defined as

$$\nabla\mathbf{T}(\mathbf{x})[\mathbf{u}] = \mathrm{D}\mathrm{T}(\mathbf{x})[\mathbf{u}]$$, a third order tensor

The *divergence* of $\mathbf{T}(\mathbf{x})$ is then defined as

$$(\nabla.\mathbf{T}).\mathbf{u} = \nabla.(\mathbf{T}^{\mathrm{T}}.\mathbf{u}), \quad \forall\mathbf{u} = \mathrm{Cte} \text{ (constant vector)}$$

– *Case of Cartesian coordinates:* the *gradient* of $\mathbf{T}(\mathbf{x})$ is written

$$\nabla\mathbf{T} = \frac{\partial \mathrm{T}_{ij}}{\partial x_k}\mathbf{e}_i \otimes \mathbf{e}_j \otimes \mathbf{e}_k$$

The divergence of $\mathbf{T}(\mathbf{x})$ is written $(\nabla.\mathbf{T}) = \frac{\partial \mathrm{T}_{ij}}{\partial x_j}\mathbf{e}_i$: indeed, from previous identity,

$$\frac{\partial \mathrm{T}_{ji}}{\partial x_i}u_j = (\nabla.\mathbf{T})_j u_j \Rightarrow (\nabla.\mathbf{T})_j = \frac{\partial \mathrm{T}_{ji}}{\partial x_i} = \mathrm{T}_{ji,i}$$

– *Curl of a vector*: it is based on the gradient, as

$$\nabla\times\mathbf{v} \rightarrow (\nabla\times\mathbf{v})\times\mathbf{u} = \left[\nabla\mathbf{v} - \nabla^{\mathrm{T}}\mathbf{v}\right].\mathbf{u}, \quad \forall\mathbf{u}$$

Observe then that $\left[\nabla\mathbf{v} - \nabla^{\mathrm{T}}\mathbf{v}\right]$ is antisymmetrical, with axial vector $\nabla\times\mathbf{v}$, thus it holds in Cartesian coordinates

$$\nabla\times\mathbf{v} = \frac{\partial v_k}{\partial x_j}\mathbf{e}_i \times \mathbf{e}_k = \varepsilon_{ijk}\frac{\partial v_k}{\partial x_j}\mathbf{e}_i$$

EXAMPLE.–

– *Curl of a second order tensor*: $\nabla\times\mathbf{T} \rightarrow (\nabla\times\mathbf{T}).\mathbf{u} = \nabla\times(\mathbf{T}^{\mathrm{T}}.\mathbf{u}), \forall\mathbf{u} = \mathrm{Cte}$

In Cartesian coordinates, we obtain the second order tensor

$$\nabla \times \mathbf{T} = \frac{\partial T_{js}}{\partial x_r}\left(\mathbf{e}_i \times \mathbf{e}_j\right) \otimes \mathbf{e}_j$$

– *Laplacian of a scalar function* $\phi(\mathbf{x})$: it is defined as $\nabla^2\phi(\mathbf{x}) = \nabla.(\nabla\phi)$

In Cartesian coordinates, it holds that $\left(\nabla^2\phi\right)_{ij} = \frac{\partial^2\phi}{\partial x_i \partial x_j}$

EXAMPLE.– A *harmonic function* is such that its Laplacian vanishes.

Laplacian of a tensor field $\mathbf{T}(\mathbf{x})$: it is defined as

$$\left(\nabla^2\mathbf{T}\right):\mathbf{H} = \nabla^2(\mathbf{T}:\mathbf{H}),\ \ \forall\mathbf{H} \to \left(\nabla^2\mathbf{T}\right)_{ij} = \frac{\partial^2 T_{ij}}{\partial x_k \partial x_k}$$

when Cartesian coordinates are used.

Recall that the evaluation of differential operators requires the consideration of the variation of the natural basis with spatial position, a general situation to be treated in the next subsection, only the specific situation of Cartesian coordinates lead to simple results due to the fact that the natural basis does not change.

EXERCISES.–

1) $\nabla \times \nabla\phi = 0$ (the curl of the gradient vanishes).

2) $\frac{1}{2}\nabla(\mathbf{u}.\mathbf{u}) = (\nabla\mathbf{u})^T.\mathbf{u}$

3) $\nabla.(\nabla\mathbf{u}.\mathbf{v}) = (\nabla\mathbf{u})^T : \nabla\mathbf{v} + \mathbf{v}.\left[\nabla(\nabla.\mathbf{u})\right]$

4) $\nabla.(\nabla\mathbf{u}) = \nabla(\nabla.\mathbf{u}) - \nabla \times (\nabla \times \mathbf{u})$

5) $\nabla.(\mathbf{u} \times \mathrm{v}) = \mathbf{v}.(\nabla \times \mathbf{u}) - \mathbf{u}.(\nabla \times \mathbf{uv})$

6) Let $\mathbf{W} \in \mathrm{Skw}$, with axial vector $\mathbf{w}$, then $\nabla \times \mathbf{W} = (\nabla.\mathbf{w})\mathbf{I} - \nabla\mathbf{w}$

1.8.3. *Useful identities in tensor analysis*

$$\mathrm{grad}(f\mathbf{u}) = \mathbf{u} \otimes \mathrm{grad} f + f \mathrm{grad}\mathbf{u}, \qquad \mathrm{div}(\mathbf{u}.\mathbf{A}) = \mathbf{A} : \mathrm{grad}\mathbf{u} + \mathbf{u}.\mathrm{div}\mathbf{A}$$

The Gauss-Green theorem (or divergence theorem) for vectors and second order tensors:

$$\int_V \mathrm{grad}\mathbf{u} dV = \int_{\partial V} \mathbf{u} \otimes \mathbf{n} dS, \quad \int_V \mathrm{grad}\mathbf{A} dV = \int_{\partial V} \mathbf{A} \otimes \mathbf{n} dS$$

with $\mathbf{n}dS$ the infinitesimal surface area element ($\mathbf{n}$ is the unit exterior normal to the surface ∂V). Taking the trace of previous identities immediately leads to the well-known relations

$$\int_V \mathrm{div}\mathbf{u} dV = \int_{\partial V} \mathbf{u}.\mathbf{n} dS, \quad \int_V \mathrm{div}\mathbf{A} dV = \int_{\partial V} \mathbf{A}.\mathbf{n} dS$$

Stokes' theorem is written (here $\mathbf{t}$ denotes the unit tangent vector to the contour $C = \partial S$ to the closed surface S)

$$\int_S \mathrm{rot}\mathbf{A}.\mathbf{n} dS = \int_C \mathbf{A}.\mathrm{t}\, dl$$

Those identities are tensor identities, thus are valid for any coordinates.

REMARK.– A more explicit notation than the nabla operator ∇ is sometimes used, such as grad or Grad. Very often, due to the symmetry in the geometry and loading, the more adequate coordinates are not Cartesian (think of polar, cylindrical, or spherical coordinates); it is then useful to understand the mechanisms for computing the gradient operator (and other differential operators) in such coordinates.

1.9. Differential operators in curvilinear coordinates

One has here to consider the variance of tensors, so that the position of indices in vector operations is important. We adopt covariant basis vectors $\{\mathbf{g}_k\}$, and the dual basis is $\{\mathbf{g}^k\}$.

1.9.1. *The natural basis*

The covariant basis vectors can be defined as the partial derivatives of the spatial position with respect to a set of curvilinear coordinates u^i, $\mathbf{g}_i := \dfrac{\partial \mathbf{OM}}{\partial u^i}$, so that each vector of this *natural basis* is tangent to the corresponding ith coordinate line $i = Ct$ (Figure 1.10). Imagine a network of lines drawn on a soft deformable sample like plasticine, which shall deform to a new configuration with distorted lines.

Note that the superscript i appearing there behaves as a subscript in the result. We assume that the mapping between Cartesian to curvilinear coordinates is invertible, so that its Jacobian is non-zero.

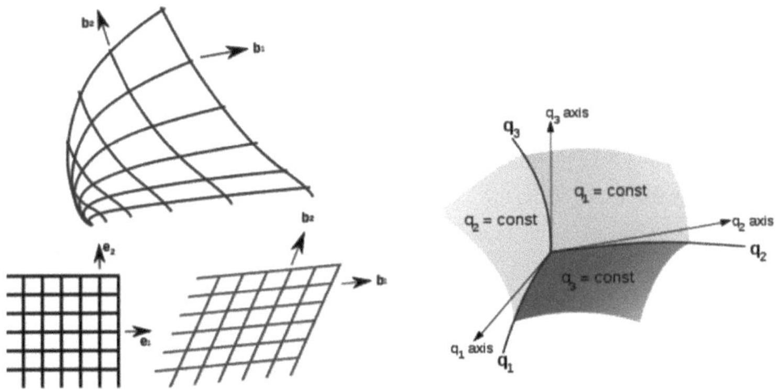

Curvilinear, Cartesian
and affine coordinates.
Coordinate surfaces, coordinate lines, and
coordinate axes of curvilinear coordinates.

Figure 1.10. *Iso-coordinate lines and natural basis. For a color version of this figure, see www.iste.co.uk/ganghoffer/biomechanics.zip*

EXAMPLE (Curvilinear Coordinates).– The Lagrangian coordinate is a label of each material point (particle), which remains attached to the particle during its motion; the Eulerian position can also be adopted as a specific curvilinear coordinate system.

The identity tensor maps the covariant basis vector $\mathbf{g}_i$ to itself, thus it decomposes as

$$\mathbf{I} := \mathbf{g}_i \otimes \mathbf{g}^i = \mathbf{g}^i \otimes \mathbf{g}_i = g_{ij}\mathbf{g}^i \otimes \mathbf{g}^j = g^{ij}\mathbf{g}_i \otimes \mathbf{g}_i = \delta_i^{.j}\mathbf{g}^i \otimes \mathbf{g}_j = \delta^i_{.j}\mathbf{g}_i \otimes \mathbf{g}^j$$

with the metric tensor **g** with components given by the internal product of the covariant base vectors, $g_{ij} = \mathbf{g}_i.\mathbf{g}_j$.

Let us then express the scalar product, the vector product and the mixed product:

– Scalar product: $\mathbf{a.b} = (a_i\mathbf{g}^i).(b^j\mathbf{g}_j) = a_i b^j \mathbf{g}^i.\mathbf{g}_j = a_i b^j \delta_i^{.j} = a_i b^i$

This is the same expression as in Cartesian coordinates, but involving the product of one covariant component with a contravariant component. Recall that the choice of covariant versus contravariant components in the decomposition of vectors **a**, **b** is immaterial.

– Vector product: $\mathbf{c} := \mathbf{a} \times \mathbf{b} = -\mathbf{b} \times \mathbf{a} = \varepsilon_{ijk} a^i b^j \mathbf{g}^k = \varepsilon^{ijk} a_i b_j \mathbf{g}_k$

– Mixed product: $[\mathbf{a}, \mathbf{b}, \mathbf{c}] = \mathbf{a}.(\mathbf{b} \times \mathbf{c}) = \varepsilon_{ijk} a^i b^j c^k = \varepsilon^{ijk} a_i b_j c_k$

A second order tensor will decompose into various bases as

$$\mathbf{S} = S_{ij}\mathbf{g}^i \otimes \mathbf{g}^j = S^{ij}\mathbf{g}_i \otimes \mathbf{g}_j = S_i^{.j}\mathbf{g}^i \otimes \mathbf{g}_j = S^i_{.j}\mathbf{g}_i \otimes \mathbf{g}^j$$

The different components of **S** are mutually related by contractions with the metric tensor. This traduces the fact that the metric measures variation (along position) of angles and distances.

We do not treat here algebraic operations on tensors, which are actually a mere extension of the relations previously written in the case of Cartesian coordinates; we shall instead focus on differential operations on tensors.

1.9.2. *Covariant derivative*

It is clear that contrary to Cartesian coordinates, the natural basis varies from point to point, which entails that the tangent space at one point is different from the tangent space in a neighboring point. This makes the comparison of vectors and tensors at different points more difficult, thus the essential problem is to define a way to connect tensor fields at different points.

Let then consider a scalar valued function: since any material point M can be parameterized by the $\left(u^i\right)$, we have $f(M)=f\left(u^i\right)$. Thus the differential

$$df = df\left(u^i\right) = \frac{\partial f}{\partial u^i} du^i$$

(here we denote the partial derivative $\partial_i = \frac{\partial}{\partial u^i}$)

For a vector field,

$$\mathbf{V} = V^i\left(u^j\right)\mathbf{g}_i \rightarrow d\mathbf{V} = dV^i\left(u^j\right)\mathbf{g}_i + V^i\left(u^j\right)d\mathbf{g}_i \equiv \frac{\partial V^i\left(u^j\right)}{\partial u^j} du^j + V^i\left(u^j\right)d\mathbf{g}_i$$

using previous result for a scalar valued function. The difficulty is to evaluate the variation of the basis vectors, term $d\mathbf{g}_i$ reflecting the variation of the basis vector with position.

Let assume that the variation of $d\mathbf{g}_i$ is linear in the variation of the curvilinear coordinates,

$$d\mathbf{g}_i = \Gamma^j_{ki}\mathbf{g}_j du^k$$

so that the variation of the base vector is linear in the variation of the coordinates and in the basis vectors; the symbols Γ^j_{ki} are called Christoffel symbols (of second kind); *they are not the components of a tensor*. They are symmetrical with respect to the lower indices, $\Gamma^k_{ij} = \Gamma^k_{ji}$ (check using the equality $\partial_{kj}\mathbf{OM} = \partial_{jk}\mathbf{OM}$). The combination of (1) and (2) leads to

$$d\mathbf{V} = \left(\partial_l V_i + \Gamma^i_{kl}V^k\right)du^l\mathbf{g}_i \equiv \frac{DV^i}{Du^l} du^l\mathbf{g}_i$$

where in we have introduced *the covariant derivative*

$$D_l V^i := \left(\partial_l V^i + \Gamma^i_{kl}V^k\right) \equiv \frac{DV^i}{Du^l}$$

so that $d\mathbf{V} = D_l V^i du^l \mathbf{g}_i \equiv DV^i\mathbf{g}_i$, with $DV^i := D_l V^i du^l$

PROOF.– The covariant derivative (contrary to the partial derivative) behaves as a contravariant vector. It is thus the right tool to define the gradient of a vector field (and all derived operators, divergence, Laplacian, curl).

The covariant derivative is the sum of the partial derivative and an additional convective term (that would vanish for Cartesian coordinates since a vector is invariant by parallel transport) traducing the way the base vectors are transported along the iso-coordinate lines; in short, they quantify the geometrical distortion of space.

From the previous relation, the Christoffel symbols can be expressed as (the dot indicates the position of indices)

$$d\mathbf{g}_i = \Gamma^j_{ki}\mathbf{g}_j du^k \rightarrow \Gamma^j_{ki} = \frac{\partial \mathbf{g}_i}{\partial u^k}.\mathbf{g}^j$$

They accordingly represent the projections of the partial derivatives of the basis vectors with respect to the curvilinear coordinates. One can further define the Christoffel symbols of the first kind by contraction with the metric,

$$\Gamma_{ijk} = g_{jl}\Gamma^l_{i.k}$$

EXAMPLE.– Sherical coordinates (r, θ, φ). The fixed Cartesian basis is denoted $(\mathbf{e}_1^0, \mathbf{e}_2^0, \mathbf{e}_3^0)$. Figure 1.11.

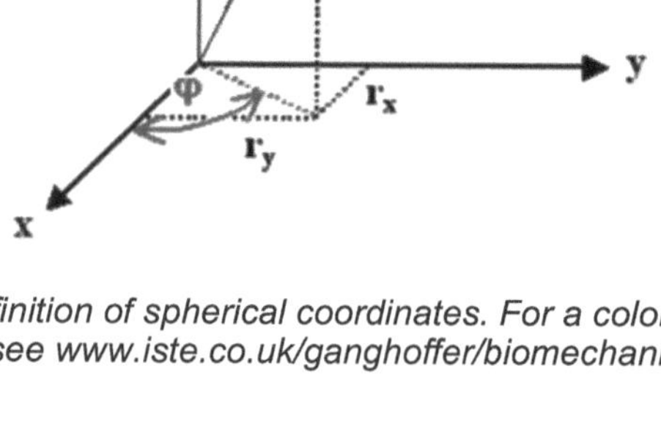

Figure 1.11. *Definition of spherical coordinates. For a color version of this figure, see www.iste.co.uk/ganghoffer/biomechanics.zip*

The natural basis is firmed by the three vectors (EXERCISE: check)

$$\begin{cases} \mathbf{g}_1 = \dfrac{\partial \mathbf{OM}}{\partial r} = \sin\theta\cos\varphi\mathbf{e}_1^0 + \sin\theta\sin\varphi\mathbf{e}_2^0 + \cos\theta\mathbf{e}_3^0 \\ \mathbf{g}_2 = \dfrac{\partial \mathbf{OM}}{\partial \theta} = r\cos\theta\cos\varphi\mathbf{e}_1^0 + r\cos\theta\sin\varphi\mathbf{e}_2^0 - r\sin\theta\mathbf{e}_3^0 \\ \mathbf{g}_3 = \dfrac{\partial \mathbf{OM}}{\partial \varphi} = -r\sin\theta\sin\varphi\mathbf{e}_1^0 + r\sin\theta\cos\varphi\mathbf{e}_2^0 + r\cos\theta\mathbf{e}_3^0 \end{cases}$$

The metric tensor components are easily calculated as (the metric tensor is here diagonal)

$$g_{11} = 1,\ \ g_{22} = r^2,\ \ g_{33} = r^2\sin^2\theta$$

This entails when mapped back into the natural basis the differentials of the base vectors

$$\begin{cases} d\mathbf{g}_1 = \dfrac{d\theta}{r}\mathbf{g}_2 + \dfrac{d\varphi}{r}\mathbf{g}_3 \\ d\mathbf{g}_2 = -rd\theta\mathbf{g}_1 + \dfrac{dr}{r}\mathbf{g}_2 + \cot\theta d\varphi\mathbf{g}_3 \\ d\mathbf{g}_3 = -r\sin^2\theta d\varphi\mathbf{g}_1 - \sin\theta\cos\theta d\varphi\mathbf{g}_2 + \left(\dfrac{dr}{r}\cot\theta d\theta\right)\mathbf{g}_3 \end{cases}$$

One immediately identifies the non-nil Christoffel symbols of the second kind

$$\Gamma_{22}^1 = -r;\ \ \Gamma_{21}^2 = \Gamma_{12}^2 = \frac{1}{r};\ \ \Gamma_{13}^1 = \Gamma_{31}^1 = \frac{1}{r};\ \ \Gamma_{33}^1 = -r\sin^2\theta;$$
$$\Gamma_{33}^2 = -\sin\theta\cos\theta;\ \ \Gamma_{23}^3 = \Gamma_{32}^3 = -\cot\theta;$$

and the Christoffel symbols of the first kind:

$$\Gamma_{212} = -r;\ \ \Gamma_{221} = \Gamma_{122} = r;\ \ \Gamma_{323} = -\left(r^2/2\right)\sin 2\theta;\ \ \Gamma_{313} = -r\sin^2\theta;$$
$$\Gamma_{133} = \Gamma_{331} = r\sin^2\theta;\ \ \Gamma_{332} = \Gamma_{233} = \left(r^2/2\right)\sin 2\theta$$

The use of the Christoffel symbols as a way to connect the basis at different points is next made more accurate.

1.9.3. *Parallel transport, connection and covariant differentiation*

We want to define a rule for transporting vector and tensor fields along iso-coordinate lines. For a vector field $\mathbf{V}(\mathrm{M})$, let us define the parallel transport of – $\mathbf{V}(\mathrm{P})$ to point Q as the vector field;

– $\tilde{\mathbf{V}}(\mathrm{Q})$, such that the following rules are satisfied:

i) $\widetilde{\left(\mathrm{V}^{\mathrm{i}}+\mathrm{W}^{\mathrm{i}}\right)}(\mathrm{M}+\Delta\mathrm{M})=\widetilde{\mathrm{V}^{\mathrm{i}}}(\mathrm{M}+\Delta\mathrm{M})+\widetilde{\mathrm{W}^{\mathrm{i}}}(\mathrm{M}+\Delta\mathrm{M})$ linearity

ii) $\widetilde{\mathrm{V}^{\mathrm{i}}}(\mathrm{M}+\Delta\mathrm{M})-\widetilde{\mathrm{V}^{\mathrm{i}}}(\mathrm{M})=\mathrm{O}(\Delta\mathrm{M})$ comparison of orders of magnitudes.

One can show that this entails the following elaboration of the covariant derivative

$$\left(\frac{\partial \mathrm{V}^{\mathrm{i}}}{\partial \mathrm{u}^{\mathrm{j}}}+\mathrm{V}^{\mathrm{k}}\Gamma^{\mathrm{i}}_{\mathrm{jk}}\right)\frac{\partial}{\partial \mathrm{u}^{\mathrm{i}}}=\lim_{\Delta\mathrm{u}^{\mathrm{j}}\to 0}\frac{\mathrm{V}^{\mathrm{i}}\left(\mathrm{u}+\Delta\mathrm{u}\right)-\widetilde{\mathrm{V}}^{\mathrm{i}}(\mathrm{u})}{\Delta\mathrm{u}^{\mathrm{j}}}\frac{\partial}{\partial \mathrm{u}^{\mathrm{i}}}$$

Here, $\frac{\partial}{\partial \mathrm{u}^{\mathrm{i}}}$ is a basis vector viewed as an operator (acting on functions on the right). The previous definition of the difference $\tilde{\mathbf{V}}(\mathrm{Q})-\mathbf{V}(\mathrm{Q})$ based on the parallel transport gives a vector, since it is built from two vectors evaluated at the same point: one has indeed transported the vector field from point P to point Q (Figure 1.12), vector $\tilde{\mathbf{V}}(\mathrm{Q})$, which is compared to the original vector field $\mathbf{V}(\mathrm{Q})$ at the same point.

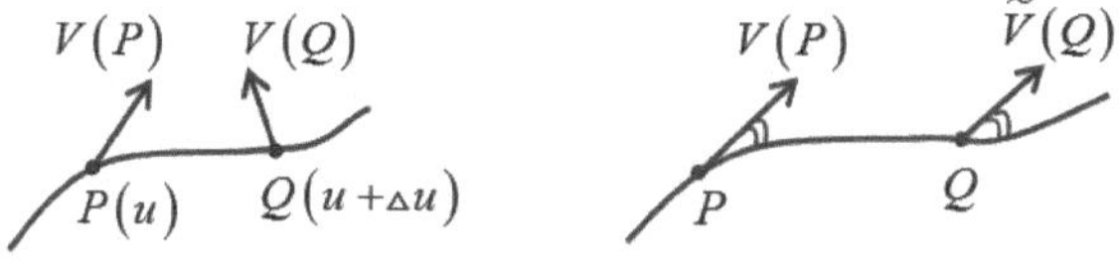

Figure 1.12. *Parallel transport of a vector field along a path (u is the curvilinear coordinate)*

NOTE (Important notation).– In the sequel, the covariant derivative will be denoted with a vertical bar (to distinguish it from the partial derivative, denoted by a comma), viz.

$$V^i_{|j} := \frac{\partial V^i}{\partial u^j} + V^k \Gamma^i_{jk}$$

The difference $\widetilde{\mathbf{V}}(Q) - \mathbf{V}(Q) \equiv d\mathbf{V}(Q)$ is the absolute differential of vector $\mathbf{V}$ at point Q. It is expressible versus the covariant derivative.

The Christoffel symbols connect the tangent spaces at different points, thus they bear the name of *connection coefficients*; one says that they define a *connection*. In metric spaces, they are expressible in terms of the metric coefficients, as

$$\Gamma^i_{kj} = g^{il}\Gamma_{klj} = \frac{1}{2} g^{il}\left(\partial_k g_{jl} + \partial_j g_{lk} + \partial_{kl} g_{kj}\right)$$

EXAMPLE.– Parallel transport on a spherical triangle (Figure 1.13).

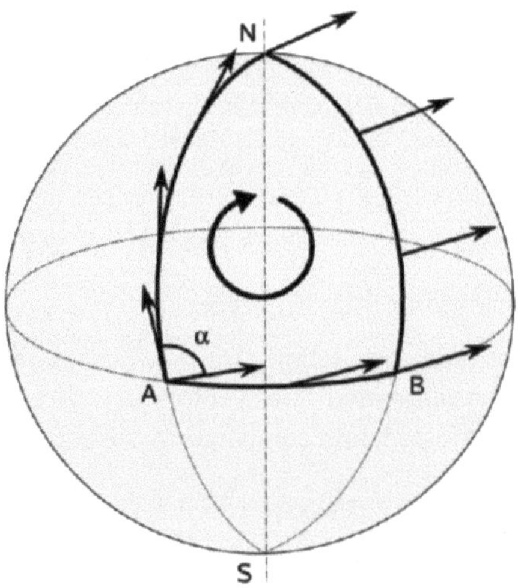

Figure 1.13. *Parallel transport of a vector field along the sphere*

The parallel transport of a vector from A to B following two different paths: the direct path along the equator and the path going through the North pole (point N) each lead to a different result (vector). The rule adopted here for the transport is that the vector makes a constant angle with the tangent vector to the trajectory. The observed twist is the sign of curvature (this angle would vanish on a flat triangle).

The covariant derivative of the basis vector $\mathbf{g}_i$ is evaluated from the relation $\mathbf{g}_{i,j} = \frac{\partial \mathbf{g}_i}{\partial \theta^j} = \Gamma^r_{ij}\mathbf{g}_r$, so that one obtains

$$\mathbf{g}_{i|j} = \mathbf{g}_{i,j} - \Gamma^r_{ij}\mathbf{g}_r = \mathbf{0}$$

This entails that the basis vectors can be considered as fixed during covariant differentiation!

1.9.4. *Application: differential operators in curvilinear coordinates*

The *Nabla operator* is defined as $\nabla = \frac{\partial}{\partial \theta^k}\mathbf{g}^k$. It entails all other differential operators:

1) The gradient of a scalar valued function of the curvilinear coordinates is evaluated as

$$\nabla F\left(\theta^i\right) = \frac{\partial F}{\partial \theta^i}\mathbf{g}^k = F_{,k}\mathbf{g}^k$$

– In case the curvilinear coordinates are Cartesian coordinates $\theta^i = x^i$, we obtain

$$\nabla F\left(x^i\right) = \frac{\partial F}{\partial x^i}\mathbf{e}_i = F_{,k}\mathbf{e}_k$$

The divergence and curl of a vector are successively given by

$$\nabla.\mathbf{u} = \frac{\partial \mathbf{u}}{\partial \theta^i}\mathbf{g}^k = \mathbf{u}_{,k}\mathbf{g}^k; \quad \nabla \times \mathbf{u} = \mathbf{g}^k \frac{\partial}{\partial \theta^i} \times \mathbf{u} = \mathbf{g}^k \times \mathbf{u}_{,k} = \text{rot}\mathbf{u}$$

The *physical meaning* of the curl of a vector is the circulation of this vector along a closed curve enclosing the point where the curl is evaluated: the limit for circuits of vanishing area of the ratio of the circulation to the area delimited by the closed circuit gives the curl at the considered point.

Note that the partial derivative of vectors and tensors coincide with the covariant derivative, due to the previous equality $\mathbf{g}_{i|j} = \mathbf{0}$; this entails the following relations

$$\text{rot } \mathbf{u} = \mathbf{g}^k \times \mathbf{u}_{,k} = \mathbf{g}^k \times \mathbf{u}_{|k} = \mathbf{g}^k \times \left(u_j \mathbf{g}^j\right)_{|k} = \mathbf{g}^k \times u_{j|k} \mathbf{g}^j$$

Note that due to the symmetry of the covariant indices of the Christoffel symbols, the curl of a vector field simply writes in terms of its covariant components as

$$(\text{curl}\mathbf{V})_{ij} = \frac{\partial V_i}{\partial u^j} - \frac{\partial V_j}{\partial u^i}$$

so that the same expression is obtained as in Cartesian coordinates.

EXERCISE.– Check previous equalities.

The gradient of a vector field $\mathbf{a} = a^k \mathbf{g}_k = a_k \mathbf{g}^k$ is evaluated as a covariant derivative based on the Christoffel symbols: the covariant derivative can either be defined with respect to the contravariant components of **a**, as

$$a^k_{|i} = \frac{\partial a^k}{\partial u^i} + \Gamma^k_{i.j} a^j$$

or with respect to its covariant components (note the minus sign)

$$a_{k|i} = \frac{\partial a_k}{\partial u^i} - \Gamma^j_{k.i} a_j$$

2) The gradient of a vector is a second order tensor, viz.

$$\text{grad}\mathbf{u} = \frac{\partial \mathbf{u}}{\partial \theta^k} \otimes \mathbf{g}^k = \mathbf{u}_{,k} \otimes \mathbf{g}^k = u_{i|k} \mathbf{g}^i \otimes \mathbf{g}^k$$

The divergence of $\mathbf{u}$ is then defined as the trace of the gradient, viz.

$$\text{div}\mathbf{u} = \nabla.\mathbf{u} = \text{tr}\left(\frac{\partial \mathbf{u}}{\partial \theta^k} \otimes \mathbf{g}^k\right) = \mathbf{u}_{,k} \cdot \mathbf{g}^k = u^k_{|k}$$

The *physical meaning* of the divergence is that of a flux across the surface enclosing an arbitrary volume element including point where div is evaluated.

3) Similarly, the gradient of a second order tensor leads to a third order tensor

$$\text{grad}\mathbf{A} = \frac{\partial \mathbf{A}}{\partial \theta^k} \otimes \mathbf{g}^k = \mathbf{A}_{ij,k} \otimes \mathbf{g}^k = \mathbf{A}_{ij|k}\mathbf{g}^i \otimes \mathbf{g}^j \otimes \mathbf{g}^k$$

Hence, the gradient increases the order by one.

The divergence of a second order tensor is defined as the trace of its gradient, so that

$$\text{div}\mathbf{A} = \text{tr}\left(\frac{\partial \mathbf{A}}{\partial \theta^k} \otimes \mathbf{g}^k\right) = \mathbf{A}_{,k}.\mathbf{g}^k = \mathbf{A}_{,k}.\mathbf{g}^k = \text{grad}\mathbf{A} : \left(\mathbf{g}_i \otimes \mathbf{g}^k\right)$$

Accordingly, the divergence reduces the order of the tensor by one.

EXERCISE.– Check the identity $\text{div}\mathbf{A} = A^{ij}{}_{|j}\mathbf{g}_i = A^{j}_{\cdot|j}$, using the identity $\mathbf{g}_{i|j} = 0$

1.9.5. *Expression of the covariant derivatives of the components of tensors*

The covariant derivative of the contravariant component S^{mn} is given by

$$S^{mn}{}_{|i} = S^{mn}{}_{,i} + \Gamma^m_{ir} S^{rn} + \Gamma^n_{ir} S^{mr}$$

In a system of curvilinear coordinates $\{\theta^i\}$, we have $\mathbf{S} = \mathbf{S}(\theta^i)$, and the partial derivative of $\mathbf{S}$ w.r.t. to θ^i is evaluated as

$$\frac{\partial \mathbf{S}}{\partial \theta^i} = S^{mn}{}_{,i}\mathbf{g}_m \otimes \mathbf{g}_n + S^{mn}\mathbf{g}_{m,i} \otimes \mathbf{g}_n + S^{mn}\mathbf{g}_m \otimes \mathbf{g}_{n,i}$$

with the partial derivatives of the basis vectors given by

$\mathbf{g}_{i,j} = \dfrac{\partial \mathbf{g}_i}{\partial \theta^j} = \Gamma^r_{ij}\mathbf{g}_r$, so that we obtain in turn

$$\mathbf{S}_{,i} := \frac{\partial \mathbf{S}}{\partial \theta^i} = \left(S^{mn}{}_{,i} + \Gamma^m_{ir} S^{rn} + \Gamma^n_{ir} S^{mr} \right) \mathbf{g}_m \otimes \mathbf{g}_n$$

Thus, the partial derivative of $\mathbf{S}$ is a tensor the components of which being the covariant derivatives of its components.

Recall that the covariant derivative of $\mathbf{S}$ is simply written as the partial derivative, viz. one has

$$\mathbf{S}_{,i} := \frac{\partial \mathbf{S}}{\partial \theta^i} = S^{mn}{}_{|i} \mathbf{g}_m \otimes \mathbf{g}_n \equiv \left(S^{mn} \mathbf{g}_m \otimes \mathbf{g}_n \right)_{|i} = S^{mn}{}_{|i} \mathbf{g}_m \otimes \mathbf{g}_n$$

EXAMPLE.– Gradient, divergence, and Laplacian in cylindrical coordinates

Cylindrical coordinates $\left(u^1 = r, u^2 = \theta, u^3 = z \right)$ are considered (Figure 1.14).

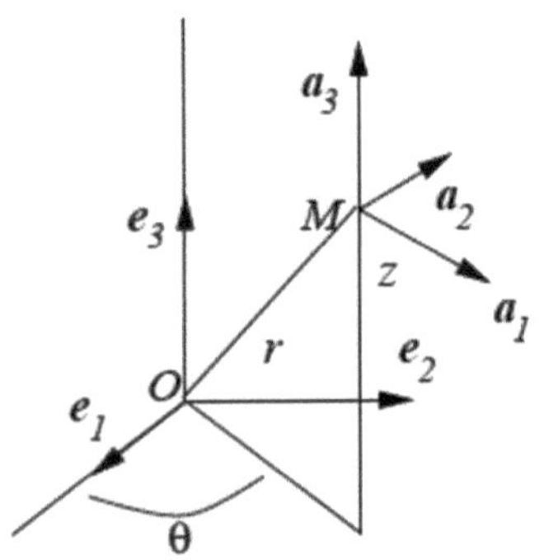

Figure 1.14. *Cylindrical coordinates*

The differential of the position vector is written

$$d(\mathbf{OM}) = dr\ \mathbf{a}_1 + d\theta\ \mathbf{a}_2 + dz\ \mathbf{a}_3$$

This entails the basis vectors

$$\mathbf{a}_1 = \frac{\partial \mathbf{OM}}{\partial r} = \begin{pmatrix} \cos\theta \\ \sin\theta \\ 0 \end{pmatrix};\ \mathbf{a}_1 = \frac{\partial \mathbf{OM}}{\partial \theta} = r \begin{pmatrix} -\sin\theta \\ \cos\theta \\ 0 \end{pmatrix};\ \mathbf{a}_1 = \frac{\partial \mathbf{OM}}{\partial z} = \begin{pmatrix} 0 \\ 0 \\ 1 \end{pmatrix}$$

The metric tensor is then evaluated in terms of its components (here it is diagonal)

$$g_{ij} = \mathbf{a}_i.\mathbf{a}_j = \begin{pmatrix} 1 & 0 & 0 \\ 0 & r^2 & 0 \\ 0 & 0 & 1 \end{pmatrix}; \; g^{ij} = \begin{pmatrix} 1 & 0 & 0 \\ 0 & 1/r^2 & 0 \\ 0 & 0 & 1 \end{pmatrix}$$

This leads to the Christoffel symbols of the first and second kind (Exercise: check)

$$\Gamma_{212} = -r, \; \Gamma_{122} = \Gamma_{221} = r, \; \Gamma_{i3j} = 0, \; \Gamma^{2}_{1.2} = 1/r = \Gamma^{2}_{2.1}, \; \Gamma^{3}_{i.j} = 0$$

The gradient of a scalar function is given by

$$\mathrm{grad}f(r,\theta,z) / \; df = \mathrm{grad}f(r,\theta,z).d\mathbf{OM} \rightarrow \mathrm{grad}f(r,\theta,z) = \begin{pmatrix} \dfrac{\partial f}{\partial r} \\ \dfrac{1}{r}\dfrac{\partial f}{\partial \theta} \\ \dfrac{\partial f}{\partial z} \end{pmatrix}$$

The gradient of a vector field

$$\mathbf{u}(r,\theta,z) = u_r(r,\theta,z)\mathbf{a}_1 + u_\theta(r,\theta,z)\mathbf{a}_2 + u_z(r,\theta,z)\mathbf{a}_3$$

is given by

$$\mathrm{grad}\mathbf{u}(r,\theta,z) = \begin{pmatrix} \partial_r u_r & \dfrac{1}{r}(\partial_\theta u_r - u_\theta) & \partial_z u_r \\ \partial_r u_\theta & \dfrac{1}{r}(\partial_\theta u_\theta + u_r) & \partial_z u_\theta \\ \partial_r u_z & \dfrac{1}{r}\partial_z u_\theta & \partial_z u_z \end{pmatrix} \quad \text{(End of example).}$$

$$\rightarrow \mathrm{div}\mathbf{u} = \mathrm{tr}(\mathrm{grad}\mathbf{u}) = \partial_r u_r + \frac{1}{r}(\partial_\theta u_\theta + u_r) + \partial_z u_z$$

1.10. Partial derivatives of function with tensor arguments

In order to derive constitutive models, it is particularly important to be able to derive scalar and tensor valued functions of tensor arguments (typically, the constitutive behavior will be prescribed by a scalar valued function of strain measures). We first define the notion of

$$F = F(\mathbf{A}) \text{ , with } \mathbf{A} \text{ a second order tensor.}$$

NOTE (Tensor functions).– We consider tensor as an argument of a function, which can be either scalar, vector, or tensor-valued, so from a generic point of view of the form

EXAMPLE (Scalar valued functions).– The scalar invariants built as $\mathrm{tr}(\mathbf{A}),\ \det(\mathbf{A}),\ F(\mathbf{A},\mathbf{B}) := \mathbf{A}:\mathbf{B}$

EXAMPLE (Vector valued functions).– $\mathbf{v}(\mathbf{A}) = \mathbf{A}.\mathbf{u}$, with $\mathbf{u}$ a given vector.

EXAMPLE (Tensor valued functions).– $\mathbf{B}(\mathbf{A}) = \mathbf{A}^k$.

1.10.1. *Differential operations on scalar valued functions with tensor arguments*

Directional derivatives of invariants: the goal here is to evaluate the derivative of a bilinear mapping $\Pi(f,g)$, such as $\phi\mathbf{v},\ \mathbf{u}.\mathbf{v},\ \mathbf{u}\otimes\mathbf{v},\ \mathbf{T}.\mathbf{v}$. One useful result is

$$D\Pi(f(\mathbf{x}),g(\mathbf{x}))[\mathbf{u}] = \Pi(f(\mathbf{x}),Dg(\mathbf{x}))[\mathbf{u}] + \Pi(Df(\mathbf{x}),g(\mathbf{x}))[\mathbf{u}]$$

The most important differential operations are listed below:

1) For a scalar valued function with tensor argument, $F = F(\mathbf{A})$, the derivative of the tensor valued function $F(\mathbf{A})$ w.r. to $\mathbf{A}$ is evaluated by taking the partial derivatives w.r. to the components of $\mathbf{A}$ (writing $\frac{\partial F}{\partial \mathbf{A}}$ does not make sense, although one is entitled to write it by an abuse of notations)

$$\frac{\partial F(\mathbf{A})}{\partial \mathbf{A}} = \frac{\partial F(\mathbf{A})}{\partial A_{kl}}\mathbf{g}_k \otimes \mathbf{g}_l \equiv B^{kl}\mathbf{g}_k \otimes \mathbf{g}_l = B^i_{.j}\mathbf{g}_i \otimes \mathbf{g}^j = B_{ij}\mathbf{g}_i \otimes \mathbf{g}_j$$

We see that the contravariant components of the partial derivative $\frac{\partial F(\mathbf{A})}{\partial A_{kl}} = B^{kl}$ are such that the subscripts of A_{kl} (a quantity in the denominator) are considered as superscripts after the evaluation of the partial derivative. The choice of the type of components $(A_{ij}, A^{ij}, \ldots)$ is immaterial; nevertheless, it is usually recognized that strain tensors tend to be covariant, whereas stress tensors tend to be contravariant.

REMARK.– In Cartesian coordinates, the variance does not matter, so that one shall have only one type of components, with here $\frac{\partial F(\mathbf{A})}{\partial \mathbf{A}} = \frac{\partial F(\mathbf{A})}{\partial A_{kl}} \mathbf{e}_k \otimes \mathbf{e}_l$ when the Cartesian basis $\{\mathbf{e}_i\}$ is adopted.

The second order partial derivative writes

$$\frac{\partial F(\mathbf{A})}{\partial(\mathbf{A}\otimes\mathbf{A})} = \frac{\partial^2 F(\mathbf{A})}{\partial\mathbf{A}\partial\mathbf{A}} = \frac{\partial^2 F(\mathbf{A})}{\partial A_{ij}\partial A_{mn}} \mathbf{g}_i \otimes \mathbf{g}_j \otimes \mathbf{g}_m \otimes \mathbf{g}_n \equiv C^{ijmn} \mathbf{g}_i \otimes \mathbf{g}_j \otimes \mathbf{g}_m \otimes \mathbf{g}_n$$

2) By extension, for a tensor valued function with tensor argument, $\mathbf{B} = \mathbf{B}(\mathbf{A})$, it holds that

$$\mathbf{B}_{,A} = \frac{\partial \mathbf{B}(\mathbf{A})}{\partial \mathbf{A}} = \frac{\partial B_{ij}}{\partial A_{kl}} \mathbf{g}^i \otimes \mathbf{g}^j \otimes \mathbf{g}_k \otimes \mathbf{g}_l = C_{ij}^{\cdot\cdot kl} \mathbf{g}^i \otimes \mathbf{g}^j \otimes \mathbf{g}_k \otimes \mathbf{g}_l$$

a fourth order tensor.

In the case where $\mathbf{A}$ is a strain and $\mathbf{B}$ a stress, then tensor $\mathbf{C}$ represents the tensor of elastic moduli. We see that the partial derivative w.r. to a second order tensor increases the order of the initial tensor by two. Partial derivatives of vector valued functions of tensor arguments can be built in the same way.

The different types of components (covariant, contravariant, mixed) of a tensor obtained from such previous partial derivatives can be mutually related by products with the metric tensor; for instance, one evaluates the fully covariant components C_{ijkl} from the mixed components as

$$C_{ijkl} = g_{kp} g_{lq} C_{ij}^{\cdot\cdot pq}$$

NOTE (Applications).– Writing the constitutive laws requires computing the partial derivatives of the invariant w.r. to tensor measures (of strain), thus we shall make use of the following derivatives

$$(\mathrm{tr}\mathbf{A})_{,A} = \mathbf{I}, \quad (\mathrm{tr}\mathbf{A})^2_{,A} = 2(\mathrm{tr}\mathbf{A})\mathbf{I}, \quad (\mathrm{tr}\mathbf{A}^2)_{,A} = (A^r_{.s}A^s_{.r})_{,A} = 2\mathbf{A}^T, \quad (\mathrm{tr}\mathbf{A}^3)_{,A} = 3(\mathbf{A}^2)^T$$

PROOF.– We leave it as an exercise, and just indicate how to compute the first previous derivative

$$\mathrm{tr}\mathbf{A} = A^k_{.k} \Rightarrow (\mathrm{tr}\mathbf{A})_{,A} = \frac{\partial A^k_{.k}}{\partial A^m_{.s}} \mathbf{g}^m \otimes \mathbf{g}_s = \delta^k_{.m}\delta^s_{.k}\mathbf{g}^m \otimes \mathbf{g}_s = \delta^s_{.m}\mathbf{g}^m \otimes \mathbf{g}_s = \mathbf{I}$$

$$(\det \mathbf{A})_{,A} = \mathrm{cof}\mathbf{A} = (\det \mathbf{A})\,\mathbf{A}^{-T}$$

EXERCISE.– Prove the last identity basing on the definition of the determinant as a multilinear alternate form.

1.10.2. *Gradient of the invariants*

Using previous definitions and statements, we can prove the following expression of the derivatives of the invariants,

$$\boxed{\frac{\partial I_1}{\partial \mathbf{T}} = \mathbf{I}, \quad \frac{\partial I_2}{\partial \mathbf{T}} = (\mathrm{tr}\mathbf{T})\mathbf{I} - \mathbf{T}^T, \quad \frac{\partial I_3}{\partial \mathbf{T}} = \mathrm{cof}\mathbf{T} = \left[I_2\mathbf{I} - I_1\mathbf{T} + \mathbf{T}^2\right]^T}$$

The derivative with respect to a tensor is here an abuse of notation; it does not make sense from the mathematical point of view, so that one can only derive w.r. to the tensor components (this defines a component to be projected in a suitable basis).

EXAMPLE.– The following derivatives are useful in continuum mechanics

$$\frac{\partial \mathbf{T}}{\partial \mathbf{T}} = \frac{\partial \mathbf{T}^T}{\partial \mathbf{T}^T} = \mathbb{I}_4, \quad \frac{\partial \mathbf{T}^T}{\partial \mathbf{T}} = \frac{\partial \mathbf{T}}{\partial \mathbf{T}^T} = \mathbb{T}$$, or equivalently in indicial notation,

$$\frac{\partial T_{ij}}{\partial T_{kl}} = \frac{\partial T_{ji}}{\partial T_{lk}} = \delta_{ik}\delta_{jl}, \quad \frac{\partial T_{ji}}{\partial T_{kl}} = \frac{\partial T_{ij}}{\partial T_{lk}} = \delta_{jk}\delta_{il}$$

– Derivation rules: let $\phi_1(\mathbf{T}), \phi_2(\mathbf{T})$ be scalar valued functions, and $\mathbf{U}(\mathbf{T}), \mathbf{V}(\mathbf{T})$ tensor valued functions. The derivative of the product $\phi_1(\mathbf{T})\phi_2(\mathbf{T})$ is given by

$$\frac{\partial\{\phi_1(\mathbf{T})\phi_2(\mathbf{T})\}}{\partial\mathbf{T}} = \phi_1(\mathbf{T})\frac{\partial\phi_2(\mathbf{T})}{\partial\mathbf{T}} + \phi_2(\mathbf{T})\frac{\partial\phi_1(\mathbf{T})}{\partial\mathbf{T}}$$

The following properties can easily be proven:

$$\frac{\partial(\phi\mathbf{U})}{\partial\mathbf{T}} = \mathbf{U}\otimes\frac{\partial\phi}{\partial\mathbf{T}} + \phi\frac{\partial\mathbf{U}}{\partial\mathbf{T}}; \qquad \frac{\partial(\mathbf{U}:\mathbf{V})}{\partial\mathbf{T}} = \left(\frac{\partial\mathbf{U}}{\partial\mathbf{T}}\right)^{\mathrm{T}}:\mathbf{V} + \left(\frac{\partial\mathbf{V}}{\partial\mathbf{T}}\right)^{\mathrm{T}}:\mathbf{U}$$

$$\frac{\partial\phi}{\partial\mathbf{T}^{\mathrm{T}}} = \mathbb{T}:\frac{\partial\phi}{\partial\mathbf{T}}; \quad \frac{\partial\mathbf{U}}{\partial\mathbf{T}^{\mathrm{T}}} = \frac{\partial\mathbf{U}}{\partial\mathbf{T}}:\mathbb{T}$$

– Partial proof: $(\mathbf{U}.\mathbf{V})_{ij} = U_{im}V_{mj} \rightarrow \left[\frac{\partial(\mathbf{U}.\mathbf{V})}{\partial\mathbf{T}}\right]_{ijkl} = U_{im}\frac{\partial V_{mj}}{\partial T_{kl}} + \frac{\partial U_{im}}{\partial T_{kl}}V_{mj} =$

$$= U_{im}\delta_{nj}\frac{\partial V_{mn}}{\partial T_{kl}} + \frac{\partial U_{nm}}{\partial T_{kl}}\delta_{in}V_{mj} = U_{im}\delta_{nj}\frac{\partial V_{mn}}{\partial T_{kl}} + \frac{\partial U_{mn}}{\partial T_{kl}}\delta_{im}V_{nj}.$$

– Previous identity reads in tensor format:

$$\frac{\partial(\mathbf{U}.\mathbf{V})}{\partial\mathbf{T}} = \left(\mathbf{U}\overline{\otimes}\mathbf{I}\right):\frac{\partial\mathbf{V}}{\partial\mathbf{T}} + \left(\mathbf{I}\overline{\otimes}\mathbf{V}^{\mathrm{T}}\right):\frac{\partial\mathbf{U}}{\partial\mathbf{T}}$$

COROLLARY.– One deduces the following useful derivatives to be used in continuum mechanics.

$$\frac{\partial(\mathbf{T}^2)}{\partial\mathbf{T}} = \mathbf{T}\overline{\otimes}\mathbf{I} + \mathbf{I}\overline{\otimes}\mathbf{T}^{\mathrm{T}} \text{ (choice } \mathbf{U} = \mathbf{T}\text{)}$$

$$\frac{\partial(\mathbf{T}^{\mathrm{T}}.\mathbf{T})}{\partial\mathbf{T}} = \left(\mathbf{I}\overline{\otimes}\mathbf{T}^{\mathrm{T}}\right)\mathbb{T} + \mathbf{T}^{\mathrm{T}}\overline{\otimes}\mathbf{I} \text{ (choice } \mathbf{U} = \mathbf{U}^{\mathrm{T}}, \mathbf{V} = \mathbf{T}\text{)}$$

$$\frac{\partial\left(\mathbf{T}^{-1}\right)}{\partial\mathbf{T}}=-\mathbf{T}^{-1}\overline{\otimes}\mathbf{T}^{-\mathrm{T}}=-\left(\mathbf{T}\overline{\otimes}\mathbf{T}^{\mathrm{T}}\right)^{-1}\ \text{(choice } \mathbf{U}=\mathbf{T}^{-1},\mathbf{V}=\mathbf{T}\text{)}$$

$$\frac{\partial\left(\mathbf{T}^{-\mathrm{T}}\right)}{\partial\mathbf{T}}=-\left(\mathbf{T}^{-\mathrm{T}}\overline{\otimes}\mathbf{T}^{-1}\right)\ \mathbb{T}=-\left(\mathbf{T}^{\mathrm{T}}\overline{\otimes}\mathbf{T}\right)^{-1}:\mathbb{T}$$

(due to previous relation)

REMARK.– The third relation can be proven as follows,

$$\mathbf{T}^{-1}.\mathbf{T}=\mathbf{I}\Rightarrow\mathbf{0}=\mathbf{T}^{-1}\overline{\otimes}\mathbf{I}+\left(\mathbf{I}\overline{\otimes}\mathbf{T}^{\mathrm{T}}\right):\frac{\partial\left(\mathbf{T}^{-1}\right)}{\partial\mathbf{T}}$$

$$\Rightarrow\frac{\partial\left(\mathbf{T}^{-1}\right)}{\partial\mathbf{T}}=-\left(\mathbf{I}\overline{\otimes}\mathbf{T}^{\mathrm{T}}\right)^{-1}:\left(\mathbf{T}^{-1}\overline{\otimes}\mathbf{I}\right)=-\left(\mathbf{I}\overline{\otimes}\mathbf{T}^{-\mathrm{T}}\right):\left(\mathbf{T}^{-1}\overline{\otimes}\mathbf{I}\right)=-\mathbf{T}^{-1}\overline{\otimes}\mathbf{T}^{-\mathrm{T}}$$

We have here used the (previous) relations:

$$\left(\mathbf{A}\overline{\otimes}\mathbf{B}\right)\left(\mathbf{C}\overline{\otimes}\mathbf{D}\right)=(\mathbf{A}.\mathbf{C})(\mathbf{B}.\mathbf{D});\quad\left(\mathbf{A}\overline{\otimes}\mathbf{B}\right)^{-1}=\mathbf{A}^{-1}\overline{\otimes}\mathbf{B}^{-1}$$

$$\mathbf{T}^{-1}.\frac{\partial\mathbf{T}}{\partial\mathbf{T}}=\mathbf{T}^{-1}.\mathbb{I}_4=\mathbf{T}^{-1}.\left(\mathbf{I}\overline{\otimes}\mathbf{I}\right)=\mathbf{T}^{-1}.\mathbb{I}_4$$

1.10.3. *Gradient of scalar and tensor valued functions with respect to symmetrical tensors*

A common situation in nonlinear elasticity is the formulation of hyperelastic constitutive models from a strain energy function (a density per unit volume of the stored elastic energy) function of a strain measure, a second order tensor, typically $\mathrm{W}=\mathrm{W}(\mathbf{C})$, with $\mathbf{C}$ a symmetrical tensor, thus one may write

$$\mathrm{W}(\mathbf{C})=\mathrm{W}\left(\frac{\mathbf{C}+\mathbf{C}^{\mathrm{T}}}{2}\right)$$

This entails the stress tensor given as the derivative

$$\mathbf{S}(\mathbf{C})=2\frac{\partial\mathrm{W}}{\partial\mathbf{C}}=\frac{\partial\mathrm{W}}{\partial\mathbf{C}}+\frac{\partial\mathrm{W}}{\partial\mathbf{C}^{\mathrm{T}}}=\frac{\partial\mathrm{W}}{\partial\mathbf{C}}+\mathbb{T}:\frac{\partial\mathrm{W}}{\partial\mathbf{C}}=\left(\mathbb{I}_4+\mathbb{T}\right):\frac{\partial\mathrm{W}}{\partial\mathbf{C}}=2\mathbb{S}:\frac{\partial\mathrm{W}}{\partial\mathbf{C}}$$

Observe that the obtained tensor is symmetrical. In indicial notation, this reads

$$S_{ij} = \frac{\partial W}{\partial C_{ij}} + \frac{\partial W}{\partial C_{ji}}$$

The tensor of elastic moduli is then elaborated as the derivative of the stress w.r. to the strain measure, viz.

$$\mathbb{C} := 2\frac{\partial \mathbf{S}}{\partial \mathbf{C}} = \frac{\partial(\mathbb{S}:\mathbf{S})}{\partial \mathbf{C}} + \frac{\partial(\mathbb{S}:\mathbf{S})}{\partial \mathbf{C}^{T}} = \frac{\partial(\mathbb{S}:\mathbf{S})}{\partial \mathbf{C}} + \frac{\partial(\mathbb{S}:\mathbf{S})}{\partial \mathbf{C}}:\mathbb{T} =$$

$$= \mathbb{S}:\frac{\partial \mathbf{S}}{\partial \mathbf{C}}:(\mathbb{I}_4 + \mathbb{T}) = 2\mathbb{S}:\frac{\partial \mathbf{S}}{\partial \mathbf{C}}:\mathbb{S} = 4\mathbb{S}\frac{\partial^2 W}{\partial \mathbf{C}\partial \mathbf{C}}:\mathbb{S}$$

The previous equation reads in indicial notation

$$\mathbb{C}_{ijkl} = \frac{1}{2}\left(\frac{\partial S_{ij}}{\partial C_{kl}} + \frac{\partial S_{ji}}{\partial C_{kl}} + \frac{\partial S_{ij}}{\partial C_{lk}} + \frac{\partial S_{ji}}{\partial C_{lk}}\right) =$$

$$\frac{\partial^2 W}{\partial C_{ij}\partial C_{kl}} + \frac{\partial^2 W}{\partial C_{ji}\partial C_{kl}} + \frac{\partial^2 W}{\partial C_{ij}\partial C_{lk}} + \frac{\partial^2 W}{\partial C_{ji}\partial C_{lk}}$$

1.11. Elements of functional analysis

We consider a vector space of dimension n, denoted E_n; we denote in general $L(X, Y)$ the set of all linear mappings from the vector space X into the vector space Y, and $L(X, Y; \mathbb{R})$ the set of bilinear applications based on $X \times Y$ into $\mathbb{R}$. An endomorphism is a linear mapping from E_n to itself, and the corresponding set is denoted $L(E_n)$. We choose a basis $\{\mathbf{e}_i\}$ in E_n.

A tensor is a multilinear form, so that a tensor of type (r, s), that is r times contravariant and s times covariant is an element of

$$L\left(E_n^*,\ldots,E_n^*,E_n,\ldots,E_n;\mathbb{R}\right),\ \text{with r times } E_n^* \text{ and s times } E_n$$

The specific case $r = 0 = s$ corresponds to scalars, elements of $\mathbb{R}$.

The space $L\left(E_n;\mathbb{R}\right)$ is the space of linear forms, and $L\left(E_n^*;\mathbb{R}\right) \equiv E_n^{**}$ the space of linear forms on E_n^*, so it is called the bidual.

The space of second order mixed tensors (once contravariant and once covariant) can be identified to $L(E_n)$.

The set of bilinear forms can be identified to the set of second order 2-covariant tensors, of the form $\mathbf{T} = T_{ij}\mathbf{e}^i \otimes \mathbf{e}^j$.

There exist isomorphisms between E_n and E_n^{**},

$$J : E_n \to E_n^{**}$$
$$\mathbf{v} \mapsto J(\mathbf{v}) \text{ such that } J(\mathbf{v})(l) = l(\mathbf{v}),\ \ \forall l \in E_n^*,\ \ \forall \mathbf{v} \in E_n$$

Elements of the bidual are derivations, such that $\partial_j.\mathbf{e}^i = \delta_j^{\,i}$: for a given $\mathbf{x} \in E_n$, the operator ∂_i is the derivation operator in the direction $\mathbf{e}_j$, so that $\partial_j = \dfrac{\partial}{\partial x_j}$.

The bidual allows an intrinsic definition of the transpose, not involving a scalar product: the transpose of a linear mapping L from E_n to itself (also called an endomorphism) is usually defined from the following equality

$$\left(L^T.\mathbf{v},\mathbf{u}\right)_g = \left(\mathbf{v}, L.\mathbf{u}\right)_g,\ \ \forall \mathbf{u} \in E_n$$

based on two scalar products in the respective vector spaces, and in which the scalar product is defined from a metric g. In given bases, the matrix of L^T is dependent on the choice of the scalar product, and in particular the notion of symmetrical mapping $L = L^T$: L is symmetrical iff

$$\left(L.\mathbf{v},\mathbf{u}\right)_g = \left(\mathbf{v}, L.\mathbf{u}\right)_g,\ \ \forall \mathbf{u} \in E_n$$

To remedy this difficulty, so to define an intrinsic notion of the transpose, one defines the adjoint mapping L^* such that

$$\left(L^*.l, \mathbf{u}\right)_{E_n^*, E_n} = \left(l, L.\mathbf{u}\right)_{E_n^*, E_n}, \quad \forall \mathbf{u} \in E_n, \quad \forall l \in E_n^*$$

using the duality brackets between E_n and E_n^*. The confusion between the transpose and the adjoint is due to the fact that one uses the Riesz representation theorem of a linear form by a vector; due to the fact that there is no natural isomorphism between E_n and E_n^*, we cannot naturally identify L^T with L^*: such an identification requires a scalar product, so it is only possible in Euclidean spaces.

The main idea behind the concept of tensor is to delineate what is intrinsic (independent from a metric or a basis); intrinsic means independent from any observer, that is from any basis or scalar product.

The following table indicates the intrinsic and non-intrinsic quantities used in tensor algebra.

Intrinsic	Not intrinsic
Vector, tensor Trace and determinant of endomorphisms (linear mappings $L(E_n)$) Scalar product (tool to measure change of length and angle) as symmetrical bilinear definite positive form. Adjoint of a linear mapping (no scalar product involved): $\left\langle L^*.l, \mathbf{u}\right\rangle_{E^*,E} := \left\langle l, L.\mathbf{u}\right\rangle_{F^*,F} := l\left(L.\mathbf{u}\right)$ Isomorphism $E^{**} := L\left(E^*; R\right) \simeq E$: identify vector fields and derivations.	Vector used as a basis vector to represent another vector. Determinant of a bilinear form (conserved only for change of orthonormal basis). Representation of a linear form by a vector (Riesz theorem): no isomorphism between E and its dual E^*. Transpose of a linear mapping L (depends on the basis) and notion of symmetrical mapping L, relative to a scalar product defined by the metric.

2

Continuum Mechanics

2.1. Motivations of nonlinear mechanics[1]

Many materials or structures are called soft in the sense that they are able to undergo large displacements / rotations, possibly in addition to large strains: rubbery materials and soft living tissues are typical examples, but truss-like structures also display such behaviors due to some internal mechanisms. Both academic and research-oriented examples will be shown to motivate the course. The mechanical response of soft materials during their service life is very seldom linear (unless restricted to the small strains regime in some specific use), and most of the time, nonlinearities are present; an intuitive idea of a nonlinear response is provided by the crushing of a soft ball (picture below), with the ball returning to its initial shape when the pressure is released.

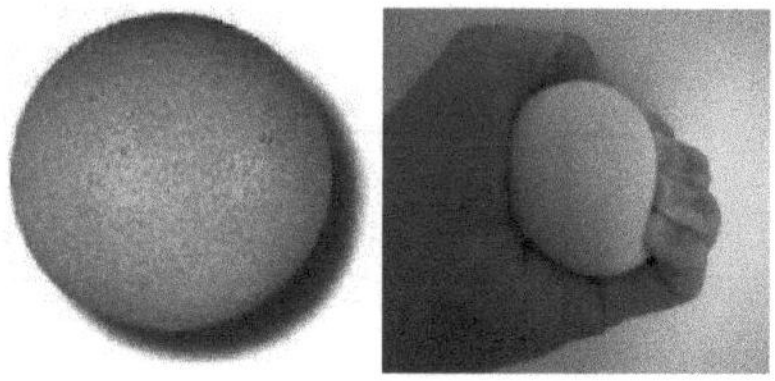

Figure 2.1. *Crushing of a soft ball: initial (left) and final configurations (right) [MOE 14]*

Nonlinearities in continuum mechanics can be classified as *geometrical nonlinearities* and *material nonlinearities*: the first type (geometrical nonlinerarities)

Chapter written by Jean-François GANGHOFFER.

1 See section 2.13 for a list of the abbreviations, notations and nomenclature used in this chapter.

occur when the displacement or rotation of material elements of the structure are large in comparison to its size; this may occur for slender structures having a small bending stiffness compared to their in-plane properties. The second type (material nonlinearities) are due to the occurrence of large strains. We provide below two examples of geometrical nonlinearities.

2.1.1. *Academic examples*

1) First example: beam under large rotations.

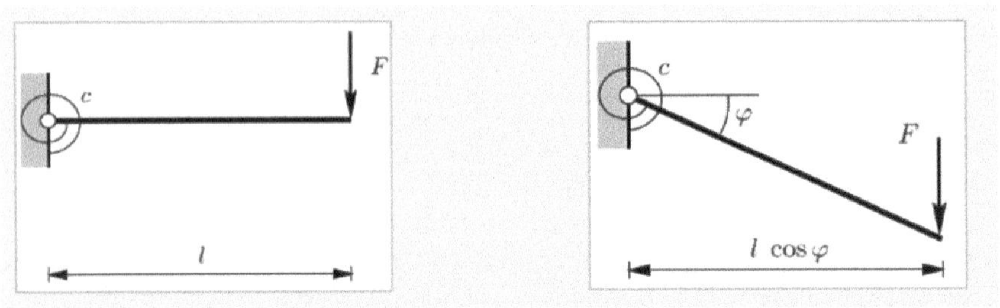

Figure 2.2. *Beam under large rotations: initial configuration (left) and deformed configuration (right)*

For flexible structures such as beams, plates or shells, the displacements/ rotations are large; geometrical nonlinearities occur for $\Delta L / L > 0.1\%$. We first write the governing equations:

– Equilibrium: $F = \dfrac{c\varphi}{l\cos\varphi}$, expressing as a nonlinear relation $F(\varphi)$.

– Linearization: in the case of small rotations, we may write

$\cos\varphi \approx 1 \rightarrow F = c\varphi / l$, which is a linear approximation of $F(\varphi)$.

By a Taylor series approximation (second order), we obtain

$\cos\varphi \approx 1 - \dfrac{\varphi^2}{2} \rightarrow F = \dfrac{c\varphi}{l\left(1-\dfrac{\varphi^2}{2}\right)}$, which is clearly a nonlinear relation.

The nonlinear response is recorded on the figure below and compared to the two developed approximations: the second order solution is valid when $\varphi < \dfrac{\pi}{3}$.

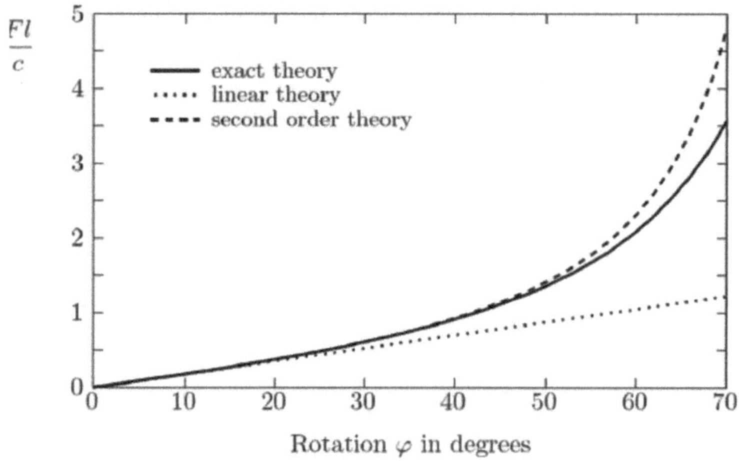

Figure 2.3. *Nonlinear response of the beam under large deflections*

2) Second example: elastic springs in large displacements (Figure 2.4).

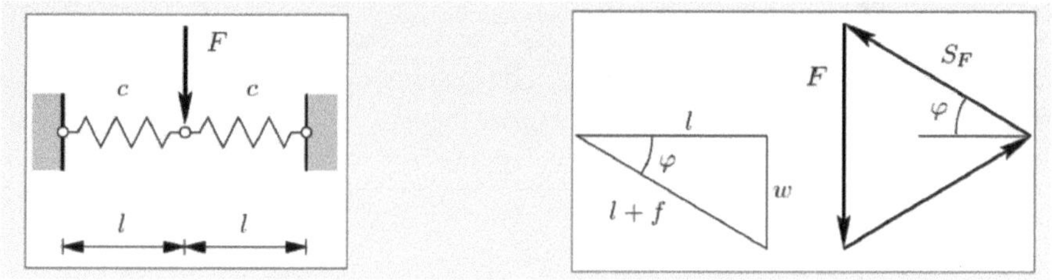

Figure 2.4. *Initial system and force diagram*

From Pythagoras rule, it holds that:

$$w^2 + l^2 = (l + f)^2 \Rightarrow f = l\left[\sqrt{1+\left(\frac{w}{l}\right)^2} - 1\right]$$

with f the spring elongation, and w the transverse displacement.

– Geometrical relation: $\sin\varphi = \dfrac{w}{l+f}$.

From the force diagram (triangle) above, it holds that

$$S_F \sin\varphi = \frac{F}{2}$$

The constitutive law of the spring writes $S_F = cf$, so that

$$\frac{w}{l}\left[1 - 1/\sqrt{1 + \left(\frac{w}{l}\right)^2}\right] = \frac{F}{2cl},$$

which is a nonlinear relation $F(w)$.

In the case of small displacements: $w/l \ll 1 \rightarrow 1/\sqrt{1+\left(\frac{w}{l}\right)^2} \approx 1 - \frac{1}{2}\left(\frac{w}{l}\right)^2$, thus

$$\frac{w}{l}\left[1 - \frac{1}{2}\left(\frac{w}{l}\right)^2\right] = \frac{F}{2cl},$$

an approximation valid for $w/l < 0.4$. The response of the structure is shown in Figure 2.5 for both the exact and second order theory.

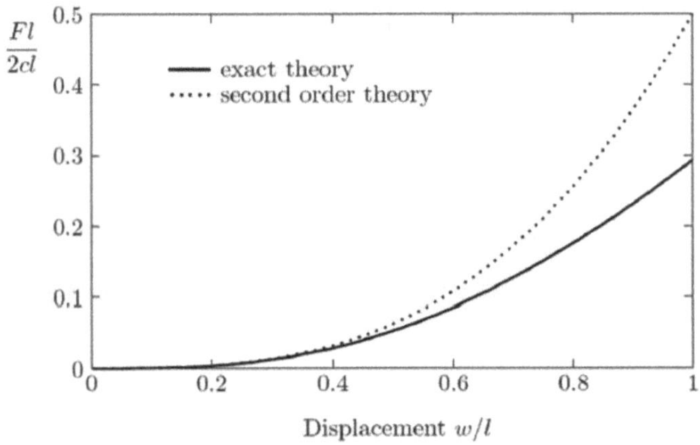

Figure 2.5. *Nonlinear response of the elastic springs*

EXAMPLE (Material nonlinearity).– Nonlinear spring with stiffness function of the spring elongation (Figure 2.6).

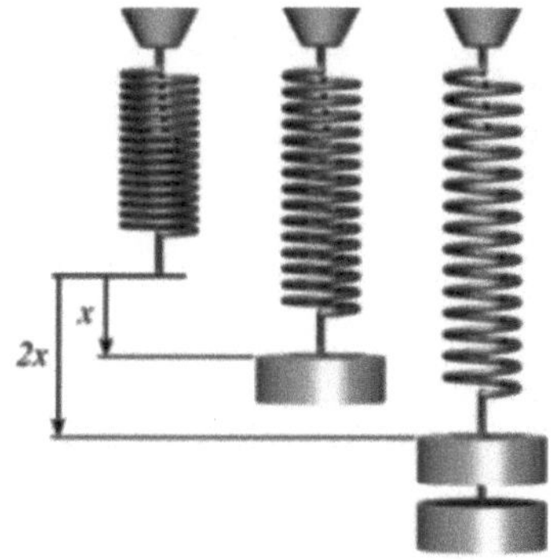

Figure 2.6. *Example of a linear spring*

– Model of linear spring: $F = k(L - L_0)$.

Model of nonlinear spring obtained by letting the spring rigidity be a function of the relative displacement, $k = k(L - L_0)$, so that $F = (L - L_0)k(L - L_0)$.

In those two academic examples concerned with structural elements, the force is at least a quadratic expression of the corresponding kinematic variable (when truncating a Taylor series expansion). Similar effects will take place in continuum mechanics, in which both the measure of strain and the stress to strain relations will possibly and very often be nonlinear.

2.1.2. *Research oriented examples*

– Mechanics of textiles: textiles are made of interwoven fibers, organized in a regular manner. They find a widespread use nowadays due to their low weight, better distribution of efforts, high resistance to external forces, impact, etc. Turbine blades of military airplanes are made of a fibrous reinforcement, which is then impregnated with a resin to make a composite material. Planar monolayer textiles such as plain weave (Figure 2.7) are made of two sets of intertwined yarns; their mechanical response is typically a multiscale problem, starting from the yarn level (microscopic scale) up to the structure (macroscopic scale).

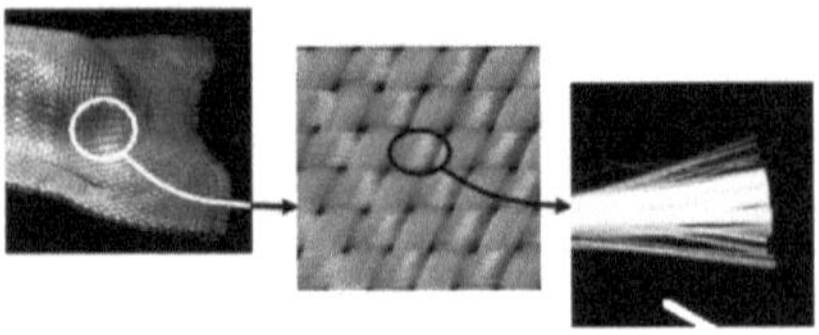

Figure 2.7. *View of a fibrous plain weave reinforcement at different scales (macroscopic, left – mesoscopic, middle – microscopic, right)*

The thick reinforcement of parts of turbine blades elaborated by SNECMA is shown in Figure 2.8.

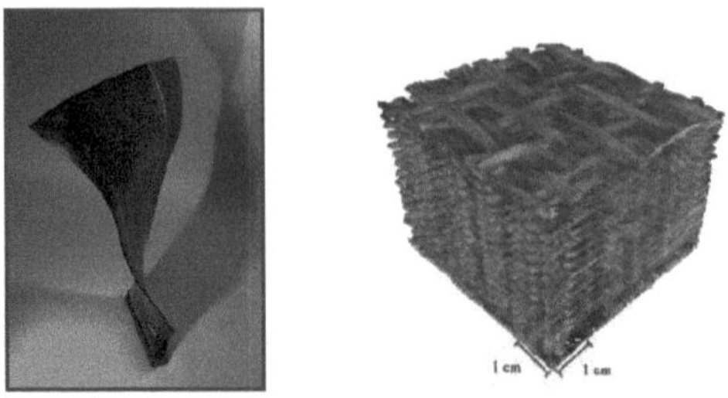

Figure 2.8. *Part of a turbine blade (left) and thick 3D reinforcement (right) [courtesy of SNECMA]. For a color version of this figure, see www.iste.co.uk/ganghoffer/biomechanics.zip*

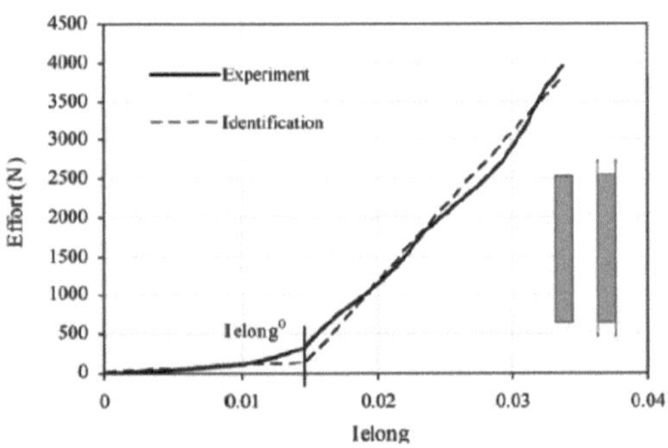

Figure 2.9. *Tensile response of the interlock sample pictured on Figure 2.8 [CHA 11]. For a color version of this figure, see www.iste.co.uk/ganghoffer/biomechanics.zip*

Both the tensile and shear responses (obtained by a bias test) are also clearly nonlinear (Figure 2.10); the model selected to describe the nonlinear response is a nonlinear elastic model (hyperelastic model).

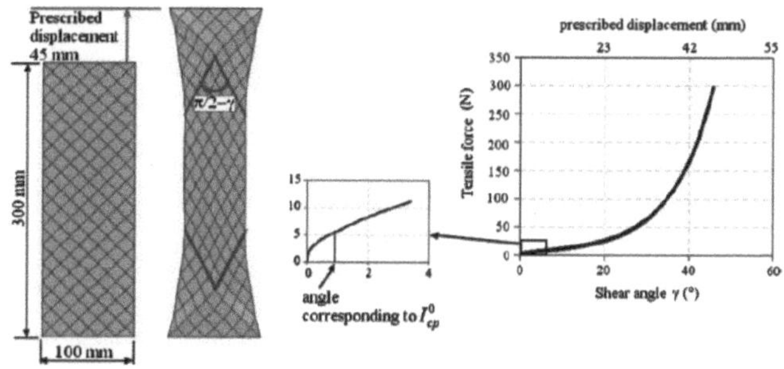

Figure 2.10. *Experimental bias extension test for a thick interlock reinforcement [CHA 11]. For a color version of this figure, see www.iste.co.uk/ganghoffer/biomechanics.zip*

Soft biological tissues such as skin, arteries (Figure 2.11), muscles and ligaments, are also prone to large deformations, since they (like textiles) consist of wavy collagen fibers immersed into a soft matrix. Healthy arteries are highly deformable composite structures and show a nonlinear stress–strain response with a stiffening effect at higher pressures (Figure 2.12). This stiffening effect is common to all biological tissues and is based on the recruitment of the embedded wavy collagen fibrils, which carry the load. Fibers give the tissue its mechanical resistance (the response is stiff in the fiber directions, and weaker in the transverse directions), so that these structures clearly behave in an anisotropic manner.

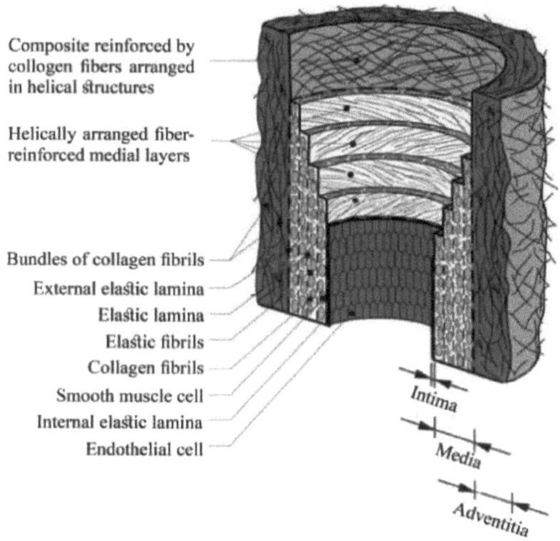

Figure 2.11. *Arteries are made of several layers including collagen fibers [HOL 00]*

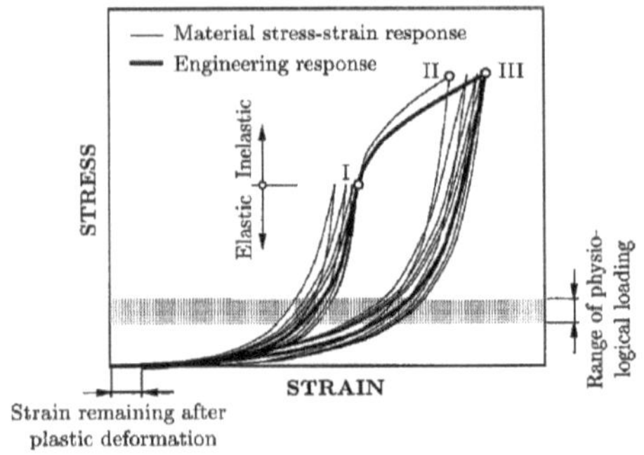

Figure 2.12. *Typical uniaxial stress–strain curves for circumferential arterial strips (from the media) in passive conditions: cyclic loading and unloading, associated with stress softening effects, lead to a pre-conditioned material which behaves (perfectly) elastically or viscoelastically (nearly repeatable cyclic behavior) – point I. Loading beyond the (visco)elastic domain up to point II leads to inelastic deformations. Additional loading and unloading cycles display stress softening again until point III is reached. Then the material exhibits (perfectly) elastic or viscoelastic response. The thick solid line indicates the (approximate) engineering response of the material [HOL 00]*

NOTE (Natural and artificial ligaments).– Tissue-engineered solutions offer a good alternative to classical surgeries in the case of anterior cruciate ligament (ACL for short) injuries (Figure 2.13).

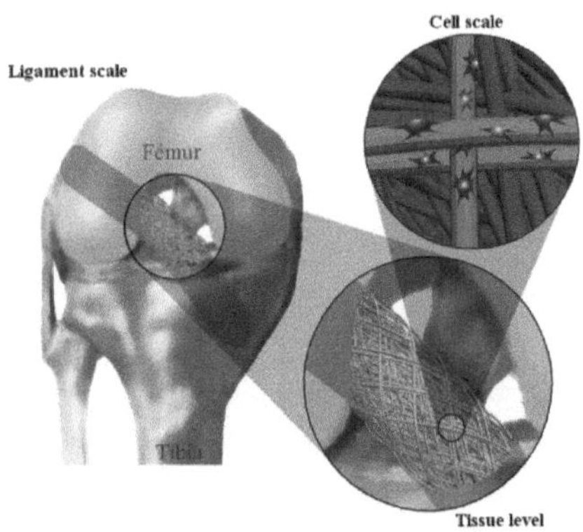

Figure 2.13. *The three scales of tissue engineering in case of the cruciate anterior ligament [LAU 11, LAU 12, LAU 13]. For a color version of this figure, see www.iste.co.uk/ganghoffer/biomechanics.zip*

Tissue engineering is based on the culture of reparative cells in a scaffold, whose properties must meet several key requirements such as material biocompatibility, biodegradability at a rate matching that of tissue formation, mechanical properties matching those of the native tissue and high porosity to allow cell infiltration. In particular, a three-dimensional porous structure flexible in the low range of strain but very stiff in the high range of strain is needed. The ideal properties to match include stiffness up to 200 N/mm, strain to failure of more than 20%, a minimum pore size of 200–250 mm for soft tissue ingrowth and a clearly convex force–displacement (or stress–strain) response. The proposed scaffold is a tailorable scaffold based on multilayer braided fibres made of copoly (lactic acid-co-(1-caprolactone)) (PLCL); cells seeded on the scaffold tend to grow new tissue (Figure 2.13), and the mechanical response of the scaffold shows a clear nonlinearity (Figure 2.14), due to both fiber alignment in the tensile load's direction in a first stage, and fiber stretch in a second stage.

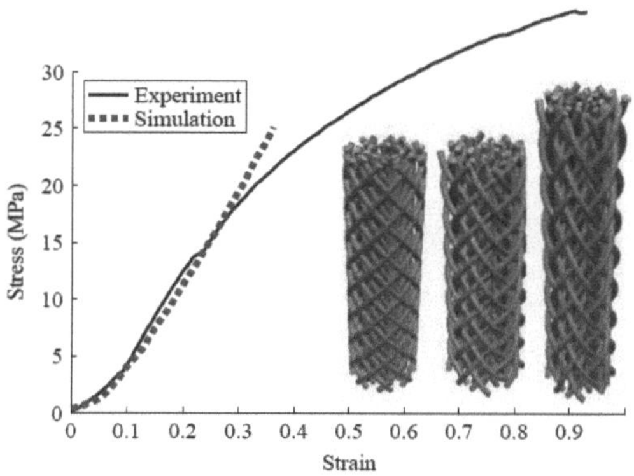

Figure 2.14. *Stress-strain response of a multilayer braided scaffold for ligament repair [LAU 11, LAU 12, LAU 13]. For a color version of this figure, see www.iste.co.uk/ganghoffer/biomechanics.zip*

2.2. Prerequisite: summary of linear elasticity

The linear elasticity theory has the following principal limitations:

– It is restricted to small displacements and rotations. Previous examples show the occurrence of large displacements and large rotations.

– There is only one configuration and thus only one system of coordinates. Accordingly, phenomena in which the configuration changes cannot be accounted for. This is the case for instance in the contact between two bodies or when the volume or shapes of the body undergo substantial modifications.

– Strains are restricted to being small, typically less than 5%. Many materials are soft, for instance rubber in the glassy state (below a certain temperature, in a state in which it behaves as a solid material).

2.2.1. *Idea of how to generalize the kinematics (analysis of motion)*

Motion	Deformation	Linearization
$\mathbf{x} = f(\mathbf{X},t)$ relates Lagrangian position $\mathbf{X}$ to spatial (physical position) $\mathbf{x}$ Trajectory: one follows a particle $M_0(\mathbf{X})$ during its motion, at fixed referential position $\mathbf{X}$ Velocity: $\mathbf{V}(\mathbf{X},\mathrm{t}) := \frac{\partial \mathbf{x}(\mathbf{X},\mathrm{t})}{\partial t} \vert X = v(\mathbf{x},t)$ Lagrangian viewpoint: we follow a material point during its motion (solid mechanics) Eulerian viewpoint: is given, hence reconstitute position $\mathbf{x} = f(\mathbf{X},t)$ by integration. Acceleration: $\gamma := \frac{d\mathbf{v}}{dt} = \frac{\partial \mathbf{v}}{\partial t} + \nabla \mathbf{v}.\mathbf{v}$ $\frac{d}{dt}$ material derivative (explicit and implicit dependencies upon time are considered).	Let $\mathbf{H} := Grad_x \mathbf{U},\ \mathbf{U} := \mathbf{x} - \mathbf{X}$ $\mathbf{F} = \mathbf{I} + \mathbf{H}$ tangent mapping between tangent spaces in reference and actual configurations, s.t. $d\mathbf{x} = \mathbf{F}.d\mathbf{X}$ Components $F_{ij} = \frac{\partial x_i}{\partial X_j}$ Lagrangian strain (vanishes for a rigid body motion): $\mathbf{E} = \frac{1}{2}(\mathbf{F}^T.\mathbf{F} - \mathbf{I})$ Allows us to compare scalar products of 'small vectors': $d\mathbf{x}'.d\mathbf{x} - d\mathbf{X}''.d\mathbf{X} =$ $= 2d\mathbf{X}'.\mathbf{E}.d\mathbf{X} = 2d\mathbf{x}'.\mathbf{e}.d\mathbf{x}$ Eulerian strain: $\mathbf{e} = \frac{1}{2}(\mathbf{I} - \mathbf{F}^{-T}.\mathbf{F}^{-1})$ Decompose $\mathbf{H} = \boldsymbol{\varepsilon} + \mathbf{w}$: $\boldsymbol{\varepsilon}$ deformation $\mathbf{w}$ rotation Volume changes: $dV = \det(\mathbf{F})dV_0 \equiv JdV_0$ Polar decomposition: $\mathbf{F} = R.\mathbf{U} = \mathbf{V}.R$	Small transformations: $\lVert\mathbf{H}\rVert << 1 \rightarrow \mathbf{E} = \boldsymbol{\varepsilon} + \frac{1}{2}\mathbf{H}^T.\mathbf{H}$ $\mathbf{E} \approx \boldsymbol{\varepsilon},\ \mathrm{U} \approx \mathbf{I} + \boldsymbol{\varepsilon},$ thus $\mathrm{R} \approx \mathbf{I} + \mathbf{w}$ $\mathbf{w} = \frac{1}{2}(\mathbf{U} \otimes \nabla + \mathbf{U} \otimes \nabla^T)$ Small strain tensor: $\varepsilon_{ij} = \frac{1}{2}\left(\frac{\partial U_i}{\partial X_j} + \frac{\partial U_j}{\partial X_i}\right)$ ε_{ii} normal deformation $\varepsilon_{ij}, j \neq i$, shear. Volume changes: $dV = tr(\boldsymbol{\varepsilon})dV_0$ *Linearization:* In the case of small displacements: $\frac{\lVert\mathbf{U}\rVert}{L} << 1$. set $f(\mathbf{x},t) = \mathrm{F}(\mathbf{X},t)$ Write $\mathrm{F}(\mathbf{X},t) = f(\mathbf{X}+\mathbf{U},t) =$ $f(\mathbf{X},t) + \nabla f.\mathbf{U} + o(\lVert\mathbf{U}\rVert)$ Hence, $\mathrm{F}(\mathbf{X},t) \approx f(\mathbf{X},t)$, thus Lagrange and Euler viewpoints coincide. *Small strains with large displacements:* Strain gradients are small, but the displacement may be large.

2.2.2. *Methodology for solving linear elasticity problems in continuum mechanics*

Let us consider a solid body Ω, a collection of material points (particles), subjected to both imposed displacements and tractions on two complementary parts of its boundary. There are two different methods for solving the field equations of linearized elasticity, based either on the displacement or on the stress field. We assume that both displacements and strains are small.

The field equations consist of the strain definition, the equilibrium equation, Hooke's law, and the kinematic compatibility of strains (automatically satisfied if strain definition is considered).

The displacement fields together with the strain and stress tensors build a set of 15 unknowns; the field equations write:

$e_{ij} = \frac{1}{2}(u_{i,j} + u_{j,i})$ strain definition;

$e_{ij,kl} + e_{kl,ij} - e_{ik,jl} - e_{jl,ik} = 0$ compatibility conditions;

$\sigma_{ij,j} + F_i = 0$ static equilibrium ($\mathbf{F}$ is a body force);

$\sigma_{ij} = (\lambda + \mu)e_{kk}\delta_{ij} + 2\mu e_{ij} \Leftrightarrow e_{ij} = \frac{1+\nu}{E}\sigma_{ij} - \frac{\nu}{E}\sigma_{kk}\delta_{ij}$ Hooke's law (two equivalent forms). The two alternative solution strategies for problems in linear elasticity are shown in Figure 2.15.

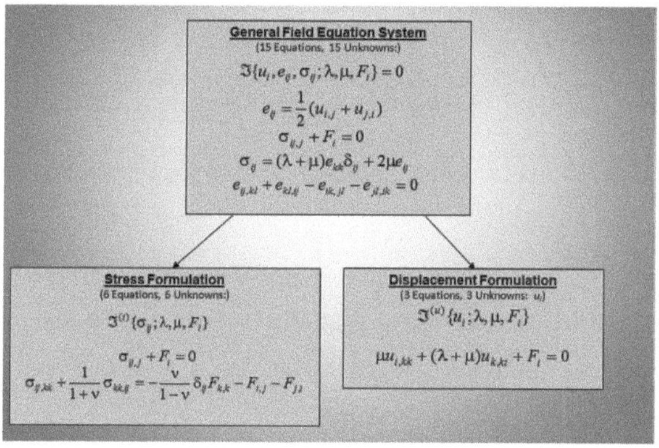

Figure 2.15. *Stress-based (Beltrami) and displacement-based (Lame-Clapeyron) formulations in linear elasticity [Course, University of Rhode Island, MCE 571]*

2.3. Introduction: notion of body in a continuum description

Although matter is discrete at a sufficiently small scale of observation (made of atoms or molecules), solid and fluid media can be considered as a continuum when viewed at a sufficiently large scale; such an approach leads to the *continuum theory*, relying on the assumption that a body can be considered as having a continuous distribution of matter in space. The notion of *particle* is here introduced as a large number of atoms or molecules, so that the notion of mass density can be introduced (instead of considering point masses in a discrete approach). The particle is thus a representative volume element upon which it makes sense to attribute fields, that is, quantities varying continuously in space and time, such as density, displacement, strain, stress, or temperature.

The *body* is defined as a collection of particles (sufficiently small volume elements that can be considered as points by a macroscopic observer) or material points; the position of each particle in the 3D Euclidean space is defined by a geometrical point, and its coordinates define a placement of this point. Each point can be clearly identified by its initial position in ambient space, denoted **X**, provided that a reference frame with origin O and basis vectors $\{\mathbf{e}_i\}$ have been defined.

DEFINITION.– The domain or region of space occupied by all material points at a given time is called the *configuration* (Figure 2.16).

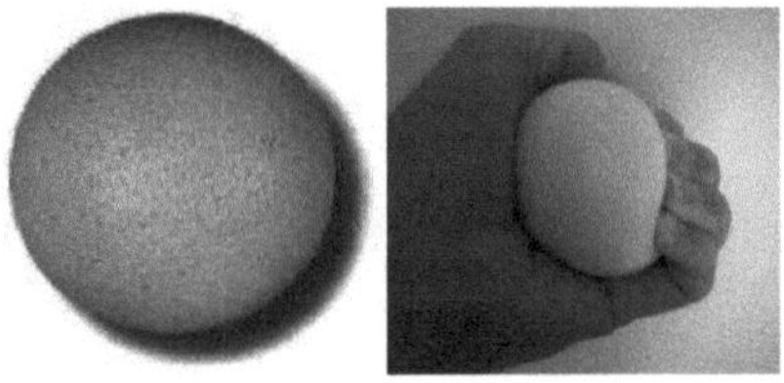

Figure 2.16. *Initial and actual configurations of a soft elastic ball [MOE 14]*

Two configurations are given some special privilege: the initial or *reference configuration* at initial time $(t=0)$, usually when the body is not deformed (although we shall see that the existence of such a stress-free configuration is problematic for biological tissues), and the *actual configuration* at time t. In this last configuration, material points are subjected to strains and stresses due to the action of external loading (body forces and contact forces).

Material points can be labeled so that particles have names.

EXAMPLE.– $\mathbf{Y}_1 = 100....0$, or $\mathbf{Y}_1 = Laura$, but it is more convenient to label particles by their initial position $\mathbf{X}$. This bypasses the need to name particles, which will be unambiguously identified by their position in the reference configuration.

2.4. Kinematics: displacement, transformation gradient, strains

During its motion, a given material point with label $\mathbf{X}$ will become a new point in space with position vector $\mathbf{x}$, function of time and initial position, so that we can define a *point mapping* from the reference to the current position

$$(\mathbf{X},t)\mapsto \mathbf{x}(\mathbf{X},t)$$

This mapping is defined for all particles (Figure 2.17), supposed to be invertible, so that we have formally

$$\mathbf{x}(\mathbf{X},t)\to \mathbf{X}=\mathbf{X}(\mathbf{x},t)$$

This presupposes that only one particle is present at one point at any time.

The reference positions $\mathbf{X}$ are called the *material coordinates*, and the corresponding vector $\mathbf{x}(\mathbf{X},t)$ the *spatial or current coordinates.*

There and in subsequent developments, the reference and actual configurations are respectively denoted by Ω_0 and Ω. Previous relation between the material and spatial positions imply in particular that both coordinates $\mathbf{X}$ and $\mathbf{x}$ shall coincide at initial time, so that one has the initial condition

$$\mathbf{X}=\mathbf{x}(\mathbf{X},0)$$

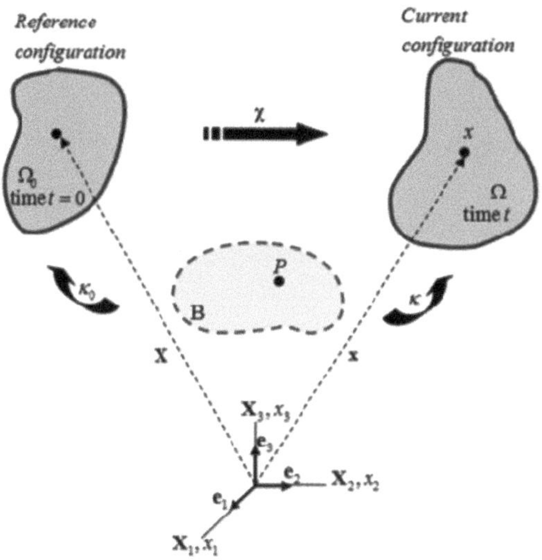

Figure 2.17. *Configuration and motion of a continuum body. For a color version of this figure, see www.iste.co.uk/ganghoffer/biomechanics.zip*

NOTE.– It is customary to denote quantities attached to the reference configuration with lower case letters, and quantities attached to the current configuration with upper case letters. This convention of notation will be adopted most of the time in the sequel.

EXAMPLE.– Rotation about $\mathbf{e}_3$ axis with angle $\alpha(t)$. The point mapping is given by

$$\begin{cases} x_1 = \cos\alpha(t) X_1 - \sin\alpha(t) X_2 \\ x_2 = \sin\alpha(t) X_1 + \cos\alpha(t) X_2 \\ x_3 = X_3 \end{cases}$$

This can be represented by the orthogonal matrix

$$[A] = \begin{pmatrix} \cos\alpha(t) & -\sin\alpha(t) & 0 \\ \sin\alpha(t) & \cos\alpha(t) & 0 \\ 0 & 0 & 1 \end{pmatrix}$$

The inverse mapping is clearly the rotation of angle $-\alpha(t)$ around axis $\mathbf{e}_3$ (this clearly saves computations), with matrix

$$\left[A^{-1}\right] = \begin{pmatrix} \cos\alpha(t) & \sin\alpha(t) & 0 \\ -\sin\alpha(t) & \cos\alpha(t) & 0 \\ 0 & 0 & 1 \end{pmatrix} \equiv \left[A^T\right]$$

This entails the inverse motion

$$\begin{cases} X_1 = \cos\alpha(t) x_1 + \sin\alpha(t) x_2 \\ X_2 = \sin\alpha(t) x_1 + \cos\alpha(t) x_2 \\ X_3 = x_3 \end{cases}$$

We note that $[A]$ does not depend on position, thus it is a *homogeneous mapping* or transformation, corresponding to a *rigid body rotation*. We can check that there is no volume change, due to the relation

$$\det[A] = 1$$

EXERCISE.– Check previous relation.

COROLLARY.– As a corollary of the continuum description, we have the following statements:

– the motion is continuous, so that two neighboring points will remain close (neighbors) during their motion;

– every connected domain in Ω_0 will remain connected in Ω;

– the boundary $\partial\Omega_0$ (set of material points) becomes the actual boundary $\partial\Omega$;

– the Jacobean of the transformation $J(\mathbf{X},t) := \det(\mathbf{F})$, is continuous w.r. to $\mathbf{X}$ and t;

– $J(\mathbf{X},t)$ keeps a constant sign during the motion, due to the initial value $J(\mathbf{X},t=0)=1$ and the physical impossibility of having $J(\mathbf{X},t)\equiv 0$, since compressing a volume to a point requires infinite energy!

DEFINITION.– Masses are represented in terms of *mass densities*, so that the (mass) density is defined by a limit process, for a volume V around point P with total mass inside $M(P)$,

$$\rho(\mathbf{x},t) = \lim_{V\to 0}\frac{M(P)}{V} \to M(P) := \int_P \rho dV \quad \text{(Eulerian representation)}$$

The referential density $\rho_0(\mathbf{X})$ is the value of $\rho(\mathbf{x},t)$ at initial time, *viz.*

$$\rho_0(\mathbf{X}) = \rho(\mathbf{x},t=0) \equiv \rho(\mathbf{x},0)$$

2.4.1. *Lagrangian versus Eulerian representation of properties*

Scalar- or tensor-valued fields attached to the material points, such as temperature, velocity, strain and stress measures, can be represented either in Lagrangian form, so that

$$F = \hat{F}(\mathbf{X},\mathbf{t})$$

or in Eulerian form, so that instead we shall write the same properties as

$$F = \tilde{F}(\mathbf{x},\mathbf{t})$$

It is clear that both previous functions are related to the motion, so that

$$\hat{F}(\mathbf{X}(\mathbf{x},\mathbf{t}),\mathbf{t}) = \tilde{F}(\mathbf{x},\mathbf{t})$$

EXAMPLE.– The temperature change for the distribution of chemical reactants within particles writes as

$$\hat{T} = atX_1 + btX_2 = \begin{pmatrix} at \\ bt \\ 0 \end{pmatrix}.\mathbf{X} \equiv \mathbf{w}.\mathbf{X}$$

The motion is the previous 2D rotation with matrix A, so that $\mathbf{x} = A.\mathbf{X}$.

– Lagrangian representation: $T = \mathbf{w}.\mathbf{X}$

– Eulerian representation:

$$\tilde{T} = \mathbf{w}.\mathrm{A}^T.\mathbf{x} = (\cos\alpha(t)x_1 + \sin\alpha(t)x_2)at + (\sin\alpha(t)x_1 + \cos\alpha(t)x_2)bt$$

(since A is an orthogonal transformation).

Temperature changes due to space-fixed radiations: the Eulerian temperature writes

$$\tilde{T} \equiv rtx_1 + stx_2 = \mathbf{z}.\mathbf{x}$$

The corresponding Lagrangian temperature is

$$\hat{T} = \mathbf{z}.A.\mathbf{X} = rt(\cos\alpha(t)X_1 - \sin\alpha(t)X_2) + st(\sin\alpha(t)X_1 + \cos\alpha(t)X_2)$$

For vector-valued functions, such as the velocity and acceleration vectors, we have from the Lagrangian viewpoint,

$$\mathbf{V}(\mathbf{X},t) := \frac{\partial \mathbf{x}(\mathbf{X},t)}{\partial t}, \quad \gamma(\mathbf{X},t) := \frac{\partial \mathbf{v}}{\partial t}$$

Both velocities are evaluated at fixed material coordinate $\mathbf{X}$. The velocity vector represents the tangent (vector) to the trajectory of a material point.

Previous partial derivatives are called *material derivatives*, that is, rate of changes holding the material particle fixed. The material derivative receives different expressions in the Lagrange or Euler representation: in a general setting, for any function F, scalar or tensor valued, the material derivative in the Lagrangian form is

$$\dot{F} = \left(\hat{F}(\mathbf{X},t)\right)^{\cdot} = \frac{\partial \hat{F}(\mathbf{X},t)}{\partial t}\dot{t} + \frac{\partial \hat{F}(\mathbf{X},t)}{\partial t}.\dot{\mathbf{X}} = \frac{\partial \hat{F}(\mathbf{X},t)}{\partial t}$$

since the particle is fixed, so that $\dot{\mathbf{X}} = \mathbf{0}$; this leads to the material gradient of F,

$$Grad\left(\hat{F}(\mathbf{X},t)\right) = \frac{\partial \hat{F}(\mathbf{X},t)}{\partial \mathbf{X}}$$

The same function when represented in Eulerian form, *viz.* $F = \tilde{F}(\mathbf{x},t)$, gives rise to the material derivative

$$\dot{F} = \left(\tilde{F}(\mathbf{x},t)\right)^{\cdot} = \frac{\partial \tilde{F}(\mathbf{x},t)}{\partial t}\dot{t} + \frac{\partial \tilde{F}(\mathbf{x},t)}{\partial \mathbf{x}}.\dot{\mathbf{x}},$$

sum of the time derivative at fixed spatial position $\mathbf{x}$, and of an additional convective term traducing the drag of the spatial gradient $\frac{\partial \tilde{F}(\mathbf{x},t)}{\partial \mathbf{x}}$ by the velocity $\mathbf{v} := \dot{\mathbf{x}}$. Thus, we are able to define the *spatial gradient* of F,

$$grad\left(\tilde{F}(\mathbf{x},t)\right) := \frac{\partial \tilde{F}(\mathbf{x},t)}{\partial \mathbf{x}}$$

EXAMPLE.– We come back to our previous example, from the Lagrangian representation, $\dot{T} = aX_1 + bX_2$. From the Eulerian representation, it results

$$\tilde{T} = \mathbf{w}.\mathrm{A}^T.\mathbf{x} = \left(\cos\alpha(t)x_1 + \sin\alpha(t)x_2\right)at + \left(\sin\alpha(t)x_1 + \cos\alpha(t)x_2\right)bt$$

$$\Rightarrow \left(\tilde{T}\right)^{\cdot} = \frac{\partial \tilde{T}}{\partial t} + grad\tilde{T}.\tilde{\mathbf{v}} = a\left(\cos\alpha(t)x_1 + \sin\alpha(t)x_2\right) + b\left(\sin\alpha(t)x_1 + \cos\alpha(t)x_2\right) \equiv aX_1 + bX_2$$

2.4.2. *Transformation gradient and strain measures*

In order to set the stage for expressing the transformation gradient in general curvilinear coordinates, we recall (see Chapter 1) the definitions of covariant and contravariant basis vectors: a system of coordinates $\left\{\theta^i\right\}$ is said to be *convective* if the material point keeps the same value of its coordinate during its motion, described by the point mapping $\mathbf{x}=\mathbf{x}\left(\theta^1,\theta^2,\theta^3,t\right)$. At a given material point M_0 in Ω_0, we define the *covariant vector basis* as the set of vectors tangent to the iso θ^i lines, *viz.*

$$\mathbf{G}_i=\mathbf{X}_{,i}:=\frac{\partial X^k}{\partial \theta^i}\mathbf{e}_i$$

Recall that the comma denotes the partial derivative and $\{\mathbf{e}_i\}$ is the fixed Cartesian basis in Euclidean space. The vectors $\{\mathbf{G}_i\}$ span the *tangent space* at point M_0, and they have dual vectors that span the cotangent space, given by

$$\mathbf{G}^i=\frac{\partial \theta^i}{\partial X^k}\mathbf{e}^i$$

The dual Cartesian basis $\{\mathbf{e}^i\}$ therein is in fact identical to the basis $\{\mathbf{e}_i\}$, but it is introduced as a matter of consistency (position of indices). One can accordingly define the following metric tensors in Ω_0

$$G_{ij}:=\mathbf{G}_i.\mathbf{G}_j,\quad \mathrm{G}^{ij}:=\mathbf{G}^i.\mathbf{G}^j$$

The identity tensor receives the following spectral decomposition,

$$\mathbf{I}=\mathbf{G}^i\otimes\mathbf{G}_i=\mathbf{G}_i\otimes\mathbf{G}^i=\mathbf{G}^{-1}$$

By analogy, one defines the same tangent and cotangent basis vectors in the actual configuration Ω

$$\mathbf{g}_i=\mathbf{x}_{,i}:=\frac{\partial x^k}{\partial \theta^i}\mathbf{e}_i,\quad \mathbf{g}^i=\frac{\partial \theta^i}{\partial x^k}\mathbf{e}^i$$

with associated metric tensor given in indicial form as

$$g_{ij}:=\mathbf{g}_i.g_j,\quad \mathrm{g}^{ij}:=\mathbf{g}^i.\mathbf{g}^j$$

and the identity tensor in Ω writes similarly as

$$\mathbf{I}=\mathbf{g}^i\otimes\mathbf{g}_i=\mathbf{g}_i\otimes\mathbf{g}^i=\mathbf{g}^{-1}$$

DEFINITION.– The *displacement* relates the material and spatial positions, so it is simply the comparison between the actual and initial position of the same particle,

$$\mathbf{u} = \mathbf{u}\left(\theta^i, t\right) = \mathbf{x} - \mathbf{X} = u_i \mathbf{g}^i = U_i \mathbf{G}^i$$

The last two equalities are the decomposition of $\mathbf{u}$ in both configurations in terms of its covariant components.

In the case where Cartesian coordinates $\left\{x^i\right\}$ are used, we shall simply write, $\mathbf{u} = \mathbf{u}\left(x^i, t\right) = \mathbf{x} - \mathbf{X} = u_i \mathbf{e}_i$.

The partial derivatives of the displacement vector shall next lead to the elaboration of the transformation gradient: let us first write,

$$\mathbf{u}_{,i} = \frac{\partial \mathbf{u}}{\partial \theta^i}\left(\theta^i, t\right) = \mathbf{x}_{,i} - \mathbf{X}_{,i} = \mathbf{g}_i - \mathbf{G}_i = u_{m|i} \mathbf{g}^m = U_{m|i} \mathbf{G}^m$$

Referring to Chapter 1 as the notion of covariant derivative (the partial derivative of a tensor coincides with its covariant derivative), this leads to

$$\mathbf{g}_i = \mathbf{G}_i + \mathbf{u}_{,i} = \left(\delta_i^m + U_{m|i}\right)\mathbf{G}^m \equiv F_{.i}^m \mathbf{G}_m$$

in which we have introduced the components $F_{.i}^m$ of a mixed tensor $\mathbf{F}$ mapping the basis vectors $\left\{\mathbf{G}_m\right\}$ to the basis vectors $\left\{\mathbf{g}_i\right\}$. Previous relation also leads to

$$\mathbf{G}_i = \mathbf{g}_i - \mathbf{u}_{,i} = \left(\delta_i^m - u^m{}_{|i}\right)\mathbf{g}_m \equiv \left(\mathbf{f}^{-1}\right)_{.i}^m \mathbf{g}_m$$

wherein the mapping $\mathbf{f}^{-1}$ with components $\left(\mathbf{f}^{-1}\right)_{.i}^m$ sends the basis vectors $\mathbf{g}_m$ to the basis vectors $\mathbf{G}_i$.

The deformation gradient is then elaborated as the dyadic product

$$\mathbf{F} = \mathbf{g}_i \otimes \mathbf{G}^i$$

with transpose and inverse successively given by

$$\mathbf{F}^T = \mathbf{G}^i \otimes \mathbf{g}_i, \quad \mathbf{F}^{-1} = \mathbf{G}_i \otimes \mathbf{g}^i$$

The reader may easily check that the inverse of the transpose is given by $\mathbf{F}^{-T} = \mathbf{g}^i \otimes \mathbf{G}_i$.

The transformation rules of covariant and contravariant vectors are accordingly

$$\mathbf{g}_i = \mathbf{F}.\mathbf{G}_i \Leftrightarrow \mathbf{G}_i = \mathbf{F}^{-1}.\mathbf{g}_i, \quad \mathbf{g}^i = \mathbf{F}^{-T}.\mathbf{G}^i \Leftrightarrow \mathbf{G}^i = \mathbf{F}^T.\mathbf{g}^i$$

Now due to the identities

$$\mathbf{F}^{-T}.\mathbf{G}^i = \left(\mathbf{g}^j \otimes \mathbf{G}_j\right).\mathbf{G}^i = \delta^i_{\ j}\mathbf{g}^j = \mathbf{g}^i, \quad \mathrm{d}\mathbf{x}{=}\mathbf{g}_i d\theta^i, \quad \mathrm{d}\mathbf{X}{=}\mathbf{G}_i d\theta^i$$

it follows the important relation

$$d\mathbf{x} = \mathbf{F}.d\mathbf{X}$$

The transformation gradient $\mathbf{F}$ thus maps the referential line element $d\mathbf{X}$ to the physical line element $d\mathbf{x}$ (imagine small vectors of infinitesimal length but with fixed directions), Figure 2.18. Observe that the infinitesimal vectors $d\mathbf{X}$ and $d\mathbf{x}$ can be conceived as the tangent vectors (in fact, they belong to the tangent spaces at points M_0 and M) of some curved lines passing through points M_0 and M.

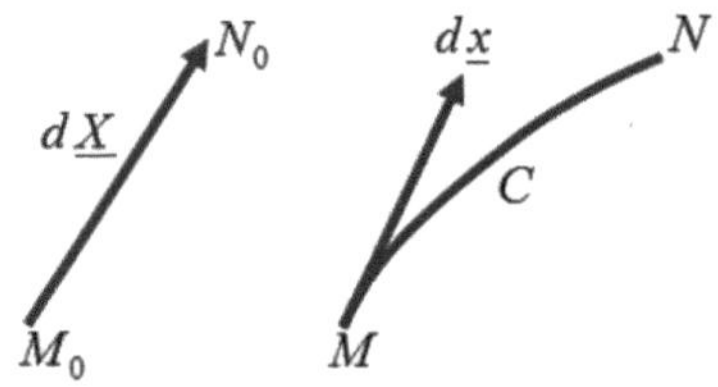

Figure 2.18. *Mapping of material line element by F*

REMARK.– The case where $\mathbf{F}$ is independent of the material point, so that $\mathbf{F} = \mathbf{F}(t)$, is referred to as an homogenous transformation. Such transformations are easy to conceive (for example, rigid body motion, rotations, etc.).

We can then write the following approximation of the point mapping:

$$\mathbf{MN} = F(X,t).\mathbf{M}_0\mathbf{N}_0 + \|\mathbf{M}_0\mathbf{N}_0\| o\left(\mathbf{M}_0\mathbf{N}_0, t\right) \text{ , with}$$

$$o\left(\mathbf{M}_0\mathbf{N}_0, t\right) \to 0 \text{ when } \|\mathbf{M}_0\mathbf{N}_0\| \to 0$$

Thus, we should not confuse the transport of vector $\mathbf{M}_0\mathbf{N}_0$ (a straight segment in Ω_0) with its transport by the motion, in general a curved arc; both notions only coincide in the case of an homogeneous motion.

REMARK.– Since $\mathbf{F}$ is attached to two different configurations Ω_0 and Ω, it is sometimes called a *two-point mapping*: as a consequence it does not really deserve the name and properties of a second order tensor. It will nevertheless be called a tensor by an abuse of vocabulary.

We have thus replaced the essentially nonlinear point mapping $\mathbf{x}(\mathbf{X},t)$ with a simpler object, which is its first order approximation by the linear mapping $\mathbf{F}$ relating differential elements.

REMARK.– Some theories in continuum mechanics involve higher order gradients, but these will not be considered in this book.

We further define the *material gradient of the spatial position*, tensor

$$Grad\mathbf{x} := \frac{\partial \mathbf{x}}{\partial X^j} \otimes \mathbf{e}^j = \frac{\partial \mathbf{x}}{\partial \theta^i}\frac{\partial \theta i}{\partial X^j}\mathbf{e}^j = \mathbf{g}_i \otimes \mathbf{G}^i = \mathbf{F}$$

We define similarly the *spatial gradient* of the material position, tensor

$$grad\mathbf{X} := \frac{\partial \mathbf{X}}{\partial x^j} \otimes \mathbf{e}^j = \frac{\partial \mathbf{X}}{\partial \theta^i}\frac{\partial \theta i}{\partial x^j}\mathbf{e}^j = \mathbf{G}_i \otimes \mathbf{g}^i = \mathbf{F}^{-1}$$

EXERCISE.– Check previous identities.

The component form of the introduced mappings is:

$$\mathbf{F} = F_{ij}\mathbf{G}^i \otimes \mathbf{G}^j = f_{ij}\mathbf{g}^i \otimes \mathbf{g}^j$$

$$\Rightarrow F_{ij} = \mathbf{G}_i.\mathbf{F}.\mathbf{G}_j,\quad \mathrm{f}_{ij} = \mathbf{g}_i.\mathbf{F}.\mathbf{g}_j,\quad \mathbf{F}^{-1} = \left(F^{-1}{}_{ij}\right)\mathbf{G}^i \otimes \mathbf{G}^j = \left(f^{-1}\right)_{ij}\mathbf{g}^i \otimes \mathbf{g}^j$$

$$\Rightarrow F_{ij} = \left(f^{-1}\right)_{ji} = \mathbf{G}_i.\mathbf{g}_j$$

Note that F_{ij} and $\left(f^{-1}\right)_{ji}$ are components defined in different bases, although they are equal.

The mapping of material line elements based on curvilinear, material and spatial coordinates is shown in Figure 2.19.

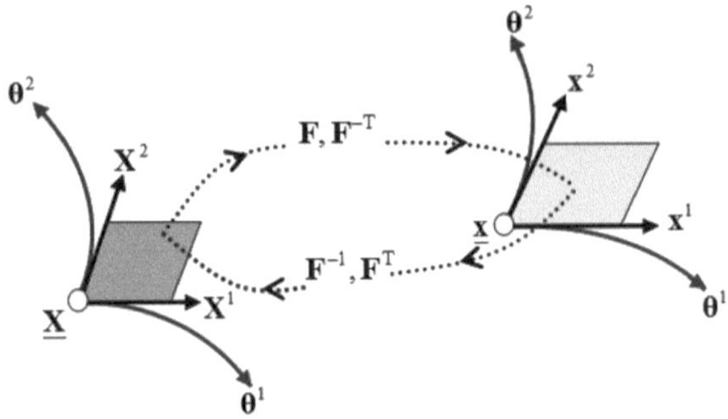

Figure 2.19. *Mapping of material line elements based on curvilinear, material and spatial coordinates. For a color version of this figure, see www.iste.co.uk/ganghoffer/biomechanics.zip*

In Cartesian coordinates, the transformation mapping expresses as

$$F_{ij} = \frac{\partial x_i}{\partial X_j} \rightarrow \mathbf{F} = \frac{\partial x_i}{\partial X_j}\mathbf{e}_i \otimes \mathbf{e}_j \qquad \text{expressed in Cartesian coordinates}$$

EXERCISE.– Check the following identities

$$F_{ij} = G_{ij} + U_{i|j},\quad \left(\mathrm{f}^{-1}\right)_{ij} = \mathbf{g}_i.\mathbf{G}_j = g_{ij} - u_{i|j},\quad \mathbf{g}_i = F_{mj}\left(\mathbf{G}^m \otimes \mathbf{G}^j\right) = .\mathbf{G}_{i|j} = F^{m}_{.i}\mathbf{G}_m = \left(\mathbf{G}^m.\mathbf{g}_i\right)\mathbf{G}_m$$

$$\mathbf{G}_i = \left(\mathrm{f}^{-1}\right)_{mj}\left(\mathbf{g}^m \otimes \mathbf{g}^j\right).\mathbf{g}_i = \left(\mathrm{f}^{-1}\right)^{m}_{.i}\mathbf{g}_m = \left(\mathbf{g}^m\mathbf{G}_i\right)\mathbf{g}_m$$

$$\mathbf{g}^i = \left(F^{-T}\right)^{m}_{.i}\mathbf{G}^m = \left(\mathbf{G}_m.\mathbf{g}^i\right)\mathbf{G}^m,\quad \mathbf{G}^i = \left(f^T\right)^{i}_{m.}\mathbf{g}^m = \left(\mathbf{g}_m.\mathbf{G}^i\right)\mathbf{g}^m$$

EXAMPLE.– Components of the transformation gradient in cylindrical coordinates (r,θ,z), with the basis set of vectors $(\mathbf{e}_r,\mathbf{e}_\theta,\mathbf{e}_z)$. The infinitesimal vector writes

$$d\mathbf{OM} = \begin{pmatrix} dr \\ rd\theta \\ dz \end{pmatrix} \equiv \begin{pmatrix} dy^1 \\ dy^2 \\ dy^3 \end{pmatrix}.$$

– For a scalar valued function,

$$f(r,\theta,z) \rightarrow df = gradf.d\mathbf{OM} \Rightarrow gradf = \begin{pmatrix} \dfrac{\partial f}{\partial r} & \dfrac{1}{r}\dfrac{\partial f}{\partial \theta} & \dfrac{\partial f}{\partial z} \end{pmatrix}$$

– For a vector-valued function

$$\mathbf{V}=\mathbf{V}(r,\theta,z)=V^i\mathbf{e}_i \rightarrow d\mathbf{V}=grad\mathbf{V}.\mathrm{d}\mathbf{OM} \rightarrow d\left(V^i\mathbf{e}_i\right)=dV^i\mathbf{e}_i+V^i\frac{\partial \mathbf{e}_i}{\partial y^k}dy^k$$

The derivatives of the vector basis $\frac{\partial \mathbf{e}_i}{\partial y_k}$ are determined: due to the relations $\mathbf{e}_r.\mathbf{e}_r=1,\ \ \mathbf{e}_r.\mathbf{e}_\theta=0$, it holds that

$$\frac{\partial \mathbf{e}_r}{\partial \theta}=\mathbf{e}_\theta,\ \ \frac{\partial \mathbf{e}_\theta}{\partial \theta}=-\mathbf{e}_r,$$ and all other derivatives vanish.

Writing $\frac{\partial \mathbf{e}_i}{\partial y^k}=\Gamma_{ij}^k\mathbf{e}_k$ formally leads to the identification of the non-nil Christoffel symbols

$$\Gamma_{12}^2=\frac{1}{r}=-\Gamma_{22}^1$$

We can thus write

$$\left(dV^i+V^p\Gamma_{pk}^i dy^k\right)\mathbf{e}_i=\left(grad\mathbf{V}\right)_{ij}dy^j\mathbf{e}_i$$

$$\Rightarrow\left(grad\mathbf{V}\right)_{ij}=\frac{\partial V^i}{\partial y^j}+\Gamma_{pk}^i V_p=\begin{pmatrix} V_{r,r} & \frac{1}{r}\left(V_{r,\theta}-V_\theta\right) & V_{r,z} \\ V_{\theta,r} & \frac{1}{r}\left(V_{\theta,\theta}+V_r\right) & V_{\theta,z} \\ V_{z,r} & \frac{1}{r}V_{z,\theta} & V_{z,z}\end{pmatrix}$$

This last tensor is the transformation gradient when $\mathbf{V}=\mathbf{x}$.

The trace of this tensor delivers the divergence as $div\mathbf{V}=V_{r,r}+\frac{1}{r}\left(V_{\theta,\theta}+V_r\right)+V_{z,z}$.

The gradient of a second order fully contravariant tensor $\mathbf{T}$ can be computed based on Christoffel symbols as

$$\left(grad\mathbf{T}\right)_k^{ij}=\frac{\partial T^{ij}}{\partial y^k}+\Gamma_{kp}^i T^{pj}+\Gamma_{pq}^j T^{iq}$$

2.4.3. *Transport of line, surface and volume elements*

Now, we determine how infinitesimal material line, surface and volume elements transform under the motion from Ω_0 to Ω; we set

$$d\mathbf{M}_0 = d\mathrm{l}_0\,\mathbf{t}_0, \quad d\mathbf{M} = d\mathrm{l}\,\mathbf{t}$$

with $\mathbf{t}_0, \mathbf{t}$ two tangent unit vectors, and $d\mathrm{l}_0, d\mathrm{l}$ the length of previous vectors. It is then clear that due to the relation $d\mathbf{x} = \mathbf{F}.d\mathbf{X}$, this results in

$$d\mathrm{l}\,\mathbf{t} = \mathbf{F}.(d\mathrm{l}_0\,\mathbf{t}_0)$$

which defines the convective transport of line elements.

Next, let us consider an elementary area of magnitude dS_0 built from the vectors $d\mathbf{X}, d\mathbf{X}'$, with

$$d\mathbf{X} \times d\mathbf{X}' = \mathbf{N}_0 dS_0$$

the infinitesimal surface element (vector), supported by the unit referential normal vector $\mathbf{N}_0$, with dS_0 as the area (Figure 2.20).

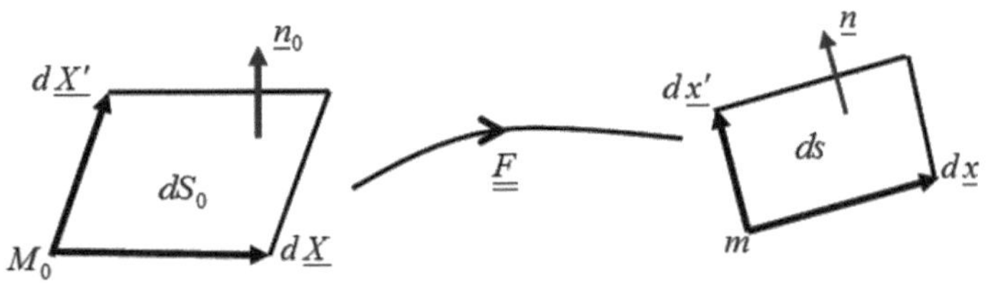

Figure 2.20. *Convective transport of surface elements. For a color version of this figure, see www.iste.co.uk/ganghoffer/biomechanics.zip*

Similarly, we have the relation

$$d\mathbf{x} \times d\mathbf{x}' = \mathbf{n}ds$$

with $\mathbf{n}$ the unit actual normal, and ds the area of the surface. Due to the relations $d\mathbf{x} = \mathbf{F}.d\mathbf{X}, \quad d\mathbf{x}' = \mathbf{F}.d\mathbf{X}'$, it holds that (check this relation)

$$n_i dS = \varepsilon_{ijk} dx_j dx'_k = \varepsilon_{ijk} F_{jq} F_{kr} dX_q dX'_r$$

Multiplying the last relation by F_{ip} and summing on index i results in

$$F_{ip} n_i dS = \varepsilon_{ijk} F_{ip} F_{jq} F_{kr} dX_q dX'_r = \varepsilon_{pqr} \det(\mathbf{F}) dX_q dX'_r = J \varepsilon_{pqr} dX_q dX'_r$$

Hence,

$$\mathbf{F}^T.\mathbf{n}ds = J d\mathbf{X} \times d\mathbf{X}' \rightarrow \mathbf{F}^T.\mathbf{n}ds = J\mathbf{N}_0 dS_0$$

Thus, finally

$$\mathbf{n}ds = J\mathbf{F}^{-T}.\mathbf{N}_0 dS_0$$

which is the transport formula of area elements (Figure 2.20), coined the *Nanson formula*. We can see that the vector $\mathbf{n}ds$ is the transport of vector $\mathbf{N}_0 dS_0$ by $\mathbf{F}$: this reflects the fact that the surface elements $\mathbf{n}ds$ and $\mathbf{N}_0 dS_0$ actually reside in the cotangent space, so they do not obey transport rules for elements of the tangent space, as line elements.

The volume of a parallelepiped built from the three elementary vectors $d\mathbf{X}, d\mathbf{X}', d\mathbf{X}''$ is their mixed product

$$dV_0 = (d\mathbf{X}, d\mathbf{X}', d\mathbf{X}'') = \varepsilon_{ijk} dX_i dX'_j dX''_k$$

From the equalities $d\mathbf{x} = \mathbf{F}.d\mathbf{X}, \quad d\mathbf{x}' = \mathbf{F}.d\mathbf{X}', \quad d\mathbf{x}'' = \mathbf{F}.d\mathbf{X}''$, we obtain

$$dv = (d\mathbf{x}, d\mathbf{x}', d\mathbf{x}'') = \det\left(\left[F_{ij}\right]\left[d\mathbf{x} \quad d\mathbf{x}' \quad d\mathbf{x}''\right]\right)$$

Denoting therein $\left[d\mathbf{x} \quad d\mathbf{x}' \quad d\mathbf{x}''\right]$, the matrix made by the three-column vector $d\mathbf{x}, d\mathbf{x}', d\mathbf{x}''$; thus we obtain the transport formula for infinitesimal volume elements

$$dv = \det(\mathbf{F}) dV_0 = J dV_0$$

All previous identities have a tensorial nature, thus they are valid in any coordinate system (Cartesian or not); it should be noted that the expression $F_{ij} = \frac{\partial x_i}{\partial X_j}$ is only valid in Cartesian coordinates.

2.5. Deformation tensors

Let us introduce the *displacement gradient* (evaluated at fixed time)

$$\mathbf{H} = Grad\mathbf{U}(\mathbf{X}, t)$$

with the material displacement given as the vector

$$\mathbf{U} = \mathbf{M_0M} = \mathbf{OM} - \mathbf{OM_0}$$

The displacement field therein is considered as a function of the Lagrangian coordinates.

The displacement gradient can itself be decomposed into symmetrical and asymmetrical parts, *viz.*

$$\mathbf{H} = \boldsymbol{\varepsilon} + \mathbf{w}, \quad \boldsymbol{\varepsilon} := \frac{1}{2}\left(\mathbf{H} + \mathbf{H}^T\right) \leftrightarrow \varepsilon_{ij} = \frac{1}{2}\left(H_{ij} + H_{ji}\right),$$
$$\mathbf{w} := \frac{1}{2}\left(\mathbf{H} - \mathbf{H}^T\right) \leftrightarrow \mathrm{w}_{ij} := \frac{1}{2}\left(H_{ij} - H_{ji}\right)$$

where in the component form has been given as well. We recognize the small strain (linearized strain) tensor $\boldsymbol{\varepsilon}$ of linear elasticity, while the asymmetrical tensor $\mathbf{w}$ obviously describes local rotations (it can be represented by its dual vector).

The transformation gradient is then given in terms of the displacement gradient as

$$\mathbf{F} = \mathbf{H} + \mathbf{I}$$

Note that since $\mathbf{F}$ is not nil for a rigid body rotation, it is not an adequate candidate for a strain measure (check this last statement), so we further proceed and define the right Cauchy-Green (right because $\mathbf{F}$ appears in the right position in the product) tensor

$$\mathbf{C} := \mathbf{F}^T.\mathbf{F} = \mathbf{U}^2 = \mathbf{I} + 2\boldsymbol{\varepsilon} + \frac{1}{2}\mathbf{H}^T.\mathbf{H}$$

This entails the *Lagrangian measure of deformation* (or strain, in other words), also called the Green Saint-Venant strain tensor.

$$\mathbf{E} := \frac{1}{2}(\mathbf{C} - \mathbf{I}) = \frac{1}{2}\left(\mathbf{H} + \mathbf{H}^T\right) + \frac{1}{2}\mathbf{H}^T.\mathbf{H} = \boldsymbol{\varepsilon} + \frac{1}{2}\mathbf{H}^T.\mathbf{H}$$

This tensor enjoys the property of vanishing for rigid body motions.

Note that $\mathbf{C}$ is a mapping from the tangent space to Ω_0 to itself.

Physical significance of $\mathbf{E}$: let us consider two infinitesimal vectors $\delta\mathbf{M}_0 = \delta\mathbf{X}, \mathrm{d}\mathbf{M}_0 = d\mathbf{X}$ in the reference configuration, being transported to the actual configuration to two new infinitesimal vectors $\delta\mathbf{M} = \delta\mathbf{x}, \mathrm{d}\mathbf{M} = d\mathbf{x}$ in the actual configuration (Figure 2.21).

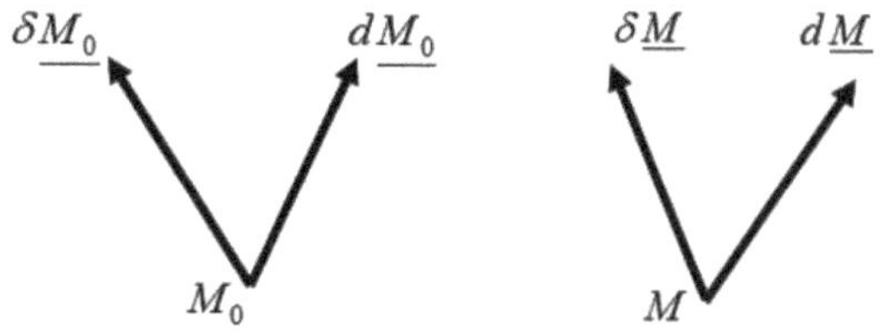

Figure 2.21. *Transport of two line elements from to Ω_0 to Ω*

The notion of strain will indicate how length and angles defined by pairs of 'small' vectors will transform under the motion; let us then compare the scalar products of two infinitesimally small vectors $\delta\mathbf{M}_0, \mathrm{d}\mathbf{M}_0$ with the similar scalar product of their spatial counterparts $\delta\mathbf{M}, \mathrm{d}\mathbf{M}$:

$$\begin{aligned}&\delta\mathbf{M}.\mathrm{d}\mathbf{M} - \delta\mathbf{M}_0.\mathrm{d}\mathbf{M}_0 = \delta\mathbf{x}.d\mathbf{x} - \delta\mathbf{X}.d\mathbf{X} = F_{i\alpha}dX_\alpha F_{i\beta}\delta X_\beta - dX_\alpha\delta X_\beta = \\ &= \left(C_{\alpha\beta} - \delta_{\alpha\beta}\right)dX_\alpha\delta X_\beta = \delta X_\beta 2E_{\alpha\beta}dX_\alpha = 2\delta\mathbf{X}.\mathbf{E}.d\mathbf{X}\end{aligned}$$

so that tensor $\mathbf{E}$ measures the variation of the scalar product of two small vectors.

It follows that

PROPOSITION.– A necessary and sufficient condition for the absence of deformation is $\mathbf{E} = \mathbf{0}$.

Let us further introduce the left Cauchy-Green tensor (observe that here $\mathbf{F}$ appears on the left in the product), also coined Finger tensor,

$$\mathbf{b} := \mathbf{F}\mathbf{F}^T = \mathbf{v}^2$$

Observe that $\mathbf{B}$ is an Eulerian mapping from the tangent space to Ω to itself, so that it should in principle be written with lower case letters as $\mathbf{b}^2$.

We now evaluate the same scalar product as before, but express it instead in Ω or rather in its tangent space: an easy calculation leads to

$$\delta\mathbf{M}.\mathrm{d}\,\mathbf{M} - \delta\mathbf{M}_0.\mathrm{d}\,\mathbf{M}_0 = \delta\mathbf{x}.d\mathbf{x} - \delta\mathbf{X}.d\mathbf{X} = 2\delta\mathbf{x}.\mathbf{e}.d\mathbf{x}$$

with the Eulerian strain tensor, called the Euler Almansi strain tensor

$$\mathbf{e} := \frac{1}{2}\left(\mathbf{I} - \mathbf{B}^{-1}\right)$$

Tensors $\mathbf{C}, \mathbf{B}, \mathbf{E}, \mathbf{e}$ are all positive definite bilinear forms when considered as acting on pairs of vectors to produce scalars: we can, for example, write the action of $\mathbf{C}$ on the small referential vector as

$$\mathbf{C}(d\mathbf{X}, d\mathbf{X}) = d\mathbf{X}.\mathbf{C}.d\mathbf{X}$$

Since they measure variations of length and angles, previous strain tensors play the role of a metric.

Variations in length and angle: the components of tensors $\mathbf{C}$ and $\mathbf{E}$ can be further interpreted in terms of changes in angle and length.

DEFINITION.– The dilatation in the direction of vector $d\mathbf{X}$ is the scalar quantity

$$\lambda(d\mathbf{X}) := \frac{\|d\mathbf{x}\|}{\|d\mathbf{X}\|} = \left(C(d\mathbf{X}, d\mathbf{X})\right)^{1/2} / \|d\mathbf{X}\|$$

EXERCISE.– Compute the dilation from the previous expression of the scalar product of two identical vectors $\delta\mathbf{X} = d\mathbf{X}$

Let us then consider the unit vectors $\mathbf{u}_0, \mathbf{u}$ such that $d\mathbf{X} = dl_0\mathbf{u}_0$, $d\mathbf{x}=dl\mathbf{u}$; we get

$$\lambda(d\mathbf{X}) := \frac{\|\mathbf{F}.d\mathbf{X}\|}{\|d\mathbf{X}\|} = \frac{\|\mathbf{F}.dl_0\mathbf{u}_0\|}{\|dl_0\mathbf{u}_0\|} = \|\mathbf{F}.\mathbf{u}_0\| \equiv \lambda(\mathbf{u}_0)$$

with $\lambda(\mathbf{u}_0) := \|\mathbf{F}(\mathbf{u}_0).\mathbf{F}(\mathbf{u}_0)\|^{1/2} = (\mathbf{u}_0.\mathbf{F}^T.\mathbf{F}.\mathbf{u}_0)^{1/2} = (\mathbf{u}_0.\mathbf{C}.\mathbf{u}_0)^{1/2} = \mathbf{C}(\mathbf{u}_0,\mathbf{u}_0)^{1/2}$

when $\mathbf{u}_0 = \mathbf{e}_i$, the basis vector, then previous identity leads to

$$\lambda(\mathbf{e}_i) = \mathbf{C}(\mathbf{e}_i,\mathbf{e}_i)^{1/2} = \sqrt{C_{ii}} \text{ called stretch ratios.}$$

Thus the square root of the diagonal coefficients of the positive definite tensor $\mathbf{C}$ represents the dilatation in this direction: the material element with length dl_0 will thus become a new material line with actual length $dl = \lambda(\mathbf{e}_i)dl_0$.

Shear of two directions: let consider two vectors making some initial angle α_0 in Ω_0, which become two vectors in Ω making the angle α (Figure 2.22).

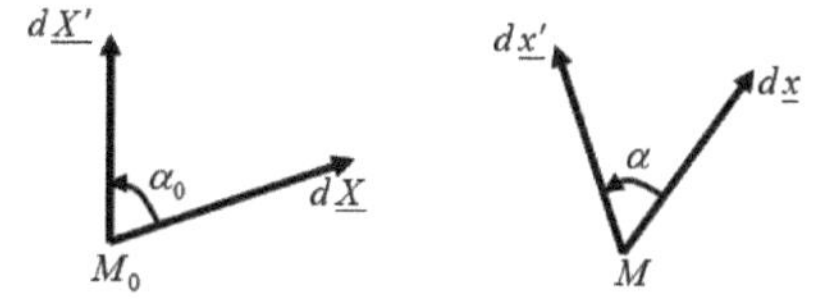

Figure 2.22. *Illustration of shear*

We easily get the scalar product $d\mathbf{x}'.d\mathbf{x} = \|d\mathbf{x}'\|\|d\mathbf{x}\|\cos\alpha \Rightarrow \cos\alpha = \dfrac{d\mathbf{x}'.d\mathbf{x}}{\|d\mathbf{x}'\|\|d\mathbf{x}\|}$.

Adopting as before the expressions $d\mathbf{X} = dl_0\mathbf{u}_0$, $d\mathbf{x}=dl\mathbf{u}$, from

$C(d\mathbf{X},d\mathbf{X}) = dl_0^2 C(\mathbf{u}_0,\mathbf{u}_0)$, $\lambda(\mathbf{u}_0) = \mathbf{C}(\mathbf{u}_0,\mathbf{u}_0)^{1/2}$, we easily obtain

$$\cos\alpha = \frac{\mathbf{C}(\mathbf{u}_0,\mathbf{u}'_0)}{\lambda(\mathbf{u}'_0)\lambda(\mathbf{u}_0)}$$

When the initial directions are orthogonal, $\alpha_0 = \frac{\pi}{2}$, we obtain the angular variation $\gamma := \frac{\pi}{2} - \alpha$, given by

$$\sin\gamma(\mathbf{u}_0, \mathbf{u'}_0) = \frac{\mathbf{C}(\mathbf{u}_0, \mathbf{u'}_0)}{\lambda(\mathbf{u'}_0)\lambda(\mathbf{u}_0)}$$

Thus the angular variations in different directions due to motion can be computed based on knowledge of the components of tensor $\mathbf{C}$; selecting the basis vectors in particular, *viz.* $\mathbf{u}_0 = \mathbf{e}_i$ and $\mathbf{u'}_0 = \mathbf{e}_j$ results in

$$\sin\gamma(\mathbf{e}_i, \mathbf{e'}_i) = \frac{C_{ij}}{C_{\underline{ii}} C_{\underline{jj}}}, \text{ valid for } i \neq j$$

Expression of strain tensors in curvilinear coordinates: the previous relation is tensorial, so it remains valid irrespective of the adopted coordinates. We have to replace the identity tensor with the corresponding metric tensor in either configuration, so that

$$\mathbf{E} = \frac{1}{2}(\mathbf{C} - \mathbf{G}), \quad \mathbf{e} = \frac{1}{2}(\mathbf{g} - \mathbf{B}^{-1})$$

$$\mathbf{e} = \mathbf{F}^{-T}.\mathbf{E}.\mathbf{F}^{-1} \leftrightarrow \mathbf{E} = \mathbf{F}^{T}.\mathbf{e}.\mathbf{F} = \frac{1}{2}\left[(\mathbf{G}^i \otimes \mathbf{g}_i).(\mathbf{g}_j \otimes \mathbf{G}^j) - G_{ij}\mathbf{G}^i \otimes \mathbf{G}^j\right]$$

$$\mathbf{C} = \mathbf{F}^{-1}.\mathbf{b}.\mathbf{F},$$

$$\text{with } \mathbf{b} = b^{ij}\mathbf{g}_i \otimes \mathbf{g}_j = \mathbf{G}^i.\mathbf{G}^j\mathbf{g}_i \otimes \mathbf{g}_j \leftrightarrow \mathbf{C} = \mathbf{g}_i.\mathbf{g}_j\mathbf{G}^i \otimes \mathbf{G}^j$$

Previous equalities entail the components of the strain tensor in curvilinear coordinates

$$E_{ij} = \frac{1}{2}(\mathbf{g}_i.\mathbf{g}_j - \mathbf{G}_i.\mathbf{G}_j) = \frac{1}{2}(g_{ij} - G_{ij}) =$$

$$= \frac{1}{2}(\mathbf{G}_i.\mathbf{u}_{,j} + \mathbf{G}_j.\mathbf{u}_{,i} + \mathbf{u}_{,i}.\mathbf{u}_{,j}) = \frac{1}{2}\left(\mathbf{U}_{i|j} + \mathbf{U}_{i|j} + \mathbf{U}_{r|i}.\mathbf{U}^r_{|j}\right)$$

When Lagrangian coordinates are considered, so that $\theta^i \equiv X^i$, this leads to

$$\mathbf{G}_i = \mathbf{e}_i, \ \mathbf{g}^i = \frac{\partial x^k}{\partial X^i}\mathbf{e}_k \to E_{ij} = \frac{1}{2}\left(\frac{\partial x^k}{\partial X^i}\frac{\partial x^m}{\partial X^j}\delta_{km} - \delta_{ij}\right)$$

in Lagrangian coordinates.

The spatial counterpart can be obtained as

$$\mathbf{e}=\frac{1}{2}\left(\mathbf{g}-\mathbf{b}^{-1}\right)=\frac{1}{2}\left(\mathbf{g}_i.\mathbf{u}_{,j}+\mathbf{g}_j.\mathbf{u}_{,i}+\mathbf{u}_{,i}.\mathbf{u}_{,j}\right)=\frac{1}{2}\left(\mathbf{u}_{i|j}+\mathbf{u}_{i|j}+\mathbf{u}_{r|i}.\mathbf{u}^r_{|j}\right)$$

$$\Rightarrow e_{ij}=\frac{1}{2}\left(\mathbf{g}_i.\mathbf{g}_j-\mathbf{G}_i.\mathbf{G}_j\right)=\frac{1}{2}\left(g_{ij}-G_{ij}\right)\equiv E_{ij}$$

This leads to the important conclusion that *both tensors* $\mathbf{E}$ *and* $\mathbf{e}$ *have the same components in their respective bases, but they do not express in the same basis.*

When adopting Eulerian coordinates, so that $\theta^i \equiv x^i$, $\mathbf{g}_i=\mathbf{e}_i$, we obtain

$$e_{ij}=\frac{1}{2}\left(\delta_{ij}-\frac{\partial X^m}{\partial x^i}\frac{\partial X^n}{\partial x^j}\delta_{mn}\right) \text{ in Eulerian coordinates.}$$

EXERCISE.– Simple shear (illustrated in Figure 2.23) is defined by the point mapping (ν is a parameter)

$$\mathbf{x}=\mathbf{X}+2\nu X_2\mathbf{e}_1 \rightarrow \begin{cases} x_1=X_1+2\nu X_2 \\ x_2=X_2 \\ x_3=X_3 \end{cases}$$

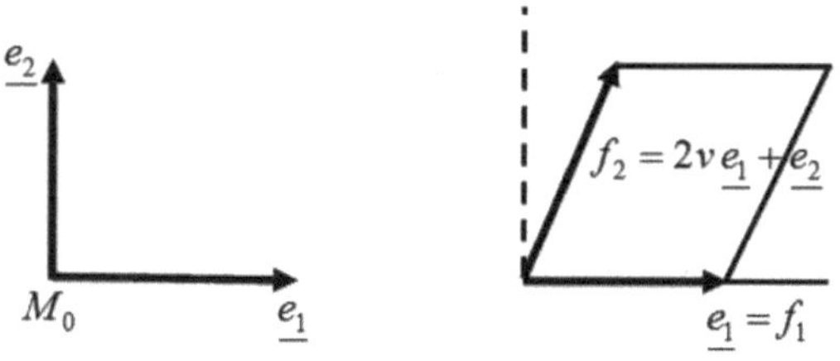

Figure 2.23. *Simple shear*

Unified definition of deformation measures:

For material tensors, $\mathbf{E}^{(m)} := \begin{vmatrix} Ln\mathbf{U} \text{ if m=0, Hencky tensor} \\ \frac{1}{m}\left(\mathbf{U}^m-\mathbf{G}\right) \end{vmatrix}$

For spatial tensors, $\mathbf{e}^{(m)} := \begin{vmatrix} Ln\mathbf{v} \text{ if m=0, Hencky spatial tensor} \\ \frac{1}{m}\left(\mathbf{v}^m - \mathbf{g}\right) \end{vmatrix}$

Due to the symmetry of tensors $\mathbf{C}$ and $\mathbf{E}$, we can write the following

PROPOSITION.– $\mathbf{C}$ is symmetrical and has positive (thus real) eigenvalues $\lambda_1, \lambda_2, \lambda_3$ called the principal dilatations. The associated eigenvectors of $\mathbf{C}$ are the principal directions of dilatation, and are also principal directions for $\mathbf{E}$ (thus $\mathbf{E}$ is diagonal in this basis). The principal directions form an orthonormal basis (this follows from the symmetry of $\mathbf{C}$ and $\mathbf{E}$).

NOTE (Interpretation of the Lagrangian strain components).– Recall the difference of the scalar product of two vectors, expression

$$\delta\mathbf{M}.\mathrm{d}\,\mathbf{M} - \delta\mathbf{M}_0.\mathrm{d}\,\mathbf{M}_0 = 2\delta\mathbf{X}.\mathbf{E}.d\mathbf{X}.$$

For the specific choice of two identical vectors, $\delta\mathbf{X} = d\mathbf{X} \rightarrow \delta\mathbf{x} = d\mathbf{x}$, hence

$$\left\|d\mathbf{x}\right\|^2 - \left\|d\mathbf{X}\right\|^2 = dl^2 - dl_0^{\ 2} = 2\mathbf{E}\left(d\mathbf{X}, d\mathbf{X}\right) = 2\mathbf{E}\left(\mathbf{u}_0, \mathbf{u}_0\right) dl_0^{\ 2}, \text{ thus}$$

$$\frac{dl^2 - dl_0^{\ 2}}{dl_0^{\ 2}} = 2\mathbf{E}\left(\mathbf{u}_0, \mathbf{u}_0\right)$$

We previously established the relations

$$\lambda^2\left(\mathbf{u}_0\right) = \left(\frac{dl}{dl_0}\right)^2 = \mathbf{C}\left(\mathbf{u}_0, \mathbf{u}_0\right) \Rightarrow \lambda\left(\mathbf{u}_0\right) = \sqrt{1 + 2\mathbf{E}\left(\mathbf{u}_0, \mathbf{u}_0\right)}$$

DEFINITION.– The scalar $\mathbf{E}\left(\mathbf{u}_0, \mathbf{u}_0\right)$ is the deformation in the direction of unit vector $\mathbf{u}_0$

In particular, the diagonal components $E_{\underline{ii}}$ represent the deformations in the directions of the vector basis $\mathbf{e}_i$.

DEFINITION.– The ratio $\delta\left(d\mathbf{X}\right) := \dfrac{\left\|d\mathbf{x}\right\| - \left\|d\mathbf{X}\right\|}{\left\|d\mathbf{X}\right\|}$ is the extension in the direction of referential vector $d\mathbf{X}$.

An easy calculation shows that

$$\delta(\mathbf{u}_0)=\frac{dl-dl_0}{dl_0}=\lambda(\mathbf{u}_0)-1\equiv\sqrt{1+2\mathbf{E}(\mathbf{u}_0,\mathbf{u}_0)}-1$$

EXERCISE.– Show that in the linearized theory, we have $\delta(\mathbf{u}_0)\approx\sqrt{1+2\boldsymbol{\varepsilon}(\mathbf{u}_0,\mathbf{u}_0)}-1\approx\boldsymbol{\varepsilon}(\mathbf{u}_0,\mathbf{u}_0)$

When $\mathbf{u}_0\equiv\mathbf{e}_i$ the unit vector basis, it yields in particular the dilatation in the direction of $\mathbf{e}_i$,

$$\delta(\mathbf{e}_i)=\lambda(\mathbf{e}_i)-1\equiv\sqrt{1+2\mathbf{E}(\mathbf{e}_i,\mathbf{e}_i)}-1=\sqrt{1+2E_{\underline{ii}}}-1$$

2.6. Polar decomposition theorem

The polar decomposition of the transformation gradient is introduced,

$$\mathbf{F}=R.\mathbf{U}=\mathbf{V}.R$$

with the goal here to separate the motion into a pure rotation R and a pure deformation (without rotation) of material elements, tensors $\mathbf{U}$ and $\mathbf{V}$, respectively called the pure right and left deformation tensors (according to their position in the product); observe that the decomposition is here multiplicative. Note the two ways of performing the decomposition (Figure 2.8), according to whether rotation is done first or last.

PROOF.– Let us consider the principal basis of the deformation tensor $\mathbf{E}$, the set of three orthonormal basis vectors $(\mathbf{B}_1 \quad \mathbf{B}_2 \quad \mathbf{B}_3)$, and let their transform under the motion (transformation gradient) be the three Eulerian vectors $(\mathbf{b}_1 \quad \mathbf{b}_2 \quad \mathbf{b}_3)$, with $\mathbf{b}_i=\mathbf{F}.\mathbf{B}_i$, thus

$$\mathbf{b}_i.\mathbf{b}_j=\mathbf{B}_i.\mathbf{C}.\mathbf{B}_j=\mu_j\mathbf{B}_i.\mathbf{B}_j=\mu_j\delta_{ij}\Rightarrow\mathbf{b}_i\perp\mathbf{b}_j\quad\text{for i}\neq\text{j},$$

with the eigenvalues of $\mathbf{C}$ given by $\mu_j=\lambda_j^2$, due to the equalities

$$\mathbf{C}.\mathbf{B}_i=\mu_i\mathbf{B}_i\ \text{(no summation on i)}\Rightarrow\mathbf{B}_i.\mathbf{C}.\mathbf{B}_i=\mu_i\mathbf{B}_i.\mathbf{B}_i=\mu_i=C_{ii}=\lambda(\mathbf{e}_i)^2$$

We thus showed that the transported vectors $(\mathbf{b}_1 \quad \mathbf{b}_2 \quad \mathbf{b}_3)$ are orthogonal; they are not unit vectors, since

$$\mathbf{b}_i = \mathbf{F.B}_i \Rightarrow \|\mathbf{b}_i\| = \|\mathbf{F.B}_i\| = \lambda_i$$

Thus, tensor $\mathbf{F}$ sends the unit vectors $\mathbf{B}_i$ to the non-unit vectors $\mathbf{b}_i$; this mapping includes a rotation transforming the direction of $\mathbf{B}_i$ to that of $\mathbf{b}_i$, and a dilatation of magnitude λ_i: we can thus decompose $\mathbf{F}$ into the product of rotation and a pure deformation (with no rotation), as pictured on Figure 2.24. This justifies the double multiplicative decomposition of $\mathbf{F}$.

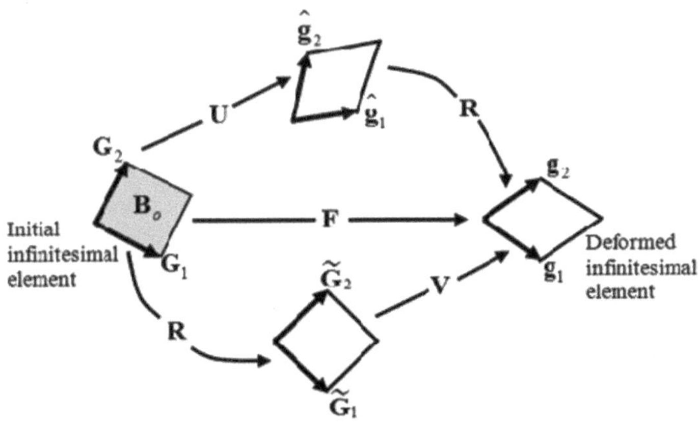

Figure 2.24. *Illustration of the polar decomposition of the t ransformation gradient [BAS 00]*

EXERCISE.– Show analytically the existence and uniqueness of tensors $\mathbf{U},\mathbf{V},R$ in the polar decomposition of $\mathbf{F}$, considering the principal basis of $\mathbf{U}$.

We accordingly decompose additively the displacement into a translation, a pure deformation and a rotation (Figure 2.25): two possible paths can be followed, according to whether the translation is done first, on path 1:

$$N_0 \underset{translation}{\rightarrow} P \underset{pure\ \text{deformation}}{\rightarrow} Q \underset{rotation}{\rightarrow} N'$$

In Figure 2.25, $\underline{U}(M_0)$ is the displacement (not to be confused with previous tensor $\mathbf{U}$) transforming $\mathbf{M_0N_0}$ into vector $\mathbf{MP}$; tensor $\mathbf{U}$ transforms $\mathbf{MP}$ into $\mathbf{MQ}$ (when

$\mathbf{M_0N_0}$ is not a principal direction of the deformation, then $\mathbf{MQ}$ is not parallel to $\mathbf{MP}$), and finally R transforms $\mathbf{MQ}$ into $\mathbf{MN'} = R.\mathbf{MQ} = R.\mathbf{U}.\mathbf{MP} = \mathbf{F}.\mathbf{M_0N_0}$.

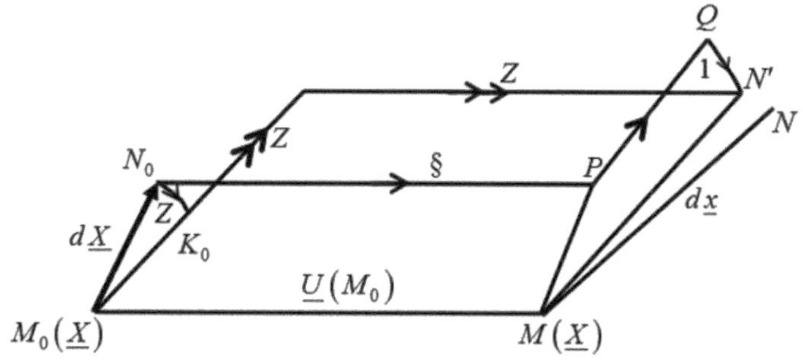

Figure 2.25. *Two possible decompositions of the motion into translation, rotation and deformation*

As an alternative, the translation is done as a last step, following path 2:

$$N_0 \underset{rotation}{\rightarrow} K_0 \underset{pure\ \text{deformation}}{\rightarrow} L_0 \underset{translation}{\rightarrow} N'$$

We then write the composition of motions

$$\mathbf{N_0N} \approx \mathbf{N_0N'} = \underline{U}(M_0) + (\mathbf{U}(M_0) - \mathbf{I}).\mathbf{M_0N_0} + (R(M_0) - \mathbf{I}).(\mathbf{U}(M_0).\mathbf{M_0N_0})$$

It is also convenient to separate the changes of volume from shape changes, by writing

$$\mathbf{F} = (J^{1/3}\mathbf{I})\overline{\mathbf{F}}$$

with $J^{1/3}\mathbf{I}$ a spherical tensor (multiple of the identity tensor) describing the volume changes, and the remaining part $\overline{\mathbf{F}}$ describing shape changes

$$\overline{\mathbf{F}} = J^{-1/3}\mathbf{F} \rightarrow \det(\overline{\mathbf{F}}) = 1$$

This induces similar decompositions of tensors $\mathbf{C}$ and $\mathbf{B}$

$$\overline{\mathbf{C}} = J^{-2/3}\mathbf{C}, \quad \overline{\mathbf{B}} = J^{-2/3}\mathbf{B}$$

Such decomposition will prove useful when writing constitutive laws for compressible and incompressible materials.

2.7. Linearization of the kinematics: small strains and small displacements

When the components of the displacement gradient $\mathbf{H} = Grad\mathbf{U}$ are small in comparison to unity, we are then able to neglect the nonlinear terms in the expressions of tensors **C** and **E**.

DEFINITION.– The assumption of small transformations means that the norm of $\mathbf{H} = Grad\mathbf{U}$ is small, so that $\|\mathbf{H}\| << 1$.

DEFINITION.– The small strains assumption means that the norm of **E** is small compared to unity, *viz.* $\|\mathbf{E}\| << 1$.

DEFINITION.– The small displacements assumption means that the norm of the displacement vector scaled by a typical length L of the medium, is small compared to unity, so that $\|\mathbf{U}\| / L \ll 1$.

These three assumptions define three specific frameworks, which have connections to each other:

From the expression of the Lagrangian strain $\mathbf{E} = \frac{1}{2}\left(\mathbf{H} + \mathbf{H}^T + \mathbf{H}^T.\mathbf{H}\right)$, it is clear that *small transformations imply small deformations.*

All other implications are unreliable and should be checked according to each specific situation.

NOTE (Especially that we may have large displacements but small strains).– This situation is called geometrical nonlinearities. The situation of large strains is referred to as material nonlinearities.

When all three previous assumptions hold, we speak of small perturbations w.r. to an initial equilibrium state.

The assumption of small displacements leads to a coincidence of the Lagrange and Euler representations (viewpoints); this assumption is valid when the reference and actual configurations are closed to each other; it then holds, for any function

$F(\mathbf{X},t)$ in the Lagrange representation, with expression $f(\mathbf{x},t)$ in the Euler representation:

$$\mathrm{F}(\mathbf{X},t) = f(\mathbf{x}(\mathbf{X},t),t) = f(\mathbf{X}+\mathbf{U}(\mathbf{X},t),t) = f(\mathbf{X},t) + Gradf(\mathbf{X},t).\mathbf{U}(\mathbf{X},t) + o(\|\mathbf{U}\|)$$

Hence, $\mathrm{F}(\mathbf{X},t) \approx f(\mathbf{X},t)$, and thus a function or quantity will be represented by the same function in both representations, that is $\mathrm{F}(\mathbf{x},t) \approx f(\mathbf{X},t)$.

There are consequences of the small transformations assumption on the form of the deformation tensors: the following approximations can easily be obtained

$$\mathbf{C} = \mathbf{I} + 2\boldsymbol{\varepsilon} + \mathbf{H}^T.\mathbf{H} \approx \mathbf{I} + 2\boldsymbol{\varepsilon}, \quad \mathbf{E}=\boldsymbol{\varepsilon} + \frac{1}{2}\mathbf{H}^T.\mathbf{H} \approx \boldsymbol{\varepsilon}$$

$$\mathbf{U}^2 = \mathbf{C} = \mathbf{I} + 2\mathbf{E} \approx \mathbf{I} + \boldsymbol{\varepsilon} \Rightarrow \mathbf{U} \approx \mathbf{I} + \boldsymbol{\varepsilon}$$

$$\mathbf{V}^2 = \mathbf{F}^T.\mathbf{F} = \left(\mathbf{I}+\mathbf{H}^T\right).\left(\mathbf{I}+\mathbf{H}\right) \approx \mathbf{I} + 2\boldsymbol{\varepsilon} \Rightarrow \mathbf{V} \approx \mathbf{I} + \boldsymbol{\varepsilon} \Rightarrow \mathbf{U} \approx \mathbf{V}$$

Let there be further linearized rotations: set $R = \mathbf{I} + r$, from the polar decomposition

$$\mathbf{F} = \mathbf{I} + \mathbf{H} = \mathbf{I} + \boldsymbol{\varepsilon} + \mathbf{w} = R.\mathrm{U} \Rightarrow \left(\mathbf{I}+r\right).\left(\mathbf{I}+\boldsymbol{\varepsilon}\right) \approx \mathbf{I} + \mathbf{r} + \boldsymbol{\varepsilon}$$

$$\Rightarrow \mathbf{I} + \boldsymbol{\varepsilon} + \mathbf{w} = \mathbf{I} + \mathbf{r} + \boldsymbol{\varepsilon} \Rightarrow \mathbf{r} \approx \mathbf{w} = \frac{1}{2}\left(\mathbf{U}\otimes\nabla + \mathbf{U}\otimes\nabla^T\right) \text{ and } \mathbf{R} \approx \mathbf{I} + \mathbf{w}$$

Thus, $R - \mathbf{I}$ is an asymmetrical tensor. We can accordingly write the following:

PROPOSITION.– For small transformations, $\boldsymbol{\varepsilon} \approx \mathbf{U} - \mathbf{I} \approx \mathbf{V} - \mathbf{I}$ characterizes the deformation, and the asymmetrical part of $\mathbf{H}$, tensor $\mathbf{w}$, characterizes displacements due to rotations, thus $\mathbf{w} \approx R - \mathbf{I}$.

In the situation of small transformations, we can decompose the displacement field as

$$\begin{aligned} \underline{U}(N_0) &\approx \underline{U}(M_0) + \boldsymbol{\varepsilon}(M_0).\mathbf{M_0N_0} + \mathbf{w}(M_0).\mathbf{M_0N_0} \\ &= \quad \textit{translation} + \textit{pure}\text{ deformation} + \text{rotation} \end{aligned}$$

due to the relation $\left(R-\mathbf{I}\right).\mathrm{U} \approx \mathbf{w}.\left(\mathbf{I}+\boldsymbol{\varepsilon}\right) \approx \mathbf{w}$.

The small transformations assumption is *justified* in the following situations:

– motions of buildings due to wind and gravity forces;

– seismic waves which usually involve small amplitudes;

– extension of a metallic sample during a tensile test in the elastic regime.

This hypothesis *cannot be adopted* in the following situations:

– deformations of a golf ball after impact;

– striction of a sample during a tensile test;

– shape forming of an aluminum can;

– fluid flow, due to the fact that the configuration is changing all the time (different particles reach the same spatial position).

2.8. Deformation velocities

The velocity is defined as the vector

$$\mathbf{v} := \mathbf{u}(\mathbf{x},t) = \frac{D\mathbf{x}}{Dt} = \frac{\partial \mathbf{x}(\mathbf{X},t)}{\partial t}.$$

In general curvilinear coordinates, the velocity gradient is elaborated as follows: the material derivative of the basis vectors is computed as

$$\dot{\mathbf{g}}_i = \dot{\mathbf{F}}.\mathbf{G}_i = \dot{\mathbf{F}}.\mathbf{F}^{-1}.\mathbf{g}_i = -\mathbf{F}.\dot{\mathbf{F}}^{-1}.\mathbf{g}_i \quad (\text{from } \mathbf{F}.\mathbf{F}^{-1} = \mathbf{I})$$

The material derivative of the contravariant basis vectors is similarly evaluated as

$$\dot{\mathbf{g}}^i = \dot{\mathbf{F}}^{-T}.\mathbf{G}^i = \dot{\mathbf{F}}^{-T}.\mathbf{F}^T.\mathbf{g}^i = -\mathbf{F}^{-T}.\dot{\mathbf{F}}^T.\mathbf{g}^i$$

In both equalities, the material derivative of the transformation gradient is given by

$$\dot{\mathbf{F}} = \dot{\mathbf{g}}_i \otimes \mathbf{G}^i$$

DEFINITION.– The velocity gradient is defined as the spatial gradient of the velocity field,

$$\mathbf{l} := grad\mathbf{v} = \mathbf{v}_{,j} \otimes \mathbf{g}^j = \dot{\mathbf{x}}_{,j} \otimes \mathbf{g}^j = \dot{\mathbf{g}}_j \otimes \mathbf{g}^j = \dot{\mathbf{F}}.\mathbf{F}^{-1} = -\mathbf{F}.\dot{\mathbf{F}}^{-1}$$

Thus, we can write

$$\dot{\mathbf{g}}_i = \mathbf{l}.\mathbf{g}_i = \mathbf{g}_i.\mathbf{l}^T; \quad \dot{\mathbf{g}}^i = \mathbf{l}.\mathbf{g}^i = -\mathbf{g}^i.\mathbf{l}$$

The velocity gradient is further decomposed into symmetrical and asymmetrical parts,

$$\mathbf{l} = \mathbf{d} + \mathbf{w}, \quad \text{with}$$

$$\mathbf{d} := \frac{1}{2}\left(\mathbf{l} + \mathbf{l}^T\right), \quad \mathbf{w} := \frac{1}{2}\left(\mathbf{l} - \mathbf{l}^T\right)$$

DEFINITION.– $\mathbf{d}$ is called the rate of *deformation tensor*, and $\mathbf{w}$ is called the *spin* or *vorticity tensor*: it quantifies local rotations .

$\mathbf{l}, \mathbf{d}, \mathbf{w}$ are all spatial tensors, since the gradient is taken w.r. to the spatial position.

The following relations are obtained for the rate of deformation and spin tensor:

$$\mathbf{d} = d_{ij}\mathbf{g}^i \otimes \mathbf{g}^j = \frac{1}{2}\frac{D}{Dt}\left(\mathbf{g}_i.\mathbf{g}_j\right)\left(\mathbf{g}^i \otimes \mathbf{g}^j\right),$$

$$\mathbf{w} = \mathrm{w}_{ij}\mathbf{g}^i \otimes \mathbf{g}^j = \frac{1}{2}\left(\mathbf{g}_i.\dot{\mathbf{g}}_j - \dot{\mathbf{g}}_i.\mathbf{g}_j\right)\mathbf{g}^i \otimes \mathbf{g}^j$$

The rate of deformation tensor can be given a geometrical interpretation:

$$\frac{D}{Dt}\left(ds^2\right) = 2d\mathbf{x}.d\dot{\mathbf{x}} = 2d\mathbf{x}.\mathbf{l}.d\mathbf{x} = 2d\mathbf{x}.\left(\mathbf{d} + \mathbf{w}\right).d\mathbf{x} \equiv 2d\mathbf{x}.\mathbf{d}.d\mathbf{x}$$

The last equality results from the asymmetry of $\mathbf{w}$ (exercise: show it); thus the rate of change of the square length ds^2 is completely described by $\mathbf{d}$. Note that the rate of deformation is the tensor generalization of the strain rate measure in 1D, $\frac{1}{l(t)}\frac{dl(t)}{dt}$, for a sample of actual length $l(t)$.

The spin tensor is given a geometrical interpretation: we can easily show that for a rigid body motion, such that when there is no deformation, $\mathbf{d} = \mathbf{0} \rightarrow d\mathbf{v} = \mathbf{w}.d\mathbf{x} = \boldsymbol{\omega} \times d\mathbf{x}$, with the vorticity vector dual to the spin, $\boldsymbol{\omega} := \frac{1}{2} curl\mathbf{v} = \frac{1}{2}\mathbf{g}^r \times \mathbf{v}_{,r}$. Thus, this justifies us calling $\mathbf{w}$ the vorticity tensor.

Connections between the rate of deformation tensor and the material derivative of the Lagrangian strain $\mathbf{E}$: we evaluate the material derivative of the difference of square length elements

$$\frac{D}{Dt}\left(ds^2\right)-\frac{D}{Dt}\left(dS^2\right)=2d\mathbf{x}.\mathbf{d}.d\mathbf{x}=2d\mathbf{X}.\dot{\mathbf{E}}.d\mathbf{X}=2d\mathbf{x}.\left(\mathbf{F}^{-T}.\dot{\mathbf{E}}.\mathbf{F}^{-1}\right).d\mathbf{x}=$$
$$=2d\mathbf{x}.\left(\mathbf{l}^T.\mathbf{e}+\dot{\mathbf{e}}+\mathbf{e}.\mathbf{l}\right).d\mathbf{x}$$

Since previous relations hold for arbitrary vector $d\mathbf{x}$, it holds that

$$\mathbf{d}=\frac{1}{2}\frac{D}{Dt}\left(\mathbf{g}_i.\mathbf{g}_j\right)\left(\mathbf{g}^i\otimes\mathbf{g}^j\right)=\mathbf{l}^T.\mathbf{e}+\dot{\mathbf{e}}+\mathbf{e}.\mathbf{l}=\mathbf{F}^{-T}.\dot{\mathbf{E}}.\mathbf{F}^{-1}\equiv L_v\mathbf{e}$$

The component form of those relations is

$$d_{ij}=\dot{e}_{ij}=\dot{E}_{ij}=\frac{1}{2}\frac{D}{Dt}\left(\mathbf{g}_i.\mathbf{g}_j\right)$$

EXERCISE.– Show previous relations between components.

REMARK.– It is then clear that the rate of deformation tensor is not the material derivative of the Lagrangian strain. In the small transformation framework, we may write

$$\mathbf{F}\approx\mathbf{I},\ \ \mathbf{E}\approx\boldsymbol{\varepsilon}\rightarrow\mathbf{d}=\dot{\mathbf{E}}\approx\dot{\boldsymbol{\varepsilon}},$$

Thus in this case, we may evaluate $\mathbf{d}$ from the small strain tensor.

The above-introduced Lie derivative operator, quantity $L_v\mathbf{e}$, is the material derivative of an arbitrary spatial tensor holding the deformed basis constant; it is expressed in absolute tensor notation versus tensor $\mathbf{l}$. It is to be observed indeed that

$$\mathbf{e}=e_{ij}\mathbf{g}^i\otimes\mathbf{g}^j\rightarrow L_v\mathbf{e}=\dot{e}_{ij}\mathbf{g}^i\otimes\mathbf{g}^j=\mathbf{l}^T.\mathbf{e}+\dot{\mathbf{e}}+\mathbf{e}.\mathbf{l}=\mathbf{F}^{-T}.\dot{\mathbf{E}}.\mathbf{F}^{-1}$$

so that the basis vectors are indeed fixed while evaluating the Lie derivative. The last equality has the sense of a push-forward of the material derivative $\dot{\mathbf{E}}$ to the spatial configuration.

These relations are next given a more profound meaning, introducing the notions of push-forward and pull-back as two opposite *transport operations* of second order tensors.

2.9. Transport operations: pull-back and push-forward

DEFINITION.– The *pull-back* transforms a spatial tensor into its material counterpart; it is denoted by the symbol ϕ^*. For a second order *spatial tensor expressed in terms of its covariant components*, it holds that

$$\mathbf{S} = S_{ij}\mathbf{g}^i \otimes \mathbf{g}^i \rightarrow \phi^*(\mathbf{S}) = \mathbf{F}^T.\left(S_{ij}\mathbf{g}^i \otimes \mathbf{g}^j\right).\mathbf{F} = \mathbf{F}^T.\mathbf{S}.\mathbf{F}$$
$$\equiv S_{ij}\left(\mathbf{F}^T.\mathbf{g}^i\right) \otimes \left(\mathbf{g}^j.\mathbf{F}\right) = S_{ij}\mathbf{G}^i \otimes \mathbf{G}^j$$

Thus, the pull-back is mapping a spatial tensor to the reference configuration, by transporting the components (which are kept fixed) and mapping the spatial basis vectors to their referential counterpart.

DEFINITION.– The *push-forward* is transporting a material (referential) second order tensor to the spatial configuration; it is denoted by the symbol ϕ_*. For a second order *spatial tensor expressed in terms of its covariant components*, it holds that

$$\mathbf{T} = T_{ij}\mathbf{g}^i \otimes \mathbf{g}^j \rightarrow \phi_*(\mathbf{T}) = \mathbf{F}^{-T}.\mathbf{T}.\mathbf{F}^{-1} = \mathbf{F}^{-T}.\left(T_{ij}\mathbf{g}^i \otimes \mathbf{g}^j\right).\mathbf{F}^{-1} = T_{ij}\mathbf{G}^i \otimes \mathbf{G}^j$$

Thus, the push-forward transports the components of $\mathbf{T}$ (which are kept fixed) from the referential to the spatial configuration in their respective bases.

The transport operations previous introduced are dependent on the nature of components: for tensors now expressed in terms of their contravariant components, it holds that

$$\mathbf{S} = S^{ij}\mathbf{g}_i \otimes \mathbf{g}_j \rightarrow \phi^*(\mathbf{S}) = \mathbf{F}^{-1}.\left(S^{ij}\mathbf{g}_i \otimes \mathbf{g}_j\right).\mathbf{F}^{-T} = \mathbf{F}^{-1}.\mathbf{S}.\mathbf{F}^{-T} \equiv S^{ij}\mathbf{G}_i \otimes \mathbf{G}_j$$

$$\mathbf{T} = T^{ij}\mathbf{G}_i \otimes \mathbf{G}_j \rightarrow \phi_*(\mathbf{T}) = \mathbf{F}.\mathbf{T}.\mathbf{F}^T = \mathbf{F}.\left(T^{ij}\mathbf{G}_i \otimes \mathbf{G}_j\right).\mathbf{F}^T = T^{ij}\mathbf{g}_i \otimes \mathbf{g}_j$$

Here also, the tensor components are mapped respectively forward and backward to the corresponding basis. Pull-back and push-forward transport operations are summarized in Figure 2.26.

It is easy to check that the push-forward and pull-back are inverse operations: for instance,

$$\phi^*(\mathbf{S}) = \mathbf{F}^T.\mathbf{S}.\mathbf{F} \rightarrow \phi_*\left(\phi^*(\mathbf{S})\right) = \mathbf{F}^{-T}.\left(\mathbf{F}^T.\mathbf{S}.\mathbf{F}\right).\mathbf{F}^{-1} = \mathbf{S}$$

Let us apply those transport operations to the right Cauchy-Green tensor, basically a covariant tensor (all deformation tensors will be expressed in covariant form, whereas the dual stress tensors to be exhibited later on will be written in contravariant form):

$$\mathbf{C} = C_{ij}\mathbf{G}^i \otimes \mathbf{G}^j = \phi^*(\mathbf{g}) = \mathbf{F}^T.\mathbf{g}.\mathbf{F} \leftrightarrow \mathbf{g} = g_{ij}\mathbf{g}^i \otimes \mathbf{g}^j = \phi_*(\mathbf{C}) = \mathbf{F}^{-T}.\mathbf{C}.\mathbf{F}^{-1}$$

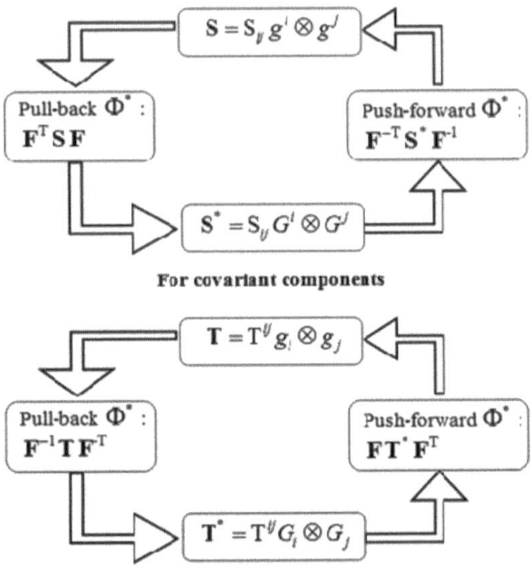

Figure 2.26. *Pull-back and push-forward operations for tensors with covariant and contravariant components [BAS 00]*

It is clear that tensors $\mathbf{C}$ and $\mathbf{g}$ have the same components, *viz.* $C_{ij} = g_{ij}$. Previous identities also show that $\mathbf{C}$ can be considered as a material metric tensor.

In the case of the Green-Lagrange tensor $\mathbf{E}$ and its spatial counterpart, the Almansi strain tensor $\mathbf{e}$, it holds that

$$\mathbf{E} = E_{ij}\mathbf{G}^i \otimes \mathbf{G}^j = \phi^*(\mathbf{e}) = \mathbf{F}^T.\mathbf{e}.\mathbf{F} \leftrightarrow \mathbf{e} = e_{ij}\mathbf{g}^i \otimes \mathbf{g}^j = \phi_*(\mathbf{E}) = \mathbf{F}^{-T}.\mathbf{E}.\mathbf{F}^{-1}$$

These relations are also due to the identity of components $E_{ij} = e_{ij}$. Recall the way $\mathbf{E}$ has been introduced as a comparison of changes of angles and lengths, so that it also can be considered as a referential metric tensor (similar to $\mathbf{C}$).

The transport operations are next involved to elaborate the *notion of Lie derivative*: let us evaluate the material derivative of the spatial strain tensor

$$\mathbf{e} = e_{ij}\mathbf{g}^i \otimes \mathbf{g}^j \rightarrow \dot{\mathbf{e}} = \frac{D\mathbf{e}}{Dt} = \dot{e}_{ij}\mathbf{g}^i \otimes \mathbf{g}^j + e_{ij}\dot{\mathbf{g}}^i \otimes \mathbf{g}^j + e_{ij}\mathbf{g}^i \otimes \dot{\mathbf{g}}^j$$

DEFINITION.– The Lie derivative is the term $L_v\mathbf{e} := \dot{e}_{ij}\mathbf{g}^i \otimes \mathbf{g}^j$; it is obtained according to the following three steps:

– Pull-back of $\mathbf{e}$: $\phi^*(\mathbf{e}) = \mathbf{F}^T.\mathbf{e}.\mathbf{F}$, a material tensor.

– Material derivative of the pull-back (in the referential configuration):

$$\frac{D}{Dt}\phi^*(\mathbf{e}) = \dot{e}_{ij}\mathbf{G}^i \otimes \mathbf{G}^j = \dot{\mathbf{F}}^T.\mathbf{e}.\mathbf{F} + \mathbf{F}^T.\dot{\mathbf{e}}.\mathbf{F} + \mathbf{F}^T.\mathbf{e}.\dot{\mathbf{F}}\,.$$

– Push-forward previous material derivative to the spatial configuration:

$$\phi_*\left(\frac{D}{Dt}\phi^*(\mathbf{e})\right) = \dot{e}_{ij}\mathbf{g}^i \otimes \mathbf{g}^j = \mathbf{F}^{-T}.\frac{D}{Dt}\phi^*(\mathbf{e}).\mathbf{F}^{-1} = \mathbf{F}^{-T}.\dot{\mathbf{F}}^{-1}.\mathbf{e} + \dot{\mathbf{e}} + \mathbf{e}.\dot{\mathbf{F}}.\mathbf{F}^{-1} = \mathbf{l}^T.\mathbf{e} + \dot{\mathbf{e}} + \mathbf{e}.\mathbf{l}$$

We thus recover a previously written expression of the Lie derivative, which is to be interpreted as the material derivative by keeping the basis vectors fixed: we see that in order to get rid of the variation of the spatial vector basis, the tensor to be varied is first pulled back to the fixed material configuration.

Case of a tensor in contravariant form, like Cauchy stress:

$$\boldsymbol{\sigma} = \sigma^{ij}\mathbf{g}_i \otimes \mathbf{g}_j \rightarrow L_v\boldsymbol{\sigma} = \dot{\sigma}^{ij}\mathbf{g}_i \otimes \mathbf{g}_j$$

The Lie derivative is obtained as previously by the following three steps:

– Pull-back of $\mathbf{e}$: $\phi^*(\boldsymbol{\sigma}) = \mathbf{F}^{-1}.\boldsymbol{\sigma}.\mathbf{F}^{-T}$, a material tensor.

– Material derivative of the pull-back (in the referential configuration):

$$\frac{D}{Dt}\phi^*(\boldsymbol{\sigma}) = \dot{\sigma}_{ij}\mathbf{G}_i \otimes \mathbf{G}_j = \dot{\mathbf{F}}^{-1}.\boldsymbol{\sigma}.\mathbf{F}^{-T} + \mathbf{F}^{-1}.\dot{\boldsymbol{\sigma}}.\mathbf{F}^{-T} + \mathbf{F}^{-1}.\boldsymbol{\sigma}.\dot{\mathbf{F}}^{-T}\,.$$

– Push-forward previous material derivative to the spatial configuration:

$$\phi_*\left(\frac{D}{Dt}\phi^*(\boldsymbol{\sigma})\right) = \mathbf{F}.\phi_*\left(\frac{D}{Dt}\phi^*(\boldsymbol{\sigma})\right).\mathbf{F}^T = \dot{\sigma}^{ij}\mathbf{g}_i \otimes \mathbf{g}_j = \mathbf{F}.\dot{\mathbf{F}}^{-1}.\boldsymbol{\sigma} + \dot{\boldsymbol{\sigma}} + \boldsymbol{\sigma}.\dot{\mathbf{F}}^{-T}.\mathbf{F}^T$$

We can show that the Lie derivative of $\boldsymbol{\sigma}$ can be expressed as

$L_v\boldsymbol{\sigma} = \dot{\boldsymbol{\sigma}} - \mathbf{l}.\boldsymbol{\sigma} - \boldsymbol{\sigma}.\mathbf{l}^T$, using the equalities $\mathbf{F}.\mathbf{F}^{-1} = \mathbf{I} \Rightarrow \dot{\mathbf{F}}.\mathbf{F}^{-1} = -\mathbf{F}.\dot{\mathbf{F}}^{-1} = \mathbf{l}$.

The velocity gradient $\mathbf{l}$ thus permits us to connect the material time-derivative of a spatial tensor and its Lie derivative. We next expose an important concept in view of the writing of constitutive laws, namely:

2.10. Isotropic tensor functions

DEFINTION.– A tensor-valued function $\mathbf{G}(\mathbf{C})$ mapping a second order tensor $\mathbf{C}$ to another second order tensor $\mathbf{G}(\mathbf{C})$ is said to be isotropic if the following relation holds

$$\mathbf{G}\left(Q.\mathbf{C}.Q^T\right) = Q.\mathbf{G}(\mathbf{C}).Q^T, \quad \forall Q \in O_3$$

the group or orthogonal transformations. Since tensor $\mathbf{C} = \mathbf{F}^T.\mathbf{F}$ now transforms under a rotation of the material basis by rotation as

$$\hat{\mathbf{G}}^i = Q.\mathbf{G}^i \rightarrow \hat{\mathbf{F}} = \mathbf{g}_i \otimes \hat{\mathbf{G}}^i = \left(\mathbf{g}_i \otimes \mathbf{G}^i\right).Q^T = \mathbf{F}Q^T$$
$$\Rightarrow \hat{\mathbf{C}} = \hat{\mathbf{F}}^T.\hat{\mathbf{F}} = Q.\mathbf{C}.Q^T$$

Thus any function of $\mathbf{C}$ is also isotropic (since it transforms as $\mathbf{C}$).

EXERCISE.–Show that integer power functions of C, $\mathbf{C}^{1/m}$ and $\mathbf{C}^{-1/m}$ are also isotropic functions of $\mathbf{C}$.

Typically, constitutive laws for isotropic materials shall involve isotropic strain tensors.

We similarly define isotropic tensor functions of spatial second order arguments of the form $\mathbf{g}(\mathbf{b})$, but now under a rotation of the spatial basis vectors

$$\mathbf{g}_i^* = Q.\mathbf{g}_i = \mathbf{F}^* = \mathbf{g}_i^* \otimes \mathbf{G}^i = Q.\mathbf{F} \Rightarrow \mathbf{b}^* = \mathbf{F}^*.\mathbf{F}^{*T} = Q.\mathbf{b}.Q^T$$

DEFINITION.– $\mathbf{g}(\mathbf{b})$ is an isotropic function of $\mathbf{b}$ when $\mathbf{g}\left(\mathbf{b}^*\right) = Q.\mathbf{g}(\mathbf{b}).Q^T, \ \forall Q \in O_3$.

We can similarly show that $\mathbf{b}^{1/m}$ and $\mathbf{b}^{-1/m}$ are isotropic functions of $\mathbf{b}$.

NOTE (Representation theorem of Truesdell and Noll (1965)).–

Any isotropic tensor-valued function of a spatial tensor $\mathbf{b}$, $\mathbf{f}(\mathbf{b})$, may be expressed by a quadratic polynomial in $\mathbf{b}$, *viz.*

$$\mathbf{f}(\mathbf{b}) = \delta_0 \mathbf{I} + \delta_1 \mathbf{b} + \delta_2 \mathbf{b}^2$$

In which $\delta_0, \delta_1, \delta_2$ are functions of the principal invariants of $\mathbf{b}$.

Similarly, any isotropic tensor-valued function of a material tensor $\mathbf{C}$, say $\mathbf{g}(\mathbf{C})$, is expressible by a quadratic polynomial in $\mathbf{C}$, *viz.*

$$\mathbf{g}(\mathbf{C}) = \gamma_0 \mathbf{I} + \gamma_1 \mathbf{C} + \gamma_2 \mathbf{C}^2$$

with $\gamma_0, \gamma_1, \gamma_2$ functions of the principal invariants of $\mathbf{C}$. This last form is in fact a consequence of the representation theorem for spatial tensors $\mathbf{f}(\mathbf{b})$, the definition of the material tensor as a pull-back of its spatial counterpart, and Cayley-Hamilton theorem (exercise).

The notion of an isotropic tensor-valued function is in fact strongly linked to the notion of *objective tensors*, a central theme in continuum mechanics: different observers must indeed see the same response of a given material in their respective referentials. We first have to define what is meant under change of observer.

DEFINITION.– A *change of observer* is defined by an orthogonal transformation in which the spatial position $\mathbf{x}$ is mapped to the new position $\mathbf{x}^* = Q.\mathbf{x} + \mathbf{a}$, $Q \in O_3$, with $\mathbf{a}$ a constant vector.

Under a change of observer, the transformation gradient is mapped into (easy exercise)

$$\mathbf{F} \mapsto \mathbf{F}^* = Q.\mathbf{F}$$

In the subsequent definitions of *objectivity*, the transformation gradient is considered as the argument of scalar valued, vector valued and tensor-valued functions successively.

DEFINITION.– A scalar-valued function of the transformation gradient $f(\mathbf{F})$ is objective iff its value is unchanged under the change of observer, so that $f(\mathbf{F}^*) = f(\mathbf{F})$. This corresponds to the strong form of covariance called invariance (two observers will measure the same scalar value).

DEFINITION.– A vector-valued function $\mathbf{v} = \mathbf{v}(\mathbf{F})$ is objective if it transforms like the basis vectors $\mathbf{g}_i^* = \mathbf{x}_{,i}^* = Q.\mathbf{g}_i$, $\mathbf{g}^{i*} = Q.\mathbf{g}^i$; thus it holds that

$$\mathbf{v}^*(\mathbf{F}) = \mathbf{v}(Q.\mathbf{F}) = Q.\mathbf{v}(\mathbf{F})$$

REMARK.– This definition is equivalent to requiring the stronger invariance of the product of the vector field $\mathbf{v}(\mathbf{F};M)$ at point M with a material vector having point M as its origin, thus $\mathbf{v}(\mathbf{F}).\mathbf{MM}'$ should be a true scalar.

DEFINITION.– A tensor-valued function $\boldsymbol{\sigma}(\mathbf{F})$ is objective iff $\boldsymbol{\sigma}(\mathbf{F}) = Q.\boldsymbol{\sigma}(\mathbf{F}).Q^T$. In fact, this corresponds to the more general form of the constitutive law for an elastic material.

REMARK.– This amounts to requiring that the product of $\boldsymbol{\sigma}(\mathbf{F})$ at point M with the vector $\mathbf{MM}'$ be an objective vector.

As a corollary of the objectivity of $\boldsymbol{\sigma}(\mathbf{F})$, it holds that

$$\sigma^{*ij} = \mathbf{g}^{*i}.\boldsymbol{\sigma}.\mathbf{g}^{*j} = \mathbf{g}^i.Q^T.\left(Q.\boldsymbol{\sigma}(\mathbf{F}).Q^T\right).Q.\mathbf{g}^j = \sigma^{ij}$$

Thus the components of $\boldsymbol{\sigma}$ remain unchanged (the two observers see the same components but in their respective bases).

The notion of objectivity defined here applies to spatial tensors; a similar notion of objectivity applies for material (referential) tensors, but adopting the following definition of the change of observer: *the material position* $\mathbf{X}$ is mapped to the new position $\mathbf{X}^* = Q.\mathbf{X} + \mathbf{a}$, $Q \in O_3$, with $\mathbf{a}$ a constant vector.

It is now interesting to examine the status of tensors with respect to objectivity, as a filter to discriminate amongst tensors those which are objective and those which are not objective; this is especially important when writing constitutive laws in rate form, since then the strain rate and stress measures have to be objective.

The transformation gradient is not objective, since $\mathbf{F} \mapsto \mathbf{F}^* = Q.\mathbf{F}$.

The right and left Cauchy Green tensors transform under a change of observer as (check)

$$\mathbf{F}^* = Q.\mathbf{F} \Rightarrow \mathbf{C}^* = \mathbf{C},\ \mathbf{B}^* = Q.\mathbf{B}.Q^T$$

Thus $\mathbf{B}$ is objective. We can check that tensor $\mathbf{C}$ is objective, according to previous definition of objectivity for material tensors relying on the adopted definition of the change of observer.

The rate of deformation tensor transforms as

$$\mathbf{l} \mapsto \mathbf{l}^* := \dot{\mathbf{F}}^*.\mathbf{F}^{*-1} = \left(\dot{Q}.\mathbf{F} + Q.\dot{\mathbf{F}}\right).\mathbf{F}^{-1}.Q^T = Q.\mathbf{l}.Q^T + \dot{Q}.Q^T = Q.\mathbf{d}.Q^T + Q.\mathbf{w}.Q^T + \dot{Q}.Q^T =$$
$$= \mathbf{d}^* + \mathbf{w}^*$$

with $\mathbf{d}^* = Q.\mathbf{d}.Q^T = \mathbf{d}^{*T}$, $\mathbf{w}^* = Q.\mathbf{w}.Q^T + \dot{Q}.Q^T$

Tensor $\mathbf{w}^*$ is antisymmetric, due to $\mathbf{w}$ being antisymmetric. It is then clear that the rate of deformation tensor is objective, whereas the spin tensor is not objective.

The stress rate will transform according to

$$\boldsymbol{\sigma}^* = Q.\boldsymbol{\sigma}.Q^T \Rightarrow \dot{\boldsymbol{\sigma}}^* = \dot{Q}.\boldsymbol{\sigma}.Q^T + Q.\dot{\boldsymbol{\sigma}}.Q^T + Q.\boldsymbol{\sigma}.\dot{Q}^T$$

Thus, the Cauchy stress rate is not objective, but the difference

$$\dot{\boldsymbol{\sigma}} + \boldsymbol{\sigma}.\mathbf{w} - \mathbf{w}.\boldsymbol{\sigma}$$

is an objective tensor, due to the relations

$$\dot{Q} = \mathbf{w}^*.Q - Q.\mathbf{w} \rightarrow \dot{Q}^T = -Q^T.\mathbf{w} + \mathbf{w}.Q^T \rightarrow \dot{\boldsymbol{\sigma}}^* + \boldsymbol{\sigma}^*.\mathbf{w}^* - \mathbf{w}^*.\boldsymbol{\sigma}^* = Q.\left(\dot{\boldsymbol{\sigma}} + \boldsymbol{\sigma}.\mathbf{w} - \mathbf{w}.\boldsymbol{\sigma}\right).Q^T$$

This means that the new stress rate $\hat{\boldsymbol{\sigma}} := \dot{\boldsymbol{\sigma}} + \boldsymbol{\sigma}.\mathbf{w} - \mathbf{w}.\boldsymbol{\sigma}$ is an objective tensor: it deserves the name Jaumann derivative or Jaumann stress rate.

2.11. Stress measures and strain–strain duality

The notion of stress has already been introduced as a motivation for using tensors (Chapter 1), see Figure 2.27; stresses are dual to deformations in the sense that their product defines the internal mechanical energy or power of internal forces. We firstly have to recall the notion of stress, here in the more general context of curvilinear coordinates, leading us to introduce the notion of physical components of the stress.

Let consider the small volume element pictured in Figure 2.27, conceived as a small RVE centered around point M.

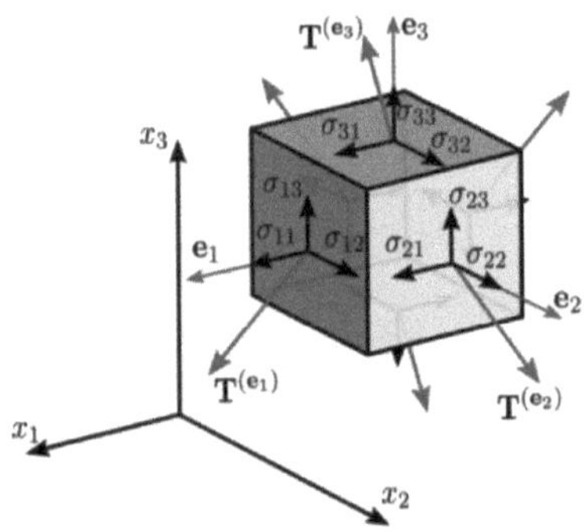

Figure 2.27. *Stress components for an infinitesimal volume element around a material point M [BAS 00]. For a color version of this figure, see www.iste.co.uk/ganghoffer/biomechanics.zip*

DEFINITION.– The stress is introduced in statics to describe contact forces between material neighboring elements. The stress tensor relates a unit-length direction vector $\mathbf{n}(M)$ defining the normal to any surface cut within the material to the stress vector $\mathbf{t}(M;\mathbf{n})$ at point M, according to

CAUCHY THEOREM.– when internal efforts are bounded, and for fixed normal $\mathbf{n}$ (at point M), the linear mapping $\mathbf{n} \mapsto \mathbf{t}(M;\mathbf{n})$ is continuous, then $\mathbf{t}(M;\mathbf{n})$ is a linear function of $\mathbf{n}$, so that there exists a second order spatial tensor called *Cauchy stress* $\boldsymbol{\sigma}$ such that

$$\mathbf{t}(M;\mathbf{n}) = \boldsymbol{\sigma}(M).\mathbf{n}(M) \rightarrow \mathbf{t}(\mathbf{x};\mathbf{n}) = \boldsymbol{\sigma}(\mathbf{x}).\mathbf{n}(\mathbf{x})$$

The proof is classical and will not be reproduced here.

The stress tensor is a measure of internal forces; it can elaborated from a physical point of view in the framework of continuum mechanics from the force $\Delta\mathbf{p}(M;\mathbf{n})$ acting on an imaginary surface with normal vector $\mathbf{n}$ and magnitude ΔA (Figure 2.28) as the limit of the ratio

$$\mathbf{t}(M) := \lim_{\Delta A \to 0} \frac{\Delta\mathbf{p}(M;\mathbf{n})}{\Delta A}$$

This defines the stress vector called traction, referring to the deformed surface element, and representing a force–per-unit deformed area in the actual configuration (vector $\mathbf{n}$ is the actual normal), Figure 2.28. The Cauchy postulate states that the stress vectors remain the same for all surfaces passing through M and sharing the same unit normal $\mathbf{n}$.

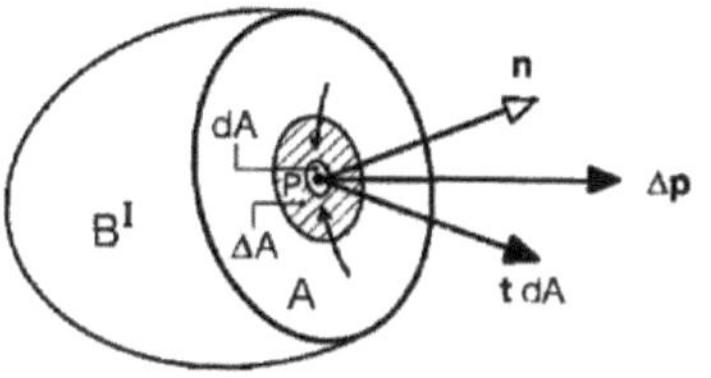

Figure 2.28. *Cauchy stress and traction vector [BAS 00]*

2.11.1. *Physical stress components*

We consider an infinitesimal parallelepiped bounded by the faces $\theta^i = Ct$ and $\theta^i + d\theta^i = Ct$, the edges of which are delimited by the vectors

$$d\mathbf{s}_i = \mathbf{g}_i d\theta^i \text{ (no summation on i)}$$

The area of any face spanned by vectors $\mathbf{g}_j d\theta^j$ and $\mathbf{g}_k d\theta^k$ is denoted by $d\mathbf{s}_i$, with

$$d\mathbf{s}_i = \mathbf{g}_j d\theta^j \times \mathbf{g}_k d\theta^k \to da_i := \left\| d\mathbf{s}_i \right\| = \left\| \mathbf{g}_j d\theta^j \times \mathbf{g}_k d\theta^k \right\| = \sqrt{g g^{ii}}\, d\theta^j d\theta^k$$

The unit exterior normal $\mathbf{n}^{\langle i \rangle}$ to the faces defined by constant curvilinear coordinates $\theta^i + d\theta^i = Ct$ coincide with the scaled basis vectors, so that

$$\mathbf{n}^{\langle i \rangle} = \mathbf{g}^i / \sqrt{g^{ii}} \text{ (since the magnitude of vector } \mathbf{g}^i \text{ is } \sqrt{g^{ii}}\text{)}$$

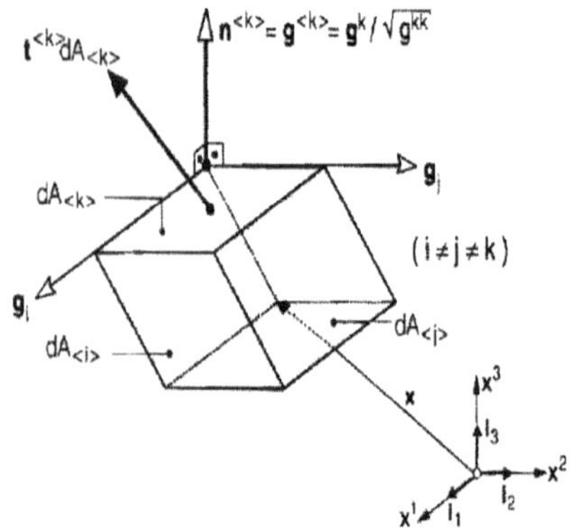

Figure 2.29. *Stress resultant acting on the surface element da_i [BAS 00]*

Let $\mathbf{t}^{\langle i\rangle}$ be the stress vector per unit area acting on the surface area da_i (Figure 2.29). The force acting on the surface element with unit normal $\mathbf{n}^i$ is then $\mathbf{t}^{\langle i\rangle} da_i$, so that the physical stress vector is defined in component form by

$$\mathbf{t}^{\langle i\rangle} = \sigma^{\langle ij\rangle}\mathbf{g}_{\langle j\rangle} = \sigma^{\langle ij\rangle}\mathbf{g}^j / \sqrt{g^{jj}}$$

This vector represents the forces per unit area of the surface element da_i, thus the set of components of the three vectors $\mathbf{t}^{\langle i\rangle}, \mathbf{t}^{\langle j\rangle}, \mathbf{t}^{\langle k\rangle}$ for three orthogonal faces with exterior normals $\mathbf{n}^i, \mathbf{n}^j, \mathbf{n}^k$ are coined the physical stress components. The first index in $\sigma^{\langle ij\rangle}$ gives the area element da_i on which $\sigma^{\langle ij\rangle}$ is acting, and the second index the direction $\mathbf{g}_j$.

The stress tensor components can next be constructed: consideration of the relations

$$da_i := \|d\mathbf{s}_i\| = \|\mathbf{g}_j d\theta^j \times \mathbf{g}_k d\theta^k\| = \sqrt{gg^{ii}}\, d\theta^j d\theta^k$$

implies the relation

$$\mathbf{t}^{\langle i\rangle} da_i = \mathbf{t}^{\langle i\rangle}\sqrt{gg^{ii}}\, d\theta^j d\theta^k = \mathbf{t}^i \sqrt{g}\, d\theta^j d\theta^k$$

introducing there in the contravariant vector $\mathbf{t}^i := \mathbf{t}^{\langle i \rangle}\sqrt{g^{ii}}$. Note that the vector $\mathbf{t}^{\langle i \rangle}$ does not behave as a contravariant vector. The tensorial stress components are then defined from Cauchy axiom as

$$\mathbf{t}^i = \sigma^{ij}\mathbf{g}_j$$

EXERCISE.– Show that the relation between the tensor components σ^{ij} and the physical stress components $\sigma^{\langle ij \rangle}$ is given by

$$\sigma^{\langle ij \rangle} = \sqrt{g_{jj} / g^{ii}}\,\sigma^{ij}$$

EXERCISE.– Show that $\{\mathbf{t}^i\}$ define contravariant vectors, using the equalities $\mathbf{t}da := \mathbf{t}^{\langle i \rangle}da_i$, defining the area da by the equality $\mathbf{n}da = \sum_i da_i\mathbf{n}^{\langle i \rangle}$ for a tetrahedron, and the definition of $\mathbf{t}^i$. We are thus led (outcome of previous exercise) to

$$\mathbf{n} = n_i\mathbf{g}^i \rightarrow \mathbf{t} = \mathbf{t}^i n_i = \sigma^{ij} n_i da$$

Thus, we may form the second order Cauchy stress tensor

$$\boldsymbol{\sigma} = \sigma^{ij}\mathbf{g}_i \otimes \mathbf{g}_j = \mathbf{t}^i \otimes \mathbf{g}_i$$

Next rewriting $\mathbf{t} = \left(\mathbf{t}^i \otimes \mathbf{g}_i\right).\mathbf{n}$, we obtain in view of previous equality the relation

$$\mathbf{t} = \boldsymbol{\sigma}.\mathbf{n}$$

which is again Cauchy theorem, which has here been proven in curvilinear coordinates.

Introduction of stress measures

Other stress tensors can be introduced, which are either artificial or natural (physical) constructs. Let us recall Nanson formula for the transport of area elements

$$\mathbf{n}da = J\mathbf{F}^{-T}.\mathbf{N}dA$$

with the Jacobian $J := \det \mathbf{F} = \dfrac{dv}{dV} = \sqrt{g/G}$ representing the ratio of two infinitesimal volume elements. Stress tensors will next be introduced via the stress-strain duality.

2.11.2. *Stress-strain duality and the principle of virtual power*

The principle of virtual power is the weak form of the conservation of momentum (equilibrium), and is the basis for introducing the various stress tensors dual to strain measures in the sense of energy. The idea behind the principle of virtual power is to evaluate forces from scalar quantities elaborated as the work performed by those forces under different motions; if we want to identify the weight of an object, $mg\mathbf{e}_z$, one way to proceed is to lift it and measure the work required, equal to $mg.\Delta z$, with Δz the change of altitude. The weight can then be reconstructed from the work, based on a model. For a force described with a vector with three independent components, we see that three independent motions are needed. In doing this, we do not have to exert real motions, but virtual motions can be imagined.

In order to give an idea of how to build the virtual power of forces acting on each point of a body, we start by writing the fundamental principle of dynamics in a Galilean reference frame (also coined inertial frame, since no fictitious forces are introduced): for a small domain $d\Omega_t$ with infinitesimal mass dm, Newton's law states that

$$d\mathbf{F} = dm\gamma = \rho\gamma d\Omega_t$$

with the total force vector $d\mathbf{F}$ decomposing into the sum of external remote forces $d\mathbf{F}_e$ (exerted by the exterior of Ω_t onto $d\Omega_t$), internal remote forces $d\mathbf{F}_{ir}$ (forces exerted by the surrounding of $d\Omega_t$ onto $d\Omega_t$), and contact forces $d\mathbf{F}_{ic}$ (forces acting through the boundary of $d\Omega_t$) due to the presence of neighboring subdomains, considered as internal contact forces, *viz.*

$$d\mathbf{F} = d\mathbf{F}_e + d\mathbf{F}_{ir} + d\mathbf{F}_{ic}$$
$$d\mathbf{F}_e = \rho\mathbf{F}_e d\Omega_t, \quad d\mathbf{F}_{ir} = \rho\mathbf{F}_{ir} d\Omega_t$$

Vectors $\mathbf{F}_e, \mathbf{F}_i$ therein denote the density per unit actual mass of the external remote forces and internal remote forces respectively (equivalently vectors $\rho\mathbf{F}_e$, $\rho\mathbf{F}_i$ are the same densities, but per unit volume).

The basic idea behind the VWP is that of a *dualisation of forces with virtual motions*: multiplying Newton's law by a virtual velocity $\tilde{\mathbf{V}}$

$$d\mathbf{F}.\tilde{\mathbf{V}} = \rho\gamma.\tilde{\mathbf{V}}d\Omega_t \rightarrow d\mathbf{F}_e.\tilde{\mathbf{V}} + d\mathbf{F}_{ir}.\tilde{\mathbf{V}} + d\mathbf{F}_{ic}.\tilde{\mathbf{V}} = \rho\gamma.\tilde{\mathbf{V}}d\Omega_t$$

we obtain a relation between scalar-valued quantities, having the physical meaning of powers.

Note that the acceleration γ is the actual acceleration, so it is *not* the material derivative of the virtual velocity (both coincide only for *real motions*).

The notion of virtual velocity requires some explanation: it can be conceived as a perturbation $\delta\mathbf{V}$ from a real velocity field $\mathbf{V}$, so that $\tilde{\mathbf{V}} = \mathbf{V} + \delta\mathbf{V}$.

We then define the linear forms of acceleration quantities, external remote forces, internal forces, and contact forces as

$$dP_a\left[\tilde{\mathbf{V}}\right] := \rho\gamma.\tilde{\mathbf{V}}d\Omega_t,\quad dP_e\left[\tilde{\mathbf{V}}\right] := d\mathbf{F}_e.\tilde{\mathbf{V}},\quad dP_{ir}\left[\tilde{\mathbf{V}}\right] := d\mathbf{F}_{ir}.\tilde{\mathbf{V}},\quad dP_{ic}\left[\tilde{\mathbf{V}}\right] := d\mathbf{F}_{ic}.\tilde{\mathbf{V}}$$

In the sequel, the internal remote forces and external remote forces will not be distinguished, so they will be collectively coined *external forces*. The density of external forces is then supposed to be described by the action of the body weight with density $\rho\mathbf{f}(x)$, thus giving rise to the virtual power

$$P_e\left[\tilde{\mathbf{V}}\right] = \int_{\Omega_t} \rho\mathbf{f}(x).\tilde{\mathbf{V}}dx$$

The virtual power of acceleration is built as a linear form of the virtual velocity, in the form of the integral over an arbitrary actual domain Ω_t, $P_a := \int_{\Omega_t} \rho\gamma.\tilde{\mathbf{V}}dx$ with $\rho\gamma$ the quantity of acceleration (a density per unit volume).

The density of contact forces is posited as a linear functional of the virtual velocity and its first order gradient,

$$P_{ic}\left[\tilde{\mathbf{V}}\right] := -\int_{c\Omega_t} \left\{\mathbf{a}.\tilde{\mathbf{V}} + \boldsymbol{\sigma} : \tilde{\mathbf{d}}\right\}dx$$

with **a** a spatial vector, and $\tilde{\mathbf{d}} := grad\tilde{\mathbf{V}}$ the virtual rate of deformation tensor. The Cauchy stress is the natural candidate for the second order tensor dual to $\tilde{\mathbf{d}}$. The condition that $P_i\left[\tilde{\mathbf{V}}\right]$ vanishes on virtual rigid body motions of the form $\tilde{\mathbf{V}} = \tilde{\mathbf{V}}_0 + \boldsymbol{\omega} \times \mathbf{OM}$, with $\tilde{\mathbf{V}}_0, \boldsymbol{\omega}$ two constant vectors, implies that **a** has to vanish, thus

$$P_{ic}\left[\tilde{\mathbf{V}}\right] := -\int_{\Omega_t} \boldsymbol{\sigma} : \tilde{\mathbf{d}} dx$$

In the sequel, in order to simplify notations, $P_{ic}\left[\tilde{\mathbf{V}}\right]$ will simply be noted $P_i\left[\tilde{\mathbf{V}}\right]$ and $d\mathbf{F}_{ic}$ as $d\mathbf{F}_c$.

EXERCISE.– Check that $P_i = P_i\left[\tilde{\mathbf{V}}\right]$ is a linear form vanishing on rigid body motions.

Transformation of this integral by the divergence theorem leads to the alternative expression

$$P_i\left[\tilde{\mathbf{V}}\right] := -\int_{\Omega_t} div\boldsymbol{\sigma} : \tilde{\mathbf{V}} dx + \int_{\partial\Omega_t} \boldsymbol{\sigma}.\mathbf{n}.\tilde{\mathbf{V}} ds$$

including the power of contact forces as the surface integral therein. The density of contact forces therein, vector **σ.n** (physically representing the cohesion of matter), is consistent with Cauchy's theorem giving the surface density of contact forces:

$$d\mathbf{F}_c = \left(\boldsymbol{\sigma}.\mathbf{n} ds\right)_{\partial d\Omega_t}$$

with the subscript $\partial d\Omega_t$ indicating that this power is restricted to the boundary of $d\Omega_t$ as its support. The virtual power of contact forces is part of the virtual power of external forces, and the previous expression of $P_i\left[\tilde{\mathbf{V}}\right]$ suggests that we consider the similar form

$$P_e\left[\tilde{\mathbf{V}}\right] := \int_{\partial\Omega_t} \mathbf{t}.\tilde{\mathbf{V}} dx + \int_{\Omega_t} \rho \mathbf{f}(x).\tilde{\mathbf{V}} dx$$

The principle of virtual power states that the virtual power of acceleration quantities is the sum of the virtual power of internal and external forces; in addition, the virtual power of internal forces shall vanish on the subset of rigid body motions (which should thus not generate any strains and stresses).

The three introduced powers are scalar-valued quantities and should be observer independent, and their construction relies on *Riesz representation theorems of linear forms*.

According to previous writings, the virtual power of acceleration $P_a = P_a\left[\tilde{\mathbf{V}}\right]$ is a linear form in the sense of the mapping

$$\tilde{\mathbf{V}} \mapsto \int_{\Omega_t} \rho\gamma.\tilde{\mathbf{V}}dx$$

Similarly, the virtual power of external forces previously written is a linear form in the sense of the mapping $P_e = P_e\left[\tilde{\mathbf{V}}\right]$,

$$\mathbf{t} \mapsto \int_{S_t} \mathbf{t}.\tilde{\mathbf{V}}ds + \int_{\Omega_t} \rho\mathbf{f}(x).\tilde{\mathbf{V}}dx$$

Summarizing, the VWP (virtual work principle) writes in Eulerian format in which all quantities are depending upon the spatial position,

$$-\int_{\Omega_t} \boldsymbol{\sigma}(\mathbf{x}):\mathbf{d}(\mathbf{x})dx + \int_{S_t} \mathbf{t}.\tilde{\mathbf{V}}dx + \int_{\Omega_t} \rho(\mathbf{x})\mathbf{f}(\mathbf{x}).\tilde{\mathbf{V}}dx = \int_{\Omega_t} \rho(\mathbf{x})\gamma(\mathbf{x}).\tilde{\mathbf{V}}dx, \ \forall\tilde{\mathbf{V}}$$

Projection of the virtual velocity $\tilde{\mathbf{V}}$ onto the subset of *real motions* leads to the kinetic energy theorem:

$$\frac{DK}{Dt} = P_e + P_i$$

with $K := \int_{\Omega_t} \frac{1}{2}\rho\mathbf{v}^2 dx$; P_e and P_i are then *real powers*, and $\mathbf{v}$ is the *real* velocity field.

A Lagrangian version of the VWP is next written, based on the following transformations: let us recall the transform formula for volume element $dx = JdX = \frac{\rho_0}{\rho}dX$, and elaborate the *first Piola-Kirchhoff stress tensor* (also called the *nominal stress*), denoted PK1 in the sequel, with associated traction vector

$$\mathbf{P} = J\boldsymbol{\sigma}.\mathbf{F}^{-T} \rightarrow \mathbf{t}_0 := \mathbf{T}.\mathbf{N}$$

involving the Lagrangian normal vector $\mathbf{N}$, related to its spatial counterpart by Nanson formula

$$\mathbf{n}ds = J\mathbf{F}^{-T}.\mathbf{N}dS$$

Accordingly, the virtual power of acceleration and external forces rewrite successively

$$P_a := \int_{\Omega_0} J\rho\boldsymbol{\gamma}.\tilde{\mathbf{V}}dX \equiv \int_{\Omega_0} \rho_0\boldsymbol{\gamma}.\tilde{\mathbf{V}}dX, \quad P_e := \int_{S_0} \mathbf{t}_0.\tilde{\mathbf{V}}dS + \int_{\Omega_0} \rho_0(\mathbf{X})\mathbf{f}(\mathbf{X}).\tilde{\mathbf{V}}dx$$

Observe that all quantities therein are now functions of the material coordinates. The virtual power of internal forces can be rewritten in terms of PK1 stress as

$$P_i := -\int_{\Omega_0} \mathbf{P} : \dot{\mathbf{F}}dx = -\int_{\Omega_0} \mathbf{S} : \dot{\mathbf{E}}dx$$

introducing therein the *second Piola-Kirchhoff stress tensor* (abbreviated as PK2)

$$\mathbf{S} := J\mathbf{F}^{-1}.\boldsymbol{\sigma}.\mathbf{F}^{-T} = \mathbf{F}^{-1}.\mathbf{P}$$

We have used the relation $\dfrac{D\mathbf{E}}{Dt} = \mathbf{F}^T.\mathbf{d}.\mathbf{F} \rightarrow \boldsymbol{\sigma} : \mathbf{d} = \dfrac{1}{J}\mathbf{S} : \dfrac{D\mathbf{E}}{Dt}$ to derive the last expression of P_i.

Observe that PK1 is not symmetrical and is lying in both configurations Ω_0 and Ω_t, whereas PK2 is a symmetrical material tensor fully living in the reference configuration.

EXERCISE.– Write the decomposition of the nominal stress and second Piola-Kirchhoff stress in their respective bases, considering that Cauchy stress is a contravariant tensor, $\boldsymbol{\sigma} = \sigma^{ij}\mathbf{g}_i \otimes \mathbf{g}_j$. Deduce that the nominal stress is attached to both configurations.

The expression of P_i shows that the double contraction of PK2 with the material derivative $\dot{\mathbf{E}} = \dfrac{D\mathbf{E}}{Dt}$ is a true scalar defining the density of virtual internal power: we say that $\mathbf{S}$ and $\dot{\mathbf{E}}$ are conjugated stress and strain measures, since we can write

$$\mathbf{S} : \dot{\mathbf{E}} = tr\left(\mathbf{S}.\dot{\mathbf{E}}\right) = \rho_0 \dot{e}$$

with $\dot{e}$ the density of internal energy-per-unit reference volume. Similarly, the nominal stress tensor $\mathbf{P}$ and $\dot{\mathbf{F}}$ are conjugated, as well as Biot stress $\mathbf{T} = \mathbf{U}.\mathbf{S} = R^T.\mathbf{P}$ (also called Jaumann stress) and $\dot{\mathbf{U}}$, so that we can write the following equalities expressing dualities between Lagrangian stresses and strains,

$$\rho_0 \dot{e} = \mathbf{S} : \dot{\mathbf{E}} = \mathbf{T} : \dot{\mathbf{U}} = \mathbf{P} : \dot{\mathbf{F}}$$

EXERCISE.– Show the relation $\rho_0 \dot{e} = \mathbf{T} : \dot{\mathbf{U}}$, with $\mathbf{U}$ the pure deformation involved in the polar decomposition of $\mathbf{F}$, using the identities $\mathbf{T} = \mathbf{U}.\mathbf{S} = R^T.\mathbf{P}$, $\mathbf{A} : (\mathbf{B}.\mathbf{C}) = \left(\mathbf{B}^T.\mathbf{A}\right).\mathbf{C} = \left(\mathbf{A}.\mathbf{C}^T\right).\mathbf{B}$, and the orthogonality of symmetrical and antisymmetrical tensors .

Adopting now an Eulerian viewpoint, the Kirchhoff stress $\boldsymbol{\tau} := J\boldsymbol{\sigma}$ is dual to the Lie derivative of the Eulerian strain $L_v \mathbf{e} = \mathbf{F}^{-T}.\dot{\mathbf{E}}.\mathbf{F}^{-1} = \mathbf{d}$ (show this last duality as an exercise), so that

$$\rho_0 \dot{e} = \boldsymbol{\tau} : L_v \mathbf{e} \equiv \boldsymbol{\tau} : \mathbf{d}$$

EXERCISE.– Deduce this last equality from the identities $\boldsymbol{\tau} = \mathbf{P}.\mathbf{F}^T = \mathbf{F}.\mathbf{S}.\mathbf{F}^T$.

2.11.3. *Physical meaning of the introduced stress tensors*

Any stress tensor can be defined as a linear mapping of a normal vector to a stress vector; this is first illustrated by introducing the pseudo stress vector $\mathbf{t}_0$ such that $\mathbf{t}_0 dS = \mathbf{t} ds \rightarrow \mathbf{t}_0 = \mathbf{t} ds / dS$; let then introduce the nominal stress $\mathbf{P}$ s.t. $\mathbf{t}_0 = \mathbf{P}.\mathbf{N}$, representing a force-per-unit reference area. Further artificial stress vectors and associated material and spatial stress tensors are shown in Figure 2.30.

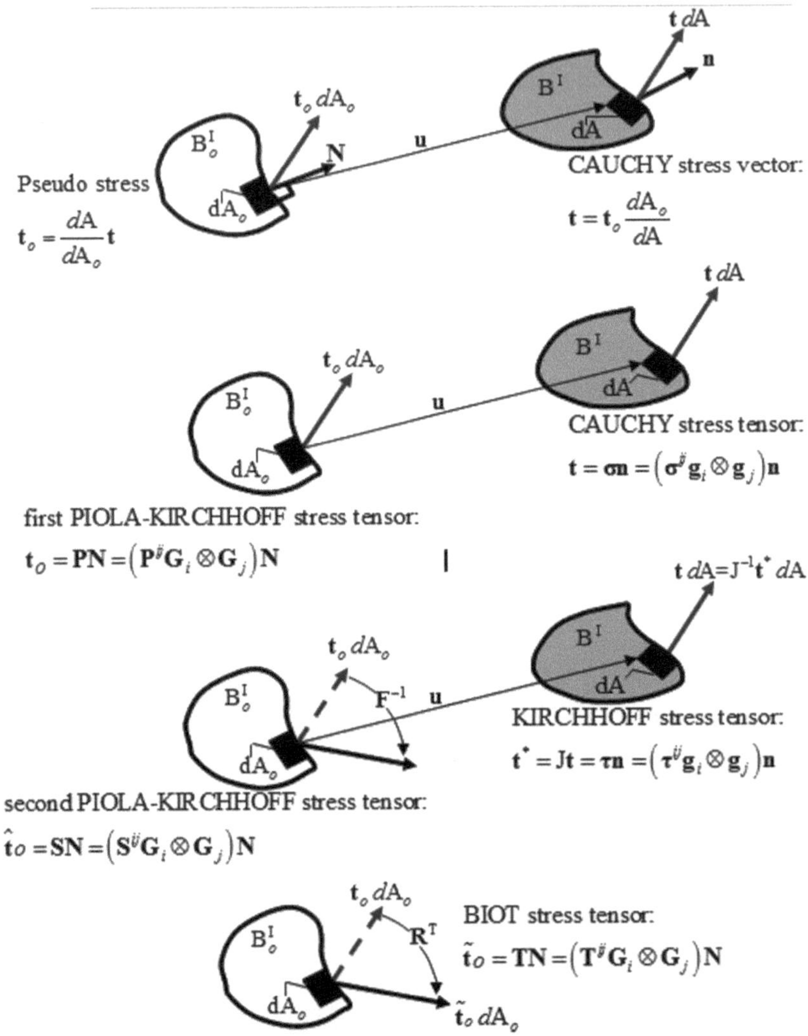

Figure 2.30. *Material and spatial stress tensors based on fictitious stress vectors [BAS 00]. For a color version of this figure, see www.iste.co.uk/ganghoffer/biomechanics.zip*

The definition of the introduced material and spatial stress tensors is summarized in Table 2.1 [BAS 00].

Notations	Definitions	Relations
Cauchy stress	$\boldsymbol{\sigma} = \sigma^{ij}\mathbf{g}_i \otimes \mathbf{g}_j$	$\boldsymbol{\sigma} = J^{-1}\boldsymbol{\tau}$
Kirchhoff stress	$\boldsymbol{\tau} = \tau^{ij}\mathbf{g}_i \otimes \mathbf{g}_j$	$\boldsymbol{\tau} = \mathbf{P}.\mathbf{F}^T = \mathbf{F}.\mathbf{S}.\mathbf{F}^T$
First Piola-Kirchhoff stress	$\mathbf{P} = P^{ij}\mathbf{G}_i \otimes \mathbf{G}_j$	$\mathbf{P} = R.\mathbf{T}$
Second Piola-Kirchhoff stress	$\mathbf{S} = S^{ij}\mathbf{G}_i \otimes \mathbf{G}_j = \mathbf{S}^T$	$\mathbf{S} = \mathbf{U}^{-1}.\mathbf{T} = \mathbf{F}^{-1}.\boldsymbol{\tau}.\mathbf{F}^{-T}$
Biot stress	$\mathbf{T} = T^{ij}\mathbf{G}_i \otimes \mathbf{G}_j$	$\mathbf{T} = \mathbf{U}.\mathbf{S}$
Important equalities	$J = \det \mathbf{F}$ $\quad S^{ij} = \tau^{ij}$ (for contravariant components)	

Table 2.1. *Summary of the definition of the introduced material and spatial stress tensors [BAS 00]*

2.11.4. *Eshelby stress in the context of configurational mechanics*

The balance of linear momentum based on the first Piola-Kirchhoff stress tensor **T** is obtained by a pull-back of the Eulerian similar balance of linear momentum

$$\rho\dot{\mathbf{v}} = Div[\boldsymbol{\sigma}] + \mathbf{f} \rightarrow J\rho\dot{\mathbf{v}} = JDiv[\boldsymbol{\sigma}] + J\mathbf{f} \Leftrightarrow \rho_0\dot{\mathbf{V}} = JDiv[\boldsymbol{\sigma}] + \mathbf{f}_0 \equiv Div[\mathbf{T}] + \mathbf{f}_0$$

Using the relation

$$Div\left(J\mathbf{F}^{-T}\right) = 0$$

tt follows:

$$\mathbf{T} = J\boldsymbol{\sigma}.\mathbf{F}^{-1} \rightarrow T^{aA} = J\sigma^{ab}F_b^{A-1}$$
$$\rightarrow T^{aA}{}_{,A} = J\sigma^{ab}{}_{,A}F_b^{A-1} = J\sigma^{ab}{}_{,i}x^i{}_{,A}F_b^{A-1} = J\sigma^{ab}{}_{,i}F^i_AF_b^{A-1} = J\sigma^{ab}_{,b} \qquad [2.1]$$

The transport on Ω_0 of the dynamical equilibrium equation

$$\rho_0\dot{\mathbf{V}} = Div[\mathbf{T}] + \mathbf{f}_0 \qquad [2.2]$$

gives according to [2.1]:

$$\mathbf{F}^T.\left(\rho_0\dot{\mathbf{V}}\right) = \mathbf{F}^T.Div[\mathbf{T}] + \mathbf{F}^T.\mathbf{f}_0 \qquad [2.2']$$

noting that all vectors are in fact covectors (covariant tensors). The local dissipation writes in terms of the free energy density $\psi=\psi(F,T,z)$ as

$$D=[\mathbf{T}-\partial_F\psi]:\dot{\mathbf{F}}-[s+\partial_T\psi]\dot{T}-\partial_z\psi.\dot{z}\geq 0$$

having as arguments a set of internal variables denoted z - which may be of scalar, vector or tensor nature. The total spatial material derivative of ψ can be elaborated as

$$d_{X^A}\psi=\frac{\partial\psi}{\partial F_I^i}F_{I,A}^i+\left(\frac{\partial\psi}{\partial X^A}\right)_{\text{expl}}\equiv T_i^I F_{I,A}^i+\left(\frac{\partial\psi}{\partial X^A}\right)_{\text{expl}}=T_i^I F_{A,I}^i+\left(\frac{\partial\psi}{\partial X^A}\right)_{\text{expl}} \quad [2.3]$$

using the compatibility condition ($\mathbf{F}$ is a field of gradient), which amounts to writing Schwartz conditions, denoting $\left(\frac{\partial\psi}{\partial X^A}\right)_{\text{expl}}$ the spatial partial derivative of ψ, which traduces the explicit dependency of ψ versus X^A (thus the notation *expl*). Such an explicit dependency may traduce the variation of the density or mechanical properties with spatial position. Combining then the identity

$$F_A^i T_{i\ ,B}^B=\left(F_A^i T_i^B\right)_{,B}-F_{A,B}^i T_i^B$$

with relation [2.3] results in

$$\begin{aligned}&F_A^i T_{i\ ,B}^B=\left(F_A^i T_i^B\right)_{,B}-d_{X^A}\psi+\left(\partial_{X^A}\psi\right)_{\text{expl}}\\&\rightarrow\mathbf{F}^T.Div[\mathbf{T}]=Div\left[\mathbf{F}^T.\mathbf{T}\right]-Div[\psi\mathbf{I}]+\left(\partial_X\psi\right)_{\text{expl}}\end{aligned}$$

The spatial gradient therein takes into account all dependencies upon the spatial variable. We then substitute the term $F^T[DivT]$ into [2.3] to reach the *purely material formulation* of equilibrium

$$\mathbf{F}^T.\rho_0\dot{\mathbf{V}}=Div\left[\mathbf{F}^T.\mathbf{T}-\psi\mathbf{I}\right]+\left(\partial_X\psi\right)_{\text{expl}}+\mathbf{F}^T.\mathbf{f}_0 \quad [2.4]$$

Under static conditions and in the absence of body forces, [2.4] becomes

$$Div[\mathbf{\Sigma}]=\left(\partial_X\psi\right)_{\text{expl}}$$

with

$$\mathbf{\Sigma} := \psi\mathbf{I} - \mathbf{F}^T.\mathbf{T}$$

the energy momentum tensor, also called Eshelby tensor. Eshelby tensor arises in the framework of configurational mechanics which also deserves the name, theory of material forces. This follows the pioneering ideas presented by Eshelby in [ESH 51, ESH 70], who introduced the so-called Maxwell-tensor of elasticity as the driving force for the motion of an inclusion in this famous Gedankenexperiment, later on called the energy-momentum tensor [ESH 70]. Eshelbian mechanics places the emphasis on the motion of particles in material space, distinctly from the motion in physical space. These ideas are in line with the intrinsic representation of continuum mechanics advanced in Noll [NOL 72]. As the name indicates, configurational mechanics provides the expression of the driving force necessary to move inhomogeneities or defects, a vector elaborated as the projection of Eshelby stress onto the normal to the boundary of the inhomogeneity. The balance law for Eshelby stress is a conservation law that can be obtained from Noether's theorem [NOE 18] expressing the translational invariance of the total potential energy of a domain enclosing an inhomogeneity. This gives Eshelby stress the meaning of the driving force (when converted into a vector by projecting it onto the normal to the moving heterogeneity) needed to move a defect in a body in its reference configuration. The concept of Eshelbian mechanics (also called configurational mechanics) is well suited to the modeling of different kinds of defects or heterogeneities in solid bodies (like interstitial atoms, dislocations, phase boundaries, cracks, interfaces, inclusions or voids), including more recently mass growth.

Eshelby stress can further be related to the domain variation, a point of view which turns out to be fruitful when considering the area of solid body growth.

Let consider as a point of departure the variation of the action functional

$$S := \int_{\Omega} L(x,u)\, d\Omega$$

the Lagrangian depending upon the independent and dependent variables x and u respectively; standard calculus of variation delivers

$$\delta S = \mu \int_{\Omega} \left(\frac{\partial L}{\partial u_k} - D_i \frac{\partial L}{\partial u_{k,i}} \right) \left(\phi_k - \xi_j u_{k,j} \right) d\Omega + \mu \int_{\partial\Omega} \left(L\xi_i + \left(\phi_k - u_{k,j}\xi_j \right) \frac{\partial L}{\partial u_{k,i}} \right) n_i d(\partial\Omega)$$

denoting by $(\xi,\phi)\equiv\mu(\delta X,\delta u)$ the components of the vector field generating the group of transformations of S. We can note that the surface contribution $\int_{\partial\Omega}\left(L\xi_i-u_{k,j}\xi_j\frac{\partial L}{\partial u_{k,i}}\right)n_i d(\partial\Omega)$ of previous surface integral expresses the domain variation (term exclusively depending on δX), and highlights the Eshelby tensor: this sets the link between domain variation and Eshelbian mechanics. The first volume integral has an integrand representing the characteristic $Q_k\equiv\phi_k-\xi_j u_{k,j}$, factor of the Euler operator applied to the Lagrangian density.

Previous variation takes the equivalent but more compact form due to Cartan:

$$L_X\omega=i_X d\omega+d\left(i_X\omega\right)$$

Allowing in turn a more compact writing of Noether's theorem: under the conditions $L_X\omega=0$ (invariance of the action integral $\int_\Omega\omega\equiv\int_\Omega Ldxdt$ under the group generated by X) and $i_X d\omega=0$ (validity of the Euler equations), the following conservation law is obtained:

$$Div\left(L\xi_i+\left(\phi_k-u_{k,j}\xi_j\right)\frac{\partial L}{\partial u_{k,i}}\right)=0$$

In particular, for a purely horizontal variation $(\delta u\equiv 0\rightarrow\phi=0)$, we recover the conservation of Eshelby tensor $\Sigma_{ij}=L\delta_{ij}-u_{k,j}\frac{\partial L}{\partial u_{k,i}}$, that is

$$Div\left(L\delta_{ij}-u_{k,j}\frac{\partial L}{\partial u_{k,i}}\right)\xi_j=0\rightarrow Div\left(L\delta_{ij}-u_{k,j}\frac{\partial L}{\partial u_{k,i}}\right)\delta X_j=0,\forall\delta X_j\rightarrow Div\Sigma=0$$

This shows that Eshelby stress is the driving force for the domain change due to the transformation of spatial coordinates – the horizontal variation – as clearly highlighted by the previous expression of the variation of S, rewritten in terms of the domain derivative $\frac{\delta S}{\delta\Omega}$:

$$\frac{\delta S}{\delta\Omega}=\int_\Omega Div\left[\left(L\delta_{ij}-u_{k,j}\xi_j\frac{\partial L}{\partial u_{k,i}}\right)\delta X_j+\left(\phi_k\frac{\partial L}{\partial u_{k,j}}\right)\delta u_j\right]d\left(\partial\Omega\right)$$

EXAMPLE.– Let us consider a bar of a homogeneous elastic material (with tensile modulus E_0 and length L) loaded in traction, clamped in $x=0$ and subjected to a tensile effort F at its right extremity $x=L$ (body forces are discarded). The solution of the linear elasticity problem compatible with the posited boundary conditions is easily obtained as

$$u(x)=F\,x/E_0$$

The bar potential energy is next written as a functional of the displacement u :

$$V[u]=\int_0^L W(x,u(x))\,dx-F.u(L)$$

With the density of elastic energy therein expressed by

$$W(x,u(x))=\frac{1}{2}E_0 u'(x)^2\equiv\frac{1}{2}E_0\left(\frac{F}{E_0}\right)^2=\frac{1}{2}\frac{F^2}{E_0}$$

This gives in elementary calculations the potential energy viewed as a function of the bar length, namely

$$V(L)=\frac{1}{2}\frac{F^2}{E_0}L-F.\left(\frac{FL}{E_0}\right)\equiv-\frac{1}{2}\frac{FL^2}{E_0}$$

Eshelby stress writes in the present small perturbations context and in 1D as the scalar

$$\Sigma=W-\sigma.\varepsilon\equiv-\frac{1}{2}E_0 u'(x)^2$$

It appears that Eshelby stress is the derivative of the total potential energy of the bar with respect to its length, *viz.*

$$\Sigma=\frac{\delta V}{\delta L}\ \text{(using the notations of domain variation,}\ \frac{\delta}{\delta\Omega}\equiv\frac{\delta}{\delta L}\text{)}$$

This shows, based on this simple 1D example, that Eshelby stress is associated with the variation of the action integral with respect to a variation of the domain.

2.12. Balance laws

Balance laws play a central role in continuum mechanics since they express the conservation of some physical quantities; they take an integral form which is consistent with the fact that measurements are made over systems having some extent in space. Localization of the balance laws is a mathematical procedure leading to their strong form as differential equations.

In order to introduce balance laws, we first determine the:

2.12.1. *Material derivatives of surface and volume integrals*

DEFINITION.– Let us consider a volume integral $A := \int_{\Omega_t} a(\mathbf{x},t)\,dx$; in order to evaluate $\frac{dA}{dt}$, we first map back A to the reference configuration with fixed domain (Ω_0 will not change in time) to evaluate the material derivative, and push forward the result to the actual configuration,

$$A := \int_{\Omega_0} a(\mathbf{x}(\mathbf{X},t),t)\,JdX \rightarrow \frac{dA}{dt} = \int_{\Omega_0} \left(\frac{da}{dt}J + a\frac{dJ}{dt} \right) dX = \int_{\Omega_0} \left(\frac{da}{dt} + a div \mathbf{v} \right) dx$$

EXERCISE.– Show the relation $\frac{dJ}{dt} = J div \mathbf{v}$ used in previous calculations, using the identities

$$J = \frac{1}{6}\varepsilon_{ijk}\varepsilon_{\alpha\beta\gamma}F_{i\alpha}F_{j\beta}F_{k\gamma}, \quad \frac{\partial F_{i\alpha}}{\partial t} = \frac{\partial v_i}{\partial x_p}F_{p\alpha}, \quad \varepsilon_{ijk}\varepsilon_{ojk} = 2\delta_{ip}.$$

An alternative expression of the material derivative is given by

$$\frac{dA}{dt} = \int_{\Omega_t} \left(\frac{\partial a}{\partial t} + div(a\mathbf{v}) \right) dx = \int_{\Omega_t} \frac{\partial a}{\partial t} dx + \int_{\partial\Omega_t} a\mathbf{v}.\mathbf{n}ds$$

The first contribution on the right-hand side traduces local variations of a, and the second surface integral the flux of a through the boundary (Figure 2.31).

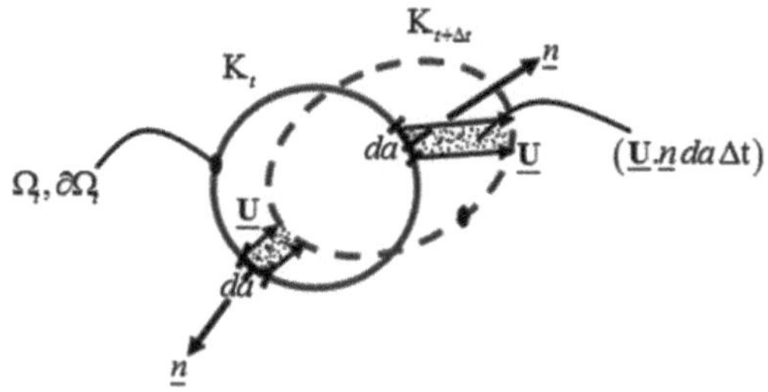

Figure 2.31. *Physical interpretation of the material derivative of a volume integral. For a color version of this figure, see www.iste.co.uk/ganghoffer/biomechanics.zip*

The material derivative of a surface integral $B := \int_{s(t)} \mathbf{b}(\mathbf{x},t).\mathbf{n}ds$ representing a flux is given by (admitted)

$$\frac{dB}{dt} = \int_{s(t)} \left(\frac{\partial b_i}{\partial t} + \frac{\partial}{\partial x_i}(b_i v_j) - b_j \frac{\partial v_i}{\partial x_j} \right) ds$$

We are now in a position to write the main balance laws in continuum mechanics.

2.12.2. *Mass conservation*

The overall mass of a set of material points in an arbitrary domain Ω_t followed in its motion, is constant, thus

$$m[\Omega_t] := \int_{\Omega_t} \rho dx \Rightarrow \frac{dm[\Omega_t]}{dt} = 0, \quad \forall \Omega_t$$

COROLLARY.– Working out the material derivative of the volume integral according to previous result leads to the following equality

$\int_{\Omega_t} \left\{ \frac{\partial \rho}{\partial t} + div(\rho \mathbf{v}) \right\} dx = 0, \quad \forall \Omega_t$ The weak form of mass conservation writes

Localization of previous condition using Dubois Reymond lemma (fundamental lemma of the calculus of variation) leads to the Eulerian form of mass conservation

$$\frac{\partial \rho}{\partial t} + div(\rho \mathbf{v}) = 0$$

the strong form of the continuity equation.

The Lagrangian form of the same balance law is easily obtained by the transformations (identity of the mass in Ω_0 and Ω_t and change of variable)

$$\int_{\Omega_t} \rho dx = \int_{\Omega_0} J\rho_0 dX = \int_{\Omega_0} \rho_0 dX \Rightarrow \rho_0 / \rho = J = \det \mathbf{F}$$

traducing mass conservation (Lagrangian form).

The continuity equation brings simplifications of the expressions of the material derivatives of volume integrals: when integrating a density per unit mass $b(\mathbf{x},t)$, we easily get

$$\frac{d}{dt}\int_{\Omega_t} b(\mathbf{x},t)\rho(\mathbf{x},t)dx = \int_{\Omega_t}\left\{\frac{\partial(\rho b)}{\partial t} + div(\rho b \mathbf{v})\right\}dx = \int_{\Omega_t} \rho \frac{db}{dt}dx$$

so that we may consider the density as constant when taking the material derivative.

Conservation of linear and angular momentum

The conservation of linear momentum is obtained in a straightforward manner from the VWP, as follows: recalling the weak form of equilibrium

$$-\int_{\Omega_t} \boldsymbol{\sigma} : \mathbf{d}dx + \int_{S_t} \mathbf{t}.\tilde{\mathbf{V}}ds + \int_{\Omega_t} \rho \mathbf{f}(\mathbf{x}).\tilde{\mathbf{V}}dx = \int_{\Omega_t} \rho \boldsymbol{\gamma}.\tilde{\mathbf{V}}dx, \quad \forall \tilde{\mathbf{V}}$$

Integrating by part the first volume integral by the divergence theorem leads to the dynamical equilibrium equations (three equations)

$$\int_{\Omega_t} \{\rho(\mathbf{x})\boldsymbol{\gamma}(\mathbf{x}) - \rho \mathbf{f} - div\boldsymbol{\sigma}\}.\tilde{\mathbf{V}}dx + \int_{S_t} (\boldsymbol{\sigma}.\mathbf{n} - \mathbf{t}).\tilde{\mathbf{V}}ds = 0, \quad \forall \tilde{\mathbf{V}}$$

The arbitrariness of $\tilde{\mathbf{V}}$ leads to (indicating the dependence of the fields upon position and time)

$$\rho(\mathbf{x},t)\boldsymbol{\gamma}(\mathbf{x},t)-\rho(\mathbf{x},t)\mathbf{f}(\mathbf{x},t)-div\boldsymbol{\sigma}(\mathbf{x},t)=\mathbf{0}, \quad \forall \mathbf{x}\in\Omega_t, \ \forall t$$

and to the static boundary conditions

$$\mathbf{t}(\mathbf{x},t)=\boldsymbol{\sigma}(\mathbf{x},t).\mathbf{n}(\mathbf{x},t), \quad \forall \mathbf{x}\in s_t, \ \forall t$$

The static boundary conditions deserve the name natural boundary conditions (since they follow from the VWP), in contrast with the kinematic boundary conditions for the displacement field – called essential, since they must be defined in the set of kinematically admissible displacements.

REMARK.– When inertia terms can be neglected, $\|\rho(\mathbf{x})\boldsymbol{\gamma}(\mathbf{x})\|\to 0$, the VWP leads to the static *equilibrium equations*.

REMARK.– The balance of linear momentum can be obtained from the equality of resultants

$$\int_{\Omega_t}\{\rho\boldsymbol{\gamma}-\rho\mathbf{f}\}dx=\int_{S_t}\boldsymbol{\sigma}.\mathbf{n}ds$$

The Lagrangian form of the VWP leads to an alternative formulation of the balance of momentum involving the second Piola-Kirchhoff stress (a Lagrangian stress), as

$$-\int_{\Omega_0}\mathbf{P}:\dot{\tilde{\mathbf{F}}}dX+\int_{S_0}\mathbf{t}_0.\tilde{\mathbf{V}}dS+\int_{\Omega_0}\rho_0(\mathbf{X})\mathbf{f}(\mathbf{X}).\tilde{\mathbf{V}}dX=\int_{\Omega_0}\rho_0\boldsymbol{\gamma}.\tilde{\mathbf{V}}dX, \quad \forall\tilde{\mathbf{V}}$$

$$-\int_{\Omega_0}\mathbf{S}:\dot{\tilde{\mathbf{E}}}dX+\int_{S_0}\mathbf{t}_0.\tilde{\mathbf{V}}dS+\int_{\Omega_0}\rho_0(\mathbf{X})\mathbf{f}(\mathbf{X}).\tilde{\mathbf{V}}dX=\int_{\Omega_0}\rho_0\boldsymbol{\gamma}.\tilde{\mathbf{V}}dX, \quad \forall\tilde{\mathbf{V}}$$

respectively involving the nominal stress $\mathbf{P}$ and the second Piola-Kirchhoff stress tensor related by

$$\mathbf{S}:=J\mathbf{F}^{-1}.\boldsymbol{\sigma}.\mathbf{F}^{-T}=\mathbf{F}^{-1}.\mathbf{P}$$

The strong form is easily obtained (check as an exercise)

$$Div\mathbf{P}+\rho_0\mathbf{f}=\rho_0\boldsymbol{\gamma}\to Div\mathbf{F}.\mathbf{S}+\rho_0\mathbf{f}=\rho_0\boldsymbol{\gamma}$$

in which Div is now the material divergence (derivatives are taken with respect to the Lagrangian coordinates).

The kinetic energy theorem in Lagrangian form then takes the form

$$\frac{d}{dt}\int_{\Omega_0}\frac{1}{2}\rho_0 V^2 dX = -\int_{\Omega_0}\mathbf{S}:\dot{\mathbf{E}}dx + \int_{S_0}\mathbf{t}_0.\mathbf{V}dS + \int_{\Omega_0}\rho_0\mathbf{f}.\mathbf{V}dx =$$
$$= -\int_{\Omega_0}\mathbf{S}:\dot{\mathbf{E}}dx + \int_{S_0}\mathbf{P}^T.\mathbf{N}.\mathbf{V}dS + \int_{\Omega_0}\rho_0\mathbf{f}.\mathbf{V}dx$$

leading to the strong form (due to the arbitrariness of the domain Ω_0):

$$\frac{1}{2}\rho_0\frac{d}{dt}V^2 = -\mathbf{S}:\dot{\mathbf{E}} + Div\left(\mathbf{P}^T.\mathbf{V}\right) + \rho_0\mathbf{f}.\mathbf{V}$$

REMARK.– There exist other formulations of the balance of linear momentum involving other stress measures (like Eshelby stress), but which will not be developed in this book.

Conservation of angular momentum: The balance of the moment of forces writes

$$\int_{\Omega_t}\mathbf{OM}\times\{\rho\boldsymbol{\gamma} - \rho\mathbf{f}\}dx = \int_{S_t}\mathbf{OM}\times\boldsymbol{\sigma}.\mathbf{n}ds$$

This writes in component form

$$\int_{\Omega_t}\{\rho\varepsilon_{ijk}x_j(\gamma_k - f_k)\}dx = \int_{S_t}\varepsilon_{ijk}x_j\sigma_{kl}n_l ds$$

EXERCISE.– Show, using the dynamical equilibrium equations, that previous condition leads to the symmetry of Cauchy stress,

$$\boldsymbol{\sigma} = \boldsymbol{\sigma}^T \rightarrow \sigma_{ij} = \sigma_{ji}$$

The symmetry of Cauchy stress entails similar relations for the other stress tensors:

$$\mathbf{P}.\mathbf{F}^T = \mathbf{F}.\mathbf{P}^T$$

This relation entails that the nominal stress and the transformation gradient are coaxial tensors, so that they share a common eigenbasis.

2.12.3. *Conservation of energy*

Let recall the expression of the material derivative:

$$\frac{d}{dt}\int_{\Omega_t} b(\mathbf{x},t)\rho(\mathbf{x},t)dx = \int_{\Omega_t} \rho\frac{db}{dt}dx$$

The first principle of thermodynamics states that energy is conserved: let us then define the total energy E of a domain Ω_t from the specific internal energy e, so that

$$E = \int_{\Omega_t} \rho e dx$$

Kinetic energy has been defined as $K := \int_{\Omega_t} \frac{1}{2}\rho^2 dx\mathbf{v}$. The *heat* received by the domain Ω_t comes from external actions (for instance inductive heating), corresponding to the density r, thus the overall quantity $\int_{\Omega_t} r dx$, and from heat conduction through $\partial\Omega_t$, the flux integral $-\int_{\partial\Omega_t} \mathbf{q}.\mathbf{n}ds$, thus the quantity of heat is

$$Q = \int_{\Omega_t} r dx - \int_{\partial\Omega_t} \mathbf{q}.\mathbf{n}ds$$

The global form of the first principle writes:

$$\frac{d}{dt}(E+K) = P_e + Q$$

The local form of this global balance equation is obtained using the dynamical equilibrium equations:

$$\rho\mathbf{f} + div\boldsymbol{\sigma} = \rho\boldsymbol{\gamma} \rightarrow \int_{\Omega_t} \rho\mathbf{f}.\mathbf{v}dx + \int_{\Omega_t} div\boldsymbol{\sigma}.\mathbf{v}dx = \int_{\Omega_t} \rho\boldsymbol{\gamma}.\mathbf{v}dx = \frac{dK}{dt}$$

We next use the identity

$$\int_{\Omega_t} div\boldsymbol{\sigma}.\mathbf{v}dx = -\int_{\Omega_t} \boldsymbol{\sigma}.\mathbf{d}dx + \int_{\partial\Omega_t} \boldsymbol{\sigma}.\mathbf{v}.\mathbf{n}ds$$

thus

$$\frac{d}{dt}(E+K)=\frac{dE}{dt}+\int_{\Omega_t}\rho\mathbf{f}.\mathbf{v}dx-\int_{\Omega_t}\boldsymbol{\sigma}.\mathbf{d}dx+\int_{\partial\Omega_t}\boldsymbol{\sigma}.\mathbf{v}.\mathbf{n}ds=P_e+Q$$

This entails

$$\frac{dE}{dt}+\int_{\Omega_t}\rho\mathbf{f}.\mathbf{v}dx-\int_{\Omega_t}\boldsymbol{\sigma}.\mathbf{d}dx+\int_{\partial\Omega_t}\boldsymbol{\sigma}.\mathbf{v}.\mathbf{n}ds=P_e+Q=\int_{\partial\Omega_t}\boldsymbol{\sigma}.\mathbf{v}.\mathbf{n}ds+\int_{\Omega_t}\rho\mathbf{f}.\mathbf{v}dx+\int_{\Omega_t}rdx-\int_{\partial\Omega_t}\mathbf{q}.\mathbf{n}ds$$

$$\Rightarrow\int_{\Omega_t}\rho\frac{de}{dt}dx-\int_{\Omega_t}\boldsymbol{\sigma}.\mathbf{d}dx=\int_{\Omega_t}rdx-\int_{\Omega_t}div\mathbf{q}dx$$

This delivers, by loalizing the last equality due to the arbitrariness of Ω_t, the local form of the first principle:

$$\rho\frac{de}{dt}=\boldsymbol{\sigma}.\mathbf{d}+r-div\mathbf{q}$$

REMARK.– Observe that the balance laws take a common format for any tensor field $\mathbf{a}(\mathbf{x},t)$ entering the spatial density, so that we evaluate the material derivative of the integral

$$\frac{D}{Dt}\int_{\Omega_t}\mathbf{a}(\mathbf{x},t)dx=\int_{\Omega_t}\boldsymbol{\Sigma}(\mathbf{x},t)dx+\int_{\partial\Omega_t}\phi(\mathbf{x},t).\mathbf{n}ds$$

involving the source term (due to internal and external sources) with density $\boldsymbol{\Sigma}(\mathbf{x},t)$ and the flux of the surface density $\phi(\mathbf{x},t)$.

EXERCISE.– Show that the balance laws previously written fit within this general writing.

2.12.4. *Entropy inequality*

Entropy inequality prescribes the direction of the dissipative process, which occurs for most materials exhibiting a local dissipation (plastic flow, creep and materials producing heat due to deformation are typical examples). The entropy is introduced as a thermodynamical potential measuring disorder, with specific density

$$s\rightarrow S:=\int_{\Omega_t}\rho s dx$$

The second law of thermodynamics is not a balance principle; it states that the material derivative of the total entropy is greater or equal than the rate of entropy input due to heat scaled by $\frac{1}{\theta}$, quantity $Q = \int_{\Omega_t} r/\theta dx - \int_{\partial\Omega_t} \frac{\mathbf{q}.\mathbf{n}}{\theta} ds$, thus

$$\frac{D}{Dt}\int_{\Omega_t} \rho s dx \geq \int_{\Omega_t} r dx - \int_{\partial\Omega_t} \mathbf{q}.\mathbf{n} ds$$

This inequality can be transformed using the divergence theorem, thus

$$\int_{\Omega_t} \left\{ \rho \frac{Ds}{Dt} - \frac{r}{\theta} + div\left(\mathbf{q}/\theta\right) \right\} dx \geq 0, \ \ \forall \Omega_t$$

The strong form is obtained as

$$\rho \frac{Ds}{Dt} - \frac{r}{\theta} + div\left(\mathbf{q}/\theta\right)$$

This inequality gives a direction for the energy transfers. The heat source term r can be eliminated using the first principle and the equality $div\left(\mathbf{q}/\theta\right) = div\mathbf{q}/\theta - \mathbf{q}.grad\theta/\theta^2$, thus

$$\rho\left(\theta \frac{Ds}{Dt} - \frac{De}{Dt}\right) \boldsymbol{\sigma} : \mathbf{d} - \frac{\mathbf{q}}{\theta}.grad\theta \geq 0$$

This inequality receives a more compact form relying on the free energy density

$\psi := e - \theta s$, *viz.*

$$\boldsymbol{\sigma} : \mathbf{d} - \rho\left(\dot{\psi} + s\dot{\theta}\right) - \frac{\mathbf{q}}{\theta}.grad\theta \geq 0$$

which is the Clausius-Duhem inequality. The left-hand side represents the local dissipation, sum of the mechanical dissipation $\boldsymbol{\sigma} : \mathbf{d} - \rho\left(\dot{\psi} + s\dot{\theta}\right)$ and the thermal dissipation, quantity $-\frac{\mathbf{q}}{\theta}.grad\theta$, which is completely transformed into heat.

It is a classical assumption to decouple the mechanical and thermal dissipations, so that for isothermal processes, it holds that

$$\boldsymbol{\sigma} : \mathbf{d} - \rho \dot{\psi} \geq 0$$

Observe that all previous derivations were done in the spatial configuration.

A similar inequality can be obtained in Lagrangian format (starting from the global writing of the dissipation and pulling back all integrals to the reference configuration), thus we arrive at

$$\mathbf{P} : \dot{\mathbf{F}} - \rho_0 \dot{\psi} \geq 0 \quad \text{or} \quad \mathbf{S} : \dot{\mathbf{E}} - \rho_0 \dot{\psi} \geq 0$$

Under isothermal conditions, the free energy density identifies with the strain energy density, which we pull back to the reference configuration:

$$\rho J \psi \equiv \rho_0 \psi(\mathbf{F}) = W(\mathbf{F}) \quad \text{or} \quad \rho_0 \psi(\mathbf{E}) = W(\mathbf{E})$$

according to which strain measure is adopted.

For a perfectly elastic material, no entropy is being produced, thus it holds that

$$\mathbf{P} : \dot{\mathbf{F}} - \rho_0 \dot{\psi} = 0 \rightarrow \mathbf{P} : \dot{\mathbf{F}} - \dot{W}(\mathbf{F}) = 0 \Rightarrow \left(\mathbf{P} - \frac{\partial W(\mathbf{F})}{\partial \mathbf{F}} \right) : \dot{\mathbf{F}} = 0, \ \ \forall \dot{\mathbf{F}} \Rightarrow \mathbf{P} = \frac{\partial W(\mathbf{F})}{\partial \mathbf{F}}$$

$$\mathbf{S} : \dot{\mathbf{E}} - \rho_0 \dot{\psi} = 0 \rightarrow \mathbf{S} : \dot{\mathbf{E}} - \dot{W}(\mathbf{E}) = 0 \Rightarrow \left(\mathbf{S} - \frac{\partial W(\mathbf{E})}{\partial \mathbf{E}} \right) : \dot{\mathbf{E}} = 0, \ \ \forall \mathbf{E} \Rightarrow \mathbf{S} = \frac{\partial W(\mathbf{E})}{\partial \mathbf{E}}$$

We thus recover, through the identification of the local dissipation, the hyperelastic constitutive models; such a procedure is called the Coleman-Noll procedure.

2.13. Abbreviations, notations and nomenclature

2.13.1. *Abbreviations*

– w.r. : with respect (to).

– s.t.: such that. Sometimes, : is used as an alternative notation.

– iff: if and only if (the logical equivalence).

2.13.2. *Notations*

Vectors and tensors are denoted as boldface symbols (although several notations may coexist). When not boldface, a tensor is considered as a linear mapping, which is a fully equivalent concept. There are two alternative ways of denoting tensors: index notation is based on components of tensors (which is convenient for proving equalities involving tensors). Absolute tensor notation is an alternative which does not rely on components, but on the essential idea that tensors are intrinsic objects, so that tensor relations are independent of any observer. It is accordingly and for this reason preferred over the indicial notation.

Einstein implicit summation rule of repeated indices in a monomial will be systematically used in order to avoid writing the summation symbol.

The transpose of a tensor $\mathbf{A}$ will be denoted $\mathbf{A}^T$. The partial derivative of a function with scalar or tensor argument $f(\mathbf{A})$ will be expressed equivalently as $f(\mathbf{A})_{,A} = \partial_A f$, denoting in both cases $\dfrac{\partial f(\mathbf{A})}{\partial A}$.

2.13.3. *Nomenclature of important symbols*

– Material and spatial positions of a material point: $\mathbf{X}, \mathbf{x}$ respectively, with

$$\mathbf{x} = \mathbf{x}(\mathbf{X}, t)$$

– Displacement and velocity fields:

$$\mathbf{u} := \mathbf{x} - \mathbf{X} \rightarrow \mathbf{v}(\mathbf{x}, t) := \frac{d\mathbf{u}}{dt} \quad \text{(Eulerian velocity)}$$

– Second order identity mapping: $\mathbf{I}$, such that $\mathbf{I}.\mathbf{x} = \mathbf{x}$

– Transformation gradient: $\mathbf{F} = Grad\mathbf{x} = \nabla_X \mathbf{x} \rightarrow d\mathbf{x} = \mathbf{F}.d\mathbf{X}$

– Right and left Cauchy-Green tensors: $\mathbf{C} = \mathbf{F}^T.\mathbf{F}, \quad \mathbf{b} = \mathbf{F}.\mathbf{F}^T$

– Lagrangian and Eulerian strains: $\mathbf{E} := \frac{1}{2}(\mathbf{C} - \mathbf{I}), \quad \mathbf{e} := \frac{1}{2}(\mathbf{I} - \mathbf{b}^{-1})$

– Transport formula for line, surface and volume elements:

$$d\mathbf{l} = \mathbf{F}.d\mathbf{l}_0\ , \ \mathbf{n}ds = J\mathbf{F}^{-T}.\mathbf{N}dS\ , \ dv = JdV$$

– Velocity gradient:

$$\mathbf{l} = grad\mathbf{v} = \mathbf{d} + \mathbf{w} \,,$$

with $\mathbf{d}, \mathbf{w}$ successively the rate of deformation (symmetrical) and spin (antisymmetrical) tensors.

3

Constitutive Models of Soft and Hard Living Tissues

3.1. Constitutive modeling

Constitutive equations are relations between stress and strain measures conjugated in the sense of energy.

DEFINITION.– An elastic material is defined as a functional relation $\boldsymbol{\sigma} = \boldsymbol{\sigma}(\mathbf{F})$, following the *principle of local state*. The state of a material point depends only on the neighborhood of this point, so that

$$\mathbf{x}(M,t) - \mathbf{x}(N,t) \approx \frac{\partial \mathbf{x}}{\partial \mathbf{X}}.d\mathbf{X} = \mathbf{F}.d\mathbf{X}$$

when M and N are close to each other. This assumption entails that the motion is described by the transformation gradient at first order with a good accuracy. More principles are needed in order to restrict the form of the constitutive law:

Principle of determinism: the stress in a body is determined by the history of its motion; this includes as special cases fluid behaviors (such as Newtonian and non-Newtonian flows) and the classical laws of elasticity, viscoelasticity, viscoplasticity, and more complex constitutive models the identification of which is the object of rheology.

Principle of material frame indifference: the frame indifference principle is a principle of objectivity stating that two observers shall see the same constitutive response in their respective frame. A change of observer may be considered as the

---

Chapter written by Jean-François GANGHOFFER.

superimposition of rigid-body motions on the current configurations; recall that a change of observer has been defined as a change of referential preserving the distances and angles between material points. This implies that the stress $\boldsymbol{\sigma}$ has to be an objective tensor, thus it should transform under an orthogonal transformation Q to the new stress

$$\mathbf{x}^* = Q.\mathbf{x} + a \rightarrow \boldsymbol{\sigma}^*(Q.\mathbf{F}) = Q.\boldsymbol{\sigma}.Q^T, \ \forall Q \in O_3$$

An important consequence is the equality of stress components in both reference systems (Figure 3.1)

$$\sigma^{*ij} = \mathbf{g}^{*i}.\boldsymbol{\sigma}.\mathbf{g}^{*j} = \mathbf{g}^i.Q^T.\left(Q.\boldsymbol{\sigma}.Q^T\right).Q.\mathbf{g}^i = \sigma^{ij}$$

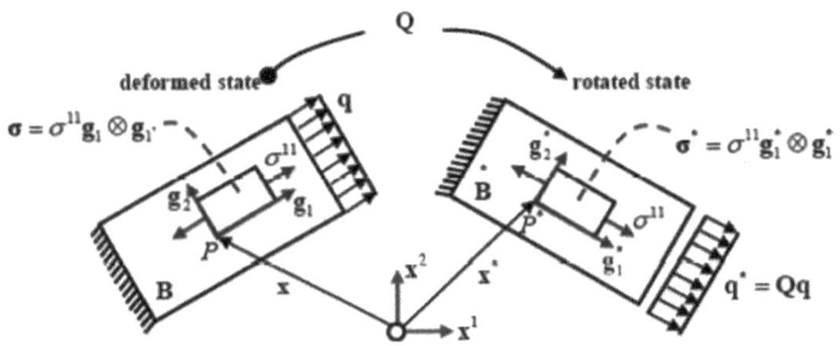

Figure 3.1. *Illustration of the principle of material frame indifference. For a color version of this figure, see www.iste.co.uk/ganghoffer/biomechanics.zip*

It can further be shown that the constitutive law can be expressed in terms of conjugated stress and strain measures: since equation $\boldsymbol{\sigma}^*(Q.\mathbf{F}) = Q.\boldsymbol{\sigma}.Q^T$ holds for an arbitrary Q, we may select $Q = R^T$, with $\mathbf{F} = R.\mathbf{U}$, leading to

$$\mathrm{R}^T.\boldsymbol{\sigma}(\mathbf{F}).\mathrm{R} = \boldsymbol{\sigma}\left(\mathrm{R}^T.\mathbf{F}\right) = \boldsymbol{\sigma}(\mathbf{U}) = \hat{\boldsymbol{\sigma}}(\mathbf{C})$$

due to (last relation) $\mathbf{U} = \mathbf{C}^{1/2}$. From the relation between stress measures,

$$\mathbf{T} = JR^T.\boldsymbol{\sigma}.\mathbf{F}^{-T} = R^T.\mathbf{P} = \mathbf{U}.\mathbf{S}$$

we further obtain

$$\mathbf{T} = JR^T.\boldsymbol{\sigma}(\mathbf{F}).\mathrm{R}.\mathbf{U}^{-1} = \det(\mathbf{U})\mathbf{U}^{-1}.\boldsymbol{\sigma}(\mathbf{U}).\mathbf{U}^{-1} \equiv \mathbf{T}(\mathbf{U})$$

This entails the expression of PK2, due to

$$\mathbf{S} = \mathbf{U}^{-1}.\mathbf{T} = \det(\mathbf{U})\mathbf{U}^{-1}.\boldsymbol{\sigma}(\mathbf{U}).\mathbf{U}^{-1} = \det\left(\mathbf{C}^{1/2}\right)\mathbf{C}^{1/2}.\boldsymbol{\sigma}(\mathbf{C}).\mathbf{C}^{1/2} \equiv \mathbf{S}(\mathbf{C})$$

Finally, we get

$$\mathbf{P} = \det(\mathbf{F})\boldsymbol{\sigma}(\mathbf{F}).\mathbf{F}^{-T} \equiv \mathbf{P}(\mathbf{F})$$

We thus recover the duality between stress and strain measures which were already shown to be conjugated in the sense of energy.

3.2. Isotropic elastic materials

We restrict our attention to materials called isotropic, meaning that the stress tensor should remain invariant when the material particle is rotated in the reference configuration, thus

$$\mathbf{X} \mapsto Q.\mathbf{X} \Rightarrow \overline{\mathbf{F}} = \mathbf{F}.Q \rightarrow \boldsymbol{\sigma}(\overline{\mathbf{F}}) = \boldsymbol{\sigma}(\mathbf{F}), \quad \forall \mathrm{Q} \in \mathrm{O}_3$$

Observe that contrary to material frame indifference, Q is here applied first (Figure 3.2).

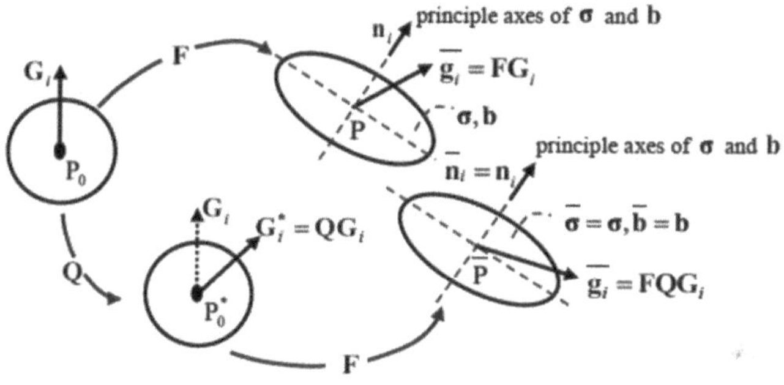

Figure 3.2. *Illustration of isotropic elasticity [BAS 00]*

The specific choice $Q = R^T$ then leads to the response of an isotropic material of the form

$$\boldsymbol{\sigma}(\mathbf{F}) = \boldsymbol{\sigma}(\mathbf{v}) = \boldsymbol{\sigma}(\mathbf{b}^{1/2}) \equiv \boldsymbol{\sigma}(\mathbf{b}), \quad \mathbf{b} = \mathbf{F}.\mathbf{F}^T$$

The objectivity requirement then imposes according to Truesdell and Noll's representation theorem

$$\boldsymbol{\sigma}(\mathbf{b}) = \varphi_0 \mathbf{I} + \varphi_1 \mathbf{b} + \varphi_2 \mathbf{b}^2$$

in which $\varphi_0, \varphi_1, \varphi_2$ are functions of the principal invariants of $\mathbf{b}$.

We then deduce the representation of the other stress measures, for instance

$$\mathbf{S} = J\mathbf{F}^{-1}.\boldsymbol{\sigma}.\mathbf{F}^{-T} = J\varphi_0\mathbf{C}^{-1} + J\varphi_1\mathbf{I} + J\varphi_2\mathbf{C}$$

with now $\varphi_0, \varphi_1, \varphi_2$ functions of the principal invariants of $\mathbf{C}$ (they will turn out to be identical to the principal invariants of $\mathbf{b}$). Going one step further using the Cayley-Hamilton theorem leads to the polynomial expansion

$$\mathbf{S} = \gamma_0\mathbf{I} + \gamma_1\mathbf{C} + \gamma_2\mathbf{C}^2$$

We may use the Lagrangian strain measure $\mathbf{E}$ instead of $\mathbf{C}$.

3.2.1. *Hypoelastic and hyperelastic isotropic materials*

There are two classes of constitutive models for nonlinear elastic materials:

1) *Hypoelastic models*: this is the weakest form of constitutive model in the sense that there is no direct relation between stress and strain, but only between stress rates and strain rates (Figure 3.3), in the form of a relation between an objective stress rate $\hat{\boldsymbol{\sigma}}$ (for instance the Jaumann derivative of $\boldsymbol{\sigma}$) and the Eulerian rate of deformation tensor,

$$\hat{\boldsymbol{\sigma}} = f(\boldsymbol{\sigma}, \mathbf{d})$$

so that we have to follow up the deformation path in order to construct the material response (in general numerically).

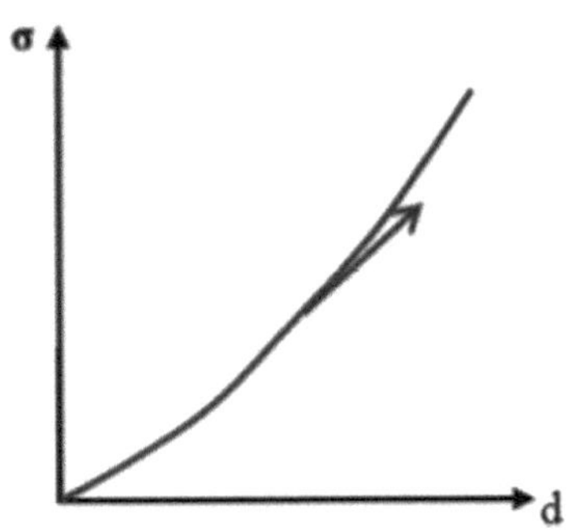

Figure 3.3. *Illustration of hypoelastic models*

EXAMPLE.– The linear relation $\hat{\boldsymbol{\sigma}} = \lambda tr(\mathbf{d})\mathbf{I} + 2\mu\mathbf{d}$, which looks similar to Hooke's law, but is NOT the material derivative of Hooke's law, due to $\mathbf{d} \neq \hat{\boldsymbol{\varepsilon}}$, unless restricted to small transformations.

Hypoelastic models suffer from some drawbacks, such as the existence of a fictitious hysteresis during a loading-unloading cycle, so that we prefer when possible to use hyperelastic models.

1) *Hyperelastic models*: while hypoelastic materials do not possess an internal energy density, hyperelastic materials do have one. We may then write the power of internal forces P_i (restricted to real motions) for those materials as the material derivative $P_i = -\frac{dE_{mech}}{dt}$, with E_{mech} the mechanical energy, expressible from a density of strain energy in the reference configuration, so that $E_{mech} = \int_{\Omega_0} W(\mathbf{E})dX$ and $P_i = -\dot{E} = -\int_{\Omega_0} \rho_0 \dot{e} dX$ lead to

$$\rho_0 \dot{e} = \mathbf{S} : \dot{\mathbf{E}} = \dot{W}(\mathbf{E}) = \frac{\partial W}{\partial \mathbf{E}} : \dot{\mathbf{E}} \rightarrow \mathbf{S} = \frac{\partial W(\mathbf{E})}{\partial \mathbf{E}} = 2\frac{\partial W(\mathbf{C})}{\partial \mathbf{C}}$$

the last relation bearing the name of *Doyle-Ericksen formula.*

Hence, the constitutive model relating PK2 to the Lagrangian strain is completely prescribed by the scalar-valued strain energy function $W(\mathbf{E})$. Alternative formulations can be similarly obtained based on the stress-strain duality as follows:

$$\mathbf{P} = \frac{\partial W(\mathbf{F})}{\partial \mathbf{F}}, \text{sym}(\mathbf{T}) = \frac{\partial W(\mathbf{U})}{\partial \mathbf{U}}$$

with $\text{sym}(\mathbf{T})$ the symmetrical part of Biot stress.

We more seldom write constitutive laws in the actual configurations, such as

$$\boldsymbol{\tau} = J\boldsymbol{\sigma} = \frac{\partial W(\mathbf{e})}{\partial \mathbf{e}}$$

due to the fact that isotropy is a property valid in the reference configuration, and that will in general be lost during motion.

The constitutive law $\mathbf{S}=\frac{\partial W(\mathbf{E})}{\partial \mathbf{E}}\equiv 2\frac{\partial W(\mathbf{C})}{\partial \mathbf{C}}W(\mathbf{E})$ is shown to be objective, since

$$\boldsymbol{\sigma}=J^{-1}\mathbf{F}.\mathbf{S}.\mathbf{F}^T=2J^{-1}\mathbf{F}.\frac{\partial W(\mathbf{C})}{\partial \mathbf{C}}.\mathbf{F}^T$$

Since under a change of observer, we have

$$\mathbf{x}\mapsto \mathbf{x}^*=Q.\mathbf{x}\Rightarrow \mathbf{F}^*=Q.\mathbf{F}\to \mathbf{C}^*=\mathbf{C}\Rightarrow \frac{\partial W(\mathbf{C}^*)}{\partial \mathbf{C}^*}=\frac{\partial W(\mathbf{C})}{\partial \mathbf{C}}$$

it holds that $\boldsymbol{\sigma}^*=\boldsymbol{\sigma}(Q.\mathbf{F})=2J^{*-1}\mathbf{F}^*.\frac{\partial W(\mathbf{C})}{\partial \mathbf{C}}.\mathbf{F}^{*T}=Q.\boldsymbol{\sigma}(\mathbf{F}).Q^T$, which concludes the proof.

EXAMPLE.– The Saint-Venant Kirchhoff material is the simplest geometrical nonlinear constitutive model in hyperelasticity, with the strain energy density given by $W(\mathbf{E})=\lambda tr(\mathbf{E})^2+2\mu tr(\mathbf{E}^2)+o(\|\mathbf{E}\|^2)$, the lowest order term $o(\|\mathbf{E}\|^2)$ indicating that it is an approximation of more complex models; this leads to the stress-strain relation

$$\mathbf{S}=\lambda tr(\mathbf{E})\mathbf{I}+2\mu\mathbf{E}+o(\|\mathbf{E}\|)$$

which resembles Hooke's law for isotropic materials, but involving the large strain measure $\mathbf{E}$, accounting for large changes of the geometry, but with small strains (the relation between stress and strain is linear). In fact, all constitutive models in hyperelasticity converge towards the Saint-Venant Kirchhoff model for small strains.

3.3. Elasticity tensors

Working out the stress-strain relation based on $W(\mathbf{E})$ leads to the stress rate (recall that strain tensors are represented as covariant tensors)

$$\dot{\mathbf{S}}=\frac{\partial^2 W(\mathbf{E})}{\partial E_{ij}\partial E_{mn}}\dot{E}_{mn}\mathbf{G}_i\otimes\mathbf{G}_j\equiv\mathbb{C}^{ijmn}\dot{E}_{mn}\mathbf{G}_i\otimes\mathbf{G}_j$$

with $\mathbb{C}^{ijmn}$ the components of the fourth order elasticity tensor, given by

$$\mathbb{C} := \frac{\partial^2 W(\mathbf{E})}{\partial \mathbf{E} \partial \mathbf{E}} \rightarrow \mathbb{C}^{ijmn} := \frac{\partial^2 W(\mathbf{E})}{\partial E_{ij} \partial E_{mn}}$$

This 4th order tensor describes the behavior of the material at a given point and for a given strain; it presents minor and major symmetries, due respectively to the symmetry of $\mathbf{E}$ and $\mathbf{S}$ and the existence of a strain energy function, viz. we have:

$\mathbb{C}^{ijmn} = \mathbb{C}^{jimn} = \mathbb{C}^{ijnm}$ *minor symmetries* (permutation of indices in each pair (I,j) and (m,n)

$\mathbb{C}^{ijmn} = \mathbb{C}^{mnij}$ *major symmetries* (permutation of the two pairs of indices)

This leads to 21 independent elastic moduli, a number that can be restricted by the consideration of *material symmetries*.

The tensor of elasticity moduli is here described in the reference configuration; it is also possible to construct a spatial 4th order tensor of moduli $\mathbf{c}$ from the rate of Kirchhoff stress

$$\dot{\tau}^{ij} = \frac{\partial^2 W}{\partial e_{ij} \partial e_{mn}} \dot{e}_{mn} \rightarrow \dot{\boldsymbol{\tau}} = \mathbf{c} : \dot{\mathbf{e}}$$

with $\quad \mathbf{c} = c^{ijmn} \mathbf{g}_i \otimes \mathbf{g}_j \otimes \mathbf{g}_m \otimes \mathbf{g}_n = \dfrac{\partial^2 W(\mathbf{e})}{\partial \mathbf{e} \partial \mathbf{e}}$

REMARK.– We can also show the relation $L_v \boldsymbol{\tau} = \mathbf{c} : L_v \mathbf{e} = \frac{1}{2} \mathbf{c} : L_v \mathbf{g}$

The components of the material and spatial tensors of elastic moduli are related next: from previous identification $E_{ij} = e_{ij}$, $\mathrm{C}_{ij} = g_{ij}$, we deduce from the definition of $\mathbb{C}$ and $\mathbf{c}$ that

$$\mathbb{C}^{ijmn} = c^{ijmn}$$

This allows us to obtain $\mathbf{c}$ as the push-forward of $\mathbb{C}$ (reciprocally, this last tensor is the pull-back of $\mathbf{c}$), thus observers located respectively in the reference and actual configurations will see the same components in their respective bases.

3.3.1. *Restrictions on the strain energy function*

The strain energy density is a non-negative scalar valued function of a tensorial strain measure. For instance, choosing the transformation gradient

$$W(\mathbf{F}) \geq 0$$

we generally assume the existence of a stress-free reference configuration (thus no residual stresses, an assumption generally incorrect for soft biological tissues), so that W should vanish on rigid body motions (not inducing any strains). Especially, we adopt a *normalization condition* (note that any potential is defined up to an arbitrary additive constant)

$$W(\mathbf{I}) = 0$$

Growth conditions shall be satisfied: $W(\mathbf{F}) \to \infty$ when either $\det(\mathbf{F}) \to 0$ or $\det(\mathbf{F}) \to +\infty$: these conditions mean the impossibility of reducing a finite volume to a point or expanding it to infinum with a finite amount of energy.

In what follows, we shall now consider specific models of hyperelastic materials.

3.4. Isotropic hyperelastic materials

Recall that for isotropic hyperelastic materials, the constitutive law is written as a quadratic expansion of the Lagrangian strain, with coefficients depending upon the principal invariants of $\mathbf{C}$: this entails that the strain energy function is a function of those principal invariants, viz.

$$W = W\left(I_1(\mathbf{C}), I_2(\mathbf{C}), I_3(\mathbf{C})\right)$$

With the three principal invariants of $\mathbf{C}$ defined by the relations:

$$\mathrm{I}_1(\mathbf{C}) = \mathrm{tr}(\mathbf{C}) = \lambda_1^2 + \lambda_2^2 + \lambda_3^2,$$
$$\mathrm{I}_2(\mathbf{C}) = \frac{1}{2}\left\{\mathrm{tr}(\mathbf{C})^2 - \mathrm{tr}(\mathbf{C}^2)\right\} = \lambda_1^2\lambda_2^2 + \lambda_2^2\lambda_3^2 + \lambda_1^2\lambda_3^2,$$
$$\mathrm{I}_3(\mathbf{C}) = \det(\mathbf{C}) = (\lambda_1\lambda_2\lambda_3)^2$$

The material stress PK2 is then formally given versus W and its derivatives from the composition rule as

$$S = 2W_{,C} = \frac{\partial W}{\partial I_1(\mathbf{C})}\frac{\partial I_1(\mathbf{C})}{\partial \mathbf{C}} + \frac{\partial W}{\partial I2(\mathbf{C})}\frac{\partial I_2(\mathbf{C})}{\partial \mathbf{C}} + \frac{\partial W}{\partial I3(\mathbf{C})}\frac{\partial I_3(\mathbf{C})}{\partial \mathbf{C}}$$

The derivatives of the invariants therein are given from Newton's identities involving the principal traces $\mathrm{tr}\left(\mathbf{C}^k\right)_{k=1..3}$:

$$I_1(\mathbf{C}) = \mathrm{tr}(\mathbf{C})$$
$$I_2(\mathbf{C}) = \frac{1}{2}\left\{I_1(\mathbf{C})^2 - \mathrm{tr}\left(\mathbf{C}^2\right)\right\}$$
$$I_3(\mathbf{C}) = \frac{1}{3}\left(I_2(\mathbf{C})I_1(\mathbf{C}) - I_1(\mathbf{C})\,\mathrm{tr}\left(\mathbf{C}^2\right) + \mathrm{tr}\left(\mathbf{C}^3\right)\right)$$

Recall that the directional derivative (or Gateaux derivative) of a scalar-valued function $f(\mathbf{A})$ of a second order tensor $\mathbf{A}$ in the direction of the second order tensor $\mathbf{X}$ is defined as

$$\frac{d}{dt}f(\mathbf{A}+t\mathbf{X})_{|t=0} = f(\mathbf{A})_{,A} : \mathbf{X}$$

The tensor $f(\mathbf{A})_{,A}$ is referred to as the gradient of the tensor function $f(\mathbf{A})$. For the derivative w.r. to a symmetric tensor, we have to symmeterize the result, so that sym(.) denotes symmetrization

$$\mathbf{A} = \mathbf{A}^T \Rightarrow f(\mathbf{Asym})_{,Asym} := \mathrm{sym}\left\{f(\mathbf{A})_{,A}\right\}$$

We have adopted the compact notation of partial derivative indicated by a comma as a subscript. The derivatives of the principal traces can then be computed

$$\frac{d}{dt}\left[\mathrm{tr}(\mathbf{A}+t\mathbf{X})^k\right]_{|t=0} = \frac{d}{dt}\left[(\mathbf{A}+t\mathbf{X})^k : \mathbf{I}\right]_{|t=0} = \frac{d}{dt}\left[(\mathbf{A}+t\mathbf{X})^k\right]_{|t=0} : \mathbf{I} =$$
$$= \frac{d}{dt}\left[(\mathbf{A}+t\mathbf{X}).(\mathbf{A}+t\mathbf{X})...(\mathbf{A}+t\mathbf{X})\right]_{|t=0} : \mathbf{I} = \frac{d}{dt}\left[\mathbf{A}^k + t\sum_{i=0}^{k-1}\mathbf{A}^i.\mathrm{X}.\mathbf{A}^{k-1-i} + t^2...\right]_{|t=0} : \mathbf{I} =$$
$$= \sum_{i=0}^{k-1}\mathbf{A}^i.\mathbf{X}.\mathbf{A}^{k-1-i} : \mathbf{I} = k\left(\mathbf{A}^{k-1}\right)^T : \mathbf{X}$$

Thus, $\mathrm{tr}\left(\mathbf{C}^k\right)_{,C} = k\left(\mathbf{C}^{k-1}\right)^T = k\mathbf{C}^{k-1}$

Inserting these equalities back into the three principal invariants delivers the derivatives

$$\mathrm{I}_1(\mathbf{C})_{,\mathrm{C}} = \mathbf{I}$$
$$\mathrm{I}_2(\mathbf{C})_{,\mathrm{C}} = \mathrm{I}_1(\mathbf{C})\mathbf{I} - \mathbf{C}$$
$$\mathrm{I}_3(\mathbf{C})_{,\mathrm{C}} = \mathbf{C}^2 - \mathrm{I}_1(\mathbf{C})\mathbf{C} + \mathrm{I}_2(\mathbf{C})\mathbf{I}$$

The last derivative receives a simpler expression since $\mathbf{C}$ is positive definite, thus invertible: from the Cayley-Hamilton theorem stating that

$$\sum_{k=0}^{3}(-1)^{3-k}\mathrm{I}_k\mathbf{C}^{3-k} = \mathbf{0} \leftarrow \mathrm{I}_{3,\mathrm{C}}.\mathbf{C}^{\mathrm{T}} = \sum_{k=0}^{2}(-1)^{i-1}\mathrm{I}_{(3-i)}\mathbf{C}^{i} = \sum_{k=0}^{3}(-1)^{i-1}\mathrm{I}_{(3-i)}\mathbf{C}^{i} + \mathrm{I}_3\mathbf{I} = \mathrm{I}_3\mathbf{I}$$

denoting $(\mathrm{I}_k)_k$ the set of three principal invariants of $\mathbf{C}$, we obtain

$$\mathrm{I}_3(\mathbf{C})_{,\mathrm{C}} = \mathrm{I}_3(\mathbf{C})\mathbf{C}^{-\mathrm{T}} = \det(\mathbf{C})\mathbf{C}^{-\mathrm{T}} = \mathrm{cof}(\mathbf{C})$$

We may easily prove the following expressions of time derivatives

$$\frac{d}{dt}\left[(\mathbf{A}+t\mathbf{X})^{T}\right]_{|t=0} = \mathbf{X}^{T} \Rightarrow \mathbf{A}^{T}{}_{,A} = \mathbb{T}$$
$$\frac{d}{dt}\left[(\mathbf{A}+t\mathbf{X})^{-1}\right]_{|t=0} = \mathbf{A}.\mathbf{X}.\mathbf{A}^{-1} \Rightarrow \mathbf{A}^{-1}{}_{,A} = -\mathbf{A}^{-1}\overline{\otimes}\mathbf{A}^{-T}$$

Summarizing, the constitutive law of an isotropic hyperelastic material is written

$$\boxed{\mathbf{S} = 2\mathrm{W}_{,\mathrm{C}} = \frac{\partial \mathrm{W}}{\partial \mathrm{I}_1(\mathbf{C})}\mathbf{I} + \frac{\partial \mathrm{W}}{\partial \mathrm{I}_2(\mathbf{C})}(\mathrm{I}_1(\mathbf{C})\mathbf{I} - \mathbf{C}) + \frac{\partial \mathrm{W}}{\partial \mathrm{I}_3(\mathbf{C})}\det(\mathbf{C})\mathbf{C}^{-1}}$$

As a consequence of the symmetry of Cauchy stress (but it also results from isotropy), we have

$$\mathrm{W}_{,\mathrm{C}} : \mathbf{C} = \mathbf{C} : \mathrm{W}_{,\mathrm{C}}$$

meaning the coaxiality of stress and strain. The spatial counterpart of the constitutive law results from the relation

$$\boldsymbol{\sigma} = J^{-1}\mathbf{F}.\mathbf{S}.\mathbf{F}^{T},\ \ \mathbf{F}.\mathbf{F}^{T} = \mathbf{b},\ \ \mathbf{F}.\mathbf{C}.\mathbf{F}^{T} = \mathbf{b}^2 \rightarrow \boldsymbol{\sigma} = 2J^{-1}\left[I_3 W_{,I_3}\mathbf{I} + \left(W_{,I_1} + I_1 W_{,I_2}\right)\mathbf{b} - W_{,I_2}\mathbf{b}^2\right]$$

3.4.1. Constitutive law in terms of principal stretches

The principal stretches (λ_i) are the eigenvalues of the pure deformation tensor $\mathbf{U}$, s.t.

$$\mathbf{U.N}_i = \lambda_i \mathbf{N}_i \rightarrow \mathbf{C.N}_i = \mathbf{U}^2.\mathbf{N}_i = \lambda_i^2 \mathbf{N}_i$$

with $\{\mathbf{N}_i\}$ the principal referential directions. Hence, we can write the spectral decomposition

$$\mathbf{U}^2 = \mathbf{C} = \lambda_i^2 \mathbf{N}_i \otimes \mathbf{N}_i$$

Let us then define the spatial principal directions, obtained by rotation of the material principal directions by the rotation $\mathrm{R} = \mathbf{v}^{-1}.\mathbf{F}$, so that $\mathbf{n}_i := \mathrm{R.N}_i$, and

$$\mathbf{v}.(\mathrm{R.N}_i) = \mathrm{R.U.N}_i = \lambda_i \mathrm{R.N}_i \Rightarrow \mathbf{b.n}_i = \mathbf{v}^2.\mathbf{n}_i = \lambda_i^2 \mathbf{n}_i$$

so that we get the spectral decomposition

$$\mathbf{v}^2 = \mathbf{b} = \lambda_i^2 \mathbf{n}_i \otimes \mathbf{n}_i$$

Thus, the components of Cauchy stress in the spatial basis $\{\mathbf{n}_i\}$ are (it is a diagonal tensor)

$$\sigma_i = \mathrm{J}^{-1} \lambda_i \mathrm{W}(\lambda_1, \lambda_2, \lambda_3)_{,\lambda_i}, \text{ with } \mathrm{J} = \lambda_1 \lambda_2 \lambda_3$$

The nominal and PK2 stress tensors are given similarly in the basis $\{\mathbf{n}_i\}$ as

$$\mathrm{P}_a = \mathrm{W}(\lambda_1, \lambda_2, \lambda_3)_{,\lambda_i}, \quad \mathrm{S}_i = \frac{1}{\lambda_i} \mathrm{W}(\lambda_1, \lambda_2, \lambda_3)_{,\lambda_i}$$

EXERCISE.– Prove previous identities. Hint: differentiate $\mathbf{C} = \lambda_i^2 \mathbf{N}_i \otimes \mathbf{N}_i$, premultiply and postmultiply $\mathbf{dC}$ with $\mathbf{N}_i$, using $\mathbf{C} = \mathbf{C}(\lambda_1, \lambda_2, \lambda_3)$ and the orthogonality property $\mathbf{N}_i.\mathrm{d}\mathbf{N}_i = 0$.

3.5. Incompressible materials

Most polymers and soft biological tissues are usually incompressible (the last category due to the large amount of water), thus those materials undergo *isochoric motions*. This entails that the Jacobian of the transformation is unity, $J = 1$. Such a material model with an internal constraint is called a *constrained material*. We then postulates a strain energy function of the form

$$W = W(\mathbf{F}) - p(J-1)$$

The scalar p is an indeterminate Lagrange multiplier, identified to a hydrostatic pressure (reaction of the continuum to the constraint $J = 1$); p can be determined thanks to the equilibrium equations and boundary conditions. The nominal stress is then easily obtained as

$$\mathbf{P} = \partial_F W(\mathbf{F}) - p\partial_F J = \partial_F W(\mathbf{F}) - pJ\mathbf{F}^{-T} \equiv \partial_F W(\mathbf{F}) - p\mathbf{F}^{-T}$$

Thus the other stress tensors

$$\mathbf{S} = \mathbf{F}^{-1}.\mathbf{P} = \mathbf{F}^{-1}.\partial_F W(\mathbf{F}) - p\mathbf{F}^{-1}.\mathbf{F}^{-T} = -p\mathbf{C}^{-1} + 2\partial_C W(\mathbf{C})$$

due to $\{\partial_F W(\mathbf{F})\}^T = 2\partial_C W(\mathbf{C}).\mathbf{F}^T$, and Cauchy stress

$$\boldsymbol{\sigma} = \mathbf{P}.\mathbf{F}^T = -p\mathbf{I} + \mathbf{F}.(\partial_F W)^T$$

Case of isotropic behavior: since $I_3(\mathbf{C}) = 1$, only the two first invariants of $\mathbf{C}$ intervene,

$$W = W(I_1(\mathbf{C}), I_2(\mathbf{C})) - \frac{1}{2}p(I_3 - 1)$$

thus

$$\mathbf{S} = 2\partial_C W(\mathbf{C}) - \partial_C[p(I_3-1)] = -p\mathbf{C}^{-1} + 2(W_{,I_1} + I_2 W_{,I_2})\mathbf{I} - 2W_{,I_2}.\mathbf{C}$$

EXERCISE.– Show last identity, using $I_{1,C} = \mathbf{I}$, $I_{2,C} = I_1\mathbf{I} - \mathbf{C}$, $I_3 = 1$.

Cauchy stress is written

$$\boldsymbol{\sigma} = J^{-1}\partial_F W.\mathbf{F}^T = J^{-1}\mathbf{F}.\mathbf{S}.\mathbf{F}^T = -p\mathbf{I} + 2\left(W_{,I_1} + I_2 W_{,I_2}\right)\mathbf{b} - 2W_{,I_2}.\mathbf{b}^2$$

EXERCISE.– Show this identity.

Alternative forms are possible, in terms of $\mathbf{b}$ or $\mathbf{b}^{-1}$: from

$\boldsymbol{\sigma} = -p\mathbf{I} + \mathbf{F}.(\partial_F W)^T$, with $(\partial_F W)^T = 2\mathbf{F}^T.\partial_b W(\mathbf{b})$ and

$$W = W(\mathbf{b}) \equiv W(\mathbf{F}) \Rightarrow \dot{W} = \mathrm{tr}\left(W_{,F}{}^T.\dot{\mathbf{F}}\right) = \mathrm{tr}\left(W_{,b}{}^T.\dot{\mathbf{b}}\right) = \mathrm{tr}\left(W_{,b}{}^T.\left(\dot{\mathbf{F}}.\mathbf{F}^T + \mathbf{F}.\dot{\mathbf{F}}^T\right)\right) =$$
$$\mathrm{tr}\left(W_{,F}.\dot{\mathbf{F}}^T\right) = 2\mathrm{tr}\left(W_{,b}.\mathbf{F}.\dot{\mathbf{F}}^T\right), \quad \forall \dot{\mathbf{F}}$$

Thus, $W_{,F}{}^T = 2W_{,b}.\mathbf{F} \Rightarrow \boldsymbol{\sigma} = -p\mathbf{I} + 2\mathbf{b}.W_{,b}$

In terms of principal stretches: if $W = W(\lambda_1, \lambda_2, \lambda_3) - p(J-1)$, with $\partial_F J = J\mathbf{F}^{-T}$, thus

$$\partial_{\lambda_i} J = J\lambda_i{}^{-1},$$

thus

$$\sigma_i = -p + \lambda_i W_{,\lambda_i}, \quad P_i = -\frac{1}{\lambda_i}p + W_{,\lambda_i}, \quad S_i = -\frac{1}{\lambda_i{}^2}p + \frac{1}{\lambda_i}W_{,\lambda_i}$$

under the constraint $\lambda_1\lambda_2\lambda_3 = 1$.

EXAMPLE.– WE determine the stress tensors in a thin elastic sheet made of an incompressible hyperelastic material, with the orthonormal basis $\mathbf{e}_1, \mathbf{e}_2, \mathbf{e}_3$ aligned with the major axis of the sheet (axis $\mathbf{e}_3$ is normal to the plane of the sheet). We assume a state of plane stress, thus $\sigma_{13} = \sigma_{23} = \sigma_{33} = 0$. Tensors $\boldsymbol{\sigma}$ and $\mathbf{b}$ are coaxial, due to the Rivlin-Ericksen representation theorem $\boldsymbol{\sigma}(\mathbf{b}) = \alpha_0\mathbf{I} + \alpha_1\mathbf{b} + \alpha_2\mathbf{b}^2$, with $\alpha_0, \alpha_1, \alpha_2$ functions of $I_1(\mathbf{b}), I_2(\mathbf{b}), I_3(\mathbf{b})$. The principal stresses are obtained as $\sigma_i = -p + 2\left(\lambda_i^2 W_{,I_1} - \frac{1}{\lambda_i^2} W_{,I_2}\right)$. Incompressibility implies $\lambda_3 = \left(\lambda_1\lambda_2\right)^{-1}$, and $\sigma_3 = 0 \Rightarrow p = 2\left(\frac{1}{(\lambda_1\lambda_2)^2} W_{,I_1} - (\lambda_1\lambda_2)^2 W_{,I_2}\right)$, thus the in-plane stresses

$$\sigma_1 = 2\left(\lambda_1^2 - (\lambda_1\lambda_2)^{-2}\right)\left(W_{,I_1} + \lambda_2^2 W_{,I_2}\right),\quad \sigma_2 = 2\left(\lambda_2^2 - (\lambda_1\lambda_2)^{-2}\right)\left(W_{,I_1} + \lambda_1^2 W_{,I_2}\right)$$

3.6. Compressible hyperelastic materials

Since the motion is isochoric for those materials, the idea here is to split the deformation into volumetric and isochoric contributions, so that we write the multiplicative decomposition

$$\mathbf{F} = \left(J^{1/3}\mathbf{I}\right)\overline{\mathbf{F}} = J^{1/3}\overline{\mathbf{F}}$$

The modified transformation gradient $\overline{\mathbf{F}}$ is associated with volume-preserving deformations. The previous definition entails the modified right Cauchy-Green tensor

$$\mathbf{C} = J^{2/3}\overline{\mathbf{C}}$$

and the modified principal stretches.

$$\overline{\lambda_i} = J^{-1/3}\lambda_i$$

DEFINITION.– A material is called nearly incompressible if dilatational changes require much higher external work compared to distortional changes.

In order to write the constitutive models, we need to compute the derivative $\frac{\partial\overline{\mathbf{C}}}{\partial\mathbf{C}}$: from $J_{,C} = J\mathbf{C}^{-1}$ since $I_{3,C} = I_3\mathbf{C}^{-1}$, with $I_3 = J^2$, $\frac{\partial J^{-2/3}}{\partial \mathbf{C}} = -\frac{1}{3}J^{-2/3}\mathbf{C}^{-1}$, thus using the identity $\partial_C\left(\phi\mathbf{A}\right) = \mathbf{A}\otimes\partial_C\phi + \phi\partial_C\mathbf{A}$, we obtain

$$\frac{\partial\overline{\mathbf{C}}}{\partial\mathbf{C}} = J^{-2/3}\left(\mathbf{I}_4 + J^{2/3}\mathbf{C}\otimes\partial_C J^{-2/3}\right) = J^{-2/3}\left(\mathbf{I}_4 - \frac{1}{3}\mathbf{C}\otimes\mathbf{C}^{-1}\right) = J^{-2/3}\mathbf{P}^T$$

with $\mathbf{P}$ the fourth order projection tensor onto the space of deviatoric second order tensors *in a Lagrangian description.* We may easily check that $\mathbf{P}$ provides the deviatoric part of a second order tensor.

We then postulate a unique decoupled representation of the strain-energy function into a volumetric density $W_{vol}(J)$ describing the volume changes, and an isochoric density $W_{iso}(\overline{\mathbf{C}})$ modeling shape changes,

$$\boxed{W(\mathbf{C}) = W_{vol}(J) + W_{iso}(\overline{\mathbf{C}})}$$

$W_{vol}(J)$ has to be a strictly convex function having a unique minimum at $J = 1$ and $\overline{\mathbf{C}} = \mathbf{I}$, thus $W_{vol}(J=1) = 0 = W_{iso}(\overline{\mathbf{C}} = \mathbf{I})$.

Constitutive equations are obtained from the expression of the local dissipation:

$$D_{int} = p_i - \dot{W} = 0 \Rightarrow p_i = \mathbf{P} : \dot{\mathbf{F}} = \mathbf{S} : \dot{\mathbf{C}}/2$$

representing the stress power per unit volume; based on previous derivations, this gives

$$D_{int} = \left(\mathbf{S} - JW_{vol,J}.\mathbf{C}^{-1} - J^{-2/3}\mathbf{P} : 2W_{iso}\right) : \frac{\dot{\mathbf{C}}}{2} = 0$$

using the identities

$$\dot{J} = J_{,C}.\dot{\mathbf{C}} = J\mathbf{C}^{-1}.\dot{\mathbf{C}}/2, \quad \dot{\overline{\mathbf{C}}} = 2\frac{\partial \overline{\mathbf{C}}}{\partial \mathbf{C}} = 2J^{-2/3}\mathbf{P}^T.\dot{\mathbf{C}}/2$$

The standard Coleman-Noll procedure then leads to

$$\mathbf{S} = \mathbf{S}_{vol} + \mathbf{S}_{iso}$$

$$\mathbf{S}_{vol} = 2W_{vol,C}(J) = Jp\mathbf{C}^{-1}, \quad \mathbf{S}_{iso} = 2W_{iso,C}(\overline{\mathbf{C}}) = J^{-2/3}\left(\mathbf{I}_4 - \frac{1}{3}\mathbf{C} \otimes \mathbf{C}^{-1}\right) : \overline{\mathbf{S}} = J^{-2/3}\mathbf{P} : \overline{\mathbf{S}}$$

with $\mathbf{P} : \overline{\mathbf{S}}$ the deviatoric part of $\overline{\mathbf{S}}$. Here, $p := W_{vol,J}(J)$ is a hydrostatic pressure that has to be prescribed by a constitutive model (contrary to the incompressible case), and $\overline{\mathbf{S}} := 2W_{iso,C}$ is a fictitious Piola-Kirchhoff stress.

The constitutive law can be rewritten in terms of Cauchy stress, with the split into volumetric and isochoric stresses

$$\boldsymbol{\sigma} = \boldsymbol{\sigma}_{\text{vol}} + \boldsymbol{\sigma}_{\text{iso}}$$
$$\boldsymbol{\sigma}_{\text{vol}} = 2\mathrm{J}^{-1}\mathbf{F}.\mathrm{W}_{\text{vol,C}}.\mathbf{F}^{\mathrm{T}} = \mathrm{J}^{-1}\mathbf{F}.\left(\mathrm{J}p\mathbf{C}^{-1}\right).\mathbf{F}^{\mathrm{T}} = p\mathbf{I}$$
$$\boldsymbol{\sigma}_{\text{iso}} = 2\mathrm{J}^{-1}\mathbf{F}.\mathrm{W}_{\text{iso,C}}.\mathbf{F}^{\mathrm{T}} = \mathrm{J}^{-5/3}\mathbf{F}.\left(\mathbf{P}:\overline{\mathbf{S}}\right).\mathbf{F}^{\mathrm{T}} = \mathrm{J}^{-1}\overline{\mathbf{F}}.\left(\mathbf{P}:\overline{\mathbf{S}}\right).\overline{\mathbf{F}}^{\mathrm{T}}$$

3.7. Isotropic compressible hyperelasticity

A decoupled representation of the strain energy function is here also adopted, as

$$\mathrm{W}(\mathbf{b}) = \mathrm{W}_{\text{vol}}(\mathrm{J}) + \mathrm{W}_{\text{iso}}\left(\overline{\mathbf{b}}\right)$$

with $\mathbf{b} = \mathbf{F}\mathbf{F}^{\mathrm{T}} = \left(\mathrm{J}^{2/3}\mathbf{I}\right).\overline{\mathbf{b}}$, and $\overline{\mathbf{b}}$ is the modified left Cauchy-Green tensor representing a volume preserving contribution to deformation, whereas $\mathrm{J}^{2/3}\mathbf{I}$ is a dilatational contribution. This entails

$$\boldsymbol{\sigma} = 2\mathrm{J}^{-1}\mathbf{b}.\mathrm{W}_{,\mathrm{b}} = 2\mathrm{J}^{-1}.\mathbf{b} = \boldsymbol{\sigma}_{\text{vol}} + 2\boldsymbol{\sigma}_{\text{iso}}$$
$$\boldsymbol{\sigma}_{\text{vol}} = p\mathbf{I}$$
$$\boldsymbol{\sigma}_{\text{iso}} = \left(\mathbf{I}_4 - \frac{1}{3}\mathbf{I}\otimes\mathbf{I}\right):\boldsymbol{\sigma} = \mathrm{Dev}\overline{\boldsymbol{\sigma}} = \mathbf{p}:\overline{\boldsymbol{\sigma}}$$

introducing $\mathbf{p} := \mathbf{I}_4 - \dfrac{1}{3}\mathbf{I}\otimes\mathbf{I}$ the projection operator onto the space of deviatoric second order tensors, but here in Eulerian format. In previous derivations, we have used the identities

$$\mathrm{J}_{,\mathrm{b}} = \frac{\mathrm{J}}{2}\mathbf{b}^{-1}.\frac{\partial\overline{\mathbf{b}}}{\partial\mathbf{b}} = \mathrm{J}^{-2/3}\left(\mathbf{I}_4 - \frac{1}{3}\mathbf{b}\otimes\mathbf{b}^{-1}\right),\quad \mathbf{A}.(\mathbf{B}\otimes\mathbf{C}).\mathbf{D} = (\mathbf{A}.\mathbf{B})\otimes(\mathbf{C}.\mathbf{D})$$

Modified strain invariants are introduced

$$\overline{\mathrm{I}}_1\left(\overline{\mathbf{C}}\right) := \mathrm{tr}\left(\overline{\mathbf{C}}\right) = \mathrm{tr}\left(\overline{\mathbf{b}}\right),\quad \overline{\mathrm{I}}_2\left(\overline{\mathbf{b}}\right) := \frac{1}{2}\left[\mathrm{tr}\left(\overline{\mathbf{b}}\right)^2 - \mathrm{tr}\left(\overline{\mathbf{b}}^2\right)\right] = \overline{\mathrm{I}}_2\left(\overline{\mathbf{C}}\right),\quad \overline{\mathrm{I}}_3 = \det\overline{\mathbf{C}} = \det\overline{\mathbf{b}}$$

to write the constitutive law, as

$$\mathrm{W} = \mathrm{W}_{\text{vol}}(\mathrm{J}) + \mathrm{W}_{\text{iso}}\left(\overline{\mathrm{I}}_1\left(\overline{\mathbf{C}}\right), \overline{\mathrm{I}}_2\left(\overline{\mathbf{C}}\right)\right) = \mathrm{W}_{\text{vol}}(\mathrm{J}) + \mathrm{W}_{\text{iso}}\left(\overline{\mathrm{I}}_1\left(\overline{\mathbf{b}}\right), \overline{\mathrm{I}}_2\left(\overline{\mathbf{b}}\right)\right)$$

Note the expression of the modified principal invariants versus their original version

$$\overline{I}_1 = J^{-2/3} I_1, \quad \overline{I}_2 = J^{-4/3} I_2, \quad \overline{I}_3 = 1$$

Thus, PK2 is expressed as

$$\mathbf{S} = 2W_{,C} = \mathbf{S}_{vol} + \mathbf{S}_{iso}$$

$$\mathbf{S}_{iso} = 2W_{iso,C}\left(\overline{I}_1\left(\overline{\mathbf{C}}\right), \overline{I}_2\left(\overline{\mathbf{C}}\right)\right) = J^{-2/3}\mathbf{P} : \overline{\mathbf{S}} = \overline{\gamma}_1 \mathbf{I} + \overline{\gamma}_2 \overline{\mathbf{C}}$$

$$\overline{\gamma}_1 = 2\left(W_{iso,\overline{I}_1} + \overline{I}_1 W_{iso,\overline{I}_2}\right), \quad \overline{\gamma}_1 = -2W_{iso,\overline{I}_2}$$

3.8. Some forms of constitutive models

We present a few classical models from the literature; the list is far from being complete and the reader is referred to the indicated references for an enhanced presentation. Those models can be classified into two categories.

3.8.1. *Incompressible materials*

1) *Ogden generic model (1972)*: developed for incompressible materials, with

$$W\left(\lambda_1, \lambda_2, \lambda_3\right) = \sum_{p=1}^{N} \frac{\mu_p}{\alpha_p}\left(\lambda_1^{\alpha_1} + \lambda_2^{\alpha_2} + \lambda_3^{\alpha_3} - 3\right)$$

and with, for instance,

$$\alpha_1 = 1.3, \quad \alpha_2 = 5, \quad \alpha_3 = -2$$

$$\mu_1 = 6.5.10^5 N/m^2, \quad \mu_2 = 0.012.10^5 N/m^2, \quad \mu_3 = -0.1.10^5 N/m^2$$

The shear modulus in a reference configuration is defined by comparison with the linear theory, giving $2\mu = \sum_{p=1}^{N} \mu_p \alpha_p$.

2) *Valanis and Landel model (1967)*: $W = \sum_{a=1}^{3} \omega(\lambda_a),\ \omega(\lambda_a) = \sum_{p=1}^{N} \frac{\mu_p}{\alpha_p}\left(\lambda_a^{\alpha_p} - 1\right)$

with the additive decomposition motivated by biaxial experiments.

3) *Mooney-Rivlin model*: it is a specific case of the Ogden model, with $N = 2,\ \alpha_1 = 2 = \alpha_2$, thus

$W = c_1(I_1 - 3) + c_2(I_2 - 3);\ I_3 = 1,\ c_1 = -c_2 = \mu/2$, with the shear modulus identified from the small strain theory as $\mu = \mu_1 - \mu_2$.

4) *The Neo-Hookean model* is a one-parameter model motivated by the statistical theory of rubber elasticity, with $N = 1,\ \alpha_1 = 2 \Rightarrow w = c_1(I_1 - 3)$.

5) *The Varga model* has density $W(\lambda_1, \lambda_2, \lambda_3) = \mu(\lambda_1 + \lambda_2 + \lambda_3 - 3)$.

6) *The Yeoh and Arruda-Boyce* model for reinforced elastomers (e.g. by carbon black) has a shear modulus given by $\mu = 2\left(W_{,I_1} + W_{,I_2}\right)$ such that $\sigma_{12} = \mu F_{12}$ under simple shear test $\mathbf{F} = \mathbf{I} + F_{12}\mathbf{e}_1 \otimes \mathbf{e}_2$. For reinforced elastomers, μ varies with deformation, which is not the case for the Mooney-Rivlin model.

7) *The Yeoh model*: $W = c_1(I_1 - 3) + c_2(I_2 - 3)^2 + c_3(I_2 - 3)^3$, thus the shear modulus is $\mu = 2c_1 + 4c_2(I_2 - 3) + 6c_3(I_2 - 3)^2 > 0$, since $W > 0 \Rightarrow I_1 > 0$.

8) *Arruda and Boyce statistical model* (sampling of rubber chain orientations into 8 directions in space): W is derived from the inverse Langevin function but can be approximated by a series $W = \mu\left\{\frac{1}{2}(I_1 - 3) + \frac{1}{20n}(I_1^2 - 9) + \frac{11}{1050n^2}(I_1^3 - 27) + \ldots\right\}$, with μ the shear modulus, and n the number of segments in a chain. We further have $\sqrt{I_1} = \sqrt{3}\lambda_{chain}$, $\lambda_{chain} = \frac{\lambda}{\lambda_0}$ as the ratio of actual to initial chain length.

3.8.2. *Compressible materials*

1) *Ogden model*: the volumetric response function is

$W_{vol}(J) = \kappa G(J) = \kappa\beta^{-2}(\beta \ln J + J^{-\beta} - 1)$, $\beta > 0$, with κ a bulk modulus in the reference configuration, with the condition $G(1) = 0 \to W_{vol}(1) = 0$. A good agreement with experimental data has been obtained.

2) *Blatz and Ko model (1962)* for foamed elastomers:

$$W(I_1, I_2, I_3) = f\frac{\mu}{2}\left[(I_1 - 3) + \frac{1}{\beta}(I_3^{-\beta} - 1)\right] + (1-f)\frac{\mu}{2}\left[\left(\frac{I_2}{I_3} - 3\right) + \frac{1}{\beta}(I_3^{\beta} - 1)\right],$$

with μ a shear modulus, and $f \in [0,1]$ an interpolation parameter. Note that this model reduces to the Mooney-Rivlin model for the choice

$$c_1 = f\frac{\mu}{2}, \quad c_2 = 1 - f\frac{\mu}{2}.$$

The determination of a suitable strain energy function for a specific material requires a sufficiently rich set of experimental data relying on different testing conditions (uniaxial, biaxial); no model can solely be determined from uniaxial tests. Experimental data are required for a proper selection of the model and in a second step for the identification of the model parameters, which is an inverse problem.

3.9. Saint-Venant Kirchhoff materials and Neo-Hookean materials through linearization

We recall the material form of the constitutive law for a hyperelastic material based on the representation theorem,

$$\mathbf{S}(\mathbf{C}) = \gamma_0\mathbf{I} + \gamma_1\mathbf{C} + \gamma_2\mathbf{C}^2$$

with $\gamma_0, \gamma_1, \gamma_2$ functions of the principal invariants of $\mathbf{C}$. The same model is written in terms of the Lagrangian strain as

$$\mathbf{S}(\mathbf{E}) = \gamma_0\mathbf{I} + \gamma_1(\mathbf{I} + 2\mathbf{E}) + \gamma_2(4\mathbf{E}^2 + 4\mathbf{E} + \mathbf{I})$$

The linearization of the tensor-valued function $\mathbf{S}(\mathbf{E})$ w.r. to $\mathbf{E}$ at point $\mathbf{X}$ is elaborated as

$$\mathrm{L}\mathbf{S}(\mathbf{E},\Delta\mathbf{E}) = \mathbf{S}(\mathbf{E}) + \Delta\mathbf{S}(\mathbf{E},\Delta\mathbf{E})$$

In which the term $\Delta\mathbf{S}(\mathbf{E},\Delta\mathbf{E}) := \frac{\mathrm{d}}{\mathrm{d}\varepsilon}\mathbf{S}(\mathbf{E}+\varepsilon\Delta\mathbf{E})_{|\varepsilon=0}$ is the Gateaux derivative or directional derivative in the direction of the variation $\Delta\mathbf{E}$. The term $\mathrm{L}\mathbf{S}(\mathbf{E},\Delta\mathbf{E})$ is obtained as

$$\begin{aligned}\mathrm{L}\mathbf{S}(\mathbf{E},\Delta\mathbf{E}) &= (\gamma_0+\gamma_1+\gamma_2)\mathbf{I} + 2(\gamma_1+2\gamma_2)\mathbf{E} + 4\gamma_2\mathbf{E}^2 + \\ &(\Delta\gamma_0+\Delta\gamma_1+\Delta\gamma_2)\mathbf{I} + 2(\gamma_1+2\gamma_2)\Delta\mathbf{E} + 4\gamma_2(\mathbf{E}.\Delta\mathbf{E}+\Delta\mathbf{E}.\mathbf{E})\end{aligned}$$

where

$$\Delta\gamma_k = \partial_{I_1}\gamma_k\Delta I_1 + \partial_{I_2}\gamma_k\Delta I_2 + \partial_{I_3}\gamma_k\Delta I_3;\ \Delta I_p = \partial_E I_p : \Delta\mathbf{E} = 2\partial_C I_p : \Delta\mathbf{E}$$

Linearization is applied at the reference position $\mathbf{x} = \mathbf{X}$ for which $\mathbf{E} = \mathbf{0}$, so that

$$\mathrm{L}\mathbf{S}(\mathbf{E}=\mathbf{0},\Delta\mathbf{E}) = \{(\gamma_0(\mathbf{I})+\gamma_1(\mathbf{I})+\gamma_2(\mathbf{I})) + (\Delta\gamma_0(\mathbf{I})+\Delta\gamma_1(\mathbf{I})+\Delta\gamma_2(\mathbf{I}))\}\mathbf{I} + 2(\gamma_1(\mathbf{I})+2\gamma_2(\mathbf{I}))\Delta\mathbf{E}$$

Herein, the coefficients $\Delta\gamma_p(\mathbf{I})$ are the directional derivatives evaluated at point **I**; those derivatives are evaluated based on the definition of the invariants:

$$\Delta I_1 = 2\mathrm{tr}\Delta\mathbf{E},\quad \Delta I_2 = 4\mathrm{tr}\Delta\mathbf{E},\quad \Delta I_3 = 2\mathrm{tr}\Delta\mathbf{E}$$

and we thus obtain

$$\Delta\gamma_k = \mathrm{tr}\Delta\mathbf{E}\left(2\partial_{I_1}\gamma_k\Delta I_1 + 4\partial_{I_2}\gamma_k\Delta I_2 + 2\partial_{I_3}\gamma_k\Delta I_3\right)(\mathbf{I}) \equiv \dot{\gamma}_k(\mathbf{I})\,\mathrm{tr}\Delta\mathbf{E},\quad k = 0,1,2$$

Introducing the abbreviation

$$\dot{\gamma}_k(\mathbf{I}) := \left(2\partial_{I_1}\gamma_k\Delta I_1 + 4\partial_{I_2}\gamma_k\Delta I_2 + 2\partial_{I_3}\gamma_k\Delta I_3\right)(\mathbf{I}),\quad k{=}0,1,2$$

and introducing the above results into $\mathrm{L}\mathbf{S}(\mathbf{E}=\mathbf{0},\Delta\mathbf{E})$ results in the linearization of the stress

$$\mathrm{L}\mathbf{S}(\mathbf{E}=\mathbf{0},\Delta\mathbf{E}) = -\pi\mathbf{I} + \lambda(\mathrm{tr}\Delta\mathbf{E}) + 2\mu\Delta\mathbf{E}$$

with the new coefficients therein

$$\pi = -(\gamma_0(\mathbf{I}) + \gamma_1(\mathbf{I}) + \gamma_2(\mathbf{I}))$$
$$\lambda = \dot{\gamma}_0(\mathbf{I}) + \dot{\gamma}_1(\mathbf{I}) + \dot{\gamma}_2(\mathbf{I})$$
$$\mu = \gamma_1(\mathbf{I}) + 2\gamma_2(\mathbf{I})$$

For sufficiently small strains (we may change $\Delta\mathbf{E}$ into $\mathbf{E}$), and assuming furthermore that the reference configuration is unstressed (thus $\pi = 0$), the linearized stress is obtained as

$$\mathbf{S}(\mathbf{E}) = \lambda(\text{tr}\mathbf{E})\mathbf{I} + 2\mu\mathbf{E}$$

This constitutive law defines the Saint-Venant Kirchhoff material representing the simplest isotropic hyperelastic material, describing a geometrically nonlinear behavior. Using the relation $S = 2\partial_C W(\mathbf{C}) = \partial_E W(\mathbf{E})$ then leads to the strain energy density

$$W(\mathbf{E}) = \frac{\lambda}{2}(\text{tr}\mathbf{E})^2 + \mu\text{tr}(\mathbf{E}^2)$$

A further linearization of the kinematics leads to the Hookean material: linearization w.r.t. the displacement field $\mathbf{u} = \mathbf{x} - \mathbf{X}$ at point $\mathbf{x} = \mathbf{X}$ (thus at $\mathbf{u} = \mathbf{0}$) leads to

$$\text{L}\mathbf{S}(\mathbf{u}, \Delta\mathbf{u}) = \{\lambda(\text{tr}\Delta\mathbf{E})\mathbf{I} + 2\mu\Delta\mathbf{E}\}_{u=0} = \lambda(\text{tr}\boldsymbol{\varepsilon})\mathbf{I} + 2\mu\boldsymbol{\varepsilon}$$

in which the small strain $\boldsymbol{\varepsilon}$ is defined as $\boldsymbol{\varepsilon} := \Delta\mathbf{E}_{|u=0}$; this last term is evaluated as follows: the first variation of the deformation gradient is computed as the Gateaux derivative

$$\delta\mathbf{F} = \frac{d}{d\varepsilon}\mathbf{F}(\mathbf{u} + \varepsilon\delta\mathbf{u})_{|\varepsilon=0} \equiv D_{\delta u}\mathbf{F} = \frac{d}{d\varepsilon}(\mathbf{F} + \varepsilon\text{Grad}\mathbf{u})_{|\varepsilon=0} = \text{Grad}\delta\mathbf{u}$$

The variation of the identity $\mathbf{F}\mathbf{F}^{-1} = \mathbf{I}$ delivers

$$\delta\mathbf{F}^{-1} = -\mathbf{F}^{-1}.\delta\mathbf{F}.\mathbf{F}^{-1} = -\mathbf{F}^{-1}.\text{Grad}\delta\mathbf{u}.\mathbf{F}^{-1} = -\mathbf{F}^{-1}.\text{grad}\delta\mathbf{u}$$

since

$$\text{grad}\delta\mathbf{u} = \text{Grad}\delta\mathbf{u}.\mathbf{F}^{-1} \leftrightarrow \frac{\partial\delta u_a}{\partial x_b} = \frac{\partial\delta u_a}{\partial X_A}\frac{\partial X_A}{\partial x_b} = \frac{\partial\delta u_a}{\partial X_A}F^{-1}_{AB}$$

Using the identities $\delta\mathbf{F} = \text{Grad}\delta\mathbf{u}$ and $(\mathbf{A}.\mathbf{B})^{\mathrm{T}} = \mathbf{B}^{\mathrm{T}}.\mathbf{A}^{\mathrm{T}}$, we elaborate the variation therein

$$\delta\mathbf{E} = \left\{\delta\left(\mathbf{F}^{\mathrm{T}}\right).\mathbf{F} + \mathbf{F}^{\mathrm{T}}.\delta\mathbf{F}\right\}/2 = \text{sym}\left(\mathbf{F}^{\mathrm{T}}.\text{Grad}\delta\mathbf{u}\right) \leftrightarrow \delta\mathrm{E}_{\mathrm{AB}} = \frac{1}{2}\left(\mathrm{F}_{\mathrm{aB}}\frac{\partial\delta\mathrm{u}_{\mathrm{a}}}{\partial\mathrm{X}_{\mathrm{A}}} + \mathrm{F}_{\mathrm{aA}}\frac{\partial\delta\mathrm{u}_{\mathrm{a}}}{\partial\mathrm{X}_{\mathrm{B}}}\right)$$

Thus, we arrive at

$$\boldsymbol{\varepsilon} := \Delta\mathbf{E}_{|\mathrm{u}=0} = \text{sym}\left(\text{Grad}\delta\mathbf{u}\right) \equiv \text{sym}\left(\text{Grad}\mathbf{u}\right)$$

replacing $\delta\mathbf{u}$ by $\mathbf{u}$ when the displacement is small enough (in norm).

The previous Hookean model considers small displacements and small strains; there is no distinction between the different stress measures (due to $\mathbf{F} = \mathbf{I}, \mathrm{J} = 1$), thus we obtain the Hookean material model

$$\boldsymbol{\sigma} = \lambda\left(\text{tr}\boldsymbol{\varepsilon}\right)\mathbf{I} + 2\mu\boldsymbol{\varepsilon}$$

with strain energy density

$$\mathrm{W}(\boldsymbol{\varepsilon}) = \frac{\lambda}{2}(\text{tr}\boldsymbol{\varepsilon})^2 + \mu\,\text{tr}\left(\boldsymbol{\varepsilon}^2\right)$$

3.10. Hyperviscoelastic models

Hyperviscoelastic materials combine a hyperelastic behavior and a viscous behavior; from Holzapfel, the strain energy of such a material takes the form

$$W\left(\mathbf{C};\Gamma_1,\ldots;\Gamma_n\right) = W_{vol}^{e}(J) + W_{iso}^{e}\left(\overline{\mathbf{C}}\right) + \sum_{\alpha=1}^{m}\Upsilon_\alpha\left(\mathbf{C};\Gamma_\alpha\right)$$

denoting therein Γ_α internal variables. The second Piola-Kirchhoff stress tensor is then obtained as

$$\mathbf{S} = \frac{\partial W\left(\mathbf{C};\Gamma_1,\ldots;\Gamma_n\right)}{\partial\mathbf{C}} = \mathbf{S}_{vol}^{e} + \mathbf{S}_{iso}^{e} + \sum_{\alpha=1}^{m}\mathbf{Q}_\alpha$$

in which the last term is the viscous stress. The isotropic part of the stress is then

$$\mathbf{S}_{iso} = \mathbf{S}^{e}_{iso} + \sum_{\alpha=1}^{m} \mathbf{Q}_{\alpha}$$

The total stress is then the sum of three contributions: in addition to the isotropic part mentioned above, there is a hydrostatic part (purely elastic) $\mathbf{S}^{e}_{vol}$ and a deviatoric part $\mathbf{S}_{iso}$ contributing to both the elastic and viscous response as part of the tensor $\sum_{\alpha=1}^{m} \mathbf{Q}_{\alpha}$.

EXERCISE.– Pioletti and Rakotomanana [PIO 00] have examined the response of biological soft tissues; the authors adopt the elastic potential in Veronda and Westmann [VER 70], given by

$$W^{e} = \frac{\alpha}{2} e^{\beta(I_1-3)} - \frac{\alpha\beta}{4}(I_2 - 3)$$

The viscous part of the strain energy is selected as

$$W^{v} = \frac{\eta}{4} tr\left(\dot{\mathbf{C}}\right)^2 (I_1 - 3)$$

The stress tensor is then written in the case of relaxation

$$\mathbf{S} = \mathbf{S}^{e}\left(\mathbf{C}(t)\right) + \mathbf{S}^{v}\left(\dot{\mathbf{C}}(t), \mathbf{C}(t)\right) + \int_{\delta}^{t} \mathbf{S}^{e}\left(\mathbf{C}(t-s)\right) \dot{M}(s)\, ds$$

with the three contributions therein traducing the immediate response, the short term response and the long term response successively, given by

$$\mathbf{S}^{e}\left(\mathbf{C}(t)\right) = 2\rho_0 \frac{\partial W^{e}}{\partial \mathbf{C}} = -p\mathbf{C}^{-1} + \alpha\beta\left(2e^{\beta(I_1-3)} - I_1\right)\mathbf{I} + \alpha\beta\mathbf{C}$$

$$\mathbf{S}^{v}\left(\dot{\mathbf{C}}(t), \mathbf{C}(t)\right) = 2\rho_0 \frac{\partial W^{v}}{\partial \mathbf{C}} = \eta(I_1 - 3)\dot{\mathbf{C}}, \quad \dot{\mathbf{C}} := \frac{\partial \mathbf{C}}{\partial t}$$

$$M(s) = \left(\sum_{k=1}^{3} a_k e^{-s/\tau_k}\right) / \left(\sum_{k=1}^{3} a_k\right)$$

The model includes 9 parameters that have to be identified: the elasticity constants α, β $[MPa]$, the short term viscosity η $[MPa.\text{s}]$ and the constant characterizing the long term viscous response, a_k, τ_k $[s]$ k=1..3 .

EXAMPLE.– Merodio and Giocolea [MER 07] proposed strain energy functions for hyper-viscoelastic incompressible media, accounting for transverse anisotropy due to the presence of a single family of fibers with unit director $\mathbf{M}$:

$$W^e(I_1, I_4) = \frac{\mu}{2}(I_1 - 3) + \frac{k_1}{2k_2}\left(e^{k_2(I_4-1)^2} - 1\right)$$

$$W^v(I_1, I_4, J_2, J_9) = \frac{\eta}{4}J_2(I_1 - 3) + \gamma J_9 \frac{k_1}{2k_2}\left(e^{k_2(I_4-1)^2} - 1\right)$$

with the invariants therein

$$I_1 = tr(\mathbf{C}),\ \mathrm{I}_2 = \frac{1}{2}\left(tr(\mathbf{C})^2 - tr(\mathbf{C}^2)\right),\ \mathrm{I}_4 = tr(\mathbf{M} \otimes \mathbf{C}.\mathbf{M}),\ J_2 = tr(\dot{\mathbf{C}}^2),\ \mathrm{J}_9 := tr(\mathbf{M} \otimes \dot{\mathbf{C}}^2.\mathbf{M})$$

and the model parameters: μ the neo-Hookean modulus of the matrix, η the viscoelastic tissue modulus, k_1, k_2 the elastic parameters of the fibers, and γ the viscosity factor of the fibers.

The model gives the following response under uniaxial traction in the fibers direction denoted 1:

Letting $\sigma := \sigma_{11}$, all other stress components vanish, and the principal stretches are written, due to transverse anisotropy,

$$\lambda_1 = \lambda,\ \lambda_2 = \lambda_3 = \lambda^{-1/2}$$

After the determination of the hydrostatic pressure, we obtain the following expression of the fiber stress

$$\sigma = \mu\left(\lambda^2 - \frac{1}{\lambda}\right) + 2k_1\left(\lambda^2 - 1\right)\lambda^2 e^{k_2\left(\lambda^2-1\right)^2} + \eta\left(\lambda^2 + 2/\lambda - 3\right)\left(2\lambda^3 + 1/\lambda^3\right)\dot{\lambda} + \gamma\frac{k_1}{k_2}\left(e^{k_2\left(\lambda^2-1\right)^2} - 1\right)4\lambda^3\dot{\lambda}$$

EXAMPLE.– In the FE code MSC Marc, the deformation potential receives two possible expressions

$$W(t) = W^{\infty} + \sum_{n=1}^{N} W^{n} e^{-t/\tau_n} = W^{0} - \sum_{n=1}^{N} W^{n}\left(1 - e^{-t/\tau_n}\right)$$

with W^0 the instantaneous elastic energy, W^{∞} the elastic energy stored after a long time, and W^n the energy characterizing the time-dependent phenomena; there is a relaxation time τ_n associated to each W^n. We further assume a proportionality relation between each W^n and the energy of instantaneous processes, so that

$W^n = \delta_n W^0$, with n a positive exponent.

For shorter times, $t \ll \tau_n$, the previous elastic energy reduces to

$$W(t) = W^{0} = W^{\infty} + \sum_{n=1}^{N} W^{n}$$

with $W^{0} = W^{\infty} + W^{0} \sum_{n=1}^{N} \delta^{n} \Rightarrow W^{\infty} = \left(1 - \sum_{n=1}^{N} \delta^{n}\right) W^{0}$

Consideration of these relations then leads to

$$W(t) = W^{0}\left(1 - \sum_{n=1}^{N} \delta^{n}\left(1 - e^{-t/\tau_n}\right)\right)$$

From this equality, a multiplicative decomposition of the deformation potential is proposed,

$$W(\mathrm{E}_{ij}, t) = W^{e}\left(E_{ij}\right).R(t)$$

With $W^{e}\left(E_{ij}\right)$ a hyperelastic potential, and $R(t)$ the relaxation function, written as a Prony series

$$R(t) = 1 - \sum_{n=1}^{N} \delta_n \left(1 - e^{-t/\tau_n}\right)$$

This can be written more specifically as

$$W(\mathbf{C},t)=\left\{W_{iso}^{e}(J)+W_{vol}^{e}\left(\overline{\mathbf{C}}\right)\right\}.R(t)$$

adopting the previous decomposition of the elastic part of the potential; the relaxation function is very often modeled using two relaxation times

$$R(t)=1-\delta_1\left(1-e^{-t/\tau_1}\right)-\delta_2\left(1-e^{-t/\tau_2}\right)$$

The relaxation is very often considered to be purely deviatoric (volume changes are deemed as purely elastic), thus we write the relaxation of the shear modulus as

$$G(t)=G^{\infty}+\sum_{n=1}^{N}G_n e^{-t/\tau_n}=G^{0}-\sum_{n=1}^{N}G_n\left(1-e^{-t/\tau_n}\right)$$

with G^{∞} the rubber modulus and G^{0} the vitreous modulus.

3.11. Anisotropic constitutive models: fiber reinforced solids, orthotropic materials

Manmade fiber-reinforced composites and soft biological tissues such as arteries or biological membranes include one or two populations (or more) of fibers immersed into a softer matrix, and give the strength of the structure (the response is stiff in the fiber direction and weaker in the transverse direction). The typical section of an artery and the mechanical response of the three main layers (intima, media, adventitia) is pictured in Figure 3.4.

The presence of fibers has two main consequences:

– Strong directional properties are given;

– The mechanical response is clearly anisotropic, so that the previous framework has to be extended.

For pedagogical reasons, we shall make a distinction between the case of one population of fibers and two populations of fibers.

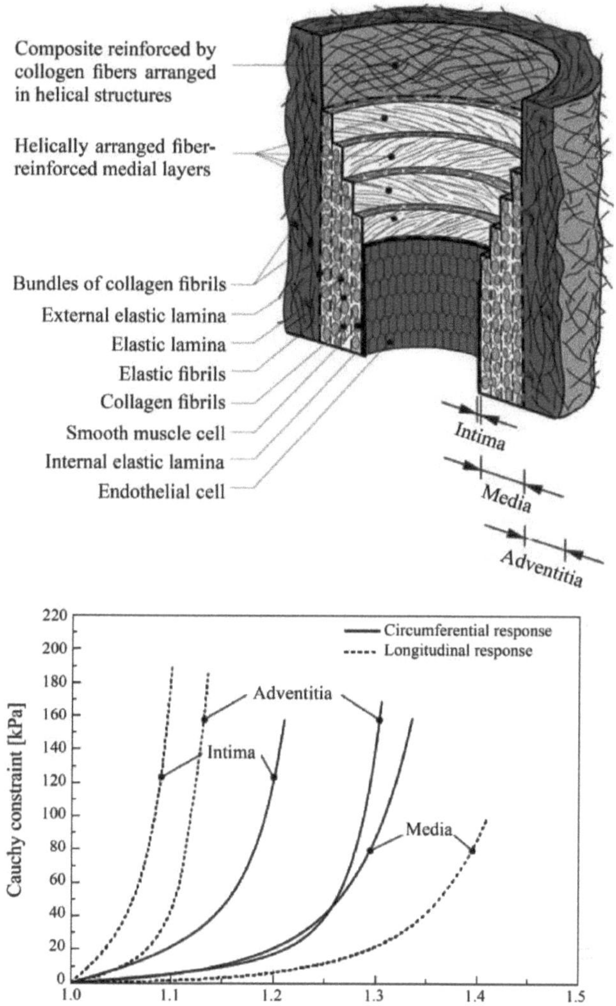

Figure 3.4. *Cross section of an artery (above) and tensile response of the three main layers [HOL 00]*

3.11.1. *Case of a single population of fibers*

We assume that the material response along directions orthogonal to the fibers is isotropic, so that the material is locally transversely isotropic. Let us denote by $\mathbf{a}_0(\mathbf{X})$ the fiber direction in the reference configuration (a unit vector), which is mapped to a vector directed along the new fiber orientation, with magnitude the fiber stretch λ:

$$\lambda \mathbf{a}(\mathbf{x},t) = \mathbf{F}.\mathbf{a}_0(\mathbf{X})$$

The fiber stretch can be evaluated from the relation

$$\|a\| = 1 \Rightarrow \lambda = \mathbf{a}_0.\mathbf{F}^T.\mathbf{F}.\mathbf{a}_0 = \mathbf{a}_0.\mathbf{C}.\mathbf{a}_0$$

Let consider a strain energy density in the reference configuration given by a function of tensor $\mathbf{C}$ as to the matrix behavior and the structural tensor $\mathbf{M} := \mathbf{a}_0 \otimes \mathbf{a}_0$ associated with the fiber, viz.

$$W = W(\mathbf{C}, \mathbf{a}_0 \otimes \mathbf{a}_0)$$

Such an energy density will model the composite material consisting of the matrix (through tensor $\mathbf{C}$ and the fibers, via tensor $\mathbf{M}$, and their interactions via mixed contributions).

Since both tensors $\mathbf{C}$ and $\mathbf{a}_0 \otimes \mathbf{a}_0$ are not affected by a rigid body rotation superimposed on the current configuration, it results that W automatically satisfies material frame-indifference, and W must remain unchanged if both matrix and fibers undergo a rotation in the reference configuration, so that

$$\mathbf{X}^* = Q.\mathbf{X} \rightarrow \mathbf{a}_0^* = Q.\mathbf{a}_0 \rightarrow \mathbf{M}^* := \mathbf{a}_0^* \otimes \mathbf{a}_0^* = Q.\mathbf{M}.Q^T$$

Due to $\mathbf{F}^* = Q.\mathbf{F} \rightarrow \mathbf{C}^* = Q.\mathbf{C}.Q^T$, W shall satisfy the condition

$$\boxed{W(\mathbf{C}, \mathbf{a}_0 \otimes \mathbf{a}_0) = W(\mathbf{C}^*, \mathbf{a}_0^* \otimes \mathbf{a}_0^*) = W(Q.\mathbf{C}.Q^T, Q.(\mathbf{a}_0 \otimes \mathbf{a}_0).Q^T), \quad \forall Q \in O_3}$$

This constraint is satisfied by forming the integrity bases of tensors $\mathbf{C}$ and $\mathbf{a}_0 \otimes \mathbf{a}_0$ [SPE 71]: we can show that the framework for isotropic materials based on invariants can be extended in the present situation by including additional invariants based on the structural tensor, thus in addition to the three principal strain invariants, we have to consider

$$I_4(\mathbf{C}, \mathbf{a}_0) := \mathbf{a}_0.\mathbf{C}.\mathbf{a}_0 = \lambda^2, \quad I_5(\mathbf{C}, \mathbf{a}_0) := \mathbf{a}_0.\mathbf{C}^2.\mathbf{a}_0$$

so that $W = W(I_1, I_2, I_3, I_4, I_5)$. The constitutive equations are then written formally as

$$\mathbf{S} = 2W_{,C} = 2\sum_{a=1}^{5} W_{,I_a}.I_{a,C}$$

We recall the derivatives of the principal invariants:

$$I_{1,C} = \mathbf{I}, \quad I_{2,C} = I_1\mathbf{I} - \mathbf{C}, \quad I_{3,C} = I_3\mathbf{C}^{-1}$$

The derivatives of the two new invariants are given by (check as an exercise)

$$I_{4,C} = \mathbf{a}_0 \otimes \mathbf{a}_0, \quad I_{5,C} = \mathbf{a}_0 \otimes \mathbf{C}.\mathbf{a}_0 + \mathbf{C}.\mathbf{a}_0 \otimes \mathbf{a}_0$$

Thus the stress tensor

$$\mathbf{S} = 2\left[\left(W_{,I_1} + I_2 W_{,I_2}\right)\mathbf{I} - W_{,I_2}\mathbf{C} + I_3 W_{,I_3}\mathbf{C}^{-1} + W_{,I_4}\mathbf{a}_0 \otimes \mathbf{a}_0 + W_{,I_5}\left(\mathbf{a}_0 \otimes \mathbf{C}.\mathbf{a}_0 + \mathbf{C}.\mathbf{a}_0 \otimes \mathbf{a}_0\right)\right]$$

A push-forward to the actual configuration gives the Cauchy stress as

$$\boldsymbol{\sigma} = 2J^{-1}\left[I_3 W_{,I_3}\mathbf{I} + \left(W_{,I_1} + I_2 W_{,I_2}\right)\mathbf{b} - W_{,I_2}\mathbf{b}^2 + I_4 W_{,I_4}\mathbf{a} \otimes \mathbf{a} + W_{,I_5}\left(\mathbf{a} \otimes \mathbf{b}.\mathbf{a} + \mathbf{b}.\mathbf{a} \otimes \mathbf{a}\right)\right]$$

3.11.2. *Case of incompressible transversely isotropic materials*

The matrix is additionally considered as incompressible, so that $I_3 = 1$. We can distinguish the two cases of extensible and inextensible fibers.

a) *Extensible fibers*: let then $W = W(I_1, I_2, I_4, I_5) - \frac{1}{2}p(I_3 - 1)$, thus the Lagrangian stress

$$\mathbf{S} = -p\mathbf{C}^{-1} + 2\left(W_{,I_1} + I_2 W_{,I_2}\right)\mathbf{I} - 2W_{,I_2}\mathbf{C} + W_{,I_4}\mathbf{a}_0 \otimes \mathbf{a}_0 + W_{,I_5}\left(\mathbf{a}_0 \otimes \mathbf{C}.\mathbf{a}_0 + \mathbf{C}.\mathbf{a}_0 \otimes \mathbf{a}_0\right)$$

EXERCISE.– Determine the Cauchy stress by a push-forward of $\mathbf{S}$.

b) *Inextensible fibers*: we have to consider the additional constraint $\lambda = 1 \Rightarrow I_4 = 1$, thus

$$W = W(I_1, I_2, I_4, I_5) - \frac{1}{2}p(I_3 - 1) - \frac{1}{2}q(I_4 - 1)$$

with $\frac{1}{2}q$ a Lagrange multiplier. The Lagrangian stress is then written

$$\mathbf{S} = -p\mathbf{C}^{-1} - q\mathbf{a}_0 \otimes \mathbf{a}_0 + 2\left(W_{,I_1} + I_2 W_{,I_2}\right)\mathbf{I} - 2W_{,I_2}\mathbf{C} + W_{,I_5}\left(\mathbf{a}_0 \otimes \mathbf{C}.\mathbf{a}_0 + \mathbf{C}.\mathbf{a}_0 \otimes \mathbf{a}_0\right)$$

with the tensor $-q\mathbf{a}_0 \otimes \mathbf{a}_0$ representing a fiber reaction stress.

3.11.3. *Case of two populations of fibers*

We consider two fibers with unit vectors $\mathbf{a}_0, \mathbf{g}_0$ in the reference configuration, and corresponding unit vectors $\mathbf{a}, \mathbf{g}$ in the actual configuration; the energy density is written

$$W = W(\mathbf{C}, \mathbf{a}_0 \otimes \mathbf{a}_0, \mathbf{g}_0 \otimes \mathbf{g}_0) \equiv W(\mathbf{C}, \mathbf{M}_1, \mathbf{M}_2)$$

Introducing the two structural tensors

$$\mathbf{M}_1 := \mathbf{a}_0 \otimes \mathbf{a}_0, \quad \mathbf{M}_2 := \mathbf{g}_0 \otimes \mathbf{g}_0$$

we can show [SPE 71, SPE 84] that the constraint

$$\boxed{W(\mathbf{C}, \mathbf{a}_0 \otimes \mathbf{a}_0, \mathbf{g}_0 \otimes \mathbf{g}_0) = W\left(Q.\mathbf{C}.Q^T, Q.(\mathbf{a}_0 \otimes \mathbf{a}_0).Q^T, Q.(\mathbf{g}_0 \otimes \mathbf{g}_0).Q^T\right), \quad \forall Q \in O_3}$$

shall be satisfied iff W is a function of the nine invariants

$$\begin{aligned}
&I_1(\mathbf{C}), I_2(\mathbf{C}), I_3(\mathbf{C}),\\
&I_4(\mathbf{C}, \mathbf{a}_0) := \mathbf{a}_0.\mathbf{C}.\mathbf{a}_0, \quad I_5(\mathbf{C}, \mathbf{a}_0) := \mathbf{a}_0.\mathbf{C}^2.\mathbf{a}_0,\\
&I_6(\mathbf{C}, \mathbf{g}_0) := \mathbf{g}_0.\mathbf{C}.\mathbf{g}_0, \quad I_7(\mathbf{C}, \mathbf{a}_0) := \mathbf{g}_0.\mathbf{C}^2.\mathbf{g}_0,\\
&I_8(\mathbf{C}, \mathbf{a}_0) := (\mathbf{a}_0.\mathbf{g}_0).\mathbf{a}_0\mathbf{C}.\mathbf{g}_0, \quad I_9(\mathbf{C}, \mathbf{a}_0) := (\mathbf{a}_0.\mathbf{g}_0)^2
\end{aligned}$$

The last invariant I_9 is a constant (independent of the deformation), thus it can be discarded. The constitutive equation is written accordingly

$$\mathbf{S}=2\begin{bmatrix}\left(W_{,I_1}+I_2W_{,I_2}\right)\mathbf{I}-W_{,I_2}\mathbf{C}+I_3W_{,I_3}\mathbf{C}^{-1}+W_{,I_4}\mathbf{a}_0\otimes\mathbf{a}_0+W_{,I_5}\left(\mathbf{a}_0\otimes\mathbf{C}.\mathbf{a}_0+\mathbf{C}.\mathbf{a}_0\otimes\mathbf{a}_0\right)+W_{,I_6}\mathbf{g}_0\otimes\mathbf{g}_0\\+W_{,I_7}\left(\mathbf{g}_0\otimes\mathbf{C}.\mathbf{g}_0+\mathbf{C}.\mathbf{g}_0\otimes\mathbf{g}_0\right)+\frac{1}{2}W_{,I_8}\left(\mathbf{a}_0.\mathbf{g}_0\right)\left(\mathbf{a}_0\otimes\mathbf{g}_0+\mathbf{g}_0\otimes\mathbf{a}_0\right)\end{bmatrix}$$

The terms $W_{,I_4}\mathbf{a}_0\otimes\mathbf{a}_0$ and $W_{,I_6}\mathbf{g}_0\otimes\mathbf{g}_0$ characterize the decoupled stress contribution of the fibers.

3.12. Case of orthotropic materials

For orthogonal fibers, viz. when $\mathbf{a}_0.\mathbf{g}_0=0$, the material is orthotropic in the reference configuration, thus the strain energy density is expressed formally as

$$W=W\left(I_1,\ldots I_7\right)$$

If, in addition, the matrix (isotropic) is incompressible, then $I_3=1$, and if the fibers are inextensible in the directions of unit vectors $\mathbf{a}_0,\mathbf{g}_0$, then $I_4=1=I_6$. Adopting both assumptions, it results in

$$W=W\left(I_1,I_2,I_5,I_7\right)-\frac{1}{2}p\left(I_3-1\right)-\frac{1}{2}q\left(I_4-1\right)-\frac{1}{2}r\left(I_6-1\right)$$

with $p/2,q/2,r/2$ Lagrange multipliers. This entails the Lagrangian stress

$$\begin{aligned}&\mathbf{S}=-p\mathbf{C}^{-1}-q\mathbf{a}_0\otimes\mathbf{a}_0-r\mathbf{g}_0\otimes\mathbf{g}_0-2W_{,I_2}\mathbf{C}+2\left(W_{,I_1}+I_2W_{,I_2}\right)\mathbf{I}+\\&2W_{,I_5}\left(\mathbf{a}_0\otimes\mathbf{C}.\mathbf{a}_0+\mathbf{C}.\mathbf{a}_0\otimes\mathbf{a}_0\right)+2W_{,I_7}\left(\mathbf{g}_0\otimes\mathbf{C}.\mathbf{g}_0+\mathbf{C}.\mathbf{g}_0\otimes\mathbf{g}_0\right)\end{aligned}$$

and the Cauchy stress

$$\begin{aligned}&\boldsymbol{\sigma}=2\mathbf{F}.\mathbf{S}.\mathbf{F}^{T}=-p\mathbf{I}-q\mathbf{a}\otimes\mathbf{a}-r\mathbf{g}\otimes\mathbf{g}+2W_{,I_2}\mathbf{b}+2\left(W_{,I_1}+I_2W_{,I_2}\right)\mathbf{I}+\\&2W_{,I_5}\left(\mathbf{a}\otimes\mathbf{b}.\mathbf{a}+\mathbf{b}.\otimes\mathbf{a}\right)+2W_{,I_7}\left(\mathbf{g}\otimes\mathbf{b}.\mathbf{g}+\mathbf{b}.\mathbf{g}\otimes\mathbf{g}\right)\end{aligned}$$

Note that the structural tensors in the actual configurations can be introduced in these two expressions, tensors $\mathbf{a}\otimes\mathbf{a}$ and $\mathbf{g}\otimes\mathbf{g}$.

Other constitutive models for hyperelastic materials and hyperviscoelastic materials will be exposed and the associated mechanical response analyzed during the exercise sessions.

3.13. Variational principles and hints to numerical solution schemes

We first summarize the field equations required to formulate a boundary value problem (BVP in short) in finite elasticity; those equations can be written either in Ω_0, or in Ω_t, and both formulations can be related by pull-back and push-forward operations.

Eulerian description: the strong form of the *initial boundary-value problem* is written

$$\begin{cases} \operatorname{div}\boldsymbol{\sigma} + \rho\mathbf{f} = \rho\dfrac{\mathrm{d}\mathbf{v}}{\mathrm{dt}} = \rho\ddot{\mathbf{u}} \quad \text{in } \Omega_t \\ \boldsymbol{\sigma} = \boldsymbol{\sigma}^T \\ \mathbf{u} = \mathbf{u}^d \quad \text{on } S_u \\ \mathbf{t} = \boldsymbol{\sigma}.\mathbf{n} = \mathbf{t}^d \quad \text{on } S_t \\ \mathbf{u}(\mathbf{x},0) = \mathbf{u}_0(\mathbf{X}) \\ \dot{\mathbf{u}}(\mathbf{x},0) = \dot{\mathbf{u}}_0(\mathbf{X}) \end{cases}$$

with $\mathbf{f}(\mathbf{x},t)$ body forces per unit mass in the actual configuration, and $\mathbf{u}^d,\mathbf{t}^d$ imposed displacement and traction vectors (they are *data*) on two complementary subparts of the external boundary of Ω_t, so that $S_u \cup S_t = \partial\Omega_t$, $S_u \cap S_t = \varnothing$. Kinematic boundary conditions (resp. static boundary conditions) are sometimes referred to as Dirichlet conditions (resp. Neumann conditions). The density $\rho = \rho(\mathbf{x},t)$ is related to the initial density through the Jacobian (itself depending on the displacement field) $\rho = \rho_0 / \det(\mathbf{F})$. This BVP has to be complemented by a constitutive relation expressing Cauchy stress in terms of a conjugated strain tensor. Since the BVP is of second order in space and time, we need two initial conditions which are the last two equations.

Observe that this BVP is formulated over the deformed configuration (the spatial divergence is involved), thus it is written in terms of the actual position $\mathbf{x}(\mathbf{X},t)$, which is precisely the unknown principal. This is a difficulty of nonlinear mechanics and a remedy consists in writing the same BVP in the fixed reference configuration.

Note that initial and boundary conditions have to be compatible so that the following conditions are satisfied:

$$\bar{\mathbf{u}}(\mathbf{x},0) = \mathbf{u}_0(\mathbf{X}), \quad \dot{\bar{\mathbf{u}}}(\mathbf{x},0) = \dot{\mathbf{u}}_0(\mathbf{X}).$$

When all data are independent of time, the BVP is referred to as *static*, so the BVP of elastostatics is written:

$$\begin{cases} \operatorname{div}\boldsymbol{\sigma} + \rho\mathbf{f} = \mathbf{0} \\ \mathbf{u} = \mathbf{u}^{d} \quad \text{on } S_u \\ \mathbf{t} = \boldsymbol{\sigma}.\mathbf{n} = \mathbf{t}^{d} \quad \text{on } S_t \end{cases}$$

The equilibrium equations have already been obtained from the VWP, which constitutes a variational principle.

The weak form of previous initial BVP is obtained by multiplying the dynamical equilibrium equation by a test function, a virtual admissible velocity field written in the form of a variation of the displacement $\delta\mathbf{u}$, and integrated over the domain Ω_t, so that we obtain the weak form of the equations of motion in the actual configuration

$$\int_{\Omega_t} \left(-\operatorname{div}\boldsymbol{\sigma}.\delta\mathbf{u} - \rho\mathbf{f}.\delta\mathbf{u} + \rho\ddot{\mathbf{u}}.\delta\mathbf{u}\right)\, dx=0$$

using the identity

$$\operatorname{div}\boldsymbol{\sigma}.\delta\mathbf{u} = \operatorname{div}\left(\boldsymbol{\sigma}.\delta\mathbf{u}\right) - \boldsymbol{\sigma} : \operatorname{grad}\delta\mathbf{u} = \operatorname{div}\left(\boldsymbol{\sigma}.\delta\mathbf{u}\right) - \boldsymbol{\sigma} : \delta\left(\operatorname{grad}\mathbf{u}\right)$$

and integrating by part (divergence th.) using the kinematic boundary condition for $\delta\mathbf{u} = \mathbf{0}$ on S_u leads to

$$\int_{\Omega_t} \left(\boldsymbol{\sigma} : \operatorname{grad}\delta\mathbf{u} - \left(\rho\mathbf{f} - \rho\ddot{\mathbf{u}}\right).\delta\mathbf{u}\right) dx - \int_{S_t} \mathbf{t}^{d}.\delta\mathbf{u}\, ds=0$$

$$\int_{\Omega_t} \mathbf{u}\left(\mathbf{x},0\right).\delta\mathbf{u}dx = \int_{\Omega_t} \mathbf{u}_0\left(\mathbf{X}\right).\delta\mathbf{u}dx$$

$$\int_{\Omega_t} \dot{\mathbf{u}}\left(\mathbf{x},0\right).\delta\mathbf{u}dx = \int_{\Omega_t} \dot{\mathbf{u}}_0\left(\mathbf{X}\right).\delta\mathbf{u}dx$$

We have written the two initial conditions for the displacement and velocity fields in weak form. The first equation is called a *variational equation*.

We recognize the VWP when we identify the previous integrals

$$P_a = \int_{\Omega_t} \rho\ddot{\mathbf{u}}.\delta\mathbf{u}\, dx=0, \quad P_e = \int_{S_t} \mathbf{t}^{d}.\delta\mathbf{u}\, ds + \int_{\Omega_t} \rho\mathbf{f}\, dx, \quad P_i = -\int_{\Omega_t} \left(\boldsymbol{\sigma} : \operatorname{grad}\delta\mathbf{u}\right) dx$$

as the virtual power of acceleration, external and internal forces respectively.

To be complete, we will show that the term $\text{grad}\boldsymbol{\delta u}$ is identified with the variation of the strain measure conjugated to Cauchy stress, namely the Almansi strain $\mathbf{e}$.

The Gateaux derivative of the Eulerian strain, the spatial tensor $\mathbf{e}$, is evaluated based on the concept of the Lie derivative: we first pull-back $\mathbf{e}$ to Ω_0, take the derivative there, and push-forward the result in Ω_t; this gives

$$\delta\mathbf{e} = \mathbf{F}^{-T}.\left(D_{\delta u}\left(\mathbf{F}^{T}.\mathbf{e}.\mathbf{F}\right)\right).\mathbf{F}^{-1} = \mathbf{F}^{-T}.\left(D_{\delta u}\mathbf{E}\right).\mathbf{F}^{-1}$$

Thus, recalling that

$$\delta\mathbf{E} = \left\{\delta\left(\mathbf{F}^{T}\right).\mathbf{F} + \mathbf{F}^{T}.\delta\mathbf{F}\right\}/2 = \text{sym}\left(\mathbf{F}^{T}.\text{Grad}\boldsymbol{\delta u}\right) \leftrightarrow \delta E_{AB} = \frac{1}{2}\left(F_{aB}\frac{\partial \delta u_a}{\partial X_A} + F_{aA}\frac{\partial \delta u_a}{\partial X_B}\right)$$

we finally arrive at

$$\delta\mathbf{e} = \frac{1}{2}\left(\mathbf{F}^{-T}.\text{Grad}^{T}\boldsymbol{\delta u} + \text{Grad}\boldsymbol{\delta u}.\mathbf{F}^{-1}\right) = \frac{1}{2}\left(\text{grad}^{T}\boldsymbol{\delta u} + \text{grad}\boldsymbol{\delta u}\right) = \text{sym}\left(\text{grad}\boldsymbol{\delta u}\right)$$

Inserting this variation back into P_i and considering the symmetry of Cauchy stress delivers

$$P_i = -\int_{\Omega_t}\left(\boldsymbol{\sigma} : \delta\mathbf{e}\right)dx$$

in which we recognize the (already written) virtual power of internal forces. Thus, the previous BVP is equivalent to the VWP principle

$$P_a = P_e + P_i$$

It is important to emphasize that the VWP does not require the existence of a potential, and that the constitutive law is not incorporated at this stage. Note that a minimization principle (a stronger requirement) does not in general exist due to loads being non-conservative or to the non-existence of a strain energy function (case of Hypoelastic materials).

For hyperelastic materials, the stress is related to the strain through a strain energy function expressed in the reference configuration, thus it is more convenient to express the VWP in the reference configuration.

Lagrangian formulation: the strong form of the initial BVP is written in terms of the nominal stress

$$\begin{cases} \mathrm{Div}\mathbf{P} + \rho_0 \mathbf{f}_0 = \rho_0 \dfrac{d\mathbf{v}}{dt} \quad \text{in } \Omega_0 \\ \mathbf{P}.\mathbf{F}^{\mathrm{T}} = \mathbf{F}.\mathbf{P}^{\mathrm{T}} \\ \mathbf{u} = \mathbf{u}^{\mathrm{d}} \quad \text{on } S_{0u} \\ \mathbf{t}_0 = \mathbf{P}.\mathbf{N} = \mathbf{t}^{\mathrm{d}} \quad \text{on } S_{0t} \\ \mathbf{u}(\mathbf{x}(\mathbf{X},t),0) = \mathbf{u}_0(\mathbf{X}) \\ \dot{\mathbf{u}}(\mathbf{x}(\mathbf{X},t),0) = \dot{\mathbf{u}}_0(\mathbf{X}) \end{cases}$$

The dynamical equilibrium can alternatively be written based on the second Piola-Kirchhoff stress, which is a symmetrical tensor, by substituting $\mathbf{P} = \mathbf{F}.\mathbf{S}$ in the first equation. This BVP is now expressed over a fixed domain, involving the Lagrangian coordinate $\mathbf{X}$; note the occurrence here of the material divergence. This BVP has to be complemented by a constitutive law expressing the stress versus a conjugated strain measure.

The weak form of the previous BVP is obtained using the same steps as for the spatial form, leading to the (already written) VWP involving Lagrangian stress measures.

EXERCISE.– Write the VWP in Lagrangian format.

For conservative loads and hyperelastic materials, the virtual power of external forces is the Gateaux derivative of a functional, thus the WVP can be rewritten as

$$\int_{\Omega_0} (\mathbf{S} : \delta\mathbf{E})\, dX - \int_{S_{0t}} \mathbf{t}_0^{\mathrm{d}}.\delta\mathbf{u} dS - \int_{\Omega_0} \rho_0 \mathbf{f}_0.\delta\mathbf{u} dX + \int_{\Omega_0} \rho_0 \ddot{\mathbf{u}}.\delta\mathbf{u} dX = 0$$

$$\rightarrow \int_{\Omega_0} \delta W(\mathbf{E})\, dX - \delta\left\{ \int_{S_{0t}} \mathbf{t}_0^{\mathrm{d}}.\mathbf{u} dS - \int_{\Omega_0} \rho_0 \mathbf{f}_0 dX \right\} + \delta\left(\frac{1}{2} \int_{\Omega_0} \rho_0 \dot{\mathbf{u}}^2 dX \right) = 0$$

$$\rightarrow \delta\left\{ \int_{\Omega_0} W(\mathbf{E})\, dX - \int_{S_{0t}} \mathbf{t}_0^{\mathrm{d}}.\mathbf{u} dS - \int_{\Omega_0} \rho_0 \mathbf{f}_0 dX + \frac{1}{2} \int_{\Omega_0} \rho_0 \dot{\mathbf{u}}^2 dX \right\} = 0$$

Thus, the introduced functional

$$V[\mathbf{u}] := \int_{\Omega_0} W(\mathbf{E})\,dX + \frac{1}{2}\int_{\Omega_0} \rho_0 \dot{\mathbf{u}}^2 dX - \int_{S_{0t}} \mathbf{t}_0^{\,d}.\mathbf{u}dS - \int_{\Omega_0} \rho_0 \mathbf{f}_0 dX$$

called the total potential energy enjoys an extremum principle; $V[\mathbf{u}]$ is the sum of the mechanical and kinetic energy (first and second terms) minus the work of external forces.

3.13.1. *Linearization of the principle of virtual work in material description*

We recall the VWP written in Lagrangian format

$$\int_{\Omega_0} (\mathbf{S} : \delta\mathbf{E})\,dX - \int_{S_{0t}} \mathbf{t}_0^{\,d}.\delta\mathbf{u}dS - \int_{\Omega_0} \rho_0 \mathbf{f}_0.\delta\mathbf{u}dX + \int_{\Omega_0} \rho_0 \ddot{\mathbf{u}}.\delta\mathbf{u}dX = 0$$

with the virtual work of internal forces

$$P_i = -\int_{\Omega_0} (\mathbf{S} : \delta\mathbf{E})\,dX \equiv -\delta W_{int}(\mathbf{u}, \delta\mathbf{u})$$

The Lagrangian strain is obviously a function of the displacement field $\mathbf{u}$. We linearized the term $\delta W_{int}(\mathbf{u}, \delta\mathbf{u})$ based on the directional derivative

$$D_{\Delta U}\delta W_{int}(\mathbf{u}, \delta\mathbf{u}) = \frac{d}{d\varepsilon}\left[\int_{\Omega_0} \left(\mathbf{S}\left(\mathbf{E}(\mathbf{u}+\varepsilon\Delta\mathbf{u})\right) : \delta\mathbf{E}(\mathbf{u}+\varepsilon\Delta\mathbf{u})\right)dX\right]_{\varepsilon=0}$$

Since differentiation and integration commute, we may write the previous derivative

$$D_{\Delta U}\delta W_{int}(\mathbf{u}, \delta\mathbf{u}) = \left[\int_{\Omega_0} \left(\mathbf{S}\left(\mathbf{E}(\mathbf{u})\right) : D_{\Delta u}\delta\mathbf{E}(\mathbf{u}) + D_{\Delta u}\mathbf{S}\left(\mathbf{E}(\mathbf{u})\right) : \delta\mathbf{E}(\mathbf{u})\right)dX\right]_{\varepsilon=0}$$

Using the chain rule, we obtain

$$D_{\Delta u}\mathbf{S}(\mathbf{E}(\mathbf{u})) = \partial_E \mathbf{S}(\mathbf{E}(\mathbf{u})) : D_{\Delta u}\mathbf{E}(\mathbf{u}) = \mathbb{C}(\mathbf{u}) : D_{\Delta u}\mathbf{E}(\mathbf{u})$$

with $\mathbb{C}(\mathbf{u})$ the fourth order tensor or elastic moduli, leading to the expression of the linear increment

$$D_{\Delta U}\delta W_{int}(\mathbf{u},\delta\mathbf{u}) = \left[\int_{\Omega_0} \left(\mathbf{S}(\mathbf{E}(\mathbf{u})) : D_{\Delta u}\delta\mathbf{E}(\mathbf{u}) + \delta\mathbf{E}(\mathbf{u}) : \mathbb{C}(\mathbf{u}) : D_{\Delta u}\mathbf{E}(\mathbf{u})\right) dX\right]_{\varepsilon=0}$$

in which we have used the minor symmetries of the elasticity tensor. Finally, inserting

$$\delta\mathbf{E}(\mathbf{u}) := D_{\delta u}\mathbf{E}(\mathbf{u}) = \mathrm{sym}\left(\mathbf{F}^T.\mathrm{Grad}\,\delta\mathbf{u}\right) \to D_{\Delta u}\delta\mathbf{E}(\mathbf{u}) = \frac{d}{d\varepsilon}\left[\mathrm{sym}\left(\mathbf{F}(\mathbf{u}+\varepsilon\Delta\mathbf{u})^T.\mathrm{Grad}\,\delta\mathbf{u}\right)\right]_{\varepsilon=0}$$

using the relations

$$\Delta\mathbf{F} = D_{\Delta u}\mathbf{F} = \mathrm{Grad}\Delta\mathbf{u}, \quad \Delta\mathbf{E}(\mathbf{u}) := D_{\Delta u}\mathbf{E}(\mathbf{u}) = \mathrm{sym}\left(\mathbf{F}^T.\mathrm{Grad}\,\Delta\mathbf{u}\right),$$

delivers the relation

$$D_{\Delta u}\delta\mathbf{E}(\mathbf{u}) = \mathrm{sym}\left(\mathrm{Grad}^T\,\Delta\mathbf{u}\mathrm{Grad}\delta\mathbf{u}\right)$$

Thus the linearization of the virtual work of internal forces leads to the increments in both tensor and indicial notations

$$D_{\Delta u}\delta W_{int}(\mathbf{u},\delta\mathbf{u}) = \int_{\Omega_0}\left(\mathrm{Grad}\delta\mathbf{u} : \mathrm{Grad}\,\Delta\mathbf{u}.\mathbf{S} + \mathbf{F}^T.\mathrm{Grad}\,\delta\mathbf{u}\delta : \mathbb{C}(\mathbf{u}) : \mathbf{F}^T.\mathrm{Grad}\,\Delta\mathbf{u}\right) dX =$$

$$\int_{\Omega_0} \frac{\partial \delta u_a}{\partial X_B}\left(\delta_{ab}S_{BD} + F_{aA}F_{bC}C_{ABCD}\right)\frac{\partial \Delta u_b}{\partial X_D} dX$$

The terms $\delta_{ab}S_{BD}$ and $F_{aA}F_{bC}C_{ABCD}$ therein represent the effective elasticity tensor, or the tangent stiffness, a symmetrical tensor, due to the fact that we can interchange the independent variations $\delta\mathbf{u}$ and $\Delta\mathbf{u}$ in $D_{\Delta U}\delta W_{int}(\mathbf{u},\delta\mathbf{u})$. The first term in $D_{\Delta U}\delta W_{int}(\mathbf{u},\delta\mathbf{u})$ represents the geometrical stress leading to the geometrical stiffness matrix, while the second term represents the material stiffness matrix.

The material form of the linearization of the virtual work principle leads to the *total Lagrangian formulation*. It is also possible to write a spatial form of the linearization of the virtual work principle (which is fully equivalent but written over the actual configuration), leading to the *updated Lagrangian formulation*. Both formulations are used in subsequent numerical solution procedures of the field equations of nonlinear elasticity using the *finite element method*; the description of those methods is however outside the scope of the volume.

4

Discrete Homogenization of Network Materials

An asymptotic homogenization technique is developed in the framework of geometrical nonlinearities to derive the large strains effective elastic response of network materials viewed as repetitive beam networks. A systematic methodology is established, allowing the prediction of the overall mechanical properties of these structures in the nonlinear regime, reflecting the influence of the geometrical and mechanical micro-parameters of the network on the overall response of the equivalent continuum. Internal scale effects of the initially discrete structure are captured by consideration of a micropolar effective continuum model. Applications to the large strain response of 3D microstructure of trabecular bone modeled with a prototype hexagonal network geometry exemplify the powerfulness of the proposed method.

4.1. Introduction

The objective of this chapter is to expose a methodology for the computation of the homogenized nonlinear response of network materials, based on an analysis performed over a representative unit cell (abbreviated as RUC); this unit cell defines the so-called mesoscopic level, intermediate between the microscopic level of the lattice and the macroscopic scale of the structure. We shall employ the discrete homogenization (DH in short) technique [MOU 03a, DOS 12] which is perfectly suited to the discrete architecture of different types of networks.

There are numerous examples of network materials, including repetitive large scale deployable structures like antenna, 3D textiles, cellular materials and especially auxetic structures with excellent damping and impact absorption

---

Chapter written by Jean-François GANGHOFFER and Khaled EL NADY.

capabilities, biological networks like biomembranes or the arterial wall, trabecular bone. Architectured materials such as polymeric and metallic foams have attracted the interest of many researchers in recent years due to their specific mechanical properties, including high crash absorbing capacity, making them suitable materials for use in the automotive industry, for instance (see [DOS 10]).

The derivation of the mechanical properties of network materials with a regular architecture (a quasi-periodic network) in relation to their topology and the mechanical properties of the inherent structural elements is especially important in order to understand the observed behavior and, possibly, to tune the internal architecture of the network to achieve certain properties at the structural level. The proper dimensioning of architectured materials regarding their topology and mechanical properties requires a good quantitative understanding of the impact of the geometry, topology and mechanical properties of the elementary unit cell at the micro level on the macroscopic effective properties under various loading conditions. Some of these network materials have been analyzed in the context of anisotropic finite-strain elasticity. This includes composites, foam-like structures, textiles in the form of 2D and 3D preforms and synthetic solids. Cellular solids for instance, by contrast to compact materials, are 2D or 3D bodies divided into cells, the walls of which are made of a solid material capable of undertaking large elastic deformations without plastic failure or fracture.

Closed form expressions of the linear behavior of the network stiffness and strength have been obtained by various authors since more than 30 years [GIB 82a, GIB 82b, HUT 06], and extensive reviews of these works can be found in [PIN 09, CHA 10]. The geometrical nonlinear response of cellular structures and network materials was extensively studied by [WAR 89, WAR 91, WAN 00], while a homogenization method was developed in [HOH 03]. Linear effective models based on beams were proposed in [JAN 03, JAN 05], and further extended in [JAN 11] to construct the strain energy function for hyperelastic cellular materials with arbitrary symmetry. An alternative approach [VIG 14] uses a computational homogenization to derive a nonlinear constitutive model for network materials.

The development of suitable micromechanical schemes for the computation of the effective mechanical response of architectured materials (such as fibrous networks) is essential in order to dispose of predictive models to analyze the overall computed response in terms of the underlying microstructure and its deformation mechanisms. When an RUC can be identified for a quasi periodic network, it is possible to develop specific homogenization schemes relying on the assumption of inherent periodicity. Although a lot of attention has been devoted to replacing large periodic networks of lattice materials by effective continuum models, less attention has been paid to the consideration of both geometrical nonlinearities and microstructure effects leading to generalized continua at the continuum level. The chief advantage of the discrete homogenization technique developed in the present work is inherent to its predictive and systematic nature, and its numerical efficiency in comparison to FE methods.

4.2. Microscopic and mesoscopic homogenization problems

The discrete homogenization method is a mathematical technique to evaluate the equivalent continuous response of repetitive discrete structures such as architectured media and network materials; it may be considered as a discrete version of the homogenization of periodic media developed thirty years ago by [SAN 80, BAK 89] and more recently applied by [WAR 02, MOU 03b, SAB 09]. The main difference lies in the fact that the initial structure to be homogenized is not modeled as a continuum, but rather as a discrete network made of beams or bars, which are then substituted by an effective continuum.

As will become apparent in the following, the chief interest of the discrete homogenization method lies in its ability to deliver the full compliance (or rigidity) matrix, thereby capturing the sometimes complex anisotropy of the so-built equivalent continuum. Regarding the issue of anisotropy, we note that the material symmetry group of the discrete network is included in the material symmetry group of the homogenized medium, as shown in [TRO 99]. The construction of constitutive laws for network materials saves the computational cost and proves convenient to perform simulations of entire network modeled as a continuum [FEY 00].

We shall present the so-called discrete homogenization method ("DH method") which is perfectly adequate for network materials comprising a periodic array of beams. The DH method is first exposed in the linear small strains regime in order to compute the initial effective moduli of the effective substitution continuum; it is then extended to the nonlinear regime in order to account for the variation of the lattice geometry under an imposed kinematic loading over an identified representative unit cell. The nonlinear response of the lattice is homogenized at the RUC level, thereby providing both stress-strain and couple stress-curvature relationships, based on an incremental scheme accounting essentially for geometrical nonlinearities.

The elastic properties of the effective continuum are first evaluated in the small strains regime, followed by an incremental computation of its nonlinear mechanical response accounting for the changes of network configuration.

4.2.1. *Determination of the initial elastic response*

The asymptotic expansion of the nodal displacement, vector $\mathbf{u}^{\varepsilon}$, is written to the second order as

$$\mathbf{u}^{\varepsilon}(\beta^{\varepsilon})=\mathbf{u}_0(\beta^{\varepsilon})+\varepsilon\ \mathbf{u}_1(\beta^{\varepsilon})+\varepsilon^2\mathbf{u}_2(\beta^{\varepsilon})$$

denoting therein ε the ratio of the unit cell size to a characteristic size of the entire network. The displacement difference $\Delta\mathbf{U}^{b\varepsilon}$ between the extremity and origin node of any beam (indexed by the letter b in the following) is expressed by a Taylor series expansion according to

$$\Delta\mathbf{U}^{b\varepsilon}=\mathbf{u}^{\varepsilon}(E(b))-\mathbf{u}^{\varepsilon}(O(b))=\varepsilon\underbrace{\left(\mathbf{u}_1^{E_{R(b)}}\left(\beta^{\varepsilon}\right)-\mathbf{u}_1^{O_{R(b)}}\left(\beta^{\varepsilon}\right)+\frac{\partial\mathbf{u}_0\left(\beta^{\varepsilon}\right)}{\partial\beta^i}\delta^{ib}\right)}_{\Delta\mathbf{U}_1^b}$$

$$+\varepsilon^2\underbrace{\left(\mathbf{u}_2^{E_{R(b)}}\left(\beta^{\varepsilon}\right)-\mathbf{u}_2^{O_{R(b)}}\left(\beta^{\varepsilon}\right)\right)}_{\Delta\mathbf{U}_2^b}$$

$$=\varepsilon\ \Delta\mathbf{U}_1^b+\varepsilon^2\ \Delta\mathbf{U}_2^b$$

with δ^{ib} therein a relative integer called the shift factor in the set $\{0,-1,1\}$. More precisely, it is nil for nodes located inside the considered cell, and non-nil for nodes belonging to a neighboring cell; the set of values of the shift factor enters the connectivity table of the selected reference unit cell.

The asymptotic expansion of the nodal microrotation $\boldsymbol{\varphi}^{n\varepsilon}$ (is here limited to the first order in ε) is defined successively at the origin and extremity of each beam as

$$\boldsymbol{\varphi}^{O(b)\varepsilon}=\boldsymbol{\varphi}_0^{O_{R(b)}}+\varepsilon\boldsymbol{\varphi}_1^{O_{R(b)}}\ ;\quad \boldsymbol{\varphi}^{E(b)\varepsilon}=\boldsymbol{\varphi}_0^{E_{R(b)}}+\varepsilon\left(\frac{\partial\boldsymbol{\varphi}_0}{\partial\beta^i}\delta^{ib}+\boldsymbol{\varphi}_1^{E_{R(b)}}\right)$$

wherein we have parameterized any point within the surface element representative of the lattice by curvilinear coordinates β^i , in order to be able to handle networks with curved material lines following the fibers (e.g. fibers in arteries) in their reference state.

The equilibrium of forces is written as the self-equilibrium (without external forces, since the reference unit cell is equilibrated under the action of neighboring cells) is conveniently written in virtual power form for the entire network and after asymptotic development as

$$\sum_{\upsilon^i \in \mathbb{Z}^3} \sum_{b \in \mathbf{B}_R} \mathbf{F}^{\varepsilon b} . \left(\mathbf{v}^{\varepsilon} \left(O(b) \right) - \mathbf{v}^{\varepsilon} \left(E(b) \right) \right) = 0 \qquad [4.1]$$

with $\mathbf{v}(\cdot)$ a virtual velocity field chosen to vanish on the unit cell edges. The force vector $\mathbf{F}^{\varepsilon b}$ decomposes into a normal and transverse contributions as

$$\mathbf{F}^{\varepsilon b} = F_x^{\varepsilon b} \mathbf{e}_x + F_y^{\varepsilon b} \mathbf{e}_y + F_z^{\varepsilon b} \mathbf{e}_z \qquad [4.2]$$

In equation [4.1], we denote $\mathbf{B}_R$ as the set of beams within the reference unit cell. The first order normal and transverse forces and the second order moment about x’, y’ and z’ at the beam extremities are successively expressed versus the kinematical nodal variables after lengthy calculations in the framework of 3D Euler-Bernoulli beams as

$$F_x^{\varepsilon b} = \frac{E_s^b A^{\varepsilon b}}{l^{\varepsilon b}} \left(\mathbf{e}_x . \left(\varepsilon \, \Delta \mathbf{U}_1^{\mathrm{b}} + \varepsilon^2 \, \Delta \mathbf{U}_2^{\mathrm{b}} \right) \right)$$

$$F_y^{\varepsilon b} = \frac{12 E_s^b I_z^{\varepsilon b}}{\left(l^{\varepsilon b} \right)^3} \left(\begin{aligned} & \mathbf{e}_y . \left(\varepsilon \, \Delta \mathbf{U}_1^{\mathrm{b}} + \varepsilon^2 \, \Delta \mathbf{U}_2^{\mathrm{b}} \right) - \\ & \frac{l^{\varepsilon b}}{2} \left(\mathbf{e}_z . \left(\boldsymbol{\varphi}_0^{O_{R(b)}} + \boldsymbol{\varphi}_0^{E_{R(b)}} + \varepsilon \left(\boldsymbol{\varphi}_1^{O_{R(b)}} + \boldsymbol{\varphi}_1^{E_{R(b)}} + \frac{\partial \boldsymbol{\varphi}_0}{\partial \beta^i} \delta^{ib} \right) \right) \right) \end{aligned} \right)$$

$$F_z^{\varepsilon b} = \frac{12 E_s^b I_y^{\varepsilon b}}{\left(l^{\varepsilon b} \right)^3} \left(\begin{aligned} & \mathbf{e}_z . \left(\varepsilon \, \Delta \mathbf{U}_1^{\mathrm{b}} + \varepsilon^2 \, \Delta \mathbf{U}_2^{\mathrm{b}} \right) + \\ & \frac{l^{\varepsilon b}}{2} \left(\mathbf{e}_y . \left(\boldsymbol{\varphi}_0^{O_{R(b)}} + \boldsymbol{\varphi}_0^{E_{R(b)}} + \varepsilon \left(\boldsymbol{\varphi}_1^{O_{R(b)}} + \boldsymbol{\varphi}_1^{E_{R(b)}} + \frac{\partial \boldsymbol{\varphi}_0}{\partial \beta^i} \delta^{ib} \right) \right) \right) \end{aligned} \right)$$

with E_s^b, G_s^b therein respectively the tensile and shear modulus of the bulk material.

In [4.2], the unit vectors $\mathbf{e}_x = \left[Cxx' \quad Cyx' \quad Czx' \right]^T$, $\mathbf{e}_y = \left[Cxy' \quad Cyy' \quad Czy' \right]^T$ and $\mathbf{e}_z = \left[Cxz' \quad Cyz' \quad Czz' \right]^T$ are the direction cosines of x', describing the transformation between the local and global coordinate system in a general 3D context (Figure 4.1), introducing the convenient notations $Cxx' = \cos\theta_x$, $Cyx' = \cos\theta_y$ and $Czx' = \cos\theta_z$. Similarly, the components of $\mathbf{e}_y$ and $\mathbf{e}_z$ are the direction cosines of the directions y' and z', respectively.

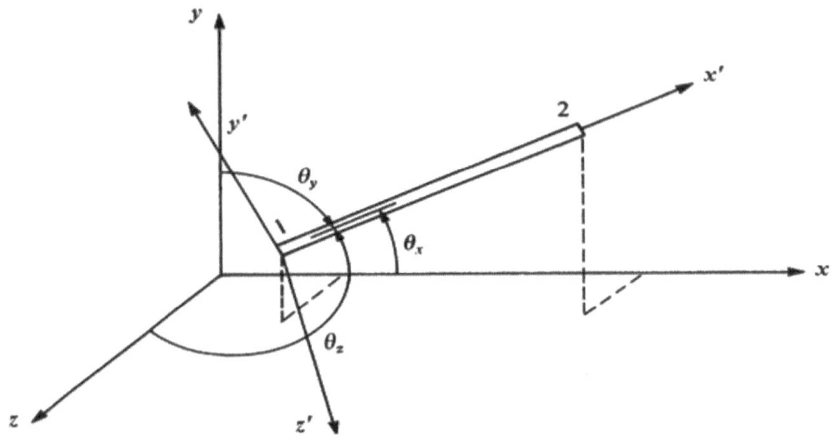

Figure 4.1. *Direction cosines associated with the x axis*

We write in parallel and to be complete the discrete equilibrium of moments, first expressed at the nodes of the unit cell in asymptotic form as

$$\sum_{\nu^i \in \mathbb{Z}^3} \sum_{b \in \mathbf{B}_R} \left(\mathbf{M}^{O(b)\varepsilon}.\mathbf{w}^{\varepsilon}\left(O(b)\right) + \mathbf{M}^{E(b)\varepsilon}.\mathbf{w}^{\varepsilon}\left(E(b)\right) \right) = 0 \qquad [4.3]$$

Another way of writing the equilibrium of moments involving the local equilibrium of each individual beam is considered for the purpose of homogenization; the moment equilibrium is expressed at the center of each beam and in asymptotic form as

$$\sum_{\nu^i \in \mathbb{Z}^3} \sum_{b \in \mathbf{B}_R} \begin{pmatrix} \mathbf{M}^{O(b)\varepsilon}.\mathbf{w}^{\varepsilon}\left(O(b)\right) + \mathbf{M}^{E(b)\varepsilon}.\mathbf{w}^{\varepsilon}\left(E(b)\right) + \\ \varepsilon l^b \left(\mathbf{e}_x \wedge \mathbf{F}^{\varepsilon b}\right).\mathbf{w}^{\varepsilon}\left(C(b)\right) \end{pmatrix} = 0 \qquad [4.4]$$

with $\mathbf{w}(\cdot)$ therein the virtual rotation velocity field and $\mathbf{M}^{O(b)}, \mathbf{M}^{E(b)}$ the moments exerted on the beam extremities, given by the expressions

$$M_x^{O(b)\varepsilon} = \frac{G_s^b J^{\varepsilon b}}{l^{\varepsilon b}} \varepsilon \left(\mathbf{e}_x . \left(\boldsymbol{\varphi}_1^{O_{R(b)}} - \left(\frac{\partial \boldsymbol{\varphi}_0}{\partial \beta^i} \delta^{ib} + \boldsymbol{\varphi}_1^{E_{R(b)}} \right) \right) \right)$$

$$M_x^{E(b)\varepsilon} = \frac{G_s^b J^{\varepsilon b}}{l^{\varepsilon b}} \varepsilon \left(\mathbf{e}_x . \left(\boldsymbol{\varphi}_1^{E_{R(b)}} - \boldsymbol{\varphi}_1^{O_{R(b)}} + \frac{\partial \boldsymbol{\varphi}_0}{\partial \beta^i} \delta^{ib} \right) \right)$$

$$M_y^{O(b)\varepsilon} = \frac{6E_s^b I_y^{\varepsilon b}}{\left(l^{\varepsilon b}\right)^2}\left(\mathbf{e}_z.\left(\varepsilon\ \Delta\mathbf{U}_1^{\mathrm{b}} + \varepsilon^2\ \Delta\mathbf{U}_2^{\mathrm{b}}\right)\right)$$

$$+\frac{E_s^b I_y^{\varepsilon b}}{l^{\varepsilon b}}\left(\mathbf{e}_y.\left(4\varphi_0^{O_{R(b)}} + 2\varphi_0^{E_{R(b)}} + \varepsilon\left(4\varphi_1^{O_{R(b)}} + 2\varphi_1^{E_{R(b)}} + 2\frac{\partial\varphi_0}{\partial\beta^i}\delta^{ib}\right)\right)\right)$$

$$M_y^{E(b)\varepsilon} = \frac{6E_s^b I_y^{\varepsilon b}}{\left(l^{\varepsilon b}\right)^2}\left(\mathbf{e}_z.\left(\varepsilon\ \Delta\mathbf{U}_1^{\mathrm{b}} + \varepsilon^2\ \Delta\mathbf{U}_2^{\mathrm{b}}\right)\right)$$

$$+\frac{E_s^b I_y^{\varepsilon b}}{l^{\varepsilon b}}\left(\mathbf{e}_y.\left(2\varphi_0^{O_{R(b)}} + 4\varphi_0^{E_{R(b)}} + \varepsilon\left(2\varphi_1^{O_{R(b)}} + 4\varphi_1^{E_{R(b)}} + 4\frac{\partial\varphi_0}{\partial\beta^i}\delta^{ib}\right)\right)\right)$$

$$M_z^{O(b)\varepsilon} = \frac{6E_s^b I_z^{\varepsilon b}}{\left(l^{\varepsilon b}\right)^2}\left(-\mathbf{e}_y.\left(\varepsilon\ \Delta\mathbf{U}_1^{\mathrm{b}} + \varepsilon^2\ \Delta\mathbf{U}_2^{\mathrm{b}}\right)\right)$$

$$+\frac{E_s^b I_z^{\varepsilon b}}{l^{\varepsilon b}}\left(\mathbf{e}_z.\left(4\varphi_0^{O_{R(b)}} + 2\varphi_0^{E_{R(b)}} + \varepsilon\left(4\varphi_1^{O_{R(b)}} + 2\varphi_1^{E_{R(b)}} + 2\frac{\partial\varphi_0}{\partial\beta^i}\delta^{ib}\right)\right)\right)$$

$$M_z^{E(b)\varepsilon} = \frac{6E_s^b I_z^{\varepsilon b}}{\left(l^{\varepsilon b}\right)^2}\left(-\mathbf{e}_y.\left(\varepsilon\ \Delta\mathbf{U}_1^{\mathrm{b}} + \varepsilon^2\ \Delta\mathbf{U}_2^{\mathrm{b}}\right)\right)$$

$$+\frac{E_s^b I_z^{\varepsilon b}}{l^{\varepsilon b}}\left(\mathbf{e}_z.\left(2\varphi_0^{O_{R(b)}} + 4\varphi_0^{E_{R(b)}} + \varepsilon\left(2\varphi_1^{O_{R(b)}} + 4\varphi_1^{E_{R(b)}} + 4\frac{\partial\varphi_0}{\partial\beta^i}\delta^{ib}\right)\right)\right)$$

The asymptotic development of the virtual velocity and rotation rates are next written; for any virtual velocity field $\mathbf{v}^{\varepsilon}(\beta)$, a Taylor series expansion leads to

$$\mathbf{v}^{\varepsilon}\left(O(b)\right) - \mathbf{v}^{\varepsilon}\left(E(b)\right) \approx \varepsilon\frac{\partial\mathbf{v}\left(\beta^{\varepsilon}\right)}{\partial\beta^i}\delta^{ib}$$

The rotation rate field is similarly expanded considering the central beam to be able to capture a change of curvature of any beam:

$$\mathbf{w}^{O(b)\varepsilon}\left(\beta\right) = \mathbf{w}\left(\beta\right);\quad \mathbf{w}^{E(b)\varepsilon}\left(\beta + \varepsilon\delta^i\right) = \mathbf{w}\left(\beta\right) + \varepsilon\frac{\partial\mathbf{w}\left(\beta\right)}{\partial\beta^i}\delta^i$$

A more detailed account of the first order homogenization method in the micropolar framework is exposed in [GOD 13a, GOD 13b, GOD 14].

In forthcoming developments, a circular section of the beams is adopted, which entails the following geometrical parameters: the cross-sectional area Aεb = π (εrb) 2, the quadratic moment of the beam $I_y^{\varepsilon b} = I_z^{\varepsilon b} = \pi \frac{(\varepsilon r^b)^4}{4}$, and the torsional constant $J^{\varepsilon b} = \pi \frac{\left(\varepsilon r^b\right)^4}{2}$.

For the homogenization of the discrete equilibrium of forces, equation [4.1] leads to the following continuous self-equilibrium for the effective continuum in virtual power form

$$\int_\Omega \mathbf{S}^i . \frac{\partial \mathbf{v}(\beta)}{\partial \beta^i} d\beta = 0$$

The stress vector $\mathbf{S}^i$ therein, which is dual to the mesoscopic transformation gradient $\frac{\partial \mathbf{v}(\beta)}{\partial \beta^i}$, decomposes into first and second order contributions versus the small parameter ε, viz. $\mathbf{S}^i = \mathbf{S}_1^i + \varepsilon \mathbf{S}_2^i$, with

$$\mathbf{S}_1^i = \sum_{b \in \mathbf{B}_R} \left(\begin{array}{l} \frac{\pi E_s^b \left(r^b\right)^2}{l^b} \left(\mathbf{e}_x . \Delta\mathbf{U}_1^b\right)\mathbf{e}_x + \left(\frac{3\pi E_s^b \left(r^b\right)^4}{\left(l^b\right)^3} \left(\mathbf{e}_y . \Delta\mathbf{U}_1^b - \frac{l^b}{2}\left(\mathbf{e}_z . \left(\boldsymbol{\varphi}_0^{O_{R(b)}} + \boldsymbol{\varphi}_0^{E_{R(b)}}\right)\right)\right)\right)\mathbf{e}_y \\ + \left(\frac{3\pi E_s^b \left(r^b\right)^4}{\left(l^b\right)^3} \left(\mathbf{e}_z . \Delta\mathbf{U}_1^b + \frac{l^b}{2}\left(\mathbf{e}_y . \left(\boldsymbol{\varphi}_0^{O_{R(b)}} + \boldsymbol{\varphi}_0^{E_{R(b)}}\right)\right)\right)\right)\mathbf{e}_z \end{array} \right) \delta^{ib}$$

$$\mathbf{S}_2^i = \sum_{b \in \mathbf{B}_R} \left(\begin{array}{l} \frac{\pi E_s^b \left(r^b\right)^2}{l^b} \left(\mathbf{e}_x . \Delta\mathbf{U}_2^b\right)\mathbf{e}_x + \left(\frac{3\pi E_s^b \left(r^b\right)^4}{\left(L^b\right)^3} \left(\mathbf{e}_y . \Delta\mathbf{U}_2^b - \frac{l^b}{2}\left(\mathbf{e}_z \left(\boldsymbol{\varphi}_1^{O_{R(b)}} + \boldsymbol{\varphi}_1^{E_{R(b)}} + \frac{\partial \boldsymbol{\varphi}_0}{\partial \beta^i} \delta^{ib}\right)\right)\right)\right)\mathbf{e}_y \\ + \left(\frac{3\pi E_s^b \left(r^b\right)^4}{\left(l^b\right)^3} \left(\mathbf{e}_z . \Delta\mathbf{U}_2^b + \frac{l^b}{2}\left(\mathbf{e}_y . \left(\boldsymbol{\varphi}_1^{O_{R(b)}} + \boldsymbol{\varphi}_1^{E_{R(b)}} + \frac{\partial \boldsymbol{\varphi}_0}{\partial \beta^i} \delta^{ib}\right)\right)\right)\right)\mathbf{e}_z \end{array} \right) \delta^{ib}$$

In analogy with previous developments, the moment equilibrium [4.4] is homogenized after inserting the asymptotic expansion of the virtual rotation rate, converting the discrete sum into a Riemann integral by passing to the limit $\varepsilon \to 0$, thus resulting in the following continuous self-equilibrium in virtual power form

$$\int_\Omega \boldsymbol{\mu}^i . \frac{\partial \mathbf{w}(\beta)}{\partial \beta^i} d\beta = 0$$

The couple stress vector $\boldsymbol{\mu}^i$ therein represent internal moments, identified on two orders, viz. $\boldsymbol{\mu}^i = \varepsilon\boldsymbol{\mu}_1^i + \varepsilon^2\boldsymbol{\mu}_2^i$, with the first and second order contributions given by

$$\boldsymbol{\mu}_1^i = \sum_{b\in\mathbf{B}_R}\left(\frac{1}{2}\left(M_{1y}^{E(b)} - M_{1y}^{O(b)}\right)\mathbf{e}_y + \frac{1}{2}\left(M_{1z}^{E(b)} - M_{1z}^{O(b)}\right)\mathbf{e}_z\right)\delta^{ib},$$

$$\boldsymbol{\mu}_2^i = \sum_{b\in\mathbf{B}_R}\left(M_{2x}^{E(b)}\mathbf{e}_x + \frac{1}{2}\left(M_{2y}^{E(b)} - M_{2y}^{O(b)}\right)\mathbf{e}_y + \frac{1}{2}\left(M_{2z}^{E(b)} - M_{2z}^{O(b)}\right)\mathbf{e}_z\right)\delta^{ib}$$

The scalar quantities therein:

$$M_{1y}^{E(b)}, M_{1y}^{O(b)}, M_{1z}^{E(b)}, M_{1z}^{O(b)}, M_{2x}^{E(b)}, M_{2y}^{E(b)}, M_{2y}^{O(b)}, M_{2z}^{E(b)}, M_{2z}^{O(b)}$$

are respectively the first and second order moment about x', y' and z'. The subscripts 1 and 2 in these expressions refer respectively to the first and second order moments.

The general form of the constitutive equations relates the Cauchy stress and couple stress tensors to the dual kinematic variables, the strain and curvature tensors (this last tensor is in general of third order when the microrotation is represented as a second order tensor); it is identified from the expressions of the homogenized stress and couple stress tensors together with $\mathbf{S}^{\mathrm{i}}$ and $\boldsymbol{\mu}^{\mathrm{i}}$ expressions as:

$$\begin{aligned}
\boldsymbol{\sigma} &= \frac{1}{g}\mathbf{S}^{\mathrm{i}} \otimes \frac{\partial \mathbf{R}}{\partial \beta^{\mathrm{i}}} = \frac{1}{g}\left(\mathbf{S}_1^{\mathrm{i}} + \varepsilon\mathbf{S}_2^{\mathrm{i}}\right) \otimes \frac{\partial \mathbf{R}}{\partial \beta^{\mathrm{i}}} \\
&= \underbrace{\frac{1}{g}\mathbf{S}_1^{\mathrm{i}} \otimes \frac{\partial \mathbf{R}}{\partial \beta^{\mathrm{i}}}}_{\left[\mathbf{K}^s\right]\{\varepsilon\}} + \underbrace{\frac{1}{g}\varepsilon\mathbf{S}_2^{\mathrm{i}} \otimes \frac{\partial \mathbf{R}}{\partial \beta^{\mathrm{i}}}}_{[\mathbf{H}]\{\chi\}} \\
\mathbf{m} &= \frac{1}{g}\boldsymbol{\mu}^{\mathrm{i}} \otimes \frac{\partial \mathbf{R}}{\partial \beta^{\mathrm{i}}} = \frac{1}{g}\left(\varepsilon\boldsymbol{\mu}_1^{\mathrm{i}} + \varepsilon^2\boldsymbol{\mu}_2^{\mathrm{i}}\right) \otimes \frac{\partial \mathbf{R}}{\partial \beta^{\mathrm{i}}} \\
&= \underbrace{\frac{1}{g}\varepsilon\boldsymbol{\mu}_1^{\mathrm{i}} \otimes \frac{\partial \mathbf{R}}{\partial \beta^{\mathrm{i}}}}_{[\mathbf{H}]\{\varepsilon\}} + \underbrace{\frac{1}{g}\varepsilon^2\boldsymbol{\mu}_2^{\mathrm{i}} \otimes \frac{\partial \mathbf{R}}{\partial \beta^{\mathrm{i}}}}_{\left[\mathbf{K}^{\mu}\right]\{\chi\}}
\end{aligned} \qquad [4.5]$$

The matrices (denoted under bracket) $\left[\mathbf{K}^s\right], \left[\mathbf{K}^{\mu}\right], [\mathbf{H}]$ are successively the classical stiffness, the micropolar stiffness and the coupling matrix. The scalar g is the Jacobian of the transformation from Cartesian to curvilinear coordinates and vector **R** is the position vector of any material point within the effective continuum.

Since for all periodical uniform structures having the property of central symmetry, the stiffness coefficients are invariant under a coordinate inversion, the pseudo-tensor $[\mathbf{H}]$ of fifth order vanishes [TRO 99]. This has the consequence that the vectors $\boldsymbol{\mu}_1^i$ and $\mathbf{S}_2^i$ in [4.5] vanish, leading to important simplification of the stress and couple stress vectors respectively,

$$\mathbf{S}^i = \mathbf{S}_1^i = \sum_{b \in \mathbf{B}_R} \left(F_{x1}^b \mathbf{e}_x + F_{y1}^b \mathbf{e}_y + F_{z1}^b \mathbf{e}_z \right) \delta^{ib}$$

and

$$\boldsymbol{\mu}^i = \boldsymbol{\mu}_2^i = \sum_{b \in \mathbf{B}_R} \left(M_{2x}^{E(b)} \mathbf{e}_x + \frac{1}{2}\left(M_{2y}^{E(b)} - M_{2y}^{O(b)} \right) \mathbf{e}_y + \frac{1}{2}\left(M_{2z}^{E(b)} - M_{2z}^{O(b)} \right) \mathbf{e}_z \right) \delta^{ib}$$

The scalar quantities F_{x1}^b, F_{y1}^b, F_{z1}^b, M_{2x}^n, M_{2y}^n, and M_{2z}^n therein are, respectively, the first order normal and transverse forces and the second order moment about x', y' and z'.

Those expressions still involve the unknown displacements $\mathbf{u}_1^n$, $\mathbf{u}_2^n$ and rotations $\boldsymbol{\varphi}_0^n$, $\boldsymbol{\varphi}_1^n$, which are determined for all nodes by solving the equilibrium equations [4.1] and [4.3].

4.2.2. *Large configuration changes of lattices by the DH method*

We next expose the extension of the DH method to the nonlinear regime, in order to account for large changes of configuration of the networks subjected to large strains. The analysis is performed over the RUC selected as the lattice unit cell; the choice of the reference unit cell is not important, as recent results [VIG 14] show that the RUC size has no effect on the predicted homogenized response in the nonlinear regime when bifurcations are prevented.

4.2.2.1. *Microscopic incremental problem over the network*

We extend the previous linear framework to account for a modification of the network geometry to a nonlinear response caused by the variation of beam director orientation and length as the kinematic loading applied over the unit cell is increased. The incremental form of the virtual work principle writes in a continuum mechanics setting as

$$\delta W_{\text{ext}} - \delta W_{\text{int}} = 0 \Rightarrow \delta V = \int_D \boldsymbol{\sigma}.\delta \mathbf{E}_G dV = \mathbf{F}_e.\delta \mathbf{u}$$

with $\delta W_{\text{ext}}, \delta W_{\text{int}}, V, \mathbf{E}_G, \mathbf{F}_e$ therein successively the virtual variation of the external and internal works, the total potential energy, the Green-Lagrange strain and the external force.

The nonlinear equilibrium problem associated to the large perturbations of the network is written in the following; the source of nonlinearity is the occurrence of large displacements and microrotations at the microscopic level of the network structural elements. A strain driven scheme is established, relying on the incremental writing of the equilibrium formation in translation and rotation, successively written as the following equations:

$$
\begin{aligned}
&\forall \mathbf{v} \in R^3, \ \sum_{b\in \mathbf{B}_R} \left(\delta F_x^{\varepsilon b(k)} \mathbf{e}_x^{b(k)} + F_x^{\varepsilon b(k)} \delta \mathbf{e}_x^{b(k)}\right).\Delta\mathbf{v} + \sum_{b\in \mathbf{B}_R} \left(F_x^{\varepsilon b(k)} \mathbf{e}_x^{b(k)}\right).\Delta\mathbf{v} \\
&+\sum_{b\in \mathbf{B}_R} \left(\delta F_y^{\varepsilon b(k)} \mathbf{e}_y^{b(k)} + F_y^{\varepsilon b(k)} \delta \mathbf{e}_y^{b(k)}\right).\Delta\mathbf{v} + \sum_{b\in \mathbf{B}_R} \left(F_y^{\varepsilon b(k)} \mathbf{e}_y^{b(k)}\right)\Delta\mathbf{v} \\
&+\sum_{b\in \mathbf{B}_R} \left(\delta F_z^{\varepsilon b(k)} \mathbf{e}_z^{b(k)} + F_z^{\varepsilon b(k)} \delta \mathbf{e}_z^{b(k)}\right).\Delta\mathbf{v} + \sum_{b\in \mathbf{B}_R} \left(F_z^{\varepsilon b(k)} \mathbf{e}_z^{b(k)}\right)\Delta\mathbf{v} = 0, \\
&\forall \mathbf{w} \in R^3, \ \sum_{b\in \mathbf{B}_R} \begin{pmatrix} \delta\left(M_x^{\varepsilon E(b)(k)} \mathbf{e}_x^{b(k)} + M_y^{\varepsilon E(b)(k)} \mathbf{e}_y^{b(k)} + M_z^{\varepsilon E(b)(k)} \mathbf{e}_z^{b(k)}\right).\mathbf{w}^{E(b)} + \\ \delta\left(M_x^{\varepsilon O(b)(k)} \mathbf{e}_x^{b(k)} + M_y^{\varepsilon O(b)(k)} \mathbf{e}_y^{b(k)} + M_z^{\varepsilon O(b)(k)} \mathbf{e}_z^{b(k)}\right).\mathbf{w}^{O(b)} \end{pmatrix} \\
&+\sum_{b\in \mathbf{B}_R} \delta\left(\varepsilon l^b \left(\mathbf{e}_x \wedge \mathbf{F}^{\varepsilon b}\right)\right).\mathbf{w}^{C(b)} + \\
&\sum_{b\in \mathbf{B}_R} \left(M^{\varepsilon E(b)(k)}.\mathbf{w}^{E(b)} + M^{\varepsilon O(b)(k)}.\mathbf{w}^{O(b)}\right) \\
&+\sum_{b\in \mathbf{B}_R} \left(\varepsilon l^b \left(\mathbf{e}_x \wedge \mathbf{F}^{\varepsilon b}\right).\mathbf{w}^{C(b)}\right) = 0
\end{aligned}
\tag{4.6}
$$

The vector $\Delta\mathbf{v} = \mathbf{v}\left(E_R(b)\right) - \mathbf{v}\left(O_R(b)\right)$ therein is the relative virtual velocity of the two extremity nodes of any beam within the network, $\mathbf{B}^b = l^b \mathbf{e}_x^b$ is the beam vector (Figure 4.2) with length $l^b = \left\|\mathbf{B}^b\right\|$ and $\mathbf{e}_x^b = \dfrac{\mathbf{B}^b}{l^b}$ $\mathbf{e}_x^b$ is the unit beam director.

For slender beams, the network is much softer in bending compared to tension, which entails that geometrical nonlinearities represents the main source of a nonlinear behaviour. We accordingly account for the variations of beam length and orientation, illustrated in Figure 4.2.

Figure 4.2. *Variation of beam orientation (left) and length (right). For a color version of this figure, see www.iste.co.uk/ganghoffer/biomechanics.zip*

For any beam $b \in \mathbf{B}_R$, the beam vector can be determined from the imposed transformation gradient denoted $\mathbf{G}^j$ and the relative position vector $\mathbf{R}^{E_R(b)} - \mathbf{R}^{O_R(b)}$ between the two extremity nodes from the relation

$$\mathbf{B}^b = \mathbf{R}^{E_R(b)} - \mathbf{R}^{O_R(b)} + \mathbf{G}^j \delta^{jb} \quad [4.7]$$

In a similar manner, the relative microrotation is elaborated versus the imposed microcurvature tensor $\mathbf{M}^j$ from the relation

$$\boldsymbol{\varphi}^{E_{R(b)}} - \boldsymbol{\varphi}^{O_{R(b)}} = \left(\boldsymbol{\varphi}_1^{E_{R(b)}} - \boldsymbol{\varphi}_1^{O_{R(b)}} + \mathbf{M}^j \delta^{jb}\right) \quad [4.8]$$

In previous two relations, we have introduced the following notation: $\mathbf{G}^j = \dfrac{\partial \mathbf{R}}{\partial \beta^j}$, $\mathbf{M}^j = \dfrac{\partial \boldsymbol{\varphi}_0}{\partial \beta^j}$ to denote the mesoscopic transformation gradient and microcurvature mappings imposed over the unit cell, and the vectors $\mathbf{R}^{E_R(b)} - \mathbf{R}^{O_R(b)}$ and $\boldsymbol{\varphi}_1^{E_R(b)} - \boldsymbol{\varphi}_1^{O_R(b)}$ are the unknown kinematic relative positions and rotations respectively that are computed incrementally, as explained in the sequel.

The variations of the beam unit vectors and lengths are obtained after few algebraic computations as

$$\begin{aligned} &\delta\mathbf{e}_x^b = \mathbf{C.P.}\delta\mathbf{B}^b / l^b, \quad \delta\mathbf{e}_y^b = \boldsymbol{\Omega}_z.\delta\mathbf{e}_x^b, \quad \delta\mathbf{e}_z^b = \boldsymbol{\Omega}_y.\delta\mathbf{e}_x^b \\ &\delta l^b = \mathbf{B}^b.[\mathbf{I} + \mathbf{C.P}].\delta\mathbf{B}^b / l^b \end{aligned} \quad [4.9]$$

with the operators $\mathbf{P}$ and $\mathbf{C}$ therein given by

$$\mathbf{P}=\left(\mathbf{I}-\mathbf{e}_x^b\otimes\mathbf{e}_x^b\right),\quad \mathbf{C}=\left(\mathbf{I}-\frac{1}{2}\mathbf{e}_x^b\otimes\mathbf{e}_x^b\right)$$

The perturbed expressions of the forces and moments express as

$$\begin{aligned}
&\delta F_x^{b(k)}\mathbf{e}_x^b=\frac{1}{l^b}\frac{dF_x^{b(k)}}{dl^b}\left(\mathbf{B}^b\otimes\mathbf{B}^b[\mathbf{I}+\mathbf{C}.\mathbf{P}]\right).\delta\mathbf{B}^b\\
&\delta F_y^{b(k)}\mathbf{e}_y^b=\frac{1}{l^b}\frac{dF_y^{b(k)}}{dl^b}\left(\mathbf{e}_y^b\otimes\mathbf{B}^b[\mathbf{I}+\mathbf{C}.\mathbf{P}]\right).\delta\mathbf{B}^b+\left(\frac{-6E_s^bI_z}{(l^b)^2}\right)\begin{bmatrix}\left(\mathbf{e}_y^b\otimes\mathbf{e}_z^b\right).\delta\boldsymbol{\varphi}^{O(b)}\\+\left(\mathbf{e}_y^b\otimes\mathbf{e}_z^b\right).\delta\boldsymbol{\varphi}^{E(b)}\end{bmatrix}\\
&\delta F_z^{b(k)}\mathbf{e}_z^b=\frac{1}{l^b}\frac{dF_z^{b(k)}}{dl^b}\left(\mathbf{e}_z^b\otimes\mathbf{B}^b[\mathbf{I}+\mathbf{C}.\mathbf{P}]\right).\delta\mathbf{B}^b+\left(\frac{6E_s^bI_y}{(l^b)^2}\right)\begin{bmatrix}\left(\mathbf{e}_z^b\otimes\mathbf{e}_y^b\right).\delta\boldsymbol{\varphi}^{O(b)}\\+\left(\mathbf{e}_z^b\otimes\mathbf{e}_y^b\right).\delta\boldsymbol{\varphi}^{E(b)}\end{bmatrix}\\
&\delta M_x^O\mathbf{e}_x^b=\left(\frac{G_s^bJ^b}{l^b}\right)\left[\left(\mathbf{e}_x^b\otimes\mathbf{e}_x^b\right).\delta\boldsymbol{\varphi}^{O(b)}-\left(\mathbf{e}_x^b\otimes\mathbf{e}_x^b\right).\delta\boldsymbol{\varphi}^{E(b)}\right]\\
&\delta M_x^E\mathbf{e}_x^b=\left(\frac{G_s^bJ^b}{l^b}\right)\left[\left(\mathbf{e}_x^b\otimes\mathbf{e}_x^b\right).\delta\boldsymbol{\varphi}^{E(b)}-\left(\mathbf{e}_x^b\otimes\mathbf{e}_x^b\right).\delta\boldsymbol{\varphi}^{O(b)}\right]\\
&\delta M_y^O\mathbf{e}_y^b=\frac{dM_y^O}{dl^b}\left(\mathbf{B}^b\otimes\mathbf{B}^b[\mathbf{I}+\mathbf{C}.\mathbf{P}]\right).\boldsymbol{\Omega}_z.\delta\mathbf{B}^b+\left(\frac{E_s^bI_y^b}{l^b}\right)\begin{bmatrix}4\left(\mathbf{e}_y^b\otimes\mathbf{e}_y^b\right).\delta\boldsymbol{\varphi}^{O(b)}\\-2\left(\mathbf{e}_y^b\otimes\mathbf{e}_y^b\right).\delta\boldsymbol{\varphi}^{E(b)}\end{bmatrix}\\
&\delta M_y^E\mathbf{e}_y^b=\frac{dM_y^E}{dl^b}\left(\mathbf{B}^b\otimes\mathbf{B}^b[\mathbf{I}+\mathbf{C}.\mathbf{P}]\right).\boldsymbol{\Omega}_z.\delta\mathbf{B}^b+\left(\frac{E_s^bI_y^b}{l^b}\right)\begin{bmatrix}2\left(\mathbf{e}_y^b\otimes\mathbf{e}_y^b\right).\delta\boldsymbol{\varphi}^{E(b)}\\-4\left(\mathbf{e}_y^b\otimes\mathbf{e}_y^b\right).\delta\boldsymbol{\varphi}^{O(b)}\end{bmatrix}\\
&\delta M_z^O\mathbf{e}_z^b=\frac{dM_z^O}{dl^b}\left(\mathbf{B}^b\otimes\mathbf{B}^b[\mathbf{I}+\mathbf{C}.\mathbf{P}]\right).\boldsymbol{\Omega}_y.\delta\mathbf{B}^b+\left(\frac{E_s^bI_z^b}{l^b}\right)\begin{bmatrix}4\left(\mathbf{e}_z^b\otimes\mathbf{e}_z^b\right).\delta\boldsymbol{\varphi}^{O(b)}\\-2\left(\mathbf{e}_z^b\otimes\mathbf{e}_z^b\right).\delta\boldsymbol{\varphi}^{E(b)}\end{bmatrix}\\
&\delta M_z^E\mathbf{e}_z^b=\frac{dM_z^E}{dl^b}\left(\mathbf{B}^b\otimes\mathbf{B}^b[\mathbf{I}+\mathbf{C}.\mathbf{P}]\right).\boldsymbol{\Omega}_y.\delta\mathbf{B}^b+\left(\frac{E_s^bI_z^b}{l^b}\right)\begin{bmatrix}2\left(\mathbf{e}_z^b\otimes\mathbf{e}_z^b\right).\delta\boldsymbol{\varphi}^{E(b)}\\-4\left(\mathbf{e}_z^b\otimes\mathbf{e}_z^b\right).\delta\boldsymbol{\varphi}^{O(b)}\end{bmatrix}
\end{aligned}\qquad [4.10]$$

Expressions [4.9] and [4.10] are then inserted into the incremental equilibrium [4.6], leading to the formulation of the incremental microscopic (at the network level) problem. This allows for the identification of the total tangent stiffness matrix, which can be additively decomposed into the stress stiffness $\left(\mathbf{K}_T^S\right)$ and the micropolar stiffness $\left(\mathbf{K}_T^m\right)$. The tangent stress stiffness itself splits into stiffness matrices, respectively the linear stiffness matrix $\mathbf{K}_{So}$, the initial displacement stiffness matrix $\mathbf{K}_u$, and the initial stress stiffness matrix $\mathbf{K}_\sigma$, explicitly given in [ELN 17, ELN 16, ELN 15], so that it holds the decomposition

$$\mathbf{K}_T^S=\sum_{b\in\mathbf{B}_R}\left(\mathbf{K}^b{}_{So}+\mathbf{K}^b{}_u+\mathbf{K}^b{}_\sigma\right)$$

A similar additive decomposition holds for the micropolar tangent stiffness matrix

$$\mathbf{K}_T^m = \sum_{b\in \mathbf{B}_R} \left(\mathbf{K}^b{}_{mo} + \mathbf{K}^b{}_{\mathrm{m}}\right)$$

with $\mathbf{K}^b{}_{mo}$ therein the initial micropolar stiffness matrix and $\mathbf{K}^b{}_{\mathrm{m}}$ the tangent micropolar stiffness matrix [ELN 16].

The unknown kinematic quantities, vectors $\mathbf{R}^{E_R(b)} - \mathbf{R}^{O_R(b)}$ and $\boldsymbol{\varphi}^{E_R(b)} - \boldsymbol{\varphi}^{O_R(b)}$, which appear in the left hand-side of equation [4.7] and [4.8]; they are determined by solving iteratively the previous incremental problem based on the modified Newton-Raphson method. For a given mesoscopic transformation gradient and microcurvature, the position and microrotation vectors are next updated at subsequent iteration $(k+1)$ from their values at previous iteration k within the reference unit cell as:

$$\mathbf{B}^{b(k+1)} = \mathbf{B}^{b(k)} + \delta\mathbf{B}^{b(k)}, \text{with } \mathbf{B}^{b(k)} = \mathbf{R}^{E_R(b)(k)} - \mathbf{R}^{O_R(b)(k)} + \mathbf{G}^j\delta^{jb}$$

$$\boldsymbol{\varphi}^{b(k+1)} = \boldsymbol{\varphi}^{b(k)} + \delta\boldsymbol{\varphi}^{b(k)}, \text{ with } \boldsymbol{\varphi}^{b(k)} = \boldsymbol{\varphi}^{E_R(b)(k)} - \boldsymbol{\varphi}^{O_R(b)(k)} + \mathbf{M}^j\delta^{jb}$$

For a given kinematic loading over the RUC, viz. the mappings $\frac{\partial \mathbf{R}}{\partial \beta^j}\delta^{jb}, \frac{\partial \boldsymbol{\varphi}}{\partial \beta^j}\delta^{jb}, j = 1,2,3,$ one can then evaluate the quantities $\mathbf{B}^b, \mathbf{e}_x^b, \mathbf{e}_y^b, \mathbf{e}_z^b, F_x^b, F_y^b, F_z^b, M_x^b, M_y^b, M_z^b$, for any beam within the reference unit cell $b \in \mathbf{B}_R$, so that the stress and the couple stress vectors, $\mathbf{S}^i$ and $\boldsymbol{\mu}^i$ respectively, can be updated.

The above formulated microscopic incremental problem over the network is then upscaled to formulate a mesoscopic incremental problem at the continuum level of the RUC.

4.2.2.2. *Mesoscopic boundary value problem over the reference unit cell*

The previously written discrete incremental equilibrium problem takes after homogenization the following similar form (self-equilibrium) written over the Lagrangian domain Ω occupied by the m nreference unit cell:

$$\sum_{\upsilon^i\in\mathbb{Z}^3}\sum_{b\in\mathbf{B}_R} \delta\mathbf{F}^b\cdot\mathbf{v}^\varepsilon + \sum_{\upsilon^i\in\mathbb{Z}^3}\sum_{b\in\mathbf{B}_R} \delta\mathbf{M}^b\cdot\mathbf{W}^\varepsilon = 0$$

$$\Rightarrow \int_\Omega \delta\mathbf{S}^{i(k)}\cdot\frac{\partial \mathbf{v}}{\partial \beta^i}d\beta + \int_\Omega \delta\boldsymbol{\mu}^{i(k)}\cdot\frac{\partial \mathbf{w}}{\partial \beta^i}d\beta = 0$$

in which $\delta \mathbf{S}^{i(\mathrm{k})}, \delta \boldsymbol{\mu}^{i(\mathrm{k})}$ denote the incremental stress and couple stress vectors.

The incremental constitutive law at the mesoscopic level is formulated in a formal manner, considering kinematic-controlled loadings applied over the unit cell; at each increment, the transformation gradient and microcurvature mappings are updated as boundary loadings over the unit cell versus a loading parameter, like stretch (if uniaxial tension is simulated) or shear angle (if a pure shear strain is simulated). We shall subsequently analyze the response of repetitive network materials under so-called elementary loadings including uniaxial tension, equibiaxial tension, simple shear, and pure bending. The kinematic load parameter is initialed at zero and then incremented, the equilibrium solution is computed at each increment.

In order to set up the nonlinear mesoscopic BVP (acronym for boundary value problem), we write the discretized version of the elastic constitutive equation relating the incremental stress $\Delta \mathbf{S}_n^{(k)}$ to the incremental strain $\Delta \mathbf{E}_{Gn}^{(k)}$, and the incremental couple stress $\Delta \mathbf{M}_n^{(k)}$ to the incremental curvature tensor $\Delta \boldsymbol{\Gamma}_n^{(k)}$ (both increments are imposed over the RUC) , as

$$\begin{aligned} \Delta \mathbf{S}_n^{(k)} &= K_{T,n}^S : \Delta \mathbf{E}_{Gn}^{(k)} \\ \Delta \mathbf{M}_n^{(k)} &= K_{T,n}^m : \Delta \boldsymbol{\Gamma}_n^{(k)} \end{aligned} \qquad [4.11]$$

The Green-Lagrange strain therein is defined as $\mathbf{E}_{\mathrm{G}} = \frac{1}{2}\left(\mathbf{F}^{\mathrm{T}}.\mathbf{F} - \mathbf{I}\right)$. The stress tangent stiffness matrix $\mathbf{K}_{T,n}^S$ is obtained from the assembly of the microscopic tangent stiffness matrices, and is computed at each increment.

The Lagrangian wryness tensor is defined (amongst many other possible choices) as the following second order tensor versus the finite rotation vector $\boldsymbol{\varphi} = \varphi_i \mathbf{e}_i$ [PIE 09]

$$\boldsymbol{\Gamma} = \nabla_X \boldsymbol{\varphi} \left[\frac{\sin\left(\|\boldsymbol{\varphi}\|\right)}{\|\boldsymbol{\varphi}\|} \mathbf{I} - \frac{1-\cos\left(\|\boldsymbol{\varphi}\|\right)}{\|\boldsymbol{\varphi}\|^2} \boldsymbol{\varphi} \times \mathbf{I} + \frac{\|\boldsymbol{\varphi}\| - \sin\left(\|\boldsymbol{\varphi}\|\right)}{\|\boldsymbol{\varphi}\|^3} \boldsymbol{\varphi} \otimes \boldsymbol{\varphi} \right]$$

with ∇_X the gradient operator with respect to the Lagrangian coordinate $\mathbf{X}$.

For a given elementary loading characterized by a continuously varying scalar loading parameter λ_n, the incremental Lagrangian strain $\Delta \mathbf{E}_{Gn}$ is related to the increment and the increment $\Delta \mathbf{F}_n$ of the transformation gradient at any increment n as [HOL 00]

$$\Delta \mathbf{E}_{Gn}(\lambda_n, \Delta\lambda_n) = sym\left(\mathbf{F}_n(\lambda_n)^T . \nabla_X \Delta \mathbf{u}_n\right) \equiv sym\left(\mathbf{F}_n(\lambda_n)^T . \Delta \mathbf{F}_n(\Delta\lambda_n)\right)$$

The discrete transformation gradient therein is given by the expression

$$\mathbf{F}_n(\lambda_n) = \mathbf{I} + \nabla_X \mathbf{u}_n \rightarrow \Delta \mathbf{F}_n(\Delta\lambda_n) = \Delta \nabla_X \mathbf{u}_n \equiv \nabla_X \Delta \mathbf{u}_n \qquad [4.12]$$

The updated Cauchy stress tensor is then obtained by a push-forward of the incremental Lagrangian stress $\Delta \mathbf{S}_n^{(k)}$ expressed in [4.11] from configuration Ω_n to Ω_{n+1}, elaborated as

$$\boldsymbol{\sigma}_{n+1}^{(k)} = \underbrace{\left(J_n^{-1}\mathbf{F}_n.\mathbf{S}_n^{(k)}.\mathbf{F}_n^T\right)}_{\boldsymbol{\sigma}_n^{(k)}} + \underbrace{\left(J_n^{-1}\mathbf{F}_n.\left(\Delta \mathbf{S}_n^{(k)}\right).\mathbf{F}_n^T\right)}_{\Delta\boldsymbol{\sigma}_n^{(k)}} \qquad [4.13]$$

with $J = \det(\mathbf{F})$ the Jacobian, defined as the determinant of the deformation gradient tensor. The subscript k is the iteration index inside the increment loop. The Cauchy stress at increment $n+1$ is accordingly given from its value at previous increment n, based on the initial stress $\mathbf{S}_n^{(k)}$ and the incremental stress $\Delta\boldsymbol{\sigma}_n^{(k)}$.

The incremental couple stress tensor is similarly evaluated as

$$\mathbf{m}_{n+1}^{(k)} = \underbrace{\left(\overline{\mathbf{R}}_n.\mathbf{M}_n^{(k)}.\mathbf{F}_n^T\right)}_{\mathbf{m}_n^{(k)}} + \underbrace{\left(\overline{\mathbf{R}}_n.\left(\Delta \mathbf{M}_n^{(k)}\right).\mathbf{F}_n^T\right)}_{\Delta\mathbf{m}_n^{(k)}} \qquad [4.14]$$

in which $\overline{\mathbf{R}}_n = \exp(spn(\boldsymbol{\varphi}))\ \overline{\mathbf{R}}_n$ is the micropolar rotation tensor at increment n, here expressed in closed form by Euler-Rodrigues formula [STE 94].

The elaborated strategy to solve the nonlinear microscopic and mesoscopic BVP over the network RUC are such that the classical and nonclassical (micropolar moduli) effective mechanical properties are evaluated for a given imposed kinematic loading over the updated unit cell configuration. The computation of the effective classical and micropolar tangent stiffness matrices entails the derivation of the characteristic bending and torsional lengths which are updated at each increment.

4.2.3. *General algorithm of the homogenization method*

The numerical evaluation of the effective nonlinear response of network materials subjected to large strains follows a stop steps procedure: the homogenized constitutive law is first evaluated in the linear framework as the ground solution valid at low strains and small displacements. In a second step, the effective nonlinear response is constructed in an incremental manner, with new mechanical properties extracted from the computed tangent compliance matrices over the new network configuration resulting from the controlled imposed kinematic loading.

The algorithm summarized in Box 4.1 has given rise to a dedicated code written in symbolic language, with the initial reference unit cell topology and mechanical properties provided as input, and delivering as an output the homogenized mechanical properties (classical and micropolar moduli) and stress-strain response for a given deformation path.

For each kinematic increments $\Delta\mathbf{E}_{Gn}^{(k)}$, $\Delta\mathbf{\Gamma}_{n}^{(k)}$;

For each iteration k inside the increment loop;

1) Compute the effective mechanical properties in the linear regime based on the linear discrete homogenization framework, equation [4.5].

2) Define the incrementally imposed strain and microcurvature applied over the RUC.

3) Compute the incremental Second Piola-Kirchhoff stress tensor and couple stress tensor, equation [4.11],

$$\Delta\mathbf{S}_{n}^{(k)} = K_{T,n}^{S} : \Delta\mathbf{E}_{Gn}^{(k)},\ \Delta\mathbf{M}_{n}^{(k)} = K_{T,n}^{m} : \Delta\mathbf{\Gamma}_{n}^{(k)}$$

4) Check convergence at iteration k; if it is attained, go to next step.

5) Compute the incremental deformation gradient, equation [4.12], and its Jacobian.

6) Update Cauchy stress and couple stress at increment ($n+1$), by a push forward of their Lagrangian couterpart from Ω_n to Ω_{n+1}, equations [4.13].and [4.14]:

$$\boldsymbol{\sigma}_{n+1}^{(k)} = J_n^{-1}\mathbf{F}_n.\left\{\mathbf{S}_n^{(k)} + \Delta\mathbf{S}_n^{(k)}\right\}.\mathbf{F}_n^T = \underbrace{\left(J_n^{-1}\mathbf{F}_n.\mathbf{S}_n^{(k)}.\mathbf{F}_n^T\right)}_{\boldsymbol{\sigma}_n^{(k)}} + \underbrace{\left(J_n^{-1}\mathbf{F}_n.\left(\Delta\mathbf{S}_n^{(k)}\right).\mathbf{F}_n^T\right)}_{\Delta\boldsymbol{\sigma}_n^{(k)}}$$

$$\mathbf{m}_{n+1}^{(k)} = \underbrace{\left(\overline{\mathbf{R}}_n.\mathbf{M}_n^{(k)}.\mathbf{F}_n^T\right)}_{\mathbf{m}_n^{(k)}} + \underbrace{\left(\overline{\mathbf{R}}_n.\left(\Delta\mathbf{M}_n^{(k)}\right).\mathbf{F}_n^T\right)}_{\Delta\mathbf{m}_n^{(k)}}$$

7) Update the network configuration from Ω_n to Ω_{n+1}.

8) Repeat steps 1-7 up to the maximum applied strain and curvature over the unit cell.

Box 4.1. *Algorithm for the nonlinear discrete homogenization of repetitive lattices*

4.3. Application to trabecular bone

Bone is a complex hierarchical material exhibiting geometrical features occurring over several length scales: a) the nanoscale of a single fiber and crystals, b) the microscale associated with the network of trabecular bone composed of random struts or plates), and c) the macroscale of the entire bone, which includes both trabecular (porous) and cortical (solid) bone types, as illustrated in Figure 4.3.

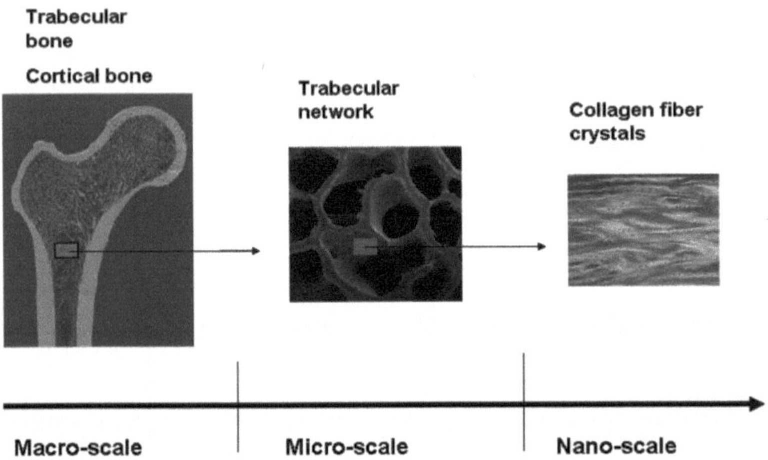

Figure 4.3. *Multi-scale organization of cortical bone [FAN 05]*

Cancellous also deserves the name trabecular bone, since it is composed of short struts of bone material called trabeculae; the trabecular network give cancellous bone a spongy appearance, thereby explaining the wording spongy bone.

Trabecular bone has a rather random and spatially heterogeneous organization, but it is nevertheless reasonable to simplify its complex geometry by modeling trabecular bone by an idealized periodic microstructure.

A mechanical analysis of a 3D anisotropic hexagonal based network is performed, with the unit cell shown in Figure 4.4 [GOD 12a, GOD 14]. The in-plane struts have length L and h and vertical struts are of length L_v (out-of-plane), all beams having a circular cross-section of diameter d. The entire network is generated from the repetition of the unit cell shown in Figure 4.4(b) thanks to three periodicity vectors $\mathbf{Y}_1, \mathbf{Y}_2, \mathbf{Y}_3$ defined in the Cartesian basis, with length $L_1 = L_2 = \sqrt{(L\cos\theta)^2 + (h + L\sin\theta)^2}$ and $L_3 = L_v$.

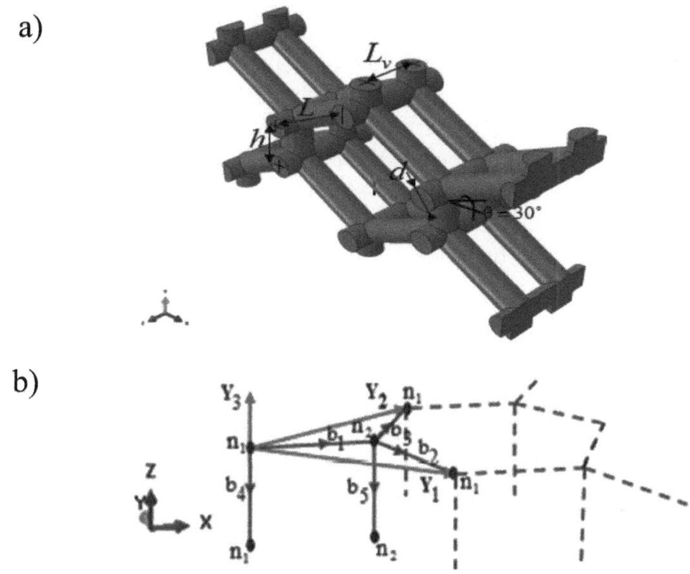

Figure 4.4. *3D hexagonal lattice. a) Geometric description of the repeatable unit cell. b) The representative unit cell*

The adopted unit cell includes five beams: three horizontal beams b_1, b_2, and b_3 and the vertical beams b_4 and b_5. Beams b_2 and b_3 have identical length L and are inclined by the angle $\theta = 30°$ towards the horizontal direction, while beam b_1 is of length h. The vertical beams b_4 and b_5 have the same length L_v. The connectivity Table 4.1 provides the numbering of beams and nodes within the chosen representative unit cell. We select the following geometrical parameters: d = 0.25 mm, L = 1 mm, L_v = 0.5 mm, h = 0.5 mm, and $\theta = 30°$; The Young's modulus and Poisson's ratio of the cell wall material are selected as E_s = 12,000 MPa, and $\nu_s = 0.3$.

Beam	1	2	3	4	5
$O(b)$	1	2	2	1	2
$O(b)$	2	1	1	1	2
δ_1	0	1	0	0	0
δ_2	0	0	1	0	0
δ_3	0	0	0	-1	-1

Table 4.1. *Connectivity array for the 3D trabecular bone unit cell*

The nonlinear elastic response of the trabecular network is evaluated in terms of the evolution of the components of Cauchy stress versus stretch for different types of loading. Figure 4.5 shows the nonlinear response in terms of the evolution of the Cauchy stress diagonal components versus stretches successively applied in the three Cartesian individual directions x,y, z. The extrapolated linear response shown as a matter of assessing the importance of nonlinear effects highlights a strong nonlinear response for tension along x, but a less pronounced nonlinear response for tension along the two other directions. The response under tension in z-direction shows a nearly initial linear behavior followed by a slight softening (Figure 4.5(c)).

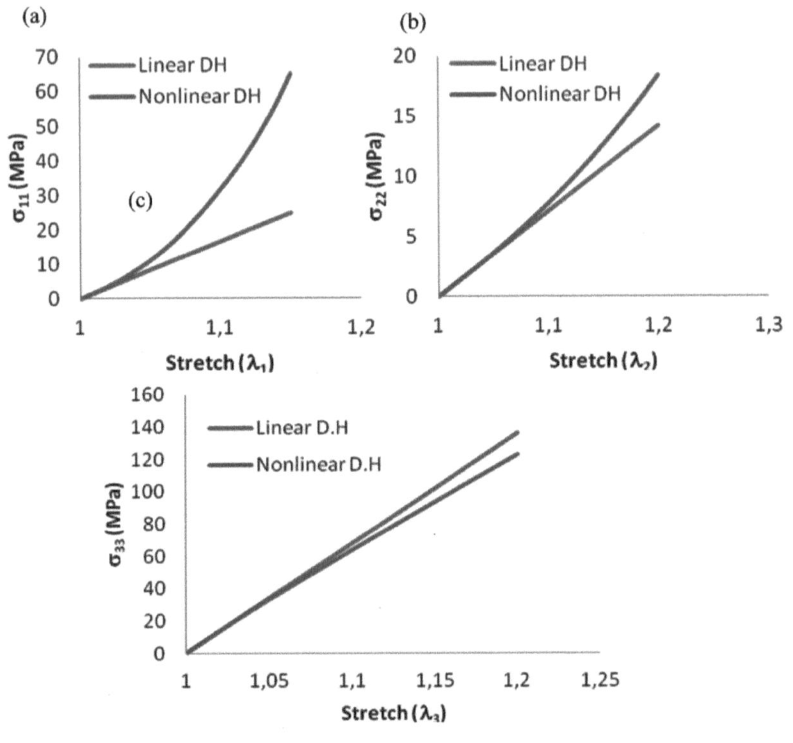

Figure 4.5. *Evolution of Cauchy stress components versus the corresponding applied stretch under uniaxial tension of the 3D hexagon in a) x-direction, b) y-direction and c) z-direction. For a color version of this figure, see www.iste.co.uk/ganghoffer/biomechanics.zip*

The responses under equi-biaxial tension present a nearly linear evolution of Cauchy stress components versus stretch in both x and y directions, Figure 4.6.

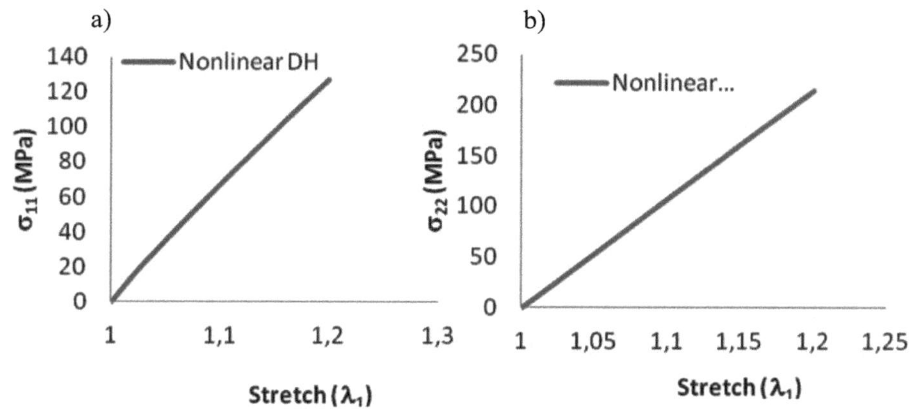

Figure 4.6. *Equi-biaxial tensile Cauchy stress versus stretch for the 3D trabecular bone network a) x-direction and b) y-direction*

The shear stress response shows important geometrical nonlinear effects beyond about 10% shear strain (Figure 4.7), due to a large change in the beam director orientation during shear, in contrast to the equi-biaxial test.

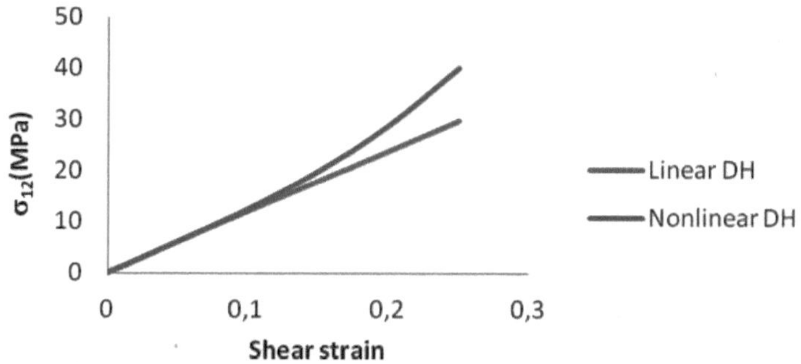

Figure 4.7. *Nonlinear in-plane shear stress response versus applied shear strain for the 3D hexagon. For a color version of this figure, see www.iste.co.uk/ganghoffer/biomechanics.zip*

A relatively good agreement has been obtained between the stress-strain responses predicted by finite elements and those calculated by discrete homogenization [ELN 16], which shows that the developed discrete homogenization method has the ability to deliver the effective response of these structures in the nonlinear regime with a good accuracy and at a low numerical cost.

The displacement distributions computed numerically over the 3D hexagonal unit cell under uniaxial tension in x, y and z, biaxial tension in x-y plane and in-plane shear are shown in Figure 4.8.

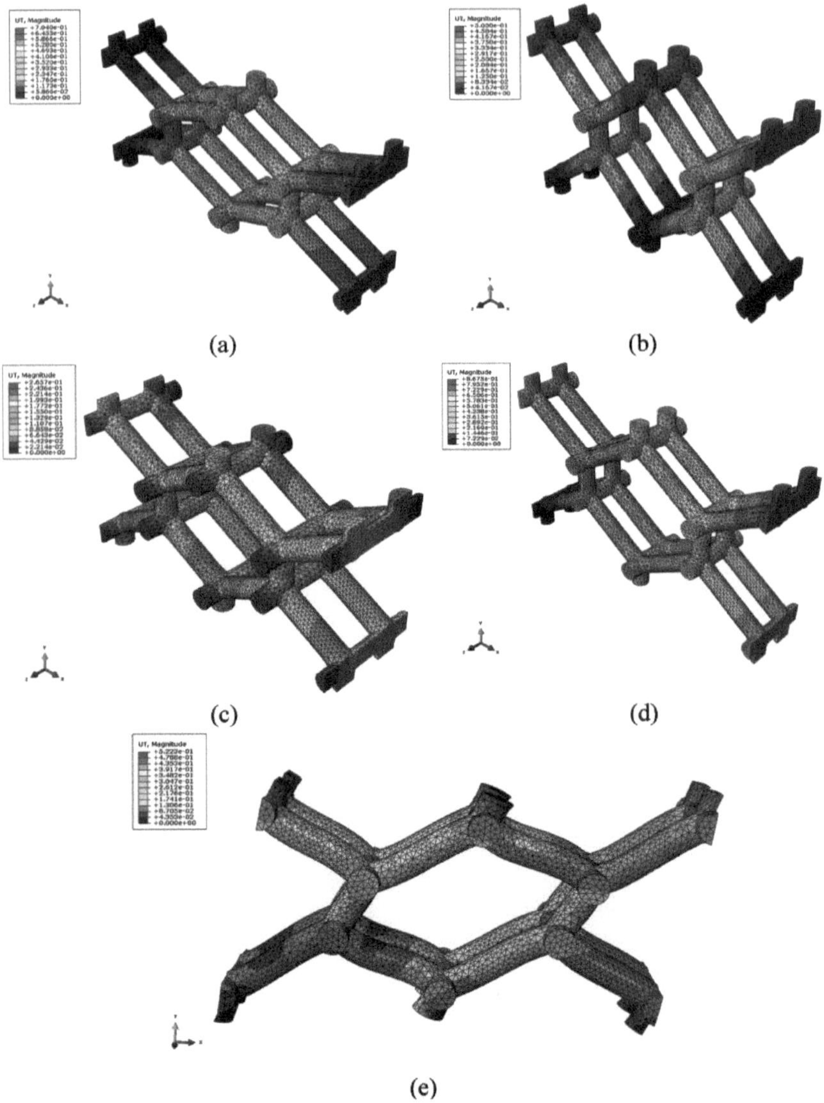

Figure 4.8. *Displacement distributions over the 3D hexagonal unit cell submitted to a uniaxial test in (a) x-direction (b), y-direction (c) z-direction (d) equi-biaxial test in x-y plane (e), and in-plane shear test. For a color version of this figure, see www.iste.co.uk/ganghoffer/biomechanics.zip*

In comparison to fully resolved FE computations, the DH method involves comparatively a much small number of degrees of freedom since the beam degrees of freedom within the unit cell are only defined at its extremity nodes, thus it saves a lot of computational cost to employ the identified homogenized model in structural computations. The proposed homogenization technique has a great versatility as to the unit cell geometry and topology, which makes it a suitable tool to explore and compute the mechanical response of repetitive architectured and network materials in 3D configurations.

5

Mechanics and Thermodynamics of Volumetric and Surface Growth

We provide in this chapter a mechanical and thermodynamical basis for the modeling of volumetric and surface growth of living tissues viewed as continua undergoing changes of mass, shape and properties due to biological evolution. It is usually recognized that soft biological tissues grow by a change of specific volume or density (volumetric growth), whereas hard tissues like bone are prone to surface growth. The field equations of continuum mechanics are extended in order to account for the existing source of mass and mass flux responsible for growth. The thermodynamics of coupled irreversible phenomena is then reviewed, and evolution laws for a growth tensor in the context of volumetric growth are formulated, considering the interactions between the transport of nutrients and the mechanical forces responsible for growth. In the last part, surface growth phenomena are introduced on the basis of preliminary aspects of the differential geometry of surfaces, and further expanded in the framework of the thermodynamics of surfaces. We conclude by a presentation of extremum thermodynamic principles for growing tissues.

5.1. Introduction

Biological evolution entered into the realm of continuum mechanics in the 1990s with attempts to incorporate into a continuum description time-dependent phenomena basically consisting of a variation of material properties, mass and shape of the solid body. One outstanding problem in developmental biology is indeed the

---

Chapter written by Jean-François GANGHOFFER.

understanding of the factors that may promote the generation of biological form, involving the processes of growth (change of mass), remodeling (change of properties), and morphogenesis (shape changes), a classification suggested by Taber [TAB 01].

The main focus in this chapter is the setting up of a modeling platform relying on the thermodynamics of surfaces [GOD 75] and configurational mechanics [CAL 85] for the treatment of surface growth phenomena in a biomechanical context. A typical situation is the external remodeling in long bones, which is induced by genetic and epigenetic factors, such as mechanical and chemical stimulations. The content of the chapter is the following: the thermodynamics of coupled irreversible phenomena is first reviewed, and evolution laws for a growth tensor (the kinematic multiplicative decomposition of the transformation gradient into a growth tensor and an accommodation tensor is adopted) in the context of volumetric growth are formulated, considering the interactions between the transport of nutrients and the mechanical forces responsible for growth. As growth deals with a modification of the internal structure of the body in a changing referential configuration, the language and technique of Eshelbian mechanics [EPS 00] are adopted and the driving forces for growth are identified in terms of suitable Eshelby stresses [AMB 02, LUB 02]. Nevertheless, and as matter of completeness, the balance equations are also formulated in physical space. Considering next surface growth, some preliminary necessary aspects of the differential geometry of surfaces are recalled, and the thermodynamics of surfaces is expounded as a basis for a consistent treatment of phenomena occurring at a growing surface, corresponding to the set of generating cells in a physiological context.

As to notations, vectors and tensors are denoted by boldface symbols. The inner product of two second order tensors is denoted $(\mathbf{A}.\mathbf{B})_{ij} = A_{ik}B_{kj}$; the material gradient is denoted ∇_R, and the actual gradient in physical space ∇. The jump and average operators over a surface of discontinuity $S(t)$ are denoted respectively $[a]_S := a^+ - a^-$ and $\bar{a} := \frac{1}{2}(a^+ + a^-)$, for any scalar or tensorial quantity a; the quantities a^+, a^- are the limit of the otherwise continuous quantity a on both sides of the interface. The unit normal to a surface in the Lagrangian (resp. physical) configuration is denoted $\mathbf{N}$ (resp. $\mathbf{n}$). The volume and surface in the actual (resp. referential) configuration are denoted Ω_t, S_t (resp. Ω_R, S_R). The convention of the summation of the repeated index in a monomial is adopted.

5.2. Thermodynamics of continuous open media: a survey

We consider multicomponent systems, mutually interacting by chemical reactions. Two alternative viewpoints shall be considered: in the first viewpoint, the system is closed, which in consideration of growth phenomena means that the nutrients are included in the overall system. The second point of view is based on the analysis of a solid body as an open system exchanging nutrients with its surrounding; hence growth shall be accounted for by additional source terms and convective fluxes.

We adopt the thermodynamic framework of open systems irreversible thermodynamics, which shall first be exposed in a general setting, and particularized thereafter for growing continuum solid bodies. Recall first that any extensive quantity A with volumetric density $a = a(\mathbf{x},t)$ satisfies a prototype balance law of the form

$$\frac{\partial a(\mathbf{x},t)}{\partial t} = -\nabla.\mathbf{J}_a(\mathbf{x},t) + \sigma_a(\mathbf{x},t) \qquad [5.1]$$

with $\mathbf{J}_a(\mathbf{x},t)$ the flux density of $a(\mathbf{x},t)$ and $\sigma_a(\mathbf{x},t)$ the local production (or destruction) of $a(\mathbf{x},t)$. The particular form of the flux and source depend on the nature of the considered extensive quantity, as shall appear in the forthcoming balance laws. We consider a system including n constituents undergoing r chemical reactions; the local variations of the partial density of a given constituent k, quantity ρ_k, obey the local balance law (Vidal *et al.*)

$$\frac{\partial \rho_k}{\partial t} = -\nabla.(\rho_k \mathbf{u} + \mathbf{J}_k) + M_k \sum_{\alpha=1..r} \nu_{\alpha k} J_\alpha \qquad [5.2]$$

with $\mathbf{u} := \frac{1}{\rho}\sum_{k=1}^{n} \rho_k \mathbf{u}_k$ the local barycentric velocity, M_k the molar mass, and $\nu_{\alpha k}$ the stoechiometric coefficients in the reaction α, such that the variation of the mass dm_k of the species k due to chemical reactions is expressed as

$$dm_k = M_k \sum_{\alpha=1..r} \nu_{\alpha k} \xi_\alpha, \quad \text{k=1..n} \qquad [5.3]$$

wherein ξ_α denotes the degree of advancement of reaction α. The molar masses M_k satisfy the global conservation law (due to Lavoisier)

$$\sum_{k=1}^{n} \nu_{\alpha k} M_k = 0, \quad \alpha = 1..r \qquad [5.4]$$

Observe that the total flux of mass is the sum of a convective flux $\rho_k\mathbf{u}$ and a diffusive flux $\mathbf{J}_k$; the mass production is identified as the contribution $M_k\sum_{\alpha=1..r}\nu_{\alpha k}J_\alpha$. In this viewpoint, the system is in fact closed, since the balance law satisfied by the global density $\rho=\sum_{k=1}^{n}\rho_k$ is written (Vidal *et al.*), accounting for the relation $\sum_{j=1}^{n}\mathbf{J}_k=\sum_{j=1}^{n}\rho_k\mathbf{u}_k=\mathbf{0}$, as

$$\frac{\partial\rho}{\partial t}=-\nabla.(\rho\mathbf{u})+\sum_{k=1}^{n}\sum_{\alpha=1..r}M_k\nu_{\alpha k}J_\alpha\equiv-\nabla.(\rho\mathbf{u}) \qquad [5.5]$$

which does not involve any source term for the total density. Instead of using the partial densities of the system constituents, one can write balance equations for the number of moles of constituent k, $n_k=m_k/M_k$, with m_k the mass of the same constituent. The molar concentration is defined as $C_k=n_k/V$, its inverse being called the partial molar volume. The partial mole number n_k satisfies the balance equation

$$\frac{\partial n_k}{\partial t}=-div\mathbf{J}_k+\frac{\partial_i n_k}{\partial t} \qquad [5.6]$$

with $\mathbf{J}_k$ the flux of species k and $\frac{\partial_i n_k}{\partial t}$ its production term, given by the De Donder definition of the rate of progress of the jth chemical reaction

$$\frac{\partial_i n_k}{\partial t}=\sum_{j=1}^{r}\nu_{kj}\dot{\xi}_j \qquad [5.7]$$

The two previous equalities enter into Gibbs relation as

$$\rho\dot{u}=\theta\rho\dot{s}_e+\theta\rho\dot{s}_i+\boldsymbol{\sigma}:\dot{\boldsymbol{\varepsilon}}-\sum_k\mu_k\frac{\rho}{M}div\mathbf{J}_k+\sum_k\mu_k\frac{\rho}{M}\sum_j\nu_{kj}\dot{\xi}_j \qquad [5.8]$$

with θ the temperature and μ_k the chemical potential of constituent k. The chemical affinity in the sense of De Donder is defined as the force conjugated to

$$A_j=-\sum_k\left(\mu_k\frac{\rho}{M}\right)\nu_{kj}=-\sum_k(\mu_k/V)\nu_{kj} \qquad [5.9]$$

Hence, Gibbs relation can be rewritten in order to highlight the variation of entropy

$$\rho\dot{s}=\frac{1}{\theta}\rho\dot{u}-\frac{1}{\theta}\boldsymbol{\sigma}:\dot{\boldsymbol{\varepsilon}}+\sum_k\frac{1}{\theta}\left(\mu_k\frac{1}{V}\right)div\mathbf{J}_k+\sum_j\frac{1}{\theta}A_j\dot{\xi}_j \qquad [5.10]$$

The local balance of internal energy expresses the first principle of thermodynamics as

$$\rho\dot{u}=-\nabla.\mathbf{J}_q+\rho\dot{w}$$

with $\mathbf{J}_q$ the heat flux, and ρw is relative to all forms of work. We shall isolate the flux-like contributions in the entropy variations, which after a few transformations , is written

$$\rho\dot{s}=\rho\dot{s}_e+\rho\dot{s}_i=\frac{1}{\theta}\nabla.\mathbf{J}_q+\sum_k\nabla.\left(\mathbf{J}_k\frac{\mu_k}{\theta}\frac{1}{V}\right)-\sum_k\mathbf{J}_k.\nabla\left(\frac{\mu_k}{\theta}\frac{1}{V}\right)+\frac{1}{\theta}(\rho\dot{w}-\boldsymbol{\sigma}:\dot{\boldsymbol{\varepsilon}})+\sum_j\frac{1}{\theta}A_j\dot{\xi}_j$$

The contribution $-\boldsymbol{\sigma}:\dot{\boldsymbol{\varepsilon}}/\theta$ (involving the virtual power of internal forces) is further decomposed into

$$\frac{1}{\theta}\boldsymbol{\sigma}:\dot{\boldsymbol{\varepsilon}}=\frac{1}{\theta}\boldsymbol{\sigma}:\nabla\mathbf{u}=\nabla.\left(\frac{1}{\theta}\boldsymbol{\sigma}:\mathbf{u}\right)-\mathbf{u}.\nabla.\left(\frac{1}{\theta}\boldsymbol{\sigma}\right)$$

Hence, the rate of the entropy density decomposes into

$$\rho\dot{s}=\rho\dot{s}_e+\rho\dot{s}_i=$$

$$-\nabla.\left(\mathbf{J}_q+\sum_k\mathbf{J}_k\frac{\mu_k}{\theta}\frac{1}{V}\right)+\mathbf{J}_q.\nabla\left(\frac{1}{\theta}\right)-\sum_k\mathbf{J}_k.\nabla\left(\frac{\mu_k}{\theta}\frac{1}{V}\right)+\frac{1}{\theta}(\rho\dot{w}-\boldsymbol{\sigma}:\dot{\boldsymbol{\varepsilon}})+\sum_j\frac{1}{\theta}A_j\dot{\xi}_j \qquad [5.11]$$

This formulation allows the identification of the divergential contribution to the exchange entropy, hence to the entropy flux

$$\mathbf{J}_s=\mathbf{J}_q+\sum_k\mathbf{J}_k\frac{\mu_k}{\theta}\frac{1}{V} \qquad [5.12]$$

and of the internal entropy production

$$\rho\dot{s}_i=\mathbf{J}_q.\nabla\left(\frac{1}{\theta}\right)-\sum_k\mathbf{J}_k.\nabla\left(\frac{\mu_k}{\theta}\frac{1}{V}\right)-\frac{1}{\theta}(\boldsymbol{\sigma}:\dot{\boldsymbol{\varepsilon}}-\rho\dot{w})+\sum_j\frac{1}{\theta}A_j\dot{\xi}_j \qquad [5.13]$$

which is due to the gradient of intensive variables (temperature, chemical potential), to the irreversible mechanical power spent and to chemical reactions.

An alternative to the previous writing of the internal entropy production bearing the name of Clausius-Duhem inequality is frequently used; as a starting point, the first principle is written as

$$\rho\dot{u} = -\nabla.\mathbf{J}_q + \boldsymbol{\sigma}:\dot{\boldsymbol{\varepsilon}} + \sum_k (\mu_k / V)\dot{n}_k \qquad [5.14]$$

We have assumed in this alternative that the mechanical power $\rho\dot{w} = \boldsymbol{\sigma}^{eq}:\nabla\mathbf{u}$ does not include a flux contribution, hence only the heat diffusion contributes to the flux of internal energy. The contribution $\boldsymbol{\sigma}:\dot{\boldsymbol{\varepsilon}} + \sum_k (\mu_k / V)\dot{n}_k$ is identified with the term $\rho\dot{w}$. The previous equality combined with the second principle, equality $\rho\dot{s} = -\nabla.\left(\frac{\mathbf{J}_q}{\theta}\right) + \rho\dot{s}_i$ (the entropy flux resolves to the sole heat flux), delivers after a few manipulations the variation of the internal energy as

$$\rho\dot{u} = \left(\theta\rho\dot{s} - \mathbf{J}_q\frac{\nabla\theta}{\theta} - T\rho\dot{s}_i\right) + \boldsymbol{\sigma}:\dot{\boldsymbol{\varepsilon}} + \sum_k (\mu_k / V)\dot{n}_k \qquad [5.15]$$

Hence, the internal entropy production is identified as

$$\theta\rho\dot{s}_i = -\rho(\dot{u} - T\dot{s}) - \mathbf{J}_q\frac{\nabla\theta}{\theta} + \boldsymbol{\sigma}:\dot{\boldsymbol{\varepsilon}} + \sum_k (\mu_k / V)\dot{n}_k \qquad [5.16]$$

which is conveniently rewritten in terms of Helmholtz free energy density $\psi := u - Ts$ as

$$\theta\rho\dot{s}_i = -\rho\left(\dot{\psi} + s\dot{\theta}\right) - \mathbf{J}_q.\frac{\nabla\theta}{\theta} + \boldsymbol{\sigma}:\dot{\boldsymbol{\varepsilon}} + \sum_k (\mu_k / V)\dot{n}_k \qquad [5.17]$$

This is at variance with the point of view adopted next, which consists in insulating a growing solid body from the external nutrients, identified as one the chemical species, but accounted for in a global manner as a source term.

5.3. General balance laws accounting for mass production due to growth

In the case of mass being created / resorbed within a solid body considered as an open system from a general thermodynamic point of view, one has to account for a source term π being produced (by a set of generating cells) at each point within the time-varying volume Ω_t ; a convective term is also added, corresponding to the transport of nutrients by the velocity field of the underlying continuum [GAN 11]. The convective flux is locally defined in terms of its surface density as $\mathbf{F}(a) = a\mathbf{v}$ for the quantity a; the overall convective flux of a across the closed surface $\partial\Omega_t$ is expressed then as

$$\phi_{conv}(a) = -\int_{\partial\Omega_t} a(\mathbf{v}-\mathbf{w}).\mathbf{n}dA \qquad [5.18]$$

(the minus accounts for the unit exterior normal $\mathbf{n}$). This diffusive flux corresponds to a macroscopic flux

$$\phi_{diff}(a) = -\int_{\partial\Omega_t} \mathbf{J}(a).\mathbf{n}dA \qquad [5.19]$$

The density of microscopic flux $\mathbf{J}(a)$ is associated to an invisible motion of molecules within a continuum description, hence must be described by a specific constitutive law. It does not depend on the velocity of the points of $\partial\Omega_t$.

The convective derivative along the vector field $\mathbf{w}$ of the field $a = a(\mathbf{x}, t)$ is written

$$\frac{\delta_w a}{\delta t} = \left(\frac{\partial a}{\partial t}\right)_x + \nabla a.\mathbf{w} \qquad [5.20]$$

In the case $\mathbf{w} = \mathbf{v}$ the velocity of the material particles, this delivers the definition of the material (or particular) derivative

$$\frac{da}{dt} \equiv \frac{\delta_v a}{\delta t} = \frac{\delta_w a}{\delta t} + \nabla a.(\mathbf{v}-\mathbf{w}) \qquad [5.21]$$

The derivative of the volume integral $A := \int_{\Omega_t} adx$ is next calculated, according to the Leibniz rule:

$$\frac{D}{Dt}\int_{\Omega_t} adx = \int_{\Omega_t} \frac{\partial a}{\partial t} dx + \int_{\partial\Omega_t} a(\mathbf{w}.\mathbf{n})\, dA \qquad [5.22]$$

with $\mathbf{w}$ the velocity field of the points on $\partial\Omega_t$, which is associated with a variation of the domain occupied by the material points of the growing solid body (Figure 5.1).

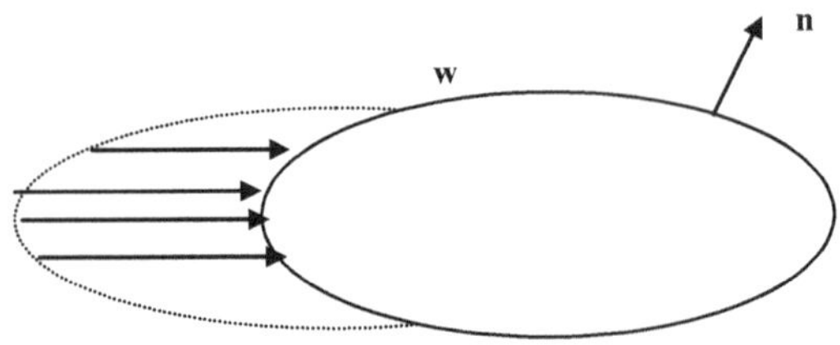

Figure 5.1. *Domain variation due to the virtual velocity field* **w**

A global balance equation can next be written, according to the natural physical rule: the balance of any quantity is the sum of the production / destruction term and of the flux; this yields

$$\frac{D}{Dt}\int_{\Omega_t} a dx = \int_{\Omega_t} \Pi dx - \int_{\partial\Omega_t} \{a(\mathbf{v}-\mathbf{w})+\mathbf{J}(a)\}.\mathbf{n} dA \qquad [5.23]$$

One can see that only the relative velocity of particles w.r.t. the surface velocity matters. Combining this identity with equation [5.5] gives

$$\int_{\Omega_t} \frac{\partial a}{\partial t} dx + \int_{\partial\Omega_t} a(\mathbf{w}.\mathbf{n}) dA = \int_{\Omega_t} \Pi dx - \int_{\partial\Omega_t} \{a(\mathbf{v}-\mathbf{w})+\mathbf{J}(a)\}.\mathbf{n} dA \qquad [5.24]$$

The corresponding local balance law is obtained after elimination of the velocity $\mathbf{w}$ according to the following steps:

$$\int_{\Omega_t} \frac{\partial a}{\partial t} dx + \int_{\partial\Omega_t} a\mathbf{v}.\mathbf{n} dA + \int_{\partial\Omega_t} a(\mathbf{w}-\mathbf{v}).\mathbf{n} dA = \int_{\Omega_t} \Pi dx - \int_{\partial\Omega_t} \{a(\mathbf{v}-\mathbf{w})+\mathbf{J}(a)\}.\mathbf{n} dA$$

$$\int_{\Omega_t} \frac{\delta_v a}{\delta t} dx + \int_{\partial\Omega_t} a\nabla.\mathbf{v} dx = \int_{\Omega_t} \Pi dx - \int_{\partial\Omega_t} \mathbf{J}(a).\mathbf{n} dA$$

Using the last equality, [5.17] is then rewritten as

$$\forall\Omega(t), \quad \int_{\Omega_t} \left(\frac{\delta_v a}{\delta t} + a div\mathbf{v}\right) dx = \int_{\Omega_t} (\Pi - div\mathbf{J}(a)) dx$$

hence giving the general balance law

$$\frac{\delta_v a}{\delta t} + a div \mathbf{v} = \Pi - div \mathbf{J}(a) \Leftrightarrow \frac{\partial a}{\partial t} + div(a\mathbf{v}) = \Pi - div \mathbf{J}(a) \quad [5.25]$$

The mass balance equation is deduced from the identification $a = \rho$, the actual density. Hence, [5.23] gives

$$\frac{D}{Dt}\int_{\Omega_t} \rho dx = \int_{\Omega_t} \pi dx - \int_{\partial\Omega_t} \mathbf{J}(\rho).\mathbf{n} dA \quad [5.26]$$

The first term on the r.h.s. corresponds to mass production, the second contribution to convection of the produced mass through the boundary $\partial\Omega(t)$, and the third contribution to diffusion through the boundary of the moving volume $\partial\Omega(t)$. The strong form of the balance law of mass is written finally

$$\frac{\delta_v \rho}{\delta t} = \Pi - div \mathbf{J}(\rho) - \rho \nabla.\mathbf{v} \quad [5.27]$$

The balance laws for mass, momentum, energy and entropy are next obtained from those general formulas for a solid body exchanging mass with its environment.

The *mass balance in Eulerian format* is given in terms of the actual density ρ by the following reasoning: we first write the general form of the balance of mass in physical space as

$$\frac{D}{Dt}\int_{\Omega_t} \rho dx = \int_{\Omega_t} \left(\frac{D\rho}{Dt} + \rho \nabla.\mathbf{v} \right) dx = \int_{\Omega_t} \pi dx + \int_{\partial\Omega_t} m ds \equiv \int_{\Omega_t} \Gamma \rho dx \quad [5.28]$$

with $\rho(\mathbf{x}, t)$ the actual density, π the physical source of mass, and $m := \mathbf{m}.\mathbf{n}$ the scalar physical mass flux across the boundary, projection of the flux (vector) $\mathbf{m}$. The previous balance law is quite general, as we account for both the variation of the integration volume through the term $\rho \nabla.\mathbf{v}$, and for the source and flux of mass reflected by the right hand side of [5.28]. Localization of the previous integral equation gives

$$\frac{D\rho}{Dt} = \pi + \nabla.\mathbf{m} - \rho \nabla.\mathbf{v} \quad [5.29]$$

with $\mathbf{v}(\mathbf{x},t) := \left(\dfrac{\partial \mathbf{x}}{\partial t}\right)_X$ the Eulerian velocity, which proves identical to [5.10]. The same balance law has been obtained in [EPS 00] starting from its Lagrangian counterpart.

In the sequel, we shall need the following expression of the material derivative of integrals of specific quantities (defined per unit mass) $a = a(\mathbf{x},t)$, obtained using the mass balance [5.29]

$$\frac{D}{Dt}\int_{\Omega_t} \rho a dx = \int_{\Omega_t} \left\{ \rho \frac{Da}{Dt} + a(\pi + \nabla.\mathbf{m}) \right\} dx \qquad [5.30]$$

The comparison of [5.27] with [5.29] gives the identification of fluxes $\mathbf{J}(\rho) = -\mathbf{m}$; the balance law is further consistent with (and equivalent to) the equation [GAN 05]

$$\dot{\rho} + \rho div(\mathbf{v}) = \Phi_\rho + \sigma_\rho$$

with $\Phi_\rho \equiv \nabla.\mathbf{m}$ the total flux of conduction and $\sigma_\rho \equiv \pi$ the volumetric source of mass. Observe the difference with the treatment of section 5.2 considering overall a closed system with no internal sources, reflected by equation [5.5]: this first point of view considers the nutrients responsible for growth as part of the system, whereas they appear as external sources in the second viewpoint.

Expressing the total mass of the domain Ω_t as $m(\Omega_t) = \int_{\Omega_t} \rho(x)dx$, the mass variation due to the transport phenomena is written as the following integral accounting for source terms, allowing the identification of the production term

$$\left(\frac{dm}{dt}\right)_{source} := \int_{\Omega_t} \pi dx = \int_{\Omega_t} dx \Rightarrow \pi = \Gamma\rho$$

The time variation of chemical concentration of nutrients is due to exchange through the boundary accounted for by the term $\int_{\partial\Omega_R} \mathbf{J}_\mu.\mathbf{N}dA$ and to a source term due to growth

$$\int_{\Omega_t} \Gamma\rho dx = \int_{\Omega_t} Tr\left(\dot{\mathbf{F}}_g.\mathbf{F}_g^{-1}\right)\rho dx$$

hence (see [5.28])

$$\frac{D}{Dt}\int_{\Omega_t}\rho dx=\int_{\Omega_t}Tr\left(\dot{\mathbf{F}}_g.\mathbf{F}_g^{-1}\right)\rho dx=\int_{\Omega_t}\rho\dot{n}_k dx+\int_{\partial\Omega_t}\mathbf{J}_k.\mathbf{N}ds$$
$$\rightarrow \rho\dot{n}_k+Div\mathbf{J}_k=\rho\left(\dot{\mathbf{F}}_g.\mathbf{F}_g^{-1}:\mathbf{I}\right)=\rho\Gamma \qquad [5.31]$$

The last equality is nothing else than $\Gamma=(\pi+\nabla.\mathbf{m})/\rho$ – a consequence of [5.28] – expressed in material format, with the identifications

$$\Gamma:=Tr\left(\mathbf{D}_g\right),\mathbf{m}_i=\mathbf{J}_i;\ \ \pi=\rho\dot{n}_k$$

The global form (first equality in

$$\frac{D}{Dt}\int_{\Omega_R}\rho dx=\frac{D}{Dt}\int_{\Omega_R}\rho JdX=\int_{\Omega_R}\frac{D}{Dt}(\rho J)\,dX=\int_{\Omega_R}\Gamma\rho JdX=\int_{\Omega_t}\Gamma\rho dx\,)$$

also fits within the general balance law for an open system [5.28].

The *Eulerian version* of the balance of momentum is written [EPS 00]

$$\frac{D}{Dt}\int_{\Omega_t}\rho\mathbf{v}dx=\int_{\Omega_t}\mathbf{f}dx+\int_{\partial\Omega_t}\mathbf{n}.\boldsymbol{\sigma}d\sigma_t+\int_{\Omega_t}\pi\mathbf{v}dx+\int_{\partial\Omega_t}\mathbf{n}.(\mathbf{m}\otimes\mathbf{v})\,d\sigma_t \qquad [5.32]$$

with $\boldsymbol{\sigma}$ Cauchy stress and $\mathbf{f}$ body forces per unit physical volume. Localizing previous balance law gives, using the mass balance [5.29],

$$\rho\frac{D\mathbf{v}}{Dt}=\mathbf{f}+div\boldsymbol{\sigma}+(\mathbf{m}.\nabla)\,\mathbf{v} \qquad [5.33]$$

For material systems that exchange work and mass with their surroundings, the material derivative of the internal energy u per unit actual volume expresses as [MUN 66]

$$\dot{u}=\nabla.\left(\mathbf{J}_q\right)-p_i+\mathbf{J}_k.\mathbf{F}_k \qquad [5.34]$$

with $\mathbf{J}_q$ the heat diffusion flux (that shall obey a constitutive relationship vs. the temperature θ), $p_i=-\boldsymbol{\sigma}:\mathbf{D}(\mathbf{u})$ the volumetric density of the power of internal forces (expressed here with an Eulerian description, with $\boldsymbol{\sigma}$ the Cauchy stress), $\mathbf{J}_k$ the diffusion flux of the k-species, and $\mathbf{F}_k$ an external force acting on the k-species.

The first law of thermodynamics for an open system has to account for the contributions to kinetic and internal energies due to the incoming material. Denoting u the specific internal energy density, one may write the energy balance in the actual configuration as

$$\begin{aligned}\frac{D}{Dt}\int_{\Omega_t}\rho\left(u+\frac{1}{2}\mathbf{v}^2\right)dx &= \int_{\Omega_t}\left\{(\mathbf{f}.\mathbf{v}+r)+\pi\left(u+\frac{1}{2}\mathbf{v}^2\right)\right\}dx \\ &+\int_{\partial\Omega_t}\mathbf{n}.\left\{\boldsymbol{\sigma}.\mathbf{v}+\mathbf{m}\left(u+\frac{1}{2}\mathbf{v}^2\right)-\mathbf{q}\right\}d\sigma(\mathbf{x})\end{aligned} \quad [5.35]$$

with r the volumetric heat supply (generated by growth), and $\mathbf{q}$ the heat flux across $\partial\Omega_t$. This writing of the energy balance can be simplified using the balance of kinetic energy with volumetric density k, obtained by multiplying [5.33] by the velocity and integrating over Ω_t, hence

$$k := \frac{1}{2}\rho\frac{D\mathbf{v}^2}{Dt} = \mathbf{f}.\mathbf{v}+\mathbf{v}.div\boldsymbol{\sigma}+\mathbf{v}.(\mathbf{m}.\nabla)\,\mathbf{v} \equiv \mathbf{f}.\mathbf{v}+\mathbf{v}.div\boldsymbol{\sigma}+\mathbf{m}.\nabla\frac{\mathbf{v}^2}{2} \Rightarrow$$

$$\int_{\Omega_t}\frac{1}{2}\rho\frac{D\mathbf{v}^2}{Dt}dx = \int_{\Omega_t}\left(\mathbf{f}.\mathbf{v}+\mathbf{v}.div\boldsymbol{\sigma}+\mathbf{m}.\nabla\left(\frac{\mathbf{v}^2}{2}\right)\right)dx$$

The left hand side of the previous equality can be expressed versus the material derivative of the total kinetic energy of the growing body, using the general equality [5.30] with $\frac{1}{2}\mathbf{v}^2$, hence

$$\begin{aligned}\frac{DK}{Dt} &= \int_{\Omega_t}\left\{\rho\frac{D\left(\frac{\mathbf{v}^2}{2}\right)}{Dt}+\frac{\mathbf{v}^2}{2}(\pi+\nabla.\mathbf{m})\right\}dx = \int_{\Omega_t}\left(\mathbf{f}.\mathbf{v}+\mathbf{v}.div\boldsymbol{\sigma}+\nabla.\left(\mathbf{m}\frac{\mathbf{v}^2}{2}\right)+\pi\frac{\mathbf{v}^2}{2}\right)dx = \\ &= \int_{\Omega_t}\left(\mathbf{f}.\mathbf{v}-\boldsymbol{\sigma}:\nabla\mathbf{v}+\pi\frac{\mathbf{v}^2}{2}\right)dx+\int_{\partial\Omega_t}\mathbf{n}.\left\{\boldsymbol{\sigma}.\mathbf{v}+\mathbf{m}\frac{\mathbf{v}^2}{2}\right\}d\sigma\end{aligned} \quad [5.36]$$

Again using [5.13] delivers similarly the material derivative of the total energy (left-hand side in [5.18]) as (the total internal energy is denoted U)

$$\frac{D}{Dt}(U+K)=\int_{\Omega_t}\rho\frac{D}{Dt}\left(u+\frac{1}{2}\mathbf{v}^2\right)dx+\int_{\Omega_t}\left(u+\frac{1}{2}\mathbf{v}^2\right)(\pi+\nabla.\mathbf{m})\,dx=$$
$$=\int_{\Omega_t}\left\{(\mathbf{f}.\mathbf{v}+r)+\pi\left(u+\frac{1}{2}\mathbf{v}^2\right)\right\}dx+\int_{\partial\Omega_t}\mathbf{n}.\left\{\boldsymbol{\sigma}.\mathbf{v}+\mathbf{m}\left(u+\frac{1}{2}\mathbf{v}^2\right)-\mathbf{q}\right\}d\sigma(x)$$

Using the balance of kinetic energy [5.19] allows us to isolate the material derivative of the internal energy

$$\frac{DU}{Dt}=\int_{\Omega_t}(-\boldsymbol{\sigma}:\nabla\mathbf{v}+r+\pi u)dx+\int_{\partial\Omega_t}\mathbf{n}.(\mathbf{m}u-\mathbf{q})d\sigma(x)$$

Its strong form is given by localization using the general equality [5.30] with the identification $a\equiv u$

$$\rho\frac{Du}{Dt}=-\boldsymbol{\sigma}:\nabla\mathbf{v}+r+\mathbf{m}\nabla.u-\nabla.\mathbf{q} \qquad [5.37]$$

The Lagrangian counterpart of previous balance law has been expressed in [EPS 00]; the dissipation inequality is written in global form as

$$\frac{D}{Dt}\int_{\Omega_t}\rho s dx\geq\int_{\Omega_t}\left(\pi s+\theta^{-1}r\right)dx-\int_{\partial\Omega_t}\mathbf{n}.\frac{\mathbf{q}}{\theta}d\sigma(\mathbf{x})\Rightarrow$$
$$\int_{\Omega_t}\left(s(\pi+\nabla.\mathbf{m})+\rho\frac{Ds}{Dt}\right)dx\geq\int_{\Omega_t}\left(\pi s+\theta^{-1}r\right)dx-\int_{\partial\Omega_t}\mathbf{n}.\frac{\mathbf{q}}{\theta}d\sigma(\mathbf{x})\Rightarrow$$
$$\int_{\Omega_t}\left(\rho\frac{Ds}{Dt}+s\nabla.\mathbf{m}\right)dx\geq\int_{\Omega_t}\left(\theta^{-1}r\right)dx-\int_{\partial\Omega_t}\mathbf{n}.\frac{\mathbf{q}}{\theta}d\sigma(\mathbf{x})$$

Hence, the local dissipation inequality is written

$$\rho\frac{Ds}{Dt}\geq\theta^{-1}r-div\left(\frac{\mathbf{q}}{\theta}\right)-s\nabla.\mathbf{m} \qquad [5.38]$$

The formulated balance laws for growing solid bodies have to be complemented by the description of the kinematics of growth, involving a specific growth mapping considered as an internal variable obeying an evolution law versus a suitable driving force identified from the writing of the local dissipation; this is the object of the next paragraph.

5.4. Growth kinematics and growth models

The kinematics of growth in a large strains formalism is elaborated from the classical multiplicative decomposition of the transformation gradient introduced in [ROD 94]

$$\mathbf{F} = \nabla_X \mathbf{x}(\mathbf{X},t) \rightarrow J := \det(\mathbf{F}) \qquad [5.39]$$

with $\mathbf{X},\mathbf{x}$ the Lagrangian and Eulerian positions respectively in the referential and actual configurations denoted Ω_R,Ω_t respectively, as the product of the growth deformation gradient $\mathbf{F}_g$ and the growth accommodation mapping $\mathbf{F}_a$

$$\mathbf{F} = \mathbf{F}_a.\mathbf{F}_g \qquad [5.40]$$

The transformation gradients $\mathbf{F}_a,\mathbf{F}_g,\mathbf{F}$ define the mappings of the tangent spaces to the various configurations. The Jacobean of the growth mapping informs us about the nature of growth:

$$J_g := \det(\mathbf{F}_g) \qquad [5.41]$$

Hence $J_g < 1$ describes growth, whereas $J_g > 1$ represents resorption. Growth essentially occurs between the referential and the actual configurations.

Adopting the framework of hyperelasticity, the first Piola-Kirchhoff stress $\mathbf{P}$ expresses from the strain energy density per unit volume in the reference configuration $W(\mathbf{F};\mathbf{X})$ (a possible explicit dependence upon the Lagrangian variable is included for heterogeneous media) as

$$\mathbf{P} := \partial_{\mathbf{F}} W(\mathbf{F}_a;\mathbf{X})$$

The balance of mass is expressed in both weak and strong forms as

$$\frac{D}{Dt}\int_{\Omega_t} \rho dx = \int_{\Omega_R} Tr\left(\dot{\mathbf{F}}_g.\mathbf{F}_g^{-1}\right)\rho J dX = \int_{\Omega_R} \rho J \dot{n}_k dX + \int_{\partial\Omega_R} \mathbf{J}_k.\mathbf{N} dA$$
$$\rightarrow \rho J \dot{n}_k + Div\mathbf{J}_k = \rho J\left(\dot{\mathbf{F}}_g.\mathbf{F}_g^{-1} : \mathbf{I}\right) = \Gamma \rho J$$

The Coleman Noll procedure then leads to the constitutive equations for the first Piola-Kirchhoff stress and the chemical potential

$$\mathbf{T} = \rho J \frac{\partial \psi}{\partial \mathbf{F}_a} . \mathbf{F}_g^{-t}; \quad \mu_k = \frac{\partial \psi}{\partial n_k}$$

The residual dissipation then splits into the sum of a *chemical dissipation* and the *intrinsic (mechanical) dissipation*

$$0 \le -\mathbf{J}_i Grad \mu_i + \rho J \left(\mathbf{F}_a^t . \frac{\partial \psi}{\partial \mathbf{F}_a} - (\psi + \mu_k)\mathbf{I} \right) : \mathbf{L}_g$$

Following Curie's principle, we are then entitled to write a general growth model as a functional relation between the growth velocity gradient $\mathbf{L}_g := \dot{\mathbf{F}}_g . \mathbf{F}_g^{-1}$ and the Eshelby stress

$$\tilde{\mathbf{\Sigma}}_a := \rho \mathbf{F}_a^t . \frac{\partial \psi}{\partial \mathbf{F}_a} - \rho(\psi + \mu_k)\mathbf{I}$$

of the general form

$$\mathbf{L}_g = f\left(\tilde{\mathbf{\Sigma}}_a\right)$$

The next sections are devoted to surface growth.

5.5. Mechanics and thermodynamics of surface growth

Surface growth is presently described as the motion of a moving interface representing physically the generating cells, separating a zone not yet affected by growth from a domain in which growth has occurred. A typical problem in mechanobiology is the evolution of the growth plate, a transition region connecting the metaphyseal bone and the epiphyseal bone, viewed as a singular surface of vanishing small thickness. Due to a flux of mass across the moving interface, a jump of density, momentum, energy and entropy occurs. A limit situation is that of a moving external surface, corresponding to external remodeling, a situation occurring in bones. The jump conditions over the moving front are expressed in both physical and material format. Surface growth can be envisaged as a production of new material inheriting the kinematics of the 'old' material, a situation of smooth growth. Material properties and density are generally changed at a material point swept by the interface; hence they undergo a jump for an interface viewed as purely geometrical entity. The writing of the jump of the internal entropy production allows the identification of a growth model relating the jump of the surface growth velocity to conjugated driving forces involving the combined jump of the free energy density and of the chemical potential.

5.5.1. *Kinematics of a growing interface and Hadamard relations*

The point of view advocated in this section is of surface growth occurring as the motion of a surface within a solid body, across which the fields experience discontinuities under a flux of nutrients leading to a mass flux. Our kinematic point of view at the base of the forthcoming developments is first exposed. Adopting a macroscopic viewpoint, a typical example in mechanobiology is the propagation of a thin transition zone called the growth plate in long bones, connecting the metaphyseal bone and the epiphyseal bone. The growth plate joins bone to cartilage, and is accordingly also called the chondro-osseous junction. This plate witnesses a competition between proliferation of chondrocytes and the ossification process, and can be considered at first sight as a singular surface moving in a stationary manner. It is clear that this picture is a simplified view, as the real growth plate has a finite thickness and is endowed with a complex microstructure.

The physical picture underlying the present vision of surface growth is the following: a solid body is swept by an internal interface delimiting a volume of the newly grown material from the volume of the material not yet affected by growth. The moving front generates new material properties and a new density, resulting from a mass flux across the front having as physical origin the transport of nutrients from the bulk. The jump conditions represent the singular part of the volumetric growth conditions (balance equations) over the propagating surface. Two cases can be considered: fixed growth surface and moving growth surface [SKA 82, SKA 97].

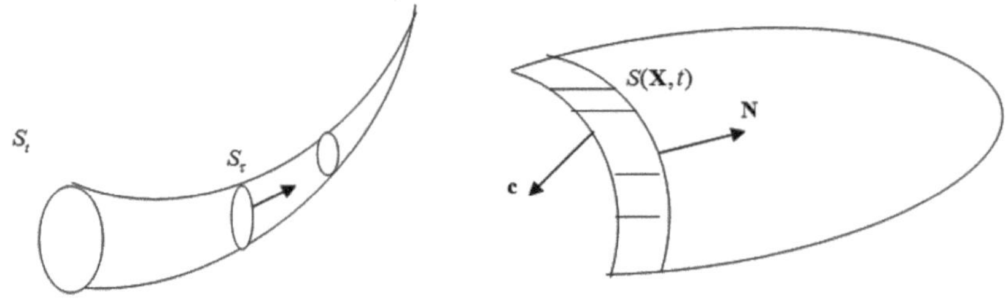

Figure 5.2. *Surface growth and geometrical domain. The (geometrical) velocity of the moving surface is denoted c. The moving interface can be viewed (right picture) as a region with a vanishing thickness.*

The surface growth (generating cells) is moving to the left (Figure 5.2) from an initial surface S_0 (at initial time); the position of the surface at time τ (denoted as S_τ) is such that it divides the total volume (at time t) into a subregion already affected by growth (in the direction of the normal on Figure 1) and a region not yet affected (in the direction of vector $\mathbf{c}$). One assumes that the newly produced material inherits the deformation and stress state of the already existing grown material, a case referred to as *smooth surface growth*. On the other hand, as the interface is a zone of discontinuity of the fields (density, kinematic and static variables) due to the mass flux it experiences, this is the source of a kinematic incompatibility generating stresses.

The growth surface can in a general situation be conveniently parameterized by convected curvilinear coordinates θ_1, θ_2, which locate a given cell on this surface. The trajectory of any particle on this surface can be generated by a third curvilinear coordinate $\theta_3(t)$, function of time, such that the growth surface has a motion described by an equation of the form

$$\theta_3(t) = f(t) \qquad [5.42]$$

A one to one mapping is supposed to exist between the Cartesian coordinates of any point on the moving surface and the curvilinear coordinates,

$$x_{Gi}(t) = x_i\left(\theta_1, \theta_2, \tau, t\right) \equiv x_i\left(\theta_1, \theta_2, \theta_3(t)\right)$$

with $\tau \le t$ a past instant in the first equality, describing the surface made of the material points generated at a time $t = \tau$. Hence, the material velocity can be written as

$$\dot{X}_{mi} = \left(\frac{\partial x_i}{\partial t}\right)_{\theta_i} \qquad [5.43]$$

wherein the curvilinear coordinates are held fixed to follow the same particle.

For a mobile interface separating two media, having a normal velocity $\mathbf{c} = c_N \mathbf{N}$, the normal to the propagating surface is defined from the equation of the surface $S(\mathbf{X}, t) = 0$, with $\mathbf{X} = \mathbf{X}\left(\theta_1, \theta_2, t\right)$ the coordinate of a material point constrained to lie on the surface. Deriving the equation $S(\mathbf{X}, t) = 0$ with respect to time shows the surface equation has a zero convected derivative (derivative of S following the vector field $\dot{\mathbf{X}} = \left(\frac{\partial \mathbf{X}}{\partial t}\right)_{\theta_1,\theta_2}$), viz.

$$\frac{\delta S}{\delta t} = \frac{\partial S}{\partial \mathbf{X}}.\dot{\mathbf{X}} + \left(\frac{\partial S}{\partial t}\right)_{\mathbf{X}} = 0 \qquad [5.44]$$

Since the vector $\frac{\partial S}{\partial \mathbf{X}}$ is orthogonal to the surface S, hence the unit normal vector to S is expressed as (this form will be made more specific in what follows, considering either the material or physical space, with specific notation for the unit normal)

$$\mathbf{N} = \frac{\partial S}{\partial \mathbf{X}} / \left\| \frac{\partial S}{\partial \mathbf{X}} \right\| \qquad [5.45]$$

Hence, the normal interface velocity $\mathbf{C}$ is given in Lagrangian format in terms of its normal component $C_N := \mathbf{C}.\mathbf{N}$, such that

$$C_N = -\frac{1}{\|Grad S\|}\left(\frac{\partial S}{\partial t}\right)_{\mathbf{X}} \qquad [5.46]$$

For any differentiable field f, the convected derivative is elaborated as

$$\frac{\delta f}{\delta t} = \frac{\partial f}{\partial t} + C_N \nabla f.\mathbf{N} \qquad [5.47]$$

Let the moving surface (in a general case) of generating cells have the velocity $\mathbf{c} = c\mathbf{n}$, with matter flowing from this surface with the (Eulerian) velocity $\mathbf{v}$, hence the growth velocity is the relative velocity of material points with respect to the moving surface, defined in Eulerian format as

$$\mathbf{v}_g := \mathbf{v} - \mathbf{c} \qquad [5.48]$$

One has to distinguish the Lagrangian from the Eulerian interface velocity, respectively denoted $\mathbf{C}$ and $\mathbf{c}$, representing the speed of propagation of the surface $S(\mathbf{X}, t) = 0$ in the reference and actual configuration.

The discontinuity of the displacement gradient results from the continuity of the displacement

$$[\mathbf{u}]_S = \mathbf{0} \rightarrow [\nabla \mathbf{u}]_S .\mathbf{e}_\alpha = \mathbf{0} \qquad [5.49]$$

with $(\mathbf{e}_\alpha)_{\alpha=1,2}$ unit tangent vectors in the tangent plane to S; this means that the discontinuity of the gradient has to satisfy

$$[\nabla \mathbf{u}]_S = \mathbf{U}(\mathbf{x}) \otimes \mathbf{N} \qquad [5.50]$$

with $\mathbf{U}(\mathbf{x})$ an arbitrary vector field. Condition [5.9] is related to the Hadamard compatibility condition, giving the jump of the velocity field in terms of the convected derivative of any function $f(\mathbf{x},t)$, defined as

$$\frac{\delta_c f}{\delta t} := \lim_{\varepsilon \to 0} \frac{f(\mathbf{x} + \varepsilon c_n \mathbf{n}, t + \varepsilon) - f(\mathbf{x},t)}{\varepsilon} \qquad [5.51]$$

Hence, we obtain

$$\frac{\delta_c ([\mathbf{u}]_S)}{\delta t} = [\mathbf{v}] + c_N [grad\mathbf{U}].\mathbf{N} = \mathbf{0} \rightarrow [\mathbf{v}] = -C_N [grad\mathbf{U}].\mathbf{N} = -C_N [\mathbf{F}.\mathbf{N}] \qquad [5.52]$$

with $\mathbf{U}$ the displacement field (the second equality is due to the continuity of the geometrical velocity of the moving front and the definition of the growth velocity). The last equality in [5.52] is recognized as the Hadamard compatibility condition.

Note that the surface gradient of any vectorial quantity is continuous over the interface. We shall herewith consider coherent propagating fronts, for which the inverse (or material) velocity $\mathbf{V}$ is continuous; recall that at all material points, the physical and material velocities are related by the well-known kinematic relation

$$\mathbf{v} + \mathbf{F}.\mathbf{V} = \mathbf{0} \qquad [5.53]$$

5.5.2. *Jump conditions in the presence of a moving surface of discontinuity*

As all balance laws involve specific quantities (defined per unit mass), one has to consider the material derivative of integrals defined from an extensive quantity with specific density a; when the considered volume $\Omega(t)$ is crossed by a discontinuity surface $S(t)$ considered as a geometrical entity (not endowed with a specific material behavior), the global balance law previously written for smooth fields not crossing such a surface of discontinuity extends to [GAN 11]

$$\frac{D}{Dt}\int_{\Omega_t}\rho a dx = \int_{\Omega_t}\left\{\frac{D}{Dt}(\rho a)+\rho a div(\mathbf{v})\right\}dx+\int_{S_t}\left[\rho a v_{gn}\right]_S dS = \\ = \int_{\Omega_t}\left\{\frac{\partial(\rho a)}{\partial t}+div(\rho a\mathbf{v})\right\}dx+\int_{S_t}\left[\rho a v_{gn}\right]_S dS \quad [5.54]$$

with $v_{gn} := \mathbf{v}_g.\mathbf{n}$ the normal growth velocity, projection of the growth velocity $\mathbf{v}_g := \mathbf{v}-\mathbf{c}$ onto the unit exterior normal $\mathbf{n}$ to the moving interface $S(t)$. For the material domain $\Omega(t)$, the general balance law [5.6] further allows us to write

$$\int_{\Omega_t}\left\{\frac{\partial(\rho a)}{\partial t}+div(\rho a\mathbf{v})\right\}dx+\int_{S_t}\left[\rho a v_{gN}\right]_S dS = -\int_{\partial\Omega_t}\mathbf{J}(a).\mathbf{n}d\sigma+\int_{\Omega_t}\pi dx \quad [5.55]$$

The flux integral $\int_{\partial\Omega}\mathbf{J}(a).\mathbf{n}d\sigma$ is further rewritten in order to account for the flux discontinuity over $S(t)$ as

$$\int_{\partial\Omega_t}\mathbf{J}(a).\mathbf{n}d\sigma = \int_{S_t}\left[\mathbf{J}(a).\mathbf{n}\right]_S d\sigma+\int_{\Omega_t}div\mathbf{J}(a)dx \quad [5.56]$$

Inserting [5.56] into [5.55] then delivers the general balance law

$$\int_{\Omega_t}\left\{\frac{\partial(\rho a)}{\partial t}+div(\rho a\mathbf{v})+div(\mathbf{J}(a))-\pi\right\}dx+\int_{S_t}\left[\rho a v_{gN}+\mathbf{J}(a).\mathbf{n}\right]_S dS = 0 \quad [5.57]$$

Identity [5.57] remains valid for any subvolume $\Omega'_t \subset \Omega_t$ enclosing the discontinuity surface $S'_t \subset S_t$; the general form [5.54] of a balance law further shows that the volume integral in [5.57] vanishes; assuming further the continuity of the jump $\left[\rho a v_{rN}+\mathbf{J}(a).\mathbf{n}\right]_{S'}$, and due to the arbitrariness of the subsurface S'_t, the singular part of [5.57] vanishes, viz.

$$\left[\rho a v_{gn}+\mathbf{J}(a).\mathbf{n}\right]_{S_t} = 0 \quad [5.58]$$

This equality means that the total flux (including diffusive and convective contributions) does not suffer any jump. Previous jump condition can be extended to vector quantities as

$$\left[\left(\rho \mathbf{a} \otimes \mathbf{v}_g + \boldsymbol{\varphi}_d\right).\mathbf{n}\right]_{S_t} = \mathbf{0} \qquad [5.59]$$

with $\boldsymbol{\varphi}_d$ the second order tensor of diffusive flux (tensorial generalization of previous $\mathbf{J}(a)$). Observe that the source terms are not involved in the jump condition.

The Lagrangian version of the jump condition is obtained from the Piola transforms of volume and area elements $dx = JdX$, $\mathbf{n}ds = J.\mathbf{F}^{-t}.\mathbf{N}dS$ (Nanson's formula), introducing the Lagrangian diffusion flux $\mathbf{J}_R(a_R) := J\mathbf{F}^{-1}.\mathbf{J}(a)$, the Lagrangian volumetric production term $\Pi := J\pi$, and the Lagrangian density $Ja := a_R$. The pull-back of the relative velocity is here introduced, viz. the vector $-\mathbf{F}^{-1}.(\mathbf{v}-\mathbf{c})$, which can further be identified to the referential inverse relative velocity [EPS 00]

$$\mathbf{V}_r(\mathbf{x},t) \equiv \left(\frac{\partial \boldsymbol{\varphi}^{-1}}{\partial t}\right)_x := -\mathbf{F}^{-1}.(\mathbf{v}-\mathbf{c}) \equiv \mathbf{V}_g \qquad [5.60]$$

Accordingly, the jump condition in Lagrangian format is then deduced by elementary linear algebra as

$$\left[\left\{\rho_R a_R \mathbf{V}_g - \mathbf{J}_R(a_R)\right\}.\mathbf{N}\right]_{S_R} = 0 \qquad [5.61a]$$

This jump condition be extended to vector quantities as

$$\left[\left(\rho_R \mathbf{a}_R \otimes \mathbf{V}_g - \boldsymbol{\varphi}_{dR}\right).\mathbf{N}\right]_{S_R} = \mathbf{0} \qquad [5.61b]$$

with $\boldsymbol{\varphi}_d$ the second order tensor of diffusive flux (the tensor generalization of previous flux $\mathbf{J}_R(a_R)$). Observe that the source terms are not involved in either jump conditions.

The balance laws for mass, momentum, energy and entropy are next obtained as the specialization of the previous general expressions, together with the corresponding jump conditions, for a solid body exchanging mass with its environment.

The jump condition associated to the mass balance in the *Eulerian version* is thus obtained using the balance law formulated in [5.54] and from the general jump condition [5.61a], with the identification

$$a = 1, \mathbf{J}(a) = -\mathbf{m} \rightarrow \left[\rho v_{gn} - \mathbf{m.n}\right]_{S_t} = 0 \leftrightarrow \left[\rho v_{gn}\right]_{S_t} = m \qquad [5.62]$$

with $m := \mathbf{m.n}$ the scalar mass flux.

The material version of the mass jump condition results from the general formula [5.61a] and the identification

$$a_R = 1, \mathbf{J}(a_R) = \mathbf{M} \rightarrow \left[\rho_R V_{gN} + \mathbf{M.N}\right]_{S_R} = 0 \Rightarrow \left[\rho_R V_{gN}\right]_{S_R} = -M \qquad [5.63]$$

with $\rho_R(\mathbf{X},t)$ the referential density, Π, M the volumetric source of mass and scalar mass flux through the boundary $\partial\Omega_R$, related to their physical counterpart by the formula

$$\mathbf{m} = J^{-1}\mathbf{F.M}, \ \pi = J^{-1}\Pi \qquad [5.64]$$

Generating cells located on the discontinuity surface delivers a flux of mass due to the growth velocity according to

$$\mathbf{M} = \rho_S \mathbf{V}_g \qquad [5.65]$$

with ρ_S the mass density within a thin layer of generating cells (it may be considered as a surface density in the limit of a very thin transition zone), with $\mathbf{C} \equiv \mathbf{V}_g$ viewed as a surface growth velocity.

The extension of the Eulerian balance of momentum with a surface of discontinuity leads to the jump relation (with the identification $a \equiv \rho\mathbf{v};\ \boldsymbol{\varphi}_d(\mathbf{v}) \equiv -(\mathbf{m}\otimes\mathbf{v}+\boldsymbol{\sigma})$) in Eulerian format

$$\left[\left(\rho\mathbf{v}\otimes\mathbf{v}_g - (\mathbf{m}\otimes\mathbf{v}+\boldsymbol{\sigma})\right).\mathbf{n}\right]_{S_t} = \mathbf{0} \qquad [5.66]$$

The Lagrangian format of the balance of momentum directly follows from the previous general treatment of discontinuities, from the identification

$$a_R := J\rho\mathbf{v}\,;\quad \mathbf{J}_R(a_R) := J\mathbf{F}^{-1}.\boldsymbol{\varphi}_d(\mathbf{v}) \equiv -J\mathbf{F}^{-1}.(\mathbf{m}\otimes\mathbf{v}+\boldsymbol{\sigma}) = -\mathbf{M}\otimes\mathbf{v}-\mathbf{P}^t$$

with $\mathbf{P} := J\mathbf{F}^{-1}.\boldsymbol{\sigma}$ the first Piola-Kirchhoff stress tensor. Application of [5.61b] gives here

$$\left[\rho_R \mathbf{v}\otimes\mathbf{V}_g + (\mathbf{P}+\mathbf{M}\otimes\mathbf{v})\right]_{S_R}.\mathbf{N} = \mathbf{0} \qquad [5.67]$$

The jump condition for the energy balance is written as

$$\left[(\rho v_{gn} - m)\left(u+\frac{1}{2}v^2\right) - \mathbf{n}^t.(\boldsymbol{\sigma}.\mathbf{v}) + \mathbf{q}.\mathbf{n}\right]_S = 0 \qquad [5.68]$$

The Lagrangian counterpart of the first principle is obtained as follows: denoting e_R the internal energy density per unit reference mass, the energy balance with respect to the reference configuration parallels the physical version [5.67] as

$$\frac{D}{Dt}\int_{\Omega_R} \rho_R\left(e_R+\frac{1}{2}v^2\right)dX = \int_{\Omega_R}\left\{(\mathbf{f}_R.\mathbf{v}+r_R)+\Pi\left(e_R+\frac{1}{2}v^2\right)\right\}dX + \int_{\partial\Omega_R}\mathbf{N}.\left\{\mathbf{P}.\mathbf{v}+\mathbf{M}\left(e_R+\frac{1}{2}v^2\right)-\mathbf{Q}\right\}d\sigma \qquad [5.69]$$

with r_R the volumetric heat supply (generated by growth), and $\mathbf{Q}$ the referential heat flux across $\partial\Omega_R$. Using again [5.61a], the jump condition associated with [5.69] results in

$$\left[(\rho_R V_{gN} + M)\left(e_R+\frac{1}{2}v^2\right) + \mathbf{N}^t.(\mathbf{P}.\mathbf{v}) - \mathbf{Q}.\mathbf{N}\right]_{S_R} = 0 \qquad [5.70]$$

traducing the jump of energy over S_R.

The internal entropy jump takes the strong form

$$\left[\rho s + \frac{\mathbf{q}.\mathbf{n}}{\theta}\right]_{S_t} \geq 0 \qquad [5.71]$$

The Lagrangian form of the Clausius-Duhem inequality receives the expression

$$\frac{D}{Dt}\int_{\Omega_R} \rho_R s_R dX \geq \int_{\Omega_R} \left(\Pi s_R + \theta^{-1} r_R\right) dX - \int_{\partial\Omega_R} \mathbf{N}.\frac{\mathbf{Q}}{\theta} d\sigma \qquad [5.72]$$

with the obvious relations

$$\mathbf{q} = J^{-1}\mathbf{F}.Q,\ \mathrm{s} = J^{-1}s_R,\ \mathrm{r} = J^{-1}r_R$$

This gives rise to the jump of the dissipation in material format

$$\left[-\rho_R s_R V_{gN} + \theta^{-1}\mathbf{Q}.\mathbf{N}\right]_{S_R} \geq 0 \qquad [5.73]$$

This inequality is next combined with the kinetic energy theorem (the virtual power of inertia forces is the sum of the virtual power of internal and external forces), written in Lagrangian format: multiplying the Lagrangian balance of momentum by the velocity and using the mass balance in referential format leads after a few rearrangements to the Lagrangian form of the kinetic energy theorem

$$\frac{\partial}{\partial t}\left(\rho_R \mathbf{v}^2/2\right) = \nabla_R.\left(\mathbf{P}.\mathbf{v} + \frac{\mathbf{v}^2}{2}\mathbf{M}\right) + \Pi\frac{\mathbf{v}^2}{2} + \mathbf{f}_{R.}\mathbf{v} - Tr\left(\mathbf{P}.\left(\nabla_R \mathbf{v}\right)^t\right) \qquad [5.74]$$

Note that the partial derivative $\frac{\partial}{\partial t}$ coincides with the total derivative $\frac{D}{Dt}$ in the present Lagrangian situation. The induced jump relation is written accordingly

$$\left[\left(\rho_R \mathbf{v}^2/2\right)V_{gN} + \mathbf{N}.\mathbf{P}.\mathbf{v} + \frac{\mathbf{v}^2}{2}M\right]_{S_R} = 0 \qquad [5.75]$$

Combining the jump condition [5.60] together with [5.75] and [5.73] then gives the positive entropy jump, assuming the continuity of temperature, as

$$\left[\left(-\rho_R s_R V_{gN} + \theta^{-1}\left(M + \rho_R V_{gN}\right)e_R\right)\right]_{S_R} \geq 0 \qquad [5.76]$$

This inequality can be rewritten using the expression of the scalar mass flux from [5.65] and the continuity of $\mathbf{V}_g \equiv \mathbf{C}$ as

$$\left\{\left[e_R\rho_R\right]_{S_R} - \theta\left[s_R\rho_R\right]_{S_R} - \left[\rho_R\right]_{S_R}\left[e_R\right]_{S_R}\right\}V_{gN} \geq 0 \qquad [5.77]$$

since the normal velocity V_{gN} is continuous across s_R. Hence, one is entitled by [5.77] to write the driving force associated with growth in the reference configuration as the scalar quantity $\left\{[e_R\rho_R]_{S_R} - \theta[s_R\rho_R]_{S_R} - [\rho_R]_{S_R}[e_R]_{S_R}\right\}$; this driving force is more complicated compared to the Hugoniot-Gibbs material force found in [MAU 95], but reduces to it in the absence of mass flux (no surface growth occurs), when $M = 0$. However, a common feature between the work of those authors and the present description is that both expressions of the driving forces do not involve inertia terms.

A more simple expression of the dissipation is obtained by considering the general form of the balance of energy and entropy discarding the heat flux; such an adiabatic situation is realistic for hard tissues such as bone which undergoes small strains, hence the heat due to the mechanical power of internal forces can be neglected:

$$\frac{DK}{dt} + \frac{D}{Dt}\int_{\Omega_t} \rho\psi dx \leq P_e + \underline{J}_k.\underline{F}_k + \Phi_m \Leftrightarrow \frac{D}{Dt}\int_{\Omega_t} \rho\psi dx \leq -P_i + \underline{J}_k.\underline{F}_k + \Phi_m \qquad [5.78]$$

with $\Phi_m := -\int_{\partial\Omega_t} \mu_i \mathbf{J}_i.\mathbf{n} d\sigma$ the flux of mass through the boundary of Ω, and $P_i = -\int_{\Omega_t} (\boldsymbol{\sigma} : grad\mathbf{v})\, dx$ the virtual power of internal forces. Inequality [5.78] traduces the fact that the flux of mechanical work and mass increases the kinetic and internal free energy of the system, the difference being dissipated. A similar form of the second principle for growing biological tissues has been obtained in [AMB 07].

The jump condition associated with [5.78] is written in Eulerian format

$$\left[-\rho\psi\mathbf{v}_g - \mu_i\mathbf{m}_i\right]_{s_t}.\mathbf{n} \geq 0 \qquad [5.79]$$

with $m_i := \mathbf{J}_i.\mathbf{n}$ the mass flux (one here considers the general case of growth due to multiple species, each labeled with the subscript i).

The relevant physical variable to characterize growth is characterized basing on the following remark: the sole motion of the interface does not generate surface growth, since one may have a nil relative velocity ($\mathbf{v}_g = \mathbf{0}$), in which case the mass flux $\mathbf{m}$ is nil, hence there is no surface growth in the sense retained in this work, but possibly only volumetric growth occurs due to the source term in the expression of the mass flux – those contributions do not intervene in the jump conditions, which give from a mathematical point of view a clear delimitation between surface and volumetric growth.

Using [5.61a], Nanson formula the limit $J\rho = \rho_R \to \rho_S$ when the volumetric density tends to the surface density of generating cells ρ_S on s_R, and considering a single phase growing continuum, together with the continuity of the velocity $\mathbf{V}_g$, the material form of the jump inequality in [5.79] is obtained as

$$\left[J\rho\psi\mathbf{V}_g - J\mu\mathbf{F}^{-1}.\mathbf{m}\right]_{s_R}.\mathbf{N} \geq 0 \Rightarrow \left[\rho_s\left(\psi_R - \mu\right)\right]_{s_R} V_{gN} \geq 0 \qquad [5.80]$$

One is therefore entitled to consider the jump of the volumetric grand potential, quantity $\rho_s\left(\psi_R - \mu\right) \equiv \rho_s\left(u_R - \theta s_R - \mu\right)$ (see [GAN 10b]) as the thermodynamic driving force for surface growth. Note that the physical body forces present may be discarded, as they represent source terms that finally will not enter the jump conditions.

The jump of the Eulerian growth velocity can then be related to the Lagrangian surface growth velocity from [5.51] and using the Maxwell-Hadamard lemma on singular surfaces as

$$\left[\mathbf{v}_g\right]_{S_t} = -\left[\mathbf{F}\right].\mathbf{V}_g = -V_{gN}\left[\mathbf{F}.\mathbf{N}\right] \qquad [5.81]$$

The field equations necessary for solving a given boundary value problem following the lines of this contribution include the bulk balance equations for mass, momentum, energy and entropy, and the associated jump conditions. Especially, the inequality [5.80] shall serve to write down a model for surface growth in terms of a relation between the geometrical velocity of the moving front and the thermodynamically conjugated driving force. Those equations have to be complemented by suitable constitutive laws for the growing tissues.

The kinetic laws for surface growth may be constructed in a phenomenological manner from [5.80], whereby the propagating interface is viewed from a macroscopic point of view as a zone of vanishing thickness. This is certainly an idealization, as the interface has a finite thickness in a true physical situation, and may instead be a transition region with a rather complex behavior. The reader is referred to the contributions in [MAU 95, MAU 10] for a clear exposition of the various pictures underlying the propagation of phase-fronts. The present point of view is nevertheless fruitful, as it provides a platform for subsequent simulations of both surface and volumetric growth phenomena occurring in hard tissues, such as bone growth.

5.6. Surface growth: a review of surface thermodynamics

Surface thermodynamics is clearly a pluridisciplinary topic, which has its origins in the study of liquids, and touches various disciplines such as metallurgy (grain boundary energy), fracture mechanics (fracture energy) mechanics (surface stress), physics of fluids (surface tension) and of solids (surface stress). Surface thermodynamic data are important parameters for specialists in each of those fields, with however a different acceptance of the term.

The thermodynamics of surfaces has a long history, going back to Gibbs; an interface exists when a thin inhomogeneous element of material forms a transition zone separating two phases of different materials (denoted α, β in what follows), as pictured in Figure 5.3. The transition zone between the bulk phases will be denoted by the Greek letter σ in what follows.

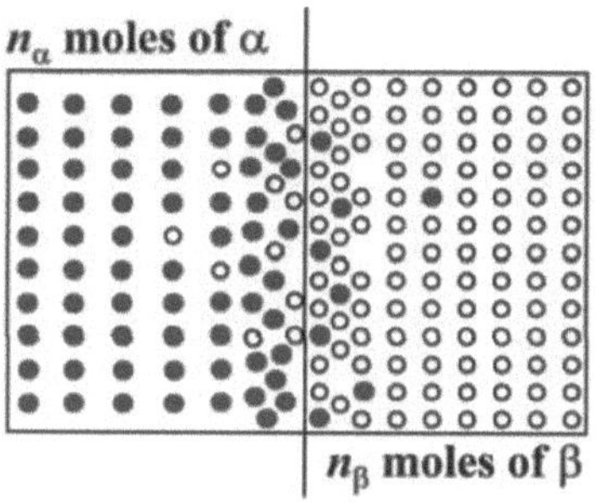

Figure 5.3. *Formation of an interface from a fixed number of moles of α and β.*

The aim of this section is not to give a detailed account in each of those fields, but rather to provide the reader with a broad overview of the basic surface thermodynamics and to review the major underlying parameters and their possible source of variation.

Different viewpoints have been considered in the literature as to the geometry of the surface (this coinage used in Linford refers to the surface, as opposed to the bulk phases): the surface phase is considered as two-dimensional by Gibbs, and named the mathematical dividing surface, as the neat separation between fluid and solid phases. Gibbs viewpoint may be called the *surface excess approach (at fixed volume)*, in which the composite system (bulk phases and the interface) is the sum of the reference system without the interface and a correction; the difference of any quantity between the actual and the reference system leads to an *interfacial excess quantity*. Important to this viewpoint is the fact that the reference and actual systems have the same volume.

Guggenheim considered the surface phase as a three dimensional body of finite small thickness and is commonly coined the *surface phase approach*. A third approach has been introduced by [GOD 69], relying on Guggenheim vision, but with the interfaces between the surface phase and the two bulk phases identified with the walls of a confining vessel. A last vision at variant with Gibbs' treatment advocates that both the actual and reference systems have the same mass, but possibly different volumes: it bears the name *Surface excess approach (at fixed mass)*, and was hardly considered in the literature, although mentioned by Gibbs in 1878. One drawback of the Guggenheim model is that the volume of the interfacial region V^{σ} is arbitrary, and has nothing to do with the volume change that occurs during the formation of the interface; this difficulty is not apparent in Gibbs approach, for which the excess volume V^{σ} is always zero.

For liquids, the situation is simple, as a single scalar parameter, the surface tension, is sufficient. Three parameters are required to characterize the thermodynamics of surfaces: the reversible work to produce the unit area of a new surface, sometimes called the *specific surface work* (the counterpart of the surface tension in liquids), the specific surface Helmholtz energy, as the change of energy of the surface region (as opposed to the bulk phases), and the surface stress tensor, defined as the reversible work required to produce a unit area of new surface by deformation. In order to avoid some existing confusion in the early literature (this is due to the oversimplified situation that prevails for liquids), these three parameters are next introduced in a distinct manner.

The thermodynamics of surfaces is based on the setting up of *excess quantities*. The reader is referred to [LIN 73] and [COU 80] for more details on the topic.

Hence, the excess (Helmholtz) free energy is defined through its differential

$$dF^{\sigma} = -S^{\sigma} dT + g dA + \mu_k dN^{\sigma}{}_k$$

The quantity g accounts for both the creation of a new surface (with a fixed number of atoms) and the elastic deformation of the surface (also with a fixed number of atoms). The addition of atoms (particles) on the surface is accounted for by the last term. Considering two phases α, β with a separating interface σ in-between, one can write the differential of the total number of particles as

$$dN = dN^{\alpha} + dN^{\beta} + dN^{\sigma}$$

the superficial excess or molar superficial concentrations are then defined as

$$n^\sigma := N^\sigma / A$$

with A the area of the interface. Any extensive quantity Z can be decomposed as

$$Z = Z^\alpha + Z^\beta + Z^\sigma = z^\alpha V^\alpha + z^\beta V^\beta + z^\sigma A$$

with $z^\sigma := Z^\sigma / A$ the *superficial excess* quantity.

Regarding surface quantities, one has to make a distinction between:

– the superficial energy γ $\left(J/m^2\right)$ – a scalar – accounting for the creation of a new surface (irreversible phenomena), with a constant number of particles;

– the purely elastic variation of the surface area, expressed by a superficial stress $\tilde{\boldsymbol{\sigma}}$, dual to an elastic surface strain $\tilde{\boldsymbol{\varepsilon}}$.

The excess total internal energy is written

$$dU^\sigma = \theta dS^\sigma + gdA + \mu_k dN^\sigma{}_k$$

For the whole system, using the previous decomposition

$$Z = Z^\alpha + Z^\beta + Z^\sigma$$

one has

$$dU = \theta dS - pdV + gdA + \mu_k dN_k$$

The variation of the free energy is

$$dF = -Sd\theta - pdV + gdA + \mu_k dN_k$$

Hence, g is defined as the partial derivative $g = \left(\dfrac{\partial F}{\partial A}\right)_{T,V,N_i}$. Combining both of the following relations

$$dF^\sigma = -S^\sigma d\theta + gdA + \mu_k dN^\sigma{}_k \,;\, F^\sigma = \gamma A + \mu_k N^\sigma{}_k$$

gives

$$S^\sigma dT + N^\sigma{}_k d\mu_k - (g - \gamma)\, dA + Ad\gamma = 0$$

This leads to the differential

$$d\gamma = -s^{\sigma} d\theta - n^{\sigma}{}_{k} d\mu_k + (g-\gamma)\, dA / A$$

expressing the variation of the superficial energy. In the case of an isothermal surface stretch with a constant chemical potential, one gets the Couchmann-Everett formula

$$g = \gamma + A\left(\frac{\partial \gamma}{\partial A}\right)_{T,\mu_i}$$

In the case of a purely elastic stretch, one obtains the following relation

$$g = \gamma + A\left(\frac{\partial \gamma}{\partial A}\right)^{el}_{T,\mu_i}.$$

The reversible work needed to form a unit area of new surface is defined at constant temperature and pressure at the partial derivative of the Gibbs free energy of the entire system (bulk phases and surface), quantity $G(P,T,n_i,A)$ with respect to the formed area A, at constant temperature T, pressure P, and number of moles of each component n_i, viz.

$$dG = -VdP - SdT + \mu_i dN_i + \gamma dA$$

whereby the multiplicative factors of the differential elements on the right-hand side of dG are defined as the partial derivatives

$$V = -\frac{\partial G}{\partial P};\ S = -\frac{\partial G}{\partial T};\ \mu_i = \frac{\partial G}{\partial N_i};\ \gamma = \frac{\partial G}{\partial A}$$

The partial derivatives are implicitly taken with all three other variables being held fixed. The specific surface work γ includes two contributions, the change of Gibbs free energy per unit area for the surface region, denoted g^{σ}, and the change per unit area of surface created from the surrounding bulk phases, evaluated as the sum over all components $\mu_i n_i^{\sigma}$,

$$\gamma = g^{\sigma} - \mu_i n_i^{\sigma}$$

with $n_i^\sigma := N_i / A$, the surface excess of the ith species. In terms of the Helmholtz energy of the whole system F, one has the similar relation involving the Helmholtz free energy per unit area f^σ, viz.

$$\gamma = f^\sigma + Pe - \mu_i n_i^\sigma$$

with e the thickness of the surface; in most cases, the parameter e is small, and one may neglect the contribution Pe, hence one has the identification $f^\sigma = g^\sigma$. The last two formulas are expressions of Gibb's adsorption equation, with the derivation due to Mullins, which is next reproduced. We consider a system with n components consisting of a solid phase σ in contact with a fluid phase and a solid phase acting as a thermal bath at temperature T and as a chemical reservoir for each component; accordingly, the components' concentrations can be adjusted to maintain the chemical potentials at fixed specified values μ_i. Imagine a modification of the temperature by dT, and the ith chemical potential by $d\mu_i$, at a fixed surface area; dn_i particles from the bulk will enter the solid phase σ from the bulk, and its change of Helmholtz free energy F^σ will be

$$dF^\sigma = -S^\sigma dT + (\mu_i + d\mu_i)\, dN_{i,\sigma} \approx -S^\sigma dT + \mu_i dN_{i,\sigma}$$

with S^σ the entropy of the phase σ. Consider next a new system for which T and μ_i are returned to their initial values, but with the surface area increased by dA^σ, and modify thereafter the temperature by dT, and the ith chemical potential by $d\mu_i$; the variation of free energy of this system of larger area is

$$dF'^\sigma \approx -S'^\sigma\, dT + \mu_i dN'_{i,\sigma}$$

Subtracting both variations of Helmholtz free energy by the unit surface area gives

$$\frac{dF'^\sigma - dF^\sigma}{dA^\sigma} \approx -\frac{\left(S'^\sigma - S^\sigma\right)}{dA^\sigma} dT + \mu_i \frac{d\left(N'_{i,\sigma} - N_{i,\sigma}\right)}{dA^\sigma}$$

Introducing therein the definitions of the specific surface Helmholtz energy

$$f^\sigma := \frac{dF'^\sigma - dF^\sigma}{dA^\sigma}$$

the specific surface entropy

$$s^{\sigma} := \frac{\left(S'^{\sigma} - S^{\sigma}\right)}{dA^{\sigma}}$$

and the surface excess

$$n_i^{\sigma} := \frac{d\left(N'_{i,\sigma} - N_{i,\sigma}\right)}{dA^{\sigma}}$$

leads to

$$df^{\sigma} = -s^{\sigma} dT + \mu_i d\Gamma_i$$

But one also can express the specific surface Helmholtz energy as

$$f^{\sigma} = \gamma + \mu_i \Gamma_i$$

Hence

$$df^{\sigma} = d\gamma + \mu_i d\Gamma_i + \Gamma_i d\mu_i = -s^{\sigma} dT + \mu_i d\Gamma_i$$

and thus finally

$$d\gamma = -s^{\sigma} dT - \Gamma_i d\mu_i$$

The same identity was derived by [GOD 69] for a one-component system using the method of Lagrange multipliers. The reversible work needed to generate a unit area of new surface by stretching at constant pressure and temperature represents the surface stress tensor, denoted $\tilde{\sigma}_{ij}$. It is related to γ by

$$\tilde{\sigma}_{ij} = \gamma \delta_{ij} + \frac{\partial \gamma}{\partial \tilde{\varepsilon}_{ij}}$$

The second order tensor $\tilde{\varepsilon}_{ij}$ is the strain (a small perturbation scheme is presently adopted) induced by the component $\tilde{\sigma}_{ij}$ acting in the jth direction per unit length of the edge normal to the ith direction, with both indices i, j lying in the plane of the surface. The previous equation is valid for an anisotropic solid, and reduces in the case of an isotropic surface to the previously written Couchmann-Everett formula, with g half the trace of the surface stress tensor. The proof of the previous formula is as follows, following Mullins: imagine a unit cube with edges parallel to the axes x, y, z, and perform two distinct operations on it:

i) Stretch the cube reversibly along the x axis by an amount dx, with the y edge fixed, but allowing the edge z to vary its length. The surface in the xz plane may then change by an inflow (or outflow) of material from the bulk, increasing (or decreasing), the cube height; denote W_0 the work expanded in this transformation. Let us next separate the stretched cube along the xy plane, requiring the work $W_2 = 2(\gamma + d\gamma)(1+dx)$, with $d\gamma$ the variation of the specific surface work γ due to the stretch dx (factor 2 arises since two surfaces are created, and the factor $(1+dx)$ since the specific surface work applies per unit surface area).

ii) Separate the original unit cube into two parts along an xy plane, requiring the work $W_3 = 2\gamma$, and stretch each half by dx in the x direction, at fixed y edge, but varying z edge. Let W_1 be the work expanded in this operation. The final configuration is the same as that obtained in the first process, hence the same total work has been expanded, hence $W_0 + W_2 = W_1 + W_3$, viz

$$W_0 + 2(\gamma + d\gamma)(1+dx) = 2\gamma + W_1$$

The difference $W_1 - W_0$ is the work expanded due to the stretching operations of both processes, and can be equalized to the x-component $\tilde{\sigma}_{xx}$ of a force in the newly formed surface times the distance $2dx$ through which this force acts, hence

$$2\tilde{\sigma}_{xx} dx = W_1 - W_0$$

The strain $\Delta\varepsilon_{xx} = dx$ (since the other side has unit length), hence

$$2\gamma + 2\gamma\Delta\varepsilon_{xx} + 2\Delta\gamma + 2\Delta\gamma\Delta\varepsilon_{xx} = 2\gamma + 2\tilde{\sigma}_{xx}\Delta\varepsilon_{xx}$$

Due to the equalities

$$\Delta\gamma\Delta\varepsilon_{xx} \approx 0$$

and

$$\Delta\gamma / \Delta\varepsilon_{xx} \approx d\gamma / d\varepsilon_{xx}$$

it finally results in

$$\tilde{\sigma}_{xx} = \gamma + \frac{d\gamma}{d\varepsilon_{xx}}$$

Similar analogous processes with the stretching replaced by shear lead to the relation [LIN 73]

$$\tilde{\sigma}_{xy} = \frac{d\gamma}{d\varepsilon_{xy}}$$

Combining the stretch and shear processes then leads to the expression of the surface stress tensor

$$\tilde{\sigma}_{ij} = \gamma\delta_{ij} + \frac{d\gamma}{d\varepsilon_{ij}}$$

represented by a 2 by 2 symmetrical matrix (3 independent components). For an isotropic material or a crystal with a threefold (or greater axis of symmetry), it follows as shown by Shuttleworth (using the principle of virtual work) that the isotropic surface stress

$$\tilde{\sigma}_{ij} = g\begin{pmatrix} 1 & 0 \\ 0 & 1 \end{pmatrix}$$

Lastly, consider a square section in the xy plane of side $\left(A^{\sigma}\right)^{1/2}$ and imagine an extension of the x edge by $\varepsilon_{xx}\left(A^{\sigma}\right)^{1/2}$; the required work is

$$W_1 = \tilde{\sigma}_{xx}\left(A^{\sigma}\right)^{1/2}\varepsilon_{xx}\left(A^{\sigma}\right)^{1/2}$$

Extend next the y edge by $\varepsilon_{yy}\left(A^{\sigma}\right)^{1/2}$, with an expanded work given by

$$W_2 = \tilde{\sigma}_{yy}\left(A^{\sigma}\right)^{1/2}\left(1+\varepsilon_{xx}\right)\varepsilon_{yy}\left(A^{\sigma}\right)^{1/2}$$

Assuming the deformation is reversible and isothermal, the total work spent is the variation of surface energy, which expresses for a high symmetry isotropic crystal as

$$d\left(A^{\sigma}\gamma\right) = W_1 + W_2 = \tilde{\sigma}_{xx}\varepsilon_{xx}A^{\sigma} + \tilde{\sigma}_{yy}\left(1+\varepsilon_{xx}\right)\varepsilon_{yy}A^{\sigma} \approx gdA^{\sigma}$$

due to the equality

$$A^{\sigma}\left(\varepsilon_{xx}+\varepsilon_{yy}\right)=dA^{\sigma}$$

Therefore, it holds the relation

$$gdA^{\sigma}=\gamma dA^{\sigma}+A^{\sigma}d\gamma \Rightarrow g=\gamma+\frac{d\gamma}{dA^{\sigma}}$$

Note that the last term vanishes for liquids; as corollary, liquid films can easily be stretched since atoms can move from the bulk to the surface without additional energy costs. The opposite situation prevails for solids, as they shear and their structure changes with an overall additional energy contribution.

The Gibbs approach towards interfacial excess quantities is as previously mentioned valid only at fixed volume; a parallel approach that is valid at fixed mass instead has been developed in (Muller and Saul), which is next exposed. The bulk phases α,β are initially separated and interface-free, and are in a thought experiment imagined to be joined along a plane to generate the α/β interface. Since mass is conserved, any change in the thermodynamic quantities of the whole system are due to the new α/β interface, coined *excess values* of the corresponding quantities, denoted with a subscript γ to distinguish them from Gibbs approach at fixed volume. The differential of the Gibbs energy of the system before and after formation of the interface is successively written (for a constant number of molecules)

$$dG_1=\left(V_{\alpha}+V_{\beta}\right)dP-\left(S_{\alpha}+S_{\beta}\right)dT$$

$$dG_2=\left(V_{\alpha}+V_{\beta}+V_{\gamma}\right)dP-\left(S_{\alpha}+S_{\beta}+S_{\gamma}\right)dT+\gamma^{*}dA$$

with γ^{*} the *reorganization surface energy*, although commonly referred to as the interfacial tension in the literature; it is a mechanical positive quantity that may depend upon interface curvature. Note that the number of atoms is the same in the reference and final states, in contrast with the Gibbs approach. Hence, the variation of the excess Gibbs free energy between states 1 and 2 for the fixed masses m_{α},m_{β} is

$$dG_{\gamma}=dG_2-dG_1=V_{\gamma}dP-S_{\gamma}dT+\gamma^{*}dA$$

which may be interpreted from an energetic point of view as follows: the term $V_\gamma dP$ is the mechanical work done against the external force field, the contribution $S_\gamma dT$ represents the heat of formation of the interface, and $\gamma^* dA$ is the mechanical work done against the internal force field of both phases α, β by motion of the molecules from the bulk to generate a new interface. The *excess free energy* of formation of the interface, potential G_γ, is the additional free energy required to form the interface from fixed masses of the pre-existing bulk phases α, β. The above equations implicitly use the conservation of mass, equation

$$n_{total} = n_\alpha + n_\beta$$

and the definition of the excess interfacial volume V_γ from the contributions to the total volume after interface formation (balance law for the volume)

$$V_{total} = V_\alpha + V_\beta + V_\gamma$$

In contrast to this treatment, Gibbs assumes a conservation of the total volume as

$$V_{total} = V_\alpha + V_\beta$$

but with addition of the new mass n^σ such that

$$n_{total} = n_\alpha + n_\beta + n^\sigma$$

as a compensation for the volume change accompanying the formation of the interface; hence, n^σ is a supply of material from outside the system, with the sense that the Gibbs volume is an open thermodynamic volume.

Due to its status as a state function, the previous differential G_γ of the allows writing relations between partial derivatives as the analogues of the bulk phase Maxwell relations

$$\left(\frac{\partial \gamma^*}{\partial T}\right)_{P,A,n_{\alpha,\beta}} = -\left(\frac{\partial S}{\partial A}\right)_{P,T,n_{\alpha,\beta}} = -S_\gamma^* \;;\; \left(\frac{\partial \gamma^*}{\partial P}\right)_{T,A,n_{\alpha,\beta}} = \left(\frac{\partial V^*}{\partial A}\right)_{P,T,n_{\alpha,\beta}} = V_\gamma$$

$$\left(\frac{\partial V_\gamma}{\partial T}\right)_{P,A,n_{\alpha,\beta}} = -\left(\frac{\partial S}{\partial P}\right)_{A,T,n_{\alpha,\beta}}$$

the introduced quantities S_γ^*, V_γ are respectively the *interfacial excess entropy* and the *specific interfacial excess volume*; the compact notation $n_{\alpha,\beta}$ stands for the two quantities $\{n_{\alpha,} n_\beta\}$.

The specific interfacial excess energy is obtained by simply integrating the differential $dG_\gamma = V_\gamma dP - S_\gamma dT + \gamma^* dA \Rightarrow \left(G_\gamma^*\right)_{n_{\alpha,\beta}} = \gamma^*$ at constant pressure and temperature, introducing the specific interfacial excess energy $G_\gamma^* := G_\gamma / A$. This last relation implies that the temperature and pressure dependence of γ^* can be determined from those of G_γ^*. The specific interfacial excess energy is obtained from a Legendre transform to $dG_\gamma = V_\gamma dP - S_\gamma dT + \gamma^* dA$ and substitution of the previous interfacial Maxwell relations, thus

$$\left(U_\gamma^*\right)_{n_{\alpha,\beta}} = \gamma^* + TS_\gamma^* - PV_\gamma^*$$

It immediately results in the specific interfacial excess enthalpy

$$\left(H_\gamma^*\right)_{n_{\alpha,\beta}} = \gamma^* + TS_\gamma^* \equiv \gamma^* - T\left(\frac{\partial \gamma^*}{\partial T}\right)_{P,A,n_{\alpha,\beta}}$$

with H_γ^* identified as the surface energy, which is the sum of the interfacial tension and the heat supplied by the surrounding for an isothermal creation of new interface.

The advantages of this last approach in comparison to Gibbs treatment is that it leads to non-nil interfacial volumes, analogues of the Maxwell relations for bulk phases can be derived, and the temperature and pressure dependence of the interfacial tension can be accessed from a comparison between simple formulae and experiments.

The reversible work to form a new surface area, parameter γ, is for a solid generally orientation dependent, although not for a liquid. This surface energy parameter has been up to now considered under the thermodynamic continuum viewpoint; we next examine two other viewpoints, the atomistic approach and Wulff plot. The atomistic approach considers the interaction between atoms to calculate the surface energy; arrangement of atoms in crystals are such that one can order atoms according to the energy required to remove atoms from the bulk: the first nearest neighbours requiring more energy compared to second and third nearest neighbours. For a crystal lattice presenting dislocations, the number of broken bonds is direction dependent, and is given by the expressions $\cos\theta / a$ and $\sin\theta / a$ in the x and y directions respectively (Figure 5.4), with a the distance to nearest neighbour (function of the type of atomic packing) and θ the inclination of the overall crystal shape resulting from the total number of steps being created.

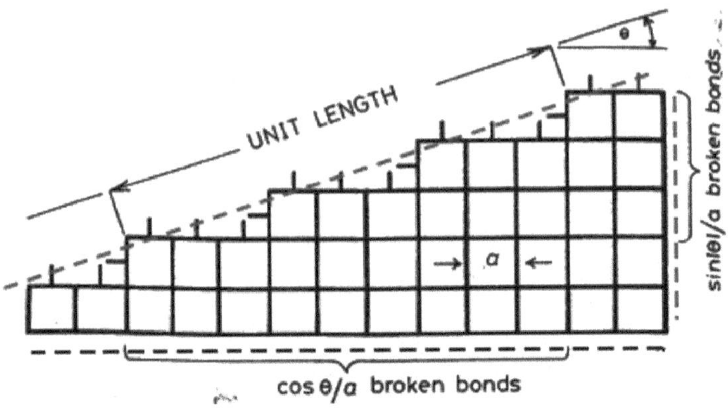

Figure 5.4. *The broken bond model for surface energy*

The surface energy is given by the expression

$$E_{surf} = (\cos\theta + \sin\theta)\varepsilon_b / (2a^2)$$

with ε_b the energy per bond. The broken bond model can be used to determine the shape of a small crystal from the minimization of the sum of surface energies γ_i over all crystal faces, a concept introduced in 1878 by J. W. Gibbs, considering constant pressure, volume, temperature and molar mass:

$$Min\sum_i A_i\gamma_i$$

at constant energy, hence adding the constraint

$$dE = 0 = \sum_i \gamma_i dA_i$$

The dependency of γ on the orientation of the crystal's surface and its equilibrium shape are condensed into a *Wulff plot*; in 1901, George Wulff stated that the length of a vector normal to a crystal face is proportional to its surface energy in this orientation. This is known as the Gibbs-Wulff theorem, which was initially given without proof, and was proven in 1953 by Conyers Herring, who at the same time provided a two step method to determine the equilibrium shape of a crystal: in the first step, a polar plot of the surface energy as a function of orientation is made, given as the so-called gamma plot denoted as $\gamma(\mathbf{n})$, with $\mathbf{n}$ the normal to the surface corresponding to a particular crystal face. The second step is Wulff construction, in which the gamma plot determines graphically which crystal faces will be present: Wulff construction of the equilibrium shape consists in drawing a plane through each point on the γ-plot perpendicular to the line connecting that point to the origin. The inner envelope of all planes is geometrically similar to the equilibrium shape (Figure 5.5).

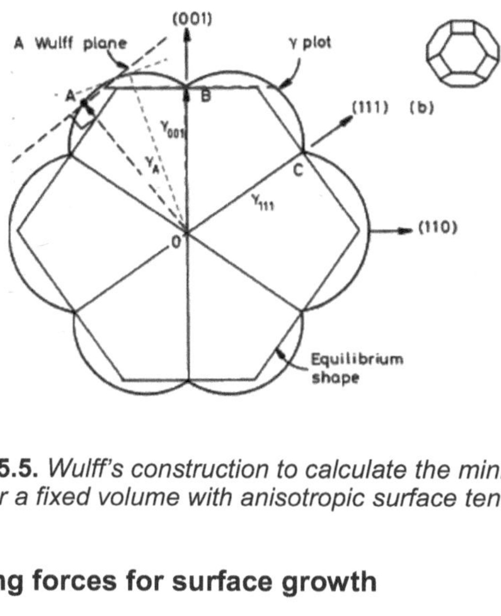

Figure 5.5. *Wulff's construction to calculate the minimizing surface for a fixed volume with anisotropic surface tension $\gamma(n)$*

5.7. Material driving forces for surface growth

In what follows, the framework for surface growth elaborated in [GAN 10] will be applied to describe bone modeling and remodeling. As a prerequiste, we recall the identification of the driving forces for surface growth. We consider a tissue element under growth submitted to a surface force field $\mathbf{f}_S$ (surface density) and to line densities p_τ, p_ν defined as the projections onto the unit vectors $\boldsymbol{\tau}_g, \mathbf{v}_g$ resp.

along the contour of the open growing surface S_g (Figure 5.6); hence, those line densities are respectively tangential and normal to the surface S_g (forces acting in the tangent plane).

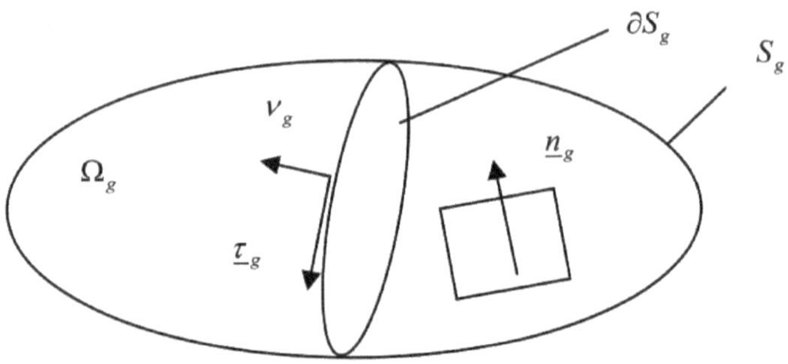

Figure 5.6. *Tissue element under growth: elements of differential geometry*

The potential energy of the tissue element is chosen as the expression

$$V = \int_{S_g} \psi^S \left(\tilde{\mathbf{F}}, \mathbf{N}; \mathbf{X}_S\right) d\sigma_g - \int_{S_g} \mathbf{f}_S . \tilde{\mathbf{x}} d\sigma_g - \int_{\partial S_g} p_\tau \tilde{\tilde{\mathbf{x}}} . \boldsymbol{\tau}_g dl_g - \int_{\partial S_g} p_\nu \tilde{\tilde{\mathbf{x}}} . \mathbf{v}_g dl_g$$

Thereby, and focusing on the surface behavior, the growing solid surface is supposed to be endowed with a surface energy, with density $\psi^S\left(\tilde{\mathbf{F}}, \mathbf{N}; \mathbf{X}_S\right)$ per unit reference surface, depending upon the surface gradient $\tilde{\mathbf{F}}$, the unit normal vector $\mathbf{N}$ to S_g, and possibly explicitly upon the surface position vector $\mathbf{X}_S$ on S_g (no tilde notation is adopted here since the support of $\mathbf{X}_S$ is strictly restricted to the surface S_g). The surface gradient $\tilde{\mathbf{F}}$ maps material lengths (or material tangent vectors) on the deformed surface; it is elaborated as the surface projection of $\mathbf{F}$ (onto the tangent plane to Ω_a),

$$\tilde{\mathbf{F}} := \mathbf{F}.\mathbf{P}$$

The material surface driving force (for surface growth) triggers the motion of the surface of the growing solid; it is identified from the material variation of V; the driving force decomposes into a tangential contribution

$$\tilde{\boldsymbol{\Upsilon}}_{gT} := \mathbf{P}.\left(\nabla_S . \tilde{\boldsymbol{\Sigma}} - \mathbf{K}' . \partial_N \psi^S\right)$$

and a normal contribution

$$\tilde{\Upsilon}_{gN} := \tilde{\boldsymbol{\Sigma}} : \mathbf{K} + \nabla_S.(\tilde{\boldsymbol{\Sigma}}.\mathbf{N})$$

in the material manifold, based on the surface Eshelby stress

$$\tilde{\boldsymbol{\Sigma}} := \tilde{\mathbf{F}}^T.\tilde{\mathbf{T}} - \psi^s \mathbf{I}_s$$

itself built from *the surface stress* $\tilde{\mathbf{T}} := \partial_{\tilde{F}} \psi^S$, and on the curvature tensor $\mathbf{K} := -\nabla_R \mathbf{N}$ in the referential configuration. The stationarity condition of the total potential energy, viz. the condition $\delta V \equiv 0$, gives by localization onto S_g the necessary equilibrium condition:

$$\nabla_S.\tilde{\boldsymbol{\Sigma}} - \mathbf{P}.\mathbf{K}'.\partial_N \psi^S + \left(\partial_{X_S} \psi^s\right)_{\exp l} + \tilde{\mathbf{F}}^T.\mathbf{f}_S = \mathbf{0}$$

5.8. Extremum principles for biological continuum bodies undergoing volumetric and surface growth

Growth of biological tissues has attracted the attention of several researchers in continuum mechanics over the last two decades [ROD 94, EPS 00, AMB 02, LUB 02, TAB 01, GAR 03, RAJ 95, RAJ 98, HUM 02]. The most advanced works treat the interactions between the mechanical equilibrium and the transport phenomena leading to growth(see [GAN 05, ATE 07] using the theory of mixtures, or [NAR 09] highlighting the coupling between reaction-transport of solutes and mechanics.

While most of the efforts have been spent on the modeling of volumetric growth, fewer works deal with surface growth, which from a conceptual point of view introduces the additional difficulty of a changing number of particles, in comparison to volumetric growth (an assumption of constant particle numbers is made there, considering that either the density or local volume change instead). A unifying framework for the treatment of both bulk and surface growth phenomena has been proposed in [DIC 05], relying on the introduction of configurational forces as the internal driving forces for growth. Configurational forces for surface growth have been identified in [GAN 11], with application to bone remodeling.

While most of the growth models are devoted to writing evolution laws for kinematic-like variables related to the growth process (growth tensor), the generalization of variational principles to account for the multiphysical phenomena occurring during growth has not been achieved so far. The main objective of this contribution is then to set up extremum principles for continuum solid bodes submitted to growth, considering successively volumetric and surface growth. Since biological tissues are open systems exchanging nutrients with their environment, it is quite natural to adopt the umbrella of the thermodynamics of open systems, allowing the incorporation of the various kinds of energies that the growing body is exchanging. We thereby advocate a novel contribution, especially the writing of new extremum principles for growing solid bodies involving generalized Eshelby stresses. This will extend the writing of nonlinear elasticity extremum principles involving the standard mechanical Eshelby stress [17]. Regarding notations, vectors and second order tensors will be denoted with boldface symbols.

5.8.1. *Extremum principle for volumetric growth*

We consider a representative tissue element (chosen at least one order of magnitude larger than the typical size of the heterogeneities, but at least one order of magnitude smaller than the macroscopic size of the body) as an open heterogeneous system exchanging work, mass and heat with its surrounding. The field variables are the absolute temperature $\theta(\mathbf{x},t)$, the entropy density $s(\mathbf{x},t)$, the transformation gradient $\mathbf{F}(\mathbf{x},t)$, the chemical potentials $\{\mu_k(\mathbf{x},t)\}_{k=1..N}$ and the number of moles (densities) of the N chemical species (nutrients) $\{n_k(\mathbf{x},t)\}_{k=1..N}$. The position $\mathbf{x}$ is introduced as a label of the material particles which occupy a finite volume in space.

As a starting point, the grand potential, a quantity used in statistical mechanics particularly for irreversible processes occurring in continuous open systems [GOO 75], is introduced in terms of its spatial density

$$\omega_{vol} = \omega_{vol}\left(\theta(x,t),\mathbf{F}(x,t),\mu_k(x,t)\right) := \int \mathbf{T} : d\mathbf{F} + \int \mathbf{F} : d\mathbf{T} \qquad [5.82]$$

ω_{vol} is the sum of the strain energy density $W_0 = W_0(\mathbf{F}) := \int \mathbf{T} : d\mathbf{F}$ and the complementary strain energy density $W_0^c(\mathbf{T}) := \int \mathbf{F} : d\mathbf{T}$, with $\mathbf{T} := \partial_F \omega$ the first Piola-Kirchhoff stress (the listed arguments of ω_{vol} will be justified later on). The previously defined densities $W_0(\mathbf{F})$ and $W_0^c(\mathbf{T})$ represent the area below and above the curve $\mathbf{T} = \mathbf{T}(\mathbf{F})$ respectively; hence, ω_{vol} is expressed as

$$\omega_{vol}[\theta,\mathbf{F},\mu_k] = \mathbf{T} : \mathbf{F} \qquad [5.83]$$

assuming the existence of a strain and stress free reference state.

Introducing next the internal energy density $e(\mathbf{F},s,n_k)$, the generalized fundamental Euler's relation is written as [16]

$$e(\mathbf{F},s,n_k)=\mathrm{T}(\mathbf{F},s,n_k):\mathbf{F}+\theta(\mathbf{F},s,n_k)s-\mu_k(\mathbf{F},s,n_k)n_k \qquad [5.84]$$

Observe that the mechanical contribution of the internal energy density, viz. the scalar $\mathrm{T}(\mathbf{F},s,n_k):\mathbf{F}$, incorporates both the strain energy and the complementary strain energy densities.

Consideration of [5.84] then leads to the rewriting of [5.83] as

$$\omega_{vol}[\theta,\mathbf{F},\mu]=e(\mathbf{F},s,\mathbf{n})-\theta(\mathbf{F},s,\mathbf{n})s-\mu_k(\mathbf{F},s,\mathbf{n}).n_k \qquad [5.85]$$

in which the convention of summation of the repeated index is adopted.

From the Gibbs and Gibbs-Duhem relations and using [5.85], the differential of the grand potential is expressed as

$$d\omega_{vol}=-sd\theta+\mathbf{T}:d\mathbf{F}-n_k d\mu_k\equiv\mathbf{T}:d\mathbf{F}+\mathbf{F}:d\mathbf{T} \qquad [5.86]$$

And is obtained from a straightforward calculation, resulting in the thermodynamic relations

$$\frac{\partial\omega_{vol}}{\partial\theta}=-s;\ \frac{\partial\omega_{vol}}{\partial\mathbf{F}}=\mathbf{T};\ \frac{\partial\omega_{vol}}{\partial\mu_k}=-n_k \qquad [5.87]$$

highlighting $\theta,\mathbf{F},\mu_k$ as the arguments of ω_{vol}; these constitutive-like relations shall be called state laws in what follows. The contributions on the r.h.s. of $d\omega_{vol}$ in [5.86] are successively the quasi-static heat flux, the incremental work of the internal stresses and the so-called quasi-static chemical work [CAL 85, p. 36].

A variational principle for volumetric growth is next constructed. From the static equilibrium equation of the tissue element submitted to body forces $\mathbf{f}_0$

$$\nabla_X.\mathbf{T}+\mathbf{f}_0=\mathbf{0} \qquad [5.88]$$

the definition of the transformation gradient $\mathbf{F} := \nabla_X \mathbf{x}$, and an integration by part, a weak formulation linking the integral of ω on a control volume V to contributions involving the applied loading (body forces and surface tractions $\mathbf{T.N}$) is obtained in a straightforward manner:

$$\int_V \omega_{vol} dV = \int_V \mathbf{f}_0.\boldsymbol{\psi} dV + \int_{\partial V} (\mathbf{T.N}).\boldsymbol{\psi} dS \qquad [5.89]$$

with $\boldsymbol{\psi}$ a kinematically admissible position field. The boundary ∂V is partitioned into a subpart S_ψ with imposed placement $\boldsymbol{\psi}^d$ and a complementary part S_t with imposed tractions $\mathbf{T}^d$, according to

$$\int_V \omega_{vol} dV = \int_V \mathbf{T} : \mathbf{F} dV = \left\{ \int_V \mathbf{f}_0.\boldsymbol{\psi} dV + \int_{S_t} \left(\mathbf{T}^d.\mathbf{N}\right).\boldsymbol{\psi} dS \right\} + \int_{S_\psi} (\mathbf{T.N}).\boldsymbol{\psi}^d dS \equiv W_T + W_\psi \qquad [5.90]$$

introducing therein the works of the imposed body forces and tractions W_T and the work of the imposed placements, viz. the quantity $W_\psi = \int_{S_\psi} (\mathbf{T.N}).\boldsymbol{\psi}^d dS$.

The Gibbs-Duhem relation allows the expression of one of the intensive variables in the triplet $(\theta, \mathbf{T}, \mu_k)$ vs. the two remaining variables. Using [5.89], a Legendre transform of ω versus the work terms gives a new potential, which identically vanishes, called for this reason the zero potential, elaborated as

$$\Omega[\theta, \boldsymbol{\psi}, \mu_k] := \int_V \omega_{vol} dV - \left\{ \int_V \mathbf{f}_0.\boldsymbol{\psi} dV + \int_{S_t} \left(\mathbf{T}^d.\mathbf{N}\right).\boldsymbol{\psi} dS \right\} - \int_{S_\psi} (\mathbf{T.N}).\boldsymbol{\psi}^d dS \equiv 0 \qquad [5.91]$$

Observe that the zero potential constructed thereabove is the tensorial generalization of the zero potential [GOO 75], accounting for the boundary conditions over the tissue element.

Since the zero potential is nil, it enjoys the property of extremality

$$\delta\Omega[\theta, \boldsymbol{\psi}, \mu_k] = 0 \qquad [5.92]$$

for the solution field, the set of primary variables $(\theta, \boldsymbol{\psi}, \mu_k)$ at thermodynamic equilibrium; this condition holds whatever the state laws. Thereby, a generalized potential energy in a thermodynamic framework has been set up, allowing for the consideration of the thermal and chemical forms of energy, in addition to the mechanical energy.

It is worth noting that the stress has been substituted by the temperature and chemical potential as new control variables: it is thus not possible to control the stress, and essential boundary conditions on the portion of surface S_t have to be prescribed in the formulation of the associated extremum principle, namely

$$\mathbf{T}.\mathbf{N} = \mathrm{t}^d \text{ on } S_{0t} \quad [5.93]$$

The material variation of the potential $\Omega[\theta, \psi, \mu_k]$ is next expressed to traduce the stationarity condition (at fixed material position: the index X is dropped in the variation, hence $\delta \equiv \delta_X$) following the steps:

1) The material variation of ω is formulated as

$$\delta\omega_{vol} = \delta\omega_{vol}(\theta, \mathbf{F} = \nabla_X \psi, \mu_k) \equiv \frac{\partial \omega_{vol}}{\partial \nabla_X \psi} : \delta(\nabla_X \psi) + \frac{\partial \omega_{vol}}{\partial \theta} \delta\theta + \frac{\partial \omega_{vol}}{\partial \mu_k} \delta\mu_k \quad [5.94]$$

2) The first contribution on the right hand side is integrated by part in the corresponding integral term in $\Omega[\theta, \psi, \mu_k]$, by a standard calculation in configurational mechanics, thereby highlighting the Eshelby stress built from the grand potential, defined as

$$\Sigma := \omega_{vol}\mathbf{I} - \mathbf{F}^T.\mathbf{T} \quad [5.95]$$

3) The variations of the spatial position, temperature and chemical potential in the previous expression are further expressed as:

$$\delta_x \psi + \mathbf{F}.\delta_x \mathbf{X} = \mathbf{0}; \; \delta\theta = \nabla_X \theta.\delta\mathbf{X}; \; \delta\mu_k = \nabla_X \mu_k.\delta\mathbf{X}$$

Assembling the calculations in those three steps delivers the material variation of the zero potential

$$\begin{aligned} \delta\Omega[\theta, \psi, \mu_k] &= \int_V Div(-\Sigma - (s\theta + n_k\mu_k)\mathbf{I}).\delta\mathbf{X}dX - \int_V (\theta Div(s\mathbf{I}) + \mu_k Div(n_k\mathbf{I})).\delta\mathbf{X}dX + \\ &+ \int_V (\partial_X W_0)_{expl}.\delta\mathbf{X}dX + \int_V \mathbf{F}^{\mathrm{T}}.\mathbf{f}_0.\delta\mathbf{X}dX + \int_{S_t} \frac{\partial \omega}{\partial \nabla_X \psi}.\mathbf{N}.\delta\psi dS - \int_{S_t} \mathbf{t}^d.\delta\psi dS - \int_{S_\psi} \delta(\mathbf{T}.\mathbf{N}).\psi^d dS \end{aligned} \quad [5.96]$$

The Eshelby stress defined in [5.95] and involved in [5.96] incorporates the density of the grand potential, the sum of the strain energy density $W_0(\mathbf{F})$ and of the complementary energy density $W_c(\mathbf{T})$; it thus differs from the classical version, which relies on $W_0(\mathbf{F})$ exclusively.

The volumetric contribution in [5.96] is then identified as

$$\delta\Omega_{vol} = 0 \Rightarrow Div\Xi + \left(\theta\nabla_X s + \mu_k \nabla_X n_k\right) + \mathbf{F}^T.\mathbf{f}_0 = \mathbf{0} \qquad [5.97]$$

introducing therein the following energy momentum tensor in a thermodynamic context

$$\Xi := -\Sigma - \left(s\theta + n_k \mu_k\right)\mathbf{I} \qquad [5.98]$$

This energy momentum tensor incorporates all forms of energies (mechanical through the classical Eshelby stress [17] $\Sigma := \mathbf{F}^T.\mathbf{T} - W\mathbf{I}$, and thermal and chemical forms in the last isotropic contribution), in line with [MAU 06a, 06b] who introduced more general internal variables in place of the chemical variables. Expressions of various energy momentum tensors including the so-called strain energy per unit stress-free volume (at fixed temperature and molar concentrations), have been proposed, (see [WU 01], equ. (45), (46); the reader is also referred to the pioneering work [ESH 70] related to those tensors). The balance law [5.97] is supplemented by the natural boundary conditions

$$t^d = \frac{\partial \omega}{\partial \nabla_X \psi}.\mathbf{N} \qquad [5.99]$$

Summarizing, a complete model of volumetric growth has been elaborated, formalized as the closed set of equilibrium equations [5.97] with boundary conditions [5.97], and constitutive behavior formally given by the dependency of the spatial density of the grand potential ω_{vol} versus its arguments, as written in the state laws [5.87]. This thermodynamic framework is further extended to account for situations in which surface growth is of importance in addition to volumetric growth.

5.8.2. *Extremum principle for combined surface and volumetric growth*

Independently of the volumetric contribution, one shall also consider an additional surface potential with (surface) density $\omega_{surf}\left[\tilde{\theta}, \tilde{\mathbf{F}}, \tilde{\mu}\right]$; this last contribution is the surface grand potential and is identified from the superficial excess quantity associated with $d\omega_{surf}$, viz.

$$d\omega_{surf} \equiv d\omega^{\sigma} = -s^{\sigma}d\theta - n_k^{\sigma}d\mu_k + \tilde{\mathbf{T}}^{el} : d\tilde{\mathbf{F}} \qquad [5.100]$$

It involves surface elastic stresses and conjugated strains, denoted as the quantities $\tilde{\mathbf{T}}^{el}, \tilde{\mathbf{F}}$ respectively. Gibbs' (1928) point of view of excess quantities at the interface between two phases has herewith been adopted (there is in Gibbs' point of view no extra volume of the interface separating the two phases). The surface excess

$$\omega^{\sigma} = \tilde{\mathbf{T}}^{el} : \tilde{\mathbf{F}} \qquad [5.101]$$

is the surface counterpart of [5.85]. Equality [5.100] shows that the true thermodynamic arguments of ω_{surf} are the (surface temperature), the superficial chemical potential and the surface strain. This last quantity may be chosen as an independent field or as the surface projection of $\tilde{\mathbf{F}}$.

The integration over a control volume (control surface resp.) of the volumetric and surface densities, quantities $\omega_{vol}[\theta, \mathbf{F}, \mu]$ ($\omega_{surf}\left[\tilde{\theta}, \tilde{\mathbf{F}}, \tilde{\mu}\right]$ resp.) further satisfies the identity

$$\int_{\Omega_g} \omega_{vol}[\theta, \mathbf{F}, \mu] dX_g + \int_{S_g} \omega_{surf}\left[\tilde{\theta}, \tilde{\mathbf{F}}, \tilde{\mu}\right] d\sigma_g = \int_{\Omega_g} \mathbf{T} : \mathbf{F} dX_g + \int_{S_g} \tilde{\mathbf{T}}^{el} : \tilde{\mathbf{F}} d\sigma_g \qquad [5.102]$$

fully expressed in the reference configuration. The right-hand side of previous equality involves the strain energy and complementary strain energy (they sum up to $\mathbf{T} : \mathbf{F}$ and $\tilde{\mathbf{T}}^{el} : \tilde{\mathbf{F}}$ regarding the volumetric and surface contributions respectively) for both the bulk and the surface.

The Gibbs adsorption equation may further be written

$$S^{\sigma} d\theta + A dg + n_i^{\sigma} du_i = 0$$

as the surface counterpart of the Gibbs-Duhem relation in classical thermodynamics [CAL 85].

The extension of the global energy balance [5.89] involving the additional surface contributions is then written

$$\int_{\Omega_g} \omega_{vol}[\theta,\mathbf{F},\mu]dX_g + \int_{S_g} \omega_{surf}\left[\tilde{\theta},\tilde{\mathbf{F}},\tilde{\mu}\right]d\sigma_g = \int_{\Omega_g} \mathbf{f}_0.\psi dX_g + \int_{\partial\Omega_g} (\mathbf{T.N}).\tilde{\psi}d\sigma_g + \int_{\partial\Omega_g} \left(\tilde{\mathbf{T}}^{el}.\mathbf{N}\right).\tilde{\psi}d\sigma_g$$
$$+\int_{S_g} \mathbf{F}_S.\tilde{\psi}d\sigma_g = \int_{\Omega_g} \mathbf{f}_0.\psi dX_g + \int_{S_{gT}} (\mathbf{T.N})^d.\tilde{\psi}d\sigma_g + \int_{S_{g\psi}} (\mathbf{T.N}).\tilde{\psi}^d d\sigma_g + \int_{S_g} \mathbf{F}_S.\tilde{\psi}d\sigma_g + \int_{\partial S_g} \left(\tilde{\mathbf{T}}^{el}.\mathbf{v}_g\right).\tilde{\tilde{\psi}}d\sigma_g \quad [5.103]$$

whereby a partition of the boundary has been made according to the nature of the boundary conditions (tractions $(\mathbf{T.N})^d$ imposed on S_{gt}; position imposed on $S_{g\psi}$, with $S_{gt} \cup S_{g\psi} = \partial\Omega_g$, $S_{gt} \cap S_{g\psi} = \varnothing$). The surface and line forces, respectively $\mathbf{F}_S$, and $\tilde{\mathbf{T}}.\mathbf{v}_g$ appear in [5.103] as dual (in the sense of the work of external forces) to the position variables $\tilde{\mathbf{x}},\tilde{\tilde{\mathbf{x}}},\tilde{\tilde{\mathbf{x}}}$ respectively, using the notations of the introductory section.

Based on the last equality in [5.103], a *zero potential* $Z[\theta,\mathbf{F},\mu]$ is elaborated as

$$Z[\theta,\mathbf{F},\mu] := \int_{\Omega_g} \omega_{vol}[\theta,\mathbf{F},\mu]dX_g + \int_{S_g} \omega_{surf}\left[\tilde{\theta},\tilde{\mathbf{F}},\tilde{\mu}\right]d\sigma_g - \int_{\Omega_g} \mathbf{f}_0.\psi dX_g -$$
$$\int_{S_{gT}} (\mathbf{T.N})^d.\tilde{\psi}d\sigma_g - \int_{S_{g\psi}} (\mathbf{T.N}).\tilde{\psi}^d d\sigma_g - \int_{S_g} \mathbf{f}_S.\tilde{\psi}d\sigma_g - \int_{\partial S_g} \left(\tilde{\mathbf{T}}.\mathbf{v}_g\right).\tilde{\tilde{\psi}}d\sigma_g \equiv 0 \quad [5.104]$$

Since the potential $Z[\theta,\mathbf{F},\mu]$ always vanishes, it is naturally endowed with an extremality property

$$\delta Z[\theta,\mathbf{F},\mu] = 0 \quad [5.105]$$

resulting in equilibrium equations following these steps:

1) Express the domain derivatives of the volume and surface energies:

$$\delta\int_{\Omega_g} \omega_{vol}[\theta,\mathbf{F},\mu]dX_g = \int_{\Omega_g} \delta\omega_{vol}[\theta,\mathbf{F},\mu]dX_g + \int_{\Omega_g} \omega_{vol}[\theta,\mathbf{F},\mu]\nabla.\delta\mathbf{x}_g dX_g,$$
$$\delta\int_{S_g} \omega_{surf}\left[\tilde{\theta},\tilde{\mathbf{F}},\tilde{\mu}\right]d\sigma_g = \int_{S_g} \delta\omega_{surf}\left[\tilde{\theta},\tilde{\mathbf{F}},\tilde{\mu}\right]d\sigma_g + \int_{S_g} \omega_{surf}\left[\tilde{\theta},\tilde{\mathbf{F}},\tilde{\mu}\right]\nabla_S.\delta\mathbf{X}_S d\sigma_g$$

Express further the variations therein

$$\delta\omega_{vol} = -s\delta\theta - n_k\delta\mu_k + \mathbf{T}:\delta\mathbf{F} \equiv \left(-s\nabla\theta - n_k\partial_{n_k}\mu_k\nabla n_k\right).\delta\mathbf{x}_g + \mathbf{T}:\delta\mathbf{F} \quad [5.106]$$

$$\delta\omega_{surf} = -s^{\sigma}\delta\theta - n_k^{\sigma}\delta\mu_k + \tilde{\mathbf{T}}:\delta\tilde{\mathbf{F}} \equiv \left(-s^{\sigma}\nabla_S\theta - n_k^{\sigma}\partial_{n_k^{\sigma}}\mu_k\nabla_S n_k^{\sigma}\right).\delta\mathbf{X}_S + \tilde{\mathbf{T}}:\delta\tilde{\mathbf{F}} \qquad [5.107]$$

2) Elaborate the contribution arising from the domain variation in step 1, viz.

$$\int_{\Omega_g}\omega_{vol}[\theta,\mathbf{F},\mu]\nabla.\delta\mathbf{x}_g dX_g = -\int_{\Omega_g}\nabla\omega_{vol}[\theta,\mathbf{F},\mu].\delta\mathbf{x}_g dX_g + \int_{\partial\Omega_g}\omega_{vol}[\theta,\mathbf{F},\mu].\delta\mathbf{X}_S.\mathbf{n}_g dX_g$$

$$\int_{S_g}\omega_{surf}\left[\tilde{\theta},\tilde{\mathbf{F}},\tilde{\mu}\right]\nabla_S.\delta\mathbf{X}_S d\sigma_g = -\int_{S_g}\nabla_S\omega_{surf}\left[\tilde{\theta},\tilde{\mathbf{F}},\tilde{\mu}\right].\delta\mathbf{X}_S d\sigma_g + \int_{\partial S_g}\omega_{surf}\left[\tilde{\theta},\tilde{\mathbf{F}},\tilde{\mu}\right]\delta\mathbf{X}_S.\mathbf{v}_g dl_g$$

using the surface divergence theorem. The volumetric and surface Eshelby stresses $\mathbf{\Sigma},\tilde{\mathbf{\Sigma}}$ are introduced (as in [GAN 10c], Appendix 2)

$$\mathbf{\Sigma} := \mathbf{F}^t.\mathbf{T} - \omega_{vol}[\theta,\mathbf{F},\mu]\mathbf{I};$$
$$\tilde{\mathbf{\Sigma}} := \tilde{\mathbf{F}}^t.\tilde{\mathbf{T}} - \omega_{surf}\left[\tilde{\theta},\tilde{\mathbf{F}},\tilde{\mu}\right]\mathbf{I}_S \qquad [5.108]$$

3) Assemble the previous variations, hence the variation of $Z[\theta,\mathbf{F},\mu]$ written finally as (dropping here the arguments of the volumetric and surface densities for simplicity)

$$\begin{aligned}\delta Z[\theta,\mathbf{F},\mu] = &\int_{\Omega_g}\left(-s\nabla\theta - n_k\partial_{n_k}\mu_k\nabla n_k + \mathbf{f}_0\right).\delta\mathbf{X}_g dX_g + \\ &+\int_{S_g}\left[\left(-s^{\sigma}\nabla_S\theta - n_k^{\sigma}\partial_{n_k^{\sigma}}\mu_k\nabla_S n_k^{\sigma}\right)\mathbf{I} + \nabla_S.\tilde{\mathbf{\Sigma}} - \mathbf{P}.\mathbf{K}'.\partial_N\omega_{surf} + \mathbf{N}.\mathbf{\Sigma}\right].\delta\mathbf{X}_S d\sigma_g \\ &+\int_{\partial\Omega_g}\omega_{vol}\mathbf{N}\delta\mathbf{X}_S dx_g - \int_{S_g}\mathbf{f}_S.\delta\mathbf{X}_S d\sigma_g - \int_{S_{gt}}(\mathbf{T}.\mathbf{N})^d.\delta\mathbf{X}_S d\sigma_g - \int_{\partial S_g}\left(\tilde{\mathbf{T}}.\mathbf{v}_g\right).\delta\tilde{\tilde{\mathbf{X}}}_S d\sigma_g + \int_{\partial S_g}\omega_{surf}\delta\tilde{\tilde{\mathbf{X}}}_S.\mathbf{v}_g dl_g \equiv 0\end{aligned} \qquad [5.109]$$

assuming a nil prescribed position field and neglecting the variation of the surface stresses (the applied loads have been considered as dead loads).

4) Express the balance equations as a result of the arbitrariness and independence of the variations $\delta\mathbf{x}_g,\delta\mathbf{X}_S$:

$$-s\nabla\theta - n_k\partial_{n_k}\mu_k\nabla n_k + \nabla.\mathbf{\Sigma} + \mathbf{f}_0 = \mathbf{0} \text{ in } \Omega_g \qquad [5.110a]$$

$$\left(\omega_{vol} - s^{\sigma}\nabla_S\theta - n_k^{\sigma}\partial_{n_k^{\sigma}}\mu_k\nabla_S n_k^{\sigma}\right)\mathbf{N} + \nabla_S.\tilde{\mathbf{\Sigma}} - \mathbf{P}.\mathbf{K}'.\partial_N\omega_{surf} + \mathbf{f}_S = \mathbf{0} \text{ on } S_g\setminus S_{gt} \qquad [5.110b]$$

with boundary conditions

$$\left(-s^{\sigma}\nabla_S\theta - n_k^{\sigma}\partial_{n_k^{\sigma}}\mu_k\nabla_S n_k^{\sigma}\right) + \nabla_S.\tilde{\mathbf{\Sigma}} - \mathbf{P}.\mathbf{K}^t.\partial_N\omega_{surf} + \omega_{vol}\mathbf{N} = (\mathbf{T}.\mathbf{N})^d \qquad [5.110c]$$

$$\omega_{surf}\mathbf{v}_g = \mathbf{T}.\mathbf{v}_g \Rightarrow \omega_{surf} = \mathbf{v}_g.\mathbf{T}.\mathbf{v}_g \text{ on } \partial S_g \qquad [5.110d]$$

The factors of the variations $\delta\mathbf{x}_g$ and $\delta\mathbf{X}_S$ in [5.109] may be identified as material driving forces for growth (in the sense of configurational mechanics [MAU 93]), since those variations represent virtual growth velocities; focusing on surface growth, the material surface driving force is then identified from [5.109] as

$$\tilde{\Upsilon} = \left(-s^{\sigma}\nabla_S\theta - n_k^{\sigma}\partial_{n_k^{\sigma}}\mu_k\nabla_S n_k^{\sigma}\right)\mathbf{I}_S + \nabla_S.\tilde{\mathbf{\Sigma}} - \mathbf{P}.\mathbf{K}^t.\partial_N\omega_{surf} - \mathbf{f}_S + \omega_{vol}\mathbf{N} \qquad [5.111]$$

As only the normal contribution can be considered as the true driving force, calculations similar to [24] lead to the normal projection of $\tilde{\Upsilon}$ as

$$\tilde{\Upsilon}_{gN} := \tilde{\Upsilon}.\mathbf{N} = \mathbf{N}.\mathbf{\Sigma}.\mathbf{N} + \tilde{\mathbf{\Sigma}} : \mathbf{K} + \nabla_S.(\tilde{\mathbf{\Sigma}}.\mathbf{N}) - \left(s^{\sigma}\nabla_S\theta + n_k^{\sigma}\partial_{n_k^{\sigma}}\mu_k\nabla_S n_k^{\sigma}\right) \qquad [5.112]$$

with the volumetric and surface Eshelby stress and curvature tensor given successively by

$$\mathbf{\Sigma} := \mathbf{F}^t.\mathbf{T} - \omega_{vol}[\theta, \mathbf{F}, \mu]\mathbf{I}\,;\, \tilde{\mathbf{\Sigma}} := \tilde{\mathbf{F}}^t.\tilde{\mathbf{T}} - \omega_{surf}\left[\tilde{\theta}, \tilde{\mathbf{F}}, \tilde{\mu}\right]\mathbf{I}_S\,;\, \mathbf{K} := -\nabla_R\mathbf{N} \qquad [5.113]$$

It is important to note that both the volumetric and surface densities, respectively the scalar functions $\omega_{vol}, \omega_{surf}$, account for all forms of energies (mechanical, thermal and chemical), as reflected in the right hand side of [5.86] and [5.100]. Especially, the surface driving force in [5.112] incorporates curvature effects (first term), a mechanical driving force (second and third term, also partly included in the bulk contribution, and a thermal and chemical contribution (part of ω_{vol} and ω_{surf}, and last parenthesis).

PART 2

Multiscale Bone Mechanics

6

Micropolar Models of Trabecular Bone

This chapter deals with the construction of Cosserat models of trabecular bone in 2D and 3D situations, based on micromechanical approaches to investigating microstructure-related scale effects on the macroscopic properties of bone. The cell walls of the bone microstructure are modeled as Timoshenko thick beams. Closed form expressions of the equivalent properties are obtained versus the geometrical and mechanical microparameters, accounting for the effects of bending, axial, and transverse shear deformations; torsion is additionally considered for a 3D geometry. The classical and micropolar effective moduli and the internal flexural and torsional lengths are identified versus the micropolar material constants. The scaling laws of the effective moduli versus density are determined in situations of low and high effective densities to assess the impact of the transverse shear deformation. Finite element models over the local architecture of the trabeculae and also over the whole lattice are performed to validate the results from homogenization. The stress distribution in a cracked 2D bone sample is also computed based on the effective micropolar model, highlighting the regularizing effect of the Cosserat continuum in comparison to a classical elasticity continuum model.

6.1. A survey of bone physiology

Bone is a particularly complex multiscale material, which is hierarchically organized to provide maximum performance with a minimum amount of material (Figure 6.1). Bone has two distinct types of tissues: cortical bone is dense and compact and forms the outer surface of bones, and it encloses the trabecular bone tissue with a honeycomb-like structure, made of trabeculae surrounded by bone marrow. The mineralized matrix of bone tissue has an organic component consisting mainly of type I collagen and an inorganic component of bone mineral made up of various salts.

Chapter written by Jean-François GANGHOFFER and Ibrahim GODA.

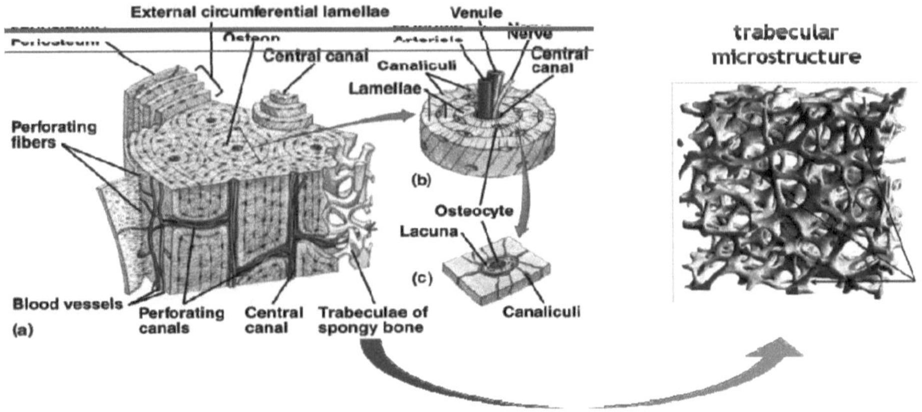

Figure 6.1. *Histological description of trabecular and cortical bone. Trabecular bone (80% of the human skeleton) has an open cell porous structure. The cortical region is compact, (20%), and contains lamellae of collagen fibers, with sensing cells trapped between the stratified layers (osteocytes), and other cells communicating by tubules through the collagen matrix. For a color version of this figure, see www.iste.co.uk/ ganghoffer/biomechanics.zip*

At the nanoscale, bone is composed of collagen fibers, apatite crystals and water; at the sub-microscale (1–10 lm), the arrangements of collagen fibers reinforced with crystals form lamellas. These are packed and organized in different orientations to form cylindrical struts called trabecula at the microscale (several hundred of micrometers). At the mesoscale, the complex and very organized network of trabeculae forms the cancellous bone. The macro scale represents the whole bone which is constituted by both the cancellous (trabecular) bone and the cortical (compact) bone. The location of these bone types in the femur is further illustrated in Figure 6.2; cortical bone is a dense material with a specific gravity of almost two in humans and a little over two in cattle; it forms most of the outer shell of a whole bone, a shell of variable thickness. Cancellous (porous) bone generally exists only within the confines of the cortical bone coverings; it is composed of short struts of bone material called trabeculae. The connected trabeculae give cancellous bone a spongy appearance, and it is consequently often called spongy bone.

Trabecular bone is very complex due to its randomness and spatial heterogeneity. In order to simplify the forthcoming analysis, one can choose to represent bone as having an idealized periodic microstructure. This work addresses the modeling of trabecular bone as hexagonal periodic structures, considering successively 2D and 3D geometries.

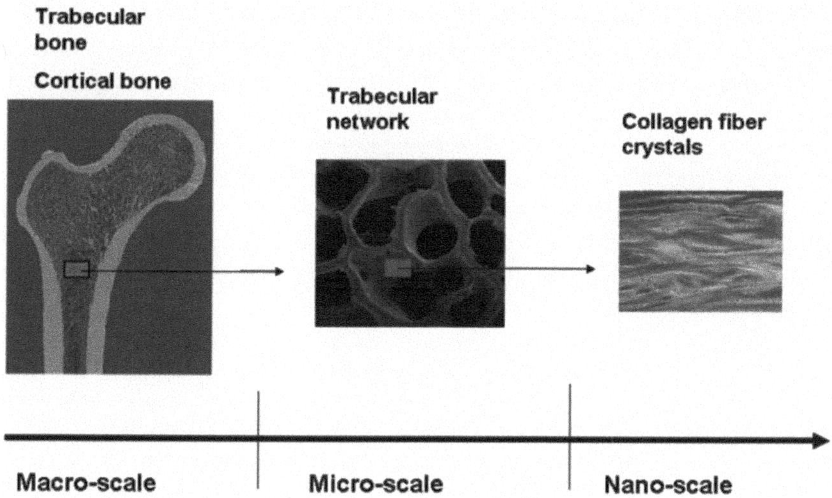

Figure 6.2. *Multiscale aspects of bone structure. For a color version of this figure, see www.iste.co.uk/ganghoffer/biomechanics.zip*

6.2. Review of trabecular bone models

The existence of couple stress was first evidenced by [YAN 81] who measured the effect of the size of a bone specimen on the apparent stiffness of cortical bone in quasi-static torsion; they further obtained the characteristic length scales for torsion and bending for cortical bone in the context of couple stress theory. Bone trabeculae are modelled as isotropic micropolar materials in [LAK 93]. Lakes and co-workers conducted a series of experiments on bone and other cellular materials, in which they observed a stiffening effect in such materials in bending and in torsion [PAR 87] and a tougher notched bone [LAK 90] than expected from classical Cauchy-type elasticity. Those last authors found that Cosserat elasticity provides better predictions of the response of bone than classical elasticity theory. [TAN 99] represented the trabeculae as beam elements in a lattice structure in which the lattice elements are rigidly interconnected to each other, for the purpose of bone remodeling. The microstructure was then embedded in the continuum in the context of the couple stress theory. The issue of application of higher-order continuum theories to mechanical analysis of bone (both cortical and cancellous) was addressed by [FAT 02], who analyzed a simplified two-dimensional bone-prosthesis configuration using a micropolar-based FE formulation. The stress and strain intensities they calculated at the bone-prosthesis interface are different from those predicted by classical elasticity. In addition, [FAT 03] identified the micropolar elastic constants of cancellous bone in the context of micromechanical analyses. In this approach, it is assumed that at the microscopic level the bone tissue is an isotropic, Cauchy-type elastic material, whereas cancellous bone behaves as a homogeneous, anisotropic micropolar-type continuum at the macroscopic level. The effective elastic constants for the micropolar continuum were determined from the response of a bone specimen, whose microstructure was obtained from micro-CT scans. This, in addition to experimental evidence reported by, e.g., [LAK 93], makes Cosserat theory a suitable candidate for the continuum modeling of cellular materials, because it contains rotations as degrees of freedom. Moreover [LAK 90, SHM 07] found that Cosserat elasticity gave better predictions of the response of bone and cellular materials than classical elasticity theory.

The evaluation of the effective mechanical properties of trabecular bone and their relationships to microstructural parameters is a critical issue when analysing stresses and strains in bone tissues and simulating their remodeling. The known limitations of experimental methods make the homogenization of "equivalent" trabecular microstructures an advantageous tool for this task. Many papers reporting results of mechanical or acoustical tests of trabecular bone have been published, e.g. [ASH 86, 89, RHO, 95, KOP 98, CEN 99] , and recent contributions [HEL 04] and [DIA 07]. However, most of the time they are limited to the identification of a few mechanical properties; in particular, anisotropic properties are difficult to access, thus pointing towards the need for predictive models of cancellous bone based on its architecture. Moreover, due to the variation of bone architecture (geometry) according to location, such predictive models will exhibit the advantage of their versatility, in the sense they are able to provide quantitative estimates of bone mechanical properties versus the bone architectural and micromechanical parameters.

This chapter focuses on the modeling of the micropolar elastic response of trabecular bone using 2D and 3D geometric models.

6.3. Construction of two-dimensional anisotropic Cosserat bone models

We adopt the viewpoint of trabecular bone as a cellular solid consisting of a quasi periodical lattice of cells with a hexagonal topology, the cell walls being modeled as thick beams. The characteristics of many periodic cellular structures are discussed in detail in the well-known contribution of [GIB 97], wherein the equivalent mechanical properties of honeycomb lattices are derived by analyzing strain and stress states in unit cells through the application (in most cases) of beam theory. Taking bending as the sole deformation mechanism, [GIB 82] derived the in-plane elastic constants for regular honeycombs with cell walls of uniform thickness; their normalized results are consequently independent of the honeycomb relative density. Taking bending and stretching as the main deformation mechanisms, [WAR 87] also developed a model to study the in-plane elastic properties of honeycombs. Considering the transverse shear as an additional deformation mechanism, [SIL 95] and [ZHU 10] obtained the closed form in-plane elastic constants of regular honeycombs. Taking bending, stretching and hinging as the deformation mechanisms, [MAS 96] theoretically analyzed the in-plane elastic constants of honeycombs. The normalized in-plane elastic constants of honeycombs generally depend on the honeycomb relative density [SIL 95, ZHU 01, ZHU 10], because the axial stretching/compression and transverse shear play an important role in the deformation.

A two-dimensional micropolar continuum model equivalent to the initial lattice is next constructed relying on discrete homogenization, and its mechanical properties are identified versus the microbeams geometry and mechanical behavior.

The adopted beam model includes transverse shear, which is necessary for thick beams. In a local coordinate systems attached to each beam element, the normal (tension) and transverse efforts adopt the expressions in asymptotic expansion forms as

$$N^{b\varepsilon} = E_s\eta\left(\mathbf{e}^b.\left(\varepsilon\,\Delta\mathrm{U}_1^{\mathrm{b}} + \varepsilon^2\,\Delta\mathrm{U}_2^{\mathrm{b}}\right)\right) = \varepsilon N_1^b + \varepsilon^2 N_2^b \qquad [6.1]$$

$$T_t^{b\varepsilon} = \frac{E_s\eta^3}{(1+\dfrac{E_s\eta^2}{(G_s k_s)})}\left(\mathbf{e}^{b\perp}.\left(\varepsilon\,\Delta\mathrm{U}_1^{\mathrm{b}} + \varepsilon^2\,\Delta\mathrm{U}_2^{\mathrm{b}}\right) - \varepsilon\frac{L^b}{2}\left(\phi_0^{O_{R(b)}} + \phi_0^{E_{R(b)}} + \varepsilon\left(\phi_1^{O_{R(b)}} + \phi_1^{E_{R(b)}} + \frac{\partial\phi_0}{\partial\lambda^i}\delta^{ib}\right)\right)\right)$$
$$= \varepsilon T_{t1}^b + \varepsilon^2 T_{t2}^b \qquad [6.2]$$

and the moments at both extremities of the beam are expressed as:

$$\mathrm{M}^{O(b)\varepsilon} = \varepsilon^2 M_1^{O(b)} + \varepsilon^3 M_2^{O(b)}\ ;\ \mathrm{M}^{E(b)\varepsilon} = \varepsilon^2 M_1^{E(b)} + \varepsilon^3 M_2^{E(b)} \qquad [6.3]$$

with the moments therein expressing as

$$M_1^{O(b)} = \frac{E_s\eta^3 L^b}{2(1+\dfrac{E_s\eta^2}{(G_s k_s)})}\left(-e^{b\perp}.\Delta\mathrm{U}_1^{\mathrm{b}}\right) + \frac{E_s\eta^3\left(L^b\right)^2}{12\left(1+\dfrac{E_s\eta^2}{(G_s k_s)}\right)}\left(\left(4+\frac{E_s\eta^2}{(G_s k_s)}\right)\phi_0^{O_{R(b)}} + \left(2-\frac{E_s\eta^2}{(G_s k_s)}\right)\phi_0^{E_{R(b)}}\right)$$

$$M_2^{O(b)} = \frac{E_s\eta^3 L^b}{2(1+\dfrac{E_s\eta^2}{(G_s k_s)})}\left(-e^{b\perp}.\Delta\mathrm{U}_2^{\mathrm{b}}\right) + \frac{E_s\eta^3\left(L^b\right)^2}{12\left(1+\dfrac{E_s\eta^2}{(G_s k_s)}\right)}\left(\left(4+\frac{E_s\eta^2}{(G_s k_s)}\right)\phi_1^{O_{R(b)}} + \left(2-\frac{E_s\eta^2}{(G_s k_s)}\right)\phi_1^{E_{R(b)}} + \left(2-\frac{E_s\eta^2}{(G_s k_s)}\right)\frac{\partial\phi_0}{\partial\lambda^i}\delta^{ib}\right)$$

$$M_1^{E(b)} = \frac{E_s \eta^3 L^b}{2(1+\frac{E_s\eta^2}{(G_s k_s)})}\left(-e^{b\perp}.\Delta \mathrm{U}_1^{\mathrm{b}}\right) + \frac{E_s\eta^3\left(L^b\right)^2}{12\left(1+\frac{E_s\eta^2}{(G_s\kappa_s)}\right)}\left(\left(2-\frac{E_s\eta^2}{(G_s k_s)}\right)\phi_0^{O_{R(b)}} + \left(4+\frac{E_s\eta^2}{(G_s k_s)}\right)\phi_0^{E_{R(b)}}\right)$$

$$M_2^{E(b)} = \frac{E_s \eta^3 L^b}{2(1+\frac{E_s\eta^2}{(G_s k_s)})}\left(-e^{b\perp}.\Delta \mathrm{U}_2^{\mathrm{b}}\right) + \frac{E_s\eta^3\left(L^b\right)^2}{12\left(1+\frac{E_s\eta^2}{(G_s k_s)}\right)}\left(\left(2-\frac{E_s\eta^2}{(G_s k_s)}\right)\phi_1^{O_{R(b)}} + \left(4+\frac{E_s\eta^2}{(G_s k_s)}\right)\phi_1^{E_{R(b)}} + \left(4+\frac{E_s\eta^2}{(G_s k_s)}\right)\frac{\partial\phi_0}{\partial\lambda^i}\delta^{ib}\right)$$

with the non-dimensional factor $\Phi_y = 12E_S I_z / G_S k_S L^2$ vanishing when transverse shear is neglected, $\mathbf{e}^{\mathrm{b}}$ the unit director for each beam, and $\mathbf{e}^{\mathrm{b}\perp}$ the transverse unit vector. The coefficients E_s, G_s are respectively the Young's modulus and shear modulus of the trabecular struts, A, L are respectively the cross-sectional area and the length of the strut, I_z is the second moment of area of the strut cross-section, and k_S is the shear correction factor.

The kinematic and static variables for any beam in the lattice are represented in the local coordinate system associated with the Timoshenko beam element in Figure 6.3.

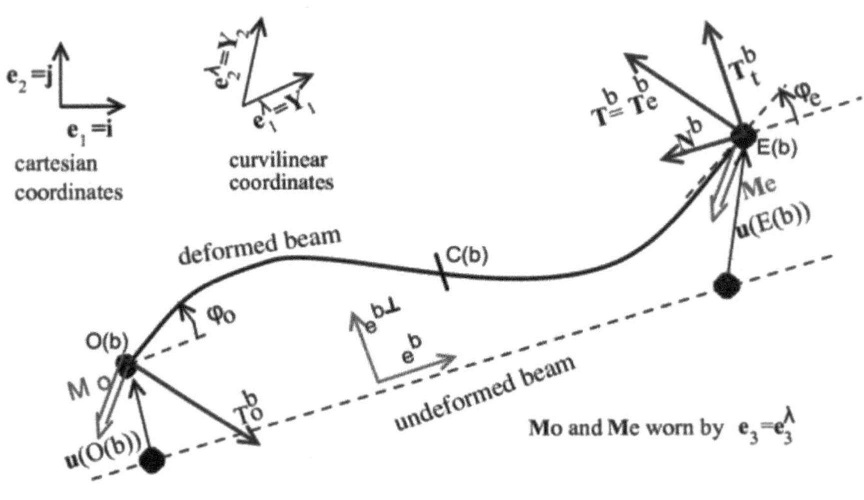

Figure 6.3. *Kinematic and static parameters of a beam element. For a color version of this figure, see www.iste.co.uk/ganghoffer/biomechanics.zip*

The constitutive behavior of the equivalent anisotropic micropolar continuum is obtained in the present 2D context in matrix format as

$$\begin{Bmatrix} \sigma_x \\ \sigma_y \\ \sigma_{xy} \\ \sigma_{yx} \\ m_{xz} \\ m_{yz} \end{Bmatrix} = \begin{bmatrix} [\mathbf{E}] & [\mathbf{B}] \\ [\mathbf{C}] & [\mathbf{K}] \end{bmatrix} \begin{Bmatrix} \epsilon_x \\ \epsilon_y \\ \epsilon_{xy} \\ \epsilon_{yx} \\ \chi_{xz} \\ \chi_{yz} \end{Bmatrix} \qquad [6.4]$$

This constitutive law can further be simplified based on symmetry properties of the studied lattices: it has indeed been shown that for centro-symmetrical lattices, the pseudo-tensors $[\mathbf{B}]$ and $[\mathbf{C}]$ vanish.

The previous form of the continuum constitutive law can presently be identified from the expressions of the homogenized stress S_1^i and couple stress μ_2^{i} vectors as

$$\boldsymbol{\sigma} = \underbrace{\frac{1}{g}\mathbf{S}_1^{\mathrm{i}} \otimes \frac{\partial \mathbf{R}}{\partial \lambda^{\mathrm{i}}}}_{[\mathbf{E}]\{\epsilon\}}, \quad \mathbf{m} = \underbrace{\frac{1}{g}\varepsilon^2 \boldsymbol{\mu}_2^{\mathrm{i}} \otimes \frac{\partial \mathbf{R}}{\partial \lambda^{\mathrm{i}}}}_{[\mathbf{K}]\{\chi\}} \qquad [6.5]$$

The homogenized stress and couple stress vectors are respectively

$$\mathrm{S}_1^i = \sum_{b \in B_R} \left(\left(\mathrm{E}_s \eta \left(\mathbf{e}^b . \Delta \mathrm{U}_1^{\mathrm{b}} \right) \mathbf{e}^b + \left(\frac{E_s \eta^3}{\left(1 + \frac{E_s \eta^2}{(G_s k_s)} \right)} \left(\mathbf{e}^{\perp b} . \Delta \mathrm{U}_1^{\mathrm{b}} - \frac{L^b}{2} \left(\phi_0^{O_{R(b)}} + \phi_0^{E_{R(b)}} \right) \right) \right) \mathbf{e}^{b\perp} \right) \right) \delta^{ib} = \sum_{b \in B_R} \left(N_1^b \mathbf{e}^b + T_{t1}^b \mathbf{e}^{b\perp} \right) \delta^{ib}$$

$$\mu_2^{\mathrm{i}} = \sum_{b \in B_R} \left(\mathrm{E}_{\mathrm{s}} \eta^3 \frac{\left(L^b\right)^2}{12} \left(\phi_1^{E_{R(b)}} - \phi_1^{O_{R(b)}} + \frac{\partial \phi_0}{\partial \lambda^i} \delta^{ib} \right) \right) \delta^{ib} = \sum_{b \in B_R} \left(\frac{1}{2} \left(M_2^{E(b)} - M_2^{O(b)} \right) \right) \delta^{ib} \qquad [6.6]$$

with N_1^b, T_{t1}^b, and, M_2^n, respectively, the first order normal and transverse effort and the second order moment. Those expressions still involve the unknown displacements u_1^n, u_2^n and rotations ϕ_0^n, ϕ_1^n (this defines the so called localization problem over the representative cell), which are determined for all nodes and as a last step by solving the equilibrium equations in translation and rotation.

6.4. Effective micropolar properties of bone

The general anisotropic hexagonal unit cell of two-dimensional honeycomb under consideration is pictured in Figure 6.4; it consists of three beams, a vertical beam of length h and two inclined beams of length l. The vertical beams have elastic and shear moduli E_{s1} and G_{s1}, respectively, while the two inclined beams have elastic and shear moduli E_s and G_s, respectively. The dimensionless parameters θ and $\alpha=h/l$ are descriptors of the cell shape, as well as the aspect ratio of the cell walls $\eta=t/l$, a parameter determining the relative density or area fraction of the solid phase. The lattice is called re-entrant when the angle θ between the two inclined beams relative to the horizontal is negative (Figure 6.5b). When this angle takes the value 30°, the classical hexagonal lattice is obtained. When $\theta < 0$, $\left(\theta \in \left[-\frac{\pi}{2}\frac{\pi}{2}\right]\right)$, an auxetic lattice [GON 08, DOS 12a, DOS 12b] having a negative Poisson's ratio is obtained. Regular unit cells correspond to the specific geometrical parameters $\alpha=1$ and $\theta=30°$ and anisotropic unit cells correspond to all other values.

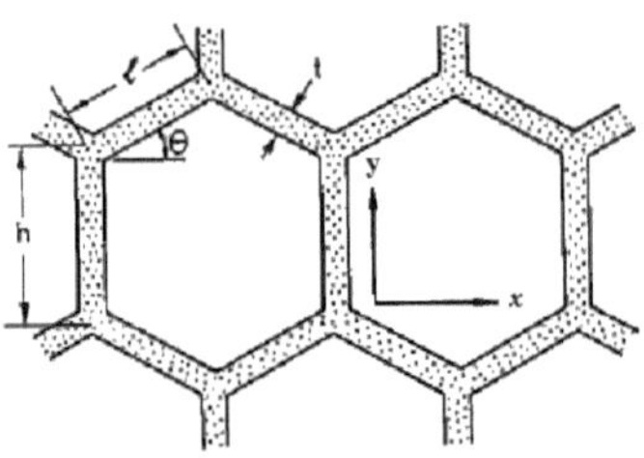

Figure 6.4. *General anisotropic unit cell for two dimensional honeycomb models. The geometrical descriptors are the cell angle (θ), the vertical cell length (h), the inclined cell length (l), and the wall thickness (t)*

The lengths of the periodicity vectors (norm of $\mathbf{Y}_1$ and $\mathbf{Y}_2$ in the Cartesian basis) are $L_1 = L_2 = \sqrt{(l\cos\theta)^2 + (h + l\sin\theta)^2}$; the other features of the lattice are collected in the connectivity Table 6.1.

Beam	1	2	3
O(b)	2	1	1
E(b)	1	2	2
δ_1	0	1	0
δ_2	0	0	1

Table 6.1. *Connectivity array for the hexagonal lattice*

The representative unit cell is easily identified for the hexagonal lattice (Figure 6.5).

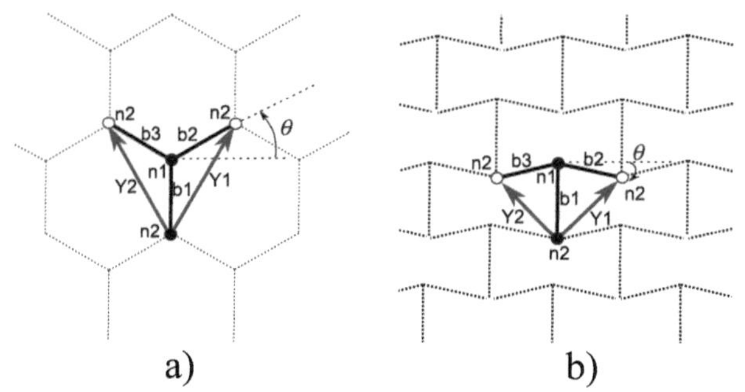

Figure 6.5. *Representative unit cell of the investigated lattices: (a) hexagonal lattice, (b) re-entrant lattice (θ < 0)*

The homogenization scheme provides the effective mechanical properties (classical and micropolar moduli) versus the geometrical and micromechanical parameters of the considered microstructure. After calculations, we extract the effective homogenized moduli of the hexagonal lattice from the equivalent stiffness matrix, expressed versus the dimensionless parameters η, α, θ, κ_s, and κ_{s1} and the elastic and shear moduli of the solid cell walls E_s, E_{s1}, G_{s1}, and G_s as follows:

$$E_x^* = -\frac{E_s\eta^3 C_t}{\left(-1+C_t^2-C_t^2\eta^2+\frac{C_t^2E_s\eta^2}{G_sk_s}-\frac{E_s\eta^2}{G_sk_s}\right)(S_t+\alpha)},\ E_y^* = \frac{(S_t+\alpha)E_sE_{s1}\eta^3}{\left(C_t^2E_{s1}+\frac{C_t^2E_sE_{s1}\eta^2}{G_sk_s}+E_{s1}\eta^2+2E_s\eta^2\alpha-C_t^2E_{s1}\eta^2\right)C_t},$$

$$G_{xy}^* = \frac{(S_t+\alpha)E_sE_{s1}\eta^3C_t}{\left(\frac{E_s\eta^2\alpha^2C_t^2E_{s1}}{G_sk_s}+\alpha^2E_{s1}C_t^2+\frac{2E_sE_{s1}\eta^2\alpha C_t^2}{G_{s1}k_{s1}}+2\alpha^3C_t^2E_s+2E_{s1}\eta^2\alpha\ S_t+E_{s1}\eta^2+\alpha^2E_{s1}\eta^2-E_{s1}\eta^2\alpha^2\ C_t^2\right)},$$

$$\nu_{xy}^* = -\frac{\left(\frac{E_s\eta^2}{G_sk_s}-\eta^2+1\right)S_tC_t^2}{\left(\frac{C_t^2E_s\eta^2}{G_sk_s}-C_t^2\eta^2+C_t^2-\frac{E_s\eta^2}{G_sk_s}-1\right)(S_t+\alpha)},\ \nu_{yx}^* = \frac{\left(\frac{E_s\eta^2}{G_sk_s}-\eta^2+1\right)S_t(S_t+\alpha)E_{s1}}{\left(\frac{C_t^2E_sE_{s1}\eta^2}{G_sk_s}+E_{s1}C_t^2+E_{s1}\eta^2+2E_s\eta^2\alpha-C_t^2E_{s1}\eta^2\right)},$$

$$K_{11} = \frac{1}{12}\frac{E_s\eta^3l^2C_t}{\alpha+S_t},\ K_{22} = \frac{1}{12}\frac{l^2E_s\eta^3E_{s1}(\alpha+S_t)}{(2E_s\alpha+E_{s1})C_t}. \qquad [6.7]$$

The characteristic length and coupling coefficient receive the following expressions

$$l_b = \frac{1}{12}\sqrt{3}\sqrt{-\frac{l^2\left(\frac{E_s\eta^2\alpha^2C_t^2E_{s1}}{G_sk_s}+E_{s1}\eta^2+\frac{2\alpha E_sC_t^2E_{s1}\eta^2}{G_{s1}k_{s1}}+2\alpha^3E_sC_t^2+2E_{s1}\eta^2S_t\alpha+\alpha^2\eta^2E_{s1}-C_t^2\alpha^2\eta^2E_{s1}+C_t^2\alpha^2E_{s1}\right)}{E_{s1}\left(-2S_t\alpha-\alpha^2-1+C_t^2\right)}},$$

$$N_{coupl}^2 = \frac{\left(\begin{array}{l}\frac{E_s\eta^2\alpha^2C_t^2E_{s1}}{G_sk_s}+E_{s1}C_t^2\alpha^2+2\alpha^3E_sC_t^2+\frac{2\alpha E_sC_t^2E_{s1}\eta^2}{G_{s1}k_{s1}}+\\ +2E_{s1}\eta^2S_t\alpha+\eta^2E_{s1}+\alpha^2\eta^2E_{s1}-C_t^2\alpha^2\eta^2E_{s1}\end{array}\right)}{\left(\begin{array}{l}4S_t\alpha E_{s1}+2E_{s1}\alpha^2+\frac{4S_t\alpha\eta^2E_sE_{s1}}{G_sk_s}+E_{s1}\alpha^2\eta^2+\frac{2E_sE_{s1}\alpha^2\eta^2}{G_sk_s}-E_{s1}\eta^2\alpha^2C_t^2-2C_t^2E_{s1}-\frac{2C_t^2E_sE_{s1}\eta^2}{G_sk_s}\\ +2E_{s1}\eta^2\alpha S_t+\frac{E_s\eta^2\alpha^2C_t^2E_{s1}}{G_sk_s}+2E_{s1}+E_{s1}C_t^2\alpha^2+\frac{2E_s\eta^2E_{s1}}{G_sk_s}+\frac{2C_t^2E_sE_{s1}\eta^2\alpha}{G_{s1}k_{s1}}+2\alpha^3E_sC_t^2+E_{s1}\eta^2\end{array}\right)}. \quad [6.8]$$

We next compare the effective properties of irregular hexagonal lattices with the values determined by [GIB 97] for a periodic array of hexagonal cells accounting for bending, axial and shear deformations of the cell walls. The homogenized properties obtained with a Euler Bernoulli beam model are plotted against cell angle θ for the purpose of assessing the impact of transverse shear (Figure 6.6 and 6.7). In order to perform this comparison, it is assumed that properties of both vertical and inclined members are the same ($E_s = E_{s1}$, $G_s = G_{s1}$, and $k_s = k_{s1}$).

All three moduli show quite similar evolutions versus the cell angle (Figure 6.6), whereas the absence of transverse shear tends to overestimate those moduli, as expected.

Both Poisson's ratios are antisymmetric function of θ (see Figure 6.7), and become negative for re-entrant configurations. The originality of the present approach is the derivation of an anisotropic micropolar model for cellular solids from homogenization of the underlying lattice. Contrary to this, many phenomenologically based generalized continuum models for bone postulate such a behavior. The homogenized properties are close to Gibson and Ashby values as pictured on previous figures with relative difference reaching a maximum of about 4% for the Young's modulus E_x^* and Poisson's ratios ν_{xy}^* and ν_{yx}^*, and about 2% for E_y^* and G_{xy}^*.

The effect of cell geometry on the effective properties is determined by considering a much lower value of $\alpha = 1/2$, leading to a concave evolution of the effective shear modulus versus θ. In this situation, we observe a noticeable difference between the homogenized shear modulus and Gibson model and this difference decreases when increasing the cell angle, as shown in Figure 6.8.

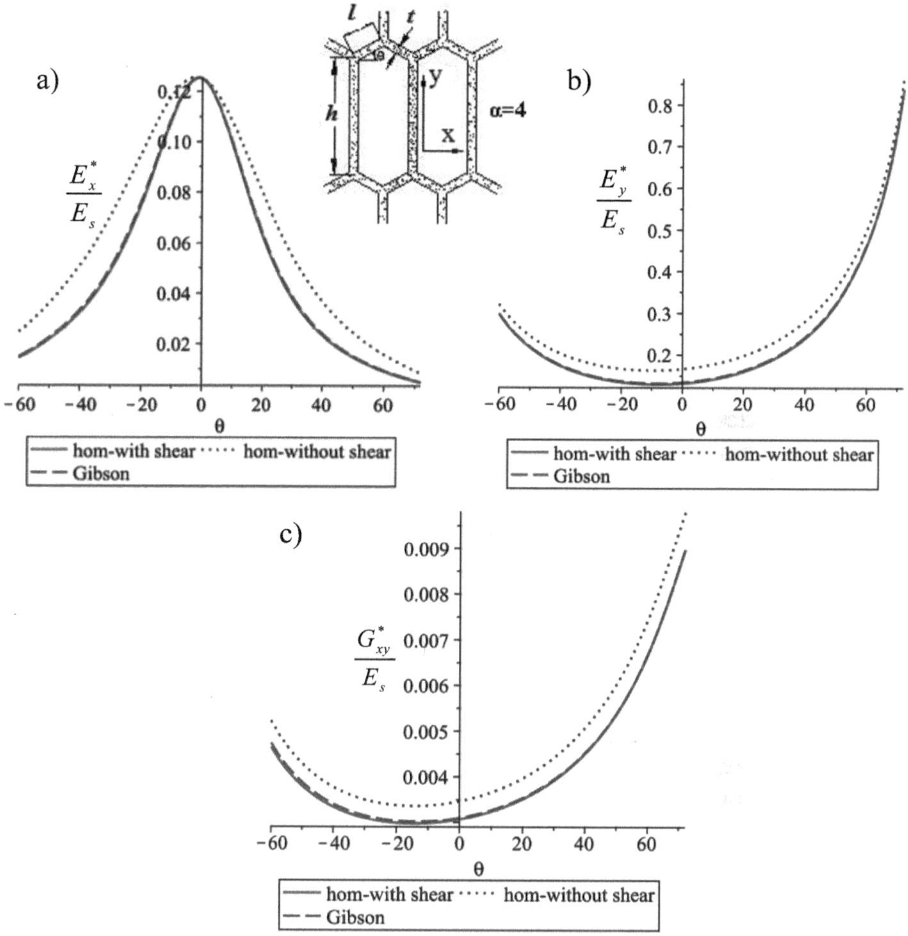

Figure 6.6. *Comparison between the homogenized elastic and shear moduli with and without considering shear effects and Gibson model for the hexagonal lattice. η=1/2, α=6. For a color version of this figure, see www.iste.co.uk/ganghoffer/biomechanics.zip*

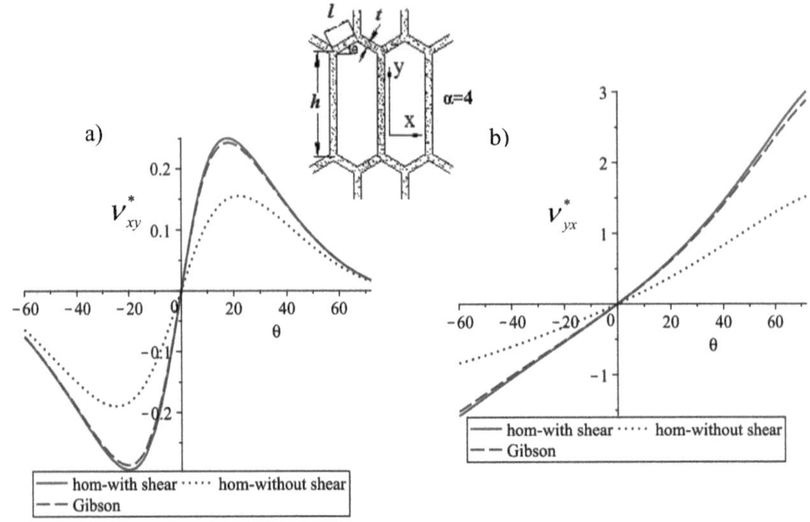

Figure 6.7. *Comparison between the homogenized Poisson's ratios v^*_{xy} and v^*_{yx} with the values obtained by Gibson for the hexagonal lattice. η=1/2, α=6. For a color version of this figure, see www.iste.co.uk/ganghoffer/biomechanics.zip*

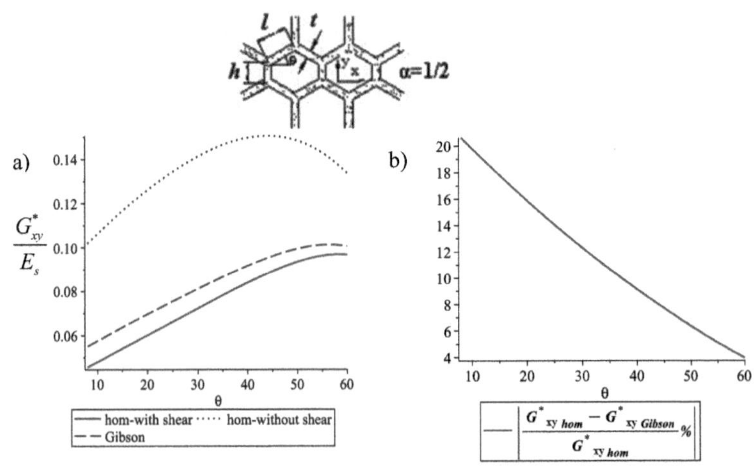

Figure 6.8. *(a) Comparison between the homogenized shear modulus G^*_{xy} / E_s (with and without shear effects) with the values obtained by Gibson for the hexagonal lattice. η=1/2, α=1/2. (b) Relative difference (in percent) between the effective modulus G^*_{xy} obtained by homogenization and Gibson model. For a color version of this figure, see www.iste.co.uk/ganghoffer/biomechanics.zip*

To demonstrate the effects of microstructural anisotropy α on the anisotropy of the elastic constants, the relative elastic and shear moduli and effective Poisson's ratios are evaluated at different values of α ranging from 0.25 to 4 (extreme value of anisotropy); they are illustrated in Figures 6.9 and 6.10. The computation is performed by setting slenderness parameter η=0.5 and θ = 30°. It is observed that the Young's and shear moduli and Poisson's ratios depend strongly on the amount of microstructural anisotropy. The relative Young's and shear moduli E_x^* / E_s and G_{xy}^* / E_s, respectively decreases (Figures 6.9(a) and 6.9(c)) whereas the relative modulus E_y^* / E_s increases (Figure 6.9(b)) when increasing the microstructural anisotropy α. Furthermore, this discrepancy increases when decreasing the cell aspect ratio α. It is observed as a noticeable difference between the homogenized shear modulus and Gibson model (see Figure 6.9(c)), which increases when decreasing the cell aspect ratio α, reaching a maximum of about 30% for α=0.25.

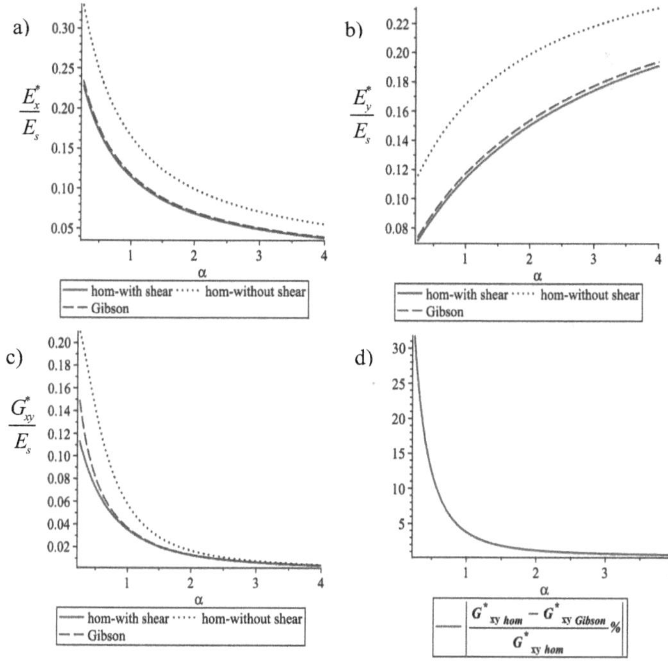

Figure 6.9. *a) and b) Evolution of the relative tensile moduli versus the geometrical parameter α. c) Relative shear modulus. d) Relative difference between the homogenized shear modulus and Gibson model. η=1/2, θ=30. For a color version of this figure, see www.iste.co.uk/ganghoffer/biomechanics.zip*

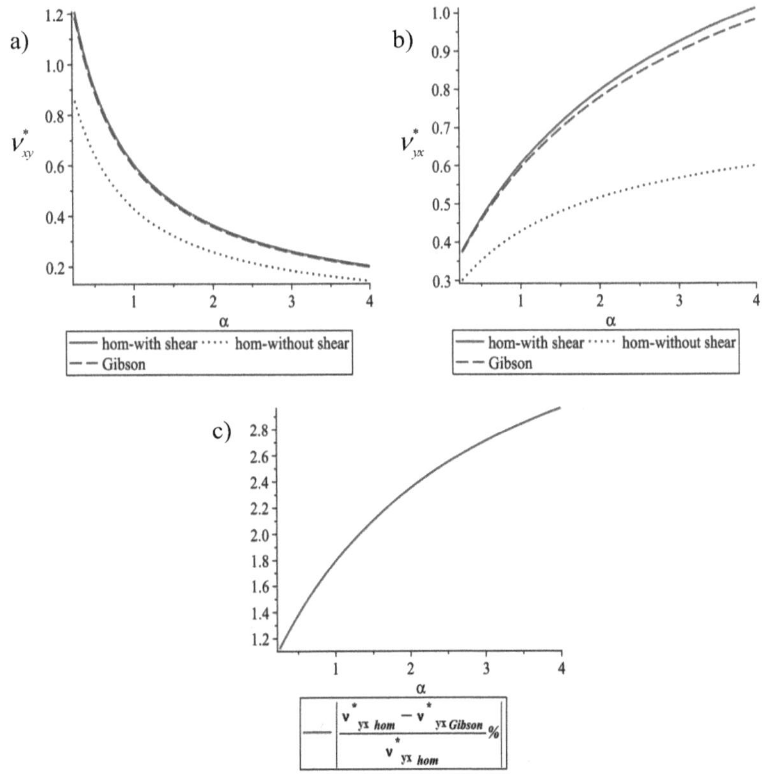

Figure 6.10. *(a) and (b) Equivalent Poisson's ratios versus the geometrical parameter α. (c) Relative difference between homogenized Poisson's ratio ν_{yx} and Gibson model. η=1/2, θ=30. For a color version of this figure, see www.iste.co.uk/ganghoffer/biomechanics.zip*

6.5. Numerical determination of the effective bone rigidities

In a 2D plane stress situation, the stress tensor has four independent components σ_x, σ_y, σ_{xy}, σ_{yx} and the couple stress tensor (or moment per unit area) has two components m_{xz}, m_{yz}. The deformation and micro-curvature components express versus the displacement gradients and micro-rotations as

$$\epsilon_x = \partial u/\partial x,\quad \epsilon_y = \partial v/\partial y,\quad \epsilon_{xy} = \partial v/\partial x - \phi,\quad \epsilon_{yx} = \partial u/\partial y + \phi$$
$$\chi_{xz} = \partial \phi/\partial x,\quad \chi_{yz} = \partial \phi/\partial y \qquad [6.9]$$

In the Cosserat theory, the rotation ϕ is an independent field, whereas in the couple-stress theory it is related to displacement gradients as in classical elasticity (equal to the local rigid rotation) $\phi=\left(\partial v/\partial x-\partial u/\partial y\right)/2$. As a result, in the couple-stress theory, the strain tensor ϵ_{ij} is symmetrical with components defined as $\epsilon_{xy}=\epsilon_{yx}=\left(\partial v/\partial x+\partial u/\partial y\right)/2$.

The equilibrium in translation and rotation, ignoring body forces and body moments, writes as the set of equations

$$\begin{aligned}&\partial\sigma_x/\partial x+\partial\sigma_{yx}/\partial y=0,\ \ \partial\sigma_{xy}/\partial x+\partial\sigma_y/\partial y=0,\\&\partial m_{xz}/\partial x+\partial m_{yz}/\partial y+\sigma_{xy}-\sigma_{yx}=0.\end{aligned}\qquad [6.10]$$

The last equation [6.10], implies that the shear stress σ_{xy} differs from σ_{yx}; Mindlin [MIN 63] suggested then resolving σ_{xy} and σ_{yx} into a symmetric part σ_S and an anti-symmetric part σ_A.

$$\sigma_S=\left(\sigma_{xy}+\sigma_{yx}\right)/2,\ \sigma_A=\left(\sigma_{xy}-\sigma_{yx}\right)/2.\qquad [6.11]$$

The symmetric part of the shear stress produces the usual shear strain ϵ_{xy}, while the anti-symmetric part tends to produce a local rigid rotation. Thus, the constitutive equation can be expressed in the following uncoupled form (for a centrally symmetric unit cell structure) as

$$\begin{Bmatrix}\sigma_x\\\sigma_y\\\sigma_S\\m_{xz}\\m_{yz}\end{Bmatrix}=\begin{bmatrix}\mathbf{E}&0\\0&\mathbf{K}\end{bmatrix}\begin{Bmatrix}\epsilon_x\\\epsilon_y\\\epsilon_{xy}\\\chi_{xz}\\\chi_{yz}\end{Bmatrix}\qquad [6.12]$$

with

$$\mathbf{E}=\begin{bmatrix}E_{11}&E_{12}&0\\E_{12}&E_{22}&0\\0&0&E_{66}\end{bmatrix},\ \mathbf{K}=\begin{bmatrix}K_{11}&0\\0&K_{22}\end{bmatrix}.\qquad [6.13]$$

The matrix **E** contains the classical moduli, while matrix **K** contains the micropolar moduli relating the two couple stress components to the corresponding curvatures.

The discrete homogenization method provides the effective homogenized moduli which are computed from the effective compliance matrix (the inverse of previous stiffness matrix). This includes Young's moduli E_x^* and E_y^*, Poisson's ratios ν_{xy}^*, ν_{yx}^*, shear modulus G_{xy}^*, and micropolar bending moduli K_{11}, K_{22}.

The numerical homogenization technique consists of determining the overall effective mechanical properties (couple-stress moduli and characteristic lengths) over a representative unit cell of trabecular bone, relying on a finite element discretization of the unit cell geometry. Each cell wall is modelled with a three-node beam element (element type B22 in ABAQUS), endowed with bending, stretching and shearing deformation mechanisms. An earlier study by [SIL 95] showed that using such a beam element to model each cell wall is sufficient for convergence.

We presently consider the honeycomb-like structural model of a trabecular bone specimen made of a 2D array of hexagonal cells and we model the trabecular architecture as a two-dimensional lattice (Figure 6.11) with a unit cell geometry of geometrical parameters given from [FOR 98] as:

$\alpha = h/l = 0.5$; $\eta = t/l = 0.25$; $l = 1\ mm$; $\theta = 60^\circ$ (Figure 6.11(a)). The bulk mechanical properties of the unit cells wall representative of the behavior of human femur [GIB 97]) are taken as $E_s = 12000$ MPa; $\nu_s = 0.3$.

In order to evaluate the effective Young's moduli E_x^* and E_y^*, Poisson's ratios ν_{xy}^* and ν_{yx}^*, and shear modulus G_{xy}^*, we conduct three finite element simulations over the whole lattice (two tensile tests and one for shear). The uniaxial tensile in the x and y-directions and shear are computed using displacement boundary conditions. In the case of the uniaxial tensile tests, the nodes forming an edge are displaced by a specified amount while the nodes on the opposite edge are prevented from translating in the direction of displacement. The node at the bottom left is also constrained from translating in y-direction when applying tension in x-direction and from translating in x-direction in the case of applying tension in y-direction. The stress–strain curves are then constructed from the reaction forces and displacements captured using the reference points in the displacement directions. The reference points do not belong to the main geometry of the lattice, and they are for adjusting movements on the lattice surfaces. The effective tensile modulus in x-direction is then estimated from the FE simulation as $E_x^* = \frac{\sigma_x}{\epsilon_x} = 74.4\ \text{MPa}$ and the Poisson's coefficient is computed as $\nu_{xy}^* = -\frac{\epsilon_y}{\epsilon_x} = 0.2$. The effective tensile modulus in y-direction is also estimated from the FE simulation as $E_y^* = \frac{\sigma_y}{\epsilon_y} = 1247\ \text{MPa}$ and the Poisson's coefficient as $\nu_{yx}^* = -\frac{\epsilon_x}{\epsilon_y} = 3.21$.

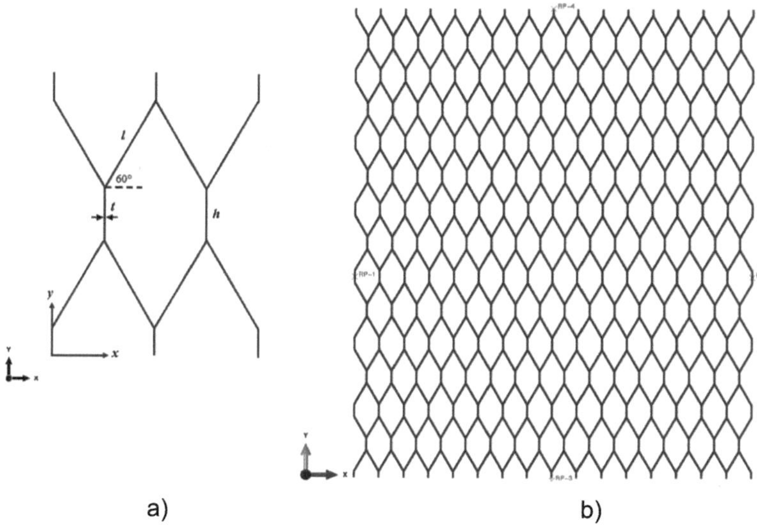

Figure 6.11. *a) Geometrical model of an elementary cell of trabecular bone. b) Hexagonal honeycomb model representing the trabecular bone lattice. For a color version of this figure, see www.iste.co.uk/ganghoffer/biomechanics.zip*

A FE simulation for simple shear on trabecular lattice is needed to be performed to construct the shear stress–strain curve, from which the shear modulus is obtained. To perform such simulation, we impose a displacement field on the top surface of the lattice along the x direction, while the bottom surface is constrained in the same direction. Both the left and right surfaces are also prevented from translating in the direction perpendicular to x. Subsequently, we capture the x-component of the resultant reaction force produced on the top surface caused by the applied displacement in the x direction. The shear strain is computed by dividing the displacement on the top surface on the length of the trabecular lattice along the y direction. The shear stress–strain curve is then constructed from the reaction force and displacement and the effective shear modulus is calculated as

$$G^*_{xy} = \frac{\sigma_{xy}}{\epsilon_{xy}} = 408.5 \text{ MPa}$$

The values of effective Young's moduli, Poisson's coefficients, and shear modulus computed from FE simulations performed over the whole trabecular lattice are summarized in Table 6.2, showing a very good agreement with those calculated by discrete homogenization. The discrepancies between the results obtained by homogenization and FE simulation is less than 1%, hence we can consider that the predicted homogenized moduli are quite accurate.

We seek now to evaluate the components of the bending stiffness matrix **K**, thus two separated bending tests shall be performed to identify the micropolar constants K_{11} and K_{22}. We employ the following equilibrium equations of a micropolar elastic beam of anisotropic material subjected to a uniform bending moment M to define the micropolar bending stiffnesses K_{11} and K_{22} as

$$\underbrace{\left(E_x^* I_z^* + K_{11} H\right)}_{K^m_{xz}} \frac{\partial \phi_z}{\partial x} = -M \;\; ; \;\; \underbrace{\left(E_y^* I_z^* + K_{22} H\right)}_{K^m_{yz}} \frac{\partial \phi_z}{\partial y} = -M \qquad [6.14]$$

with E_x^* and E_y^* the effective homogenized moduli of the macroscopic equivalent beam, I_z^* the quadratic moment of the macroscopic beam along z, given by $I_z^* = H^3/12$.

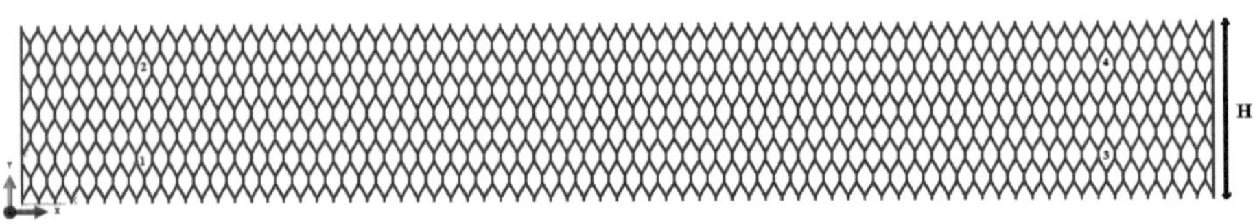

Figure 6.12. *Hexagonal honeycomb beam model representing the trabecular bone lattice submitted to pure bending moment. For a color version of this figure, see www.iste.co.uk/ganghoffer/biomechanics.zip*

We calculate the bending stiffness of the beam K_{xz}^{m} (exhibting micropolar effect), with the microrotation determined from the displacements of nodes 1-4 (Figure 6.12). The horizontal and vertical distance separating the points 1-4 are respectively 58 mm and 5.96 mm. The microcurvature of this macrobeam is defined as the relative variation of the microrotation, calculated from the two intermediate variables as

$$\phi_{left} = -\left(U_{2x} - U_{1x}\right)/dist\left(2-1\right);\ \phi_{right} = -\left(U_{4x} - U_{3x}\right)/dist\left(4-3\right) \qquad [6.15]$$

Therefore the gradient of the microrotation and the bending stiffness are computed successively as

$$\frac{\partial \phi}{\partial x} = \left(\phi_{left} - \phi_{right}\right)/dist\left(3-1\right);\ K_{xz}^{m} = M \Big/ \frac{\partial \phi}{\partial x} \qquad [6.16]$$

The calculation of the response of the trabecular lattice under pure bending is achieved by applying a concentrated moment = 0.2 N.mm on the lateral edges. The simulation results are shown in Figure 6.13. The micropolar constant K_{11} is computed from the relation in [6.14] and given in Table 6.2.

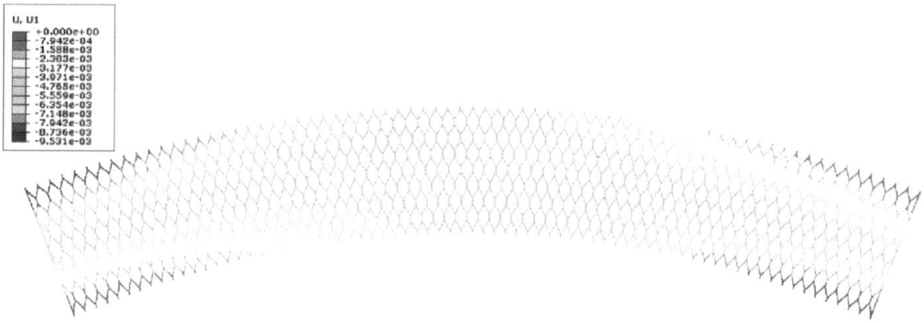

Figure 6.13. *Iso-displacements along x-axis under the bending moment. For a color version of this figure, see www.iste.co.uk/ganghoffer/biomechanics.zip*

For the planar orthotropic couple-stress model considered in this work, there are four characteristics lengths which are defined by the following expressions in terms of the engineering constants [BOU 02, ROV 07]

$$l_{b1} = \sqrt{\frac{K_{11}}{4E_{66}}}, \quad l_{b2} = \sqrt{\frac{K_{22}}{4E_{66}}},$$
$$l_{Ex} = \sqrt{\frac{(1+\nu^*_{yx})K_{11}}{2E^*_x}}, \quad l_{Ey} = \sqrt{\frac{(1+\nu^*_{xy})K_{22}}{2E^*_y}} \qquad [6.17]$$

The components of the effective elastic constants of trabecular bone evaluated by discrete homogenization and FE simulation are compared in Table 6.2. They are in a very good agreement.

	Discrete homogenization	Finite element simulation
E^*_x (MPa)	75.2	76.4
E^*_y (MPa)	1255	1247
G^*_{xy} (MPa)	407.5	408.5
ν^*_{xy}	0.197	0.2
ν^*_{yx}	3.28	3.21
K_{11} (N)	5.7	6.0
K_{22} (N)	21.3	–
l_{b1} (μm)	59.0	60.6
l_{b2} (μm)	116.0	–
l_{Ex} (μm)	402.7	412.0
l_{Ey} (μm)	100.8	–

Table 6.2. *Couple-stress elastic constants and characteristics lengths of trabecular bone unit cell*

In the next section, the established anisotropic micropolar constitutive model for bone is applied to investigate the expected regularizing effect of such an enhanced continuum around a crack present in the bone structure.

6.6. Application to bone fracture

Simulations of a square bone sample including a crack are performed, using the previously obtained homogenized anisotropic micropolar model. The results are compared to those obtained by a classical four node finite element called CPS4 within the Abaqus environment.

6.6.1. *Four node 2D Cosserat finite element*

A four node Cosserat finite element has been developed and implemented in the industrial code Abaqus in order to simulate the two dimensional response of the trabecular bone including a crack. The effective trabecular bone behavior has been obtained by the discrete asymptotic homogenization technique described in the previous sections. A simple finite element has been constructed, the four node quadrilateral with three degrees of freedom per node u_{1j}, v_{2j}, ϕ_j at each vertex (there are four nodes: j=1,2,3,4), as shown on Figure 6.14.

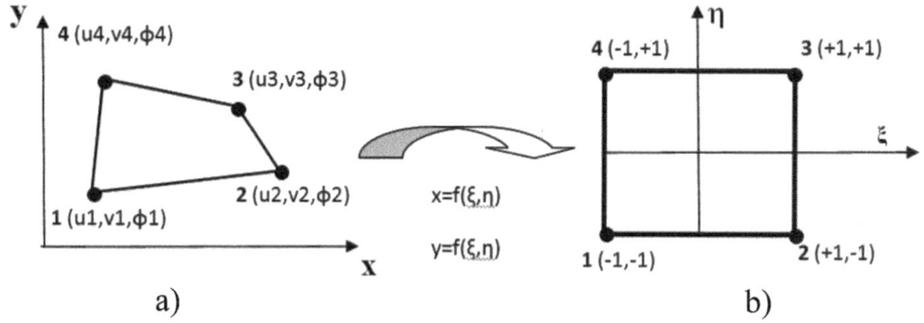

Figure 6.14. *Four node linear quadrangle Cosserat FE. (a) Arbitrary four-node quadrilateral FE with three degrees of freedom (u,v,ϕ). (b) Reference element*

The geometrical transformation of the element is given by the following relations:

$$x(\xi,\eta)=\sum_{i=1}^{4} N_i(\xi,\eta)x_i;\ y(\xi,\eta)=\sum_{i=1}^{4} N_i(\xi,\eta)y_i;\ \phi(\xi,\eta)=\sum_{i=1}^{4} N_i(\xi,\eta)\phi_i; \quad [6.18]$$

In those expressions, ξ, η are the coordinates of any point of the reference element, $x(\xi,\eta)$, $y(\xi,\eta)$ and $\phi(\xi,\eta)$ successively the position and rotation of one point of the physical element; x_i, y_i and ϕ_i are respectively the coordinates and rotation of node i of the real element and N_i are the shape functions.

The stress-strain relation in a two dimensional plane stress situation can be written as follows:

$$\begin{Bmatrix} \sigma_x \\ \sigma_y \\ \sigma_{xy} \\ \sigma_{yx} \\ m_{xz} \\ m_{yz} \end{Bmatrix} = [\mathbf{D}] \begin{Bmatrix} \epsilon_x \\ \epsilon_y \\ \epsilon_{xy} \\ \epsilon_{yx} \\ \chi_{xz} \\ \chi_{yz} \end{Bmatrix} = [\mathbf{D}] \begin{Bmatrix} \dfrac{\partial u}{\partial x} \\ \dfrac{\partial v}{\partial y} \\ \dfrac{\partial v}{\partial x} - \phi \\ \dfrac{\partial u}{\partial y} + \phi \\ \dfrac{\partial \phi}{\partial x} \\ \dfrac{\partial \phi}{\partial y} \end{Bmatrix} \qquad [6.19]$$

with $[\mathbf{D}]$ therein the effective stiffness matrix of the bone, obtained by the discrete asymptotic homogenized method. Introducing the stress and couple stress vector, the constitutive equation for an anisotropy centro-symmetric unit cell can be written in the form given by [6.4]. The kinematic relations [6.19] are conveniently written in the matrix form

$$\{\epsilon\} = [B]\{u\} \qquad [6.20]$$

The displacement-rotation mixed vector therein is defined as $\{u\} = \{u_1 \; v_1 \; \phi_1 \; u_2 \; v_2 \; \phi_2 \; u_3 \; v_3 \; \phi_3 \; u_4 \; v_4 \; \phi_4\}^T$ and $[B]$ is the matrix defined as:

$$[B] = \begin{bmatrix} B_{11} & 0 & 0 & B_{14} & 0 & 0 & B_{17} & 0 & 0 & B_{110} & 0 & 0 \\ 0 & B_{22} & 0 & 0 & B_{25} & 0 & 0 & B_{28} & 0 & 0 & B_{211} & 0 \\ 0 & B_{32} & B_{33} & 0 & B_{35} & B_{36} & 0 & B_{38} & B_{39} & 0 & B_{311} & B_{312} \\ B_{41} & 0 & B_{43} & B_{44} & 0 & B_{46} & B_{47} & 0 & B_{49} & B_{410} & 0 & B_{412} \\ 0 & 0 & B_{53} & 0 & 0 & B_{56} & 0 & 0 & B_{59} & 0 & 0 & B_{512} \\ 0 & 0 & B_{63} & 0 & 0 & B_{66} & 0 & 0 & B_{69} & 0 & 0 & B_{612} \end{bmatrix}$$

The components of this matrix are as follows:

$$B_{11}=\frac{1}{4}j_{11}(-1+\eta)+\frac{1}{4}j_{12}(-1+\xi),\ B_{14}=\frac{1}{4}j_{11}(1-\eta)+\frac{1}{4}j_{12}(-1-\xi),$$

$$B_{17}=\frac{1}{4}j_{11}(1+\eta)+\frac{1}{4}j_{12}(1+\xi),\ B_{110}=\frac{1}{4}j_{11}(-1-\eta)+\frac{1}{4}j_{12}(1-\xi),$$

$$B_{22}=\frac{1}{4}j_{21}(-1+\eta)+\frac{1}{4}j_{22}(-1+\xi),\ B_{25}=\frac{1}{4}j_{21}(1-\eta)+\frac{1}{4}j_{22}(-1-\xi),$$

$$B_{28}=\frac{1}{4}j_{21}(1+\eta)+\frac{1}{4}j_{22}(1+\xi),\ B_{211}=\frac{1}{4}j_{21}(-1-\eta)+\frac{1}{4}j_{22}(1-\xi),$$

$$B_{32}=B_{11},\ B_{33}=\frac{1}{4}(-1+\xi+\eta-\xi\eta),\ B_{35}=B_{14},\ B_{36}=\frac{1}{4}(-1-\xi+\eta+\xi\eta),$$

$$B_{38}=B_{17},\ B_{39}=\frac{1}{4}(-1-\xi-\eta-\xi\eta),\ B_{311}=B_{110},\ B_{312}=\frac{1}{4}(-1+\xi-\eta+\xi\eta),$$

$$B_{41}=B_{22},\ B_{43}=\frac{1}{4}(1-\xi-\eta+\xi\eta),\ B_{44}=B_{25},\ B_{46}=\frac{1}{4}(1+\xi-\eta+\xi\eta),\ B_{47}=B_{28},$$

$$B_{49}=\frac{1}{4}(1+\xi+\eta+\xi\eta),\ B_{410}=B_{211},\ B_{412}=\frac{1}{4}(1-\xi+\eta-\xi\eta),$$

$$B_{53}=B_{11},\ B_{56}=B_{14},\ B_{59}=B_{17},\ B_{512}=B_{110},$$

$$B_{63}=B_{22},\ B_{66}=B_{25},\ B_{69}=B_{28},\ B_{612}=B_{211},$$

and j_{ij}, i,j = 1,2 are the coefficients of the Jacobean matrix.

6.6.2. *Simulation results*

The stress distribution across the crack is compared with the corresponding distribution obtained for a classical Cauchy continuum, relying on the developed special finite element. Especially, the anisotropy reflected in the equivalent stiffness matrix is shown to act on both the classical and the micropolar parts of the overall constitutive behavior; the directions of anisotropy are here dictated by the underlying cellular structure, which is distributed over the considered bone sample, although represented in an average sense in this specific application. The principal directions of anisotropy (here orthotropy) are supposed to be aligned with the fiber directions within the selected bone sample, which by definition coincide with the X and Y directions of the representative volume element (the unit cell).

The specimen dimensions correspond to 1mm width and 1 mm height. The effective material properties calculated by the discrete homogenization technique are summarized in the following matrix:

$$[\mathbf{D}]=\begin{bmatrix}212.7 & 698.3 & 0 & 0 & 0 & 0\\ 698.3 & 3547.4 & 0 & 0 & 0 & 0\\ 0 & 0 & 436.2 & 378.8 & 0 & 0\\ 0 & 0 & 271.6 & 543.4 & 0 & 0\\ 0 & 0 & 0 & 0 & 5.7 & 0\\ 0 & 0 & 0 & 0 & 0 & 21.3\end{bmatrix} \quad [6.21]$$

The two characteristic lengths of this anisotropic micropolar behavior are calculated from the rigidities in [6.21], as $l_{b1}=\sqrt{K_{55}/2(K_{33}+K_{34})}=60\mu m$ and $l_{b2}=\sqrt{K_{66}/2(K_{43}+K_{44})}=115\mu m$.

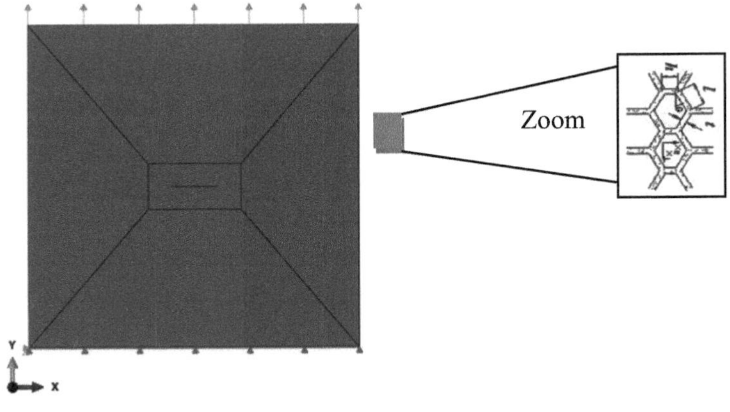

Figure 6.15. *Applied loading and boundary conditions of the 2D bone specimen with bone fracture. Zoom of the material specimen on two hexagonal unit cells. For a color version of this figure, see www.iste.co.uk/ganghoffer/biomechanics.zip*

A horizontal crack representing 14% of the total specimen width is introduced to represent longitudinal fracture, as viewed in Figure 6.15. The bottom edge of the specimen (1 x 1 mm) is simply supported in Y-direction (the corner node at the origin is fixed), and the top edge is submitted to a uniaxial displacement magnitude equal to 0.25 mm in the vertical direction.

In these simulations, three different meshes referred to as Coarse, Medium and Fine are used; those meshes contain respectively 128, 489 and 1188 elements. We checked that using a finer mesh (than the mesh referred to as Fine) does not lead to different stress distributions. The spatial distribution of the in-plane stress components S_{11}, S_{12}, and $\frac{\partial\phi}{\partial x}=\left(\phi_{left}-\phi_{right}\right)/dist(3-1)$ is pictured below for those three mesh sizes, in Figures 6.16, 6.17, 6.18, corresponding successively to the coarse, medium and fine meshes for both the Cauchy and the Cosserat anisotropic continuum models for the purpose of assessing the effect of enhancement of the Cauchy continuum by microrotations. As a matter of clarity of the presentation of the results and despite the existence of two orthogonal symmetry planes, we choose to plot stress distributions over the complete sample including the whole crack. We observe as a general tendency that the peak stress is reduced by the enhancement of the Cauchy model by micropolar degrees of freedom, for all three mesh sizes, at the exception of the vertical component S_{22} in the case of the coarse mesh (Figures 6.16e and f). One observes that a stress peak is obtained for both models close to the crack tip, a fact which is well known in the field of fracture mechanics [BRO 74]. However, the amplitude of the peak stress is significantly reduced when the Cosserat continuum model is employed; the percentage of reduction is evaluated for each stress component (and each mesh size) in Table 6.3. It reaches up to 50% for the Fine mesh, a value in agreement with the amount of reduction of the stress concentration factor obtained for sharp cracks using a linear isotropic Cosserat behavior in [LAK 90]. Note that the reduction of the stress concentration depends on the Cosserat characteristic length, which is fixed in present analysis, since it is a byproduct of the effective moduli.

We further plot the isovalues of the two in-plane displacement components, U_1 and U_2 (Figure 6.23); we observe a more diffuse displacement field when the Cosserat model is employed, in comparison to Cauchy continuum, for which steep stress gradients are obtained. This confirms the regularizing effect of the Cosserat continuum model.

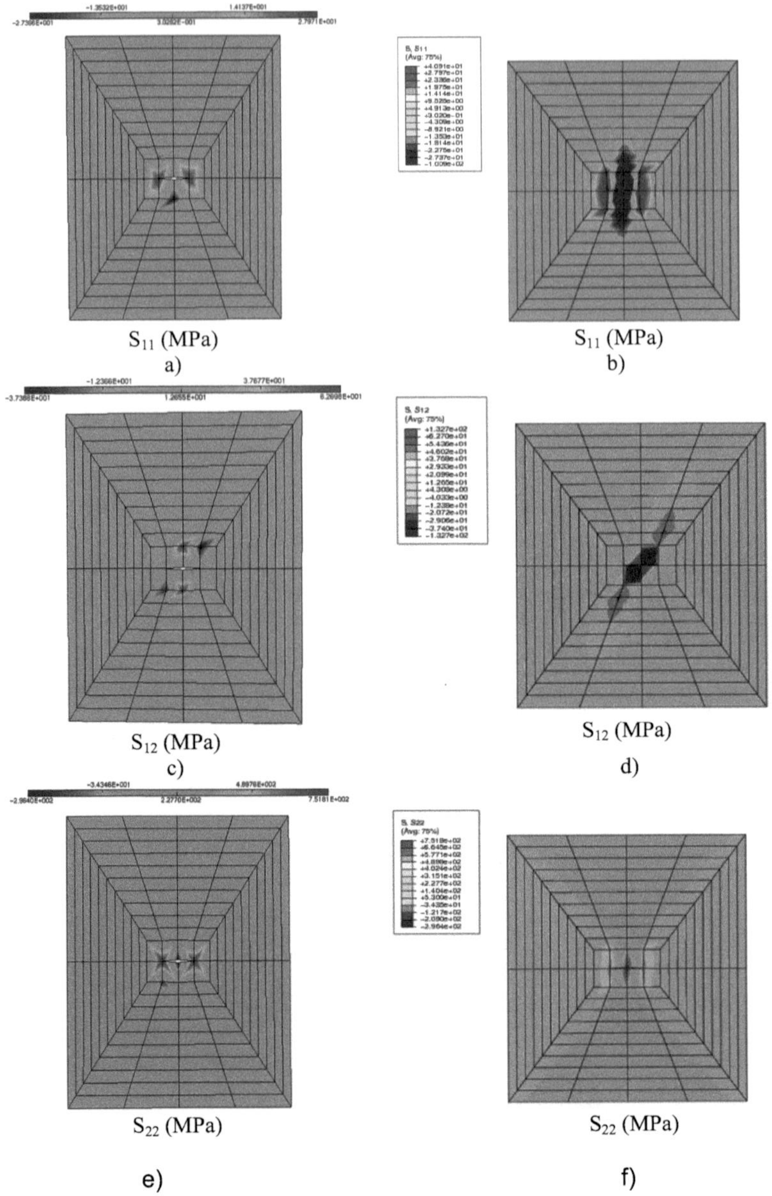

Figure 6.16. *Spatial stress distribution obtained by extrapolation of stresses from the integration points to the nodal points. a), c), and e) correspond to Cosserat finite element results and b), d), and f) correspond to the classical four node finite element. Coarse mesh (128 elements). For a color version of this figure, see www.iste.co.uk/ganghoffer/biomechanics.zip*

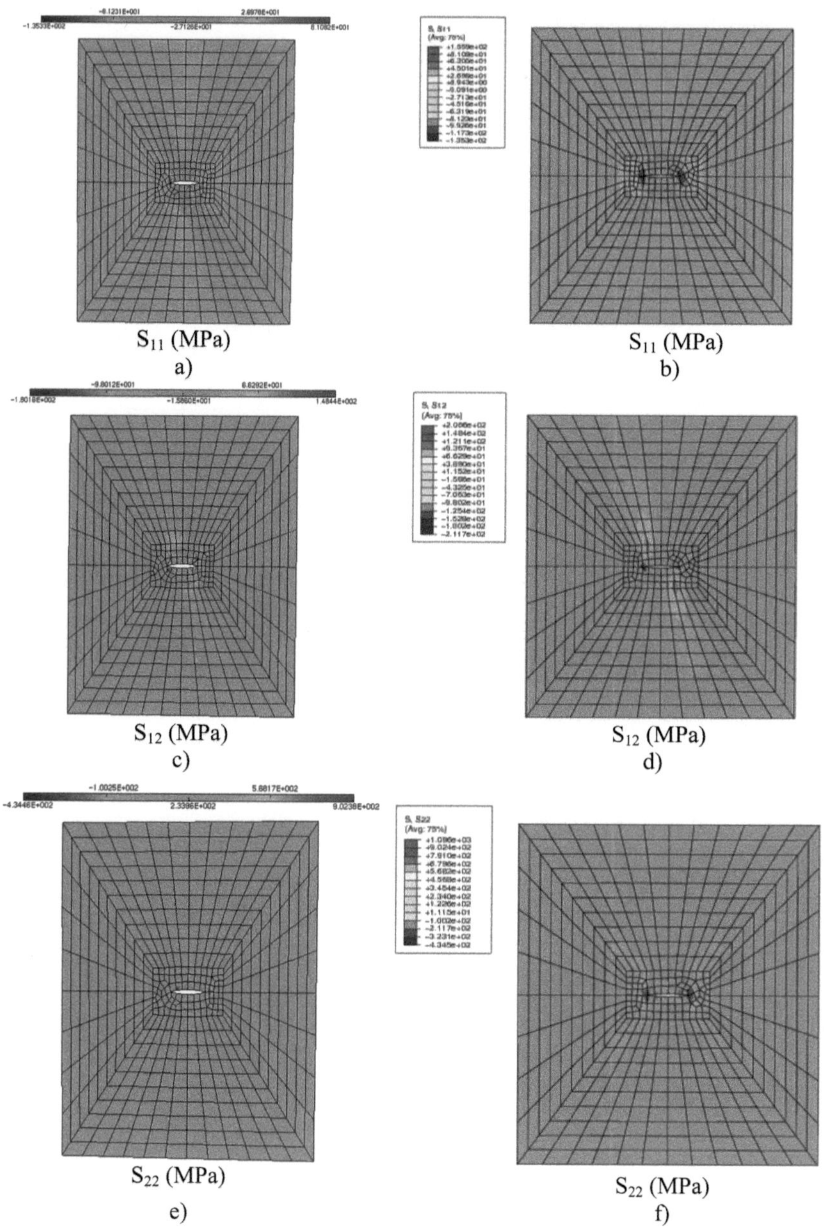

Figure 6.17. *Spatial stress distribution obtained by extrapolation of the stresses from integration points to the nodal points. a), c), and e) correspond to Cosserat finite element results and b), d), and f) correspond to the classical four node finite element. Medium mesh (489 elements). For a color version of this figure, see www.iste.co.uk/ganghoffer/biomechanics.zip*

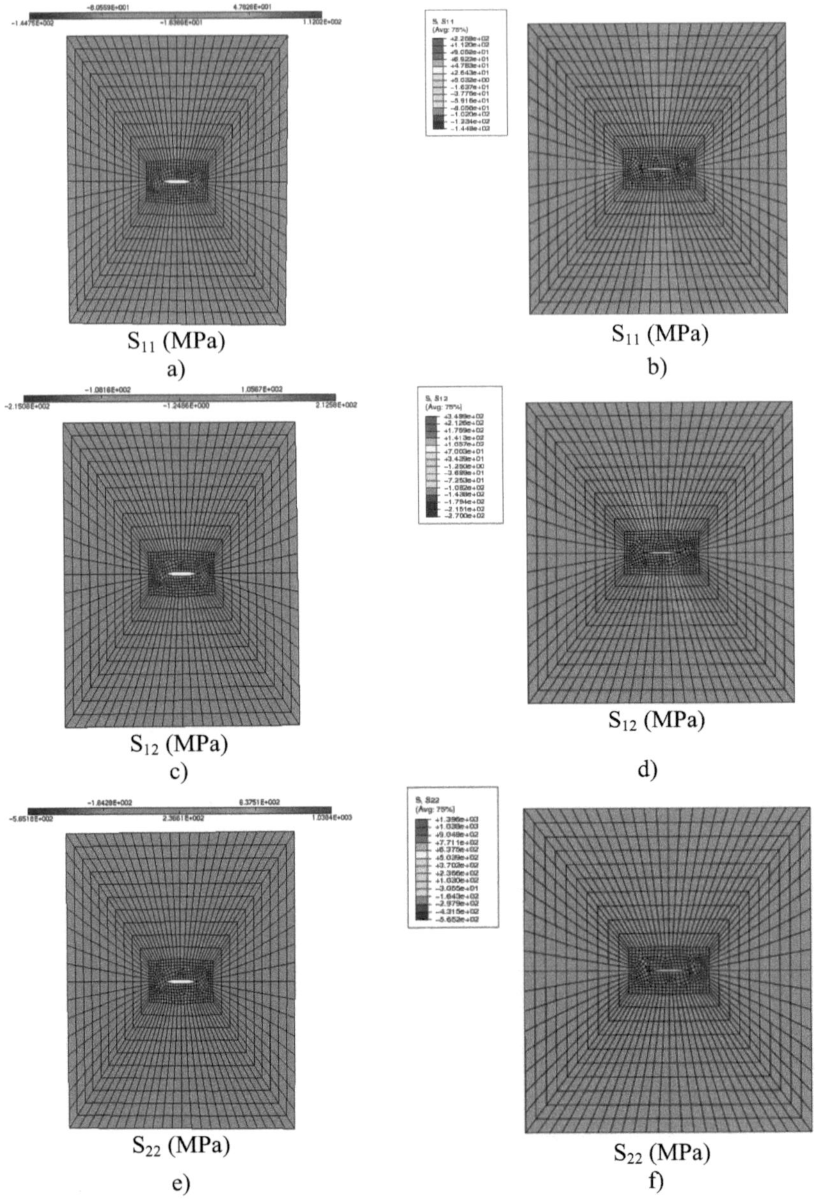

Figure 6.18. *Spatial stress distribution obtained by extrapolation of stresses from the integration points to nodal points. a), c), and e) correspond to Cosserat finite element results and b), d), and f) correspond to the classical four node finite element. Fine mesh (1188 elements). For a color version of this figure, see www.iste.co.uk/ganghoffer/biomechanics.zip*

	Cosserat finite element			CPS4 finite element (Abaqus)			% of stress reduction $\frac{S_{Cauchy}-S_{Cosserat}}{S_{Cauchy}}\times 100$		
	S11	S12	S22	S11	S12	S22	S11	S12	S22
Coarse	28	63	752	41	133	752	31.7	52.6	0
Medium	81	148	902	156	206	1096	48.1	28.2	17.7
Fine	112	213	1038	227	350	1396	50.7	39.2	25.6

Table 6.3. *Comparison of peak stress values for the two models (Cauchy and micropolar)*

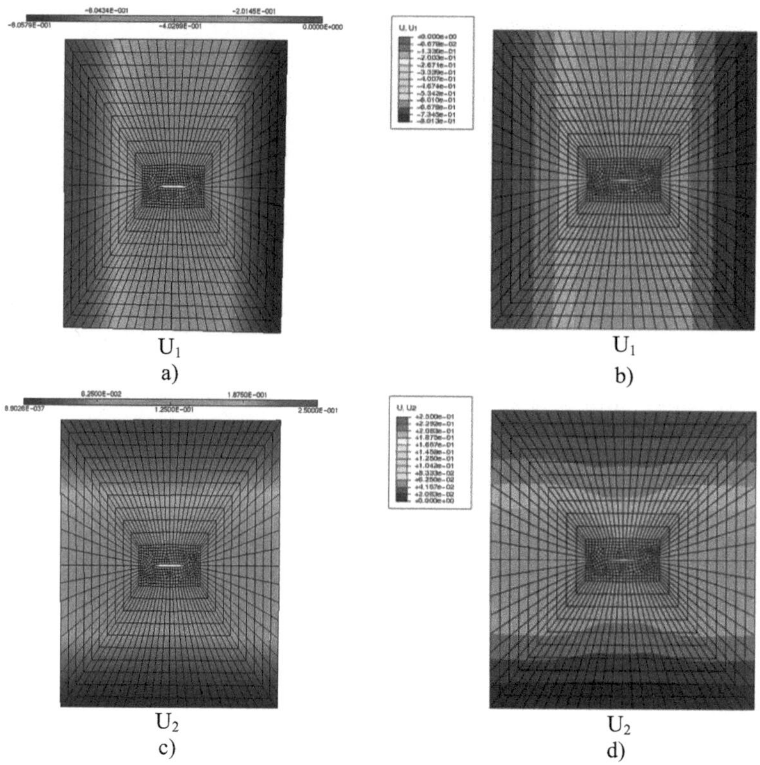

Figure 6.19. *Displacement distribution. a), c) correspond to Cosserat finite element results and b), d) correspond to the classical four node finite element (CPS4 from Abaqus). Fine mesh (1188 elements). For a color version of this figure, see www.iste.co.uk/ganghoffer/biomechanics.zip*

6.7. Effective mechanical properties of 3D vertebral trabecular bone

Trabecular bone, or spongy bone, is a highly porous biological tissue with an open-celled cellular structure, which accounts for up to 20% of the bone mass in humans. It is found in vertebrae; Figure 6.20 shows lumbar vertebra in superior and lateral view with x and y as the lateral directions while z is oriented in the superior direction. The individual trabeculae that make up the trabecular tissue tend to have one of two basic shapes, either slender rods (or beam like struts) or relatively thin and flat plates. These struts and plates form a three-dimensional scaffolding within the cortical shell. Structurally, trabecular bone helps to distribute loading from the articular surfaces to the cortical shell. Trabecular tissue is also very important for the stability and fixation of joint implants and other fixation devices placed into bone. At microscopic scale, trabecular bone contains small trabecular struts connecting in a way that generates a three-dimensional porous structure. In the vertebral body, cancellous bone appears to contain a somewhat regular array of trabecular struts (Figure 6.20), and thus can be idealized as a cellular solid. It is reported that the predominant arrangement of vertebral bone is that of vertically oriented columns with reinforcing horizontal struts, and the structure can be assumed to be columnar with cylindrical symmetry [GIB 85, OVE 97, KIM 02]. Therefore, we consider the idealized shape of trabecular bone as a hexagonal cellular structure comprised of a doubly prism shape unit cell, as will be described later on.

In order to be specific to the considered application to vertebral trabecular bone, a general 3D anisotropic hexagonal lattice with horizontal struts of lengths L and h (lengths of in-plane struts) and vertical struts of length L_v (out-of-plane) is considered, both struts being endowed with uniform circular cross-section of diameter d. The whole lattice is generated from the repetition of the unit cell shown in Figure 6.21 thanks to three periodicity vectors defined in the Cartesian basis. The considered hexagonal unit cell is composed of five beams; three horizontal beams b_1, b_2, and b_3 and the vertical beams b_4 and b_5. The horizontal beams b_2 and b_3 have the same length L and are inclined by the angle θ with the horizontal direction, while b_1 has length h. The two vertical beams b_4 and b_5 have the same length L_v.

This geometry is representative of the unit cell of vertebral trabecular bone. The lengths of the periodicity vectors (the norm of the periodicity vectors $\mathbf{Y}_1$, $\mathbf{Y}_2$, and $\mathbf{Y}_2$ in the Cartesian basis) are $L_1 = L_2 = \sqrt{(L\cos\theta)^2 + (h + L\sin\theta)^2}$ and $L_3 = L_v$. The connectivity Table 6.4 gives the numbering of beams and nodes within the chosen representative unit cell.

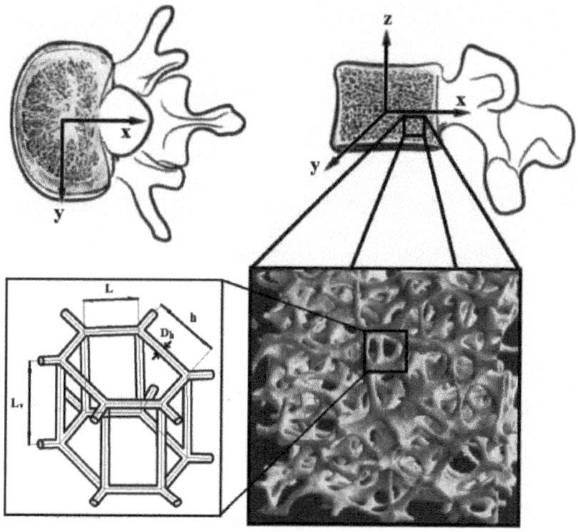

Figure 6.20. *General anisotropic unit cell for 3D hexagonal models. The geometric parameters include the cell angle θ, the in-plane strut lengths (L, h), and the vertical length L_v*

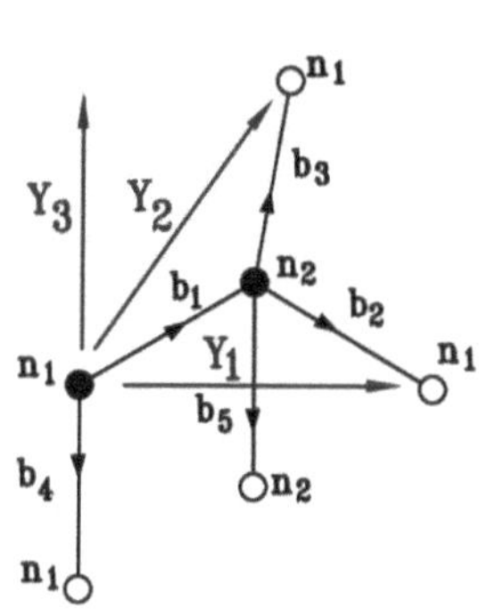

Figure 6.21. *Reference unit cell of vertebral trabecular bone and its periodicity vectors Y_1, Y_2, and Y_3*

Beam	1	2	3	4	5
O(b)	1	2	2	1	2
E(b)	2	1	1	1	2
δ_1	0	1	0	0	0
δ_2	0	0	1	0	0
δ_3	0	0	0	-1	-1

Table 6.4. *Connectivity array for the 3D hexagonal lattice*

The lattice is called re-entrant when the angle θ between the horizontal direction and the strut with length L (in-plane) is negative. Regular unit cells correspond to geometrical parameters $L = h$ and $\theta = 30°$ and anisotropic unit cell correspond to all other values.

The relative density of the general 3D anisotropic hexagonal is computed in terms of the geometrical parameters L, h, L_v, r and θ as

$$\frac{\rho^*}{\rho_s} = \frac{1}{2} \frac{\pi r^2 (h + 2L + 2L_v)}{LL_v \cos\theta (h + L\sin\theta)} \qquad [6.22]$$

where ρ^* is the equivalent density of the lattice, and ρ_s the density of the bulk material. After homogenization, the equivalent orthotropic stiffness matrix for the 3D hexagonal lattice is finally obtained in terms of the geometrical and mechanical lattice parameters by rearranging the constitutive equation in matrix format as

$$\mathbf{D} = \begin{bmatrix} \mathbf{E} & 0 \\ 0 & \mathbf{K} \end{bmatrix} = \begin{bmatrix}
E_{11} & E_{12} & E_{13} & 0 & 0 & 0 & 0 & 0 & 0 & 0 & 0 & 0 & 0 & 0 & 0 & 0 & 0 & 0 \\
E_{21} & E_{22} & E_{23} & 0 & 0 & 0 & 0 & 0 & 0 & 0 & 0 & 0 & 0 & 0 & 0 & 0 & 0 & 0 \\
E_{31} & E_{32} & E_{33} & 0 & 0 & 0 & 0 & 0 & 0 & 0 & 0 & 0 & 0 & 0 & 0 & 0 & 0 & 0 \\
0 & 0 & 0 & E_{44} & E_{45} & 0 & 0 & 0 & 0 & 0 & 0 & 0 & 0 & 0 & 0 & 0 & 0 & 0 \\
0 & 0 & 0 & E_{54} & E_{55} & 0 & 0 & 0 & 0 & 0 & 0 & 0 & 0 & 0 & 0 & 0 & 0 & 0 \\
0 & 0 & 0 & 0 & 0 & E_{66} & E_{67} & 0 & 0 & 0 & 0 & 0 & 0 & 0 & 0 & 0 & 0 & 0 \\
0 & 0 & 0 & 0 & 0 & E_{76} & E_{77} & 0 & 0 & 0 & 0 & 0 & 0 & 0 & 0 & 0 & 0 & 0 \\
0 & 0 & 0 & 0 & 0 & 0 & 0 & E_{88} & E_{89} & 0 & 0 & 0 & 0 & 0 & 0 & 0 & 0 & 0 \\
0 & 0 & 0 & 0 & 0 & 0 & 0 & E_{98} & E_{99} & 0 & 0 & 0 & 0 & 0 & 0 & 0 & 0 & 0 \\
0 & 0 & 0 & 0 & 0 & 0 & 0 & 0 & 0 & K_{11} & K_{12} & K_{13} & 0 & 0 & 0 & 0 & 0 & 0 \\
0 & 0 & 0 & 0 & 0 & 0 & 0 & 0 & 0 & K_{21} & K_{22} & K_{23} & 0 & 0 & 0 & 0 & 0 & 0 \\
0 & 0 & 0 & 0 & 0 & 0 & 0 & 0 & 0 & K_{31} & K_{32} & K_{33} & 0 & 0 & 0 & 0 & 0 & 0 \\
0 & 0 & 0 & 0 & 0 & 0 & 0 & 0 & 0 & 0 & 0 & 0 & K_{44} & K_{45} & 0 & 0 & 0 & 0 \\
0 & 0 & 0 & 0 & 0 & 0 & 0 & 0 & 0 & 0 & 0 & 0 & K_{54} & K_{55} & 0 & 0 & 0 & 0 \\
0 & 0 & 0 & 0 & 0 & 0 & 0 & 0 & 0 & 0 & 0 & 0 & 0 & 0 & K_{66} & K_{67} & 0 & 0 \\
0 & 0 & 0 & 0 & 0 & 0 & 0 & 0 & 0 & 0 & 0 & 0 & 0 & 0 & K_{76} & K_{77} & 0 & 0 \\
0 & 0 & 0 & 0 & 0 & 0 & 0 & 0 & 0 & 0 & 0 & 0 & 0 & 0 & 0 & 0 & K_{88} & K_{89} \\
0 & 0 & 0 & 0 & 0 & 0 & 0 & 0 & 0 & 0 & 0 & 0 & 0 & 0 & 0 & 0 & K_{98} & K_{99}
\end{bmatrix} \qquad [6.23]$$

The rigidity constants therein are lengthy expressions of the lattice descriptors and microbeams moduli. From the effective compliance matrix (the inverse of previous stiffness matrix), one is able to identify the classical and micropolar effective moduli as closed form expressions of the lattice geometrical and mechanical parameters. The effective elastic constants are given in terms of the geometrical parameters L, h, L_v, r and θ , the shear correction k_s , and the elastic moduli of the solid cell walls E_s, G_s , as follows:

$$E_x^* = -\frac{3\pi r^4 C_t G_s k_s E_s}{L_v \begin{pmatrix} -3C_t^2 r^2 G_s k_s L S_t - S_t L^3 G_s k_s - 3S_t E_s r^2 L \\ +S_t L^3 C_t^2 G_s k_s + 3S_t E_s r^2 C_t^2 L + L^2 C_t^2 h G_s k_s \\ -3C_t^2 r^2 h G_s k_s - L^2 h G_s k_s - 3h E_s r^2 + 3C_t^2 h E_s r^2 \end{pmatrix}}; E_y^* = \frac{3\pi r^4 (h + L S_t) G_s k_s E_s}{L_v L C_t \begin{pmatrix} 3L C_t^2 r^2 E_s + L^3 C_t^2 G_s k_s + 3r^2 L G_s k_s \\ +6r^2 h G_s k_s - 3L r^2 C_t^2 G_s k_s \end{pmatrix}};$$

$$E_z^* = \frac{\pi r^2 E_s}{L C_t (h + L S_t)};$$

$$G_{xy}^* = \frac{E_{45} + E_{44}}{2} = \frac{E_{54} + E_{55}}{2} = \frac{3\pi r^4 C_t (h + L S_t) G_s k_s E_s}{L_v \begin{pmatrix} 6LhS_t r^2 G_s k_s + 3h^2 C_t^2 E_s r^2 + L^2 h^2 C_t^2 G_s k_s + 3r^2 h^2 G_s k_s \\ +2C_t^2 h^3 L G_s k_s + 6C_t^2 L h r^2 E_s + 3L^2 r^2 G_s k_s - 3C_t^2 h^2 r^2 G_s k_s \end{pmatrix}};$$

$$G_{yz}^* = \frac{E_{66} + E_{67}}{2} = \frac{E_{76} + E_{77}}{2} = \frac{3\pi r^4 (h + L S_t) G_s k_s E_s}{L C_t \begin{pmatrix} 2LhL_v^2 S_t G_s k_s + 6LhS_t E_s r^2 + L^3 G_s k_s L_v + 3h^2 E_s r^2 + 3L^2 E_s r^2 + 3E_s r^2 L L_v \\ +2h^3 L_v G_s k_s + 6h E_s r^2 L_v + h^2 L_v^2 G_s k_s + L^2 L_v^2 G_s k_s - 3L^2 C_t^2 E_s r^2 - L^2 C_t^2 L_v^2 G_s k_s \end{pmatrix}};$$

$$G_{xz}^* = \frac{E_{88} + E_{89}}{2} = \frac{E_{98} + E_{99}}{2} = \frac{3\pi r^4 C_t G_s k_s E_s}{(h + L S_t)\left(3C_t^2 L r^2 E_s + C_t^2 L_v^2 L G_s k_s + L^2 G_s k_s L_v + 3E_s r^2 L_v\right)};$$

$$\nu_{xy}^* = -\frac{\left(L^2 G_s k_s + 3r^2 E_s - 3r^2 G_s k_s\right) L C_t^2 S_t}{\begin{pmatrix} -3r^2 C_t^2 G_s k_s L S_t - L^3 S_t G_s k_s - 3L S_t r^2 E_s + S_t L^3 C_t^2 G_s k_s + 3S_t E_s r^2 C_t^2 L \\ +L^2 C_t^2 h G_s k_s - 3r^2 C_t^2 h G_s k_s - L^2 h G_s k_s - 3h E_s r^2 + 3C_t^2 h E_s r^2 \end{pmatrix}};$$

$$\nu_{yx}^* = \frac{S_t (h + L S_t)\left(L^2 G_s k_s + 3r^2 E_s - 3r^2 G_s k_s\right)}{\left(3C_t^2 r^2 L E_s + L^3 C_t^2 G_s k_s + 3r^2 L G_s k_s + 6r^2 h G_s k_s - 3C_t^2 L r^2 G_s k_s\right)}; \qquad [6.24]$$

$$\nu_{xz}^* = \nu_{zx}^* = \nu_{yz}^* = \nu_{zy}^* = 0.$$

E_x^* and E_y^* are the in-plane effective Young's moduli; E_z^* is the out-of-plane effective Young's modulus; G_{xy}^* is the in-plane effective shear modulus; G_{xz}^* and G_{yz}^* are the out-of-plane effective shear modului; ν_{xy}^* and ν_{yx}^* are the in-plane effective Poisson's ratios and $\nu_{xz}^*, \nu_{zx}^*, \nu_{yz}^*, \nu_{zy}^*$ the out-of-plane effective Poisson's ratios. C_t and S_t stand for $\cos\theta$ and $\sin\theta$, respectively.

The present 3D micropolar model substantiates additional material constants. The micropolar shear constant κ couples the rotation of particles to shear stresses;

if $\kappa = 0$ (coupling number *(N]* = 0) the Cauchy stress does not depend on the rotational degree of freedom. The limit $\kappa \to \infty$ is a condition energetically admissible, similar to "incompressibility" in classical elasticity, and corresponds to $N = 1$ (upper bound) which is called "couple stress theory". The modulus γ sets the intensity of the couple stresses and is proportional to the characteristic lengths for bending and torsion, quantities l_b and l_t respectively. In Cauchy solids, the internal characteristic length is of the order of the atomic distance and moments of forces and consequently do not produce any macroscopic effect. However, in microstructured solids such as hard biological tissues possessing a cellular structure where an intrinsic internal length of the order of several microns may be detected, couple stresses may influence the macroscopic behaviour. In the limit situation $\gamma = 0$, Cauchy elasticity is recovered.

The micropolar moduli κ and μ^* are further identified from the effective stiffness matrix [6.23] as closed form expressions of the geometrical and micromechanical parameters of the hexagonal unit cell as follows:

$$\kappa_{xy} = \frac{3\pi r^4 C_t E_s}{L^2 L_v\left(\frac{3S_t r^2 E_s}{G_s k_s L} + LS_t + h + \frac{3hr^2 E_s}{G_s k_s L^2}\right)};\ \kappa_{yx} = \frac{3\pi r^4 S_t E_s}{L^3 L_v C_t\left(1 + \frac{3r^2 E_s}{G_s k_s L^2}\right)};$$

$$\kappa_{yz} = \frac{3\pi r^4 S_t E_s}{L^3 L_v C_t\left(1 + \frac{3r^2 E_s}{G_s k_s L^2}\right)};\ \kappa_{zy} = \frac{3\pi r^4 E_s}{LL_v^2 C_t\left(\frac{3S_t r^2 LE_s}{G_s k_s L_v^2} + LS_t + h + \frac{3hr^2 E_s}{G_s k_s L_v^2}\right)};$$

$$\kappa_{zx} = \frac{3\pi r^4 E_s}{LL_v^2 C_t\left(h + LS_t\right)\left(1 + \frac{3r^2 E_s}{G_s k_s L_v^2}\right)};\ \kappa_{xz} = \frac{3\pi r^4 C_t E_s}{L^2 L_v\left(h + LS_t\right)\left(1 + \frac{3r^2 E_s}{G_s k_s L^2}\right)};$$

$$\mu_{xy}^* = -\frac{3}{2}\frac{\pi r^4 C_t E_s\left(\begin{array}{l} -4S_t hL^3 + C_t^2 L^2 h^2 - \frac{12S_t Lhr^2 E_s}{G_s k_s} + 6S_t r^2 hL + 3L^2 r^2 + 2C_t^2 Lh^3 + \frac{3C_t^2 h^2 r^2 E_s}{G_s k_s} + \frac{6C_t^2 hLr^2 E_s}{G_s k_s} \\ +2C_t^2 L^4 - 2L^2 h^2 - \frac{6L^2 r^2 E_s}{G_s k_s} + \frac{6L^2 C_t^2 r^2 E_s}{G_s k_s} - 2L^4 - 3C_t^2 h^2 r^2 + 3h^2 r^2 - \frac{6h^2 r^2 E_s}{G_s k_s} \end{array}\right)}{L^2 L_v\left(1 + \frac{3r^2 E_s}{G_s k_s L^2}\right)\left(\begin{array}{l} 9S_t r^2 Lh^2 - 3Lh^2 r^2 S_t C_t^2 + 2S_t h^3 L^2 C_t^2 + C_t^2 S_t L^3 h^2 + 3S_t L^3 r^2 + \frac{6S_t L^2 hC_t^2 r^2 E_s}{G_s k_s} + \frac{3S_t Lh^2 C_t^2 r^2 E_s}{G_s k_s} \\ -3C_t^2 h^3 r^2 + \frac{6h^2 LC_t^2 r^2 E_s}{G_s k_s} + 9hL^2 r^2 + 2h^4 LC_t^2 - 6hL^2 r^2 C_t^2 + 3h^3 r^2 + \frac{3h^3 C_t^2 r^2 E_s}{G_s k_s} + h^3 L^2 C_t^2 \end{array}\right)};$$

$$\mu^*_{yx}=\frac{3}{2}\frac{\pi r^4E_s\left(\begin{array}{l}-\frac{6S_tC_t^2hLr^2E_s}{G_sk_s}-\frac{3C_t^2S_th^2r^2E_s}{G_sk_s}+\frac{6C_t^2S_tL^2r^2E_s}{G_sk_s}-2C_t^2S_tLh^3-L^2C_t^2S_th^2-3L^2r^2S_t\\ +3h^2r^2C_t^2S_t+2C_t^2S_tL^4-3S_th^2r^2+6C_t^2r^2hL-6Lhr^2+2L^3C_t^2h+\frac{6LC_t^2hr^2E_s}{G_sk_s}\end{array}\right)}{L^3L_vC_t\left(1+\frac{3r^2E_s}{G_sk_sL^2}\right)\left(\begin{array}{l}6S_tr^2Lh+C_t^2L^2h^2+\frac{3C_t^2h^2r^2E_s}{G_sk_s}+3h^2r^2\\ +\frac{6C_t^2hLr^2E_s}{G_sk_s}+2C_t^2Lh^3+3L^2r^2-3C_t^2h^2r^2\end{array}\right)};$$

$$\mu^*_{yz}=\frac{3}{2}\frac{\pi r^4E_s\left(\begin{array}{l}S_tL^3-\frac{6S_thE_sr^2}{G_sk_s}+\frac{3S_tLE_sr^2}{G_sk_s}+S_tC_t^2L^2L_v-2S_th^3+\frac{3S_tL^2C_t^2E_sr^2}{L_vG_sk_s}-S_tL_vL^2-\frac{3S_th^2E_sr^2}{L_vG_sk_s}\\ -\frac{3S_tL^2E_sr^2}{L_vG_sk_s}-L_vS_th^2+2L^2h+2C_t^2L_vLh+\frac{6C_t^2LhE_sr^2}{L_vG_sk_s}+\frac{6hE_sr^2}{G_sk_s}-\frac{6hLE_sr^2}{L_vG_sk_s}-2L_vhL\end{array}\right)}{L^3L_vC_t\left(1+\frac{3r^2E_s}{G_sk_sL^2}\right)\left(\begin{array}{l}-C_t^2L_vL^2-\frac{3L^2C_t^2E_sr^2}{L_vG_sk_s}+\frac{3Lr^2E_s}{G_sk_s}+2LhL_vS_t+\frac{6LhS_tE_sr^2}{L_vG_sk_s}\\ +\frac{6hE_sr^2}{G_sk_s}+\frac{3h^2E_sr^2}{L_vG_sk_s}+L^2L_v+L^3+\frac{3L^2E_sr^2}{L_vG_sk_s}2h^3+h^2L_v\end{array}\right)};$$

$$\mu^*_{zy}=-\frac{3}{2}\frac{\pi r^4E_s\left(\begin{array}{l}-2LL_vhS_t-\frac{6LhS_tE_sr^2}{L_vG_sk_s}+2h^3-\frac{3L^2E_sr^2}{L_vG_sk_s}+\frac{3LE_sr^2}{G_sk_s}+L^3\\ +\frac{6hE_sr^2}{G_sk_s}-L^2L_v-\frac{3h^2E_sr^2}{L_vG_sk_s}-L_vh^2+\frac{3L^2C_t^2E_sr^2}{L_vG_sk_s}+L_vL^2C_t^2\end{array}\right)}{LL_v^2C_t\left(1+\frac{3r^2E_s}{G_sk_sL_v^2}\right)\left(\begin{array}{l}-\frac{3S_tC_t^2L^3E_sr^2}{L_vG_sk_s}-\frac{9C_t^2hL^2E_sr^2}{L_vG_sk_s}-S_tC_t^2L^3L_v-3C_t^2hL_vL^2+\frac{3S_tL^3E_sr^2}{L_vG_sk_s}\\ +\frac{9S_th^2E_sr^2L}{L_vG_sk_s}+\frac{9hL^2E_sr^2}{L_vG_sk_s}+\frac{3h^3E_sr^2}{L_vG_sk_s}+2S_tLh^3+S_tL^3L_v+3hL_vL^2+3S_th^2L_vL\\ +\frac{3hLE_sr^2}{G_sk_s}+2h^4+\frac{6S_thLE_sr^2}{G_sk_s}+hL^3+S_tL^4+\frac{6h^2E_sr^2}{G_sk_s}+\frac{3S_tL^2E_sr^2}{G_sk_s}+L_vh^3\end{array}\right)};$$

$$\mu^*_{zx}=-\frac{3}{2}\frac{\pi r^4E_s\left(-\frac{3C_t^2r^2E_s}{G_sk_sL_v}-C_t^2L_v+\frac{3E_sr^2}{G_sk_sL}+L\right)}{L_v^2LC_t\left(h+LS_t\right)\left(\frac{9r^4E_s^2}{LL_v^2G_s^2k_s^2}+C_t^2L_v+\frac{3E_sr^2L}{G_sk_sL_v^2}+\frac{6C_t^2r^2E_s}{G_sk_sL_v}+\frac{9C_t^2E_s^2r^4}{G_s^2k_s^2L_v^3}+L+\frac{3r^2E_s}{LG_sk_s}\right)};$$

$$\mu^*_{xz}=\frac{3}{2}\frac{\pi r^4C_tE_s\left(-\frac{3C_t^2r^2E_s}{G_sk_sL_v}-C_t^2L_v+\frac{3E_sr^2}{G_sk_sL}+L\right)}{L_vL^2\left(h+LS_t\right)\left(\frac{3C_t^2r^2E_s}{L_vG_sk_s}+C_t^2L_v+\frac{9C_t^2E_s^2r^4}{G_s^2k_s^2L^2L_v}+\frac{3C_t^2L_vr^2E_s}{G_sk_sL^2}+L+\frac{6r^2E_s}{LG_sk_s}+\frac{9r^4E_s^2}{L^3G_s^2k_s^2}\right)}. \qquad [6.25]$$

The characteristic lengths for bending and torsion are next identified from the homogenized stiffness matrix as

$$l_{b_{xy}}^2 = \frac{K_{99}}{2(E_{45}+E_{44})};\ l_{b_{yx}}^2 = \frac{K_{66}}{2(E_{54}+E_{55})};\ l_{b_{yz}}^2 = \frac{K_{55}}{2(E_{67}+E_{66})};$$
$$l_{b_{zy}}^2 = \frac{K_{88}}{2(E_{76}+E_{77})};\ l_{b_{zx}}^2 = \frac{K_{77}}{2(E_{89}+E_{88})};\ l_{b_{xz}}^2 = \frac{K_{44}}{2(E_{98}+E_{99})};\qquad [6.26]$$

$$l_{t_{xy}}^2 = \frac{K_{99}+K_{98}}{E_{44}+E_{45}};\ l_{t_{yx}}^2 = \frac{K_{66}+K_{67}}{E_{54}+E_{55}};\ l_{t_{yz}}^2 = \frac{K_{55}+K_{54}}{E_{66}+E_{67}};$$
$$l_{t_{zy}}^2 = \frac{K_{88}+K_{89}}{E_{76}+E_{77}};\ l_{t_{zx}}^2 = \frac{K_{77}+K_{76}}{E_{88}+E_{89}};\ l_{t_{xz}}^2 = \frac{K_{44}+K_{45}}{E_{98}+E_{99}}.$$

One additional micropolar constant, the coupling number N, determines the strength of the micropolar behavior; it is calculated for the three planes x-y, y-z and x-z as

$$N_{xy}^2 = \frac{E_{44}-E_{45}}{2E_{44}};\ N_{yx}^2 = \frac{E_{55}-E_{54}}{2E_{55}};\ N_{yz}^2 = \frac{E_{66}-E_{67}}{2E_{66}};$$
$$N_{zy}^2 = \frac{E_{77}-E_{76}}{2E_{77}};\ N_{zx}^2 = \frac{E_{88}-E_{89}}{2E_{88}};\ N_{xz}^2 = \frac{E_{99}-E_{98}}{2E_{99}}.\qquad [6.27]$$

The micropolar material constants, the characteristic lengths in bending and torsion l_b and l_t, respectively, and the coupling number N, indicate the nature and significance of non-classical phenomena in the response of a medium with microstructure. The coupling number $0 \le N \le 1$ indicates the mechanisms producing micro-rotations and the contribution of the skew-symmetric part of the stress tensor or non classical effects [ERI 99]. The two extreme cases $N = 0$ and $N = 1$ denote classical-elasticity behavior and the coincidence of micro-rotations with macro-rotations respectively. The latter is known as couple-stress or Mindlin theory [ERI 99], and it is a special case of micropolar elasticity with constrained micro-rotations which occurs when $\gamma \to \infty$. A value of N between zero and unity ($0 < N < 1$) denotes a micropolar behavior [COW 70].

6.8. Results and discussion

The anisotropic mechanical behavior of vertebral trabecular bone is largely determined by its trabecular architecture. The precise determination of the mechanical behavior requires the determination of the geometrical and mechanical features of the micro-architecture of bone. Although detailed architectural measurements are not available for three-dimensional specimens, we suggest from visual inspection of several representative cross sections that the range of $\theta \in$ [10° - 80°] is reasonable to represent, on average, the relatively steep angle at which the oblique trabeculae lies relative to the horizontal. By contrast to the parameter θ, which could be reasonably estimated by visual inspection of the strut alignment, the geometrical model parameter h / L could not be reasonably estimated (from simple visual inspection). However, this parameter does affect the degree of anisotropy of the material. Therefore, to determine an appropriate value of h / L, we choose a value that results in a degree of material anisotropy consistent with previous measurements for the type of trabecular bone presented in [FOR 96]); specifically, we choose $h/L = 0.5$. Hence, a combination of $h/L = 0.5$ and $L_v / L = 1$ is used to simulate the anisotropic vertebral trabecular bone. We consider all struts (trabeculae) with identical cross-section geometry of diameter d = 0.25mm and strut lengths L=1mm [MIL 05]). The mechanical parameters of the trabeculae building structure are: $G_s = 0.5 * E_s / (1+\nu_s)$, with Poisson's ratio ν_s=0.3 [YOO 06]). The shear correction coefficient is $k_s = 9/10$. We plot the calculated homogenized elastic properties previously obtained for vertebral trabecular bone versus the geometrical parameter θ (Figures 6.22, 6.273, and 6.24), with a range of variation chosen in the interval [10°, 80°]; this range is also supported by [LIU 09], who scanned specimens of vertebral trabecular bone using microcomputed tomography. Here and in the sequel, the ratio of the effective properties to the bulk value is recorded.

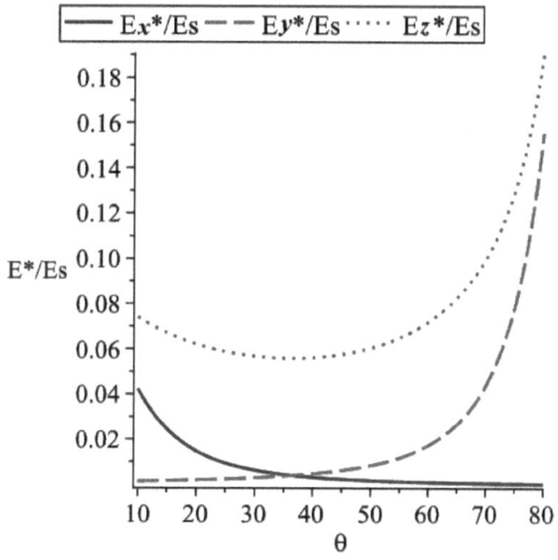

Figure 6.22. *Effective elastic moduli E_x^*, E_y^*, and E_z^* of vertebral trabecular bone versus the geometrical parameter θ. For a color version of this figure, see www.iste.co.uk/ganghoffer/biomechanics.zip*

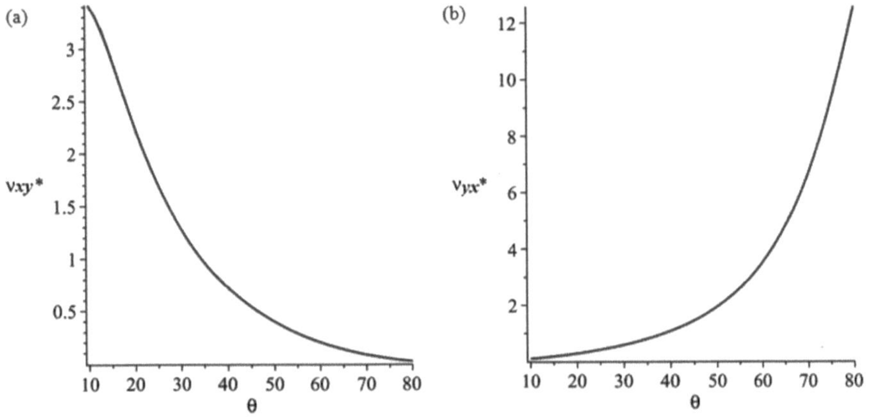

Figure 6.23. *Effective in-plane Poisson's ratios of vertebral trabecular bone versus θ a) V^*_{xy} and b) V^*_{yx}*

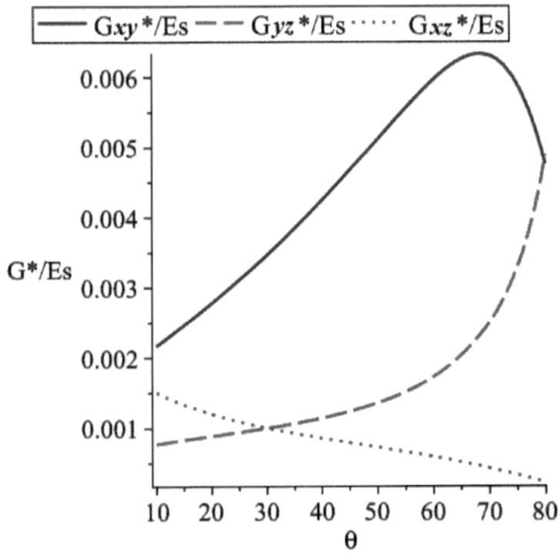

Figure 6.24. *Effective shear moduli G^*_{xy}, G^*_{yz}, and G^*_{xz} of vertebral trabecular bone versus angle θ. For a color version of this figure, see www.iste.co.uk/ganghoffer/biomechanics.zip*

As can be seen on Figure 6.24, the equivalent tensile modulus E^*_x exhibits a maximum for θ close to 10° and a minimum at 80°; it shows strong variations around those two cell angle values. It is obvious that the Young's modulus E^*_y

increases gradually with increasing θ, contrary to E_x^*; it exhibits a maximum for θ close to 80° and a minimum at 10° maximum for θ close to 10° . Those evolutions can be explained as follows: when the inclination angle θ gets closer to zero (θ is nil), the bone architecture becomes close to a tetragonal topology, thus the equivalent tensile modulus along the x-direction (E_x^*) exhibits a maximum (the tensile modulus along the y-direction shows a minimum). The opposite situation occurs when the angle θ gets closer to 90°. At about $\theta = 35°$, we note that the bone structure appears as isotropic $(E_x^* = E_y^*)$. In addition, the Young's modulus E_z^* decreases slightly with θ to reach a minimum at about 40° and then further increases up to 80°. We note from Figure 6.28 that the effective in-plane shear modulus G_{xy}^* gradually increases with the inclination angle up to a maximum at about 70°, and then rapidly decreases. The out-of-plane shear moduli G_{yz}^*, G_{xz}^* have behaviours similar to those of E_y^* and E_x^*.

In order to reflect the impact of transverse shear deformations on the effective properties of vertebral trabecular structure, the scaling behavior of the equivalent elastic and shear moduli as well as of Poisson's ratios versus effective density is determined, considering the following geometrical parameters: $L_v = 0.5L$; $h = 0.5L$; θ=30° Figures 6.25, 6.26, and 6.27. The obtained results show that the absence of transverse shear leads to overestimate Young's moduli (E_x^* and E_y^*) and the shear moduli (G_{xy}^*, G_{yz}^*, and G_{xz}^*) of vertebral trabecular bone; there is no such effect of transverse shear on E_z^*. We also note that neglecting shear deformation leads to underestimating the in-plane Poisson's ratios ν_{xy}^* and ν_{yx}^*.

The equivalent properties are obviously dependent on the relative densities of vertebral trabecular bone, as the following results show. The elastic (E_x^* and E_y^*) and shear moduli (G_{xy}^*, G_{yz}^*, and G_{xz}^*) exhibit a power law dependency on the relative density (Figures 6.25 and 6.26 respectively); there is a linear dependence for E_z^*. Observe that the effective density takes high values (up to 0.3) due to the consideration of thick beams.

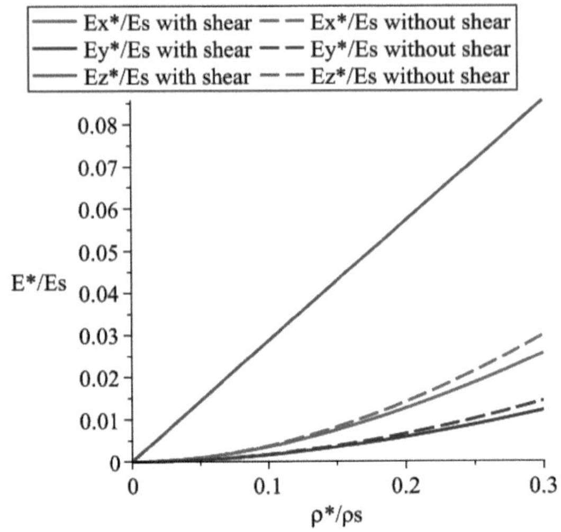

Figure 6.25. *Effective elastic moduli E_x^*, E_y^*, and E_7^* of vertebral trabecular bone versus the relative density ρ / ρ_S and impact of transverseshear. For a color version of this figure, see www.iste.co.uk/ganghoffer/biomechanics.zip*

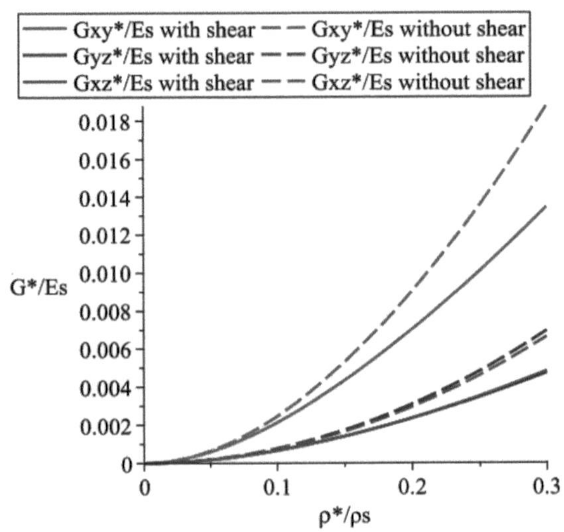

Figure 6.26. *Effective shear moduli G_{xy}^*, G_{yz}^*, and G_{xz}^* of vertebral trabecular bone versus the relative density ρ / ρ_s and impact of transverse shear. For a color version of this figure, see www.iste.co.uk/ganghoffer/biomechanics.zip*

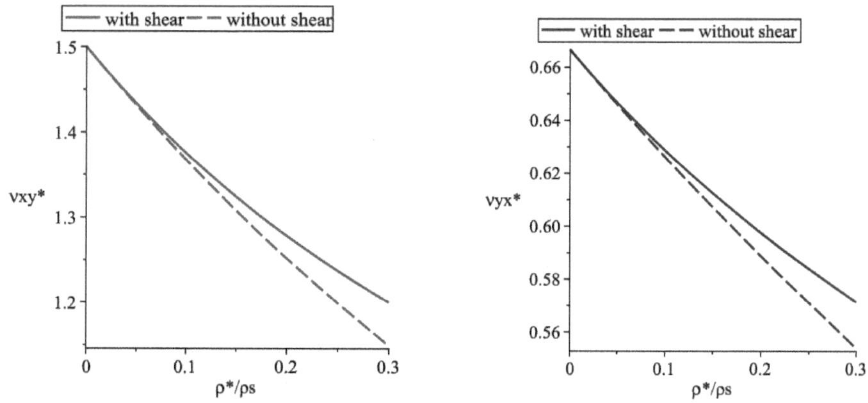

Figure 6.27. *Effective Poisson's ratios v_{xy}^* and v_{yx}^* of vertebral trabecular bone versus the relative density ρ / ρ_s and impact of transverse shear. For a color version of this figure, see www.iste.co.uk/ganghoffer/biomechanics.zip*

Two main aspects of the homogenization method and effective micropolar bone model shall be analyzed in the chapters that follow: the justification of the micropolar effects and the validation of the method in terms of the accuracy of the predicted effective properties.

6.9. Relevance of the choice of a micropolar effective continuum

In order to assess the relevance of a micropolar continuum model, the characteristic lengths and coupling number for the micropolar behaviour respectively described by equations [6.26] and [6.27], are evaluated versus the angle θ. The ratio of the characteristic internal micropolar lengths for bending and torsion, quantities l_b and l_t, to the characteristic unit cell size, the factor $2L\cos(\theta)$, are plotted to assess the strength of the micropolar effect (Figures 6.28a-c). It appears that those ratio reach values close to unity (and higher than unity for $l_{b_{zx}}, l_{t_{zx}}$), indicating that micropolar effects indeed impact the continuum behavior.

Furthermore, the coupling number N is plotted against the angle θ, to point out the nature and significance of non-classical phenomena in the response of bone considered as a medium with a microstructure (Figure 6.28d). It is recognized that values of the coupling number between zero and one ($0 < N < 1$) indicate a micropolar behaviour: this appears to be the case for the three planes, and for all values of the cell angle θ in the considered range.

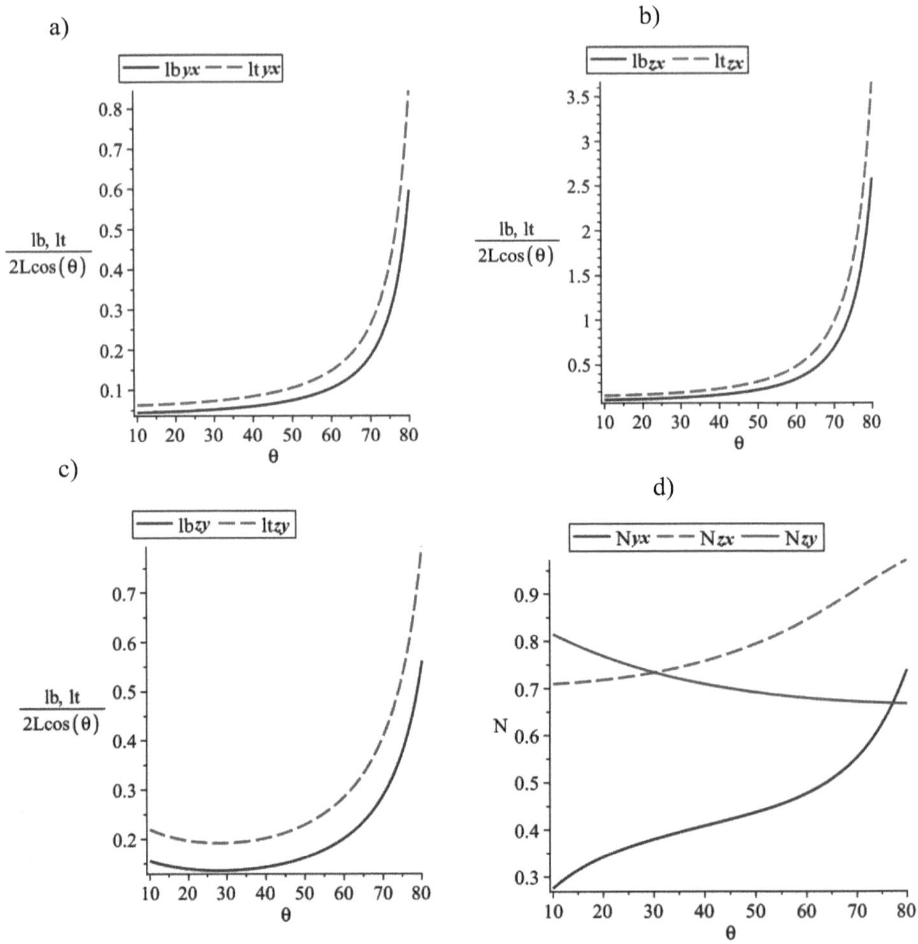

Figure 6.28. *a), b), c) Ratio of the characteristic micropolar lengths in bending and torsion to the unit cell size of vertebral trabecular bone and d) Micropolar coupling number versus the angle $\theta \in$ [10° - 80°]. For a color version of this figure, see www.iste.co.uk/ganghoffer/biomechanics.zip*

As mentioned earlier, the classical and micropolar shear moduli, respectively quantities G^* and μ^*, are related by $G^* = \mu^* + \kappa / 2$; the coefficient $2\mu^* + \kappa$ in micropolar elasticity replaces the term $2G^*$ in classical elasticity. The case $\kappa = 0$

corresponds to a decoupling of the rotational and translational degrees of freedom, hence the solid becomes classically elastic; the coupling number N then vanishes and the shear modulus μ^* is identical to the classical continuum theory ($\mu^* = G^*$). One observes (Figure 6.29) a correlative increase of the micropolar shear constant κ and of the coupling number N. We also notice the negative values of μ^*_{yz} / G^*_{yz} in the range $\theta = 60° \sim 80°$ in which the coupling number is greater than 0.7. As mentioned by [JEO 08a, JEO 08b], values of the coupling number larger than $1/\sqrt{2}$ lead to negative μ^*. Subsequently, in order to preserve the positive definiteness of strain-energy density, upper and lower bounds for the coupling number should be chosen such that $0.1 \le N \le 1/\sqrt{2}$.

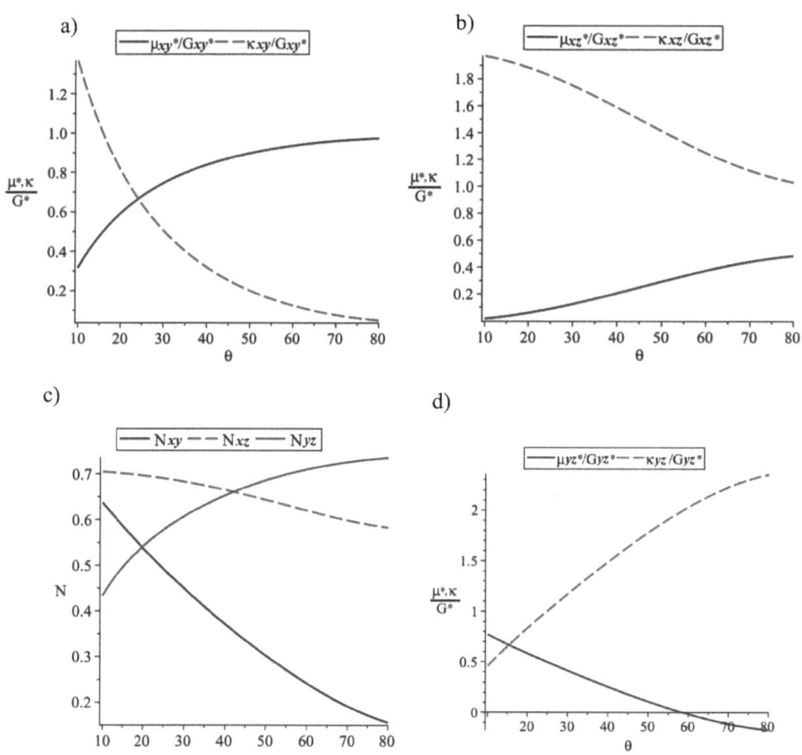

Figure 6.29. *Non dimensional micropolar anisotropic linear media material constants of vertebral trabecular bone: μ^* / G^* and k / G variations and its corresponding coupling number N versus the geometrical parameter $\theta \in$ [10° - 80°]. For a color version of this figure, see www.iste.co.uk/ganghoffer/biomechanics.zip*

In order next to assess the accuracy of the calculated effective properties, numerical simulations are performed to calculate those properties from a micromechanical model of vertebral trabecular bone.

6.10. Finite element validation

In this subsection, a numerical homogenization approach is applied to evaluate the accuracy of the effective properties predicted by the homogenization scheme. The finite element simulations presented here are performed over a representative volume element of the bone structure, equivalent to the unit cell of vertebral trabecular bone previously used in the analytical homogenization. The moduli are compared with those issued from the discrete homogenization analysis developed in the previous sections. Only the classical moduli will be computed and compared with those obtained from discrete homogenization.

The developed finite element model consists of three basic steps. Firstly, appropriate periodic boundary conditions are applied to the representative unit cell. Subsequently, the non-homogenous strain fields obtained from the analysis are integrated to define volume averaged strains. Finally, the effective elastic coefficients of the vertebral trabecular bone structure are obtained as the ratio of the average stress to the average strain. The average stresses and strains in the unit cell are defined by

$$\tilde{\sigma}_{ij} = \frac{1}{V_u}\int_{V_u} \sigma_{ij} dV_u \;;\quad \tilde{\epsilon}_{ij} = \frac{1}{V_u}\int_{V_u} \epsilon_{ij} dV_u \qquad [6.28]$$

where V_u is the volume of the periodic unit cell, and σ_{ij} and ϵ_{ij} are the local (microscopic) stresses and strains (in the unit cell). The average strain $\tilde{\epsilon}_{ij}$ can be related to the periodic boundary displacements of the unit cell by means of the divergence theorem [LAI 93] as follows:

$$\tilde{\epsilon}_{ij} = \frac{1}{V_u}\int_{V_u} \epsilon_{ij} dV_u = \frac{1}{2V_u}\int_S \left(u_i n_j + u_j n_i\right) dS \qquad [6.29]$$

where S is the boundary surface of the unit cell, u_i the displacement and n_j the unit exterior normal to S. Since the analysis is based on the RVE, periodical boundary conditions should be applied in the model to obtain a reasonable stress distribution. Two continuity conditions must be satisfied at the boundary surfaces of the neighbouring unit cells: the displacements must be continuous, and the traction distribution at the opposite parallel boundaries of the unit cell must be uniform.

As previously mentioned, the equivalent vertebral trabecular microstructure model consists of repeatable cells within the 3D space; Figure 6.30 depicts the geometry of vertebral trabecular unit cell. Since the microstructure is repeatable, only one cell needs to be analysed, with periodic boundary conditions. A condition of continuity of the displacement has been prescribed to the boundaries of the unit cell, in terms of periodic displacement boundary conditions. Hence, opposite surfaces of the unit cell are constrained to have an equivalent deformation state.

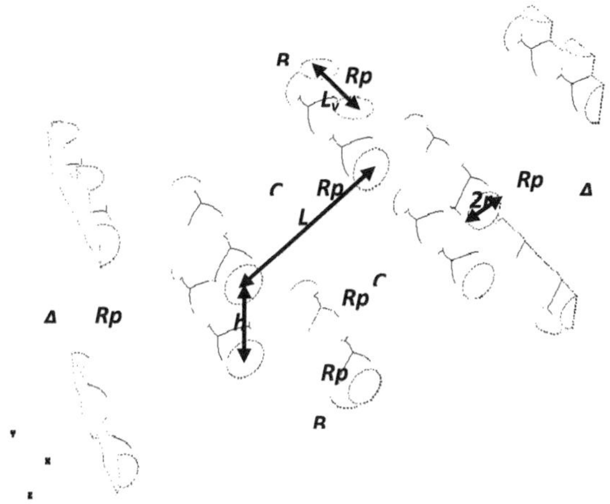

Figure 6.30. *Geometry of a repeatable vertebral trabecular structural unit cell with periodic boundary conditions. For a color version of this figure, see www.iste.co.uk/ganghoffer/biomechanics.zip*

The periodic boundary conditions must be applied on the boundary of the unit cell model shown in Figure 6.30, to ensure that there is no separation or overlap between the neighbouring unit cells. We apply the nodal displacement constraint equations [6.11] in the finite element procedure directly as the boundary conditions.

To this end, the unit cell has been meshed by the so called C3D10 element, a 10-node quadratic tetrahedral finite element, used in ABAQUS environment and the analysis has been performed using a linear perturbation analysis. We have performed simulations with different number of elements; results are here presented for the finer meshes (88,574 elements and 137,091 nodes). In order to apply the constraint equations [6.11], the six side surfaces are defined as sets, which are actually the boundaries and then the nodal degree of freedom on each surface (set) is constrained by an equation written for the degree of freedom of a reference point (six reference points are defined, labelled Rp1, Rp2, ... Rp6). The reference point does not belong to the main geometry of the unit cell for adjusting movements on the unit cell surface (to implement the periodic and loading boundary conditions). This mechanical constraint enforces all the nodes on the plane to have the same displacement in the specified direction, which means those nodes remain in the same plane. Because of the different kinematic conditions for applying the loading on the unit cell in axial and shear tests, separate sets of interactions will be defined. However, it should be noted that both of these interaction sets are designed in a way to satisfy the periodic boundary conditions when applying the macroscopic displacement field on the unit cell. The coupling here is quite similar to the equation constraint defined for the axial loadings.

In order to calculate the effective properties of vertebral trabecular bone, the general strain–stress relationship for an orthotropic material is expressed in matrix format as

$$\left[\tilde{\epsilon}_{ij}\right]=\left[S_{ij}^{eff}\right]\left[\tilde{\sigma}_{ij}\right] \qquad [6.30]$$

where $\left[S_{ij}^{eff}\right]$ is the effective compliance matrix, written as

$$\left[S_{ij}^{eff}\right]=\begin{bmatrix} S_{11}^{eff} & S_{12}^{eff} & S_{13}^{eff} & 0 & 0 & 0 \\ S_{12}^{eff} & S_{22}^{eff} & S_{23}^{eff} & 0 & 0 & 0 \\ S_{13}^{eff} & S_{23}^{eff} & S_{33}^{eff} & 0 & 0 & 0 \\ 0 & 0 & 0 & S_{44}^{eff} & 0 & 0 \\ 0 & 0 & 0 & 0 & S_{55}^{eff} & 0 \\ 0 & 0 & 0 & 0 & 0 & S_{66}^{eff} \end{bmatrix} \qquad [6.31]$$

According to the relationship between the engineering elastic constants and the compliance matrix S_{ij}, nine independent effective elastic constants of 3D vertebral trabecular bone can be calculated as

$$\begin{aligned} E_x^* &= \frac{1}{S_{11}^{eff}}; \nu_{xy}^* = -\frac{S_{12}^{eff}}{S_{11}^{eff}}; G_{xy}^* = \frac{1}{S_{44}^{eff}}; \\ E_y^* &= \frac{1}{S_{22}^{eff}}; \nu_{xz}^* = -\frac{S_{13}^{eff}}{S_{11}^{eff}}; G_{xz}^* = \frac{1}{S_{55}^{eff}}; \\ E_z^* &= \frac{1}{S_{33}^{eff}}; \nu_{yz}^* = -\frac{S_{23}^{eff}}{S_{22}^{eff}}; G_{yz}^* = \frac{1}{S_{66}^{eff}}. \end{aligned} \qquad [6.32]$$

A Python programming subroutine has been written within the ABAQUS environment, in order to get the effective properties automatically.

By assuming a set of the global strain, $\tilde{\epsilon}_{ij}$, and applying the periodic boundary conditions in the form of equation [6.11], we obtain a unique stress distribution of the unit cell. Then the global stress, $\tilde{\sigma}_{ij}$, corresponding to this set of global strain, is obtained; the effective compliance matrix S_{ij} is evaluated from equation [6.30].

We consider to be specific a unit cell of vertebral trabecular bone with the following geometrical parameters: r=0.125mm, L=1mm, L_v=0.5mm, h=0.5mm, and θ=30°; the bulk modulus of the trabeculae is chosen as E_s =12000 MPa, Poisson's ratio ν_s=0.3, and density 2000 kg/m^3 [GIB 05]). We do not take into account the heterogeneity of properties inside the trabeculae, although researchers showed that properties of trabecular bone at the tissue level are heterogeneous (e.g. [HAR 08]). Also, some experimental data indicates anisotropy in the properties of trabeculae; bone tissue stiffness depends on direction (e.g. [VAN 04]). We follow here instead most of the works which assume a homogeneous and isotropic behaviour of the trabeculae, e.g. [KIM 02, GIB 05, YOO 06], and [LIU 09].

As shown in Table 6.5, a relatively good agreement is obtained between the moduli predicted by finite elements and those calculated by discrete homogenization. In Figures 6.31a-c, the contour plots of the von Mises effective stress within the unit cells are shown for uniaxial tensile loadings in x, y directions, as well as for shear loading in the y-z plane.

Non-dimensional elastic constants	ABAQUS	Discrete homogenization	$\% \left\lvert \dfrac{\text{Abaqus} - \text{Hom}}{\text{Abaqus}} \right\rvert$
E_x^* / E_s	0.014	0.013	7.1
E_y^* / E_s	0.0067	0.0059	11.9
E_z^* / E_s	0.059	0.057	3.3
G_{xy}^* / E_s	0.0056	0.0069	23.2
G_{xz}^* / E_s	0.0030	0.0024	20.0
G_{yz}^* / E_s	0.0027	0.0023	16.8
ν_{xy}^*	1.27	1.28	0.8
ν_{yx}^*	0.57	0.59	3.5

Table 6.5. *Normalized elastic constants predicted by homogenization and FE simulations*

The relative density takes the value $\rho^* / \rho_S = 0.19$, so the apparent density is 380 kg/m^3.

The more pronounced differences (up to 23%) between both moduli occur under shear (the homogenized tensile moduli are predicted with a better accuracy) can be explained by the occurrence of stress gradients appearing at the articulation of beams within the unit cell.

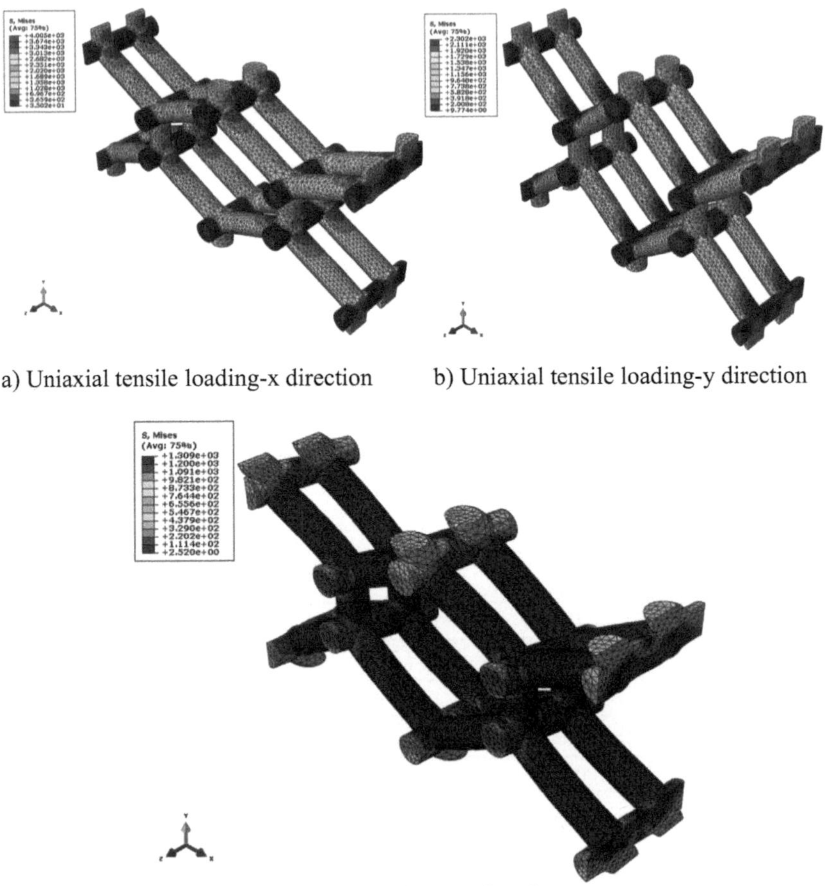

a) Uniaxial tensile loading-x direction

b) Uniaxial tensile loading-y direction

c) Shear loading-yz direction

Figure 6.31. *Contour plot of von Mises effective stress after uniaxial tensile (a) and b)) and shear c) loading. For a color version of this figure, see www.iste.co.uk/ganghoffer/biomechanics.zip*

6.11. Discussion and limitations

There is a limitation of our microstructural models that should be discussed when extrapolating the results to real trabecular bone: simple observations indeed reveal that the trabeculae are most of the time not perfectly uniform, but have a certain degree of non uniformity due to the natural variance in trabecular thickness for real vertebral bones. Models for such a variation of cross-section have been proposed such as the double-tapered shape model in [KIM 02]), with a cross-section area minimal at the center and maximum at the base (Figure 6.36, right view). A correction of this approximation by the trabeculae thickening could recover the part of the reduction due to considering the idealized uniform cross-sectional area. Detailed information on the strut morphology near the strut joints is not available in the literature and experimental measurements of the variation in strut thickness are lacking as well. [GEF 08] reported that the maximal thickness at the ends of the bases of a trabecula d_{max} (at the junctions of the lattice) is proportional to the minimal thickness at the center of a trabecula $d_{min,}$ adopting the relation $d_{\max} = 0.9922 d_{\min} + 60.174\ (\mu m)$. This linear proportion is observed by measuring the d_{max} and d_{min} dimensions of the trabeculae of canine cancellous bone from digital optical micrographs.

We next consider an extension of our model in order to compensate the thickening near the strut joints by an additional thickness of the strut. The diameter of the strut midway between large and small diameters shall be adopted as the average thickness of an equivalent uniform trabecula, $d = (d_{\max} + d_{\min})/2$, as pictured on Figure 6.32.

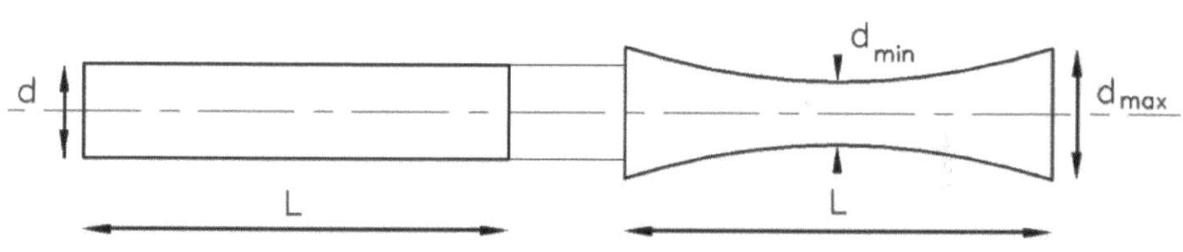

Figure 6.32. *Geometric dimensions of rod-like trabecula: Right: base thickness (d_{max}) and minimum thickness at the centre (d_{min}). Left: Equivalent uniform strut with diameter* $d = (d_{\max} + d_{\min})/2$

The mechanical properties of trabecular bone according to this enhanced geometrical model will be modified, due to the increase of the relative density of trabecular bone. The scaling behavior of the equivalent elastic and shear moduli versus the effective relative density is next obtained to highlight the effect of increasing trabeculae thickness; we adopt the following geometrical microparameters: $d_{\min} = 0.25\text{mm}$; $L_v = 0.5L$; $h = 0.5L$; θ=30° (see Figures 6.37–6.39). The relative elastic and shear moduli vary as power law functions of the relative density ρ^* / ρ_s (except for E_z^* / E_s varies linearly) for both geometric dimensions of the initial and extended model, based respectively on dmin and $d = (d_{\max} + d_{\min}) / 2$, given in Table 6.6.

	d_{min} (at the centre)		$d = (d_{\max} + d_{\min}) / 2$	
Young's moduli	$E_x^* / E_s = 0.26(\rho^* / \rho_s)^{1.89}$	R^2=0.99	$E_x^* / E_s = 0.38(\rho^* / \rho_s)^{1.87}$	R^2=0.99
	$E_y^* / E_s = 0.13(\rho^* / \rho_s)^{1.91}$	R^2=0.99	$E_y^* / E_s = 0.19(\rho^* / \rho_s)^{1.89}$	R^2=0.99
	$E_z^* / E_s = 0.285(\rho^* / \rho_s)$	R^2=1	$E_z^* / E_s = 0.357(\rho^* / \rho_s)$	R^2=1
Shear Moduli	$G_{xy}^* / E_s = 0.13(\rho^* / \rho_s)^{1.81}$	R^2=0.99	$G_{xy}^* / E_s = 0.17(\rho^* / \rho_s)^{1.78}$	R^2=0.99
	$G_{yz}^* / E_s = 0.048(\rho^* / \rho_s)^{1.87}$	R^2=0.99	$G_{yz}^* / E_s = 0.068(\rho^* / \rho_s)^{1.85}$	R^2=0.99
	$G_{xz}^* / E_s = 0.05(\rho^* / \rho_s)^{1.89}$	R^2=0.99	$G_{xz}^* / E_s = 0.07(\rho^* / \rho_s)^{1.88}$	R^2=0.99
Poisson's ratio	$\nu_{xy}^* = -0.098(\rho^* / \rho_s) + 1.48$	R^2=0.99	$\nu_{xy}^* = -1.15(\rho^* / \rho_s) + 1.47$	R^2=0.98
	$\nu_{yx}^* = -0.32(\rho^* / \rho_s) + 0.66$	R^2=0.99	$\nu_{yx}^* = -0.37(\rho^* / \rho_s) + 0.66$	R^2=0.99

Table 6.6. *Relationships between the relative Young's and shear moduli as well as Poisson's ratios versus relative density for two types of trabecular geometry*

The Young's moduli (E_x^*, E_y^*) and shear moduli $(G_{xy}^*, G_{yz}^*, G_{xz}^*)$ are approximately 50% higher than those of trabeculae with minimum thickness at the centre, while E_z^* shows an increase by about 25%, as shown on Figures 6.33 and 6.34.

There is no comprehensive experimental evidence to fully verify the predicted homogenized mechanical behavior by comparison with real cancellous bone, due to an insufficient amount of experimental data: in most cases, only the longitudinal stiffness modulus is determined. Experimental studies generally neglect the issue of principal directions of orthotropy: the specimens are cut in arbitrary directions and the measured data are thus not sufficient to determine the principal directions. A number of authors have measured the elastic modulus of vertebral trabecular bone. [KOP 02] tested human vertebral trabecular bone and obtained a mean elastic modulus of 319±189 MPa with apparent density in the range 0.09-0.38 g/cm^3. [KEA 97] measured the Young's modulus of human vertebral bone by compression tests as 165±110 MPa with a mean value of apparent density of 0.14±0.06 g/cm^3. [KOP 98] reported the elastic moduli from compression of human vertebral specimens to be 291±113 MPa, with a wet apparent density of 0.17±0.04 g/cm^3. Moreover, [CEN 99] measured by compression tests an average Young's modulus of human lumbar cancellous bone of 134±81 MPa, for a bone volume fraction of 17.77±5.65%. [BAN 02] measured the average elastic modulus of human vertebral cancellous bone of 352±145 MPa and the mean apparent density of the samples as 0.174±0.052 g/cm^3, while [MOR 01] measured a similar value of 344±148 MPa from compression tests of human vertebral trabecular bone specimens, with a mean apparent density of 0.18±0.05 g/cm^3. [MOR 03] reported power-law regressions between modulus E^* (in MPa) and apparent density ρ^* (in g/cm^3) as $E^* = 4730\rho^{*1.56}$, from tension and compression tests of human vertebral trabecular bone; the mean measured value of apparent density is 0.18 g/cm^3. [WOL 10] performed non-destructive macroscopic mechanical compression and tension tests on human vertebral bones, reporting apparent Young's moduli of 424 MPa in compression and 439 MPa in tension; morphological analysis of the samples yielded volume fraction of 11±0.03%. [HAV 10] reported a mean elastic modulus of 813.10±419.28% MPa for vertebral trabecular bone of baboons, with bone volume fraction of 29.93±8.16%.

[HU 02] performed nondestructive compression tests in the lumbar vertebral specimens of dogs in the anteroposterior (AP) and left-right (LR) directions, and destructive compression tests in the cephalocaudal (CC) direction: the measured Young's moduli are respectively 209±54 MPa, 269±96 MPa, and 967±83 MPa in (AP), (LR), and (CC), with an apparent density of 0.35±0.03 g/cm^3, a tissue density of 2.05±0.01 g/cm^3 and bone volume fraction of 19±2%; those moduli are normalized with respect to the mean value E_s=19.9 ± 2.5 GPa, computed by [JOR 02] from a canine distal femur using a 1 GHz acoustic microscope.

Shear properties of trabecular bone received limited attention, perhaps due to the experimental and conceptual difficulties encountered when testing such highly anisotropic porous structures. The shear moduli for cancellous bone are rarely measured; e.g. in [ASH 86]), the mean shear modulus is measured using an ultrasonic technique in the range 58-89 MPa. [WOL 10] reported an apparent shear modulus of 85 MPa by performing non-destructive macroscopic mechanical torsion tests on human vertebral trabecular samples with volume fraction of 11±0.03%. [KAS 07] performed torsion testing on ovine vertebral cancellous bones, reporting shear modulus of 282±81MPa and a mean apparent density of 0.69±0.07 g/cm^3; the shear is normalized with respect to the Young's modulus of a single trabecula for sheep, measured by [LOR 06] as 15.0±1.4 GPa. Moreover, [WAN 05] measured the mean shear modulus as 183±75 MPa by conducting torsion tests on bovine tibial trabecular bone samples of volume fraction 26.3±0.041%.

We have plotted on Figures 6.33 and 6.34 the variation of the effective mechanical properties obtained by our model, together with previously mentioned measurements. The computed values of Young's and shear moduli compare well with the reported experimental values; however, measurements are dispersed around our anisotropic computed values, partially due to the arbitrary directions in which specimens are cut (the principal directions are undetermined). Another source of discrepancy is the variation of the geometry (the tested struts have different lengths and thicknesses, which are in most cases not described) and location of the tested samples.

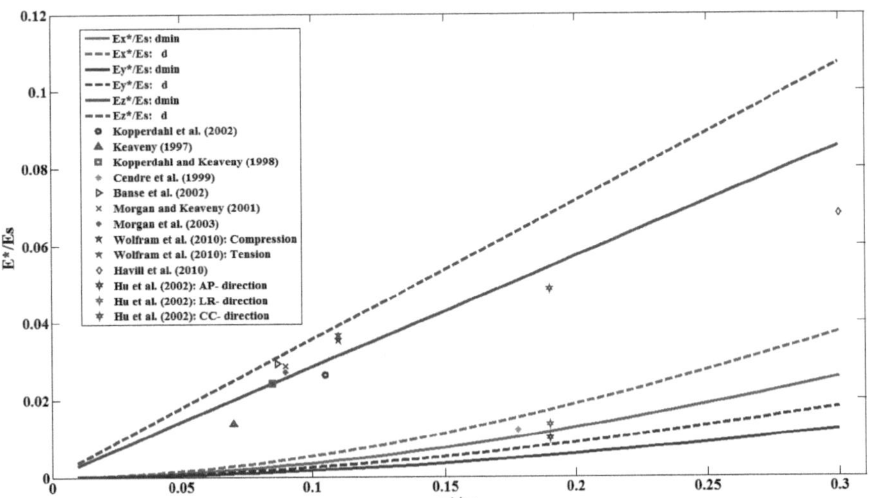

Figure 6.33. *Normalized homogenized effective Young's moduli of vertebral trabecular bone and experimental values from literature plotted against relative density. For a color version of this figure, see www.iste.co.uk/ganghoffer/biomechanics.zip*

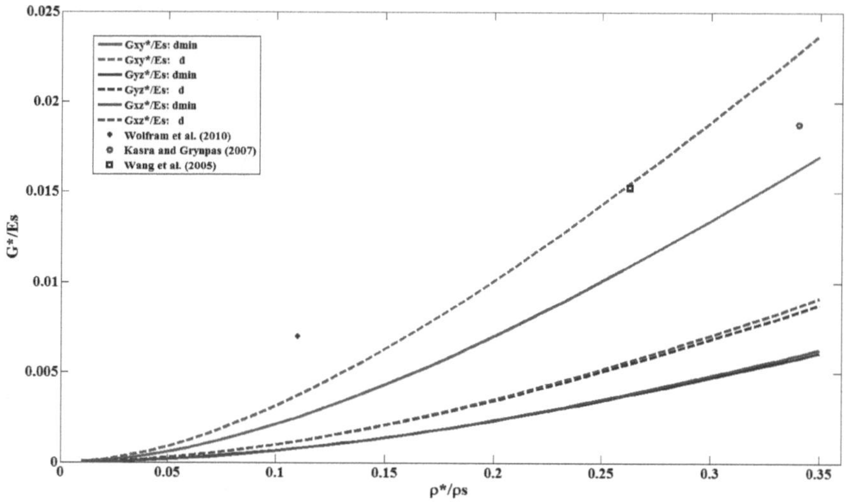

Figure 6.34. *Normalized homogenized effective shear moduli of trabecular bone and experimental values from literature plotted against relative density. For a color version of this figure, see www.iste.co.uk/ganghoffer/biomechanics.zip*

Finally, the anisotropic Poisson coefficients are never reported to be investigated experimentally. The calculated Poisson's ratios are shown to only weakly depend on relative density (Figure 6.35).

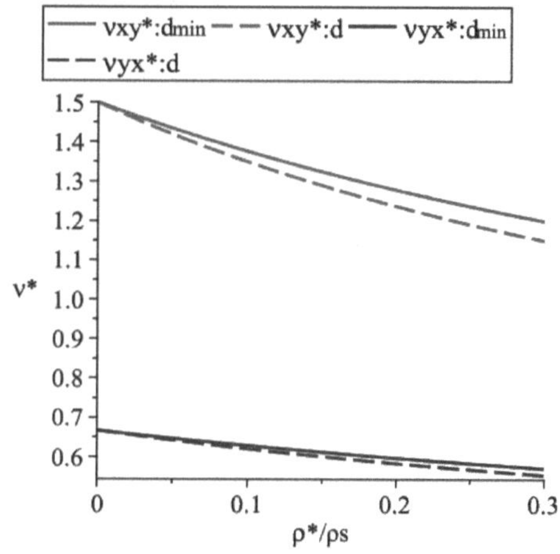

Figure 6.35. *In-plane Poisson's ratios of vertebral trabecular bone plotted against relative density. For a color version of this figure, see www.iste.co.uk/ganghoffer/biomechanics.zip*

Such micromechanically based enhanced continuum models developed at the mesoscopic scale pave the way towards realistic simulations of the mechanical response of complete bone structures submitted to complex loadings.

The developed homogenization model constitutes an interesting tool for an efficient virtual experimentation of the mechanical behavior of lattices endowed with diverse topologies, and can therefore in the future be adapted to model many repetitive representative cells of bone by which the main structural features of the trabecular microstructure can be characterized, such as tapered hexagonal and tetrakaidecahedral morphologies, or topologies acquired from microcomputed tomography.

7

Size-dependent Dynamic Behavior of Trabecular Bone

In order to investigate microstructure-related scale effects on macroscopic bone properties, Cosserat models of vertebral trabecular bone are constructed, based on micromechanical approaches. A micropolar equivalent continuum model is firstly constructed, the effective mechanical properties of which are expressed versus the geometrical and mechanical microparameters, accounting for bending, axial, transverse shear deformations, and torsion. The static and dynamic effective behaviors of vertebral trabecular bone in the form of beam structures are next analyzed, in terms of the deflection, torsion and eigenfrequencies of deformations. The governing differential equations of static and dynamic bending and torsion of trabecular bone are derived using variational principles based on non-classical theory, and explicit solutions are derived, accounting for length scale effects. The static bending and torsion behaviors developed by the non-classical theory show significant differences with those obtained by the classical theory when the ratio of the characteristic beam size to the internal material length scale parameter is small, or for small specimen sizes. The identified static and dynamical effective properties of bone are correlated to physiological factors, such as the age of patient, the effective bone density, and pathologies leading to a modification of the internal architecture.

7.1. Introduction

Classical continuum theory based on the Cauchy-Boltzmann approach is the simplest continuum model in which the stress at a material point depends only on the strain at the same point. As a consequence, it cannot incorporate microstructural

---

Chapter written by Jean-François GANGHOFFER, Ibrahim GODA and Rachid RAHOUADJ.

size effects and internal lengths, which occur in micron- and sub-micron-scale structures. However, real materials such as biological tissues often exhibit a number of important length scales, which must be included in any realistic model. Although there have been many continuum models of bone developed over the last two decades on the basis of classical elasticity [BOW 98, TAY 02, TAY 07], those models ignore microstructure-related scale effects on the macroscopic mechanical properties. The structural hierarchy of materials in microstructures such as bones plays an important role in determining their macroscopic mechanical behavior as well as the stress and strain distribution. This issue is presently investigated by studying generalized continuum mechanics theories which account for the influence of microstructure-related scale effects on the macroscopic properties of bone. Bone is a strongly heterogeneous material with microstructural features, requiring generalized continuum mechanics theories when the macroscopic length scale (identified as the smallest length scale of the deformation pattern) becomes comparable or smaller than the typical microstructural length scale, such as the size of trabeculae in cancellous bone. Especially, the classical assumptions inherent to classical elasticity are no more valid under such conditions.

The generalized continuum mechanics like micro-dilatation theory [NUN 79], micro-polar or so-called Cosserat theory [ERI 01], microstretch theory [ERI 01], micro-strain theory [FOR 06] and micromorphic theory [ERI 01]) have the inherent capacity to bridge micro and macro scales using their explicit size effect features. These theories were widely applied for modeling purposes of foams [DIE 03], cellular materials and bone [LAK 86] as well as heterogeneous materials [FOR 98, FOR 99] like sand, soil [MAJ 04, KAR 10] and rock [STE 10]. The existence of couple stress in bones was first evidenced by [YAN 81], who measured the effect of the size of a bone specimen on the apparent stiffness of cortical bone in quasi-static torsion; they further obtained the characteristic length scales for torsion and bending for cortical bone in the context of couple stress theory. Bone trabeculae are modelled as isotropic micropolar materials in [LAK 93]. Lakes and co-workers conducted a series of experiments on bone and other cellular materials, in which they observed a stiffening effect in such materials in bending and in torsion [PAR 87] and a tougher notched bone [LAK 90] than expected from classical Cauchy-type elasticity. Those last authors found that Cosserat elasticity provides better predictions of the response of bone than classical elasticity theory.

The dynamics of cellular solids obeying a Cosserat behavior such as bones don't appear to have received much attention in the literature. [WAN 01] used a micropolar theory to calculate displacements and stresses generated by a fundamental boundary value problem in the dynamics of elastic continua; namely, a time-harmonic line force acting at a point on the edge of a two-dimensional honeycomb half-space. [BAK 98] studied the effect of impact and energy absorption behaviour of cellular materials. [BAN 05] developed a numerical scheme for reducing the computational expense associated with the free vibration and the response calculations of the cellular structures. Moreover, [BAN 09] investigated the applicability and the limitations of the effective medium theory for dynamics of cellular structures.

7.2. Age-related change of vertebral bone architecture

We model the beam-like architecture of vertebral bone as a three-dimensional lattice with a unit cell geometry shown in Figure 7.1b. The architecture of the model (thick vertical columns and thinner horizontal struts) is based on studies of samples taken from the central part of vertebral bodies from normal individuals aged from 30 to 90 years [MOS 88, MOS 89]. Four indices are used to describe the morphology of the unit cell: the horizontal trabecular thickness (D_h), the vertical trabecular thickness (D_V), and the mean distance between the horizontal trabeculae (L_h) as well as between the vertical trabeculae (L_v). From those morphological indices, an idealized model of trabecular microarchitecture is constructed – the prismatic shape. The age-related change of vertebral bone architecture (the changes in trabecular thickness and spacing that occur with aging) has been described by [MOS 88, MOS 89], using the relations in Table 7.1.

The whole structure is generated from the repetition of the reference unit cell Figure 7.1a by means of three periodicity vectors defined with a Cartesian basis. Note that both vertical and horizontal trabeculae are idealized as a circular beam with diameter D_v and D_h, respectively.

Parameter	Relation
Horizontal trabeculae thickness (μm)	$D_h = -1.03 \times \text{Age} + 189$
Vertical trabeculae thickness (μm)	$D_v = 0.14 \times \text{Age} + 208$
Distance between horizontal trabeculae (μm)	$L_h = 13.74 \times \text{Age} + 288$
Distance between vertical trabeculae (μm)	$L_v = 6.74 \times \text{Age} + 456$

Table 7.1. *Micro-structural parameters for vertebral bone based on morphology–age relationships by [MOS 88], with age measured in years*

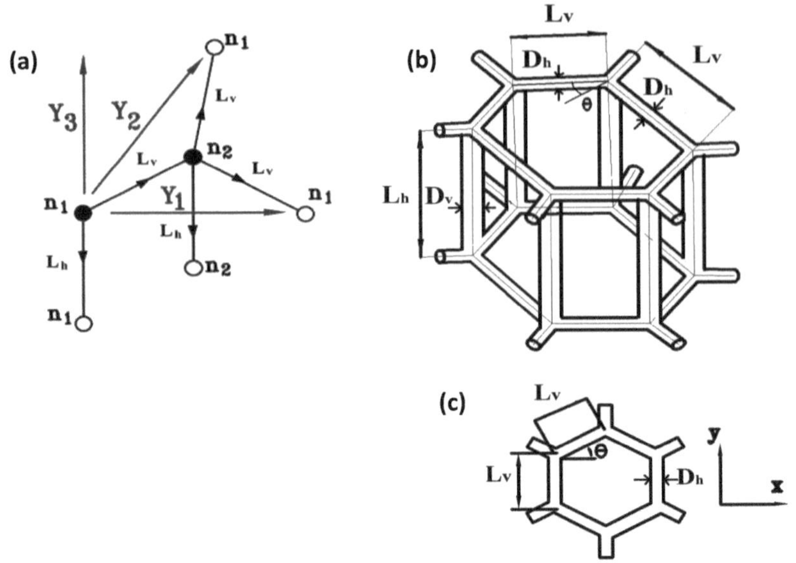

Figure 7.1. *Representative model of vertebral trabecular bone and its parameters. (a) Reference unit cell and its periodicity vectors Y_1; Y_2; and Y_3 and, (b) 3D hexagonal unit cell with thick vertical trabeculae and thinner horizontal ones rendering human vertebral bone, (c) 2D hexagonal lattice*

A 3D geometric model of a regular hexagonal structure is considered as an idealized representation of vertebral trabecular bone, for which the micropolar elastic constants at the macroscopic level are identified from a micromechanical analysis. The considered hexagonal unit cell is composed of five beams; three horizontal beams have the same length L_v and diameter D_h (two lengths of in-plane struts are inclined by the angle $\theta = 30°$ with the horizontal direction) and two vertical struts of length L_h and diameter D_v (out-of-plane). The lengths of the periodicity vectors are $L_1 = L_2 = \sqrt{3}L_v$ and $L_3 = L_h$.

The relative density of the 3D hexagonal is computed in terms of the geometrical parameters of vertebral trabecular bone architecture $L_v, L_h, D_h,$ and D_v as

$$\frac{\rho^*}{\rho_s} = \frac{1}{18}\frac{\pi\left(3L_vD_h^2 + 2L_hD_v^2\right)\sqrt{3}}{L_v^2L_h} \qquad [7.1]$$

The effective density decreases with age, a fact in agreement with physiology (Figure 7.2).

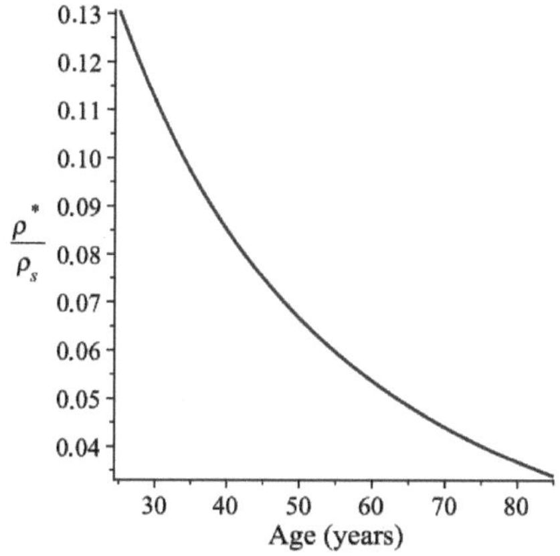

Figure 7.2. *Evolution with age of the effective normalized density of trabecular bone*

In the forthcoming sections, microstructure-dependent beam models for vertebral trabecular bone are developed for the purpose of evaluating the bending and torsional behaviors in both the static and dynamic regimes. The derivations are based on the micropolar model and Hamilton's principle. The models contain material length scale parameters and are able to capture size effects, unlike classical beam theory.

7.3. Size dependent static and dynamic bending behaviors

In this section, we identify the static and dynamic response of vertebral cancellous bone beam samples subjected to bending, in relation to their microstructural properties.

7.3.1. *Governing equations of the motion of micropolar beams in bending*

We consider a straight beam made of a micropolar continuum representative of the homogenized bone architecture, with cross section of height H and width W, subjected to a pure bending moment around the z axis, as shown in Figure 7.3. According to the micropolar theory, the strain energy density is a function of both strain (conjugated to stress) and curvature (conjugated to couple stress). It then follows that the strain energy U in a deformed isotropic linear elastic material occupying a fixed domain Ω is given by

$$U = \frac{1}{2}\int_{\Omega}\left(\sigma_{ij}\epsilon_{ij} + m_{ij}\chi_{ij}\right)\mathrm{d}V \qquad [7.2]$$

The total length (or span) of the beam member is L_m. According to the basic assumptions of Bernoulli–Euler beams and the one-dimensional beam theory, the strain energy writes

$$U = \frac{1}{2}\int_{\Omega}\left(\sigma_{xx}\epsilon_{xx} + m_{xz}\chi_{xz}\right)\mathrm{d}V \qquad [7.3]$$

with

$$\sigma_{xx} = E_x^* y\frac{\partial^2 v(x,t)}{\partial x^2},\ \epsilon_{xx} = y\frac{\partial^2 v(x,t)}{\partial x^2},\ \text{and}\ m_{xz} = \gamma_{xz}\frac{\partial^2 v(x,t)}{\partial x^2},$$

Hence

$$U = \frac{1}{2}\int_0^{L_m}\int_A\left(E_x^* y^2 + \gamma_{xz}\right)\left(\frac{\partial^4 v(x,t)}{\partial x^4}\right)dA dx = \frac{1}{2}\int_0^{L_m}\left(E_x^* I_z^* + \gamma_{xz}A\right)\left(\frac{\partial^4 v(x,t)}{\partial x^4}\right)dx \qquad [7.4]$$

where I_z^* is the second moment of cross-sectional area defined by $I_z^* = \int_A y^2 dA$ and A the cross-sectional area of the macroscopic beam. The rotation angle of the centroidal axis of the beam is related to the deflection as $\phi_z(x,t) = \dfrac{\partial v(x,t)}{\partial x}$.

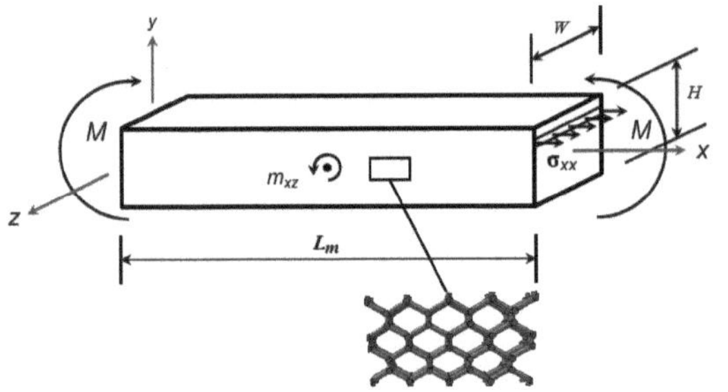

Figure 7.3. *Configuration of the micropolar micro-structured beam of the vertebral trabeculae*

The previous expression [7.4] of the strain energy can be written with the characteristic micropolar length (internal material length scale parameter) $l_{b_{xy}}$ as

$$U=\frac{1}{2}\int_0^{L_m}\left(E_x^*I_z^*+2l_{b_{xy}}^2\left(2\mu_{xy}^*+\kappa_{xy}\right)A\right)\left(\frac{\partial^4 v(x,t)}{\partial x^4}\right)dx \qquad [7.5]$$

On the other hand, the kinetic energy can be written as

$$KE=\frac{1}{2}\int_0^{L_m}\rho^*(x)A(x)\left(\frac{\partial^2 v(x,t)}{\partial t^2}\right)dx \qquad [7.6]$$

with $\rho^*(x)$ the effective density of the beam material (for trabecular bone this is equivalent to the volume fraction (BV/TV) or relative density multiplied by the tissue density).

By neglecting the body force and the body couple, the work done by the external forces in the form of lateral forces $q(x,t)$ can be written as

$$W^{ex}=\int_0^{L_m}q(x,t)v(x,t)dx \qquad [7.7]$$

The governing equation of motion of the micropolar beam can then be determined thanks to Hamilton's principle

$$\delta\int_{t_1}^{t_2}\left(KE-U+W^{ex}\right)dt=0 \qquad [7.8]$$

where δ denotes the variation symbol. Substituting equations [7.5]–[7.7] into [7.8] then leads to

$$\delta\int_{t_1}^{t_2}\int_0^{L_m}\left(\frac{1}{2}\rho^*A\left(\frac{\partial v(x,t)}{\partial t}\right)^2-\frac{1}{2}\left(E_x^*I_z^*+2l_{b_{xy}}^2\left(2\mu_{xy}^*+\kappa_{xy}\right)A\right)\left(\frac{\partial v^2(x,t)}{\partial x^2}\right)^2+q(x,t)v(x,t)\right)dxdt=0 \qquad [7.9]$$

The variation in equation [7.9] can be evaluated using integration by parts to obtain

$$\begin{aligned}&\int_{t_1}^{t_2}\int_0^{L_m}\left(-\rho^*A\frac{\partial v^2(x,t)}{\partial t^2}-\left(E_x^*I_z^*+2l_{b_{xy}}^2\left(2\mu_{xy}^*+\kappa_{xy}\right)A\right)\frac{\partial v^4(x,t)}{\partial x^4}+q(x,t)\right)\delta v dx dt+\int_0^{L_m}\rho^*A\frac{\partial v(x,t)}{\partial t}\delta v dx\Bigg|_{t_1}^{t_2}\\&-\int_{t_1}^{t_2}\left(E_x^*I_z^*+2l_{b_{xy}}^2\left(2\mu_{xy}^*+\kappa_{xy}\right)A\right)\frac{\partial v^2(x,t)}{\partial x^2}\delta\frac{\partial v(x,t)}{\partial x}dt\Bigg|_0^{L_m}+\int_{t_1}^{t_2}\left(E_x^*I_z^*+2l_{b_{xy}}^2\left(2\mu_{xy}^*+\kappa_{xy}\right)A\right)\frac{\partial v^3(x,t)}{\partial x^3}\delta v dt\Bigg|_0^{L_m}=0\end{aligned} \qquad [7.10]$$

The governing differential equation of motion and boundary conditions follows from the previous equation: the Euler–Lagrange equation coincides with the dynamical equation of motion

$$\left(E_x^* I_z^* + 2 l_{b_{xy}}^2 \left(2\mu_{xy}^* + \kappa_{xy}\right) A\right) \frac{\partial^4 v(x,t)}{\partial x^4} + \rho^* A \frac{\partial^2 v(x,t)}{\partial t^2} = q(x,t) \qquad [7.11]$$

Since δv is zero at time t_1 and time t_2, the boundary conditions follow from [7.10] as

$$\frac{\partial v^2}{\partial x^2} \delta \frac{\partial v}{\partial x} = 0; \quad \frac{\partial v^3}{\partial x^3} \delta v = 0, \; x = 0 \text{ and } x = L_m \qquad [7.12]$$

Equation [7.11] shows that the bending deformation of the beam has two contributions: one associated with the normal stress component σ_{xx}, and the second one due to the couple stress component m_{xz}. The bending rigidity of the beam, quantity $E_x^* I_z^* + 2 l_{b_{xy}}^2 \left(2\mu_{xy}^* + \kappa_{xy}\right) A$, explicitly depends on the internal bending length $l_{b_{xy}}$, itself related to the underlying microstructure of the beam material. This internal bending length enables the incorporation of the material microstructural features in the micropolar model and accounts for the size effect, as will be proven in the next section. Clearly, when the microstructural effect is suppressed by letting $l_{b_{xy}} = 0$, the model defined by equation [7.11] reduces to the classical Bernoulli-Euler beam theory.

7.3.2. *Size-dependent static bending behavior*

The static size-dependent behavior of micro-structured beams is analyzed by considering a specific case. Figure 7.4 illustrates a micro-structured cantilever with uniform rectangular cross section with height H and width W, subjected to a moment M at its end. From equation [7.11], the static form of the governing (equilibrium) equation can be obtained as

$$\left(E_x^* I_z^* + 2 l_{b_{xy}}^2 \left(2\mu_{xy}^* + \kappa_{xy}\right) A\right) \frac{\partial^4 v(x)}{\partial x^4} = q(x) \qquad [7.13]$$

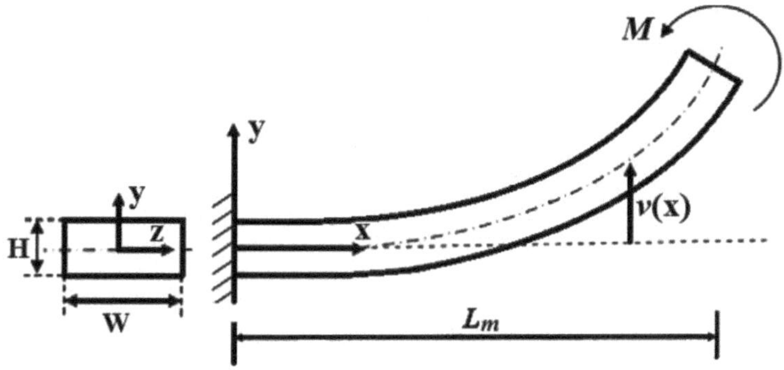

Figure 7.4. *Schematic diagram of a micro-structured cantilever beam with rectangular cross section exposed to a moment at its free end*

The boundary conditions are prescribed in terms of the displacement and rotation, variables v and ϕ_z, at x=0 and x=L_m as

$$v(0)=0,\ \phi_z(0)=\frac{\partial v(0)}{\partial x}=0,\ \frac{\partial \phi_z^2(L_m)}{\partial x}=\frac{\partial^3 v(L_m)}{\partial x^3}=0,$$
$$\left(E_x^* I_z^* + 2l_{b_{xy}}^2\left(2\mu_{xy}^* + \kappa_{xy}\right)A\right)\frac{\partial \phi_z(L_m)}{\partial x},\frac{\partial^2 v(L_m)}{\partial x^2}=M \qquad [7.14]$$

Integrating the governing equation [7.13] four times with q(x) = 0, gives

$$\left(E_x^* I_z^* + 2l_{b_{xy}}^2\left(2\mu_{xy}^* + \kappa_{xy}\right)A\right)v(x)=\frac{C_1}{6}x^3+\frac{C_2}{2}x^2+C_3x+C_4 \qquad [7.15]$$

Applying the boundary conditions [7.14] then leads to the four constants

$$C_1 = C_3 = C_4 = 0,\ C_2 = M \qquad [7.16]$$

Summarizing, the following expression for the static deflection is obtained

$$\left(E_x^* I_z^* + 2l_{b_{xy}}^2\left(2\mu_{xy}^* + \kappa_{xy}\right)A\right)v(x)=\frac{M}{2}x^2 \qquad [7.17]$$

The closed-form expression for the static deflection of the micropolar beam is then obtained as a function of the geometrical and mechanical micro-parameters (mechanical: E_s, G_s; geometrical: L_v, L_h, D_h, D_v, and k_s) of vertebral trabeculae as well as a function of geometrical parameters of the macroscopic beam (*W* and *H*):

$$v(x)=\frac{32\sqrt{3}L_hL_v\left(3E_sD_h^2+4G_sk_sL_v^2+9D_h^2G_sk_s\right)Mx^2}{\pi D_h^4WHE_s\left(16G_sk_sH^2+3E_sD_h^2+4G_sk_sL_v^2+9D_h^2G_sk_s\right)} \quad [7.18]$$

The previous equation can be rewritten as a function of the internal material length scale, a parameter describing microstructure-dependent size effects, as

$$v(x)=\frac{2L_vL_h\sqrt{3}\left(3E_sD_h^2+4G_sk_sL_v^2+3D_h^2G_sk_s\right)\left(4G_sk_sL_v^2+3E_sD_h^2+9D_h^2G_sk_s\right)Mx^2}{\pi D_h^4E_sG_sk_sWH\left(3E_sD_h^2H^2+36E_sD_h^2l_{b_{xy}}^2+3D_h^2G_sk_sH^2+108l_{b_{xy}}^2G_sk_sD_h^2+48l_{b_{xy}}^2G_sk_sL_v^2+4H^2G_sk_sL_v^2\right)} \quad [7.19]$$

If the microstructural effect is neglected, by letting $l_{b_{xy}}=0$, the static deflection formula predicted by the classical theory is recovered as

$$\left(E_x^*I_z^*\right)\bar{v}(x)=\frac{M}{2}x^2 \quad [7.20]$$

7.3.3. *Size-dependent free flexural vibration behavior*

In this subsection, the size-dependent flexural vibration behavior of a micro-structured micropolar beam is analyzed. Many papers have appeared in the literature in the last 10 years that develop microstructure-dependent beam models based on non-local theories; those models are however by essence phenomenological. For instance, a Timoshenko beam model was considered in [MA 08] to investigate dynamical bending behavior; the authors employ a modified couple stress theory which however does not result from a micromechanical analysis of an underlying microstructure.

According to Hamilton's principle, the dynamic governing equation for the free flexural vibrations of a micro-structured beam in terms of v(x, t) is determined by equation [7.11] as

$$\left(E_x^*I_z^*+2l_{b_{xy}}^2\left(2\mu_{xy}^*+\kappa_{xy}\right)A\right)\frac{\partial^4v(x,t)}{\partial x^4}+\rho^*A\frac{\partial^2v(x,t)}{\partial t^2}=0 \quad [7.21]$$

To be specific, we consider the dynamic analysis of a cantilever beam to demonstrate the size dependency of the beams' natural frequencies. The problems of a cantilever beam of length L_m with rectangular cross-sectional shape are solved next; the corresponding boundary conditions are written as

$$v(0,t)=0,\quad \frac{\partial v(0,t)}{\partial x}=0,\quad \frac{\partial^2 v(L_m,t)}{\partial x^2}=0,\quad \frac{\partial^3 v(L_m,t)}{\partial x^3}=0 \qquad [7.22]$$

Using variable separation, the substitution of the Ansatz $v(x,t)=V(x)\underbrace{\left[A\sin(\omega_b t)+B\cos(\omega_b t)\right]}_{F(t)}$ into equation [7.21] leads to the frequency equation

$$\cos(\zeta L_m)\cosh(\zeta L_m)=-1 \qquad [7.23]$$

where $\zeta^4=\dfrac{\rho^* A\omega_b^{\,2}}{E_x^* I_y+2l_{b_{xy}}^2\left(2\mu_{xy}^*+\kappa_{xy}\right)A}$.

The bending natural frequency therein is denoted by ω_b. The solution of equation [7.23] is written

$$(\zeta L_m)_n=1.875,\ 4.694,\ 7.855,\ 10.996,\ 14.137,\ldots\ n=(1,\ 2,\ 3,\ 4,\ 5,\ldots) \qquad [7.24]$$

The bending natural frequencies are then derived as

$$(\omega_b)_n=(\zeta L_m)_n^2\sqrt{\frac{E_x^* I_y+2l_{b_{xy}}^2\left(2\mu_{xy}^*+\kappa_{xy}\right)A}{\rho^* A L_m^4}} \qquad [7.25]$$

By replacing the parameters involved in equation [7.25] by their expression from Appendix B, a closed form solution for the circular bending natural frequency ω_b due to micropolar effects is obtained in terms of the micro-architecture geometrical parameters of vertebral trabecular bone, quantities L_v, L_h, D_h, and D_v, the shear correction k_s, the bulk density ρ_s, the bulk elastic modulus E_s, the shear modulus

G_s of the solid tissue struts of vertebral bone, and the macro-size parameters of the macrobeam H, L_m as follows:

$$(\omega_b)_n = (\zeta L_m)_n^2 \frac{\sqrt{6}}{8L_m^2}\sqrt{\frac{D_h^4 E_s\left(16G_s k_s H^2 + 3E_s D_h^2 + 4G_s k_s L_v^2 + 9D_h^2 G_s k_s\right)L_v}{\rho_s\left(3E_s D_h^2 + 4G_s k_s L_v^2 + 9D_h^2 G_s k_s\right)\left(3L_v D_h^2 + 2L_h D_v^2\right)}} \qquad [7.26]$$

The previous equation can be written in terms of the characteristic micropolar length $l_{b_{xy}}$ as

$$(\omega_b)_n = (\zeta L_m)_n^2 \frac{\sqrt{6}}{2L_m^2}\sqrt{\frac{G_s k_s E_s D_h^4\left(3E_s D_h^2 H^2 + 36E_s D_h^2 l_{b_{xy}}^2 + 3D_h^2 G_s k_s H^2 + 108 l_{b_{xy}}^2 G_s k_s D_h^2 + 48 l_{b_{xy}}^2 G_s k_s L_v^2 + 4H^2 G_s k_s L_v^2\right)L_v}{\rho_s\left(3L_v D_h^2 + 2L_h D_v^2\right)\left(3E_s D_h^2 + 4G_s k_s L_v^2 + 3D_h^2 G_s k_s\right)\left(4G_s k_s L_v^2 + 3E_s D_h^2 + 9D_h^2 G_s k_s\right)}} \qquad [7.27]$$

Flexural natural frequency in cycles per second can be obtained by the relation $\left(f_b\right)_n = \left(\omega_b\right)_n / 2\pi$.

7.3.4. *Numerical results*

To illustrate the derived solution for the static deflection and the free flexural vibration of the fixed-free beam based on micropolar theory, a few numerical results are presented, focusing on the first two natural frequencies. For the purpose of illustration, the beam considered here and thereafter is taken to be made of vertebral cancellous human bone; the material (tissue) properties used in the calculations are selected as $E_s = 12000$ MPa, $G_s = 0.5 * E_s / (1 + \nu_s)$, Poisson's ratio $\nu_s = 0.3$, and density $\rho_s = 2000 \text{ kg/m}^3$ following [GIB 05]. The shear correction coefficient is $k_s = 9/10$.

The aforementioned properties hold at the microlevel, while the cross-sectional shape of the macroscopic beam is kept as W=2H, with the length L_m=15H (let H=1mm). Figures 7.5 and 7.6 show the change with age (in the range 25–85 years) of the bending deflection at the free end of the beam and of the first two natural frequencies. In the static deflection analysis, the moment at the end of the beam is set to 0.01 N.mm.

Figure 7.7 compares the deflection and rotation caused by an applied moment M = 0.01 N.mm at the free end of the cantilevered beam, predicted by the micropolar model and by the classical theory. The geometrical parameters

L_v, L_h, D_h, and D_v of the trabeculae building structure are selected (from Table 7.1) for a person at the age of 40. It is clearly observed from Figure 7.7 that the deflection and rotation predicted by the current microstructure-dependent model are always smaller than those predicted by the classical model, along the entire length of the cantilever beam. Hence, the micro-structured beam is predicted from the micropolar theory as stiffer in bending in comparison to the classical beam theory.

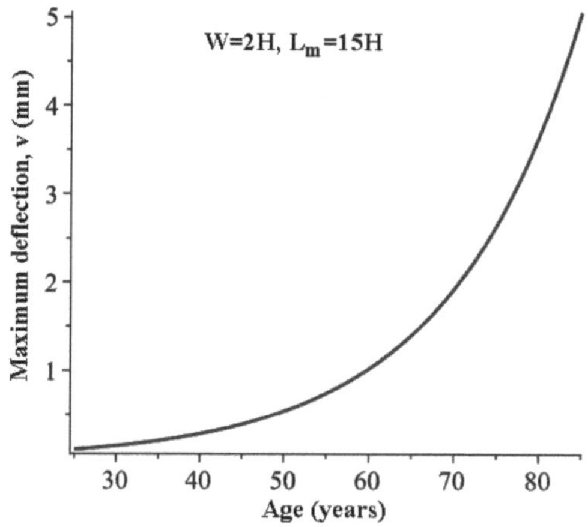

Figure 7.5. *Variation of the maximum deflection of the micro-structured cantilever beam with age*

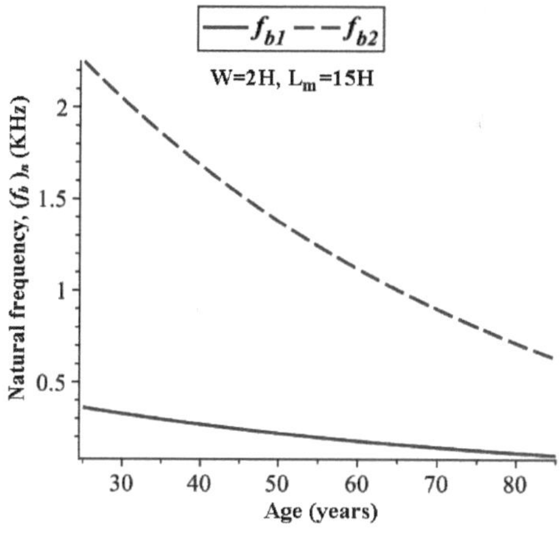

Figure 7.6. *Variation of the first two flexural natural frequencies of the micro-structured cantilever beam with age*

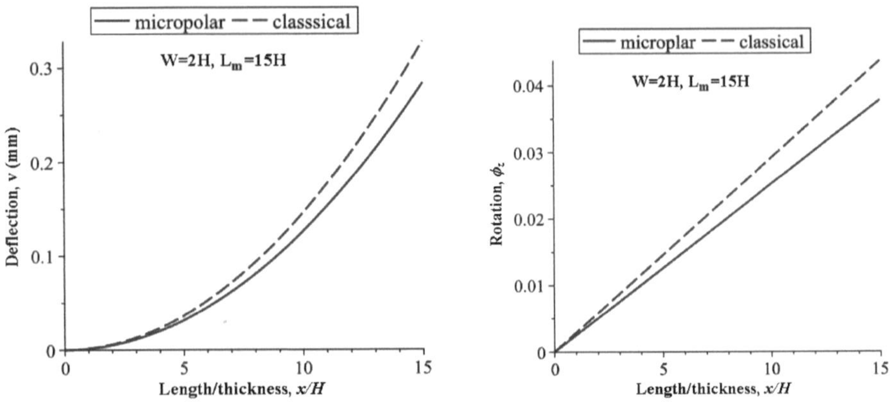

Figure 7.7. *Deflection and rotation of the micro-structured cantilever beam*

In order to assess the size-dependent static and vibration behaviours of micro-structured beams, a few numerical case studies are analyzed for both static and vibration situations, based on the equations established in the previous sections. In order to study the dependence of the size effect on the geometric dimensions of the cantilever beam, we set the definitions of the normalized static deflection (or rotation) as $v/\overline{v}$ (or $\phi_z/\overline{\phi}_z$) and the normalized natural frequency as $(\omega_b)_n/(\overline{\omega}_b)_n$, where $\overline{v}$ and $\overline{\omega}_b$ are respectively the static deflection and bending natural frequency that do not consider the size effect ($l_{b_{xy}}=0$) as evaluated by the classical theory, and v and ω_b are respectively the bending deflection and natural frequency incorporating the size effect. By this definition, for certain length-scale parameters, the influence of the size effect on mechanical properties is more substantial when the normalized value is smaller.

For the static case, the normalized static deflection of the micro-structured cantilever versus the ratio of the beam thickness to the length scale parameter $H/l_{b_{xy}}$ is depicted in Figure 7.8. The difference between the static deflection predicted by the classical beam theory and by the micropolar theory increases as the macroscopic beam thickness H approaches the characteristic length $l_{b_{xy}}$. As a specific case, for $H/l_{b_{xy}}=10$, the static deflection of the micro-cantilever predicted by the classical beam theory is about 1.15 greater than the static deflection estimated by micropolar theory. For large ratios, typically $H/l_{b_{xy}}>20$, the results predicted by classical beam theory are close to the predictions of the micropolar theory.

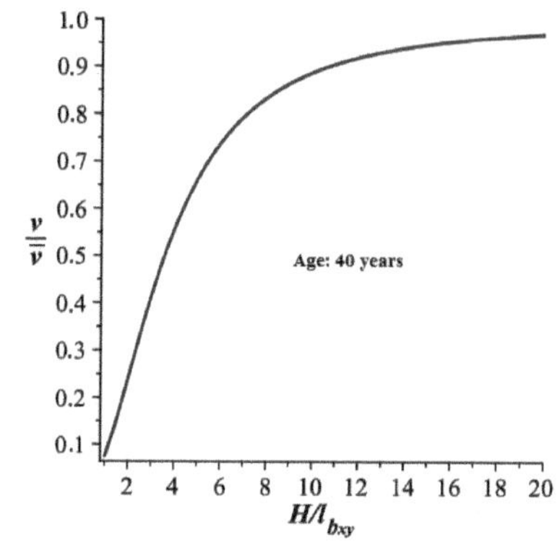

Figure 7.8. *Normalized static deflection of a micro-structured cantilever versus the ratio of the cantilever thickness to the material length scale parameter of trabecular bone*

For the vibration case, the normalized size-dependent natural frequency of the considered micro-structured beam is recorded versus the ratio of the beam characteristic size (thickness) to the internal material length scale parameter $H / l_{b_{xy}}$ in Figure 7.9. For large ratios $H / l_{b_{xy}} > 15$, the classical and micropolar beam theories predict approximately the same results. It is notable that for all values of $H / l_{b_{xy}}$, the natural frequencies predicted by micropolar theory are greater than those predicted by classical beam theory, thus revealing that the micropolar beam is stiffer than the classical beam.

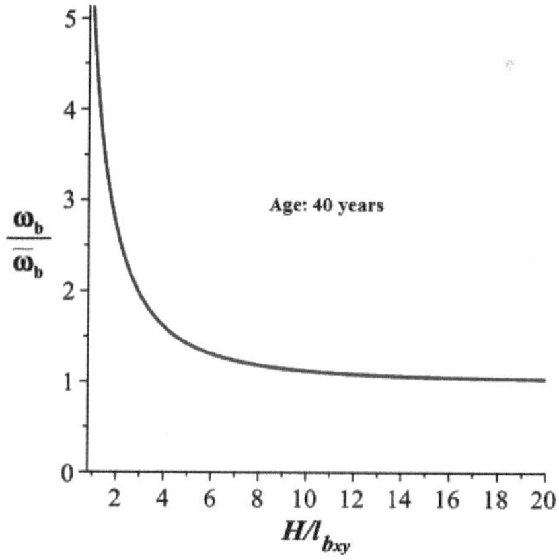

Figure 7.9. *Size effect on the flexural natural frequencies with respect to various sizes for micro-structured cantilevered beam*

The numerical results show that the size effect on static deflection and free vibration characteristics is pronounced for micro-structured beams with a ratio less than $H / l_{b_{xy}} < 15$. The static deflection behavior of micro-structured beams is much more affected by the size effect than its natural frequencies; as the length scale parameter $H / l_{b_{xy}}$ decreases, the deflection decreases due to an increase of bending stiffness, whereas the natural frequencies increase.

We conclude from previous results that the difference between the normalized static deflection and natural frequencies predicted by the micropolar theory with those of the classical theory is significant when the ratio of the beam characteristic size to the internal material length scale parameter is small. The classical theory underestimates the bending stiffness for relatively small thicknesses, due to the absence of any internal length; hence, the use of non-classic theories such as micropolar theory seems to be essential for the analysis in these cases.

In the forthcoming section, the torsion of vertebral trabecular bone in beam samples is formulated based on Cosserat reduced theory. The governing differential equation of torsion is derived employing Hamilton's principle (variational approach) and an explicit analytical solution is derived for the static torsion. The characteristic equation giving the natural frequencies is derived and solved analytically for the free torsional vibrations of beams.

7.4. Size-dependent static and dynamic torsional behaviors

7.4.1. *Governing equations of motion of a twisted micro-structured beam*

The problem of a twisted micro-structured fixed-free beam of total length L_m, thickness H and width W is considered (Figure 7.10). For pure torsion, the components of the displacement of the micro-structured beam can be expressed as [RAO 07]

$$u_x(x,t) = 0; \; u_y(x,t) = -z\phi(x,t); \; u_z(x,t) = y\phi(x,t) \qquad [7.28]$$

where u_x, u_y, and u_z denote the displacement along x, y and z axes, respectively; the field $\phi(x,t)$ stands for the rotation angle about the center of the twist (x-axis).

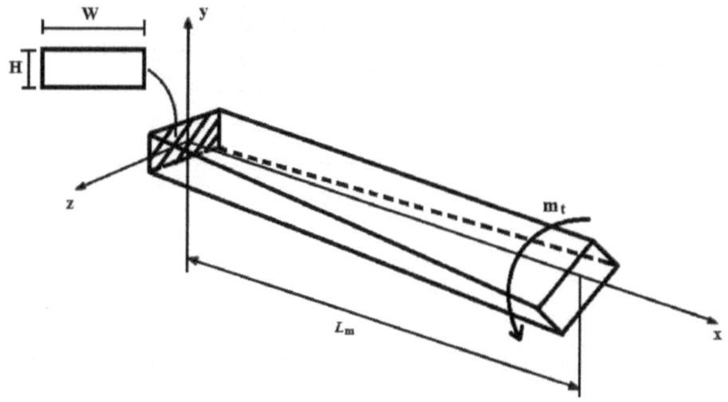

Figure 7.10. *Schematic of the micro-structured beam: geometry, loading, and coordinate system*

In the Cosserat reduced or couple-stress theory, the rotation is related to displacement gradients (equal to the local rigid rotation) as:

$$\phi_x = \left(\partial u_z / \partial y - \partial u_y / \partial z\right)/2;\ \phi_y = \left(\partial u_x / \partial z - \partial u_z / \partial x\right)/2;\ \phi_z = \left(\partial u_y / \partial x - \partial u_x / \partial y\right)/2$$

As a result, the strain tensor ϵ_{ij} is symmetrical, with components depending upon the displacements gradients as

$$\epsilon_{xy} = \epsilon_{yx} = \left(\partial u_y / \partial x + \partial u_x / \partial y\right)/2; \qquad \epsilon_{xz} = \epsilon_{zx} = \left(\partial u_z / \partial x + \partial u_x / \partial z\right)/2;$$
$$\epsilon_{yz} = \epsilon_{zy} = \left(\partial u_z / \partial y + \partial u_y / \partial z\right)/2.$$

Subsequently, the components of the rotation vector are derived as follows:

$$\phi_x = \phi(x,t);\ \phi_y = -\frac{y}{2}\frac{\partial \phi(x,t)}{\partial x};\ \phi_z = -\frac{z}{2}\frac{\partial \phi(x,t)}{\partial x}$$

The components of the strain tensor are then obtained as,

$$\epsilon_{xx} = \epsilon_{yy} = \epsilon_{zz} = \epsilon_{yz} = 0;\ \epsilon_{xy} = \epsilon_{yx} = -\frac{z}{2}\frac{\partial \phi(x,t)}{\partial x};\ \epsilon_{xz} = \epsilon_{zx} = \frac{y}{2}\frac{\partial \phi(x,t)}{\partial x} \qquad [7.29]$$

and the non-zero components of the symmetrical part of the curvature (or rotation of gradient) tensor $\chi_{ij} = 1/2\left(\phi_{i,j} + \phi_{j,i}\right)$ are written as

$$\chi_{xx} = \frac{\partial \phi(x,t)}{\partial x}; \chi_{yy} = -\frac{1}{2}\frac{\partial \phi(x,t)}{\partial x}; \chi_{zz} = -\frac{1}{2}\frac{\partial \phi(x,t)}{\partial x}; \chi_{xy} = \chi_{yx} = -\frac{y}{4}\frac{\partial^2 \phi(x,t)}{\partial x^2}; \chi_{xz} = \chi_{zx} = -\frac{z}{4}\frac{\partial^2 \phi(x,t)}{\partial x^2} \quad [7.30]$$

By substituting the strain components from equation [7.29] into the force stresses σ_{ij}, the non-zero components of the stress tensor are obtained as

$$\sigma_{xy} = \sigma_{yx} = -\frac{z}{2}\left(2\mu_{xy}^* + \kappa_{xy}\right)\frac{\partial \phi(x,t)}{\partial x}; \ \sigma_{xz} = \sigma_{zx} = \frac{y}{2}\left(2\mu_{xz}^* + \kappa_{xz}\right)\frac{\partial \phi(x,t)}{\partial x} \quad [7.31]$$

Proceeding in a similar manner, by substituting equation [7.30] into the couple stresses m_{ij}, we obtain the non-zero components of the couple stress tensor as

$$\begin{aligned} m_{xx} &= E_{tor-x}^*\frac{\partial \phi(x,t)}{\partial x}; \ m_{yy} = -\frac{1}{2}E_{tor-y}^*\frac{\partial \phi(x,t)}{\partial x}; \ m_{zz} = -\frac{1}{2}E_{tor-z}^*\frac{\partial \phi(x,t)}{\partial x}; \\ m_{xy} &= -\frac{y}{2}\gamma_{xy}\frac{\partial^2 \phi(x,t)}{\partial x^2}; \ m_{yx} = -\frac{y}{2}\beta_{yx}\frac{\partial^2 \phi(x,t)}{\partial x^2}; \ m_{xz} = -\frac{z}{2}\gamma_{xz}\frac{\partial^2 \phi(x,t)}{\partial x^2}; \ m_{zx} = -\frac{z}{2}\beta_{zx}\frac{\partial^2 \phi(x,t)}{\partial x^2}. \end{aligned} \quad [7.32]$$

where $E_{tor-x,y,z}^*$ are the homogenized torsional elastic moduli in directions x, y, and z respectively. The additional bending and torsion effects present in the micropolar continuum can be characterized by characteristic lengths for bending and torsion involved in the following micropolar coefficients

$$\gamma_{xy} = 2\left(2\mu_{xy}^* + \kappa_{xy}\right)l_{b_{xy}}^2, \ \gamma_{xz} = 2\left(2\mu_{xz}^* + \kappa_{xz}\right)l_{b_{xz}}^2,$$

$$\beta_{yx} = \left(2\mu_{yx}^* + \kappa_{yx}\right)\left(l_{t_{yx}}^2 - 2l_{b_{yx}}^2\right)$$

$$\beta_{zx} = \left(2\mu_{zx}^* + \kappa_{zx}\right)\left(l_{t_{zx}}^2 - 2l_{b_{zx}}^2\right).$$

This entails the expression of the strain energy of the micro-structured beam

$$\begin{aligned} U &= \frac{1}{2}\int_\Omega \left(\sigma_{xy}\epsilon_{xy} + \sigma_{yx}\epsilon_{yx} + \sigma_{xz}\epsilon_{xz} + \sigma_{zx}\epsilon_{zx} + m_{xx}\chi_{xx} + m_{yy}\chi_{yy} + m_{zz}\chi_{zz} + m_{xy}\chi_{xy} + m_{yx}\chi_{yx} + m_{xz}\chi_{xz} + m_{zx}\chi_{zx}\right)dV \\ &= \frac{1}{2}\int_\Omega \left(\begin{aligned} &\frac{z^2}{2}\left(2\mu_{xy}^* + \kappa_{xy}\right)\frac{\partial^2 \phi(x,t)}{\partial x^2} + \frac{y^2}{2}\left(2\mu_{xz}^* + \kappa_{xz}\right)\frac{\partial^2 \phi(x,t)}{\partial x^2} + \left(E_{tor-x}^* + \frac{E_{tor-y}^*}{4} + \frac{E_{tor-z}^*}{4}\right)\frac{\partial^2 \phi(x,t)}{\partial x^2} \\ &+ \frac{y^2}{8}\gamma_{xy}\frac{\partial^4 \phi(x,t)}{\partial x^4} + \frac{y^2}{8}\beta_{yx}\frac{\partial^4 \phi(x,t)}{\partial x^4} + \frac{z^2}{8}\gamma_{xz}\frac{\partial^4 \phi(x,t)}{\partial x^4} + \frac{z^2}{8}\beta_{zx}\frac{\partial^4 \phi(x,t)}{\partial x^4} \end{aligned}\right)dV \\ &= \frac{1}{2}\int_0^{L_m}\int_A \left(\begin{aligned} &\left(\frac{1}{2}\left(z^2\left(2\mu_{xy}^* + \kappa_{xy}\right) + y^2\left(2\mu_{xz}^* + \kappa_{xz}\right)\right) + \left(E_{tor-x}^* + \frac{E_{tor-y}^*}{4} + \frac{E_{tor-z}^*}{4}\right)\right)\frac{\partial^2 \phi(x,t)}{\partial x^2} \\ &+ \frac{1}{8}\left(\left(\gamma_{xy} + \beta_{yx}\right)y^2 + \left(\gamma_{xz} + \beta_{zx}\right)z^2\right)\frac{\partial^4 \phi(x,t)}{\partial x^4} \end{aligned}\right)dA\,dx \end{aligned} \quad [7.33]$$

with the moment of inertia about the z-axis and y-axis defined as $I_z^* = \int_A y^2 dA$; $I_y^* = \int_A z^2 dA$; hence,

$$U = \frac{1}{2}\int_0^{L_m} \left(\begin{array}{l} \left(\frac{1}{2}\left(\left(2\mu_{xy}^* + \kappa_{xy}\right) I_y^* + \left(2\mu_{xz}^* + \kappa_{xz}\right) I_z^* \right) + \left(E_{tor-x}^* + \frac{E_{tor-y}^*}{4} + \frac{E_{tor-z}^*}{4} \right) A \right) \left(\frac{\partial \phi(x,t)}{\partial x} \right)^2 \\ \frac{1}{8}\left(\left(\gamma_{xy} + \beta_{yx}\right) I_z^* + \left(\gamma_{xz} + \beta_{zx}\right) I_y^* \right) \left(\frac{\partial^2 \phi(x,t)}{\partial x^2} \right)^2 \end{array} \right) dx \quad [7.34]$$

The total kinetic energy of the micro-structured beam can be expressed as

$$\begin{aligned} KE &= \frac{1}{2}\int_\Omega \rho^*(x) \left(\frac{\partial^2 u_x(x,t)}{\partial t^2} + \frac{\partial^2 u_y(x,t)}{\partial t^2} + \frac{\partial^2 u_z(x,t)}{\partial t^2} \right) dV = \frac{1}{2}\int_0^{L_m} \int_A \rho^*(x)\left(y^2 + z^2\right) \left(\frac{\partial^2 \phi(x,t)}{\partial t^2} \right) dx \\ &= \frac{1}{2}\int_0^{L_m} \rho^*(x) I_x^* \left(\frac{\partial \phi(x,t)}{\partial t} \right)^2 dx \end{aligned} \quad [7.35]$$

with the polar moment of inertia about the x-axis $I_x^* = \int_A \left(y^2 + z^2\right) dA$.

The work done by the distributed torque-per-unit length about the x-axis, $m_t\left(x,t\right)$ can be represented as

$$W^{ex} = \int_0^{L_m} m_t\left(x,t\right) \phi\left(x,t\right) dx \quad [7.36]$$

The governing equation of motion and the boundary conditions for a microstructured beam undergoing torsional deformation under an externally applied torque are similarly obtained from Hamilton's principle

$$\int_{t_1}^{t_2} \left(\delta KE - \delta U + \delta W^{ex} \right) dt = 0 \quad [7.37]$$

with t_1 and t_2 the times at which the configuration of the system is assumed to be known.

The first variation or virtual change of the strain energy of the beam in the time interval [t₁, t₂] is written

$$\delta U=\int_0^{L_m}\left(\begin{array}{l}\left(\frac{1}{2}\left(\left(2\mu_{xy}^*+\kappa_{xy}\right)I_y^*+\left(2\mu_{xz}^*+\kappa_{xz}\right)I_z^*\right)+\left(E_{tor-x}^*+\frac{E_{tor-y}^*}{4}+\frac{E_{tor-z}^*}{4}\right)A\right)\frac{\partial\phi(x,t)}{\partial x}\delta\frac{\partial\phi(x,t)}{\partial x}\\ \frac{1}{8}\left(\left(\gamma_{xy}+\beta_{yx}\right)I_z^*+\left(\gamma_{xz}+\beta_{zx}\right)I_y^*\right)\frac{\partial^2\phi(x,t)}{\partial x^2}\delta\frac{\partial^2\phi(x,t)}{\partial x^2}\end{array}\right)dx \quad [7.38]$$

Similarly, we obtain the variations of the kinetic energy and external work

$$\delta KE=\int_0^{L_m}\rho^*(x)I_x^*\frac{\partial\phi(x,t)}{\partial t}\delta\frac{\partial\phi(x,t)}{\partial t}dx \quad [7.39]$$

$$\delta W^{ex}=\int_0^{L_m}m_t(x,t)\,\delta\phi(x,t)dx \quad [7.40]$$

The variation $\int_{t_1}^{t_2}\delta KEdt$ in equation [7.37] can be evaluated by integration by parts with respect to time:

$$\begin{aligned}\int_{t_1}^{t_2}\delta KEdt&=\int_0^{L_m}\left(\int_{t_1}^{t_2}\rho^*(x)I_x^*\frac{\partial\phi(x,t)}{\partial t}\delta\frac{\partial\phi(x,t)}{\partial t}dt\right)dx=\int_0^{L_m}\left(\rho^*(x)I_x^*\frac{\partial\phi(x,t)}{\partial t}\delta\phi\Big|_{t=t_1}^{t=t_2}-\int_{t_1}^{t_2}\rho^*(x)I_x^*\frac{\partial^2\phi(x,t)}{\partial t^2}\delta\phi dt\right)dx\\&=-\int_{t_1}^{t_2}\int_0^{L_m}\rho^*(x)I_x^*\frac{\partial^2\phi(x,t)}{\partial t^2}\delta\phi dxdt\end{aligned} \quad [7.41]$$

It is assumed that the initial and final configurations are known, in terms of the data $\phi(x,t_{1,2})$. Thus, $\delta\phi(x,t_1)=\delta\phi(x,t_2)=0$, and the integrated terms on the RHS of equation [7.41] vanish.

Similarly, for the strain energy, integrating by parts with respect to x leads to

$$\int_{t_1}^{t_2}\delta Udt=\int_{t_1}^{t_2}\left(\begin{array}{l}\left(\left(\frac{1}{2}\left(\left(2\mu_{xy}^*+\kappa_{xy}\right)I_y^*+\left(2\mu_{xz}^*+\kappa_{xz}\right)I_z^*\right)+\left(E_{tor-x}^*+\frac{E_{tor-y}^*}{4}+\frac{E_{tor-z}^*}{4}\right)A\right)\frac{\partial\phi(x,t)}{\partial x}-\frac{1}{8}\left(\left(\gamma_{xy}+\beta_{yx}\right)I_z^*+\left(\gamma_{xz}+\beta_{zx}\right)I_y^*\right)\frac{\partial^3\phi(x,t)}{\partial x^3}\right)\delta\phi\Big|_{x=0}^{x=L_m}\\ -\int_0^{L_m}\left(\frac{1}{2}\left(\left(2\mu_{xy}^*+\kappa_{xy}\right)I_y^*+\left(2\mu_{xz}^*+\kappa_{xz}\right)I_z^*\right)+\left(E_{tor-x}^*+\frac{E_{tor-y}^*}{4}+\frac{E_{tor-z}^*}{4}\right)A\right)\frac{\partial^2\phi(x,t)}{\partial x^2}\delta\phi dx\\ +\frac{1}{8}\left(\left(\gamma_{xy}+\beta_{yx}\right)I_z^*+\left(\gamma_{xz}+\beta_{zx}\right)I_y^*\right)\frac{\partial^2\phi(x,t)}{\partial x^2}\delta\frac{\partial\phi(x,t)}{\partial x}\Big|_{x=0}^{x=L_m}+\int_0^{L_m}\frac{1}{8}\left(\left(\gamma_{xy}+\beta_{yx}\right)I_z^*+\left(\gamma_{xz}+\beta_{zx}\right)I_y^*\right)\frac{\partial^4\phi(x,t)}{\partial x^4}\delta\phi dx\end{array}\right)dt \quad [7.42]$$

Substituting the aforementioned expressions in equation [7.37], one obtains the governing equation and boundary conditions in weak form as

$$\begin{aligned}
&\int_{t_1}^{t_2}\int_0^{L_m}\left(\frac{1}{8}\left(\left(\gamma_{xy}+\beta_{yx}\right)I_z^*+\left(\gamma_{xz}+\beta_{zx}\right)I_y^*\right)\frac{\partial^4\phi(x,t)}{\partial x^4}-\left(\begin{array}{l}\frac{1}{2}\left(\left(2\mu_{xy}^*+\kappa_{xy}\right)I_y^*+\left(2\mu_{xz}^*+\kappa_{xz}\right)I_z^*\right)\\+\left(E_{tor-x}^*+\frac{E_{tor-y}^*}{4}+\frac{E_{tor-z}^*}{4}\right)A\end{array}\right)\frac{\partial^2\phi(x,t)}{\partial x^2}+\rho^*(x)I_x^*\frac{\partial^2\phi(x,t)}{\partial t^2}-m_t(x,t)\right)\delta\phi\,dxdt\\
&+\int_{t_1}^{t_2}\left(\left(\frac{1}{2}\left(\left(2\mu_{xy}^*+\kappa_{xy}\right)I_y^*+\left(2\mu_{xz}^*+\kappa_{xz}\right)I_z^*\right)+\left(E_{tor-x}^*+\frac{E_{tor-y}^*}{4}+\frac{E_{tor-z}^*}{4}\right)A\right)\frac{\partial\phi(x,t)}{\partial x}-\frac{1}{8}\left(\left(\gamma_{xy}+\beta_{yx}\right)I_z^*+\left(\gamma_{xz}+\beta_{zx}\right)I_y^*\right)\frac{\partial^3\phi(x,t)}{\partial x^3}\right)\delta\phi\Bigg|_{x=0}^{x=L_m}dt\\
&+\int_{t_1}^{t_2}\frac{1}{8}\left(\left(\gamma_{xy}+\beta_{yx}\right)I_z^*+\left(\gamma_{xz}+\beta_{zx}\right)I_y^*\right)\frac{\partial^2\phi(x,t)}{\partial x^2}\delta\frac{\partial\phi(x,t)}{\partial x}\Bigg|_{x=0}^{x=L_m}dt=0
\end{aligned}\qquad [7.43]$$

From [7.43], the governing equation of motion of torsion is obtained in terms of $\phi(x, t)$ as

$$\begin{aligned}
&\frac{1}{8}\left(\left(\gamma_{xy}+\beta_{yx}\right)I_z^*+\left(\gamma_{xz}+\beta_{zx}\right)I_y^*\right)\frac{\partial^4\phi(x,t)}{\partial x^4}-\left(\begin{array}{l}\frac{1}{2}\left(\left(2\mu_{xy}^*+\kappa_{xy}\right)I_y^*+\left(2\mu_{xz}^*+\kappa_{xz}\right)I_z^*\right)\\+\left(E_{tor-x}+\frac{E_{tor-y}}{4}+\frac{E_{tor-z}}{4}\right)A\end{array}\right)\frac{\partial^2\phi(x,t)}{\partial x^2}\\
&+\rho^*(x)I_x^*\frac{\partial^2\phi(x,t)}{\partial t^2}=m_t(x,t)
\end{aligned}\qquad [7.44]$$

7.4.2. *Size-dependent static torsional behavior*

The static torsion of the micro-structured beam is next analyzed; from the relations $\frac{\partial\phi(x,t)}{\partial t}=0$ and $\frac{\partial\phi(x,t)}{\partial x}=\frac{d\phi}{dx}$, equation [7.44] reduces to

$$\underbrace{\frac{1}{8}\left(\left(\gamma_{xy}+\beta_{yx}\right)I_z^*+\left(\gamma_{xz}+\beta_{zx}\right)I_y^*\right)}_{K_{t1}}\frac{d^4\phi}{dx^4}-\underbrace{\left(\frac{1}{2}\left(\left(2\mu_{xy}^*+\kappa_{xy}\right)I_y^*+\left(2\mu_{xz}^*+\kappa_{xz}\right)I_z^*\right)+\left(E_{tor-x}^*+\frac{E_{tor-y}^*}{4}+\frac{E_{tor-z}^*}{4}\right)A\right)}_{K_{t2}}\frac{d^2\phi}{dx^2}=m_t(x)\qquad [7.45]$$

Letting $K_t=\frac{K_{t2}}{K_{t1}}=\frac{\left(\frac{1}{2}\left(\left(2\mu_{xy}^*+\kappa_{xy}\right)I_y^*+\left(2\mu_{xz}^*+\kappa_{xz}\right)I_z^*\right)+\left(E_{tor-x}^*+\frac{E_{tor-y}^*}{4}+\frac{E_{tor-z}^*}{4}\right)A\right)}{\frac{1}{8}\left(\left(\gamma_{xy}+\beta_{yx}\right)I_z^*+\left(\gamma_{xz}+\beta_{zx}\right)I_y^*\right)}$,

one can rewrite [7.45] as

$$\frac{d^4\phi}{dx^4}-K_t\frac{d^2\phi}{dx^2}=\frac{1}{K_{t1}}m_t(x)\qquad [7.46]$$

This equation is a non-homogenous linear ordinary differential equation with constant coefficients that can be solved analytically; a double integration of equation [7.46] results in:

$$\frac{d^2\phi}{dx^2} - K_t\phi = (1/K_{t1})\int\int m_t(x)\,dxdx + C_1x + C_2 \qquad [7.47]$$

The general solution of [7.47] decomposes into

$$\phi(x) = \underbrace{C_3e^{\sqrt{K_t}x} + C_4e^{-\sqrt{K_t}x}}_{\phi_h(x)} + \phi_p(x) \qquad [7.48]$$

in which $\phi_h(x)$ represents the complementary solution (or homogeneous solution) and $\phi_p(x)$ the particular (or non-homogeneous solution)

$$\phi_p(x) = e^{\sqrt{K_t}x}\int\frac{e^{-\sqrt{K_t}x}\left((1/K_{t1})\int\int m_t(x)\,dxdx + C_1x + C_2\right)}{2\sqrt{K_t}}dx - e^{-\sqrt{K_t}x}\int\frac{e^{\sqrt{K_t}x}\left((1/K_{t1})\int\int m_t(x)\,dxdx + C_1x + C_2\right)}{2\sqrt{K_t}}dx \qquad [7.49]$$

The general solution of [7.47] can then be written as

$$\phi(x) = C_3e^{\sqrt{K_t}x} + C_4e^{-\sqrt{K_t}x} + e^{\sqrt{K_t}x}\int\frac{e^{-\sqrt{K_t}x}\left(m_tx^2/2K_{t1} + C_1x + C_2\right)}{2\sqrt{K_t}}dx - e^{-\sqrt{K_t}x}\int\frac{e^{\sqrt{K_t}x}\left(m_tx^2/2K_{t1} + C_1x + C_2\right)}{2\sqrt{K_t}}dx \qquad [7.50]$$

After simplification, we obtain

$$\phi(x) = -\frac{1}{2}\frac{2K_tK_{t1}C_2 + 2m_t + 2xK_tK_{t1}C_1 + m_tx^2K_t}{K_t^2K_{t1}} + C_3e^{\sqrt{K_t}x} + C_4e^{-\sqrt{K_t}x} \qquad [7.51]$$

with C_1,...C_4 constants that can be obtained by applying the fixed-free boundary conditions, that is from [7.43]:

$$\phi(0) = \frac{d\phi(0)}{dx} = 0;\quad \frac{d^2\phi(L_m)}{d^2x} = 0;\quad \mathrm{K}_t\frac{d\phi(L_m)}{dx} - \frac{d^3\phi(L_m)}{d^3x} = 0 \qquad [7.52]$$

This delivers the constants

$$C_1 = -\frac{m_t L_m}{K_{t1}}; C_2 = -\frac{m_t \left(e^{-\sqrt{K_t}L_m} + e^{\sqrt{K_t}L_m} - 2 - \sqrt{K_t}e^{\sqrt{K_t}L_m}L_m + L_m\sqrt{K_t}e^{-\sqrt{K_t}L_m}\right)}{\left(e^{-\sqrt{K_t}L_m} + e^{\sqrt{K_t}L_m}\right)K_t K_{t1}};$$
$$C_3 = -\frac{m_t\left(L_m\sqrt{K_t}e^{-\sqrt{K_t}L_m} - 1\right)}{K_t^2 K_{t1}\left(e^{-\sqrt{K_t}L_m} + e^{\sqrt{K_t}L_m}\right)}; C_4 = \frac{m_t\left(1 + \sqrt{K_t}e^{\sqrt{K_t}L_m}L_m\right)}{K_t^2 K_{t1}\left(e^{-\sqrt{K_t}L_m} + e^{\sqrt{K_t}L_m}\right)}. \quad [7.53]$$

The torsion angle of the micro-structured beam is then obtained as

$$\phi(x) = -\frac{1}{2}\frac{-\frac{2m_t\left(e^{-\sqrt{K_t}L_m} + e^{\sqrt{K_t}L_m} - 2 - \sqrt{K_t}e^{\sqrt{K_t}L_m}L_m + L_m\sqrt{K_t}e^{-\sqrt{K_t}L_m}\right)}{e^{-\sqrt{K_t}L_m} + e^{\sqrt{K_t}L_m}} + 2m_t - 2xK_t m_t L_m + m_t x^2 K_t}{K_t^2 K_{t1}}$$
$$-\frac{m_t\left(L_m\sqrt{K_t}e^{-\sqrt{K_t}(-x+L_m)} - e^{\sqrt{K_t}x} - e^{-\sqrt{K_t}x} - L_m\sqrt{K_t}e^{\sqrt{K_t}(-x+L_m)}\right)}{K_t^2 K_{t1}\left(e^{-L_m\sqrt{K_t}} + e^{L_m\sqrt{K_t}}\right)} \quad [7.54]$$

The maximum torsion angle of the beam occurs at its end (when $x = L_m$), given by

$$\phi(x)_{\max} = \frac{1}{2}\frac{\frac{2m_t\left(e^{-\sqrt{K_t}L_m} + e^{\sqrt{K_t}L_m} - 2 - \sqrt{K_t}e^{\sqrt{K_t}L_m}L_m + L_m\sqrt{K_t}e^{-\sqrt{K_t}L_m}\right)}{e^{-\sqrt{K_t}L_m} + e^{\sqrt{K_t}L_m}} + L_m^2 K_t m_t}{K_t^2 K_{t1}} \quad [7.55]$$

The torsion angle can be derived on the basis of the classical theory from equation [7.54] as

$$\bar{\phi}(x) = \frac{x m_t}{2\left(G_{xy}^* I_y^* + G_{xz}^* I_z^*\right)}\left(2L_m - x\right) \quad [7.56]$$

7.4.3. *Size-dependant free torsional vibration behavior*

The governing equation of the free torsional vibration of the micro-structured beam can be written from equation [7.44], considering $m_t\left(x,t\right) = 0$, as

$$\underbrace{\frac{1}{8}\left(\left(\gamma_{xy} + \beta_{yx}\right)I_z^* + \left(\gamma_{xz} + \beta_{zx}\right)I_y^*\right)}_{K_{t1}}\frac{\partial^4\phi(x,t)}{\partial x^4} - \underbrace{\left(\frac{1}{2}\left(\left(2\mu_{xy}^* + \kappa_{xy}\right)I_y^* + \left(2\mu_{xz}^* + \kappa_{xz}\right)I_z^*\right) + \left(E_{tor-x}^* + \frac{E_{tor-y}^*}{4} + \frac{E_{tor-z}^*}{4}\right)A\right)}_{K_{t2}}\frac{\partial^2\phi(x,t)}{\partial x^2}$$
$$+\rho^*(x)I_x^*\frac{\partial^2\phi(x,t)}{\partial t^2} = 0 \quad [7.57]$$

To solve this vibrational problem, we use the separation of variables method, writing

$$\phi(x,t)=\varphi(x)\left[A\sin(\omega_T t)+B\cos(\omega_T t)\right] \qquad [7.58]$$

Here, ω_T denotes the torsional natural frequency of the micro-structured beam. By substituting equation [7.58] into equation [7.57], we obtain the homogeneous linear ordinary differential equation with constant coefficients

$$K_{t1}\frac{d^4\varphi}{dx^4}-K_{t2}\frac{d^2\varphi}{dx^2}-\rho^* I_x^* \omega_T^2\varphi=0 \qquad [7.59]$$

The analytical solution of [7.59] is given by

$$\varphi(x)=C_1\cosh S_1x+C_2\sinh S_1x+C_3\cos S_2x+C_4\sin S_2x \qquad [7.60]$$

where

$$S_1=\sqrt{\frac{K_{t2}}{2K_{t1}}+\sqrt{\frac{K_{t2}^{\;2}}{4K_{t1}^{\;2}}+\frac{\rho I_x^*\omega_T^{\;2}}{K_{t1}}}};\quad S_2=\sqrt{-\frac{K_{t2}}{2K_{t1}}+\sqrt{\frac{K_{t2}^{\;2}}{4K_{t1}^{\;2}}+\frac{\rho I_x^*\omega_T^{\;2}}{K_{t1}}}} \qquad [7.61]$$

with C_1-C_4 constants that can be defined by applying appropriate boundary conditions. In addition, S_1 and S_2 are the solutions of the characteristic equation of [7.59].

Applying the boundary conditions mentioned in [7.52] for a fixed-free beam to the general solution, equ. [7.60] gives a homogeneous system of equations for the constants C_1-C_4 expressed in matrix format as:

$$\begin{bmatrix} 1 & 0 & 1 & 0 \\ 0 & S_1 & 0 & S_2 \\ S_1^2\cosh(S_1L_m) & S_1^2\sinh(S_1L_m) & -S_2^2\cos(S_2L_m) & -S_2^2\cos(S_2L_m) \\ (K_{t2}S_1-K_{t1}S_1^3)\sinh(S_1L_m) & (K_{t2}S_1-K_{t1}S_1^3)\cosh(S_1L_m) & -(K_{t2}S_2+K_{t1}S_2^3)\sin(S_2L_m) & (K_{t2}S_2+K_{t1}S_2^3)\cos(S_2L_m) \end{bmatrix}\begin{Bmatrix} C_1 \\ C_2 \\ C_3 \\ C_4 \end{Bmatrix}=0 \qquad [7.62]$$

Therefore, for non-trivial solutions to exist, we require the vanishing of the determinant of system [7.62], leading to the relations

$$S_2L_m\approx(n-0.5)\pi \quad \text{n=1,2,3,...} \qquad [7.63]$$

By inserting equation [7.63] in [7.61] and after some easy manipulations, the nth natural torsional frequency of the fixed-free micro-structured beam is obtained as

$$\begin{aligned}(\omega_T)_n &= \frac{(n-0.5)\pi}{L_m}\sqrt{\frac{K_{t2}+K_{t1}\left(\frac{(n-0.5)\pi}{L_m}\right)^2}{\rho^* I_x^*}} \\ &= \frac{(n-0.5)\pi}{L_m}\sqrt{\frac{\left(\frac{1}{2}\left((2\mu_{xy}^*+\kappa_{xy})I_y^*+(2\mu_{xz}^*+\kappa_{xz})I_z^*\right)+\left(E_{tor-x}^*+\frac{E_{tor-y}^*}{4}+\frac{E_{tor-z}^*}{4}\right)A\right)+\frac{1}{8}\left((\gamma_{xy}+\beta_{yx})I_z^*+(\gamma_{xz}+\beta_{zx})I_y^*\right)\left(\frac{(n-0.5)\pi}{L_m}\right)^2}{\rho^* I_x^*}}\end{aligned} \qquad [7.64]$$

where K_{t1} and K_{t2} are identified from the micropolar constants of vertebral trabecular bone; those coefficients are explicitly written in terms of the geometrical and mechanical microparameters of trabeculae as well as the geometrical macrostructure of the beam as

$$\begin{aligned}K_{t1} &= \frac{1}{8}\left(\begin{aligned}&\frac{\sqrt{3}}{384}\frac{\pi D_h^4 E_s\left(3D_h^4E_s^2D_v^2+18D_h^4E_sD_v^2G_s+24L_h^2D_h^4G_s^2k_s+24E_sD_v^4L_vL_hG_sk_s+4L_h^2D_h^4E_sG_sk_s+16L_vG_s^2k_sD_v^4L_h\right)}{L_hL_v\left(6D_h^4E_sD_v^2G_s+3D_h^4E_s^2D_v^2+8L_h^2D_h^4G_s^2k_s+4L_h^2D_h^4E_sG_sk_s+16E_sD_v^4L_vL_hG_sk_s\right)}\\ &-\frac{\sqrt{3}}{384}\frac{\pi D_h^4E_s\left(3D_h^4E_sD_v^2+24D_v^4L_vL_hG_sk_s+4L_h^2D_h^4G_sk_s\right)(E_s-2G_s)}{L_hL_v\left(6D_h^4E_sD_v^2G_s+3D_h^4E_s^2D_v^2+8L_h^2D_h^4G_s^2k_s+4L_h^2D_h^4E_sG_sk_s+16E_sD_v^4L_vL_hG_sk_s\right)}\end{aligned}\right)I_z^*+\frac{1}{1152}\frac{\sqrt{3}\pi D_v^4E_sI_y^*}{L_v^2};\\ K_{t2} &= \frac{1}{4}\frac{\sqrt{3}\pi D_h^4E_sG_sk_sI_y^*}{L_vL_h\left(3E_sD_h^2+4G_sk_sL_v^2+3D_h^2G_sk_s\right)}+\frac{D_h^4D_v^4\sqrt{3}\pi E_sG_sk_sI_z^*}{L_v\left(9D_h^4D_v^2E_sL_v+12L_h^2D_h^4G_sk_sL_v+12D_v^4E_sD_h^2L_h+16D_v^4L_v^2L_hG_sk_s\right)}\\ &+\frac{1}{24}\frac{\sqrt{3}\pi D_h^4E_sG_s}{(3E_s+2G_s)L_vL_h}+\frac{1}{96}\frac{\left(3E_sD_v^4L_v+\frac{3}{4}\frac{D_h^4E_sD_v^2}{L_hk_s}+L_hD_h^4G_s\right)\sqrt{3}\pi G_sE_sD_h^4}{L_vL_h\left(6E_sL_vG_sD_v^4+E_s^2D_v^4L_v+\frac{9}{4}\frac{D_h^4E_s^2D_v^2}{L_hk_s}+3L_hD_h^4E_sG_s+\frac{3}{2}\frac{D_h^4G_sE_sD_v^2}{L_hk_s}+2L_hD_h^4G_s^2\right)}+\frac{1}{288}\frac{\sqrt{3}\pi G_sD_v^4}{L_v^2}.\end{aligned} \qquad [7.65]$$

An examination of equation [7.64] shows that the classical torsional frequency can be recovered by letting the micropolar constants vanish (couple stress constants)

$$(\bar{\omega}_T)_n = \frac{(n-0.5)\pi}{L_m}\sqrt{\frac{G_{xy}^* I_y^* + G_{xz}^* I_z^*}{\rho^* I_x^*}} \qquad [7.66]$$

The torsional natural frequency in cycles per second can be obtained by the following relation $(f_T)_n = (\omega_T)_n / 2\pi$.

7.4.4. *Numerical results*

We consider here the geometry of a micro-structured cantilever beam with circular cross-sectional shape (diameter (d) = 1 mm, L_m =15d) for the purpose of numerical illustration. The maximum torsion angle and the first torsion natural frequency versus age (25-85 years) are depicted in Figures 7.11 and 7.12 respectively. The torque acting on the end of the beam in the static case is taken as 0.0001 N.mm.

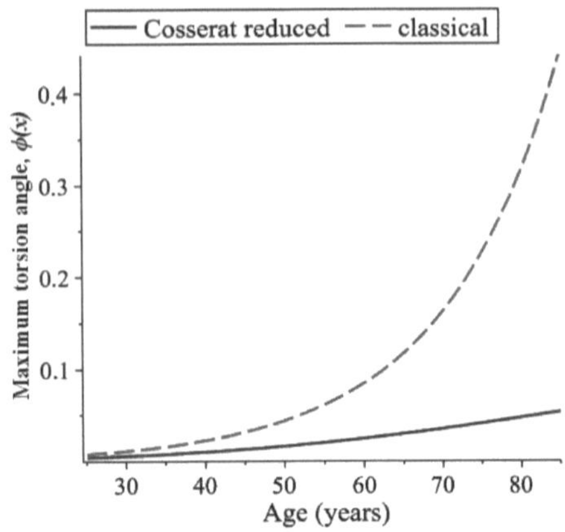

Figure 7.11. *Variation with age of the maximum static torsion angle of the micro-structured cantilever beam. For a color version of this figure, see www.iste.co.uk/ganghoffer/biomechanics.zip*

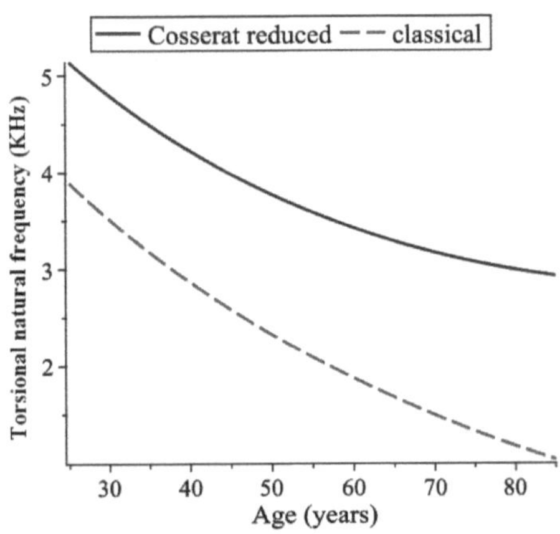

Figure 7.12. *Variation with age of the first torsional natural frequencies of the micro-structured cantilever beam For a color version of this figure, see www.iste.co.uk/ganghoffer/biomechanics.zip*

It appears in Figure 7.11 that the maximum static torsional angle increases gradually with age, contrary to the torsional natural frequency. One can also observe from Figure 7.11 that the static torsion angle predicted by the current microstructural model is always smaller than that predicted by the classical model; this confirms that the micro-structured beams modeled by Cosserat theory are stiffer than those modeled by the classical theory. From the responses recorded in Figure 7.12, it can be seen that the first torsional natural frequencies predicted by the classical theory are about (1.5–2.5) times lower than that predicted by the Cosserat reduced theory.

In Figure 7.13, we examine the static torsion angle versus the diameter of vertebral trabecular bone specimen of age 40 years; the difference between the Cosserat reduced model and the classical theory becomes significant for specimens of small diameters. The normalized maximum torsion angle of the vertebral trabecular bone beam specimen of age 40 years, variable $\phi / \overline{\phi}$, is recorded versus the beam diameter in Figure 7.13(b); it is observed that the static torsional behavior of the beam is size-dependent.

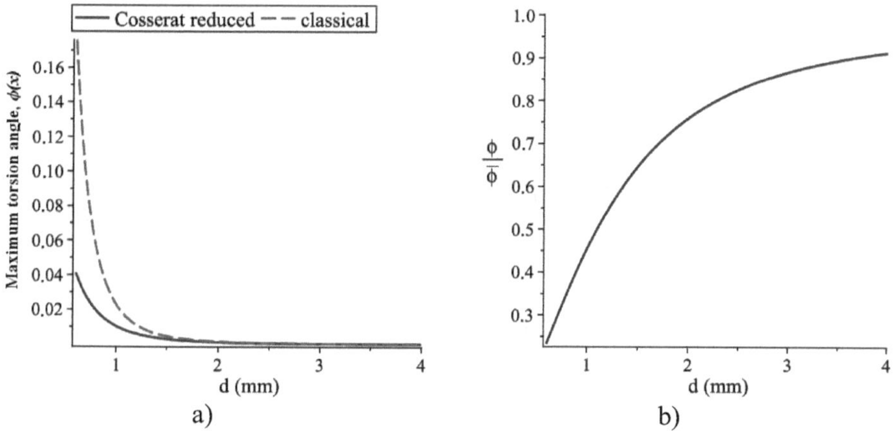

Figure 7.13. *a) Effect of specimen size on the maximum torsion angle; b) the normalized maximum torsion angle versus beam diameter. Age is 40 years. For a color version of this figure, see www.iste.co.uk/ganghoffer/biomechanics.zip*

Numerical results for the free torsional vibration of size-dependent behavior are depicted in Figure 7.14, which compares the torsional natural frequencies predicted by the present model and by the classical theory. Differences are significant especially for small beam diameters; this indicates that the size dependence of the material properties increases the torsional stiffness and hence it increases the torsional frequencies. The normalized torsional frequency of the vertebral trabecular bone beam specimen of age 40 years, variable $\omega_T / \overline{\omega_T}$, is recorded versus the beam diameter in Figure 7.14(b); an increase in the beam dimensions leads to a decreasing size effect which become evanescent for large diameters.

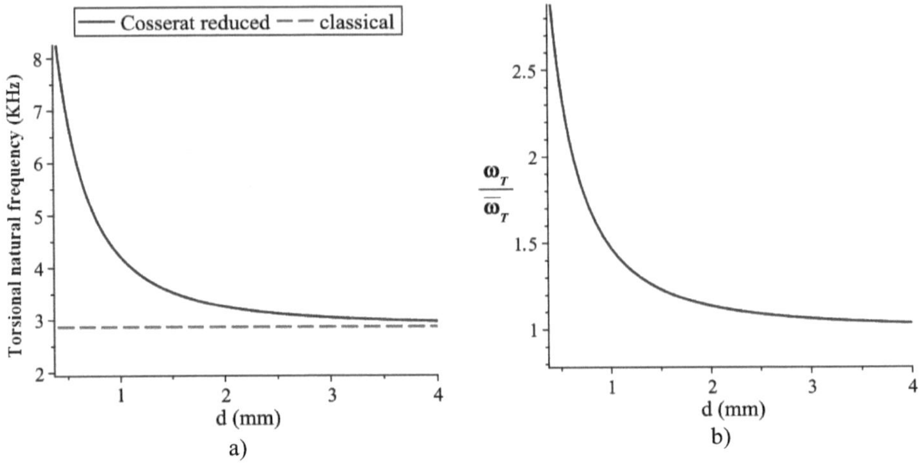

Figure 7.14. *a) Effect of specimen size on the torsional natural frequency; b) the first normalized torsional natural frequency versus beam diameter. Age is 40 years. For a color version of this figure, see www.iste.co.uk/ganghoffer/biomechanics.zip*

The bending and torsion of the vertebral trabecular bone beam specimens have been formulated based on the Cosserat theory, a non-classical theory capable of capturing the size effects. Employing the Hamilton principle, the governing equations of motions for both bending and torsion have been obtained. The results obtained for the static bending problem reveal that both the deflection and rotation predicted by the micropolar model are smaller than those predicted by the classical model. Those differences are very large when the beam thickness is small. Similar trends are observed for the free vibration problem, where it is shown that the natural frequency predicted by the micropolar model is higher than the frequency resulting from the classical model, the difference being significantly large only for very thin beams. This work further outlines that the static and dynamic torsional behavior of vertebral bone is size-dependent, with the non-classical theory predicting a stiffer behavior than does the classical theory, especially when the specimen dimensions are reduced.

8

Prediction of Size Effects in Bone Brittle and Plastic Collapse

The overall plastic yield and brittle failure behaviors of three-dimensional trabecular bone are investigated by a microstructural modeling approach based on the homogenization of the initially discrete architecture. The multiaxial yield behavior is analyzed to formulate a multiaxial plastic yield criterion. Furthermore, the brittle fracture of the lattice is modeled under triaxial stress states to construct the failure surfaces, defined in the tension–tension quadrant. In plastic yielding, the analyses are performed assuming an elastic perfectly plastic lattice, and a micromechanical model based on homogenization scheme is applied to a representative unit cell to determine the macroscopic plastic yield surfaces in stress space. This general framework is applied to evaluate the yield and failure properties of trabecular bone, which are of key interest in understanding and predicting the fracture of bones and bone implant systems. The effective strength of trabecular bone is evaluated in the two situations of fully brittle (fracture with no tissue ductility) and fully ductile failure (yield with no tissue fracture) of the trabecular tissue.

An adaptation and extension of the discrete homogenization method towards a micropolar effective medium has been introduced in order to construct the plastic yield surfaces for which the material point of the effective continuum supports couple stresses in addition to Cauchy-type stresses. The size effects in the ductile fracture mode have been addressed by considering a micropolar behavior, reflecting the influence of additional degrees of freedom and internal bending length effects on the initiation of plasticity. It is observed that when the characteristic size of the microscale structure is comparable to the bending length, a significant difference is

---

Chapter written by Jean-François GANGHOFFER and Ibrahim GODA.

shown between the results based on the non-classical theory and those obtained by the classical theory.

8.1. Introduction

Nowadays, cellular solids and network materials such as foams, lattice truss materials and grid materials are widely used in a variety of commercial and military applications such as automotive industry, aerospace or marine engineering, due to their many advantages, including high mechanical properties and strength, energy absorption capacities, thermal and acoustic insulation properties, and lightweight structural components. Cellular lattice structures endowed with a specific mechanical behavior due to the presence of an inherent microstructure are encountered in many applications such as light-weight industrial components, trabecular bone and bone scaffolds. Such structures are a complex of cells made of an interconnected network of edges. They are usually categorized as 2D and 3D cellular solids, and coined honeycombs and foams respectively. Many man-made and biological structures present a discrete topology, such as fibrous materials (textiles, collagen fiber networks, biological membranes), with a more or less complex organization of the fibrous microstructure; the membrane of biological cells is viewed as an assembly of filaments which are linked together as part of a network. Since those structures consist of many repetitive elements, a need arose to develop mechanical models for the prediction of their deformation behavior.

Among their notable mechanical properties, cellular structures are of more importance, since the microstructure of the lattice can be adjusted so that desired mechanical properties can be achieved at the mesoscopic or macroscopic level. In view of exploiting the full potential of cellular solids, their constitutive mechanical response has to be entirely understood and modeled. Recently, researchers worldwide have widely used additive manufacturing methods to fabricate lattice structures. Although these fabrication methods are capable of producing lattice structures with tunable porosity and pore sizes and with high repeatability, they are costly and time consuming even for fabricating small sized parts. Accordingly, developing numerical prediction models with the capability to predict the mechanical properties of lattice structures can reduce the required experimental measurements in addition to the manufacturing cost.

The material strength of architectured materials has been the topic of several studies related to either their elastic strength, or to their non-linear elastoplastic behavior. The choice of method of analysis of such lattices in the plastic range is especially guided by the nature and type of lattice under consideration. Several classification methods exist in the literature: Deshpande proposed to classify lattices as either stretching dominated or bending dominated lattices [DES 01a], considering that beams are working either in tension–compression, also known as "direct action mechanism", or in bending [CHR 00, MOH 05]. The elastic strength of beam lattices has been considered by various authors: [GIB 97] analysed foams (bending dominated) with the relative density as the dominant criterion, [DEM 07, SUL 08] and [KIM 02] with a more numerical or analytical approach, [FLO 06] with an energy homogenization method, [DES 01b] and [WAN 05] with various stretching dominated lattices and [DOY 06] studied the octet lattice. The stretching dominated lattices are proved to be much stronger than bending governed lattices in [DES 01a]; this raises the interest of this material as a substitute for metallic foams in lightweight structures. More recently, the initial yield surface for 2D truss-lattice materials under biaxial loading was investigated by [ALK 09], based on FE analyses and analytical techniques relying on an energy criterion for orthotropic materials. The extended finite element method is used in [ZAN 10] for the elastoplastic analysis of periodic truss materials in the small strain regime. Multiscale base functions are constructed to capture the small scale heterogeneities of the unit cells; this local information is then brought to the upper macroscopic scale to perform structural calculations. The mechanical properties of micro-lattice structures subjected to normal stresses are evaluated in [USH 13], based on an analytical method relying on classical beam theory. The yield surface is determined under an external biaxial loading state.

The investigation of the stress–strain relationships of beam lattices in the plastic range is more involved. The classical criteria of continuum mechanics do indeed not allow us to describe the nonlinearities in the plastic range [FAN 08]. The effective behavior of three different lattice materials endowed with cubic symmetry has been studied by means of analytical and numerical techniques in [PAR 10]. A multiscale finite element method was developed by [ZAN 10] to analyse the elastoplastic small strain behavior of 2D periodic lattices. A continuum mechanism-based multi-surface plasticity model has been introduced by [MOH 05] to simulate the mechanical behavior of 2D or 3D stretching dominated lattices. This method was later extended by [FAN 08]; this model however relies on an underlying hypothesis of uniform deformation of the cells. This hypothesis is not necessary true in the case of a lattice with internal nodes (in the unit cell), even if the lattice is stretching dominated.

Those networks are endowed with a specific mechanical behavior because of the presence of an inherent microstructure. The prediction of the effective mechanical behavior of such lattices in relation to the geometrical and mechanical parameters at the micro level, is an especially significant issue. The setting up of such predictive models allows us to fully understand the microstructural origin of the mechanical behavior and the lattice architecture required to achieve optimized properties at the structural level. The suitable size materials, the lattice topology and mechanical properties can be chosen relying on a quantitative understanding of the macroscopic impact of the microstructural parameters.

An essential category of such network materials consists of lattices such as trabecular bone having a discrete kinematics and topology, showing size effects at the macroscopic scale of description. Such size effects have been proven to be important when sample dimensions are comparable with the size of the cell [LAK 86]. In other words, the effective mechanical properties show size dependence if the dimension of a specimen or a structure is in a close order to the unit cell size. This behavior called "size effects" [ONC 01, TEK 08] designates the effect of the macroscopic sample size, relative to the unit cell size, on the mechanical behavior. However, these effects are not easily accessible from a direct analysis at the macroscopic structure scale. Consequently, the motivation of such micromechanical-inspired constitutive models is to increase understanding of the yielding behavior of those structures in different loading situations.

A key issue in bone biomechanics is the influence of the microarchitecture strut ductility on the overall strength of trabecular bone. With aging and disease, the individual trabeculae become more brittle, so the question arises: how does this influence the strength of the apparent trabecular bone? In the present work, we analyze the structural behavior of trabecular bone failure by investigating two extreme behaviors, known as fully ductile and fully brittle. For both strut level behaviors, we investigate failed tissue amount and reactive strength evaluated at the trabecular bone scale at yield or ultimate point, respectively for ductile or brittle tissue level material behavior. The yield point corresponds to the stress at which a material begins to deform plastically, and the ultimate point corresponds to the highest point of the stress-strain curve. For the entirely ductile case, we consider tissue level failure by plastic yielding, whereas in the brittle case the yielding is not allowed (break without significant deformation or failed by fracture). Furthermore, when determining the plastic yield strength of trabecular bone structure, bone struts are assumed to be made of elastic perfectly plastic material [GIB 85, KEA 94, MCD 10]. Basically, once the ultimate yielding (plastic collapse) occurs, the structure cannot bear extra loads any more.

The objective here is then to develop an adequate two-dimensional and three-dimensional model for describing the multiaxial yield and failure behavior of network materials like trabecular bone, and to set up criteria for the brittle and ductile collapse based on micromechanical approaches. The discrete homogenization technique is presently developed as a convenient micromechanical approach to construct the plastic yield surfaces of 2D and 3D trabecular architecture. The initial network is replaced by an effective Cauchy continuous medium at an intermediate scale, endowed with effective properties representative of an identified representative unit cell within the structure.

The effective elastoplastic response of general beam networks is additionally obtained thanks to an adaptation and extension of the discrete asymptotic homogenization method exposed in Chapter 3. The 3D homogenization model is applied to evaluate the yield and failure properties of trabecular bone, which are of key interest in understanding and predicting the fracture of bones and bone implant systems. The effective strength of trabecular bone is evaluated in the two situations of fully brittle (fracture with no tissue ductility) and fully ductile failure (yield with no tissue fracture) of the trabecular tissue. A size-dependent non-classical plastic yield criterion is finally developed relying on the Cosserat theory to capture the size-dependency of the trabecular bone structures. Accounting for the moments allows the homogenization towards the more complete micropolar framework, incorporating a microrotation in addition to the displacement as kinematic descriptors at the continuum level. Such extension and adaptation towards a micropolar effective medium leads to constructing the plastic yield surfaces for which the material point supports couple stresses in addition to Cauchy-type stresses. As a consequence, the plastic yield surfaces of trabecular bone would be determined under the externally applied couple stresses.

8.2. Determination of the plastic yield and brittle fracture surfaces in stress space

Two different types of analyses will be performed in the so-called ductile and brittle analysis in order to assess effects of strut level material behavior on the apparent behavior. For this purpose, we expose a sequence of equations written in stress space based on homogenization method in order to determine the onset of plastic yielding of lattice structure and the fracture surface under multiaxial loading conditions, relying on a micromechanical analysis.

8.2.1. *Microscopic stresses versus the macroscopic deformation tensor*

In this subsection, the homogenization method previously described is used to analytically determine the matrix relating the microscopic stress tensor to the apparent (effective) stress tensor. We first construct a vector of the forces that gather all resultants (axial and transverse) and moments acting on each beam b of the reference unit cell. Let $\left(F_x^b, F_{y,z}^b, M_{y,z}^{O(b)}, M_{y,z}^{E(b)}\right)$ denote the microscopic stress components of beam b, $b = 1,2,3,...$ Recall that F_x^b is the axial force, $F_{y,z}^b$ the shearing forces and $M_{y,z}^{O(b)}$, $M_{y,z}^{E(b)}$ the bending moments with respect to y and z at both beam extremities. Therefore, the vector of the forces is written lengthily as

$$[F] = \begin{bmatrix} F_x^1 \\ F_y^1 \\ F_z^1 \\ M_y^{O(1)} \\ M_y^{E(1)} \\ M_z^{O(1)} \\ M_z^{E(1)} \\ F_x^2 \\ .... \end{bmatrix} \qquad [8.1]$$

Previous expressions of resultants and moments still involve the unknown displacements u^n and rotations ϕ^n, which are determined for all nodes by solving the equilibrium of efforts and moments, equations [8.1] and [8.2]. As a consequence, these expressions are functions of the macroscopic strain tensor $[\mathbf{E}]$; then we can construct a matrix $[Q_e]$ relating the force vector on the unit cell to the homogenized strain, so that

$$[F] = [Q_e][\mathbf{E}] \qquad [8.2]$$

Additionally, we can relate the effective homogenized strain and stress tensors to the following relation $[\mathbf{E}] = [S][\boldsymbol{\Sigma}]$, leading in turn to a relation between the stresses at the micro (forces and moments) and macro levels,

$$[F] = [Q_e][S][\boldsymbol{\Sigma}] \qquad [8.3]$$

The previous expression of the microscopic stress is next involved in defining the criteria of onset of plastic yielding and fracture surface for 3D lattice under triaxial and shear loadings.

8.2.2. *Plastic collapse and yield surface*

Yielding in the struts of such cellular materials is concentrated in small zones. When a section becomes completely plastic, it is usually referred to as formation of plastic hinges. The initial yield strength is based on the first cell strut(s) to reach the fully plastic limit moment of the cell struts, representing complete loss to carry additional load.

We here consider a uniform strut thickness of unit cell; hence, the plastic hinge occurs at a cross-section of maximum bending moment. Based on the plastic hinge concept, lattice collapses plastically when the bending moment exerted by the loads on the individual struts reaches the fully plastic moment, creating plastic hinges at the corners. The fully plastic moment for a perfectly plastic beam under combined bending moment and extensional stress receives the expression [STR 93]

$$\left|M^b\right| = \sigma_{ys} Z_{\lambda b}\left(1-\left(\frac{\sigma_a^b}{\sigma_{ys}}\right)^2\right) \qquad [8.4]$$

with M^b the bending moment at the beam extremity nodes with respect to y or z axis, σ_{ys} the initial yield strength of the individual struts (bulk material) and $Z_{\lambda b}$ the plastic section modulus at the plastic hinge section of the struts, which is adopted as $Z_{\lambda b} = D_b^3/6$ assuming a circular cross sectional area of the struts. We here consider a 3D geometry of the circular struts, hence the extensional stress acting on them is elaborated as $\sigma_a^b = F_x^b / A^b$. Therefore, the criterion that defines the failure surface for plastic yield is described from [8.4] as

$$\frac{6\left|M^b\right|}{\sigma_{ys} D_b^3} + \left(\frac{F_x^b}{A^b \sigma_{ys}}\right)^2 = 1 \qquad [8.5]$$

with M^b the bending moments and F_x^b the axial forces within struts arising from the external applied normal ($\Sigma_x, \Sigma_y, \Sigma_z$) and shear stresses ($\Sigma_{xy}, \Sigma_{yz}, \Sigma_{xz}$) acting on the unit cell.

Introducing the homogenized stress components of $[\Sigma]$ after normalization by the initial yield strength of the bulk material σ_{ys} into equation [8.3] leads to the following relation

$$\left[F/\sigma_{ys}\right]=\left[F^N\right]=\begin{bmatrix} F_x^1 \\ F_y^1 \\ F_z^1 \\ M_y^{O(1)} \\ M_y^{E(1)} \\ M_z^{O(1)} \\ M_z^{E(1)} \\ F_x^2 \\ \cdots \end{bmatrix}^N =[Q_e][S]\left[\Sigma/\sigma_{ys}\right]=[Q_e][S]\begin{bmatrix} \Sigma_x/\sigma_{ys} \\ \Sigma_y/\sigma_{ys} \\ \Sigma_z/\sigma_{ys} \\ \Sigma_{xy}/\sigma_{ys} \\ \Sigma_{yz}/\sigma_{ys} \\ \Sigma_{xz}/\sigma_{ys} \end{bmatrix} \qquad [8.6]$$

As a final step, the system of previous equations in [8.5] then rewrites for the set of struts b that belong to the reference unit cell by which the whole structure is generated as

$$\frac{6\left|M^{bN}\right|}{D_b^3}+\left(\frac{F_x^{bN}}{A^b}\right)^2=1 \qquad [8.7]$$

The plastic collapse strength domain of the structure is then the surface delimited by previous equality in the normalized stress space. As a consequence, the criteria of plastic collapse are derived in three dimensional stress states; addressing the criteria for in-plane and out-of-plane yielding: in-plane stresses Σ_x and Σ_y combined with out-of-plane normal stress Σ_z in addition to in-plane Σ_{xy} and out-of-plane shears Σ_{yz}, . Σ_{xz} . are considered in the analysis. Note that the normalized axial forces and moments, quantities F_x^{bN} and M^{bN} are extracted from $\left[F^N\right]$ in equation [8.6].

8.2.3. *Brittle failure and fracture surfaces*

Brittle failure occurs when the maximum stress in a cell strut of lattice subjected to both bending and axial loads exceeds the modulus of rupture σ_{fs} of the cell materials. This criterion has been used in [GIB 97]. The moment that causes the maximum stress writes as

$$\sigma_{\max}=M^b/S_E^b \qquad [8.8]$$

with S_E^b the elastic section modulus adopted as $S_E^b = \pi D_b^3 / 32$ for circular struts. Therefore, failure occurs when the previous stress cumulated to the axial stress exceeds σ_{fs},

$$\sigma_{fs} = \sigma_a^b + M^b / S_E^b \qquad [8.9]$$

where σ_a^b is the axial stress in the cell strut adopted as $\sigma_a^b = \dfrac{F_x^b}{A^b}$, M^b the bending moment at the strut ends either with respect to y or z axis. The brittle failure condition now becomes

$$\frac{F_x^b}{\sigma_{fs} A^b} + \frac{M^b}{\sigma_{fs} S_E^b} = 1 \qquad [8.10]$$

Inserting the macroscopic stress components $[\Sigma]$ normalized by the modulus of rupture of cell strut σ_{fs} into equation [8.3] yields the following relation

$$[F/\sigma_{fs}] = [F^N] = \begin{bmatrix} F_x^1 \\ F_y^1 \\ F_z^1 \\ M_y^{O(1)} \\ M_y^{E(1)} \\ M_z^{O(1)} \\ M_z^{E(1)} \\ F_x^2 \\ \cdots \end{bmatrix}^N = [Q_e][S][\Sigma/\sigma_{fs}] = [Q_e][S] \begin{bmatrix} \Sigma_x / \sigma_{fs} \\ \Sigma_y / \sigma_{fs} \\ \Sigma_z / \sigma_{fs} \\ \Sigma_{xy} / \sigma_{fs} \\ \Sigma_{yz} / \sigma_{fs} \\ \Sigma_{xz} / \sigma_{fs} \end{bmatrix} \qquad [8.11]$$

It is obvious that the combination of applied normal and shear stresses required to cause tensile failure in the cell strut is found from the applied moments and the axial stresses acting on the cell struts.

Finally, based on the system of previous equations [8.10], the failure surface of lattices under multiaxial stress conditions accounting for anisotropy is described by a brittle failure criterion condition rewritten for the set of struts *b* that belong to the reference unit cell as

$$\frac{F_x^{bN}}{A^b} + \frac{32 M^{bN}}{\pi D_b^3} = 1 \qquad [8.12]$$

A dedicated code has been constructed from the previous developments for the calculation of the plastic yield and brittle fracture surfaces of 3D porous open-celled structure in the homogenized stress space. The code uses an input file including the topology and mechanical properties within a selected unit cell, and delivering as an output the plastic collapse and brittle fracture stresses.

Although the developed models have the ability to describe the plastic and brittle failure within the microstructure of cellular materials in a broad sense subjected to multiaxial loadings, we are focusing on vertebral trabecular bone.

8.3. 3D plastic collapse and brittle fracture surface of trabecular bone

Trabeculae form a 3D porous lattice in cancellous bone, whose micro-architecture determines its mechanical performance including the macroscopic strength. Therefore, a 3D geometric model of a hexagonal structure is considered as an idealized representation of vertebral trabecular bone, for which the yield domain and fracture surface at the macroscopic level are identified from a micromechanical analysis. The whole structure is generated from the repetition of this unit cell using the three periodicity vectors defined in a Cartesian basis as shown in Figure 8.1(a). This study examines the plastic collapse and brittle failure of a hexagonal model of rod-like columnar structure considered as a prototype topology of vertebral trabecular bone. The model is comprised of struts of length L_v with diameter D_h for the horizontal ones and length L_h with diameter D_v for the vertical struts (Table 8.1).

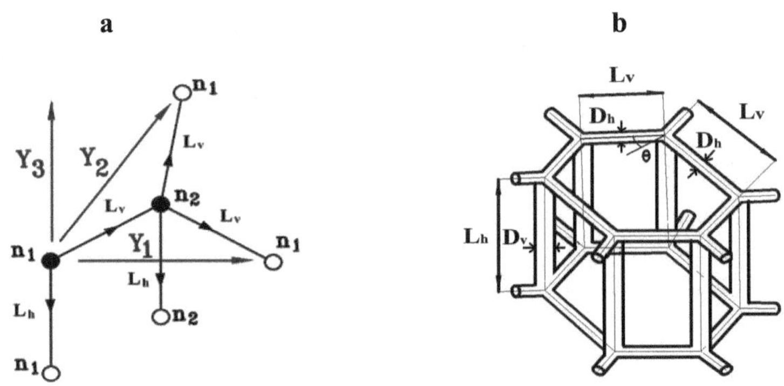

Figure 8.1. *Representative model of vertebral trabecular bone and its parameters. a) Reference unit cell and its periodicity vectors Y_1, Y_2, and Y_3, b) 3D hexagonal unit cell with thick vertical trabeculae and thinner horizontal ones rendering of human vertebral bone. For a color version of this figure, see www.iste.co.uk/ganghoffer/biomechanics.zip*

Parameter	Relation
Horizontal trabeculae thickness (μm)	$D_h = -1.03 \times \text{Age} + 189$
Vertical trabeculae thickness (μm)	$D_v = 0.14 \times \text{Age} + 208$
Distance between horizontal trabeculae (μm)	$L_h = 13.74 \times \text{Age} + 288$
Distance between vertical trabeculae (μm)	$L_v = 6.74 \times \text{Age} + 456$

Table 8.1. *Micro-structural parameters for vertebral bone based on morphology–age relationships by [MOS 88], with age measured in years*

Two different types of analysis can be performed in the so-called ductile and brittle analysis in order to assess effects of tissue level material behavior on the apparent trabecular bone behavior. In the ductile case, tissue failure occurs by yielding, while in the brittle case, tissue failure occurs via brittle fracture.

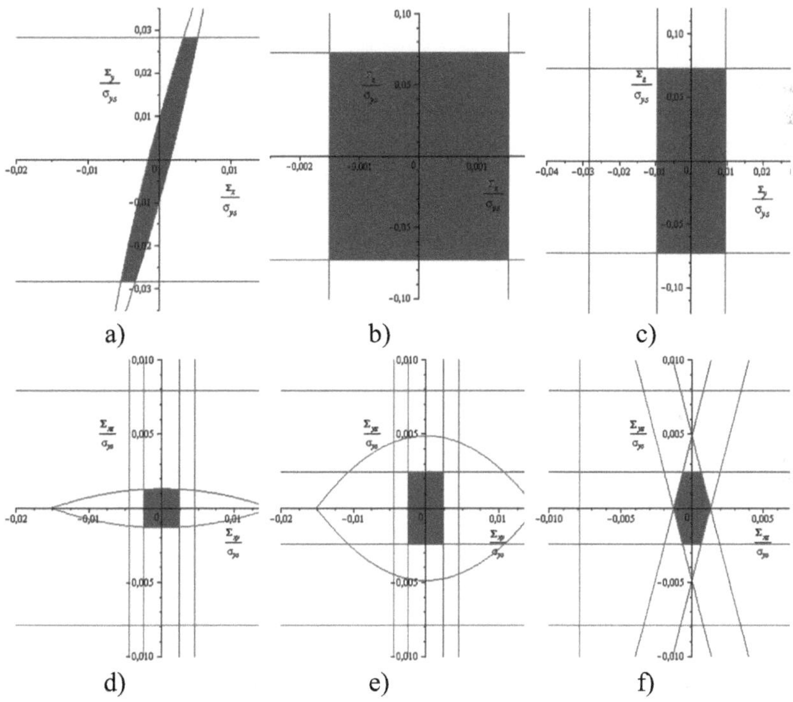

Figure 8.2. *Plastic yield surfaces in the stress spaces of (a) Σ_x - Σ_y, (b) Σ_x - Σ_z, (c) Σ_y - Σ_z, (d) Σ_{xy} - Σ_{xz}, (e) Σ_{xy} - Σ_{yz}, and (f) Σ_{xz} - Σ_{yz} , all the stresses are normalized with respect to the yield strength of the bulk material. Age is 40 years with relative density ρ^*/ρ_s = 0.12. For a color version of this figure, see www.iste.co.uk/ganghoffer/biomechanics.zip*

The overall collapse surface in the macroscopic stress space consists of intersecting collapse surfaces which are associated with particular collapse modes. The initial yield strength of periodic trabecular structure is based on the first cell wall(s) reaching the fully plastic limit moment of the cell walls in either tension or compression. The plastic yield surface is the inner envelope of the intersecting surfaces for the plastic collapse mechanism. We define the onset of plastic yielding under multiaxial state of stress for vertebral trabecular bones relying on the homogenization model established in section 8.3 with the plasticity criterion defined by equality equation [8.7]. The micro-architecture parameters of the trabeculae are selected for a person at the age of 40, with an inclination angle of the oblique trabeculae $\theta = 60°$, corresponding to a relative density of about 12%. The combination of the applied normal and shear stresses which causes plastic collapse, plotted on the stress axes Σ_{ij} (i = x, y, z), is a closed surface called the yield surface for the vertebral trabecular bone, as illustrated in Figure 8.2. Initial yield surfaces are plotted according to equation [8.7] in various stress sub-spaces, normalized by the trabecula yield strength σ_{ys}.

From the results illustrated in Figure 8.2, some key points shall be addressed: the plastic yield domains are closed, anisotropic, and consist of several convex curved surfaces, with some planar facets. All the plastic failure envelopes are symmetrical with respect to tensile and compressive loadings. This may derive from the fact that the initial yield is based on the first cell strut to collapse in bending or reaching yield by extension. If we would consider the post-buckling behavior past initial yield, the behavior would exhibit a tension–compression asymmetry (in this study, we assume that the first compressive buckling load is greater than the initial yield limit of any strut within the lattice, so that no buckling will arise).

At very low bone volume fraction (high slenderness ratios of trabeculae), trabecular bone may fail in a brittle mode. Additionally, at low slenderness ratios dry individual trabeculae can also fracture in a brittle way (fracture with no tissue ductility); the stress reaches the ultimate point very quickly (only a small amount of plastic strain can be sustained, and consequently only a very small amount of load is enough to break the whole specimen). We consider trabecular architectures corresponding to a typical lattice for a person aged 85 years. Such an old trabecular lattice is selected since it has a low relative density of about 5%, and at very low bone relative density trabecular bone can fail in a brittle manner. Relying on brittle failure criterion equation [8.12], we evaluate in Figure 8.3 the brittle failure of this lattice under a multiaxial loading by determining the brittle fracture surfaces, to provide the complete failure envelope.

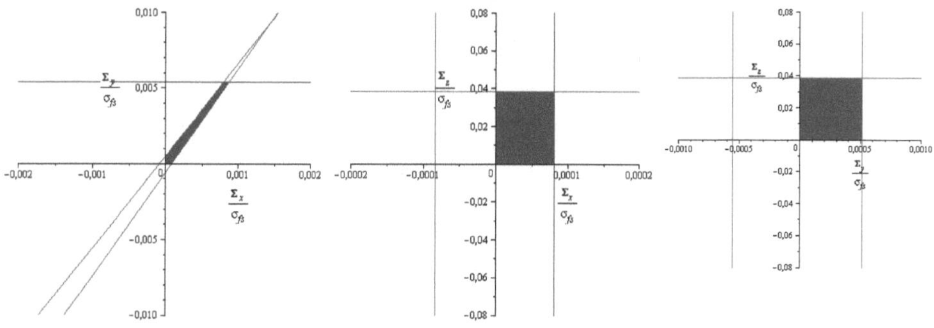

Figure 8.3. *Brittle fracture surfaces in the stress spaces of* **(a)** Σ_x - Σ_y, **(b)** Σ_x - Σ_z, *and* **(c)** Σ_y - Σ_z *for a person aged 85 years with bone relative density* ρ/ρ_s = 0.05. *The tension–tension quadrant is colored. For a color version of this figure, see www.iste.co.uk/ganghoffer/biomechanics.zip*

From the evaluated brittle fracture surfaces (Figure 8.3), some key points should be noted; from in-plane stress state (Figure 8.3(a)), it can be seen that the fracture surface is elongated in the direction of tension and it is truncated in the biaxial tension zone because of the brittle fracture of in-plane vertical trabeculae under axial stress Σ_y. Moreover, the fracture surfaces in stress spaces Σ_x - Σ_z, and Σ_y - Σ_z are composed of several planar facets; they exhibit anisotropic properties and are closed in the tension–tension zone.

8.4. Plastic yield surface based on Cosserat theory

Several works from the literature show the importance of considering microstructure-related scale effects on the macroscopic properties of bone [GOD 12b, GOD 13a]. This motivates us to extend previous analyses of collapse mechanisms to micropolar behaviors focusing on the plastic yield behavior in 2D situations.

Accounting for the moments would allow the homogenization towards more general effective media, such as micropolar continua, a more complete equivalent continuum incorporating a microrotation in addition to the displacement as kinematic descriptors at the continuum level. This entails that the material can transmit couple stresses in addition to tractions (classical stresses); those couple stresses develop internal work in the variation of microcurvatures, defined as the spatial gradients of the microrotation. As a consequence, the second order development of the displacement is taken into account in this specific theory to capture the micropolar effect and hence the asymptotic expansion of the rotation is limited to the first order. The moments are then expressed at both ends as:

$$M^{O(b)} = \frac{6E_s^b I_z^b}{\left(L^b\right)^2 (1+\Phi)} \left(-e_y.\left(\Delta \mathbf{U}_1^b + \Delta \mathbf{U}_2^b\right)\right) +$$

$$\frac{E_s^b I_z^b}{L^b(1+\Phi))} \left((4+\Phi))\phi_0^{O(b)} + (2-\Phi))\phi_0^{E(b)} + \left((4+\Phi))\phi_1^{O(b)} + (2-\Phi))\phi_1^{E(b)} + (2-\Phi))\frac{\partial \phi_0}{\partial \lambda^i}\delta^{ib} \right)\right)$$

$$M^{E(b)} = \frac{6E_s^b I_z^b}{\left(L^b\right)^2 (1+\Phi)} \left(-e_y.\left(\Delta \mathbf{U}_1^b + \Delta \mathbf{U}_2^b\right)\right) +$$

$$\frac{E_s^b I_z^b}{L^b(1+\Phi)} \left((2-\Phi)\phi_0^{O(b)} + (4+\Phi)\phi_0^{E(b)} + \left((2-\Phi)\phi_1^{O(b)} + (4+\Phi)\phi_1^{E(b)} + (4+\Phi)\frac{\partial \phi_0}{\partial \lambda^i}\delta^{ib} \right)\right)$$

with $\Delta \mathbf{U}_2^b$ the second order displacement difference between the ends of each beam, defined as $\Delta \mathbf{U}_2^b = \mathrm{u}_2^{E(b)} - \mathrm{u}_2^{O(b)}$. The unknown displacements u_1^n, u_2^n and rotations φ_0^n, φ_1^n, are determined for all nodes by solving the equilibrium equations for the reference unit cell. The solutions lead to u_1^n and φ_0^n expressed versus the displacement gradient or deformation tensor, while φ_1^n and u_2^n express versus the gradient of micro-rotation or the micro-curvature tensor.

The homogenization of the discrete equilibrium of forces leads to the construction of the stress vectors therein $\mathrm{F}^b = \mathrm{F}_x^b + \mathrm{F}_y^b$ in the 2D case; this in turn leads to the homogenized Cauchy stress tensor. In order to get the couple stress components, the local equilibrium of each individual strut is considered for the purpose of homogenization, written at the centre of each strut. After some developments and simplifications, the moment equilibrium equation after homogenization takes the following virtual power form

$$\begin{aligned}\int_\Omega \boldsymbol{\mu}^i.\frac{\partial \mathbf{w}}{\partial \lambda^i} d\lambda = \int_\Omega \boldsymbol{\mu}^i.\left(\nabla_x \mathbf{w}.\frac{\partial \mathbf{R}}{\partial \lambda^i}\right) d\lambda = \int_\Omega \left(\boldsymbol{\mu}^i \otimes \frac{\partial \mathbf{R}}{\partial \lambda^i}\right) : \frac{1}{g}(\nabla_x \mathbf{w})\, dx \\ = \int_\Omega (\mathbf{m}.\nabla_x).\mathbf{w} dx = 0\end{aligned} \qquad [8.13]$$

with $\mathbf{m}$ the macroscopic (effective) couple stress and $\mathbf{w}$ the virtual rotational velocity. The identification of the couple stress vectors $\boldsymbol{\mu}^i$ leads in turn to the expression of the homogenized couple stress according to

$$\mathbf{m} = \frac{1}{g}\boldsymbol{\mu}^i \otimes \frac{\partial \mathbf{R}}{\partial \lambda^i} \qquad [8.14]$$

with μ^i therein expressed as the following sum over all struts forming the lattice

$$\mu^i = \sum_{b \in B_R} \frac{1}{2}\left(M_2^{E(b)} - M_2^{O(b)}\right)\delta^{ib} \qquad [8.15]$$

After solving for all unknown displacements and rotations, we obtain the effective constitutive equations, written in a 2D context in matrix format

$$\underbrace{\begin{Bmatrix} \sigma_x \\ \sigma_y \\ \sigma_{xy} \\ m_{xz} \\ m_{yz} \end{Bmatrix}}_{\Sigma} = [K] \underbrace{\begin{Bmatrix} \varepsilon_x \\ \varepsilon_x \\ \varepsilon_{xy} \\ \kappa_{xz} \\ \kappa_{yz} \end{Bmatrix}}_{\mathbf{E}} \qquad [8.16]$$

where $[K]$ is the effective (homogenized) stiffness matrix, σ_{ij} the symmetric part of Cauchy stress tensor and m_{ij} the couple stress tensor.

We consider here the reduced Cosserat theory, where the rotation is related to the displacement gradients $\phi = \left(\partial u_y / \partial x - \partial u_x / \partial y\right)/2$. As a result, the strain tensor ε_{ij} is symmetrical, with components defined as $\varepsilon_{xy} = \varepsilon_{yx} = \left(\partial u_y / \partial x + \partial u_x / \partial y\right)/2$. The couple stress tensor (or moment per unit area) has two components m_{xz}, m_{yz} and the corresponding micro-curvatures express versus the micro-rotations as $\kappa_{xz} = \partial\phi / \partial x,\ \kappa_{yz} = \partial\phi / \partial y$.

We can next construct the matrix relating the microscopic stress tensor to the macroscopic stress tensor; we first elaborate the force vector gathering all resultants and moments for all struts of the reference unit cell

$$[F] = \left[F_x^1 \quad F_y^1 \quad M^{O(1)} \quad M^{E(1)} \quad F_x^2 \quad \ldots\right]^T \qquad [8.17]$$

After solving for all kinematic unknowns, we can obtain the individual expressions for all forces and moments. These expressions involve the macroscopic deformation tensor $[\mathbf{E}]$, allowing us to build the following relation between the microscopic stress and the macroscopic deformation

$$[F] = [Q_e][\mathbf{E}] \qquad [8.18]$$

This in turn entails the following relation between the microscopic stress and the macroscopic stress

$$[F]=[Q_e][K]^{-1}[\Sigma] \tag{8.19}$$

On the basis of the previous calculation of the microscopic stress, we can use the criterion of plasticity initiation in equation [8.5] to define the boundary of the elastic domain of the micropolar continuum medium.

By inserting the macroscopic Cauchy and couple stresses Σ^N normalized by the initial yield strength of strut material σ_{ys} into equation [8.19], the following relation is obtained

$$\left[F^N\right]=\begin{bmatrix} F_x^1 \\ F_y^1 \\ M^{O(1)} \\ M^{E(1)} \\ F_x^2 \\ \ldots \end{bmatrix}^N=[Q_e][K]^{-1}\left[\Sigma^N\right]=[Q_e][K]^{-1}\begin{bmatrix} \sigma_x/\sigma_{ys} \\ \sigma_y/\sigma_{ys} \\ \sigma_{xy}/\sigma_{ys} \\ m_{xz}/\sigma_{ys} \\ m_{yz}/\sigma_{ys} \end{bmatrix} \tag{8.20}$$

The plastic yield surface domain of the trabecular lattice is then the surface delimited by the yield criterion; the system of previous rewrites for the set of struts that belong to the reference unit cell forming the lattice as

$$\left(\left(F_x^{bN}\right)^2+4\left|M^{bN}\right|\right)\Big/t^2=1 \tag{8.21}$$

We next plot the plastic yield surfaces of trabecular bone lattice adopting the micro-architecture parameters of bone cell as follows: the trabecular separation and trabecular thickness are respectively 0.726 and 0.148 mm. To evaluate the overall plastic yield strength relying on micropolar theory, we plot the plastic yield surfaces under the shear stress combined with the couple stress in x-z plane and the couple stress state in both x-z and y-z planes as well (Figure 8.4). The plastic yield surface consists of the intersection surfaces of collapse of vertical and inclined struts.

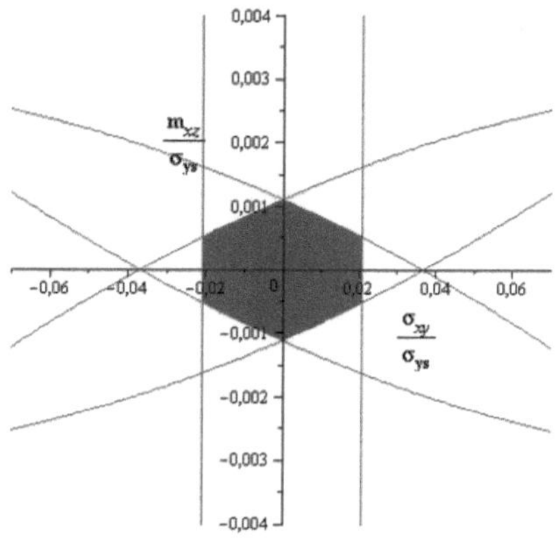

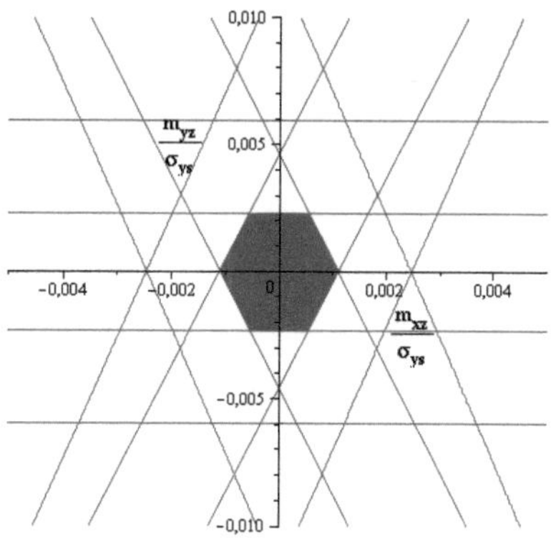

Figure 8.4. *Plastic yield surfaces under combined shear σ_{xy} and couple stresses state m_{xz} and m_{yz}. For a color version of this figure, see www.iste.co.uk/ganghoffer/biomechanics.zip*

8.4.1. *Size dependent plastic yield criterion*

As a matter of fact, the classical yield criterion does not describe the yield size dependency that has been observed in microstructured structures, and thus it may underestimate the yield loads of structures. So, we would suggest a specific size-dependent plastic yield criterion relying on micromechanical model on the basis of the micropolar theory. In order to develop the micropolar yield criterion, the deviatoric part of the strain energy density of the microscale structure including both classical and non-classical parts is calculated based on this theory, and equated to the deviatoric strain energy of a macro-sample subjected to its yielding tensile load.

The strain energy density consists of one part due to dilation (volume changes) and a second part accounting for distortion (change of shape). The former is called the dilatational strain energy (hydrostatic part) and the latter the distortional energy (deviatoric part). From this decomposition, the hydrostatic and deviatoric energy densities based on micropolar theory take the forms

$$U = \frac{1}{2}\left(\sigma_{ij}\epsilon_{ij} + m_{ij}\chi_{ij}\right)\Big|_H + \frac{1}{2}\left(\sigma_{ij}\epsilon_{ij} + m_{ij}\chi_{ij}\right)\Big|_D \qquad [8.22]$$

The hydrostatic stress tensor writes as follows $\sigma_{ij}\big|_H = \frac{1}{3}\delta_{ij}\sigma_{kk}$ and the corresponding hydrostatic strain will be $\epsilon_{ij}\big|_H = \frac{1}{3}\delta_{ij}\epsilon_{kk} = \frac{(1-2\nu)}{3E}\delta_{ij}\sigma_{kk}$. Here, the hydrostatic components of the couple stress and curvature tensors will vanish; however, they will appear later on in the distortional part. Accordingly, the dilatational part of the strain energy can be obtained as

$$U_H = \frac{1-2\nu}{6E}(\sigma_{kk})^2 \quad [8.23]$$

The deviatoric stress tensor can be obtained by subtracting the hydrostatic stress from the total stress tensor as $\sigma_{ij}\big|_D = \sigma_{ij} - \frac{1}{3}\delta_{ij}\sigma_{kk}$. In the same manner, the deviatoric strain is obtained as

$$\epsilon_{ij}\big|_D = \epsilon_{ij} - \frac{1}{3}\delta_{ij}\epsilon_{kk} = \frac{1}{E}\big((1+\nu)\sigma_{ij} - \nu\delta_{ij}\sigma_{kk}\big) - \frac{(1-2\nu)}{3E}\delta_{ij}\sigma_{kk} = \frac{(1+\nu)}{E}\underbrace{\left(\sigma_{ij} - \frac{1}{3}\delta_{ij}\sigma_{kk}\right)}_{\sigma_{ij}|_D}$$

The non-classical part of the strain energy due to the couple stress and curvature tensors will have a deviatoric nature as will be shown next; therefore, the distortional part of the strain energy can be evaluated by subtracting the dilatational term from the total strain density as follows

$$\begin{aligned} U_D &= \frac{1}{2}\left(\sigma_{ij}\epsilon_{ij} + m_{ij}\chi_{ij}\right) - \frac{1-2\nu}{6E}(\sigma_{kk})^2 \\ &= \frac{1}{2}\left(\frac{\sigma_{ij}}{E}\big((1+\nu)\sigma_{ij} - \nu\delta_{ij}\sigma_{kk}\big) + m_{ij}\chi_{ij}\right) - \frac{1-2\nu}{6E}(\sigma_{kk})^2 \end{aligned} \quad [8.24]$$

According to von Mises's yield theory, a ductile solid will yield when the distortional energy density reaches a critical value for that material; this critical value of the distortional energy can be estimated from a uniaxial tensile test. At the instance of yielding, $\sigma_{xx} = \sigma_Y$ (yield stress), all other normal and shear stresses will vanish; therefore, the distortion energy density associated with yielding is

$$U_D\big|_Y = \frac{1+\nu}{3E}\sigma_Y^2 \quad [8.25]$$

We then equate the distortional strain energy density of the micro-level structure which incorporates both the classical and non-classical terms to the distortional strain energy stored in a macro-level sample at the yield point during tensile test, as follows

$$\left(\sigma_{ij}\sigma_{ij} - \frac{(\sigma_{kk})^2}{3}\right) + \frac{E}{\underbrace{1+\nu}_{2G}} m_{ij}\chi_{ij} = \frac{2}{3}\sigma_Y^2 \quad [8.26]$$

Consequently, yielding will occur when the stresses and couple stresses produced by loadings satisfy the previous equation. We can also define an equivalent stress σ_{eq} based on the previous relation to show the failure condition under which yield occurs, $\sigma_{eq} \geq \sigma_Y$ as

$$\sigma_{eq} = \sqrt{\frac{1}{2}\left(3\sigma_{ij}\sigma_{ij} - (\sigma_{kk})^2\right) + 3Gm_{ij}\chi_{ij}} \quad [8.27]$$

It is obvious that when the non-classical term in equation [8.27] vanishes, the classical von Mises yield criterion will be recovered.

Consistent with the basic assumptions of Bernoulli–Euler beam and the one-dimensional beam theory (Figure 11), the non-zero components of the static and kinematic variables write

$$\sigma_{xx} = E_x^* y \frac{\partial^2 v(x,t)}{\partial x^2}, \ \chi_{xz} = \frac{\partial \phi_z(x,t)}{\partial x} = \frac{\partial^2 v(x,t)}{\partial x^2}, \ m_{xz} = \gamma_{xz} \frac{\partial^2 v(x,t)}{\partial x^2} \quad [8.28]$$

where v represents the lateral deflection of the beam in y-direction, x and y refer respectively to the longitudinal and lateral coordinates, and γ_{xz} is the homogenized micropolar bending constant. The rotation angle of the centroidal axis of the beam is related to the deflection as: $\phi_z(x,t) = \frac{\partial v(x,t)}{\partial x}$.

Upon substitution of non-zero components of stress, couple stress, and curvature in equation [8.28] into equation [8.27], the equivalent or yield stress of the microstructured beam writes

$$\sigma_Y^* = \sqrt{\left(E_x^* y \frac{\partial^2 v(x,t)}{\partial x^2}\right)^2 + 3G_{xy}\gamma_{xz} \frac{\partial^4 v(x,t)}{\partial x^4}} \quad [8.29]$$

The external bending moment acting on a beam based on the micropolar theory can be written as

$$M = \int_A \left(E_x^* y^2 \frac{\partial^2 v(x,t)}{\partial x^2} + \gamma_{xz} \frac{\partial^2 v(x,t)}{\partial x^2} \right) dA = \left(E_x^* I_z^* + \gamma_{xz} A\right) \frac{\partial^2 v(x,t)}{\partial x^2} \quad [8.30]$$

I_z^* refers to the second moment of cross-sectional area defined by $I_z^* = \int_A y^2 dA$, with A the cross-sectional area of the macroscopic beam and y the perpendicular distance to the neutral axis.

By substituting the second displacement gradient from equation [8.30] into equation [8.29], we can obtain, after simplification, a relation for the yielding load in the following form

$$\left(\sigma_Y^*\right)^2 = \frac{M^2\left(\left(E_x^* y\right)^2 + 3G_{xy}\gamma_{xz}\right)}{\left(E_x^* I_z^* + \gamma_{xz} A\right)^2} \qquad [8.31]$$

The size-dependent yielding behavior of the micro-structured beams is analyzed by considering a micro-structured cantilever with uniform rectangular cross-section with height H and width W, subjected to a concentrated force F acting on the free end as shown on Figure 8.5. For this beam, the maximum bending moment happens at the clamped end based on the relation $M_{\max} = FL$, and the maximum distance on the beam section from the neutral axis is $y_{\max} = H/2$. Hence, the yield moment is determined as $M_{\max} = FL$.

The yielding moment of the beam can be written with the characteristic micropolar bending length l_b using the relation $l_b = \sqrt{(\gamma_{xz})/2(2\mu^* + k)}$, as

$$M_Y = \frac{\sigma_Y^*\left(E_x^* I_z^* + 4l_b^2 G_{xy} A\right)}{\sqrt{\left(E_x^* y\right)^2 + 12G_{xy}^2 l_b^2}} \qquad [8.32]$$

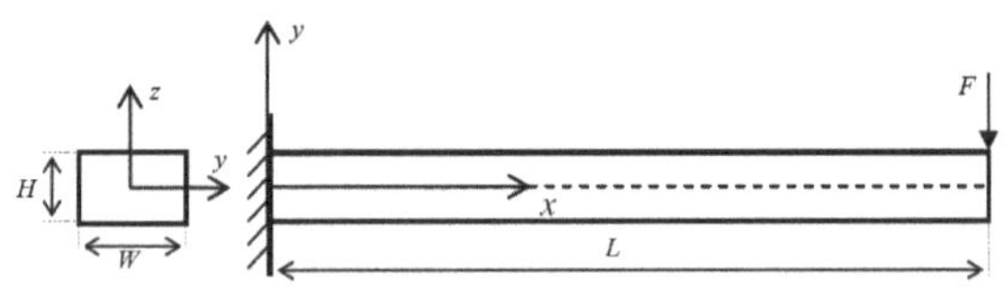

Figure 8.5. *Schematic diagram of a micro-structured cantilever beam with rectangular cross-section exposed to a concentrated force at its free end*

We here focus on the fully plastic moment of a yielded section of a beam, wherein the plastic moment M_P is always greater than the yielding moment. Rewriting equation [8.32] leads to the fully homogenized plastic moment of a beam with rectangular cross-section as

$$M_P = \frac{3}{2}\frac{\left(\frac{E_x^* H^3}{12} + 4l_b^2 G_{xy} H\right)}{\sqrt{\left(\frac{E_x^* H}{2}\right)^2 + 12G_{xy}^2 l_b^2}}\sigma_Y^* \qquad [8.33]$$

We note that when the length scale bending parameter l_b is negligible, the plastic yield load of a classical cantilever beam can be obtained from equation [8.33] as $\frac{FL}{H^2}=\frac{M_P}{H^2}=\frac{1}{4}\sigma_Y^*$. It should be noted that σ_Y^* refers to the effective homogenized initial yield strength of the beam material. It is determined based on the micromechanical model using the maximum elastic bending moment criteria.

In order to assess the size-dependent plastic yield of micro-structured beams, we consider the microbeams to be made of human trabecular bone; the geometrical parameters of 2D trabecular lattice are selected as follows: the trabecular separation and trabecular thickness are respectively 0.726 and 0.148 mm, and the inclined angle of the hexagonal trabecular cell is 30°

The plastic moment M_P / H^2 normalized by the initial yield stress of bulk material σ_{ys} is evaluated on the basis of the developed size effect plastic yield criterion and compared to the classical plastic moment (Figure 8.6). We observe that, contrary to the classical plastic yield criterion, the present non-classical yield criterion is able to capture the size dependency. It is indicated that when the characteristic size of the microscale structure is comparable to the length scale of the structure material, there will be a considerable gap between the results of the classical criterion and the results of the micropolar based criterion. It is also worth noting that, when the size of the beam samples increases, we recover the plastic yielding moment predicted by the classical criterion.

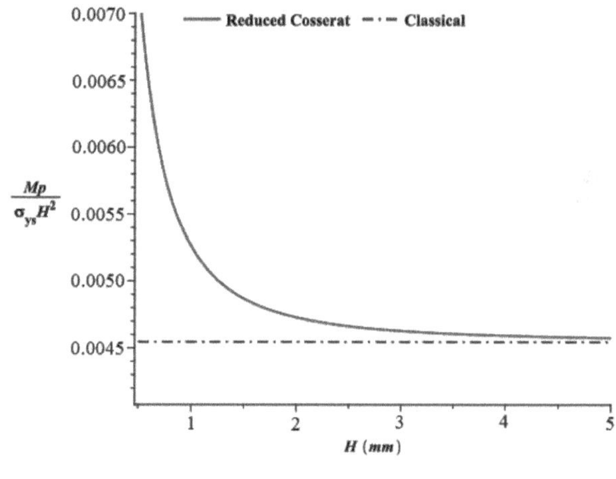

Figure 8.6. *Effect of bone specimen size on the plastic yield moment of the beam.* $l_b = 0.12$ mm *. For a color version of this figure, see www.iste.co.uk/ganghoffer/biomechanics.zip*

8.5. 3D homogenization of trabecular bone lattices

Classical continuum mechanics may not be the most appropriate macroscopic description of such porous structures because higher order effects - absent in classical theories - are required to afford the transition from the discrete to an equivalent continuum description. New measures of deformation, which involves some internal length, such as the curvature tensor, are needed in a more complete continuum theory [HAD 11]. In comparison with a classical continuum, a micropolar continuum is obtained by adding a rotation to each point of the continuum; it is based on the assumption that micromoments exist at each point of the continuum (e.g. [GOD 12c, GOD 13b]). As a consequence, such a theory will require the introduction of micropolar couple-stresses. This idea, as formalized in the micropolar theory, hypothesizes the material micro-rotation to be independent of the continuum mechanical macro-rotation (defined as one half of the curl of the displacement field). The gradient of the micro-rotation vector is used as a curvature tensor. The micropolar couple-stress tensor is non-symmetric and is energetically conjugate to the non-symmetrical curvature tensor. The appealing aspect of the micropolar couple-stress is that a physically meaningful link can be made between the microscopic couple stresses (or micromoments) and the homogenized curvatures, which in turn leads to the existence of a relationship between the couple stress at the micro (at each trabeculae) and macro levels; this is done via the homogenization approaches wherein the effective micropolar couple stress rigidity tensor is determined.

In the forthcoming development, we focus on the homogenization of the initially discrete bone lattice towards an equivalent 3D micropolar couple-stress continuum.

The homogenized constitutive equation is written with reference to the continuum relation of linear couple stress elasticity in matrix format as

$$[\mathbf{m}] = \frac{1}{g}\boldsymbol{\mu}^{i} \otimes \frac{\partial \mathbf{R}}{\partial \lambda^{i}} = [\mathbf{D}][\chi] \qquad [8.34]$$

with $[\mathbf{D}]$ the equivalent (homogenized) stiffness matrix of couple stress continuum.

The micro-curvature tensor χ in [8.34] can be decomposed into two parts related to the torsional curvature and the bending curvature:

The torsional microcurvature field is elaborated as

$$\left.\begin{aligned}\chi_{xx} &= \frac{\partial \phi_x}{\partial x}\\ \chi_{yy} &= \frac{\partial \phi_y}{\partial y}\\ \chi_{zz} &= \frac{\partial \phi_z}{\partial z}\end{aligned}\right\} \rightarrow \chi^{tor} = \begin{bmatrix} \chi_{xx} & 0 & 0 \\ 0 & \chi_{yy} & 0 \\ 0 & 0 & \chi_{zz} \end{bmatrix}$$

and the bending microcurvature field is

$$\left.\begin{aligned}\chi_{xy} &= \frac{\partial \phi_y}{\partial x} & \chi_{yx} &= \frac{\partial \phi_x}{\partial y}\\ \chi_{xz} &= \frac{\partial \phi_z}{\partial x} & \chi_{zx} &= \frac{\partial \phi_x}{\partial z}\\ \chi_{yz} &= \frac{\partial \phi_z}{\partial y} & \chi_{zy} &= \frac{\partial \phi_y}{\partial z}\end{aligned}\right\} \rightarrow \chi^{bend} = \chi - \chi^{tor} = \begin{bmatrix} 0 & \chi_{xy} & \chi_{xz} \\ \chi_{yx} & 0 & \chi_{yz} \\ \chi_{zx} & \chi_{zy} & 0 \end{bmatrix}$$

We also define a decomposition of the couple stress tensor $\mathbf{m}$ as: $\mathbf{m} = \mathbf{m}^{\text{tor}} + \mathbf{m}^{\text{bend}}$, where

$$\mathbf{m}^{\text{tor}} = \begin{bmatrix} m_{xx} & 0 & 0 \\ 0 & m_{yy} & 0 \\ 0 & 0 & m_{zz} \end{bmatrix} \text{ and } \mathbf{m}^{\text{bend}} = \begin{bmatrix} 0 & m_{xy} & m_{xz} \\ m_{yx} & 0 & m_{yz} \\ m_{zx} & m_{zy} & 0 \end{bmatrix}$$

are the couple stress tensor of torsion and the couple stress tensor of bending, respectively.

Figure 8.7 shows the representation of couple stress measures in 3D; the first subscript of the couple stress (or moment stress) refers to the surface on which the moment stress acts, while the second subscript denotes the axis around which it rotates.

Accordingly, the constitutive equation for the equivalent couple stress continuum can be written in 3D matrix format as

$$\begin{aligned}[\mathbf{m}] &= \left[m_{xx}, m_{yy}, m_{zz}, m_{xy}, m_{yx}, m_{yz}, m_{zy}, m_{zx}, m_{xz}\right]^T \\ &= [\mathbf{D}]\left[\frac{\partial \phi_x}{\partial x}, \frac{\partial \phi_y}{\partial y}, \frac{\partial \phi_z}{\partial z}, \frac{\partial \phi_y}{\partial x}, \frac{\partial \phi_x}{\partial y}, \frac{\partial \phi_z}{\partial y}, \frac{\partial \phi_y}{\partial z}, \frac{\partial \phi_x}{\partial z}, \frac{\partial \phi_z}{\partial x}\right]^T\end{aligned} \qquad [8.35]$$

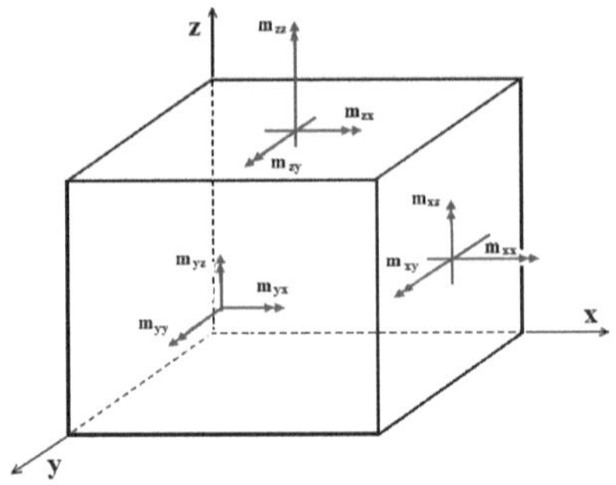

Figure 8.7. *3D representation of couple stress measures. For a color version of this figure, see www.iste.co.uk/ganghoffer/biomechanics.zip*

We can then build the corresponding compliance matrix from the previous expression

$$[\mathbf{C}] = [\mathbf{D}]^{-1} \qquad [8.36]$$

8.6. Plastic collapse and yield surface models under micropolar couple stresses

The plastic yield surface equations for three-dimensional cellular solid of vertebral trabecular bone subjected to combined actions of macroscopic bending moments and torques are derived based on asymptotic homogenization of moment equilibriums over the trabecular unit cell which in turn leads us to construct an equivalent micropolar couple-stress continuum medium.

8.6.1. *Formulation of micromoments within the trabecular cell*

In this section, the homogenization method is used to analytically determine the matrix relating the microscopic couple stress tensor to the apparent (effective) couple stress tensor. We first construct a vector of the moments which gathers all moments (bending and torsional) acting on each beam b of the reference unit cell considered as an idealized representation of the vertebral trabecular network.

Let $\left(M_x^{O(b)}, M_x^{E(b)}, M_{y,z}^{O(b)}, M_{y,z}^{E(b)}\right)$ denote the microscopic couple-stress components (or micromoments) of beam b, $b = 1,2,3,...$ Recall that $M_{x,y,z}^{O(b)}$ and

$M_{x,y,z}^{E(b)}$ are the micromoments with respect to x, y and z at the origin and beam extremity. Therefore, the vector of the micromoments is written next, replacing M_x^O and M_x^E with the more standard notation T^O and T^E

$$[M]=\begin{bmatrix} T^{O(1)} \\ T^{E(1)} \\ M_y^{O(1)} \\ M_y^{E(1)} \\ M_z^{O(1)} \\ M_z^{E(1)} \\ T^{O(2)} \\ \ldots \end{bmatrix} \qquad [8.37]$$

Previous expressions still involve the unknown displacements $\mathbf{u}^n$ and rotations ϕ_1^n, which are determined for all nodes by solving the equilibrium of forces and moments over the entire lattice. As a consequence, these expressions are functions of the macroscopic curvature tensor $[\chi]$; we can then construct a matrix $[Q_m]$ relating the moment vector on the unit cell to the homogenized curvature, so that

$$[M]=[Q_m][\boldsymbol{\chi}] \qquad [8.38]$$

In addition, we can relate the effective homogenized curvature and micropolar couple-stress tensors by the following relation $[\boldsymbol{\chi}]=[\mathbf{C}][\mathbf{m}]$, which in turn leads to the existence of a relationship between the couple stress at the micro (torsions and moments at each beam extremity) and macro levels

$$[M]=[Q_m][\mathbf{C}][\mathbf{m}] \qquad [8.39]$$

The previous expressions of the microscopic couple stresses are next involved to define the criteria for the onset of plastic yielding for vertebral bone under macroscopic torsion and moment loadings.

8.6.2. *Plastic yield criterion under a combined action of bending and torsion*

Among the key issues in bone biomechanics is the influence of the tissue-level ductility on the overall strength of trabecular bone. In the present work, we seek to analyze the structural behavior of trabecular bone failure by investigating its fully ductile behavior. We investigate the failed tissue amount and the impact of strength evaluated at the scale of individual trabeculae on the overall bone behavior, for a full plastic deformation of the ductile trabecular tissue. Note that the yield point corresponds in general to the moment stress at which a material begins to deform plastically, and the ultimate point corresponds to the highest point of the moment stress-curvature response; we shall focus on the latter in this contribution. For the entirely plastic case, we consider tissue level failure by fully plastic deformation, at which it reaches the fully plastic limit moments and/or torque yielding. Basically, once the ultimate yielding (plastic collapse) occurs, the structure cannot bear extra loads any more. It is demonstrated that after the onset of initial yielding, the unit cell of the bone structure is still capable of withstanding further loading, and continuously increasing the loads results in at least one more pair of struts also yielding; accordingly, the structural unit cell will become a mechanism and ultimately collapse. We here consider a uniform strut thickness of trabecular bone; hence, the plastic hinge occurs at a cross-section of maximum bending or twisting moment. Based on the plastic hinge concept, trabecular bone collapses plastically when the bending moment and/or twisting moment exerted by the external micropolar couple-stress loads on the individual trabeculae reaches the fully plastic moments and/or torque, creating plastic hinges at the corners.

The post-yield behavior of trabecular bone has been shown to exhibit remarkable similarities to that of open-celled plastic foams [CAR 76, HAY, 79, GIB 97]. In the case of wet trabecular bone at a lower slenderness ratio, the tissue fails in a fully ductile manner; it collapses plastically when loaded beyond the linear-elastic regime [BUR 75, GIB 85]. Yielding in the struts of such cellular materials type is concentrated in small zones. When a section becomes completely plastic, it is usually referred to as a formation of plastic hinges. For all vertebral trabecular bones used in this study, fully plastic collapse dominates, for which the structure reaches its ultimate load-carrying capacity, either in bending or in torsion.

In order to consider the interaction between bending and torsion, the following circular interaction plastic yield criterion that defines the failure surface is adopted

$$\left(\frac{M^b}{M_p}\right)^2 + \left(\frac{T^b}{T_p}\right)^2 = 1 \qquad [8.40]$$

with M^b the bending moment at the beam extremity nodes with respect to y or z axis, T^b the torque at the beam extremities, M_p the fully plastic bending moment of the cross-section (plastic moment capacity of a section subjected to bending alone), and T_P the fully plastic torque for pure torsion of the cross section (plastic torsional moment capacity of a section subjected to torsion alone).

Since bone trabeculae with circular cross-sectional area will be considered, further insight into the plastic behavior of a shaft or rod in torsion and bending shall be obtained by considering the idealized case of a solid circular shaft made of an elastoplastic material. As long as the shearing stress does not exceed the yield strength τ_Y, Hooke's law applies, and the stress distribution across the section is linear. As the torque increases, the shear stress eventually reaches the shear yield strength τ_Y, and we can then obtain the value of the torque at the onset of yield for solid circular section, as: $T_Y = \frac{\pi D^3}{16}\tau_Y$. This expression is referred to as the maximum elastic torque, since it is the largest torque for which the deformation remains fully elastic. As the torque is further increased, a plastic region develops in the shaft, around an elastic core; the plastic region further expands until, in the limit, the deformation is fully plastic: T_P is given by $T_P = \frac{4}{3}T_Y = \frac{\pi D^3}{12}\tau_Y$. Thus, only a one-third increase in torque is required after yielding to bring the rod to its ultimate load-carrying capacity. This value of the torque, which corresponds to a fully plastic deformation, is called the plastic torque (ultimate torque) of the considered solid circular rod. Figure 8.8 shows the behavior of T against the twist ϕ for an idealized elastoplastic material.

Let us denote ϕ_Y the angle of twist at the onset of yield: we can then express the torque T as a function of the twist angle ϕ,

$$T = \frac{4}{3}T_Y\left(1 - \frac{1}{4}\frac{\phi_Y^3}{\phi}\right)$$

where T_Y and ϕ_Y represent, respectively, the torque and the angle of twist at the onset of yield. Note that previous relation may be used only for values of ϕ larger than ϕ_Y. We note that, as ϕ increases indefinitely, T approaches the limiting value $T_P = \frac{4}{3}T_Y$ corresponding to the case of a fully developed plastic zone. However, the value T_P cannot actually be reached, but it is rapidly approached as ϕ increases.

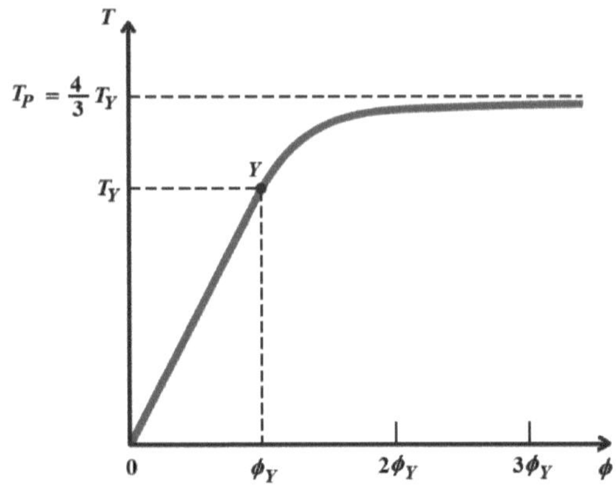

Figure 8.8. *Torque rotation relation for elastoplastic material. For a color version of this figure, see www.iste.co.uk/ganghoffer/biomechanics.zip*

In the same way, in order to gain a better insight into the plastic behavior of a member in bending, let us consider the case of a member made of an elastoplastic material and having a circular cross-section. As long as the normal stress does not exceed the tensile yield strength σ_Y, Hooke's law applies, and the stress distribution across the section is linear. As the bending moment increases, the normal stress eventually reaches the yield strength σ_Y, and the value of the bending moment at the onset of yield is given by [GER 04] as $M_Y = \frac{\pi D^3}{32}\sigma_Y$. The moment M_Y is referred to as the maximum elastic moment, since it is the largest moment for which the deformation remains fully elastic. As the bending moment further increases, plastic zones develop at the outside of the member with an elastic core subsisting between the plastic zones. If we consider the effects of increasingly larger moment, then the plastic moment $M_P = Z_\lambda \sigma_Y$ will be reached, at which point the member has yielded fully. This value of the bending moment, which corresponds to a fully plastic deformation, is called the plastic moment of the member. Note that Z_λ stands for the plastic section modulus of the trabeculae, adopted as $Z_{\lambda b} = D^3 / 6$ [GER 04].

8.6.3. *Determination of the plastic yield surfaces in couple stress space*

Although Young's modulus of the trabecular tissue is determined experimentally in many studies, its strength is rarely reported. In one of the few experimental studies in which trabecular tissue yield strength is measured using four-point bending tests, a static strength of about 160 MPa is reported [CHO 92]. In fact, there is also no experimental data available at tissue level for the shear yield strength of trabecular bone, perhaps due to the experimental and conceptual difficulties at this level. Since we do not have experimental data for the shear yield strength, we rely on shear stress failure theory, that is, shear yield strength τ_Y is equal to "yield strength / $\sqrt{3}$".

After substitution of the expressions of the full plastic bending and torsion (M_P, T_P) of the circular trabeculae section in equation [8.40] and adopting a shear yield strength of $\tau_Y = \sigma_Y / \sqrt{3}$ (according to the usual von-Mises yield criterion), the criterion that defines the failure surface for plastic yield in torsion and bending can be written as

$$\left(\frac{6M^b}{\sqrt{3}D_b^3\tau_Y}\right)^2 + \left(\frac{12T^b}{\pi D_b^3\tau_Y}\right)^2 = 1 \qquad [8.41]$$

with M^b the bending moments (M_y^b, M_z^b) and T^b the twisting moments (torques) within the trabecular struts, arising from the torsion couple stresses (m_{xx}, m_{yy}, m_{zz}) and bending couple stresses ($m_{xy}, m_{yx}, m_{yz}, m_{zy}, m_{zx}, m_{xz}$) externally applied on the unit cell.

Inserting the couple stress components of $[\mathrm{m}]$ after normalization by the initial shear yield strength of the trabeculae material τ_Y into equation [8.39] delivers the following relation

$$[M/\tau_Y] = \left[M^N\right] = \begin{bmatrix} T^{O(1)} \\ T^{E(1)} \\ M_y^{O(1)} \\ M_y^{E(1)} \\ M_z^{O(1)} \\ M_z^{E(1)} \\ T^{O(2)} \\ \ldots \end{bmatrix}^N = [Q_m][\mathbf{C}][\mathbf{m}/\tau_Y] = [Q_m][\mathbf{C}] \begin{bmatrix} m_{xx}/\tau_Y \\ m_{yy}/\tau_Y \\ m_{zz}/\tau_Y \\ m_{xy}/\tau_Y \\ m_{yx}/\tau_Y \\ m_{yz}/\tau_Y \\ m_{zy}/\tau_Y \\ m_{zx}/\tau_Y \\ m_{xz}/\tau_Y \end{bmatrix} \qquad [8.42]$$

As a final step, the system of previous equations in [8.41] then rewrites, for the set of struts b that belong to the reference elementary cell by which the whole vertebral bone structure is generated, as

$$\left(\frac{6M^{bN}}{\sqrt{3}D_b^3}\right)^2+\left(\frac{12T^{bN}}{\pi D_b^3}\right)^2=1 \qquad [8.43]$$

in which M^{bN} and T^{bN} are the normalized bending and torsion moments.

The plastic collapse strength domain of the lattice is then the surface delimited by the previous equality in the normalized couple stress space. As a consequence, the criterion of plastic collapse for hexagonal cellular geometry considered as a morphological model of vertebral trabecular is derived in the three dimensional couple stress states; this criterion incorporates twisting moment, in-plane and out-of-plane bending moment yielding stresses. Note that the normalized torques and bending moments, quantities T^{bN} and M^{bN}, are extracted from $\left[M^N\right]$ in equation [8.42].

It should be noted that the plastic yield strength of the periodic trabecular structure is based on the cell wall(s) reaching the fully plastic limit moments and/or torque of the cell walls in either bending or torsional couple stress. We develop a yield surface descriptive of the plastic collapse mechanism of vertebral bone in three dimensional couple stress space using a circular plastic yield criterion for the description of torsional/bending interaction. The plastic yield surface is the inner envelope of the intersecting surfaces for the plastic collapse mechanism; such a mechanism is characterized by a failure surface, namely a surface in couple stress-space describing the combination of bending and torsional couple stresses causing failure via plastic collapse.

Plastic failure models of three-dimensional reticulated porous materials of vertebral bone are next established, and the failure modes are further analyzed for the corresponding porous components under several typical loads, which include torsion and bendings in three-dimensional space. The failure modes cover the plastic failure due to pure twisting moment about x, y, and z and bending moments in xy, yx, xz, zx, yz and zy planes. The mathematical formulations, which characterize different plastic failure modes, have been derived for the trabecular strut failure resulting from loading for these porous components under the above loading conditions.

We are now in a position to determine the complete multiaxial bending and torsional yield envelopes for human trabecular bone. In the forthcoming section, the proposed plastic failure criterion is used to compute the plastic failure surfaces for trabecular bone under multiaxial couple stress states.

8.7. Numerical results

It is convenient to summarize the main steps of the computational procedure leading up to the construction of the plastic collapse surfaces for microstructures submitted to combined bending and torsion moments.

Steps for the determination of the plastic yield surfaces based on micropolar theory from homogenization of 3D lattice

1) Definition of the coordinates' transformation function $\mathbf{R}$ such that $x \xrightarrow{R} x(\lambda)$, with λ the curvilinear coordinates associated with the unit basis vectors.

2) For each beam in the elementary cell $b \in B_R$, define the following:

a) The origin and end node of the beam respectively $O(b)$, $E(b)$

b) The non-dimensional transverse shear factor $\Phi^b_{y,z} = 12E^b_s I^b_z / G^b_s A^b k_s \left(L^b\right)^2$

c) The displacement difference $\Delta\mathbf{U}^{b\varepsilon}$ between the extremity and origin node of each beam

$$\Delta\mathbf{U}^{b\varepsilon} = \mathbf{u}^{\varepsilon}(E(b)) - \mathbf{u}^{\varepsilon}(O(b)) = \varepsilon^2\left(\mathbf{u}^{E(b)}\left(\lambda^{\varepsilon}\right) - \mathbf{u}^{O(b)}\left(\lambda^{\varepsilon}\right)\right)$$

d) The nodal microrotations $\phi^{O(b)\varepsilon} = \varepsilon\phi_1^{O(b)}$, $\phi^{E(b)\varepsilon} = \varepsilon\left(\dfrac{\partial\phi}{\partial\lambda^i}\delta^{ib} + \phi_1^{E(b)}\right)$

e) The beam resultant force vector $\mathbf{F}^b$

f) The moments at the beam extremities $\mathbf{M}^{O(b)}$ and $\mathbf{M}^{E(b)}$

3) Solve the systems of equilibrium equations over the unit cell in rotation and translation for the kinematic unknowns $\mathrm{u}^n = \left(u_x^n \;\; u_y^n \;\; u_z^n\right)$, $\phi_1^n = \left(\phi_x^n, \phi_y^n, \phi_z^n\right)$; these kinematic variables are then expressed versus the micro-curvature tensor $[\chi]$.

4. Construct a vector gathering all twisting and bending moments for all beams of the unit cell, such that $[M] = \left[T^{O(1)} \;\; T^{E(1)} \;\; M_y^{O(1)} \;\; M_y^{E(1)} \;\; M_z^{O(1)} \;\; M_z^{E(1)} \;\; T^{O(2)} \;\; \ldots.\right]^T$, with T denoting the twisting moment, and M_y, M_z bending moments in y and z directions.

5) Construct the matrix $[Q_m]$ that relates the micromoments on the unit cell $[M]$ to the homogenized micro-curvature tensor $[\boldsymbol{\chi}]$ as $[M]=[Q_m][\boldsymbol{\chi}]$.

6) Construct the couple stress vectors $\boldsymbol{\mu}^i$

7) Construct the homogenized couple stress tensor $[\mathbf{m}]=\frac{1}{g}\boldsymbol{\mu}^i\otimes\frac{\partial\mathbf{R}}{\partial\lambda^i}$

8) Construct the constitutive equation for the homogenized couple stress continuum, such that $[\mathbf{m}]=[\mathbf{D}][\chi]$. In 3D format, it writes

$$[\mathbf{m}]=\left[m_{xx},m_{yy},m_{zz},m_{xy},m_{yx},m_{yz},m_{zy},m_{zx},m_{xz}\right]^T=[\mathbf{D}]\left[\frac{\partial\phi_x}{\partial x},\frac{\partial\phi_y}{\partial y},\frac{\partial\phi_z}{\partial z},\frac{\partial\phi_y}{\partial x},\frac{\partial\phi_x}{\partial y},\frac{\partial\phi_z}{\partial y},\frac{\partial\phi_y}{\partial z},\frac{\partial\phi_x}{\partial z},\frac{\partial\phi_z}{\partial x}\right]^T$$

with the homogenized stiffness matrix $[\mathbf{D}]$ given by

$$\begin{bmatrix} D_{11} & D_{12} & D_{13} & 0 & 0 & 0 & 0 & 0 & 0 \\ D_{21} & D_{22} & D_{23} & 0 & 0 & 0 & 0 & 0 & 0 \\ D_{31} & D_{32} & D_{33} & 0 & 0 & 0 & 0 & 0 & 0 \\ 0 & 0 & 0 & D_{44} & D_{45} & 0 & 0 & 0 & 0 \\ 0 & 0 & 0 & D_{54} & D_{55} & 0 & 0 & 0 & 0 \\ 0 & 0 & 0 & 0 & 0 & D_{66} & D_{67} & 0 & 0 \\ 0 & 0 & 0 & 0 & 0 & D_{76} & D_{77} & 0 & 0 \\ 0 & 0 & 0 & 0 & 0 & 0 & 0 & D_{88} & D_{89} \\ 0 & 0 & 0 & 0 & 0 & 0 & 0 & D_{98} & D_{99} \end{bmatrix}$$

9) Construct the compliance matrix $[\mathbf{C}]=[\mathbf{D}]^{-1}$

10) Construct a relation between the homogenized curvature tensor and macroscopic couple stress tensor as $[\boldsymbol{\chi}]=[\mathbf{C}][\mathbf{m}]$

11) Construct a relation between the microscopic moment vector and the macroscopic couple stress tensor, such that $[M]=[Q_m][\mathbf{C}][\mathbf{m}]$

12) Normalize the equation in step 11 by the initial shear yield strength of the beam material, so that

$[M^N]=[M/\tau_Y]=[Q_m][\mathbf{C}][\mathbf{m}/\tau_Y]$. It reads in explicit form

$$\left[M^N\right]=\left[M/\tau_Y\right]=\begin{bmatrix} T^{O(1)} \\ T^{E(1)} \\ M_y^{O(1)} \\ M_y^{E(1)} \\ M_z^{O(1)} \\ M_z^{E(1)} \\ T^{O(2)} \\ \ldots \end{bmatrix}^N =[Q_m][\mathbf{C}][\mathbf{m}/\tau_Y]=[Q_m][\mathbf{C}]\begin{bmatrix} m_{xx}/\tau_Y \\ m_{yy}/\tau_Y \\ m_{zz}/\tau_Y \\ m_{xy}/\tau_Y \\ m_{yx}/\tau_Y \\ m_{yz}/\tau_Y \\ m_{zy}/\tau_Y \\ m_{zx}/\tau_Y \\ m_{xz}/\tau_Y \end{bmatrix}$$

13. The failure surface for plastic yield in torsion and bending is defined by writing for each beam $b \in B_R$ in the unit cell by the following criterion

$$\left(\frac{M^b}{M_p}\right)^2+\left(\frac{T^b}{T_p}\right)^2=1 \Rightarrow \left(\frac{6M^{bN}}{\sqrt{3}D_b^3}\right)^2+\left(\frac{12T^{bN}}{\pi D_b^3}\right)^2=1$$

with M^{bN} and T^{bN} the normalized bending and torsion moments acting on each beam, M_p and T_p the fully plastic bending and twisting moments of the beam cross section. The failure envelope delineating the plastic yield is then the intersection of the surfaces of all beams defined by previous equality in the normalized couple stress space.

As a heterogeneous porous cellular solid, trabecular bone has anisotropic mechanical properties that depend on the porosity of the specimen as well as the architectural arrangement of the individual trabeculae. Its apparent (whole-specimen) level properties depend on the tissue-level material properties of the individual trabeculae. It is well-known that trabecular bone consists of an interconnected network of rods. The micro-architecture of trabecular bone makes it extremely difficult to characterize. Trabecular bone experiences complex multiaxial loads which can lead to entire bone fracture. The knowledge of the failure properties of trabeculae under multiaxial stress state is quite limited, especially under bending and twisting moment-induced stresses. The bending and torsional strengths measure the ability of trabecular bone to withstand bending and twisting loads; they represent the ultimate strength of a material subjected to bending or torsional loading, and are the maximum bending or torsional stresses that a material sustains before rupture. Analytical models of trabecular bone based on microstructural model can significantly enhance our understanding of multiaxial failure behavior under such loadings. Microstructural models relying on homogenization approaches serve to relate deformation and failure mechanisms at the trabecular level to the macroscopic material behaviors, and thus provide a framework to interpret and extend the existing experimental observations.

The overall collapse surfaces in the macroscopic couple stress space consist of intersecting collapse surfaces which are associated with particular collapse modes. In this section, we detail the calculations of the collapse surfaces for nine essential practical combinations of the macroscopic couple stress loadings. The plastic yield surfaces are the inner envelopes of the intersecting surfaces describing the combination of non-symmetrical micropolar couple stresses causing failures via plastic collapses.

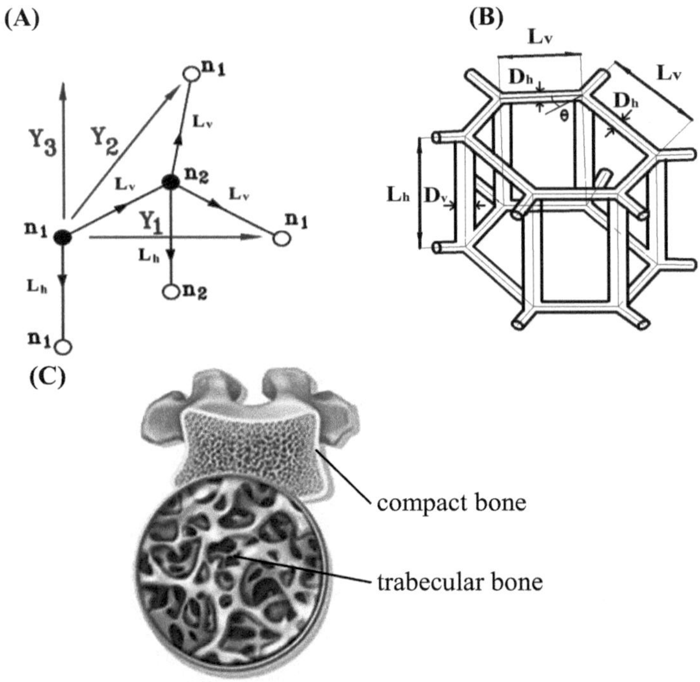

Figure 8.9. *Geometrical model of vertebral trabecular bone and its micro-architecture parameters. A) Reference unit cell and its periodicity vectors Y_1, Y_2 and Y_3. B) Three-dimensional hexagonal elementary cell with thick vertical trabeculae and thinner horizontal ones, rendering of human vertebral bone. C) View of human vertebral bone (trabecular and cortical). For a color version of this figure, see www.iste.co.uk/ganghoffer/biomechanics.zip*

At the microscopic scale, trabecular bone contains small struts connected in a way that it generates a three-dimensional porous structure. In the vertebral body, trabecular bone contains a somewhat regular array of trabecular struts (Figure 8.9 (C)), and thus can be idealized as a cellular solid. Therefore, we model the rod-like architecture of vertebral bone as a three-dimensional lattice with a unit cell geometry shown in Figure 8.9 (B). The whole structure is generated from the repetition of the reference unit cell (Figure 8.9 (A)) by means of three periodicity vectors defined in a Cartesian basis. The architecture of the model (thick vertical columns and thinner horizontal struts) is based on studies of samples taken from the central part of vertebral bodies from normal individuals aged from 30 to 90 years. Four morphological indices are used to describe the morphology of the unit cell: the horizontal trabecular thickness (D_h), the vertical trabecular thickness (D_V) and the mean distance between the horizontal trabeculae (L_h) as well as between the vertical trabeculae (L_v). The architecture of human trabecular bone and the changes in trabecular thickness and spacing that occur with aging are described by [MOS 88, MOS 89], using the relations in Table 8.1.

Parameter	Relation
Horizontal trabeculae thickness (μm)	$D_h = -1.03 \times \text{Age} + 189$
Vertical trabeculae thickness (μm)	$D_v = 0.14 \times \text{Age} + 208$
Distance between horizontal trabeculae (μm)	$L_h = 13.74 \times \text{Age} + 288$
Distance between vertical trabeculae (μm)	$L_v = 6.74 \times \text{Age} + 456$

Table 8.2. *Micro-architecture parameters for vertebral bone based on morphology–age relationships; age is measured in years*

The apparent density is generally the most important factor affecting the failure behavior of such cellular materials. The apparent density (ρ^*) is the density of a volume of the porous material, including voids, whereas the material density (ρ_s) is the density of the structural matrix itself. For the plastic analysis, there is a strong correlation between strength and bone relative density. The plastic strength of bone will be expressed later on in terms of the relative density to give insight on the bone relative density dependence. The vertebral plastic failures under couple stresses will be related to the relative density of vertebral trabecular bone expressed in terms of the trabecular bone architecture, itself changing with age. The relative density is then computed in terms of geometrical parameters of the vertebral bone unit cell as

$$\frac{\rho^*}{\rho_s} = \frac{1}{8} \frac{\pi\left(3L_v D_h^2 + 2L_h D_v^2\right)}{L_v^2 L_h \cos\theta\left(1+\sin\theta\right)} \qquad [8.44]$$

We next seek to define the plastic yielding under multiaxial couple state of stress for vertebral trabecular bones relying on the developments presented in the previous sections with the plasticity criterion defined by equality equation [8.43]. The micro-architecture parameters L_v, L_h, D_h and D_v of the trabecular are selected for a person at the age of 40, with an inclination angle of the oblique trabeculae $\theta = 60°$, hence a relative density of about 12% is received. The combination of the applied torsion and bending stresses which causes plastic collapse, plotted on couple stress axes m_{ij} (i,j = x, y, z), is a closed surface called the plastic yield surface for the vertebral trabecular bone, as illustrated in Figure 5. The failure envelopes for biaxial couple stress states are obtained as the intersection of the plastic yield surfaces of the struts forming the vertebral bone unit cell, in various stress sub-spaces, considering all couple stress components normalized by shear yield strength of the solid cell strut material. Plastic yield surfaces are plotted according to equations [8.43] in various couple-stress sub-spaces, with all twisting and bending moment stress components; (m_{ij}, $i = j$) corresponding to torsion, namely m_{xx}/τ_Y , m_{yy}/τ_Y , m_{zz}/τ_Y , and (m_{ij}, $i \neq j$) corresponding to bending, namely m_{xy}/τ_Y , m_{yx}/τ_Y m_{xz}/τ_Y , m_{zx}/τ_Y , m_{yz}/τ_Y , m_{zy}/τ_Y . They are normalized by τ_Y, the trabecular shear yield strength. In Figure 5(A), (B), (C), (D), (E), (F), (G), (H), (I), (J), (K) and (L), the plastic yield surfaces are pictured in the subspaces

$$\left(m_{xx}, m_{yy}\right)\Big|_{m_{zz}, m_{xy}, m_{yx}, m_{xz}, m_{zx}, m_{yz}, m_{zy}=0}, \left(m_{xx}, m_{zz}\right)\Big|_{m_{yy}, m_{xy}, m_{yx}, m_{xz}, m_{zx}, m_{yz}, m_{zy}=0},$$

$$\left(m_{yy}, m_{zz}\right)\Big|_{m_{xx}, m_{xy}, m_{yx}, m_{xz}, m_{zx}, m_{yz}, m_{zy}=0}, \left(m_{xx}, m_{xy}\right)\Big|_{m_{yy}, m_{zz}, m_{yx}, m_{xz}, m_{zx}, m_{yz}, m_{zy}=0},$$

$$\left(m_{xx}, m_{xz}\right)\Big|_{m_{yy}, m_{zz}, m_{xy}, m_{yx}, m_{zx}, m_{yz}, m_{zy}=0}, \left(m_{xy}, m_{yy}\right)\Big|_{m_{xx}, m_{zz}, m_{yx}, m_{xz}, m_{zx}, m_{yz}, m_{zy}=0}$$

$$\left(m_{xy}, m_{zz}\right)\Big|_{m_{xx}, m_{yy}, m_{yx}, m_{xz}, m_{zx}, m_{yz}, m_{zy}=0}, \left(m_{xy}, m_{xz}\right)\Big|_{m_{xx}, m_{yy}, m_{zz}, m_{yx}, m_{zx}, m_{yz}, m_{zy}=0},$$

$$\left(m_{xy}, m_{yz}\right)\Big|_{m_{xx}, m_{yy}, m_{zz}, m_{yx}, m_{xz}, m_{zx}, m_{zy}=0}, \left(m_{yx}, m_{zy}\right)\Big|_{m_{xx}, m_{yy}, m_{zz}, m_{xy}, m_{xz}, m_{zx}, m_{yz}=0},$$

$$\left(m_{xz}, m_{yz}\right)\Big|_{m_{xx}, m_{yy}, m_{zz}, m_{xy}, m_{yx}, m_{zx}, m_{zy}=0}, \text{ and } \left(m_{zx}, m_{zy}\right)\Big|_{m_{xx}, m_{yy}, m_{zz}, m_{xy}, m_{yx}, m_{xz}, m_{yz}=0}$$

respectively.

Since trabecular bone experiences highly complex multiaxial loads which can lead to entire bone fracture, the determination of its multiaxial failure behavior is of great clinical importance as it determines its fracture strength (namely the bone's ability to withstand multiaxial loadings). The determination of its plastic failure properties under combinations of macroscopic couple stresses including all twisting and bending moments is quite limited or has never been studied. The multiaxial ductile state failure behavior of trabecular bone in couple-stress space is important for modeling the entire bone fracture and can provide insight into structure–function relationships. There is currently no study on the multiaxial plastic yield criterion for trabecular bone under couple stresses. We thus derive three-dimensional states of couple-stress plastic failure for trabecular bone based on homogenization of bone elementary cells. Formulations of plastic failure criterion are plotted under biaxial loadings in alternative combinations of the macroscopic couple stress components (see Figure 8.10). From the results depicted in Figure 8.10, some key points shall be addressed: the plastic yield domains are closed, anisotropic, and consist of several convex curved surfaces, with some planar facets. All the plastic collapse envelopes are symmetrical with respect to positive and negative torsion and bending loadings. For twisting, the positive torsion moment indicates a clockwise rotation, and a negative torsion a counter clockwise rotation. In the case of bending, positive bending moment occurs when the deflected structure is concave upward and negative bending moment causes the structure to be concave downward.

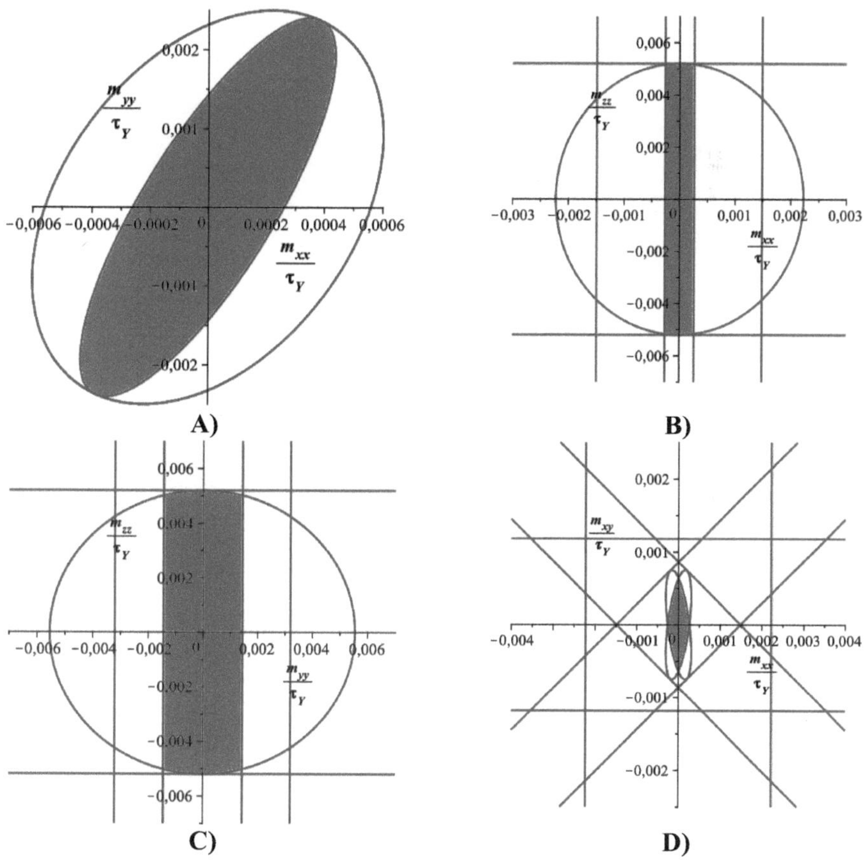

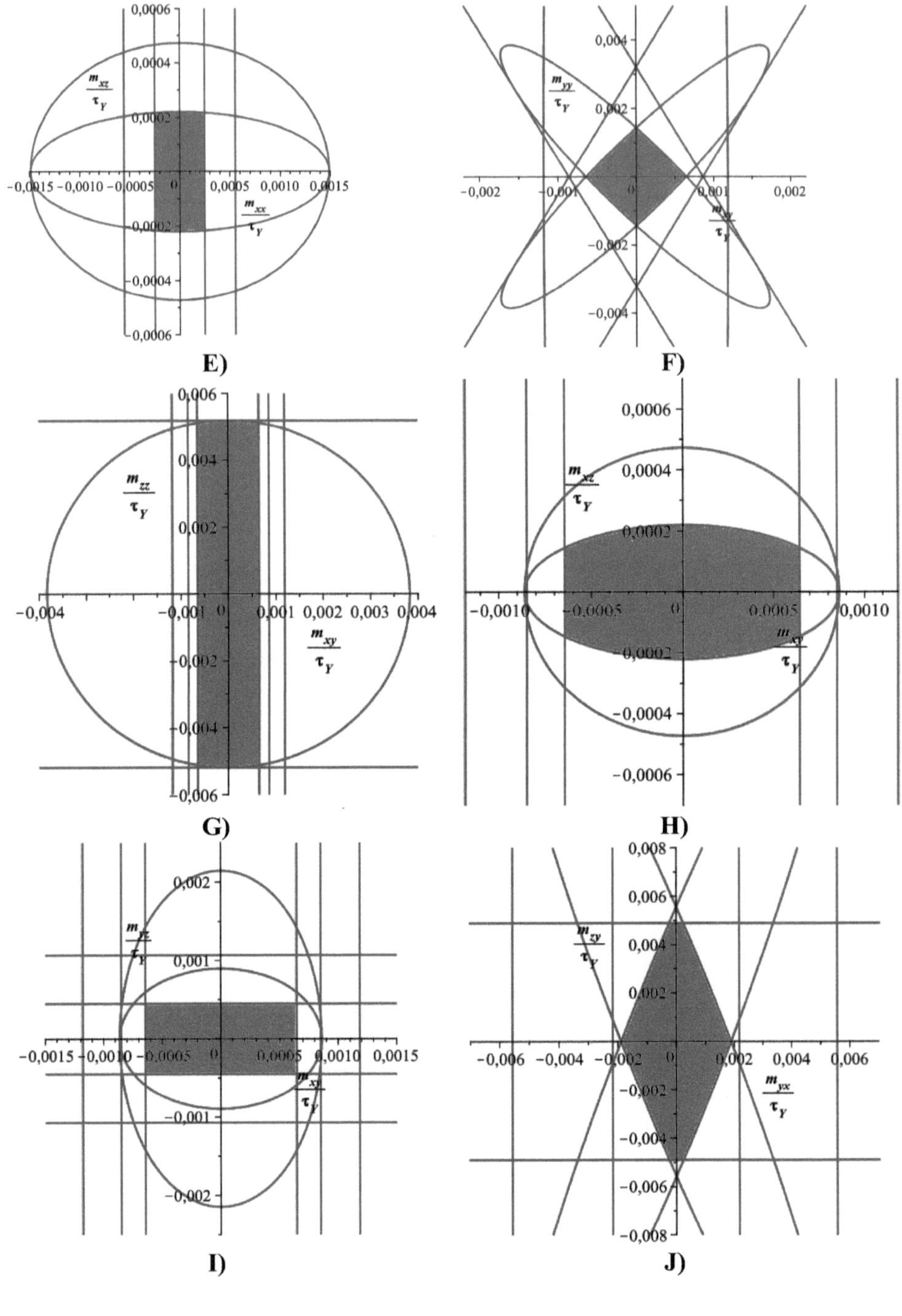

$\frac{m_{xz}}{\tau_Y}$
$\frac{m_{xx}}{\tau_Y}$
E)
$\frac{m_{yy}}{\tau_Y}$
$\frac{m_{xy}}{\tau_Y}$
F)
$\frac{m_{zz}}{\tau_Y}$
$\frac{m_{xy}}{\tau_Y}$
G)
$\frac{m_{xz}}{\tau_Y}$
$\frac{m_{xy}}{\tau_Y}$
H)
$\frac{m_{yz}}{\tau_Y}$
$\frac{m_{xy}}{\tau_Y}$
I)
$\frac{m_{zy}}{\tau_Y}$
$\frac{m_{yx}}{\tau_Y}$
J)

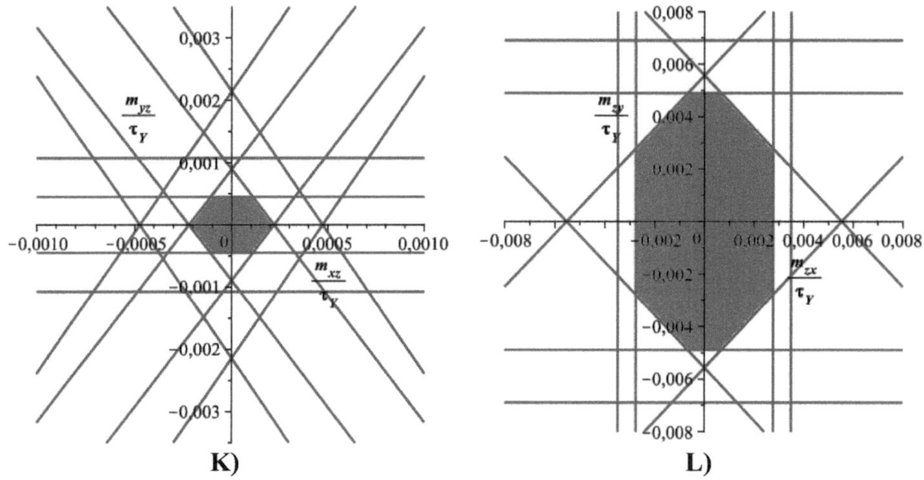

Figure 8.10. *Plastic collapse surfaces in the couple stress spaces of A) m_{xx} - m_{yy}, B) m_{xx} - m_{zz}, C) m_{yy} - m_{zz}, D) m_{xx} - m_{xy}, E) m_{xx} - m_{xz}, F) m_{xy} - m_{yy}, G) m_{xy} - m_{zz}, H) m_{xy} - m_{xz}, I) m_{xy} - m_{yz}, J) m_{yx} - m_{zy}, K) m_{xz} - m_{yz}, L) m_{zx} - $m_{z;}$; all the couple stresses are normalized with respect to the shear yield strength of bone bulk material. Age is 40 years with chosen relative density ρ^*/ρ_s = 0.12. For a color version of this figure, see www.iste.co.uk/ganghoffer/biomechanics.zip*

We shall further examine the separate effect of individually applying each of the macroscopic couple stress components m_{ij} on the periodic vertebral elementary cell, in order to determine the plastic collapse strength (i.e. the value of each of these couple stresses which produces full plasticity at any section); we summarize the plastic collapse criteria defined by the collapse criterion adopted for bone aged 40 years in Table 8.2.

When we come across the analysis of the previous results, we observe that when vertebral bone structure is subjected to the remote external twisting moment stress m_{xx}, plastic collapse is characterized by two modes: collapse of inclined and vertical trabecular struts, depending on which strut fails. Consistent with the plastic collapse criteria in Table 8.2, failure does not occur in the vertical strut, but in the slanted struts because the plastic collapse stress of the vertical struts exceeds the collapse stress of the inclined struts. The plastic collapse modes under twisting moment stress m_{yy} are characterized by three states (collapse of the in-plane vertical, inclined, and vertical struts), depending on which strut fails. Relying on Table 8.2, we find that the plastic twisting moment m_{yy} causes collapse of vertebral bone of inclined struts. Under the in-plane bending couple stresses m_{xy} and m_{yx}, there will be also three collapse modes: the plastic collapse occurs in the in-plane vertical, inclined, and vertical struts, also depending on which strut fails. The results show that failure of the inclined struts occurs first, based on the expressions of the plastic collapse criterion written previously.

Category	Plastic yield strength
In-plane vertical struts	$718\left(\frac{m_{yy}}{\tau_Y}\right)^2=1;\ 713704\left(\frac{m_{xy}}{\tau_Y}\right)^2=1;\ 212415\left(\frac{m_{yx}}{\tau_Y}\right)^2=1;$ $130598\left(\frac{m_{zx}}{\tau_Y}\right)^2=1;\ 5003672\left(\frac{m_{yz}}{\tau_Y}\right)^2=1;$
Inclined struts	$14636556\left(\frac{m_{xx}}{\tau_Y}\right)^2=1;\ 485487\left(\frac{m_{yy}}{\tau_Y}\right)^2=1;\ 2412703\left(\frac{m_{xy}}{\tau_Y}\right)^2=1;$ $279501\left(\frac{m_{yx}}{\tau_Y}\right)^2=1;\ 20675986\left(\frac{m_{xz}}{\tau_Y}\right)^2=1;\ 32649.7\left(\frac{m_{zx}}{\tau_Y}\right)^2=1;$ $1250918\left(\frac{m_{yz}}{\tau_Y}\right)^2=1;\ 32226.5\left(\frac{m_{zy}}{\tau_Y}\right)^2=1;$
Vertical struts	$201924\left(\frac{m_{xx}}{\tau_Y}\right)^2=1;\ 32487\left(\frac{m_{yy}}{\tau_Y}\right)^2=1;\ 37069.7\left(\frac{m_{zz}}{\tau_Y}\right)^2=1;$ $68191.8\left(\frac{m_{xy}}{\tau_Y}\right)^2=1;\ 55529.8\left(\frac{m_{yx}}{\tau_Y}\right)^2=1;\ 81976.8\left(\frac{m_{zx}}{\tau_Y}\right)^2=1;$ $41757\left(\frac{m_{zy}}{\tau_Y}\right)^2=1;$

Table 8.3. *Fully plastic collapse criteria for 3D vertebral trabecular bone (aged 40 years) unit cell*

Considering next the effect of an applied macroscopic out-of-plane twisting moment stress m_{zz}, we find that the out-of-plane twisting m_{zz} produces a torque (no bending moment) on the vertical struts in the periodic unit cell, so when the torque of the vertical struts lies above the plastic torsional moment capacity of the strut cross section, the vertical struts will yield torsionally under m_{zz}. Furthermore, the out-of-plane bending couple stress m_{xz} applied to the structure causes the collapse of the inclined struts only, with no effect on other struts.

The plastic collapses under out-of-plane bending couple stress m_{yz} are characterized by two collapse modes (in-plane vertical and inclined) depending on which strut fails: according to Table 8.2, failure does not occur in the inclined strut, but in the in-plane vertical ones, because the collapse stress of inclined struts exceeds the collapse stress of the in-plane vertical struts. The plastic collapses under out-of-plane bending couple stress m_{zx} are characterized by three states including collapses in inclined, in-plane vertical and vertical struts, depending on which strut fails. The results highlight that the in-plane vertical struts fail first, followed sequentially by the vertical and inclined struts.

Lastly, under bending couple stress m_{zy}, there will be two collapse modes: the plastic collapses in the inclined and in the vertical struts, also depending on which strut fails. The results emphasize that the vertical struts fail first, followed by the inclined ones.

Since the initial yielding is generally premonitory of structural collapse, understanding of such trabecular lattice yielding behavior is an essential aspect. We therefore seek to calculate the initial yield surfaces together with plastic collapse surfaces for vertebral bone structure. The initial yield surface gives the couple-stress range where all the material points stay in the elastic regime, whereas the plastic collapse surface gives the couple stress range before total collapse of the structure occurs. The initial yield surfaces of vertebral bone in three-dimensional couple-stress space are constructed using the yield criterion in equation [8.40] for the description of the initial yielding for all beams under macroscopic torsional and bending couple stresses. The initial yield of the periodic trabecular is based on the cell wall(s) reaching the maximum elastic moment (M_Y) and/or torque (T_Y) of the cell walls. Note that, while we compute the initial yield surfaces, the fully plastic moment, M_P and torque, T_P are replaced by the maximum elastic moment, M_Y and torque, T_Y, respectively. Initial yield surfaces (blue lines) are plotted together with plastic collapse surfaces (red lines) in various couple-stress sub-spaces: (m_{xx}, m_{yy}), (m_{yy}, m_{zz}), (m_{xy}, m_{yy}) and (m_{zx}, m_{zy}), respectively (Figure 8.11). The initial yield surface is the inner envelope of the intersecting surfaces (blue) in couple stress-space, describing the combination of bending and torsional couple-stresses causing onset of yield.

In addition, for the plastic collapse strength there is a strong correlation between the strength of vertebral trabecular bone, bone age and its relative density as well. In order to provide an insight into the age dependence of the couple-stress characteristics, for the representative vertebral trabecular lattice, we derive the effective torsional and bending plastic yield strengths normalized by the initial trabecula shear strength as a function of the bone age or bone relative density: Figures 8.18 and 8.19 show the plastic yield strengths as a function of bone age and/or relative density. These results are obtained by applying nine loading conditions (three twisting moments in Figure 8.12, and six bending moments which include in-plane and out-of-plane bending moment loadings in Figure 8.13) to determine the effective plastic yield strength. It is obvious from the plots that the effective torsion and bending plastic yield strengths decrease with age and increase with the relative density of the vertebral trabecular lattice.

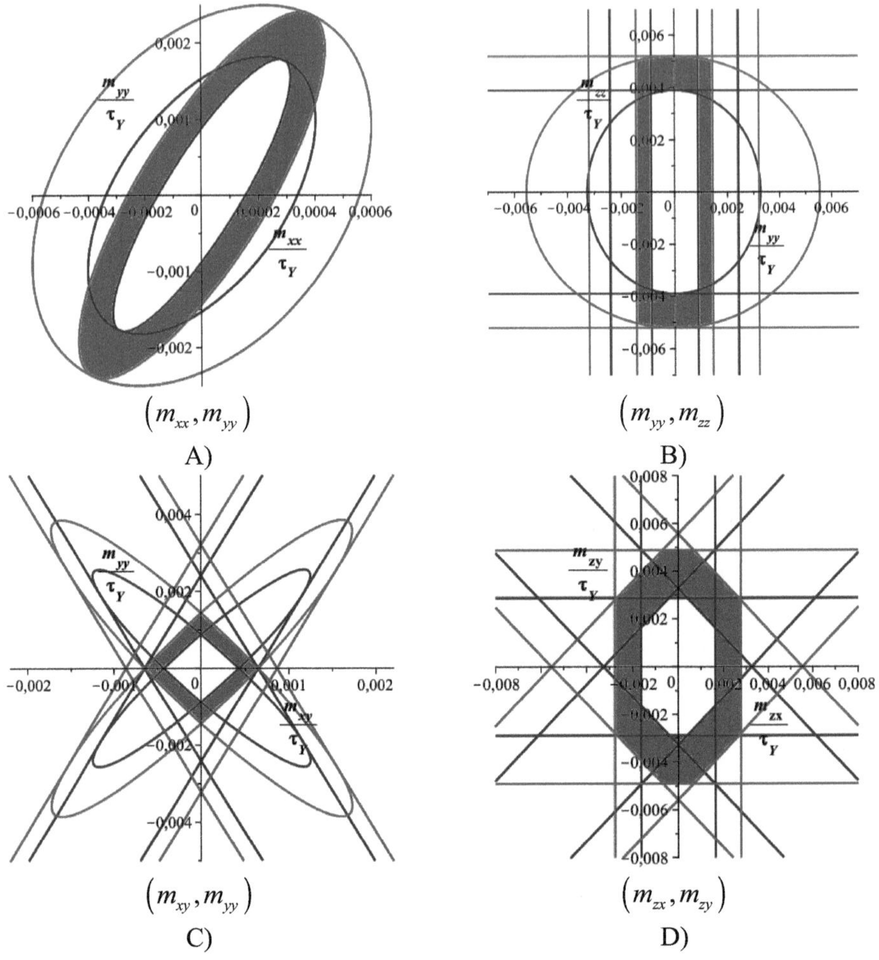

Figure 8.11. *Initial and plastic yield surfaces of vertebral trabecular bone under combinations of twisting and bending couple stresses. Age is 40 years. The blue lines denote the initial yield surfaces, while red lines denote plastic collapse surfaces. For a color version of this figure, see www.iste.co.uk/ganghoffer/biomechanics.zip*

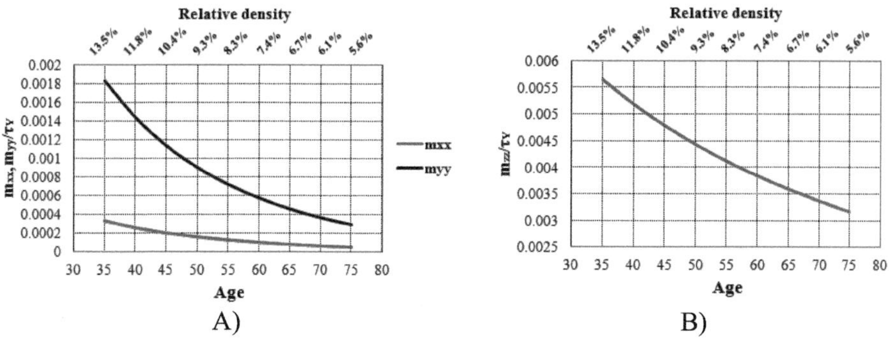

Figure 8.12. *Plastic torsional couple stresses versus vertebral bone age or bone relative density. For a color version of this figure, see www.iste.co.uk/ganghoffer/biomechanics.zip*

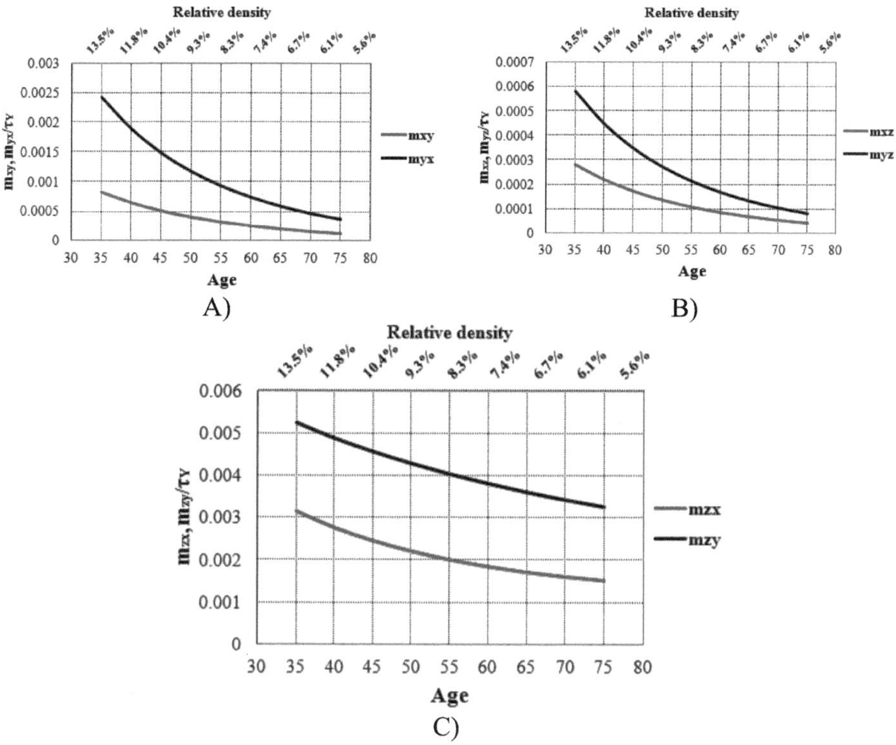

Figure 8.13. *Plastic bending couple stresses versus vertebral bone age or bone relative density. For a color version of this figure, see www.iste.co.uk/ganghoffer/biomechanics.zip*

The characterization of trabecular bone tissue material and the determination of its failure characteristics and role in trabecular bone mechanical behavior is of significant clinical importance. In this regard, we have established a complete strength criterion under multiaxial loads for human trabecular bone based on relationships between the morphology and multiaxial failure properties of bone.

The present work could investigate large deformation aspects of the bone structure, exploiting more detailed models for beams in large deformations [GRE 15, GAB 14, GAB 15], especially in order to build hyperelastic models. Since bone properties overall may degrade due to the development of either plastic flow or brittle damage, this last mechanism shall be given some consideration in future works, relying for instance on the methods exposed in [CIA 07].

A potentially important development of the results presented in the present chapter could be the application of the proposed approach to bone biosubstitutes used for bone reconstruction [GIO 15, GIO 16], including microstructure design to get tailored mechanical properties.

9

Multiscale Aspects of Bone Internal and External Remodeling

Models for the internal and external bone remodeling are developed in this chapter in the framework of the thermodynamics of irreversible processes. Internal remodeling is essentially accounted for by an evolution of the internal bone density versus the trace of the Eshelby stress. The growing surface is endowed by a specific mechanical behavior elaborated from a surface potential, depending upon the elastic surface strain and the external normal. The surface remodeling velocity is related to the driving force for growth identified as the surface divergence of Eshelby stress. The developed formalism is applied to simulate bone internal and external remodeling for 2D geometries, showing the influence of external mechanical stimuli on the evolution of the external shape of bone. Simulations of the evolution of the shape and density of bone structures are carried out successively at mesoscopic, microscopic and macroscopic levels, the former representing cellular level simulations of the trabecular unit cell. At the macroscopic level, a linear elastic constitutive law for trabecular bone is derived, relying on the discrete homogenization approach, whereby the continuum parameters such as stiffness evolve with morphology as remodeling occurs within bone. The developed numerical platform is able to simulate combined bone internal and external remodeling at both trabecular and macroscopic scales.

9.1. Introduction: biological mechanics of bone remodeling

Bone tissue is an adaptable living structure, having the capacity to adapt its internal structure and external shape to variations in its mechanical environment. Bone tissue adapts its density and hence its mechanical properties in such a way that

Chapter written by Jean-François GANGHOFFER and Ibrahim GODA.

its structure is optimized in some sense. It aims at preserving the mechanical properties and adapting its structure in response to the mechanical and physiological environments it experiences. This adaptive capacity of bone tissue to the mechanical forces and their history is called bone remodeling, with a classification as either internal or external remodeling [COW 79]. External or surface remodeling results in a change of the external shape of the overall bone structure, and occurs by the resorption or apposition of bone material on the surfaces. On the contrary, internal remodeling refers to the resorption or deposition of bone material only, accompanied by the removal and densification of the architecture of cancellous bone.

Obviously, since these processes have a tremendous impact on the overall behavior and health of the entire body, bone remodeling simulations are of great importance. The need for bone-remodeling models is especially marked in applications dealing with bone adaptivity, such as bone implants and scaffold design (see [SAN 09, FER 02]), and also to predict the outcome of dental or orthodontic treatment (see [FIE 12, WAN 12]). Bone remodeling plays an important role in the processes of formation, maintenance and repair of bone tissues. It largely determines the internal microstructure of the bone and has an essential influence on its strength and mechanical properties.

9.1.1. *Bone internal and external remodeling*

Bone is a dynamic living tissue, with the ability to modify its internal architecture and external shape in response to its environment in order to maintain its structural integrity as the support of the body and its physiological functions, especially as a reservoir of nutrients [PAA 94, SOM 01].

In adult bones, the skeletal development is the result of growth, modeling and remodeling processes. Growth means the change of overall bone mass and is due to both genetic and epigenetic factors, including mechanical physiological loads supported by the body. Modeling occurs in a concomitant manner to growth, as it modulates the bone architecture during growth in response to a variation of the mechanical conditions. It accordingly controls the shape, size and strength of bones by the addition or resorption of bone material at the surfaces of trabeculae. Remodeling involves the replacement of old bone by new lamellar bone by resorption processes; bone remodeling as an integrated process incorporates the mechanisms through which living bone adapts its internal structure to changes of the external load. A golden standard for bone remodeling is the BMU, an acronym for the discrete packets of Bone Remodeling Units or Bone Multicellular Units [FRO 01].

Both cortical and trabecular bones undergo remodeling, the difference between them being in the geometrical progression: the BMU has the ability to move along the surface of the trabeculae, whereas it penetrates the compact cortical bone by digging a tunnel. The temporal sequence of remodeling steps on a trabeculum is illustrated in Figure 9.1.

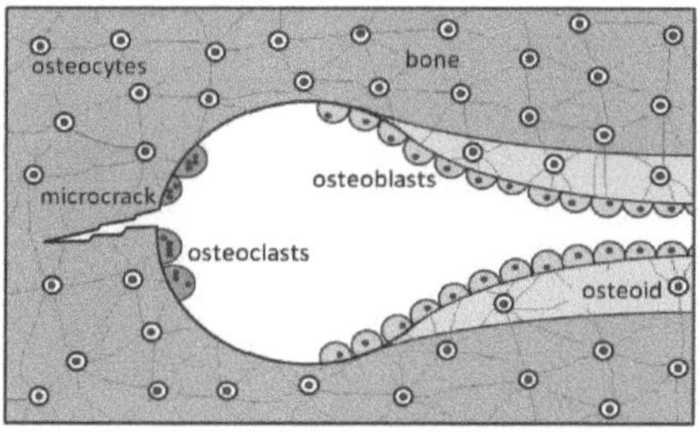

Figure 9.1. *Schematic view of a BMU moving along a microcrack within trabecular bone. Osteoclasts resorb bone in the form of a cutting cone. Osteoblasts subsequently fill the resorbed space with osteoid (the new bone matrix), and differentiate into sensing cells called osteocytes [GAN 16]*

Bone remodeling plays an important role in the processes of formation, maintenance and repair of bone tissues; it aims at shaping the skeleton and repairing bone fractures by adapting the microstructure in response to mechanical and physiological stimuli: following a phase of growth and development, bone undergoes a continuous cycle of repair, renewal and optimization in which mechanical stimuli play a fundamental role. Bone remodeling can also be perturbed in a pathological manner, due to age-related factors, changes in physical activity, drugs, secondary diseases, menopause-associated hormonal changes, all leading to an unbalance of the extracellular matrix remodeling and local angionesis.

Bone undergoes a continuous and repetitive remodeling cycle involving the sequence of activation, resorption of old bone (modeling), inversion and formation of new bone (remodeling) by specialized cells (osteoclasts/osteoblasts/osteocytes), as illustrated on the synthetic Figure 9.2. In the remodeling process, specialized cells called osteoclasts resorb bone tissue, whereas other cells called osteoblasts deposit new bone tissue. When osteoblasts become trapped within the bone matrix they secrete, they differentiate into osteocytes, playing the role of load sensing cells. Osteoblasts increase bone mass under high loading conditions, by producing hydroxyapatite, the mineral forming the new bone, together with collagen I, an organic matrix. Bone tissue is removed by osteoclasts under low loading conditions; the overall cell activity aims to reach an optimized bone architecture. The interaction between the activity of osteoblasts and osteoclasts is responsible for bone remodeling at the level of the BMU.

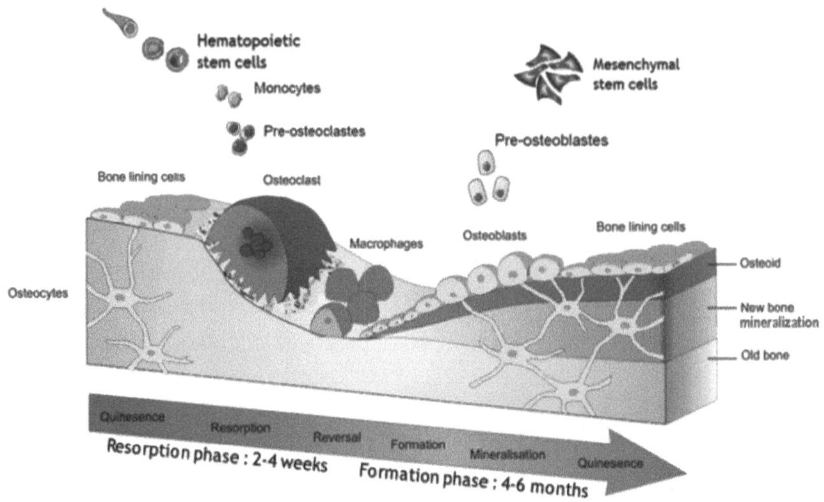

Figure 9.2. *Bone remodeling cycle (from Biomedical Tissue Research, University of York. www.york.ac.uk/res/bonefromblood; figure completed by the delivery of cells). For a color version of this figure, see www.iste.co.uk/ganghoffer/biomechanics.zip*

Two types of bone functional adaptation may then occur simultaneously: internal adaptation and external adaptation. Most of the efforts have been paid to internal bone adaptation in the literature: indeed, numerous models are developed to simulate bone density evolution, where various mechanical stimuli have been proposed as triggers for bone adaptation, including stresses [BEA 90, VAN 93, CAR 96, DOB 02, GAR 14], strains [COW 76, FER 99], strain rate [NEG 07], strain energy density [WEI 92, HUI 00, KUH 03, FER 10, KAC 11], and tissue damage [MAC 92, PRE 94, MAR 95, LEV 98, RAM 99, RAM 01]. Conversely, only a few models are developed for external adaptation [HUI 87, HUI 00, COW 81]. It is clear that most of the efforts have been driven to study internal remodeling, and fewer one to external remodeling. This difference of interest maybe due to the external remodeling (change of shape) occurring more slowly and for being less important than internal remodeling [FRI 88]; this is true especially for adult persons. However, external bone remodeling is significant in the bone healing process, and it is essential to predict the bone external shape in the design of artificial bone grafts [CHE 07a].

Bone undergoes morphological changes in response to applied mechanical loads by modifying its external geometry as well as its internal structure, i.e. external and internal remodeling occurs simultaneously. Remodeling processes take place in order to adjust the amount of tissue and its geometrical shape.

Incorporating in a coupled approach both external and internal remodeling into an integrated model including both processes is a challenging task we propose to address in the present contribution. We accordingly propose a two-dimensional mathematical model of trabecular bone remodeling at both microscopic and macroscopic scales that simulates the surface-based addition or removal of bone tissue material, combined with bone density alterations induced by the mechanical loads it experiences.

We call internal remodeling the reorganization of trabecular bone resulting in changes of bone density, while external remodeling refers to the processes occurring at the external surface of bone resulting in geometrical changes. Total remodeling denotes both processes; in addition to the redistribution of bone mineral content, the external remodeling or the change in shape (surface remodeling) is of great interest in order to understand the adjustment of bone to a new mechanical environment. Therefore, the objective of the present study is to develop a numerical model capable of simulating both internal and external bone adaptation.

The function of internal remodeling is the renewal of the tissue with alteration of the bone mechanical properties either at the strut level or at macroscopic level. Bone formation occurs in overloaded conditions, in order to reinforce bone where needed, while resorption occurs when disuse is observed. This resorption tends to reduce the amount of bone at locations where it is not needed from a mechanical viewpoint. Bone therefore adapts its density in such a way as to attain a homeostatic state of a given mechanical stimulus. Besides density changes, remodeling also occurs to modify bone topology, mainly in trabecular tissue, since the trabeculae tend to align along the load directions.

External bone adaptation at macro and micro-levels is defined as the bone capacity to form or destroy bone surface in response to mechanical loadings. Moreover, throughout this study, the external geometries of the proximal femur and trabecular strut are considered to be variable, so that changes in the external shape are allowed during bone local adaptation (surfaces move freely). As we shall demonstrate later, the divergence and trace of Eshelby stress are adopted as feedback control variables to determine the shape and bone density adaptations. In the case of femur bone remodeling, both types of bone remodeling (internal and external) occur simultaneously, but the most important mechanism which determines the outer geometry and the rate of boundary movement is external bone remodeling. Therefore, in order to determine the rate of bone remodeling, external bone remodeling is mathematically modeled and simulated jointly with internal remodeling.

The integrity and strength of bone microstructure is determined by the balance between the volume of resorbed bone and that of newly formed bone

[CHR 01, SEE 06]. One can classify the models developed in the literature aiming at simulating the functional adaptation of bone into three main classes: optimization theories considering bone as a mechanical structure undergoing an evolutionary process, phenomenological models and mechanistic models, some of which in this third category incorporating the chemical and biological processes responsible for the modification of bone architecture. More details relative to each of these classes of models is given in [GAN 11].

The need for bone-remodeling models and simulations is especially marked in applications dealing with bone adaptivity, such as bone implants and scaffold design [SAN 09]. This contribution focuses on the bone internal remodeling processes occurring at the scale of individual trabeculae, in order to refine the more external remodeling approaches [GAN 10, GAN 14, GAN 16].

As proposed in [HAM 14], remodeling models can be classified into several categories: (i) models based on the global optimality criterion [HOL 94, BAG 00, TOV 06, JAN 08, JAN 10]; (ii) models relying on maintaining a homeostatic state of mechanical energy [CAR 86, ADA 10] and references therein; (iii) models based on damage [PRE 94, NAM 07] and references therein; (iv) models including both mechanical and metabolic factors in the remodeling cycle [HER 00, AZN 05] (and references therein); (v) one model considering the interstitial fluid flow [TSU 09]; (vi) and models based on a reaction–diffusion and including mechanical stress but without biophysical activities of osteoblasts and osteoclasts [MAT 03, TEZ 05]. To complete the picture, one shall also consider mathematical models describing the activity of BMUs and based on the work of [KOM 03] focusing on the non-linear autoregulation between osteoblasts and osteoclasts [PIV 08] and references therein. The interactions with biological factors highlighted by [PIV 08] were completed by [BON 11] who considered the influence of mechanical stimulus effects on cell dynamics. More recently, [HAM 14] proposed an approach which combined the work of [KOM 03] and [BON 11] to include fatigue damage growth and repair, mineralization, porosity, bone properties evolution and cellular accommodation during the load history.

9.1.2. *Bone remodeling cycle*

Following a phase of growth and development, bone undergoes a continuous cycle of repair, renewal and optimization in which mechanical stimuli play a fundamental role. The various stages of remodeling are summarized in Figure 9.1, showing the corresponding activity of specialized cells. The first stage of the cycle is activation: under the action of repeated stimuli, such as strain energy, stress concentrations, high strain levels, and existing microcracks, mechanical signals are detected by the osteocytes forming a network within the available space between successive bone lamellae. The osteocytes then send a signal to the osteoclasts, which then erode the matrix to resorb old bone. This is followed by the formation of osteoid, a non-mineralized organic portion of the bone matrix forming prior to the maturation of bone tissue. Mineralization resulting from the activity of osteoblasts then proceeds by the synthesis of an extracellular matrix, composed of 90% collagen fibers, 10% proteoglycans and non-collagen proteins involved in the regulation of mineralization. New bone consists of hydroxyapatite, the mineral phase, the dissolution of which obeys the following chemical reaction (bone formation corresponds to the inverse reaction), which condenses a more complicated set of (up to 14) chemical reactions

$$[Ca_3(Po_4)_2]_3Ca(OH)_2 + 8H^+ \rightarrow 10Ca^{2+} + 6HPO_4^{2-} + 2H_20$$

Because of the complexity of bone remodeling mechanisms, we shall consider a small representative volume of material from the standpoint of physicochemical / biochemical mechanisms and mechanical phenomena, as illustrated in the simplified view depicted in Figure 9.3. In the present work, we shall consider mechanical factors as the main stimuli triggering bone remodeling: due to the formation and development of microcracks under mechanical stresses, strain energy builds up, leading to stress concentration around the crack tips (Figure 9.3). The existing bone needs first to be eroded to remove the stress concentration zones (modeling process), before new bone can be formed to shape the eroded trabeculae close to their original trabecular shape (remodeling process).

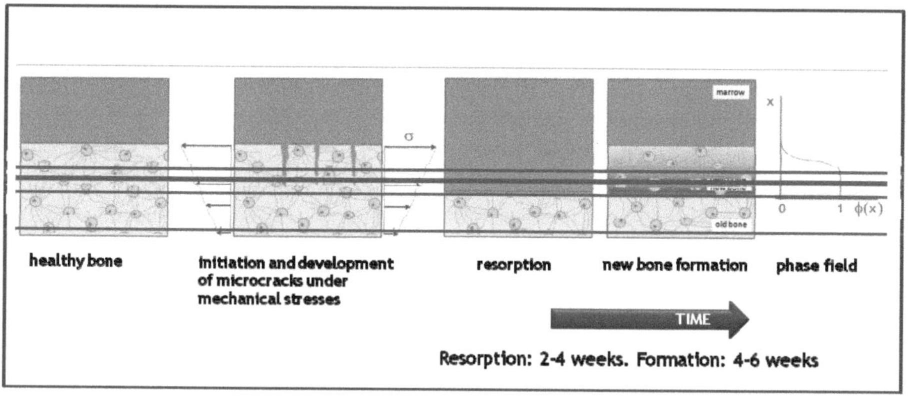

Figure 9.3. *Schematic remodeling cycle including 4 main step: healthy bone, initiation and development of microcracks under mechanical stresses, resorption, and new bone formation. The evolution of the phase order parameter is represented in 1D as a function of variable x [GAN 16]. For a color version of this figure, see www.iste.co.uk/ganghoffer/biomechanics.zip*

9.1.3. *Connection between cell activity and bone production*

Osteocytes are cells which form an interconnecting network in the bone tissue, and differentiate into osteoblasts. The location and propagation of information signaling the mechanical damage and stress concentrations take place through this network. Osteoclasts are derived from the differentiation of stem cells in the bone marrow and can live for about ten days. They resorb mineralized bone tissue while moving on the surface of the bone matrix. A key stimulator for osteoclast differentiation and activation is a molecule called RANK. Osteoblasts are also differentiated from stem cells in the bone marrow, and express the RANKL messenger molecule and its decay receptor OPG (osteoprotegerin). There are many biological factors involved in the regulation of the bone remodeling process [9, 10], including autocrine and paracrine signaling molecules, systemic hormones and extracellular matrix components that affect cell to-cell communication, migration, adhesion, proliferation and differentiation. The interactions between osteoclasts and osteoblasts are very complex, since osteoblasts can regulate the activity of osteoblasts by specialized molecules called RANK and RANKL, see [11] for a review.

Once osteoclasts have resorbed bone, they recruit osteoblasts which fulfill the previously resorbed cavity to form osteoid, organic part of the bone tissue. The bone formation takes about 10 times longer than resorption, and the final phase corresponds to the mineralization of the matrix. Then, osteoblasts differentiate into osteocytes after two weeks and are inserted into the newly formed bone or die.

Coordination of osteoclasts, osteoblasts and osteocytes in a BMU (Figure 9.4) is performed through a complex exchange of information which passes through autocrine signaling pathways (between cells of the same type) and paracrine (between cells of different types) and various couplings. Among the messengers involved, RANKL and OPG play a critical role both in the physiological bone remodeling and in the development of diseases. RANKL has a stimulating effect on the differentiation of osteoclast precursors and subsequent activation of mature osteoclasts in resorption of the active cells. OPG molecule produced by mature osteoblasts, acts as a decoy receptor for RANKL, inhibiting the RANK-RANKL bond formation and thereby the stimulation of osteoclasts. A high ratio of RANKL/OPG promotes bone resorption phenomenon while a low ratio leads to decrease osteoclast activity. On the other hand, it seems that canalicular fluid flow is responsible for osteoclast activity and their ability to move in the BMU.

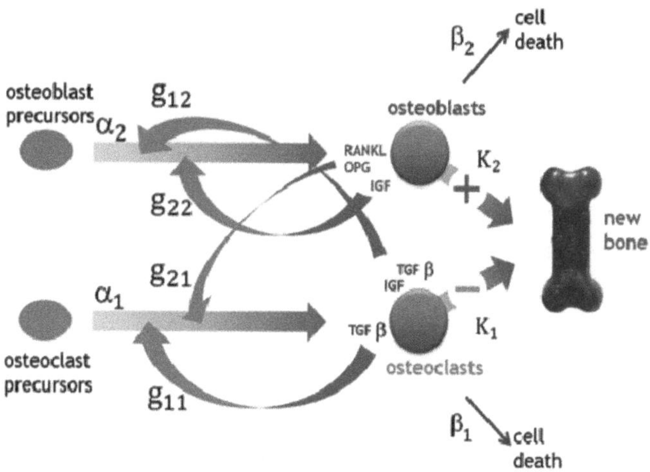

Figure 9.4. *Dynamical system illustrating cell activity and the different interactions leading to the formation of new bone, or bone resorption [KOM 03]. For a color version of this figure, see www.iste.co.uk/ganghoffer/biomechanics.zip*

As underlined in [HAM 14], predictive models of bone remodeling generally ignore the coupled activity of the specialized cells (osteoclasts, osteoblasts and osteocytes), which combines with the mechanical response of bone. The cell activity involves many interactions between families of cells and their precursors, as illustrated in Figure 9.4. We shall here consider the well-known model of [KOM 03] which synthetically describes these phenomena. Here, we remember that the osteocytes sense the mechanical stimulus and transmit the information to the other cells located in their vicinity. The scheme adopted by previous authors describes the effects of osteoblasts and osteoclasts precursors on the mature cells and includes the feedback effect and the interaction between osteoblasts producing material and osteoclasts responsible for resorption. The various produced biochemical factors (RANKL, OPG, IGF and IGF TGF) are at the origin of these effects. The corresponding dynamic system of equations writes

$$\dot{n}_{oc} = \alpha_1 \, n_{oc}^{g_{11}} \, n_{ob}^{g_{21}} - \beta_1 \, n_{oc} \quad [9.1]$$

$$\dot{n}_{ob} = \alpha_2 \, n_{oc}^{g_{12}} \, n_{ob}^{g_{22}} - \beta_2 \, n_{ob} \quad [9.2]$$

where $n_{ob}(t)$ and $n_{oc}(t)$ represent the total numbers of osteoblasts and osteoclasts, respectively (they reach equilibrium values denoted (n_{oc}^{eq} and n_{ob}^{eq}), and the exponents $g_{11}, g_{12}, g_{21}, g_{22}$ traduce the transduction phase by the osteocytes, which sense mechanical signals and activate osteoblasts and osteoclasts in response of it.

Previous system describes the temporal changes of the osteoclasts and osteoblasts populations including the autocrine and paracrine regulations between both populations. The 8 parameters of this nonlinear system represent the mutual effects, interaction and feedback among cell populations. Thus, parameters α_1 and α_2 reflect the effect of precursors of osteoclasts and osteoblasts on the osteoclasts and osteoblasts proliferation; in contrast, parameters β_1, β_2 are the speed of the cell degradation process. The equilibrium points of previous dynamical system ($\dot{n}_{oc}^{eq} = \dot{n}_{ob}^{eq} = 0$) corresponding to the stability of the cell populations are easily computed as

$$n_{oc}^{eq} = \left(\frac{\beta_1}{\alpha_1}\right)^{\frac{1-g_{22}}{\Gamma}} \left(\frac{\beta_2}{\alpha_2}\right)^{\frac{g_{21}}{\Gamma}} \quad [9.3]$$

$$n_{ob}^{eq} = \left(\frac{\beta_1}{\alpha_1}\right)^{\frac{g_{12}}{\Gamma}} \left(\frac{\beta_2}{\alpha_2}\right)^{\frac{1-g_{11}}{\Gamma}} \quad [9.4]$$

with $\Gamma := g_{12}\, g_{21} - (1 - g_{11})(1 - g_{22})$.

Depending on the sign of $\Phi = \beta_1(g_{11} - 1) + \beta_2(g_{22} - 1)$, the model predicts different types of dynamical behavior: for $\Phi = 0$, one obtains a periodic solution (as illustrated in Figure 9.4), the case $\Phi < 0$ leads to damped oscillations converging to n_{oc}^{eq} and n_{ob}^{eq} , and the last situation $\Phi > 0$ to unstable oscillations.

Following the model developed in [BON 11], the sensing activity of osteocytes may be connected to mechanical stimuli through the deformation energy of the bone at the vicinity of the region of high stress concentration such as microcracks. So, one may assume that

$$g_{ij} = g_{ij}\ (\Delta Y) \quad [9.5]$$

with the following mechanical signal obtained as a volume integral over -a characteristic zone D_{ocy} including a crack (Figure 9.5) and submitted to a stress field corresponding to a combined tension and bending (i.e. corresponding the mode 1 of crack opening),

$$\Delta Y(x) = \int_{D_{ocy}} dV\ f(x,y)\left(\frac{\omega(y)}{\rho(y)} - w_0\right) = \left(\int_{D_{ocy}} dV\ f(x,y)\frac{\omega(y)}{\rho(y)}\right) - S_0 \quad [9.6]$$

including the specific strain energy $\frac{\omega(y)}{\rho}$, with $\omega(y) = \frac{1}{2}\sigma(y):\varepsilon(y)$.

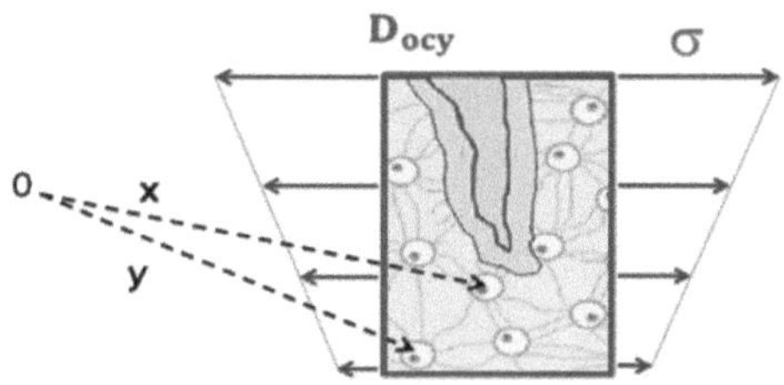

Figure 9.5. *Perception of the mechanical stimulus by the network of osteocytes [GAN 16]. For a color version of this figure, see www.iste.co.uk/ganghoffer/biomechanics.zip*

The spatial influence function may be defined as the following fading exponential

$$f(x,y) = \left| \begin{array}{ll} 0 & if\ d > R_{OC} \\ e^{\frac{-d(x,y)}{R_{OC}}} & if\ d \leq R_{OC} \end{array} \right. \quad [9.7]$$

in which $d(x,y) = |x - y|$ is the distance between the considered osteocyte and pre-osteoclasts or pre-osteoblasts. According to [BON 11], it is estimated that no signal will be received from the outside of this zone of influence. Finally, the signal received by osteoclasts and osteoblasts influences the autocrine and paracrine factors through the exponents g_{ij}.

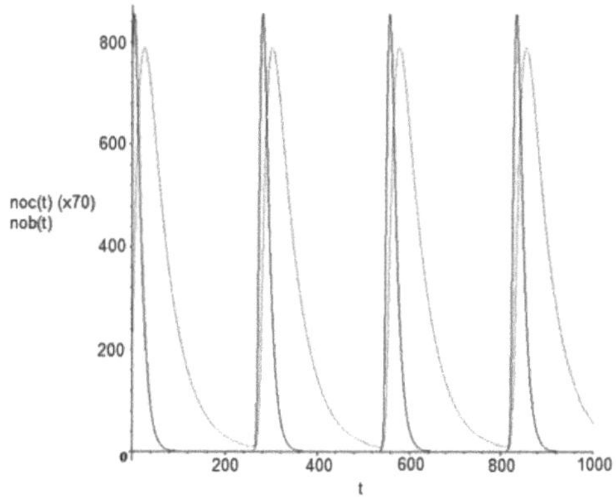

Figure 9.6. *Stable regulated oscillatory changes in bone cell populations and bone mass. Red line: osteoclasts (x70); blue line: osteoblasts. Calculations were performed with the following parameters: $\alpha_1 = 3$, $\alpha_2 = 4$, $\beta_1 = 0.2$, $\beta_2 = 0.02$, $g_{11} = 1.1$, $g_1 = 1$, $g_{21} = -0.5$, $g_{22} = 0$. Initial number of cells: 11 osteoclasts and 212 osteoblasts [KOM 03]. For a color version of this figure, see www.iste.co.uk/ganghoffer/biomechanics.zip*

9.2. Volumetric and surface growth model

We consider a solid body Ω with external boundary $\partial\Omega$; in the continuum macroscopic formulation of growth, those boundaries are only external, whereas internal boundaries are likely to exist when adopting a micromechanical viewpoint.

Part of the external boundary of the solid body is submitted to an evolution due to new mass being added or removed, leading to an overall shape change of the body. This process is called external remodeling in the literature, as illustrated on Figure 9.7. Internal remodeling generally also occurs during the life time of a bone, due to the dynamical equilibrium between bone resorbing units (specialized cells called osteoclasts) and bone generating units (the osteoblasts), leading to a morphological evolution of the trabecular structure.

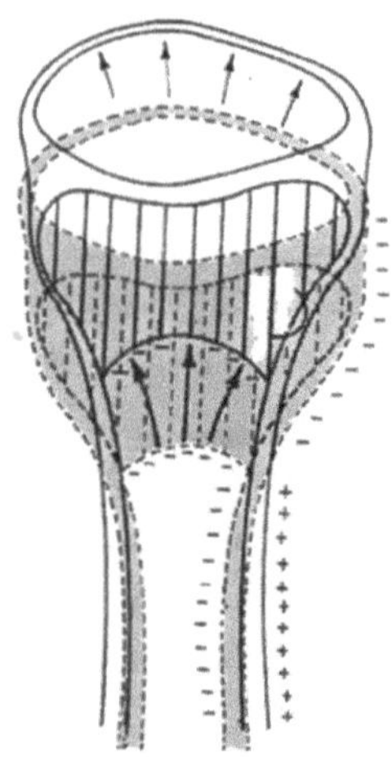

Figure 9.7. *External remodeling of a bone. + (resp. –) zones indicate bone apposition (resp. resorption). From [WEI 88]*

Regarding notations, any field $\mathrm{A}(\mathbf{x})$ (scalar valued or tensorial valued) having its domain within the growing volume Ω_g shall have a counterpart (that can be envisaged as its trace) on the growing surface S_g, using the same variable with a tilde $\tilde{\mathrm{A}}$, and a line trace on the edge of S_g $\partial\mathrm{S}_g$, when the surface is not closed), noted with a double tilde $\tilde{\tilde{A}}$. The transpose of a second order tensor is denoted $\mathbf{A}^{\mathrm{T}}$. The curvature tensor is defined as $\mathbf{K}_r = -\nabla\mathbf{N}$, with $\mathbf{N}$ the unit exterior normal to the considered surface. The double contraction of two second order tensors is defined as $\mathbf{A}:\mathbf{B}:=A_{ij}B_{ij}$ in a Cartesian coordinate system. The dot over any quantity denotes its material derivative, viz. $\dot{a}:=Da/Dt$.

9.2.1. *Bulk growth model*

In order to set the stage for the construction of the bulk growth model, we shall first express the local dissipation in order to highlight the duality between the growth velocity gradient and a conjugated driving force for bulk growth. Although bone mechanics involves small strains, one has to consider large shape changes that will be induced by growth, thus it is convenient to adopt in full generality a large strains framework.

Since the number of particles is the same in any configuration, a motion that connects all configurations can be defined: a multiplicative decomposition of the total transformation gradient $\mathbf{F}$ is adopted, in line with the pioneering work of [ROD 94], relating the reference, intermediate and actual configurations, respectively the set of material points $\Omega_0,\Omega_a,\Omega_t\equiv\Omega$ as:

$$\mathbf{F}=\mathbf{F}_a\mathbf{F}_g \qquad [9.8]$$

with $\mathbf{F}_g$ the growth transformation and $\mathbf{F}_a$ the mapping accounting for growth accommodation (Figure 9.8); observe that neither of these two mappings are compatible in general. The kinematic decomposition [9.8] induces the similar decomposition of the Jacobean

$$J_g=\text{Det}\left(\mathbf{F}_g\right);\ J_a=\text{Det}\left(\mathbf{F}_a\right)\rightarrow J=J_aJ_g \qquad [9.9]$$

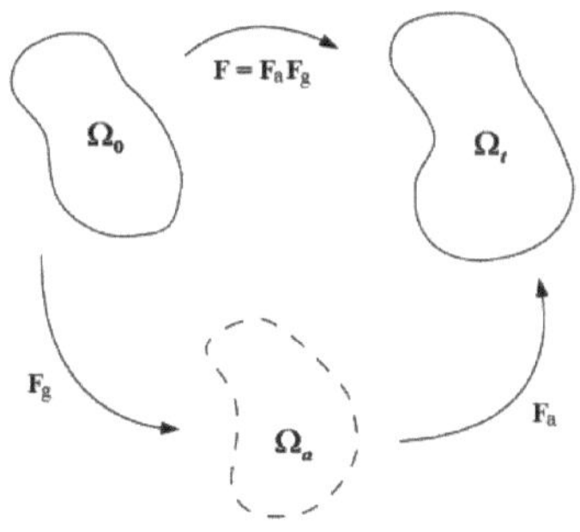

Figure 9.8. *Schematic representation of the multiplicative decomposition of deformation gradient into its elastic and growth parts*

Assuming mass is preserved between the natural (intermediate) and the actual configuration, and that the density does not change from the reference to the natural configuration (following the theory of multiple natural configurations), and in line with [AMB 07], the continuity equation is obtained as

$$\rho_0 = J_a \rho \quad [9.10]$$

The local dissipation expresses in the presence of transport of chemical species as follows, recalling the inequality obtained in [GAN 05] in the initial configuration, and considering first in full generality a large strains framework:

$$J\rho\sigma_s = J\boldsymbol{\sigma} : \mathbf{D} - \frac{d}{dt}(\rho_0 \psi) \geq 0 \quad [9.11]$$

with σ_s the intrinsic dissipation. Here, $\psi = \psi(\mathbf{F}_a; \mathbf{X})$ is the energy density per unit mass in the reference configuration, related to the referential volumetric strain energy density $W_0 = W_0(\mathbf{F}_a; \mathbf{X})$ by the relation $W_0(\mathbf{F}_a; \mathbf{X}) = \rho_0 \psi(\mathbf{F}_a; \mathbf{X})$; the second order tensor $\mathbf{D}$ is the symmetrical part of the velocity gradient $\mathbf{L} := \text{grad}\mathbf{V}$. The assumed dependencies of the potential $W_0 = W_0(\mathbf{F}_a)$ for a supposedly uniform body (we can then drop the dependency on the coordinate $\mathbf{X}$) then lead to the writing of the second law, dropping the thermal and chemical dissipation to focus on the purely mechanical dissipation:

$$J\boldsymbol{\sigma} : (\dot{\mathbf{F}}.\mathbf{F}^{-1}) - \frac{dW_0(\mathbf{F}_a)}{dt} \geq 0 \Rightarrow \left(J_a \boldsymbol{\sigma}.\mathbf{F}_a^{-T} - \frac{\partial W_0(\mathbf{F}_a)}{\partial F_a} \right) : \delta\mathbf{F}_a + \left(J_a \mathbf{F}_a^{T}.\boldsymbol{\sigma}.\mathbf{F}_a^{-T} - \rho_0 \psi I \right) : \left(\delta\mathbf{F}_g.\mathbf{F}_g^{-1} \right) \geq 0 \quad [9.12]$$

substituting therein variations to the material derivatives, and dividing both members of the inequality by ρJ, using the relation $\frac{1}{\rho} = \frac{J_a}{\rho_0}$. The arbitrariness of the variations $\delta\mathbf{F}_a$ then leads to the vanishing of the multiplicative factor therein, viz.

$$J_a \boldsymbol{\sigma}.\mathbf{F}_a^{-T} - \frac{\partial W_0(\mathbf{F}_a)}{\partial F_a} = 0 \rightarrow \frac{\partial W_0(\mathbf{F}_a)}{\partial F_a} = J_a \boldsymbol{\sigma}.\mathbf{F}_a^{-T} \quad [9.13]$$

Inserting this expression into the previous local dissipation then gives

$$\left(\mathbf{F}_{\mathbf{a}}^{\mathrm{T}}.\frac{\partial \mathrm{W}_0(\mathbf{F}_{\mathrm{a}})}{\partial \mathrm{F}_{\mathrm{a}}} - \mathrm{W}_0(\mathbf{F}_{\mathrm{a}})\mathbf{I}\right):\left(\delta\mathbf{F}_{\mathrm{g}}.\mathbf{F}_{\mathrm{g}}^{-1}\right) \geq 0 \qquad [9.14]$$

thereby identifying the Eshelby stress attached to the intermediate configuration, second order tensor

$$\boldsymbol{\Sigma}_{\mathbf{a}} := \mathrm{W}_0(\mathbf{F}_{\mathrm{a}};\mathbf{X})\mathbf{I} - \mathbf{F}_{\mathbf{a}}^{\mathrm{T}}.\frac{\partial \mathrm{W}_0(\mathbf{F}_{\mathrm{a}};\mathbf{X})}{\partial \mathrm{F}_{\mathrm{a}}} \qquad [9.15]$$

with the second order tensor **I** therein denoting the identity mapping.

The nominal stress $\mathbf{P}_{\mathrm{a}} := \frac{\partial \mathrm{W}_0}{\partial \mathbf{F}_{\mathrm{a}}}$ (adopting identical notations as in previous paragraph) appearing in Eshelby stress is dual to the accommodation (elastic) mapping $\mathbf{F}_{\mathbf{a}}$. From the dependency $\mathrm{W}_0(\mathbf{F}_{\mathrm{a}})$, previous derivations lead to the expression of Cauchy stress in the actual configuration as

$$\boldsymbol{\sigma} = \rho\frac{\partial \mathrm{W}_0(\mathbf{F}_{\mathrm{a}})}{\partial \mathbf{F}_{\mathrm{a}}}.\mathbf{F}_{\mathrm{a}}^{\mathrm{T}} \qquad [9.16]$$

Hence, the dissipation rewrites in terms of the highlighted Eshelby stress in [9.15] as

$$\mathrm{J}\rho\sigma_{\mathrm{s}} = -\boldsymbol{\Sigma}_{\mathrm{a}}:\left(\delta\mathbf{F}_{\mathrm{g}}.\mathbf{F}_{\mathrm{g}}^{-1}\right) \geq 0 \Leftrightarrow -\boldsymbol{\Sigma}_{\mathrm{a}}:\mathbf{L}_{\mathrm{g}} \geq 0 \qquad [9.17]$$

defining therein the growth velocity gradient

$$\mathbf{L}_{\mathrm{g}} := \dot{\mathbf{F}}_{\mathrm{g}}.\mathbf{F}_{\mathrm{g}}^{-1} \qquad [9.18]$$

Evolution laws for the flux variables can be written from the formulation of the second law: simple linear growth models may be obtained, in line with [AMB 07], considering a quadratic dissipation (versus $\mathbf{L}_{\mathrm{g}}$), whence

$$\mathbf{L}_{\mathrm{g}} = -\mathrm{A}\boldsymbol{\Sigma}_{\mathrm{a}},\ \mathrm{A} \geq 0 \qquad \textit{A being a non-negative constant} \qquad [9.19]$$

In view of the application to bone structures, the small strain version of previous dissipation inequality can be adopted for bone due to the fact that peak strains are of the order of 2–3%; it clearly writes:

$$-\boldsymbol{\Sigma}_a : \dot{\boldsymbol{\varepsilon}}_g \geq 0 \qquad [9.20]$$

with $\dot{\boldsymbol{\varepsilon}}_g$ the growth strain rate and the small strains version of Eshelby stress expressing as

$$\boldsymbol{\Sigma}_a := W_0(\boldsymbol{\varepsilon}_a)\mathbf{I} - \boldsymbol{\varepsilon}_a{}^T . \frac{\partial W_0(\boldsymbol{\varepsilon}_a)}{\partial \boldsymbol{\varepsilon}_a} \qquad [9.21]$$

Here, $W_0(\boldsymbol{\varepsilon}_a)$ is the strain energy density per unit volume depending upon the small elastic strains $\boldsymbol{\varepsilon}_a$, the form of which will be obtained later on from the discrete homogenization of the trabecular architecture considered as a repetitive unit cell of prismatic shape.

The small strains kinematics here relies on an additive decomposition of the total strain into a grown strain and an elastic strain necessary to restore the kinematic integrity at each material point of the evolving body; it writes in rate form:

$$\dot{\boldsymbol{\varepsilon}} = \dot{\boldsymbol{\varepsilon}}_a + \dot{\boldsymbol{\varepsilon}}_g \qquad [9.22]$$

with $\dot{\boldsymbol{\varepsilon}}_a$ given by the elastic constitutive law, and the growth strain rate $\dot{\boldsymbol{\varepsilon}}_g$ satisfying the positive character of the local dissipation in equation [9.20]; a sufficient condition for this writes

$$\dot{\boldsymbol{\varepsilon}}_g = -A\boldsymbol{\Sigma}_a, \; A \geq 0 \qquad [9.23]$$

with the small strains version of the Eshelby stress $\boldsymbol{\Sigma}_a$.

Note that the intermediate configuration is needed since the grown strains are in fact cumulative, thus they induce a change of the initial reference configuration, from which elastic strains have to be measured. The introduced intermediate configuration can be viewed as the configuration obtained by unloading any material point from the physical configuration.

Integrating the previously written local dissipation on an arbitrary volume Ω_g and considering that the growth strain rate arises from the volumetric growth velocity field $\mathbf{V}_g$ further entails

$$-\int_{\Omega_g} \boldsymbol{\Sigma}_a : \nabla\mathbf{V}_g \geq 0 \Rightarrow \int_{\Omega_g} \nabla.\boldsymbol{\Sigma}_a.\mathbf{V}_g - \int_{\partial\Omega_g} \mathbf{N}.\boldsymbol{\Sigma}_a.\mathbf{V}_g \geq 0 \qquad [9.24]$$

Although the growth velocity $\mathbf{V}_g$ is normally not globally integrable (compatible), one assumes it is locally an integrable field, thus permitting integration by part in [9.24]; this is further allowed since we consider both internal bulk growth and surface growth as fully irreversible phenomena.

Thereby, and focusing on the volume integral in previous inequality, a sufficient condition for the local dissipation to be non-negative is

$$\mathbf{V}_g = K_g \nabla.\boldsymbol{\Sigma}_a,\ K_g \geq 0 \qquad [9.25]$$

The involvement of the fully material Eshelby stress tensor relies on the belief that configurational changes due to internal or external remodeling occur in the reference configuration, and have to be distinguished from the motion problem *per se* [GAN 13].

Since the newly apposed or resorbed material on the surface of bone leads from a kinematic point of view to a purely irreversible growth deformation, one has to update the shape continuously since the variation of the amount of material on the surface of the solid body under external remodeling leads to a permanent change of shape. Since apposition or removal of bone material is presently considered as being fully irreversible, it means that the removal of the external load will lead to the shape induced by the growth velocity field and associated growth velocity gradient computed from the previous step of the growth sequence. In a second stage, one has to redistribute the existing stresses and (elastic) strains to the new shape of the (new) bone sample subjected to the same external loads, by interpolating the mechanical fields to the new domain (a remeshing is possibly done based on an indicator of mesh distortions). This evolution in two successive steps (a growth phase followed by an elastic computation) is achieved thanks to some operator splitting method explained into more details in the forthcoming sections.

9.2.2. *Evolution of bulk density*

Two alternative growth schemes can be considered from a kinematic point of view: constant-density growth is usually coupled with the assumption of an incompressible tissue, so that I_z^* and thus density does only change due to growth, but it remains constant when elastic accommodation is considered. For constant-volume growth, the change in volume from the initial body to the unstressed body is neglected, so that $I_z^* = H^3/12$. Bone tissue growth then occurs by densification only, a viewpoint advocated in this contribution; accordingly, the balance of mass expresses the rate of change of the material density versus the growth mapping

$$\frac{\mathrm{d}\rho}{\mathrm{d}t} = \Gamma^g \rho \qquad [9.26]$$

In a biological context, the growth rate Γ^g appearing in the mass balance equation receives the following expression versus the growth mapping

$$\Gamma^g = \mathrm{Tr}\left(\dot{\mathbf{F}}_g \mathbf{F}_g^{-1}\right) \qquad [9.27]$$

In the present context, various authors (e.g. [GAN 10, GAN 12, AMB 07]) argued based on thermodynamic principles that the driving force behind growth is the Eshelby stress. [GAN 10] introduced general balance laws for solid bodies undergoing growth phenomena in the framework of irreversible thermodynamics of open systems. The growth is viewed as a multiphysical phenomenon coupling the transport of nutrients with mechanical factors promoting a local variation of the density, at constant number of particles. The growing continuum body is endowed with a general hyperelastic constitutive behavior encapsulated into a strain energy function $\mathrm{W}(\mathrm{u})$, expressed per unit referential volume.

Relying on that the growth rate depends on the Eshelby-like tensor, we propose a constitutive relation for the mass source, wherein the growth rate is proportional to the trace of the Eshelby tensor

$$\Gamma^g = \frac{\mathrm{B_c}}{\rho^2} \mathrm{Tr}\left(\boldsymbol{\Sigma}_a(\boldsymbol{\varepsilon}_a) - \boldsymbol{\Sigma}_0\right) \qquad [9.28]$$

where $\mathrm{B_c}$ is a proportionality constant measuring the rate of the adaptation process (it governs the velocity of the adaptation process), $\boldsymbol{\Sigma}_\mathrm{a}(\boldsymbol{\varepsilon}_\mathrm{a})$ is the actual volumetric Eshelby stress tensor receiving the expression

$$\boldsymbol{\Sigma}_\mathrm{a}(\boldsymbol{\varepsilon}_\mathrm{a}) = \mathrm{W}_0(\boldsymbol{\varepsilon}_\mathrm{a})\mathbf{I} - \boldsymbol{\varepsilon}_\mathrm{a} . \frac{\partial \mathrm{W}_0(\boldsymbol{\varepsilon}_\mathrm{a})}{\partial \boldsymbol{\varepsilon}_\mathrm{a}} \qquad [9.29]$$

and $\boldsymbol{\Sigma}_0$ the threshold value for the volumetric Eshelby stress stimulus (reference value).

The evolution equation of the density [9.26] combined with the definition of the mass source Γ^g in [9.28] then delivers

$$\frac{d\rho}{dt} = \frac{B_c}{\rho} Tr\left(\Sigma_a(\varepsilon_a) - \Sigma_0\right) \qquad [9.30]$$

This equation describes bone adaptation, and more precisely whether the stress stimulus within bone tissue will induce bone formation, bone resorption or equilibrium, according to the sign and value of the rate of change of the apparent bone density. Equilibrium is reached when the Eshelby stress attains the homeostatic value Σ_0, supposed to be known at each point. The present model describes density changes at constant volume, so that $J_g = 1$.

It is noteworthy that the trace of Eshelby stress is the strain energy density

$$Tr(\Sigma_a(\varepsilon_a)) = 3W_0(\varepsilon_a) - \varepsilon_a : \sigma(\varepsilon_a) = 3W_0(\varepsilon_a) - 2W_0(\varepsilon_a) = W_0(\varepsilon_a) \qquad [9.31]$$

Accordingly, the volumetric strain energy density (or the trace of Eshelby stress) is the mechanical stimulus that triggers bone adaptation locally, a scalar quantity that incorporates on an equal footing local stress and strain tensors. We then propose a formulation for internal bone adaptation utilizing the strain energy density function as the remodeling signal representing the deformational energy at any point, taking the following form:

$$\frac{d\rho}{dt} = B_c\left(\frac{W_0(\varepsilon_a)}{\rho} - \frac{W_t}{\rho_m}\right) \equiv f(t,\rho) \qquad [9.32]$$

This modified evolution equation ensures a good stability behavior of the time-dependent remodeling. The driving mechanism for adaptive activity is then the deviation between the actual energy per unit mass $G_s = 0.5E_s/(1+\nu_s)$ from a site-specific homeostatic equilibrium $\nu_s = 0.3$; the scalar M_P^*/H^2 therein is the maximum density. The internal remodeling process in [9.32] is considered to have converged when the rate σ_{ys} tends to zero (equilibrium is then reached) or when the density secures a minimal or maximal value. The baseline value of strain energy, quantity $l_b = 0.12$ mm, is determined based on the relation $\sigma_{ij}\big|_D$, wherein a constant reference value for bone formation of the order of 2000 microstrain (variable $\rho_0 = J_a\rho$) is considered [NAM 07]). The remodeling rate B_c is chosen from [CHE 07a] to be 1.0 (kg m^{-3})^2/(Pa.time) in the simulation at the macroscopic level.

The rate of change of the local density of the trabecular bone at a particular location, quantity $J\rho\sigma_s = J\boldsymbol{\sigma}:\mathbf{D} - \frac{d}{dt}(\rho_0\psi) \geq 0$, introduced in the last equation is evaluated by the finite difference method. The improved Euler's method known as Crank-Nicolson scheme is a modified method used instead of Euler forward method in order to improve the solution of the differential equation satisfied by density, using the average of the slopes of the solution curves; it is presently involved to estimate the density of bone tissue at the considered region, based on a first order non-linear ordinary differential equation [9.32]; the internal adaptation is updated according to

$$J\boldsymbol{\sigma}:\left(\dot{\mathbf{F}}.\mathbf{F}^{-1}\right) - \frac{dW_0(\mathbf{F}_a)}{dt} \geq 0 \Rightarrow \left(J_a\boldsymbol{\sigma}.\mathbf{F_a}^{-T} - \frac{\partial W_0(\mathbf{F}_a)}{\partial F_a}\right):\delta\mathbf{F_a}$$
$$+\left(J_a\mathbf{F_a}^T.\boldsymbol{\sigma}.\mathbf{F_a}^{-T} - \rho_0\psi I\right):\left(\delta\mathbf{F}_g.\mathbf{F}_g^{-1}\right) \geq 0 \qquad [9.33]$$

involving the function $J_a\boldsymbol{\sigma}.\mathbf{F_a}^{-T} - \frac{\partial W_0(\mathbf{F}_a)}{\partial F_a} = 0 \rightarrow \frac{\partial W_0(\mathbf{F}_a)}{\partial F_a} = J_a\boldsymbol{\sigma}.\mathbf{F_a}^{-T}$ defined in equation [9.32], evaluated at the discrete point

$$\left(\mathbf{F_a}^T.\frac{\partial W_0(\mathbf{F}_a)}{\partial F_a} - W_0(\mathbf{F}_a)\mathbf{I}\right):\left(\delta\mathbf{F}_g.\mathbf{F}_g^{-1}\right) \geq 0.$$

This scheme contains the implicit terms, $\boldsymbol{\Sigma}_\mathbf{a} := W_0(\mathbf{F}_a;\mathbf{X})\mathbf{I} - \mathbf{F_a}^T.\frac{\partial W_0(\mathbf{F}_a;\mathbf{X})}{\partial F_a}$; in order to make it explicit, we approximate ρ_{n+1} using a forward Euler method, so that $\mathbf{P}_a := \frac{\partial W_0}{\partial \mathbf{F}_a}$, with $\mathbf{F_a}$ the time step size (iteration).

A surface growth model is next elaborated, in a manner that mimics the developed bulk growth model.

9.3. Surface growth model

Surface growth is based on the framework elaborated in [GAN 10, GAN 12]; hence we only provide the main thrust of the kinematics, referring the reader to those two contributions for more details.

From a kinematic viewpoint, and adopting unless otherwise stated a large deformation formalism, the surface deformation gradient is split multiplicatively into a surface growth mapping $\tilde{\mathbf{F}}_g$ and an accommodation mapping $\tilde{\mathbf{F}}_a$ needed to restore the kinematic compatibility, hence

$$\tilde{\mathbf{F}} = \tilde{\mathbf{F}}_a.\tilde{\mathbf{F}}_g \qquad [9.34]$$

The surface mapping $\tilde{\mathbf{F}} := \mathbf{F}.\mathbf{\Pi}$ represents the projection of the bulk transformation gradient onto the tangent plane of $\partial\Omega$; it is obtained by projecting the transformation gradient onto the local tangent plane $\mathbf{N}$, with $\mathbf{\Pi} = \mathbf{I}_S - \mathbf{N} \otimes \mathbf{N}$ the corresponding projector, with the second order tensor $\mathbf{I}_S$ denoting the surface identity tensor, mapping any vector in the tangent plane to the surface to itself. A generic point on the boundary is denoted $\mathbf{X}_S$. Note that $\tilde{\mathbf{F}}$ is in general not invertible, but it has a pseudo inverse $\tilde{\mathbf{f}}$ such that

$$\tilde{\mathbf{F}}.\tilde{\mathbf{f}} = \mathbf{i}_s;\ \tilde{\mathbf{f}}.\tilde{\mathbf{F}} = \mathbf{I}_s \qquad [9.35]$$

This multiplicative decomposition mimics a similar one that holds for volumetric growth; it induces a decomposition of the Jacobean of the total transformation, viz.

$$\tilde{J} = \tilde{J}_a \tilde{J}_g \qquad [9.36]$$

with $\tilde{J}_a := \det\left(\tilde{\mathbf{F}}_a\right); \tilde{J}_g := \det\left(\tilde{\mathbf{F}}_g\right)$ respectively the growth and accommodation Jacobean, such that the surface densities in the reference and actual configurations, quantities ρ_{0S}, ρ_S respectively, are related by

$$\rho_{0S} = \tilde{J}_a \rho_S \qquad [9.37]$$

One here assumes that mass is preserved between the intermediate and actual configurations (growth of mass only occurs between the initial and intermediate configuration), thereby following a similar assumption made classically in the case of volumetric growth.

The mechanical equilibrium is obtained by requiring the vanishing of the first variation of the elastic energy of the growing surface, described from a surface potential energy $W^S\left(\tilde{\mathbf{F}},\mathbf{N}\right)$ defined per unit reference surface; the surface energy potential is built from the surface part of the transformation gradient and the normal to the surface under remodeling. The dependency on the normal accounts for curvature effects not being present in the bulk growth model. The elastic surface energy density is simply the projection of the bulk energy density onto the surface under external remodeling.

The surface nominal stress is defined as the partial derivative $\tilde{\mathbf{T}} := \partial_{\tilde{F}} W^S$. The surface is submitted to a surface force field $\mathbf{f}_S$, a driving force for surface growth. The stationnarity condition of the total potential energy writes [GAN 10]:

$$\nabla_S.\tilde{\mathbf{\Sigma}}_a - \mathbf{\Pi}.\mathbf{K}_r^t.\partial_N W^S + \left(\partial_{X_S} W^s\right)_{\exp l} + \tilde{\mathbf{F}}^T.\mathbf{f}_S = \mathbf{0} \qquad [9.38]$$

with $\mathbf{K}_r := -\nabla_R \mathbf{N}$ the curvature tensor of the surface. Previous balance equation represents mechanical equilibrium in terms of the surface Eshelby stress

$$\tilde{\mathbf{\Sigma}}_a := W^s \mathbf{I}_s - \tilde{\mathbf{F}}_a^{\ t}.\tilde{\mathbf{T}} \qquad [9.39]$$

with the second order surface tensor $\mathbf{I}_S$ therein denoting the identity mapping with support included in S_g.

One further shows that the tangential projection of Eshelby stress is obtained as the introduced surface Eshelby stress $\tilde{\mathbf{\Sigma}}_a$. Due indeed to the involution of the projection operator $\mathbf{\Pi}$ and its symmetry, on obtains the projection of the volumetric Eshelby stress on the tangent plane at any point on the boundary

$$\mathbf{\Sigma}_a.\mathbf{\Pi} := W_0(\mathbf{F}_a;\mathbf{X})\mathbf{\Pi} - \mathbf{F}_a^{\ T}.\frac{\partial W_0(\mathbf{F}_a;\mathbf{X})}{\partial F_a}.\mathbf{\Pi} \equiv W_0(\mathbf{F}_a.\mathbf{\Pi};\tilde{\mathbf{X}})\mathbf{\Pi} - (\mathbf{F}_a^{\ T}.\mathbf{\Pi}).\left(\frac{\partial W_0(\mathbf{F}_a;\mathbf{X})}{\partial F_a}.\mathbf{\Pi}\right) \qquad [9.40]$$

in which the strain energy density $W_0(\mathbf{F}_a.\mathbf{\Pi};\tilde{\mathbf{X}})$ has to be evaluated for boundary points $\tilde{\mathbf{X}}$ and for the projected elastic mapping $\tilde{\mathbf{F}}_a := \mathbf{F}_a.\mathbf{\Pi}$.

Note that Eshelby stress is a purely material stress attached to the reference (initial) configuration. It can be obtained by a pull-back of the balance of linear momentum (equilibrium equation) involving the purely spatial Cauchy stress. It also represents the driving force for the change of domain for a general functional. In shape optimization, the shape derivative of the total potential energy is the jump of Eshelby stress over the moving interface. Both last aspects and derivations can be found in [GAN 14].

The surface position and its variation on S_g are denoted $\mathbf{X}_S$ and $\delta\mathbf{X}_S$ resp. For a surface with local normal $\mathbf{N}$ (to the locally defined tangent plane), the operators $\nabla_S := \mathbf{\Pi}.\nabla$ and $\mathrm{div}_S(.) := \mathrm{Tr}(\nabla_S(.))$ successively denote the surface gradient and the surface divergence.

The surface mass balance equation expresses as

$$\dot{\rho}_S = \Gamma^g \rho_S \qquad [9.41]$$

with ρ_S the surface density of nutrients (HA in the case of bone), and Γ^g the rate of mass growth, given by the trace of the surface growth velocity gradient as

$$\Gamma^g = \mathrm{Tr}\left(\nabla \tilde{\mathbf{V}}_g\right) \qquad [9.42]$$

with $\tilde{\mathbf{V}}_g$ the surface growth velocity, bringing an irreversible contribution to the kinematics.

As only the normal component of the surface growth velocity induces a change of shape, we assume a normal surface growth velocity proportional to the normal driving force for growth (to be identified later on), quantity $\tilde{\Upsilon}_{gN}$

$$\tilde{V}_N\left(\tilde{\mathbf{X}}, t\right) = \tilde{f}\left(\tilde{\Upsilon}_{gN}\right) \qquad [9.43]$$

The kinematic picture of this surface growth model is then such that mass growth only occurs through surfaces, generating in turn shape changes of the growing body. The new material accumulated (or resorbed) on those surfaces (with a density different from that of the parent material) induces in turn a volumetric growth process, although no internal material point is submitted to growth. This kinematic viewpoint is believed to be physically consistent with the way growth remodeling occurs in bones through a mineralization process localized close to the surface: the thickness of the erosion zone due to the osteoclasts activity is about 60 µm in trabecular bone and 100 µm in cortical bone; hence it is realistic to consider this process as surface growth.

The surface growth velocity can be related to driving forces for surface growth identified from the writing of the local surface dissipation (Clausius-Duhem inequality) for growth processes involving simultaneously transport of mass in addition to irreversible deformations due to growth, as

$$\begin{aligned}\tilde{J}\tilde{\boldsymbol{\sigma}}:\left(\dot{\tilde{\mathbf{F}}}.\tilde{\mathbf{F}}^{-1}\right)-\frac{dW^S\left(\tilde{\mathbf{F}}_a\right)}{dt}\geq 0 \Rightarrow \left(\tilde{J}_a\tilde{\boldsymbol{\sigma}}.\tilde{\mathbf{F}}_a^{-T}-\frac{\partial W^S}{\partial \tilde{\mathbf{F}}_a}\right):\delta\tilde{\mathbf{F}}_a \\ +\left(\tilde{J}_a\tilde{\mathbf{F}}_a^{T}.\tilde{\boldsymbol{\sigma}}.\tilde{\mathbf{F}}_a^{-T}-W^S\mathbf{I}_S\right):\left(\delta\tilde{\mathbf{F}}_g.\tilde{\mathbf{F}}_g^{-1}\right)\geq 0\end{aligned} \qquad [9.44]$$

substituting variations to the material derivatives, and dividing both members of the inequality by $\rho_S \tilde{J}$, using the relation $\frac{1}{\rho_S} = \frac{\tilde{J}_a}{\rho_{0S}}$. The arbitrariness of the variations $\delta\tilde{\mathbf{F}}_a$ in previous inequality then leads to the expression of surface Cauchy stress

$$\boldsymbol{\sigma} = \rho \frac{\partial W_0(\mathbf{F}_a)}{\partial \mathbf{F}_a}.\mathbf{F}_a^{T} \qquad [9.45]$$

Hence the surface dissipation (dividing the previous expression by factor J_a) resumes to

$$J\rho\sigma_s = -\boldsymbol{\Sigma}_a : \left(\delta\mathbf{F}_g.\mathbf{F}_g^{-1}\right) \geq 0 \Leftrightarrow -\boldsymbol{\Sigma}_a : \mathbf{L}_g \geq 0 \qquad [9.46]$$

thereby recognizing the surface Eshelby stress built from the surface energy

$$\tilde{\boldsymbol{\Sigma}}_a := W^S \mathbf{I}_S - \tilde{\mathbf{F}}_a^{T}.\frac{\partial W^S}{\partial \tilde{\mathbf{F}}_a} \qquad [9.47]$$

The surface growth velocity gradient in previous inequality is defined by

$$\tilde{\mathbf{L}}_g := \dot{\tilde{\mathbf{F}}}_g.\tilde{\mathbf{F}}_g^{-1} \qquad [9.48]$$

According to Curie principle, couplings between variables are possible only for variables of the same tensor nature; it follows from the local dissipation that the growth velocity gradient $\tilde{\mathbf{L}}_g$ is coupled to the sole Eshelby stress, viz. $\tilde{\mathbf{L}}_g = \mathbf{f}\left(\tilde{\boldsymbol{\Sigma}}_a\right)$, thereby defining the general format of the growth model.

One may go one step beyond, assuming that the surface growth velocity is fully irreversible, due to the fact that the source of shape changes of the growing body is the mass flux (hence it will correspond to situations of either growth or resorption); this means that the surface growth velocity gradient can be expressed as the surface gradient of the surface velocity. This does not preclude the existence of elastic surface strains induced by surface growth, thus the surface velocity gradient splits into a growth part and an elastic contribution, which is nevertheless assumed to be negligible.

In the case of a completely irreversible surface growth, $\tilde{\mathbf{L}}_g$ represents the surface (Eulerian) gradient of the surface growth velocity field $\tilde{\mathbf{V}}_g$, viz.

$$\tilde{\mathbf{L}}_g = \nabla_S \tilde{\mathbf{V}}_g \qquad [9.49]$$

Hence, considering the case of a purely mechanical dissipation as a special case of the surface dissipation inequality, integration by part followed by localization of the corresponding weak statement leads to

$$\mathbf{L}_g := \dot{\mathbf{F}}_g . \mathbf{F}_g^{-1} \qquad [9.50]$$

Note that the surface divergence $\nabla_S.(.)$ is involved in previous inequality as the tensors are attached to the surface. The line (boundary) terms in previous inequality vanish in case of a closed surface.

The last inequality then leads in view of the arbitrariness of the considered surface S_g to the following linear surface growth model

$$\tilde{V}_g = \tilde{K}_g \mathrm{div}_S\left(\tilde{\Sigma}_a\right), \ \tilde{K}_g \geq 0 \qquad [9.51]$$

The positive constant $\tilde{K}_g$ is chosen such that the local surface dissipation is nonnegative. Thereby, a linear relation between the surface growth velocity and the surface divergence of the (surface) Eshelby stress shall be adopted for simplicity reason in the sequel.

Note that as for bulk growth, the integration by parts presupposes that one is entitled to derive the surface growth velocity. However, note that the system of coordinates is in general non-integrable (surface growth strains are compatible) in the intermediate configuration in which the derivation is performed; we nevertheless assume that is possible to evaluate such a gradient locally in the neighborhood of a material point.

The bone adaptation algorithm relies on the solution of a linear elasticity problem in the bulk bone, coupled to the surface growth problem. These two problems are solved in a staggered manner; the solution of bulk problem provides at each time step the stress field on each point in the domain including the surface, leading in turn to an evaluation of the remodeling velocity given versus the divergence of Eshelby stress.

9.4. A coupled bone internal and external remodeling approach

9.4.1. *Strong formulation*

The previous set of equations is now specialized to treat the growth of hard tissues like bones [GAN 16c]; in this case, deformations are limited in both the bulk and on the growing surface; hence one may restrict to a small strains framework, leading to the specialization of previous BVP to this simplified context. The algorithms incorporating both internal and external remodeling mechanisms of bone will later on be performed at different mechanical levels.

Following the representation theorems for isotropic scalar valued functions of tensorial arguments, the mechanical part of the surface strain energy density $W^S(\tilde{\boldsymbol{\varepsilon}}, \mathbf{N})$ can in general be selected as a function of the invariants of the curvature tensor (the mean and Gaussian curvatures), the invariants of the small strain surface deformation tensor $\tilde{\boldsymbol{\varepsilon}}$ and of its square, and the unit exterior normal to the growing surface, vector $\mathbf{N}$. We choose for simplicity reasons a form linearly dependent on the second invariant of the linearized part of Cauchy Green strain, viz. $\tilde{\mathbf{C}} \cong \mathbf{I} + 2\tilde{\boldsymbol{\varepsilon}}$, and quadratic in the normal to the growing surface, hence

$$W^S(\tilde{\boldsymbol{\varepsilon}}, \mathbf{N}) = \frac{\lambda(\rho)}{2} \mathrm{Tr}(\tilde{\boldsymbol{\varepsilon}})^2 + \mu(\rho) \mathrm{Tr}(\tilde{\boldsymbol{\varepsilon}}^2) + C\mathbf{N}.\mathbf{K}_r.\mathbf{N} \qquad [9.52]$$

with $\mathbf{L}_g = -A\boldsymbol{\Sigma}_a$, $A \geq 0$ the surface strain (induced by the existing volumetric strain), and $-\boldsymbol{\Sigma}_a : \dot{\boldsymbol{\varepsilon}}_g \geq 0$ the surface mechanical properties, expressing versus the surface density and the maximum value of the traction modulus according to

$$\boldsymbol{\Sigma}_a := W_0(\boldsymbol{\varepsilon}_a)\mathbf{I} - \boldsymbol{\varepsilon}_a^T . \frac{\partial W_0(\boldsymbol{\varepsilon}_a)}{\partial \boldsymbol{\varepsilon}_a} \qquad [9.53]$$

and C a constant associated to the energy of geometrical curvature of the growing surface; the second order tensor $\mathbf{K}_r = -\nabla\mathbf{N}$ in previous expression is the curvature tensor, defined as the gradient of the unit exterior normal $\mathbf{N}$. Interactions terms between curvature and mechanical deformation are presently not considered.

The elastic surface stress then results from previous expression of the surface potential in [9.52] as

$$\tilde{\mathbf{T}}^{el} \equiv \tilde{\boldsymbol{\sigma}}^{el} := \frac{\partial W^S}{\partial \tilde{\boldsymbol{\varepsilon}}} = \lambda(\rho)\,\mathrm{tr}(\tilde{\boldsymbol{\varepsilon}})\mathbf{I}_S + 2\mu(\rho)\tilde{\boldsymbol{\varepsilon}} \qquad [9.54]$$

An orthotropic linear elastic constitutive behavior for the bulk may alternatively be considered as a more general model, in agreement with the literature.

Regarding the kinematics, and additive split of the total surface strain into a growth part $\tilde{\boldsymbol{\varepsilon}}_g$ and an elastic part $\tilde{\boldsymbol{\varepsilon}}_a$ needed to restore kinematic compatibility on the surface is performed, according to

$$\tilde{\boldsymbol{\varepsilon}} = \tilde{\boldsymbol{\varepsilon}}_a + \tilde{\boldsymbol{\varepsilon}}_g \qquad [9.55]$$

The elastic part $\tilde{\boldsymbol{\varepsilon}}_a$ enters into the constitutive law expressed in [9.52], [9.54]. The field variables can be separated into the set of surface variables and the set of volumetric variables.

As a summary of the field equations of the surface growth of hard tissues, the following BVP can be written on the domain $\Omega \cup S_u \cup S_t \cup S_g$, denoting therein S_u (resp. S_t) the portion of the external boundary with kinematic (resp. static) boundary conditions, with $S_u \cap S_t = \varnothing$, and $S_g \subset S_u$ the subsurface on which external remodeling occurs in the sequel:

Surface kinematic variables

$$\begin{cases} \tilde{\boldsymbol{\varepsilon}} = \tilde{\boldsymbol{\varepsilon}}_a + \tilde{\boldsymbol{\varepsilon}}_g = \boldsymbol{\varepsilon}.\boldsymbol{\Pi} = \boldsymbol{\varepsilon}.(\mathbf{I}_S - \mathbf{N}\otimes\mathbf{N}) \\ \tilde{\boldsymbol{\varepsilon}} = \mathrm{Sym}(\nabla\tilde{\mathbf{U}}) \\ \dot{\tilde{\boldsymbol{\varepsilon}}}_g = \frac{1}{2}(\nabla_S\tilde{\mathbf{V}}_g + \nabla^T{}_S\tilde{\mathbf{V}}_g) \end{cases}$$

Mass preservation between intermediate and current configuration

$$\rho_{0S} = (\mathrm{Tr}\tilde{\boldsymbol{\varepsilon}}_a)\rho_S \text{ on } S_g$$

Stationnarity condition of the surface potential energy:

$$\begin{cases} \nabla_S.\tilde{\boldsymbol{\Sigma}}_a - \boldsymbol{\Pi}.\mathbf{K}_r^t.\frac{\partial W^S}{\partial N} + \tilde{\mathbf{F}}^T.\mathbf{f}_S = \mathbf{0} \text{ on } S_g, \text{ surface gradient } \nabla_S := \boldsymbol{\Pi}.\nabla \\ \mathbf{f}_S = \mathbf{0} \text{ on } S_g / S_t \end{cases}$$

Constitutive laws from surface energy and Eshelby stress

$$\begin{cases} W^S(\tilde{\varepsilon}_a) = \frac{\lambda(\rho_S)}{2}\mathrm{Tr}(\tilde{\varepsilon}_a)^2 + \mu(\rho_S)(\tilde{\varepsilon}_a : \tilde{\varepsilon}_a) \to \tilde{\mathbf{T}}^{el} \equiv \tilde{\boldsymbol{\sigma}}^{el} := \frac{\partial W^S(\tilde{\varepsilon}_a)}{\partial\tilde{\varepsilon}_a} = \lambda(\rho_S)\mathrm{tr}(\tilde{\varepsilon}_a)\mathbf{I}_S + 2\mu(\rho_S)\tilde{\varepsilon}_a \\ \tilde{\boldsymbol{\Sigma}}_a := W^S(\tilde{\varepsilon}_a)\mathbf{I}_S - \tilde{\boldsymbol{\varepsilon}}_a{}^T.\frac{\partial W^S}{\partial\tilde{\varepsilon}_a} \quad \text{on } S_g \end{cases}$$

Surface growth velocity

$$\tilde{\mathbf{V}}_g = \tilde{K}_g \mathrm{div}_S(\tilde{\boldsymbol{\Sigma}}_a), \ \tilde{K}_g = \mathrm{Cte} \text{ on } S_g$$

Continuity equation

$$\dot{\rho} = \Gamma^g\rho \text{ on } S_g;\ \Gamma^g := \mathrm{Tr}(\dot{\mathbf{F}}_g.\mathbf{F}_g^{-1}) = \frac{B_c}{\rho^2}\mathrm{Tr}(\boldsymbol{\Sigma}_a(\boldsymbol{\varepsilon}_a) - \boldsymbol{\Sigma}_0);\ \rho(\mathbf{X}, t=0) = \rho_0(\mathbf{X})$$

Equilibrium equation of bulk stress

$$\begin{cases} \mathrm{div}\boldsymbol{\sigma} + \rho\mathbf{g} = 0 \text{ in } \Omega \\ \boldsymbol{\Sigma}_a := W_0\mathbf{I} - \boldsymbol{\varepsilon}_a{}^T.\boldsymbol{\sigma} \\ \mathbf{U} = \mathbf{0} \text{ on } S_u; \boldsymbol{\sigma}.\mathbf{N} = \mathbf{f}_S \text{ on } S_t \\ \boldsymbol{\sigma} = 2\mu(\rho)\boldsymbol{\varepsilon}_a + \lambda(\rho)\mathrm{Tr}(\boldsymbol{\varepsilon}_a)\mathbf{I} \text{ in } \Omega \end{cases}$$

The tensor $\tilde{\boldsymbol{\Sigma}}_{\mathbf{a}} := W^S(\tilde{\boldsymbol{\varepsilon}}_a)\mathbf{I}_S - \tilde{\boldsymbol{\varepsilon}}_{\mathbf{a}}^T . \dfrac{\partial W^S(\tilde{\boldsymbol{\varepsilon}}_a)}{\partial \tilde{\boldsymbol{\varepsilon}}_a}$ represents the small strain Eshelby stress, the surface divergence of which drives surface growth. Note that in the considered small strains framework, the surface transformation gradient may be approximated to first order by the surface identity mapping, viz. one has $\tilde{\mathbf{F}} \cong \mathbf{I}_S$. The reversible surface strain $\tilde{\boldsymbol{\varepsilon}}_a$ coincides with the surface elastic strain $\tilde{\boldsymbol{\varepsilon}}_e$, obtained from a projection of the volumetric elastic strain onto the tangent plane to the growing surface.

The growth velocity field on S_g coincides with the total velocity field, which accordingly imposes a kinematic boundary conditions; S_t is the subsurface on which tractions $\mathbf{t}^d$ are applied. Note that the curvature of the surface intervenes in the momentum balance equation through the curvature tensor, and that Eshelby stress $\tilde{\boldsymbol{\Sigma}}_a$ is involved in both the growth model and in the equilibrium equation. Since the surface growth velocity $\tilde{\mathbf{V}}_g$ is imposed on the growing subsurface S_g (in fact as the total surface velocity), it determines a kinematic boundary condition in terms of a velocity field. This field is however not known a priori, as it is related to Eshelby stress $\tilde{\boldsymbol{\Sigma}}_a$. Note that the newly generated material on the growing surface accumulates in the bulk (it is "pushed" in some sense to the bulk by the continuously appearing material on the surface), and it does in fact generate volumetric growth: it generates a new domain with new properties (density, elastic moduli), thereby giving rise to internal stresses. Due to this kinematic picture, the newly generated material is "transferred" to the bulk with a purely elastic strain, which defines the total bulk strain (contrary to surface strain that splits into a growth part and an elastic contribution).

The bone structure is defined by its planar geometry, and a preload is applied to the whole structure, in order to mimic physiological conditions of exercise. This determines an initial state of stress (which is elastic as no growth has yet occurred), from which the surface Eshelby stress $\tilde{\boldsymbol{\Sigma}}_a$ can be evaluated.

The solution of the complete BVP involves the knowledge of the kinematic boundary condition on the growing surface (the velocity field on the growing surface is the total velocity field), which itself requires as a first step to solve the surface BVP. The surface and bulk variables are coupled by the growth model, since the surface growth velocity depends on the trace of the elastic deformation (a tangential field).

The numerical solution of previous BVP requires setting up an incremental solution scheme, which involves the following steps:

1) Time discretization of the interval of interest $[0,T]=\cup_{n=0}^{N}[t_n,t_{n+1}]$: the time derivatives arising in the weak form are accordingly replaced by suitable algorithmic approximations, hence leading to the formulation of a global time-stepping algorithm, which is independent from the constitutive models.

2) At a particular point in $\Omega\cup S_g\cup S_u\cup S_t$ (typically a quadrature point of a finite element), the goal is to compute an approximation of the field variables appearing in the weak form.

3) At the point of interest, the incremental volume and surface displacements successively denoted by $\{\Delta\mathbf{U}^{n+1},\Delta\tilde{\mathbf{U}}^{n+1}\}$ are regarded as given.

4) At time t_n, the state at any point in the closed set $\Omega^n\cup S_g^n\cup S_u^n\cup S_t^n$ characterized by the set of variables

$\{\rho^n,E^n,\lambda^n,\mu^n,W_0{}^n,\mathbf{U}^n,\tilde{\mathbf{U}}^n,\boldsymbol{\varepsilon}^n,\tilde{\boldsymbol{\varepsilon}}_g^n,\tilde{\boldsymbol{\varepsilon}}_a^n,\tilde{\mathbf{V}}_g^n,\boldsymbol{\Sigma}_a{}^n,\tilde{\boldsymbol{\Sigma}}_a{}^n,\Omega^n\cup S_g^n\cup S_u^n\cup S_t^n\}$ is given and is assumed to be in equilibrium, i.e. it satisfies previous equilibrium equation.

5) One then has to produce an update of the state variables at any point to the new values

$\{\rho^{n+1},E^{n+1},\lambda^{n+1},\mu^{n+1},W_0{}^{n+1},\mathbf{U}^{n+1},\tilde{\mathbf{U}}^{n+1},\boldsymbol{\varepsilon}^{n+1},,\tilde{\boldsymbol{\varepsilon}}_g^{n+1},\tilde{\boldsymbol{\varepsilon}}_a^{n+1},\tilde{\mathbf{V}}_g^{n+1},\boldsymbol{\Sigma}_a{}^{n+1},\tilde{\boldsymbol{\Sigma}}_a{}^{n+1},\Omega^{n+1}\cup S_g^{n+1}\cup S_u^{n+1}\cup S_t^{n+1}\}$

at time step t_{n+1} in a manner consistent with the local BVP.

One can set a separate algorithm for the update of the field variables on the growing surface as follows:

The surface force field $\mathbf{f}_S$ is prescribed on the growing surface, and it is thought that different choices will lead to different grown shapes. The update of the surface velocity field $\tilde{\mathbf{V}}_g$ leads to a shape evolution during the time increment resulting in an updated domain written as

$$\Omega_{n+1}=\left(I+\Delta t\tilde{\mathbf{V}}_g{}^n\right)\Omega_n \qquad [9.56]$$

A new geometry is then generated, and the whole domain splits into the old domain and the new domain generated by the material point created on the growing surface. The density of the material points of the new domain is the surface density, and new material properties are also given for the newly generated material.

1. Initialization of the data

$\underline{t = t_0}$: $\rho_0, \nu, \Delta t, g, \mathbf{f}_S, B_c, K, \mathbf{U}^0, \tilde{\mathbf{U}}^0, \tilde{\mathbf{V}}_g^0 = \mathbf{0}, \Omega^0 \cup S_g^0 \cup S_u^0 \cup S_t^0$ data

2. Initialization of the variables at time t_n

$\underline{t = t_n}$: $\rho^n, E^n, \lambda^n, \mu^n, \mathbf{U}^n, \tilde{\mathbf{U}}^n, W_0^n, \tilde{\boldsymbol{\Sigma}}_a^n, \mathbf{V}_g^n, \tilde{\boldsymbol{\varepsilon}}_a^n, \tilde{\boldsymbol{\varepsilon}}_g^n, \tilde{\boldsymbol{\varepsilon}}^n, \Omega^n \cup S_g^n \cup S_u^n \cup S_t^n$

2.1. Computation of lame's coefficients

$$\lambda^n(\rho) = \frac{E^n(\rho)\nu}{(1-2\nu)(1+\nu)};\ \mu^n(\rho) = \frac{E^n(\rho)}{2(1+\nu)}$$

2.2. Set up the constitutive law based on strain energy density

$$W^S = W^S\left(\tilde{\boldsymbol{\varepsilon}}_a^n\right) = \frac{\lambda^n}{2}\mathrm{Tr}(\tilde{\boldsymbol{\varepsilon}}_a^n)^2 + \mu^n\left(\tilde{\boldsymbol{\varepsilon}}_a^n : \tilde{\boldsymbol{\varepsilon}}_a^n\right) + C\mathbf{N}^n.\mathbf{K}^n.\mathbf{N}^n$$

2.3. Set up the constitutive laws for Cauchy and Eshelby stress

$$\to \tilde{\boldsymbol{\sigma}}^n := \frac{\partial W^S\left(\tilde{\boldsymbol{\varepsilon}}_a^n\right)}{\partial \tilde{\boldsymbol{\varepsilon}}_a} = 2\mu^n \tilde{\boldsymbol{\varepsilon}}_a^n + \lambda^n \mathrm{tr}\left(\tilde{\boldsymbol{\varepsilon}}_a^n\right)\mathbf{I}_S \to \tilde{\boldsymbol{\Sigma}}_a^n := W^S\left(\tilde{\boldsymbol{\varepsilon}}_a^n\right)\mathbf{I}_S - \tilde{\boldsymbol{\varepsilon}}_a^n.\frac{\partial W^S\left(\tilde{\boldsymbol{\varepsilon}}_a^n\right)}{\partial \tilde{\boldsymbol{\varepsilon}}_a}$$

3. Update variables

$\underline{t_{n+1} = t_n + \Delta t}$: update variables according to

3.1. Equilibrium equation for the bulk elasticity

$$\begin{cases} \mathrm{Div}\boldsymbol{\sigma}\left(\boldsymbol{\varepsilon}_a\left(\mathbf{U}^{n+1}\right)\right) + \rho\mathbf{g} = 0 \text{ in } \Omega^n \\ \boldsymbol{\sigma}^{n+1} = 2\mu^{n+1}\boldsymbol{\varepsilon}_a^{n+1} + \lambda^{n+1}\mathrm{Tr}\left(\boldsymbol{\varepsilon}_a^{n+1}\right)\mathbf{I} \text{ in } \Omega^n \\ \mathbf{u}^{n+1}\big|_{S_u} = \mathbf{0} \text{ on } S_u \\ \mathbf{U}^{n+1}\big|_{S_g} = \tilde{\mathbf{U}}^{n+1} \\ \boldsymbol{\sigma}^{n+1}.\mathbf{N}^{n+1} = \mathbf{f}_S \text{ on } S_t \end{cases}$$

$$\tilde{\boldsymbol{\sigma}}^{n+1} = 2\mu^{n+1}\tilde{\boldsymbol{\varepsilon}}_a^{n+1} + \lambda^{n+1}\mathrm{tr}\left(\tilde{\boldsymbol{\varepsilon}}_a^{n+1}\right)\mathbf{I}_S$$

3.2. Balance of surface momentum

$$\text{On } S_g^n : \begin{cases} \nabla_S.\tilde{\boldsymbol{\Sigma}}_a^{n+1} - \boldsymbol{\Pi}.\left(\mathbf{K}^{n+1}\right)^T.\frac{\partial \psi^S}{\partial \mathbf{N}} + \mathbf{f}_S = \mathbf{0},\ \ \mathbf{f}_S = \mathbf{0} \text{ on } S_g / S_t \\ \mathbf{K}^n := -\nabla\mathbf{N}^n \text{ curvature tensor} \end{cases}$$

3.2.1 Update surface growth velocity and density

$$\text{On } S_g^n : \begin{cases} \tilde{\mathbf{V}}_g^{n+1} = \tilde{K}\left(\nabla_S.\tilde{\boldsymbol{\Sigma}}_a^n\right) \to \tilde{\mathbf{U}}^{n+1} = \tilde{\mathbf{U}}^n + \mathbf{V}_g^{n+1}\Delta t \\ \rho_S^{n+1} - \rho_S^n = \Gamma^S \Delta t \rho_S^n \to \rho_S^{n+1} \\ \Gamma^S = \frac{B_c}{\left(\rho_S^n\right)^2}\mathrm{Tr}\left(\boldsymbol{\Sigma}_a^n(\boldsymbol{\varepsilon}_a) - \boldsymbol{\Sigma}_0\right) \end{cases}$$

3.2.2 Update surface kinematic variables

$$\text{On } S_g^n : \begin{cases} \dot{\tilde{\boldsymbol{\varepsilon}}}_g^n = \mathrm{sym}\left(\nabla_S\left(\mathbf{V}_g^n\right)\right) \\ \tilde{\boldsymbol{\varepsilon}}^{n+1} = \mathrm{Sym}\left(\mathrm{Grad}_S\tilde{\mathbf{U}}^{n+1}\right) \\ \tilde{\boldsymbol{\varepsilon}}_g^{n+1} = \tilde{\boldsymbol{\varepsilon}}_g^n + \dot{\tilde{\boldsymbol{\varepsilon}}}_g^{n+1}\Delta t \\ \tilde{\boldsymbol{\varepsilon}}_a^{n+1} = \tilde{\boldsymbol{\varepsilon}}^{n+1} - \tilde{\boldsymbol{\varepsilon}}_g^{n+1} \\ \mathbf{K}^{n+1} = -\nabla\mathbf{N}^{n+1} \end{cases}$$

3.4. Update domain and transport all variables to new mesh

$$\Omega_{n+1} = \left(I + \Delta t \mathbf{V}_g^{\ n}\right)\Omega_n \to S_g^{\ n+1} = \left(I + \Delta t \mathbf{V}_g^{\ n}\right)S_g^{\ n}$$

4. Loop again

5. Stop when $\rho_S^{n+1} = \rho_{SMax}$

End

The growth process is arrested when a criterion is satisfied (for instance a maximum or minimum density $\rho_{SMax}(\tilde{\mathbf{X}},t)$ has been reached); it is clear from the formulation of bone internal remodeling that this process shall stop when the strain energy density (the driving force for internal remodeling) reaches a homeostatic value. A similar condition applies to bone external remodeling, although no homeostatic value of the driving force appears in the model for the surface growth velocity. It is nevertheless clear from the mathematical formulation that external remodeling will stop when the bone material reaches a state of homogeneous stress (or elastic strain).

Remodeling and the computation of a new internal stress distribution is done in a two steps staggered scheme, whereby new bone is being apposed (or resorbed) – based on mechanical stimulus - and both strains and stresses are redistributed in the domain by a subsequent elastic computation. The external load is kept fixed during the remodeling process.

9.4.2. *Variational formulation of the growth boundary value problem*

In most cases, it is not possible to find an analytical solution for these equations (strong-form of the solution) and thus one has to recourse to an approximation method. In view of the finite element discretization, these equations need to be rewritten weak form, i.e. an integrated variational formulation.

The variational formulation of surface equilibrium (balance of surface momentum) is written for the purpose of finite element discretization by multiplicating the strong form by a test function $\mathbf{v}$ chosen to vanish on the domain boundary, viz. $\mathbf{v} = \mathbf{0}$ on $\partial\Omega$. The balance of surface momentum is conveniently recalled in order to build the variational form

$$\nabla_S.\tilde{\boldsymbol{\Sigma}}_{\mathbf{a}} - \boldsymbol{\Pi}.\mathbf{K}_r^t.\frac{\partial W^S}{\partial N} + \mathbf{f}_S = \mathbf{0} \qquad [9.57]$$

is further transformed using the selected expression of the surface energy and elaborating the contribution

$$\boldsymbol{\Pi}.\mathbf{K}_r^t.\frac{\partial W^S}{\partial N} = \boldsymbol{\Pi}.\mathbf{K}_r^t.\mathrm{grad}_S W^S.\mathbf{N} = -(\mathrm{grad}\mathbf{N})^T.\frac{\partial W^S}{\partial N} = -\mathrm{grad}W^S \qquad [9.58]$$

This delivers the following balance equation

$$\nabla_S.\tilde{\boldsymbol{\Sigma}}_{\mathbf{a}} + \mathrm{grad}W^S + \mathbf{f}_S = \mathbf{0} \qquad [9.59]$$

The surface divergence of Eshelby stress is further elaborated as

$$\nabla_S.\tilde{\boldsymbol{\Sigma}}_{\mathbf{a}} = \mathrm{div}_S\left(W^S\mathbf{I}\right) - \mathrm{div}\left(\boldsymbol{\varepsilon}.\frac{dW^S}{d\boldsymbol{\varepsilon}}\right) \equiv \mathrm{grad}_S W^S - \mathrm{div}\left(\boldsymbol{\varepsilon}.\frac{dW^S}{d\boldsymbol{\varepsilon}}\right) \quad [9.60]$$

Hence, the surface equilibrium writes as the following nonlinear differential equation

$$-\mathrm{div}_S\left(\boldsymbol{\varepsilon}.\frac{dW^S}{d\boldsymbol{\varepsilon}}\right) + 2\mathrm{grad}_S W^S + \mathbf{f}_s = \mathbf{0} \quad [9.61]$$

Its weak form is obtained as

$$\begin{aligned} 0 &= \int_\Omega -\mathrm{div}\left(\boldsymbol{\varepsilon}.\frac{dW}{d\boldsymbol{\varepsilon}}\right).\mathbf{v}dx + 2\int_\Omega \mathrm{grad}(W).\mathbf{v}dx + \int_\Omega \mathbf{f}.\mathbf{v}dx \\ &\equiv \int_\Omega \boldsymbol{\varepsilon}.\frac{dW}{d\boldsymbol{\varepsilon}}:\boldsymbol{\varepsilon}(\mathbf{v})dx - 2\int_\Omega W\mathrm{div}\mathbf{v}dx - \int_\Omega \mathbf{f}.\mathbf{v}dx \end{aligned} \quad [9.62]$$

For the specific form of the energy density, previous writing becomes

$$0 = \int_\Omega \boldsymbol{\varepsilon}.\left[\lambda(\rho)\mathrm{Tr}(\boldsymbol{\varepsilon})\mathrm{I} + 2\mu(\rho)\boldsymbol{\varepsilon}\right]\boldsymbol{\varepsilon}(\mathbf{v})dx - 2\int_\Omega \left\{\frac{\lambda(\rho)}{2}\{\mathrm{Tr}(\boldsymbol{\varepsilon})\}^2 + \mu(\rho)\boldsymbol{\varepsilon}:\boldsymbol{\varepsilon}\right\}\mathrm{div}\mathbf{v}dx - \int_\Omega \mathbf{f}.\mathbf{v}dx = 0 \quad [9.63]$$

This nonlinear problem is solved by a relaxation method: consider a given initial strain $\boldsymbol{\varepsilon}_0$, given as the solution of the elasticity problem in the bulk material: the idea is to freeze the actual strain $\boldsymbol{\varepsilon} = \boldsymbol{\varepsilon}_0$, and to solve the resulting linear problem, which writes

$$0 = \int_\Omega \boldsymbol{\varepsilon}.\left[\lambda(\rho)\mathrm{Tr}(\boldsymbol{\varepsilon}_0)\mathrm{I} + 2\mu(\rho)\boldsymbol{\varepsilon}_0\right]\boldsymbol{\varepsilon}(\mathbf{v})dx - 2\int_\Omega \left\{\frac{\lambda(\rho)}{2}\{\mathrm{Tr}(\boldsymbol{\varepsilon}_0)\}^2 + \mu(\rho)\boldsymbol{\varepsilon}_0:\boldsymbol{\varepsilon}\right\}\mathrm{div}\mathbf{v}dx - \int_\Omega \mathbf{f}.\mathbf{v}dx = 0 \quad [9.64]$$

The strain is then updated, thus $\boldsymbol{\varepsilon}_0 = \boldsymbol{\varepsilon}$, and the loop is continued up to satisfaction of the arrest criterion, for instance when $\left\|\frac{\boldsymbol{\varepsilon}_0 - \boldsymbol{\varepsilon}}{\boldsymbol{\varepsilon}_0}\right\| \leq \mathrm{eps}$, with eps a prescribed tolerance.

This nonlinear problem is solved by a relaxation method: consider a given initial strain $\boldsymbol{\varepsilon}_0$, given as the solution of the elasticity problem in the bulk material: the idea is to freeze the actual strain $\boldsymbol{\varepsilon} = \boldsymbol{\varepsilon}_0$, and to solve the resulting linear problem, which writes

The variational formulation of the bulk elasticity equilibrium equation is further written for the purpose of the finite element discretization in weak form by multiplying the strong form of equilibrium by a test (vector) function $\mathbf{v}$ chosen to vanish on the domain boundary, viz. $\mathbf{v} = \mathbf{0}$ on $\partial\Omega$. The weak form then writes as follows: we seek the displacement field $\mathbf{u}$ satisfying

$$\int_{\Omega} \boldsymbol{\sigma}(\mathbf{u}) : \boldsymbol{\varepsilon}(\mathbf{v}) dx - \int_{\Omega} \rho \mathbf{g}.\mathbf{v}\ dx = 0, \quad \forall \mathbf{v} /\ \mathbf{v} = \mathbf{0} \text{ on } \partial\Omega \qquad [9.65]$$

After substitution in [9.65] of the constitutive law, this delivers the following variational equation

$$\int_{\Omega} \lambda(\rho) \nabla.\mathbf{u} \nabla.\mathbf{v} + 2\mu(\rho) \boldsymbol{\varepsilon}(\mathbf{u}) : \boldsymbol{\varepsilon}(\mathbf{v}) dx - \int_{\Omega} \rho \mathbf{g}.\mathbf{v}\ dx = 0 \qquad [9.66]$$

9.5. Applications to hard tissues like bone

Trabecular bone is characterized by struts or plates of bone, or both, connected into a lattice-like network; interstitial spaces are filled with marrow. For trabecular bone of low apparent density, the structure is composed entirely of small struts with a thickness of about 250 μm and variable arrangement of lengths of 700–2,000 μm. In the forthcoming sections, the previous algorithms of the coupled bone internal and external adaptation shall be applied to hard tissues like trabecular bone at different mechanical levels. The external bone adaptation process at micro and macro-level is defined as the bone capacity to form or destroy bone surface in response to the mechanical loadings. Moreover, throughout this study, the external geometries of the proximal femur and trabecular strut are considered to be variable, i.e., changes in the external shape are allowed during bone local adaptation. Internal remodeling processes aim at regulating the local trabecular bone density. The divergence and trace of Eshelby stress are adopted as feedback control variables to determine the shape and bone density adaptations.

9.5.1. *Micromechanical simulations of the density and morphological changes of the trabecular structure*

We elaborate in this section a two-dimensional mathematical model of trabecular bone remodeling at meso- and microscale levels to simulate the surface-based addition and removal of material together with density evolution inside the trabeculae. It is essential to remark here that external bone remodeling always goes together with internal bone remodeling. The model at the microscopic scale is based on a finite element representation of individual trabecular struts in which the material properties of the subtrabecular elements are inconstant (they change over time). The remodeling stimulus is based on Eshelby stress, which is developed in the framework of the thermodynamics of irreversible processes. It is also assumed that the marrow spaces surrounding trabeculae do not contribute any stiffness or structural integrity to the region being modeled. The apposition or removal of bone material is described by modifying the external boundary as a function of time, and then the geometry is adapted.

In fact, when determining the stiffness of trabecular bone tissue, the non-uniformity of the mineral distribution within the trabeculae is usually not taken into account. The assumption that the trabecular bone tissue has uniform mechanical properties is not necessarily valid, since the bone tissue stiffness is strongly dependent on the degree of mineralization [VAN 01, VAN 04]. The individual trabeculae of trabecular bone are composed of packets of remodeled bone of different ages and thus different degrees of mineralization. Furthermore, the bone remodeling rate varies and is supposed to be relevant, for instance, to the amount of mechanical loading it experiences. These differences lead to a non-uniform mineral distribution within the trabeculae, which entails a variation of trabecular tissue stiffness.

The surface layer of trabeculae is frequently renewed, but tissue in the middle of any trabeculae can escape from this remodeling [BOY 93], and become older than the surface layer; this results in a large difference in mineral content between the core and the surface layer. Such non-uniform mineral distribution has been shown with a variety of techniques (e.g. [GAD 00, ROS 97]).

Although tissue mineral density (TMD) has been used to quantify changes in bone tissue as related to a variety of factors (see [BOY 93, MUL 08, MIL 07]), TMD represents only one constituent of bone (i.e., the mineral phase) and it ignores the contributions of the organic and water phases. Recently, alternative micro-scale approaches based on tissue composition have been proposed by [WAG 11] relating tissue mass densities to elastic properties. Such approaches would allow for the accurate prediction of the mechanical behavior of trabeculae. These authors present a method for estimating the tissue density and the isotropic elastic modulus from micro-computed tomography (microCT) derived tissue mineral density (gHA/cm^3). The derived relationship between micro-computed tomography and tissue density utilizes four parameters: the density of the three bone constituents (water, ash, and organics) and the volume fraction of the organic constituents The tissue density (g/cm^3) as a function of microCT (mineral) density (gHA/cm^3) is then given by the following relation

$$\rho_t = 0.6667\rho_{HA} + 1.144 \qquad [9.67]$$

The tissue-level bone elastic modulus (GPa) is related to the microCT (mineral) density (gHA/cm^3) as

$$\log_{10}(E_s) = -8.58 + 4.05\log_{10}\left(400/(1 + 0.504/\rho_{HA})\right) \qquad [9.68]$$

Combining the two previous relations yields the following relation between tissue elastic modulus (GPa) and tissue density (g/cm^3)

$$E_s = e^{4.05\ln\left(\frac{400(1.49\rho_t - 1.71)}{1.49\rho_t - 1.21}\right) - 19.75} \qquad [9.69]$$

The theoretical basics of the current numerical models of bone remodeling are derived from the observations that the bone architecture is related to the mechanical loads experienced and controlled by a regulatory mechanism.

The coupled internal and external growth model is presently applied to simulate the growth of the individual trabeculae of cancellous bone; Figure 8.3a represents the architecture of a two-dimensional trabecular unit cell. Such coupling model takes into account the adapted bone density to adjust the mechanical response of tissues with the external remodeling phenomenological model. Trabecular bone is assumed to be homogeneous, linear elastic and isotropic at the trabecular level, with mechanical parameters identified by adjusting the previous relation by a second order polynomial, relating tissue elastic modulus (GPa) to tissue density (g/cm^3) as

$$E_s = 6.5\rho_t^2 + 12.3\rho_t - 26.5 \qquad [9.70]$$

with a shear modulus also depending on trabeculae material density relying on the relation $G_s = E_s/2(1 + \nu_s)$; Poisson ratio is selected as $\nu_s = 0.3$. This 2nd order polynomial equation [9.70] is used within the tissue density range $1.5 \leq \rho_t \leq 2$; this defines the minimum and maximum values of trabeculae bone density over all elements in the design domain. Equation [9.70] determines the change in elastic properties versus bone tissue density, itself varying as a function of a given mechanical stimulus according to [9.32].

In order to simplify the representation of the trabeculae geometry, simulations at the scale of the mesoscopic will be performed, considering a repetitive unit cell (with periodic boundary conditions). At the mesoscale, the complex and very organized network of trabeculae forms the cancellous bone. Relying on microstructural modeling approaches, we address the modeling of trabecular bone as having an idealized periodic network of hexagonal cells. The considered hexagonal unit cell is composed of assembly of inclined and vertical trabeculae of lengths 1 mm with uniform trabeculae thickness 0.25 mm and an inclination angle $\theta = 30°$.

We next carry out a series of growth simulations at the scale of trabecular bone cell and then at the scale of the triple junction of the individual trabeculae to investigate the adaptation responses under different mechanical loading conditions.

Simulations are done with the open source finite element (FE) software Freefem++, relying on the variational formulation of the equilibrium equations. Let Ω be the region occupied by the trabecular bone material enclosed by a boundary $\Lambda = \partial\Omega$. The border is decomposed into four disjoint parts Λ_1, Λ_2, Λ_3, and Λ_4, where Λ_1 is part of the boundary along which the trabecular cell is fixed (bottom edges), and Λ_2 the part of the boundary (top edges) along which the different loading conditions P (P1, P2) are applied (Figures 8.3–8.5). Periodic boundary conditions are then imposed at the left and right edges (Λ_3 and Λ_4) of the hexagonal trabecular unit cell.

In order to illustrate the application of bone adaptation in numerical computations using the FE method, bone internal and external remodeling at trabecular level is predicted, starting from an initial geometry with a uniform density distribution of 1500 kg/m^3. We consider a static loading as a mechanical stimulus for internal and external bone remodeling. Three loading conditions are then investigated: case I, a compressive load (P1 = 25 MPa), case II, a shear load (P2 = 10 MPa), and case III, a combined compressive and shear loadings (P1 = 5 MPa, P2 = 10 MPa).

The two proportionality constants B_c and $\tilde{K}_g$ that play the role of gain in the adaptation process are chosen as follows: the remodeling rate B_c governing the velocity of the local adaptation process is taken as 0.01 (kg m^{-3})^2/(Pa.timestep) and the constant $\tilde{K}_g$ measuring the rate of the surface adaptation process is selected as 1.e-16 (m^4/N.timestep). The growth time step size is selected small enough, $\Delta t = 1$. After 50 iteration time steps, the growth simulation results are illustrated sequentially in Figures 8.3–8.5, for compressive, shear, and combined compressive to shear loadings.

It is important to note that during the bone remodeling process, the external mechanical load is transmitted in the form of stress/strain to the local bone site, which are sensed by osteocytes, which subsequently activate the regulating cells (osteoclasts and osteoblasts) causing changes of bone masses and mechanical properties. External mechanical loads provoke local strains in the bone tissue, and there is a consensus in the literature that the remodeling stimulus depends on the strain magnitude.

The effective strain (which is a scalar value summarizing the strain tensor) of trabecular bone is then computed using the following expression

$$\varepsilon_{eq} = \sqrt{\frac{2}{3}\varepsilon_{ij}\varepsilon_{ij}} \qquad [9.71]$$

in which ε_{ij} are the strain tensor components.

The distributions of equivalent strain in the adapted bone are shown in Figures 9.9(a)–9.9(b). In general, one can clearly see that high surface strains give locally rise to significant bone formation (growth). It is also remarkable that when the trabecular structure at the representative unit cell is subjected to alternative load directions, strains in the trabeculae surfaces become very high at the conjunction of the trabecular struts (Figures 9.9(b)–9.11(b)); this induces rapid bone formation along those trabecular edges. Thus, we deduce that bone formation most likely occurs at sites of high mechanical strain.

An additional simulation is performed over the hexagonal trabecular unit cell to give insights into different loading conditions. In this loading case (V), the part of the boundary along which the trabecular cell is fixed is selected to be the left edge, and the part of the boundary along which the loads (P1 = 5 MPa, P2 = 7 MPa) are applied is the right edge; periodic boundary conditions are then imposed at bottom and top edges. Under those combined compressive and shear loadings, the adaptive response of bone is presented on Figure 9.12 for a fixed number of growth timesteps. The adaptation simulations are also carried out over a maximum of 20 timesteps.

The external bone adaptation process at this level can then be defined as the bone capacity to form or destroy bone surface in response to mechanical stimuli. The tissue response is shown by modifications of the bone shape or its microstructure in addition to a slight reorganization of its material properties. Additionally, as a general remark in simulations of trabecular bone adaptation, one can notice the increasing trabecular thickening near the strut joints. Simple observations indeed reveal that the trabeculae are most of the time not perfectly uniform, but have a certain degree of non-uniformity, due to the natural variance in trabecular thickness for real trabecular bones. This in turn illustrates the importance of our obtained results of trabecular thickening near the strut joints; this will enhance the possibility of improving the representation of the topology of the bone unit cell structure. These computed variations of the trabeculae thickness and information on the strut morphology near the strut joints may indeed help to consider more realistic bone topology prototypes rather than the idealized uniform cross section of the trabeculae.

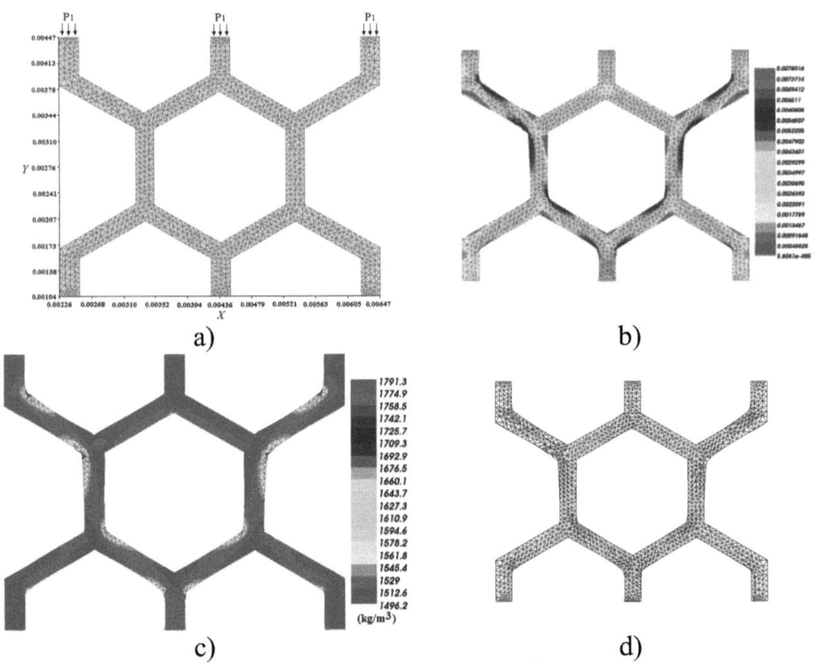

Figure 9.9. *Combined internal and external trabecular bone adaptation under compressive load at the top edge: a) initial geometry with uniform density of 1500 kg/m^3 submitted to compressive load; b) strain distribution; c) bone density distribution; and d) initial and new grown geometry meshes. The adaptations lasted 50 time steps. For a color version of this figure, see www.iste.co.uk/ganghoffer/ biomechanics.zip*

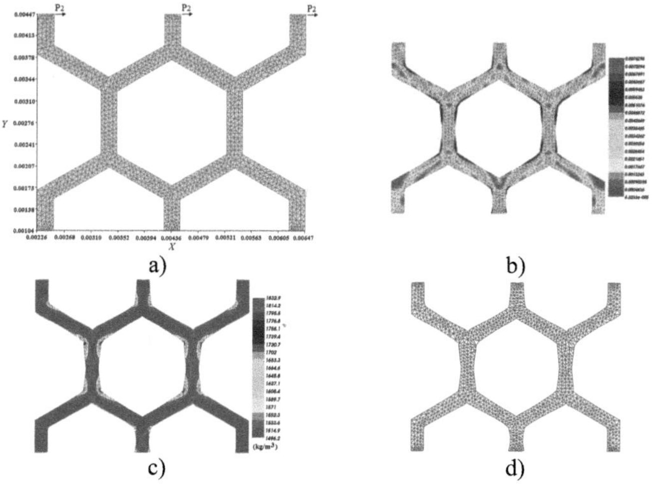

Figure 9.10. *Combined internal and external trabecular bone adaptation under shear load at the top edge: a) initial geometry with uniform density of 1500 kg/m^3 submitted to shear load; b) strain distribution; c) bone density distribution; and d) meshes of the new grown geometry. The adaptations lasted 50 timesteps. For a color version of this figure, see www.iste.co.uk/ganghoffer/biomechanics.zip*

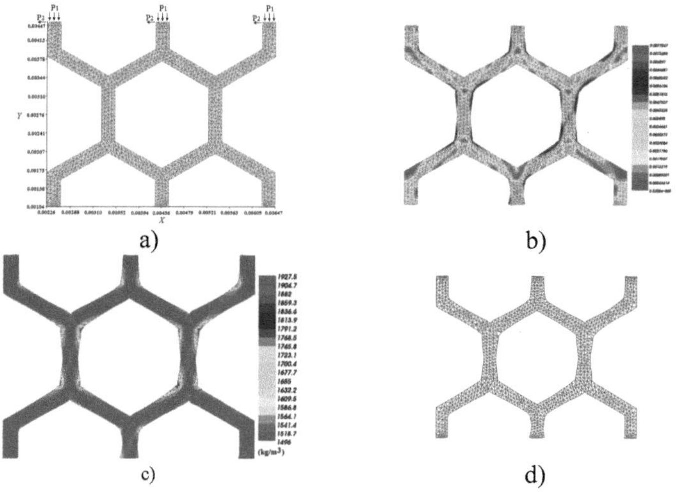

Figure 9.11. *Combined internal and external trabecular bone adaptation under compressive and shear loads at the top edge: a) initial geometry with uniform density of 1500 kg/m^3 submitted to combined compressive and shear loads; b) strain distribution; c) bone density distribution; and d) meshes of the new grown geometry. The adaptations lasted 50 timesteps. For a color version of this figure, see www.iste.co.uk/ganghoffer/biomechanics.zip*

In order to quantitatively investigate the morphological changes in trabecular bone, we next compute the bone volume fraction or bone relative density. In such porous open-celled structures, the apparent density is generally the most important factor affecting the effective mechanical behavior. The apparent density is the density of a volume of the porous material, including voids, whereas the material density is the density of the structural matrix itself. The relative density is equivalent to BV/TV (bone volume fraction), which stands for bone volume over total volume respectively. The bone volume fraction is presently computed before submitting the trabecular bone to mechanical stimulus and after stimulation to give insights about how much bone volume fraction has been changed due to growth (evolution of the external shape). In order to quantify the external adaptation response, we compute the initial bone volume fraction (original shape geometry) and compare it with the bone volume fraction of the grown geometry; this has been computed for all loading cases. The original bone volume fraction (before evolution occurs) is equal to 0.24. In loading case I, the volume fraction is found equal to 0.25, which means that about 4% increase of bone volume is attained as a result of external growth under compressive load exerted at the top edge. Similarly, in the loading cases II, III, the volume fraction of the grown bone is very similar, about 0.265; this means that about 10.5% of bone volume increase is attained due to the shear and combined compressive and shear loads. As a final growth simulation comparison (case V), it is found that the volume fraction of bone is equal to 0.27; this implies that approximately 12.5% of bone has been added (in terms of volume) to the original geometry.

The developed growth model is next applied to simulate the growth of the individual trabeculae of cancellous bone. Simulations at the scale of the triple junction of the individual trabeculae will be performed, considering the same geometrical (strut lengths and thicknesses) and mechanical microparameter as in previous simulations. We simulate separately two static loadings as a mechanical stimulus for external bone adaptation, a compressive load (P1 = 10 MPa) and a shear load (P2 = 5 MPa); in both cases, the nodes are fixed on the bottom edges.

Bone undergoes morphological changes in response to applied mechanical loads by modifying its external geometry as well as its internal structure, i.e., external and internal remodeling occur simultaneously. At this scale nevertheless, bone remodeling is nearly a purely surface phenomena, traduced by an evolution of the external shape of the individual trabeculae (density variations have been shown to be negligible).

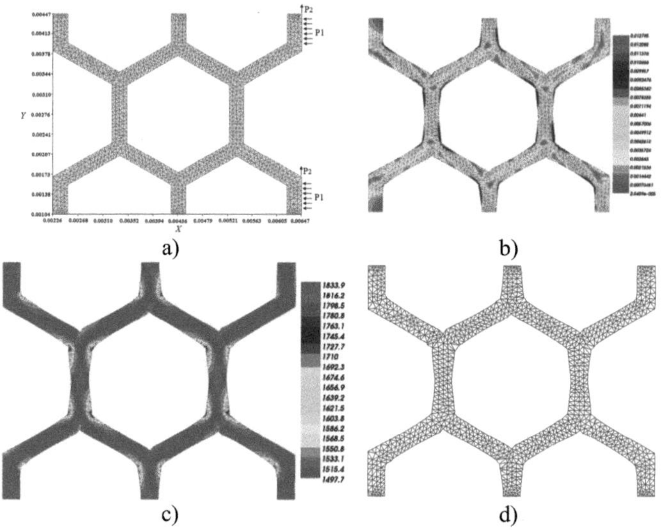

Figure 9.12. *Combined internal and external trabecular bone adaptation under compressive and shear loads at the right edg: a) original bone geometry with uniform density of 1500 kg/m^3 submitted to combined compressive and shear loads; b) strain distribution; c) bone density distribution; and d) plot of the mesh of the newly grown bone geometry. Bone adaptations are shown after 20 growth-timesteps. For a color version of this figure, see www.iste.co.uk/ganghoffer/biomechanics.zip*

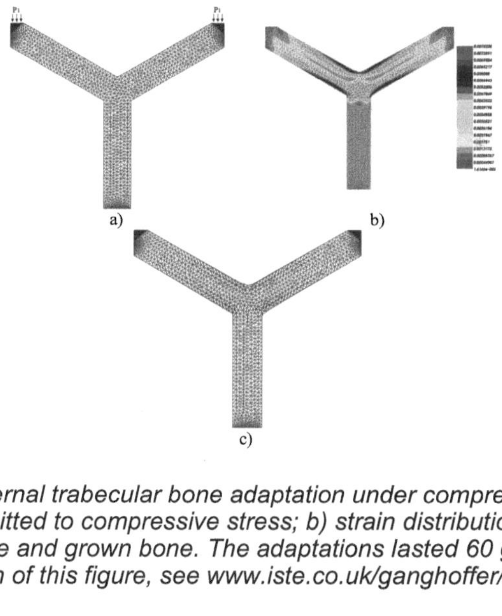

Figure 9.13. *External trabecular bone adaptation under compressive: a) original configuration submitted to compressive stress; b) strain distribution; c) overlay plots of the original bone and grown bone. The adaptations lasted 60 growth-timesteps. For a color version of this figure, see www.iste.co.uk/ganghoffer/biomechanics.zip*

We plot the new external geometrical shape, as indicative of the trend of the external remodeling process. The proportionality constant K_g governing the rate of surface adaptation process in [9.51] is taken as 10^{-11} (m^4/N.timestep). The geometry is evolved as a function of time (external remodeling) using the FE method: the adapting process leads to the mesh nodes repositioning without modifying the connectivity or the definition of additional elements (Figures 9.13 and 9.14).

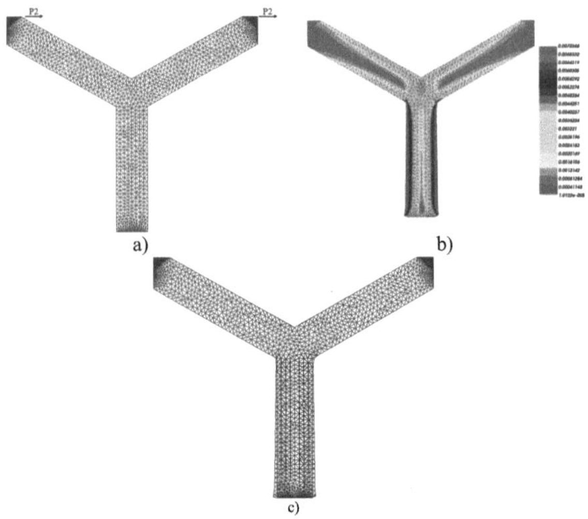

Figure 9.14. *External trabecular bone adaptation under shear stress: a) original configuration submitted to shear stress; b) strain distribution; c) overlay plots of the original bone and grown bone meshes. The adaptations lasted 60 growth-timesteps. For a color version of this figure, see www.iste.co.uk/ganghoffer/biomechanics.zip*

9.6. Macroscopic viewpoint

We next perform structural remodeling computations of the evolution of the external shape jointly with adaptation of the apparent internal density for a sample of the human femur subjected to physiological loads, in order to illustrate the powerfulness of the developed methodology. Actually, this type of bone remodeling is physiologically caused by the remodeling of the trabecular (spongy) bone.

In order to model surface growth over time, it is necessary to account for internal bone remodeling occurring in the entire bone. Internal remodeling is considered as the mechanism for the redistribution of the apparent trabecular bone density under the action of the physiological loading. The trace of Eshelby stress which identifies to the energy density is adopted as previously as the biomechanical signal that controls internal remodeling of trabecular bone. For external remodeling, the outer geometry of the femur bone adapts due to the applied loads by allowing surfaces to move freely according to a constitutive relation based on the divergence of Eshelby stress. In our structural model, trabecular bone is represented as a cellular structure with the cell walls modeled as a bulk material. Internal and external bone growth shall be expressed in algorithmic format from a variational formulation of the strong form of the field equations, leading to a numerical implementation for 2D situations.

9.6.1. *Homogenization scheme for trabecular bone*

Bone is a particularly complex multiscale material, which is hierarchically organized to provide maximum performance with a minimum amount of material. At the nanoscale, bone is composed of collagen fibers, apatite crystals and water; at the sub-microscale (1–10 lm), the arrangements of collagen fibers reinforced with crystals form lamellas. These latest are packed and organized in different orientations to form cylindrical struts called trabeculae at the microscale (several hundred of micrometers). At the mesoscale, the complex and very organized network of trabeculae forms the trabecular bone. The macro scale represents the whole bone which is constituted by both the trabecular bone and the cortical bone. Bone is a heterogeneous and anisotropic material normally present into two major forms of bone tissue at the macroscopic level, cortical bone (at the outer surface) and trabecular bone (Figure 9.15). Trabecular bone is very complex due to its randomness and spatial heterogeneity. In order to simplify the forthcoming analysis, one can choose to represent bone as having an idealized periodic microstructure: in this work, we model bone as having a periodic prismatic structure in a two-dimensional context.

Discrete homogenization is a tool utilized to determine the effective macroscopic material properties of microstructured materials. Trabecular bone defines the microstructure of bones, mainly located within the confines of the cortical bone coverings. Homogenization aims then at identifying the material properties of trabecular bone at the apparent level; we assume the architecture of trabecular bone consists of an elastically isotropic material.

Following the procedures of discrete homogenization introduced in Chapter 4, the constitutive equation for the equivalent Cauchy continuum writes

$$\{\sigma_x,\sigma_y,\sigma_{xy}\}=[C]\left\{\frac{\partial u_1}{\partial x},\frac{\partial u_2}{\partial y},\frac{1}{2}\left(\frac{\partial u_2}{\partial x}+\frac{\partial u_1}{\partial y}\right)\right\} \qquad [9.72]$$

with u_1, u_2 the two components of the planar displacement field, and $[C]$ the equivalent rigidity matrix written for an orthotropic material in terms of Young moduli (E_x, E_y), the shear modulus G_{xy} and the two Poisson's ratios (ν_{xy} and ν_{yx}) as

$$[C] = \frac{1}{1-\nu_{xy}\nu_{yx}} \begin{bmatrix} E_x & \nu_{yx}E_x & 0 \\ \nu_{xy}E_y & E_y & 0 \\ 0 & 0 & 2G_{xy}\left(1-\nu_{xy}\nu_{yx}\right) \end{bmatrix} \qquad [9.73]$$

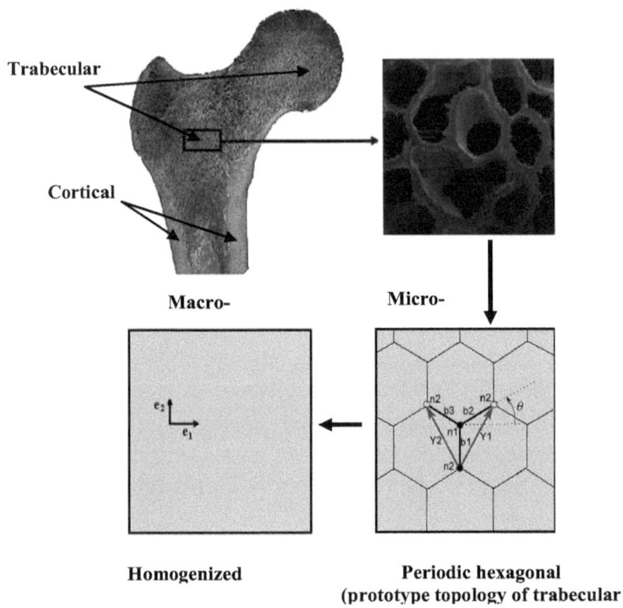

Figure 9.15. *Trabecular bone and microstructure homogenization towards an effective continuum*

A 2D geometrical model of a hexagonal structure is then considered as an idealized representation of trabecular bone, for which the Young's modulus, shear modulus, are identified at the mesoscopic level from a micromechanical analysis. The model geometry consists of struts of length L for the inclined beams and h for the vertical beams with uniform trabeculae thickness t. The struts of length L are inclined with the angle $\theta = 30°$.

The bony structure has been further assumed to be described by a linear isotropic behavior; this simplifies the previous effective constitutive law in [9.73], which only involves two independent mechanical moduli.

Homogenization results provide the evolution of the effective Young's and shear moduli of trabecular bone versus apparent density. The apparent tensile and shear moduli of trabecular bone are obtained from discrete homogenization as power law functions of the apparent density (kg/m^3),

$$E=72.5\rho^{2.4} \text{, } G=36.25\rho^{2.4}/(1+\nu) \qquad [9.74]$$

with the effective Poisson ratio assumed to be independent of ρ , $\nu(\rho)=\nu=0.3$.

We next perform structural remodeling computations of the evolution of the combined internal and external adaptation of bone for a sample of the human femur subjected to physiological load.

9.6.2. *Simulation of joined internal and external remodeling*

The shape of part of the human femur is presented by B-splines, and then meshed; simulations are done with the open source FE software Freefem++, adopting the variational formulation of the equilibrium equations. Let Ω be the region occupied by the effective homogenized isotropic bone material enclosed by a boundary $\Lambda=\partial\Omega$. The border is decomposed into two disjoint parts Λ_1 and Λ_2, where Λ_1 is the part of the boundary along which the femur is fixed, and Λ_2 the part of the boundary along which the loads P (P1, P2) are applied (Figure 9.16). The mechanical load environment for bones is complex, with force components resulting from joint contact, muscles, tendons, and soft tissue, and the load magnitudes and directions vary constantly as a person moves about. The load values and directions are usually corresponding to a static load case representing the stance phase of gait for a person weighing 80 kg. As a representative loading condition, the bone structure is subject to three load cases (one-legged stance, abduction, and adduction). Remodeling simulations are currently conducted under a single-loading condition. We consider static loading as a mechanical stimulus with the force corresponding to the instant when the foot touches the floor (case 1), which is assumed to be predominant (higher force magnitude than the other load cases). Load values are taken from [BEA 90] and loads (distributed forces) are applied at the head of the femur and as the reaction of the muscles located at the great trochanter (Figure 9.16). The first load condition P1 = 2317 N is assigned at an angle of $\theta_1=24°$ and the second load P2 = 703 N is positioned at an orientation of $\theta_2=28°$ (the angles are given with respect to the vertical axis).

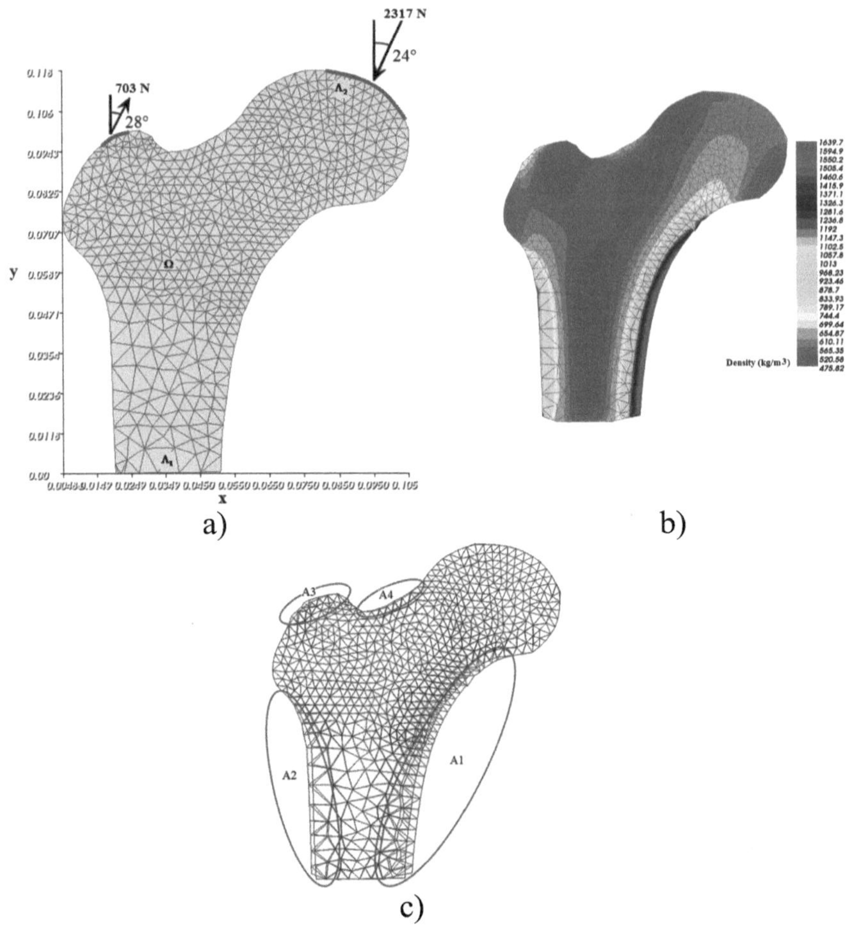

Figure 9.16. *Simulations of internal and external bone growth at the macroscopic level: (a) loading conditions and original structure with uniform distribution at the beginning of simulation: initial density= 800 kg/m^3; (b) bone density distribution after 140 growth-timesteps; (c) overlay plots of the original bone and grown bone meshes after 140 growth-timesteps. For a color version of this figure, see www.iste.co.uk/ganghoffer/biomechanics.zip*

Bone growth results show that the external shape of the femur bone and its internal morphology simultaneously adapt to the surrounding mechanical environment (Figure 9.16). Figure 9.16(a) shows the geometry of a 2D frontal plane of a proximal femur, which is taken as the initial configuration in our structural internal and external remodeling analysis. The process starts with an assumed uniform density distribution chosen as 800 kg/m^3. The time step size (iteration) is selected as $\Delta t = 1$. The combinative algorithms of both adaptive internal and external remodeling lead to changes of both the local bone density and external shape geometry. The adaptation simulations are carried out over a maximum of 140 timestep growth. At the end of the adaptation process, one can observe from the bone density distribution (Figure 9.16(b)) the formation of two high density external layers along the diaphysis of the femur: a zone of low density corresponding to the medullar channel inside the diaphysis and a complex distribution of density of the femoral head and trochanter including three different density zones with the head surrounded by a lower density zone (low density region on the distal part of the head).

In addition to bone density adaptation and based on surface remodeling, we generated a computational framework for simulating the movement of the surface of femur bone based on the divergence of Eshelby stress. Thereby, the evolution of the whole bone geometry and its shape due to surface growth is simulated. After a maximum of 140 growth timesteps, we observe that most parts of the initial bone show geometrical change during bone remodeling, with however only the periosteal surface on the diaphysis presenting a noticeable change (location A1 and A2). The overlay plots of the original bone and grown bone meshes (Figure 9.16(c)) show that much of the diaphysis is formed by appositional bone growth. From the assessment of results on Figure 9.16(c), we compare the bone outline of the initial structure with that of the final structure. Bone remodeling changes the shape slightly in most regions, except the periosteal surfaces on the diaphysis; growth is more significant on the right surface than on the left. Slight changes occur in regions of great trochanter (A3) and femoral neck cortex (A4).

In conclusion, in this macroscopic point of view, we developed a computational model, based on mechanical loading-induced bone remodeling, which can simulate the changes of internal density (to account for trabecular morphology variations) as well as the simultaneous overall shape changes, especially on the periosteal surfaces in the diaphysis. Simulation results show that the external shape of the femur bone and the internal morphology of trabecular bone simultaneously adapt to the surrounding mechanical environment. This type of external remodeling simulation platform could be useful for situations like implant surgery, and fracture healing where more important modifications of the external geometry occur.

10

Integrated Remodeling to Fatigue Damage Model of Bone

Bone is a living tissue that continually undergoes damage under normal everyday mechanical loadings. It maintains its structural integrity by a continuous process of damage detection and repair, and by adaptation to changes in the mechanical environment. These adaptive changes are regulated by the so-called process of bone remodeling. The aim of the current chapter is to develop an integrated remodeling-to-fracture model to simulate the bone remodeling process considering the anisotropy of trabecular bone. For this purpose, models for density of continuum solid bodies are developed in the framework of the thermodynamics of irreversible processes, considering bone elasticity with damage properties, in which the damage stimuli is incorporated. The asymptotic homogenization technique is involved to derive the effective anisotropic continuum mechanical properties of the trabecular bone from an initially discrete structure representative of femur bone microstructure, consisting of a piling up of hexagonal planar cells. As a consequence, relationships are established between the apparent density of bone and its effective anisotropic properties, that is the elastic and shear moduli and Poisson's ratio. Damage is directly incorporated into the effective density, the evolution of which is computed versus time. This formalism is involved to simulate bone remodeling for 2D geometries in the context of damaged anisotropic bone elasticity, showing the influence of the external mechanical stimuli on the evolution of the density of bone.

10.1. Introduction

As a result of daily loading activities, fatigue and fracture may be present and grow within bone. On the other hand, bone is a living tissue having the ability to

---

Chapter written by Jean-François GANGHOFFER and Ibrahim GODA.

sense damage and repair itself within a certain level of damage. Bone is also able to adapt its mechanical properties to the varying mechanical and physiological environments in a process commonly coined bone remodeling. Accordingly, the analysis of bone remodeling considering bone fatigue and damage is an important issue in biomechanics.

Bone is a living tissue, a remarkable and extremely adaptable substance in the human body. It continuously undergoes processes of growth, reinforcement, and resorption which occur in living situation; these processes are collectively termed bone remodeling. Additionally, bone has the ability to adapt its external shape as well as its internal structure to variations in its mechanical environment. The adaptive response of bones to changes in load history is called bone remodeling [WOL 92], with a classification as either internal or external remodeling [COW 79]. External or surface remodeling results in a change of the external geometry of the overall bone structure, and occurs by the resorption or deposition of bone material on the surfaces. To the contrary, internal remodeling refers to the resorption or deposition of bone material only, accompanied by the removal and densification of the cancellous bone, but no change in its overall shape.

Bone remodeling provides a mechanism to repair damage as it is well known that daily activities damage bone; this damage is normally repaired in a continuous remodeling process. During physiological daily activities, bone is regularly subjected to repetitive fatigue loading and when this load goes beyond a certain level of strain or stress, damage is induced into the bone tissue. The presence of such damage causes a reduction in mechanical properties of the bone material such as stiffness and strength, due to the formation of microcracks. These microcracks have been implicated in physiological phenomena including remodeling [MOR 93], and bone adaptation [PRE 94, LEE 99].

A few researchers considered crack damage and fatigue as the inducer of bone remodeling [CAR 87, PRE 94] hypothesized that fatigue damage acts to produce a bone maintenance stimulus. [PRE 94] proposed the hypotheses required to develop damage-adaptive remodeling laws and derived a remodeling law to predict the time course of bone adaptation. Furthermore, [MAR 92] derived an analytical expression for the remodeling equilibrium state of damage based on osteonal repair of cracks. [PRE 95] proposed a microstructural finite element analysis to explore the relationship between damage formation and local strain of osteocyte containing lacunae for various types of damage. [RAM 02] proposed an extension of the adaptive-elasticity theory in order to include the contribution of bone microdamage as a stimulus. [TAY 07] proposed that bone tissue is a continuous system incorporating the growing and repairing of microcracks. [MCN 07] showed that by incorporating both microdamage and strain as mechano-regulatory stimuli, the creation and refilling of a resorption cavity along the surface of bone trabeculae can be simulated. [WAN 11] proposed a numerical model based on the theory of bone remodeling to predict the evolution of the trabecular bone architecture within the vertebral body. Those authors assumed that the multicellular units activation probability depends on the state of damaged bone tissue (damage accumulation), which is evaluated as to bone fatigue damage.

The resistance of bones to damage by self-repair mechanisms and adaptation to the environmental conditions are the most important characteristics of bone. Such adaptive changes are regulated by adaptive physiological process called bone remodeling. A better understanding of the process requires applying the elastic damage theory under the hypothesis of small strain to the bone structure and investigating its mechanical response. From the standpoint of biomechanics, there is a significant need to attempt to put some of the theoretical framework in which a hypothetical statement can be made regarding the damage consequence on the process of adaptation of bone. To reach this goal, we use the framework of continuum thermodynamics to introduce and model the presence of microcracks. Thus, the main purpose of this chapter is to introduce a strategy for modeling the interaction of damage with the remodeling process in bones.

Most models (except for a few, including [WAT 04, KOW 10]) consider bone as an isotropic structural material; this assumption is clearly a shortcoming of the treatment of trabecular bone, ignoring the importance of anisotropic bone material behavior in the remodeling process [BEA 90, WEI 92].

In this chapter, we couple the material density and bone anisotropy; it is assumed that at the microscopic level, the bone trabeculae consists of an isotropic, Timoshenko-type elastic cellular material, whereas cancellous bone behaves as a homogeneous, anisotropic continuum at the macroscopic level. Anisotropic properties of trabecular bone are difficult to access by observations, thus pointing toward the need for predictive models of cancellous bone, based on its architecture and microstructural properties. Moreover, due to the variation of bone architecture (geometry) according to location, homogenization models will exhibit the advantage of their versatility, in the sense that they are able to provide quantitative estimates of bone mechanical properties versus the bone architectural and micromechanical parameters. The main purpose of this chapter is accordingly to set up a model which combines an anisotropic internal remodeling rule with fatigue damage, based on continuum damage theory. We propose to account for the trabecular structure in the proximal femur by using an orthotropic material model to represent the effective properties of the trabecular bone in a 2D framework on the basis of discrete homogenization. In our structural model, trabecular bone is represented as a cellular structure with the cell walls modeled as a Timoshenko beam. Consequently, at the macroscopic level, the constitutive laws for trabecular bone shall account for morphological effects, including anisotropy as well as damage. Internal growth of bone shall be expressed in algorithmic format from a variational formulation of the strong form of the field equations, leading to a numerical implementation for 2D situations.

10.2. Balance laws and constitutive equations of density growth

Tissue growth associated with changes in volume and mechanical properties (the so-called local evolution of materials) can be interpreted as local material transformations or local deformation measures. The evolution of internal bone density formulated in Chapter 9 is recalled:

$$\frac{\mathrm{d}\rho}{\mathrm{d}t} = \Gamma^g \rho \qquad [10.1]$$

with the growth rate Γ^g therein receiving the following expression versus the growth mapping and its rate:

$$\Gamma^g = \mathrm{Tr}\left(\dot{\mathbf{F}}_g \mathbf{F}_g^{-1}\right) \qquad [10.2]$$

We have further assumed that an incompressible tissue, so that $\mathrm{J}_{\mathrm{a}} = 1$ and that bone tissue growth then occurs by densification only, so that $\mathrm{J}_g = 1$. Relying on that the growth rate depends on the Eshelby-like tensor, we propose a constitutive relation for the mass source, wherein the growth rate is proportional to the trace of the Eshelby tensor

$$\Gamma^g = \frac{\mathrm{B}_{\mathrm{c}}}{\rho^2} \mathrm{Tr}\left(\boldsymbol{\Sigma}(u) - \boldsymbol{\Sigma}_0\right) \qquad [10.3]$$

where B_{c} is a proportionality constant measuring the rate of the adaptation process (it governs the velocity of the adaptation process), $\boldsymbol{\Sigma}(\mathbf{u})$ is the actual volumetric Eshelby stress tensor receiving the expression

$$\boldsymbol{\Sigma}(\mathbf{u}) = \mathrm{W}(\mathbf{u})\mathbf{I} - \epsilon(\mathbf{u})\frac{\partial \mathrm{W}(\mathbf{u})}{\partial \epsilon} \qquad [10.4]$$

and $\boldsymbol{\Sigma}_0$ the threshold value for the volumetric Eshelby stress stimulus (reference value). The identity tensor is therein denoted $\mathbf{I}$. The evolution equation of the density, equation [10.4], combined with the definition of the mass source Γ^g in equation [10.3] then delivers the following evolution law for internal bone density

$$\frac{d\rho}{dt} = \frac{B}{\rho}\mathrm{Tr}\left(\boldsymbol{\Sigma}(\mathbf{u}) - \boldsymbol{\Sigma}_0\right) \qquad [10.5]$$

This equation describes bone adaptation, and more precisely whether the tissue stress stimulus will induce bone formation, bone resorption or equilibrium, according to the sign and value of the rate of change of the apparent bone density. Equilibrium is reached when the Eshelby stress attains the homeostatic value $\boldsymbol{\Sigma}_0$, supposed to be known at each point.

It is worth noting that the trace of Eshelby stress is the strain energy density. Accordingly, the volumetric strain energy density (or the trace of Eshelby stress) is the mechanical stimulus that triggers bone adaptation locally, as a scalar quantity that incorporates local stress and strain tensors. We then propose a formulation for bone adaptation utilizing the strain energy density function as the remodeling signal representing the deformational energy at any point. In particular, the remodeling stimulus is obtained by the calculation of strain energy per unit density (strain energy stored per unit mass of bone) taking the following form:

$$\frac{d\rho}{dt} = B_c\left(\frac{W(u)}{\rho} - \frac{W_t}{\rho_m}\right) \equiv f(t,\rho) \qquad [10.6]$$

This modified evolution equation ensures a good stability behavior of the time-dependent remodeling. The driving mechanism for adaptive activity is then the deviation between the actual energy per unit mass $W(u)/\rho$ from a site-specific homeostatic equilibrium W_t/ρ_m; the scalar ρ_m therein is the maximum density. The internal remodeling process in equation [10.6] is considered to have converged when the rate $d\rho/dt$ tends to zero (equilibrium is then reached) or when the density secures a minimal (10 kg/m^3) or maximal value (1740 kg/m^3). The remodeling rate B_c is chosen from [CHE 07a] to be 1.0 (kg m^{-3})^2/(Pa.time).

Although there are many studies [HAZ 01, HE 00] which consider a constant reference value for bone adaptation of 1000 $\mu\epsilon$ (this last notation means microstrain, so that one computes the strain ε_0 based on the number of loading cycles), [QIN 98] showed, however, that the reference value of remodeling strain depends on the daily loading cycles. Based on experimental fitted curves, a continuous nonlinear relation between strain and daily loading cycle is obtained by previous author as

$$\epsilon_0 = 10^{2.28}\left(5.6 - \log_{10} N\right)^{1.5} \qquad [10.7]$$

where N represents the daily loading cycle number and ϵ_0 is the resultant micro-strain magnitude ($\mu\epsilon$). Hence, the reference value of W_t can be obtained using relation [10.6], with $W_t(u) = 1/2\sigma_0\epsilon_0$.

10.3. Elastic damage bone remodeling process

Bone adaptability allows for efficient repair, which in turn helps to prevent fractures. Bone can be remodeled in response to external loads to give denser bone matrix in regions that are subjected to higher levels of stress. These external loads induce changes in the mechanical fields and damage, such as macrocracks or microcracks [TAY 07]. A damage-adaptive bone remodeling model is presently developed, wherein microdamage is incorporated as a stimulus.

Bone continuously checks for the presence of cracks and other types of damage within its microstructure, and deliberately replaces damaged regions with new material through remodeling processes. In addition to removing fatigue damage, the remodeling process leads to an adjustment of bone mechanical properties in response to imposed loads. The overall aim is thus to numerically simulate the remodeling of femur bone using an elastic-damage theory in the context of small displacements and small strains. In the theory of elastic damage mechanics, the mechanical properties may degrade gradually as damage progresses; combining damage definition and mechanical properties evolution in framework of bone remodeling, the elastic and shear moduli at each location will be calculated from density relying on a micromechanical scheme. Subsequently, we apply the theory of elastic-damage mechanics to the bone remodeling process.

The continuum behavior of bone material is modeled based on the concept of continuum damage mechanics, as an effective elastic anisotropic behavior coupled to fatigue damage. We consider the following rational elastic-damage potential energy based on the physically appealing effective stress concept [LEM 78, LEM 84, JU 91]

$$W = \frac{1}{2}\left(1 - D_e\right)\epsilon : C^0 : \epsilon \qquad [10.8]$$

This equation describes the strain energy density function coupled to the fatigue damage, described by the scalar variable D_e (ranging from 0 to 1, $D_e = 0$ corresponding to the undamaged state while $D_e = 1$ physically means the local rupture state or local failure), ϵ is the elastic strain tensor, C^0 the virgin (undamaged) elasticity stiffness tensor.

The effective stress σ incorporates damage as

$$\sigma = (1 - D_e) C^0 : \epsilon \qquad [10.9]$$

This last relation entails that the overall elastic-damage stiffness tensor C^D takes the form

$$C^D = (1 - D_e) C^0 \qquad [10.10]$$

For simplicity reasons, a homogenized measure of isotropic damage variable D_e is retained coupled with the orthotropic bone behavior that will be obtained later on from a micromechanical analysis of the bone microstructure.

In order to describe the degree of damage D_e, we shall adopt the non-linear damage model for high cycle fatigue under purely elastic strains proposed by [HAM 11], expressing as

$$D_e = 1 - \left(1 - \left(\frac{N}{N_f}\right)^{1/(1-\alpha_f)}\right)^{1/(1+\beta_f)} \qquad [10.11]$$

where $\alpha_f = 0.2$ and $\beta_f = 0.4$ represent the material fatigue parameters of bone [HAM 11], and N / N_f is the proportion of fatigue life with N the cycles per day and N_f the number of cycles to failure.

The model expressed in [10.11] is plotted in Figure 10.1; it shows the nonlinear increase of the damage variable (D_e) with the normalized number of cycles (N/N_f); close to final failure, the damage variable increase significantly, representing the critical damage.

The strain energy density, considered as the mechanical stimulus controlling the bone adaptation in the presence of damage, is pictured in Figure 10.2. Note that when the damage density reaches its maximum value ($D_e = 1$), the stimulus is set to zero while for a virgin bone tissue ($D_e = 0$), the stimulus is maximum because the stiffness of the virgin material is not yet reduced.

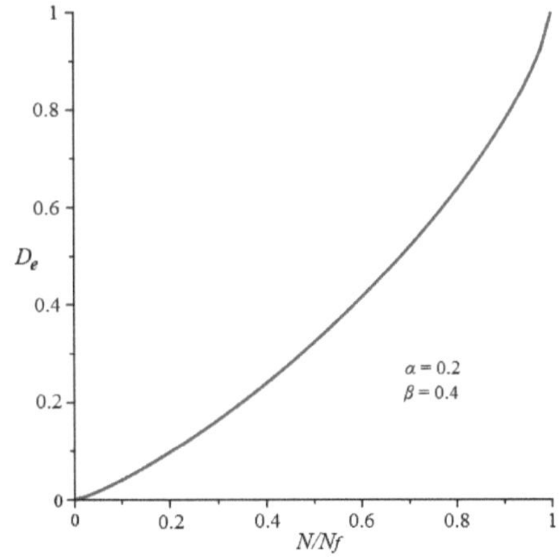

Figure 10.1. *Evolution of damage variable (D_e) versus the normalized number of cycles (the ratio of the number of cycles N to the number of cycles to failure N_f)*

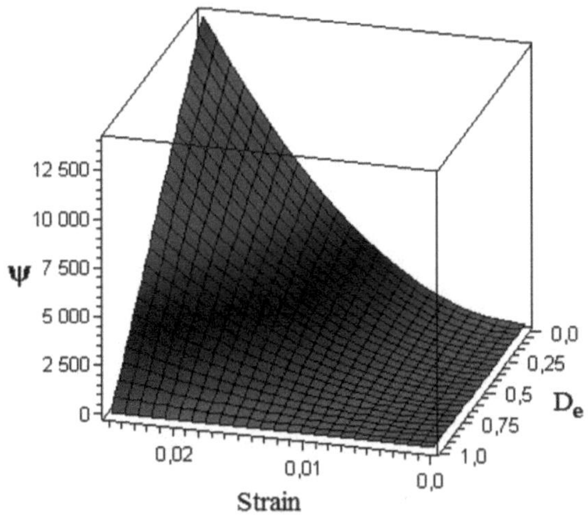

Figure 10.2. *3D plot of strain energy density coupled with damage, function* $W = \frac{1}{2}(1-D_e)\sigma\epsilon = \frac{1}{2}(1-D_e)E^{*}\epsilon^{2}$ *. For a color version of this figure, see www.iste.co.uk/ganghoffer/biomechanics.zip*

Under plane stress conditions, it holds that $\sigma_{zz} = \sigma_{zx} = \sigma_{zy} = 0$; Hooke's law for an orthotropic material coupled to fatigue damage then takes the following form, expressed in terms of the homogenized engineering constants (that will be subsequently determined from the trabecular bone architecture), the two Young moduli (E_x^*, E_y^*), the shear modulus G_{xy}^* and the two Poisson's ratios (ν_{xy}^* and ν_{yx}^*).

$$\begin{bmatrix} \sigma_x \\ \sigma_y \\ \sigma_{xy} \end{bmatrix} = \underbrace{(1-D_e)C^0}_{C^D} : \epsilon = \frac{(1-D_e)}{1-\nu_{xy}^*\nu_{yx}^*} \begin{bmatrix} E_x^* & \nu_{yx}^* E_x^* & 0 \\ \nu_{xy}^* E_y^* & E_y^* & 0 \\ 0 & 0 & 2G_{xy}\left(1-\nu_{xy}^*\nu_{yx}^*\right) \end{bmatrix} \begin{bmatrix} \epsilon_x \\ \epsilon_y \\ \epsilon_{xy} \end{bmatrix} \quad [10.12]$$

These engineering parameters will be computed later on, separately from the proposed micromechanical scheme introduced in Chapter 6.

As the apparent density of bone tissue changes during the process of remodeling, the elastic and shear moduli change accordingly. Next, we establish the relationships between the apparent density and the anisotropic mechanical properties of trabecular bone using a discrete homogenization method. As a consequence, an anisotropic homogenization model of cancellous bone, considered as a cellular solid (modeled as two-dimensional lattices of articulated beams), shall be constructed in order to derive its effective mechanical properties from the trabecular bone microstructural parameters.

Homogenization aims to identify the material properties of trabecular bone at the apparent level. We adopt the viewpoint of trabecular bone as a cellular solid consisting of a quasi periodical lattice of cells having a hexagonal topology. We assume the architecture of trabecular bone consists of an elastically isotropic material.

Following the procedures present in section 10.3, the constitutive equation for the equivalent continuum of Cauchy stresses is written

$$\left\{\sigma_x, \sigma_y, \sigma_{xy}\right\} = \left[C^0\right] \left\{ \frac{\partial u_1}{\partial x}, \frac{\partial u_2}{\partial y}, \frac{1}{2}\left(\frac{\partial u_2}{\partial x} + \frac{\partial u_1}{\partial y} \right) \right\} \quad [10.13]$$

where $\left[C^0\right]$ is the equivalent undamaged rigidity matrix, and u_1, u_2 the two components of the planar displacement field.

A 2D geometrical model of a hexagonal structure is considered as an idealized representation of trabecular bone, for which the Young's modulus, shear modulus, and Poisson's ratio are identified at the mesoscopic level. The model geometry consists of struts of length L for the inclined beams and h ($h = 0.5\ L$) for the vertical beams with uniform trabeculae thickness t. The struts of length L are inclined with the angle $\theta = 30°$. The tensile and shear moduli of trabecular bone express as power law functions of apparent density (kg/m^3), whereas the Poisson's ratio is linearly related to density, viz.

$$E_x^* = 7.8\rho^{2.72},\quad E_y^* = 2.7\rho^{2.76},\quad G_{xy}^* = 12.3\rho^{2.55},$$
$$\nu_{xy}^* = -0.0005\rho + 1.57;\quad \nu_{yx}^* = -0.0002\rho + 0.69 \qquad [10.14]$$

Those relations are then inserted in [10.13], leading to a formulation of Hooke's law for the orthotropic plane-stress bone material in terms of apparent bone density ρ and fatigue damage D_e.

Furthermore, combining the damage definition and elastic and shear modulus evolutions in the framework of bone remodeling involves the determination of the damaged elastic and shear modulus at each location, based on the knowledge of the density. Note that the state of anisotropy is evolving over the course of remodeling, due to the evolution of internal density and the distinct scaling laws of the moduli versus density.

In the following we present structural remodeling computations of the evolution of the internal density of anisotropic bone for a sample of the human femur subjected to physiological load, in order to illustrate the powerfulness of the developed methodology.

10.4. Simulation of internal bone remodeling with damage

The shape of part of the human femur (shown in the top-left of Figure 10.3) is described by B-splines, and then meshed; simulations are done with open source finite element software Freefem++ [HEC 05], adopting the variational formulation of equilibrium equations. Let Ω be the region occupied by the effective homogenized anisotropic bone material enclosed by a boundary $\Gamma = \partial\Omega$. The border is decomposed into two disjoint parts Γ_1 and Γ_2, where Γ_1 is the part of the boundary along which the femur is fixed, and Γ_2 the part of the boundary along which the loads P (P1, P2) are applied (Figure 10.3). Indeed, the load environment for bones is complex with force components resulting from joint contact, muscles, tendons, and soft tissue, and the load magnitudes and directions vary constantly as a person moves about. The load values and directions usually correspond to a static load case representing the stance phase of gait for a person weighing 80 kg. As a representative daily loading condition, the bone structure is subject to three load cases (one-legged stance, abduction and adduction). Remodeling simulations is presently conducted under single-loading condition. We consider static loading as a mechanical stimulus in bone remodeling with the force corresponding to the time when the foot touches the floor (case 1) which is assumed to be predominant (higher force magnitude than the other load cases). Load values are taken from [BEA 90], and they are applied at the head of the femur and as the reaction of the muscles located at the great trochanter (Figure 10.3(a)). The first load condition $P_1 = 2317$ N is assigned at an angle of $\theta_1 = 24°$ and the second load $P_2 = 703$ N is positioned at an orientation of $\theta_2 = 28°$ (the angles are given with respect to the vertical axis) with a fixed number of N = 6000 cycles/day. In the present work, the cycle at failure N_f is assumed to be 15,000, then the fraction of fatigue lifetime cycle is $N / N_f = 0.4$, thus the fatigue damage D_e (which is defined as the loss of stiffness) is computed from the relation in equation [10.13] to be equal to 0.25.

We recall that the linearized strain tensor is given by $\epsilon(\mathrm{u}) = \frac{1}{2}\left((\nabla \mathrm{u}) + (\nabla \mathrm{u})^T\right)$. The linear elastic deformation of anisotropic bone material is described in terms of the stress tensor $\boldsymbol{\sigma}(\mathrm{u})$, the strain tensor $\epsilon(\mathrm{u})$, the displacement vector $\mathbf{u}(\mathrm{u}_1, \mathrm{u}_2)$, the traction vector P, and the body force vector $\rho \mathrm{g}_a$ (g_a is the acceleration due to gravity). Therefore, the governing field equations of the problem are the following

$$\begin{cases} \mathrm{Div}(\boldsymbol{\sigma}(\mathrm{u})) + \rho \mathrm{g}_a = 0 \text{ in } \Omega \\ \mathbf{u} = \mathbf{0} \qquad \text{on } \Gamma_1 \\ \boldsymbol{\sigma}.\mathbf{N} = \mathrm{P} \qquad \text{on } \Gamma_2 \end{cases} \qquad [10.15]$$

The constitutive law of stress-strain relation of elasticity based damage mechanics of the anisotropic bone behavior is expressed as

$$\boldsymbol{\sigma}_{ij}(\mathrm{u}) = (1 - \mathrm{D}_e)\mathrm{C}^0_{ijkl}(\rho)\epsilon_{kl}(\mathrm{u}) \qquad [10.16]$$

The variational formulation of the equilibrium equation is written for the purpose of the finite element discretization in weak form by multiplying the strong form of equilibrium by a test (vector) function **v** chosen to vanish on the domain boundary, viz. $\mathbf{v} = \mathbf{0}$ on $\partial\Omega$. The weak form then writes as follows: we seek the displacement field $\mathbf{u}$ satisfying

$$\int_{\Omega} \sigma(\mathbf{u}) : \epsilon(\mathbf{v}) dx - \int_{\Omega} \rho g_a . \mathbf{v} \, dx = \int_{\Gamma_2} P.\mathbf{v} \, ds, \quad \forall \mathbf{v}, \; \mathbf{v} = \mathbf{0} \text{ on } \Gamma_1 \qquad [10.17]$$

After the substitution of equation [10.16], the following variational equation is obtained

$$\int_{\Omega} (1-D_e) \begin{pmatrix} \left(C_{11}^0 \frac{\partial u_1}{\partial x} + C_{12}^0 \frac{\partial u_2}{\partial y} \right) \frac{\partial v_1}{\partial x} + \left(C_{21}^0 \frac{\partial u_1}{\partial x} + C_{22}^0 \frac{\partial u_2}{\partial y} \right) \frac{\partial v_2}{\partial y} \\ + \frac{C_{33}^0}{2} \left(\frac{\partial u_1}{\partial y} + \frac{\partial u_2}{\partial x} \right) \left(\frac{\partial v_1}{\partial y} + \frac{\partial v_2}{\partial x} \right) \end{pmatrix} dx - \int_{\Omega} \rho g_a . v \, dx = 0 \qquad [10.18]$$

which is the basis for the subsequent FE computations.

In order to illustrate the application of anisotropic internal bone adaptation based on continuum damage mechanics in numerical computations, the proximal femur bone remodeling is predicted, considering an initial geometry with a uniform density distribution of 300 kg/m^3. The rate of change of the apparent density of the bone at a particular location, quantity $d\rho / dt$, shall be evaluated by the finite difference method.

The improved Euler's method is a modified method used instead of Euler forward method in order to improve the solution of the differential equation satisfied by density, using the average of the slopes of the solution curves; it is presently involved to estimate the apparent density of bone tissue at the considered region, based on a first order non-linear ordinary differential equation [10.6]; the density is updated according to

$$\rho_{n+1} = \rho_n + \frac{\Delta t}{2} \left[f(t_n, \rho_n) + f(t_{n+1}, \rho_{n+1}) \right] \qquad [10.19]$$

involving the function $f(t_n, \rho_n)$ defined in equation [10.6], evaluated at the discrete point (t_n, ρ_n). The second scheme contains the implicit terms, $f(t_{n+1}, \rho_{n+1})$; in order to make the second scheme explicit, we approximate ρ_{n+1} using a forward Euler, so that $\rho_{n+1}=\rho_n+\Delta t f(t_n,\rho_n)$, with $\Delta t = t_{n+1}-t_n$ the time step size (iteration). Consequently, the improved Euler scheme rewrites

$$\rho_{n+1}=\rho_n+\underbrace{\frac{\Delta t}{2}\left[f(t_n,\rho_n)+f(t_{n+1},\rho_n+\Delta t f(t_n,\rho_n))\right]}_{d\rho} \qquad [10.20]$$

The mechanical stimulus is computed from the finite element analysis according to [10.18] which rewrites in extended form as

$$W=\frac{1}{2}\left(C_{11}^{D}\frac{\partial u_1}{\partial x}+C_{12}^{D}\frac{\partial u_2}{\partial y}\right)\frac{\partial u_1}{\partial x}+\frac{1}{2}\left(C_{12}^{D}\frac{\partial u_1}{\partial x}+C_{22}^{D}\frac{\partial u_2}{\partial y}\right)\frac{\partial u_2}{\partial y}+\frac{1}{4}C_{33}^{D}\left(\frac{\partial u_1}{\partial y}+\frac{\partial u_2}{\partial x}\right)^2 \qquad [10.21]$$

Based on the evaluation of expression [10.21], the density increment $d\rho$ for each element can be calculated from equation [10.19], and the element density is then updated as $\rho_{n+1} = \rho_n + \alpha\rho$. As the density changes for each element, the stiffness (elastic and shear modulus and Poisson's ratio) of the continuum is updated at each new iteration as well. At the end of each iteration, the mechanical properties of each element are calculated according to equation [10.14].

In any subsequent iteration, the mechanical stimulus and its proximity to the threshold value are evaluated for each element, with the overall structural response of bone based on the biomechanically adapted stiffness computed from the previous iteration. This iterative process continues until all bone mass is in remodeling equilibrium. The remodeling solution is converged when all elements have reached one of three states: the element reaches the maximum density value, i.e. bone formation; the element density reaches the minimum value, i.e. all bone tissue has fully resorbed, or when the strain energy per unit mass of bone is equal to the reference stimulus, in which case equilibrium is reached.

It should be noted that the aforementioned models of elastic damage bone remodeling are phenomenological in nature; they describe the stimulus and the response, without taking into the very biological mechanisms taking place internally in the bone tissue.

Figure 10.3(a) shows the geometry and physiological static surface loads of a 2D frontal plane of a proximal femur, which is taken as the initial model in our structural remodeling analysis (the bone sample is simultaneously subjected to its own weight). The process starts with the solution of the elasticity problem, adopting a uniform density distribution chosen as 300 kg/m^3. The time step is selected small enough, $\Delta t = 1$, in order not to affect the temporal sequence of the mechanical fields within the bone sample in a significant way. The combinative algorithms of both adaptive and damage remodeling change the local bone density (see Figure 10.3). After 100 iteration steps (Figure 10.3(c)), one can observe the formation of two high density external layers along the diaphysis of the femur, a zone of low density corresponding to the medullar channel inside the diaphysis, and a complex distribution of density of the femoral head and trochanter including three different density regions with the head surrounded by a lower density zone (low density region on the distal part of the head).

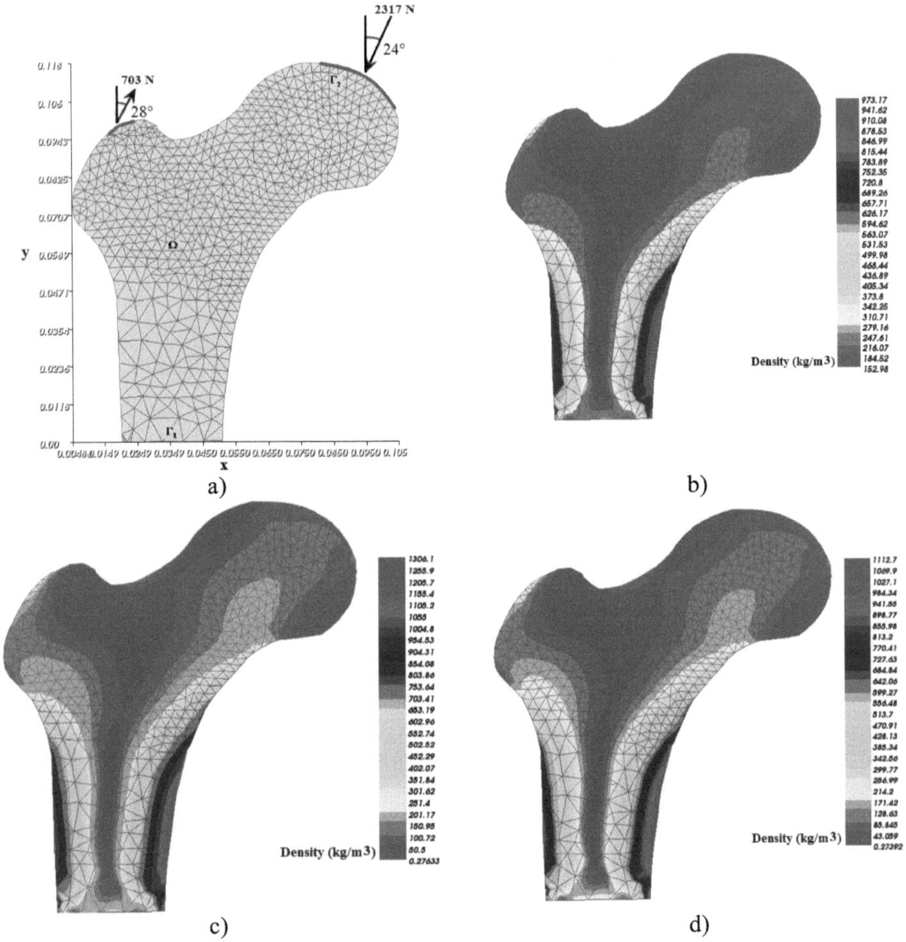

Figure 10.3. *Bone density distribution of 2D model of the proximal femur: a) loading conditions and uniform distribution at the beginning of simulation: initial density= 300 kg/m^3;b) bone response to simulated physiological loading after 50 days with D_e = 0.25; c) after 100 days with D_e =0.25; (d) after 100 days with D_e =0. For a color version of this figure, see www.iste.co.uk/ganghoffer/biomechanics.zip*

Density increases with the level of damage, due to the fact that the components of the rigidity matrix will be reduced, resulting in a strain increase, thus to an increase of the mechanical stimulus (strain energy), increasing in turn the driving force for bone adaptation and therefore the density. This is reflected qualitatively in equation [10.18], since damage and density act in opposite manner on the effective mechanical properties, which means that the density increase generated by damage acts as a repair mechanism at the scale of existing microcracks. By analyzing the bone density distribution after a simulation of 100 remodeling days (Figures 10.3(c) and (d)), we outline the fact that damaged and undamaged bone density distributions show similar patterns and approximately produce the same spatial distribution; however, they do not show identical values of density. The overall density is increased with the level of damage, e.g., the maximum is 1306 kg/m^3 for the damaged bone and 1112 kg/m^3 for the virgin material. This indicates that the density of undamaged bone never reaches the density of a damaged bone.

Additionally, we observe that the bone density distribution evolves with time; comparing results after 50 days (Figure 10.3(b)) of application of the loading with the results after 100 remodeling days (Figure 10.3(c)) an increase of density is observed after 100 days remodeling, with formation of new zones of density as well. Therefore, bone remodeling redistributes the bone material in response to the applied load, leading to new bone morphology: the bone density is increased wherever the bone is highly loaded (high mechanical stimulus) and is decreased where its presence is not necessary due to less internal stress. The development of models of distributed damage coupled to internal bone remodeling is one important perspective of this work.

PART 3

Mechanics of Soft Biological Tissues: The Intervertebral Disk, Biological Networks, Ligaments and Tendons

11

Micromechanics of the Intervertebral Disk

11.1. Introduction

Intervertebral disks are key elements of the spine as presented in Figure 11.1. They allow spine motion like flexion extension, lateral bending and axial rotation dissipating mechanical loadings during all these motions [COS 07, WOO 10].

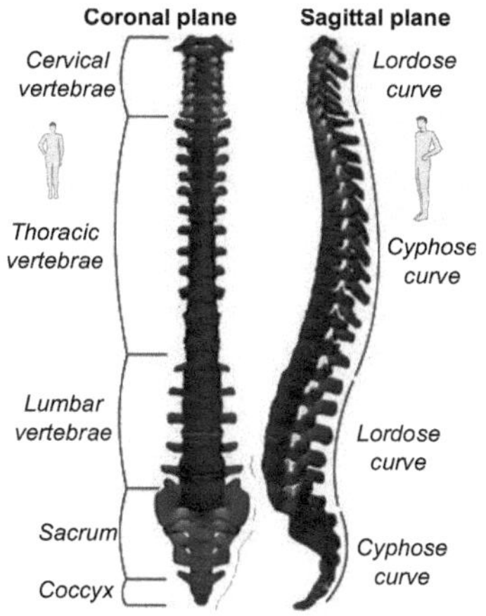

Figure 11.1. *Spine regions and curvatures adapted from [BAL 13]*

The intervertebral disk (IVD) is a fibro-cartilage bonding two adjacent vertebrae. It is divided into two main parts: the nucleus pulposus and the annulus fibrosus as illustrated on Figure 11.2. The vertebral endplate can also be considered as a third element but it is common to both intervertebral disk and vertebrae.

Chapter written by Adrien BALDIT.

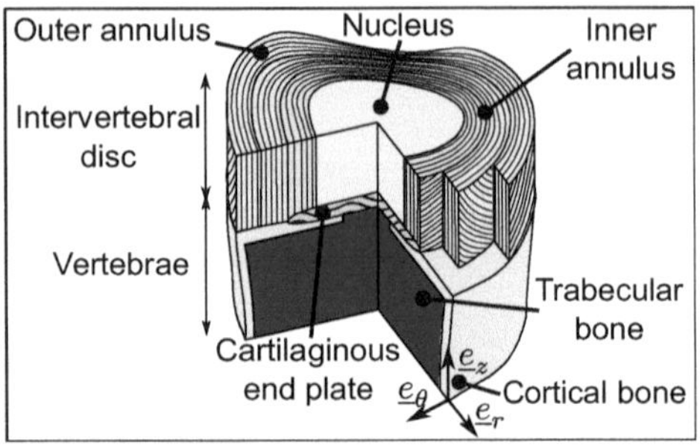

Figure 11.2. *Intervertebral disk drawing with the bottom adjacent vertebra (adapted from [BAL 13]). For a color version of this figure, see www.iste.co.uk/ganghoffer/biomechanics.zip*

The IVD size decreases from the cervical level to the lumbar region (Figure 11.1) where height and diameter are respectively around 10 to 15 mm and 30 mm [CAM 07, RAN 04]. This region is very often studied due to chronic pain and complex treatments [BRE 06, KRA 09]. IVD cylindrical shape allows using cylindrical coordinates to describe the directions of this joint as presented in Figure 11.2.

11.2. Intervertebral disk components and structure

Water, collagen fibers and proteoglycans molecules represent 90 to 95% of IVD components [BIB 01]. The rest is composed of minor molecules and cells maintaining and renewing the IVD tissues. With 6000 cell/mm3, the IVD has a really low cell density compared to other joints cartilages where it is usually around 14000 to 15000 cell/mm3 [BIB 01, MAR 75, RAN 00, RAN 04]. Within the IVD the cell distribution is related to cell nutrients and metabolic wastes routes. Being avascular the IVD cell's environment is renewed thanks to blood capillaries surrounding the annulus fibrosus and blood marrow through vertebral endplates [HOL 81, RAN 04, SHI 10]. The cell population is mainly composed of:

– fibroblasts responsible for collagen type I production especially within outer part of the annulus fibrosus [BIB 01];

– chondrocytes that produce collagen type II and proteoglycans in the whole disk.

These cells are responsible for the IVD tissue homeostasis-producing extracellular matrix of which the predominant components are collagen, elastin and proteoglycan molecules [BIB 01, KRA 09].

11.2.1. *Intervertebral disk nucleus pulposus*

The ovoid shape of the nucleus pulposus represents 30 to 50% of IVD volume [FER 04, RAN 00]. Due to its soft and homogeneous consistency, the nucleus pulposus is considered as a highly hydrated gel with nearly 90% water content for young people and 70% for adults. During physiological motions, it has been observed by nucleography that the nucleus pulposus position remains the same [BOL 89]. Being surrounded by annulus fibrosus, the interface in between is not clearly defined; instead it is a gradient of both tissues' components distribution [KRA 09, INO 81, RAN 04].

11.2.2. *Intervertebral disk annulus fibrosus*

The annulus fibrosus is made of 7 to 15 concentric layers clearly visible at macroscopic scale as presented on Figure 11.3. Usually lamellae tend to be thinner in posterior regions than in anterior ones. This tissue is the link in between two adjacent vertebrae with one to two thirds of the lamellae fixed to the vertebral endplates [MAG 09, SHI 10]. The annulus fibrosus is assimilated to a hyaline cartilage with 60 to 85% of water content [ACA 95, RAN 04]. As a fiber-oriented cartilage, from inner to outer IVD the fiber orientation decreases in regards to the circumferential direction (see Figure 11.2) from 45° to 25° with an average value around 28° [AMB 09, MAL 12, NOA 05, SHI 10].

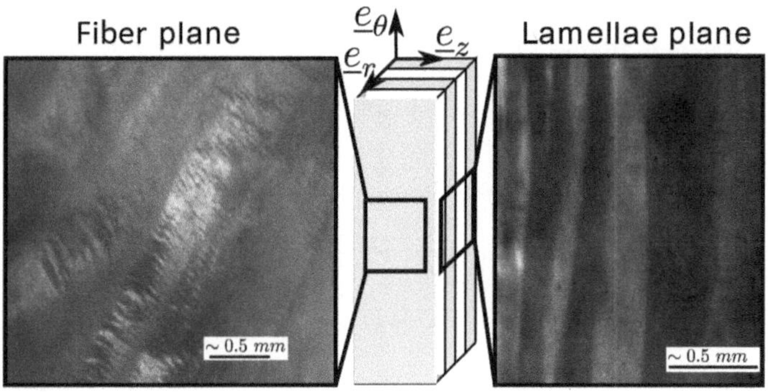

Figure 11.3. *Macroscopic pictures of the annulus fibrosus oriented fiber structure adapted from [BAL 13]*

Made of collagen type I, this structure gives an anisotropic behavior to this tissue as a composite material. Being soft, these fibers only play a role undergoing tensile loadings [AMB 09, BAL 13].

11.3. Intervertebral disk's roles

Daily motions induce mechanical loadings that are mainly absorbed by the IVD dampers at all spine stages.

11.3.1. *Physiological loadings*

Within the spine physiological motions induce large deformations mainly undergone by intervertebral disks as the softest material of this musculoskeletal segment. Bearing the upper body part load daily, IVD tissues have to undergo large strains with a recoverable ability to maintain the structure as its original shape.

The mechanical loadings applied to an IVD are complex, combining compression traction and torsion. They depend on the disk location and the considered motion but in literature mainly compression loadings ranged from 350 to 2000 N are found [MAL 12, SCH 10, SHI 10].

11.3.2. *Mechanical loadings within intervertebral disk*

During those solicitations, the load is transferred from one vertebra to another through the IVD where the highly hydrated nucleus pulposus translates the loadings in hydrostatic pressures contained by the annulus fibrosus and the adjacent vertebra endplates as shown in Figure 11.4.

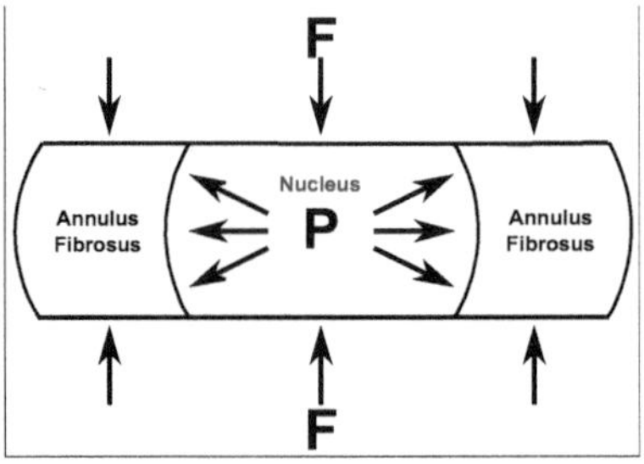

Figure 11.4. *Drawing of the load transfer through an intervertebral disk. The vertical force F is transferred by the vertebrae and transformed by the nucleus pulposus in a hydrostatic pressure P contained by the annulus fibrosus*

Therefore, the annulus fibrosus is mainly solicited in circumferential tension. Moreover, the biphasic character of IVD tissues has a strong influence on the loading recovery process. While IVD deforms due to mechanical loadings, fluid flows take place within the disk and its periphery. This leads to a height loss around

1 cm for adults in between morning and evening measurements. During the sleeping time (full night lying) the mechanical loading is changed and thanks to negatively charged proteoglycan molecules within tissues an osmotic effect induces IVD rehydration [COR 11, HUY 10, MAG 09]. This phenomenon has been illustrated by immerging in pure water nucleus pulposus tissue of which the volume increased by 200% [FRI 97].

11.4. Intervertebral disk's experimental characterization

Many techniques allow the characterization of soft tissues and only a few that reveal the previously described phenomena will be presented in this section. The descriptions will be mainly focused on the annulus fibrosus tissue due to its complex anisotropic behavior.

11.4.1. *Mechanical characterization*

Mechanical test are mainly uni- or bi-axial tensile and compression tests. Often they start with a conditioning test to skip the transient behavior and to do measurements only in permanent states, reducing viscous effects to limit the study to the elastic or hyper-elastic framework. The uni-axial tensile test is the easiest one to carry out on annulus fibrosus tissue, allowing the soliciting of the oriented fiber's structure to characterize its anisotropy [AMB 09, GRE 12, HOL 05, OCO 12]. The stress strain curves in cyclic tests or relaxation tests illustrate its nonlinear, viscous behavior as presented in the following Figures 11.5(a) and 11.5(b).

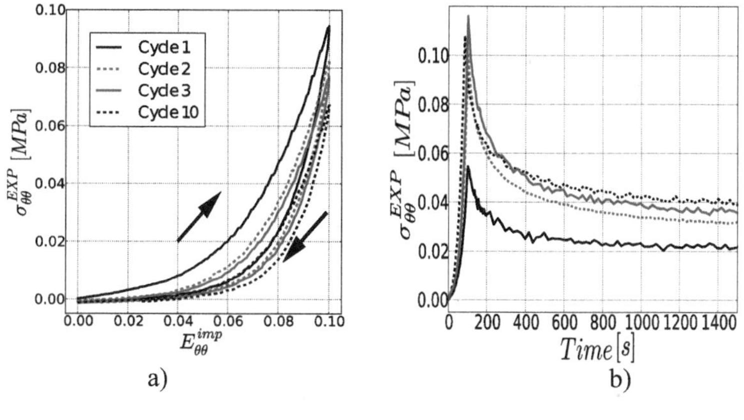

Figure 11.5. *Annulus fibrosus a) cyclic tensile curve and b) relaxation tensile curve adapted from [BAL 13]. The curves represent the second Piola Kirchhoff stress tensor circumferential component ($\sigma_{\theta\theta}^{EXP}$) as function of a) imposed strain ($E_{\theta\theta}^{imp}$) and b) time. The tests were done on pig samples*

Bi-axial tensile tests are more complex to realize due to the sample size and the complexity to fix them in the machine clamps [BAS 04, BRU 04, OCO 12]. To compensate the loss of data using uni-axial tensile tests, image analyses are performed to obtain surface strain measurements while loading [MIC 10, HAN 12, WAG 04].

Tensile tests perfectly fit with physiological loadings applied on annulus fibrosus, especially the bi-axial one, but for nucleus pulposus they do not. Actually, as a gel it cannot be easily fixed on a tensile test machine. Moreover, physiologically it bears compression that can be reproduced as presented on Figure 11.6.

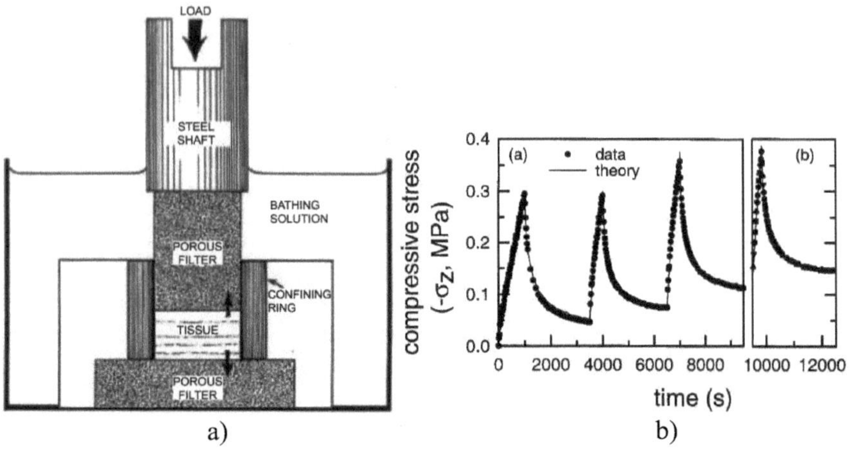

Figure 11.6. *a) Confined compression test [BES 94] and b) evolution of the vertical stress (σ_z) as function of the time through four loadings [IAT 98]*

This kind of test is also used on annulus fibrosus to characterize the collagen matrix embedding the oriented fiber structure but do not solicit it in a physiological way. All those tests highlight the viscous behavior of the tissue that can be treated within the mechanic behavior law or thanks to a coupling between the fluid and solid phase of the tissue studied.

11.4.2. *Poro-mechanical characterization*

The expression "poro-mechanical" aims to describe viscous effects observed on tissues thanks to their biphasic composition that has a paramount influence on cell environment quality. Actually, while IVD are loaded they generate fluid flows that renew cell nutrients and expel metabolic wastes. As well as cyclic tests, creep (see Figure 11.6b) and relaxation tests are used to characterize tissue permeability as presented in Figure 11.7.

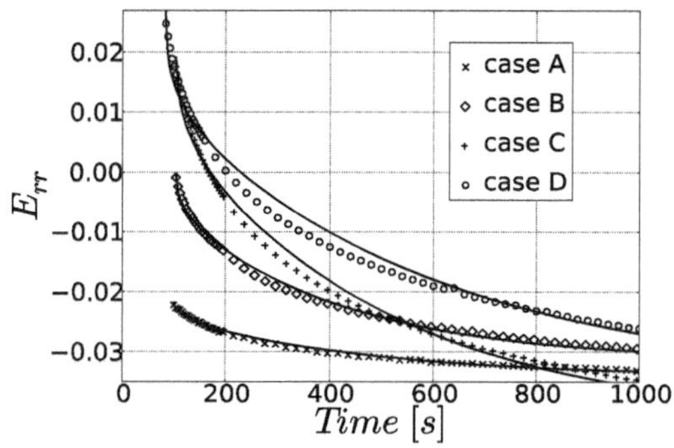

Figure 11.7. *Permeability characterization on strain measurements realized during tensile loads [BAL 14a]. A poro-mechanical model is fitted on radial strain (E_{rr}) / time measurements during a relaxation test*

This characterization has to be done in adequacy with the physiological loadings therefore a confined compression test well suits nucleus pulposus [BES 94] and tensile tests fit annulus fibrosus solicitation [BAL 14a]. Although, due to the complexity of the material, even those dedicated tests provide data with coupled phenomena that cannot be easily differentiated. An alternative is to perform direct permeability measurements as presented in Figure 11.8.

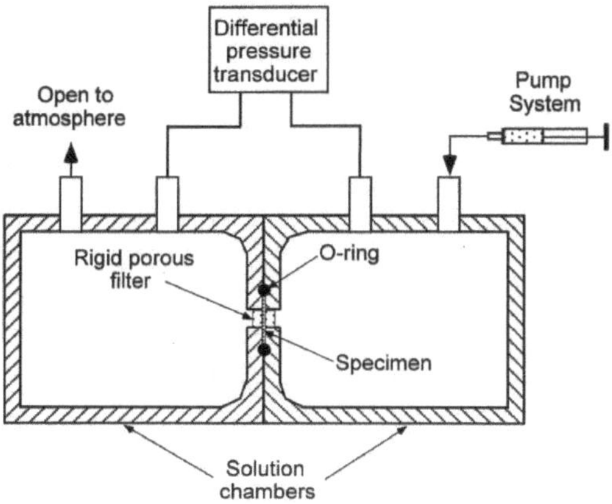

Figure 11.8. *Direct permeability measurements [GU 99]*

Not considering the difficulty to fix the sample into the experimental set up, this method does not take into account the effect of the mechanical loading on the permeability that is often defined as strain dependent [MAL 11].

11.4.3. *Poro-chemo-mechanical characterization*

When seeking to understand fluid and chemical component transfers through the IVD, poro-chemo-mechanical couplings have to be characterized. Therefore, mechanical tests are performed while the sample's chemical environment (often [NaCl] concentration) is changed [BAL 13, DRO 95, FRI 97]. This test allows the quantification of the tissue chemo-sensitivity plus its reversible behavior as shown in Figure 11.9.

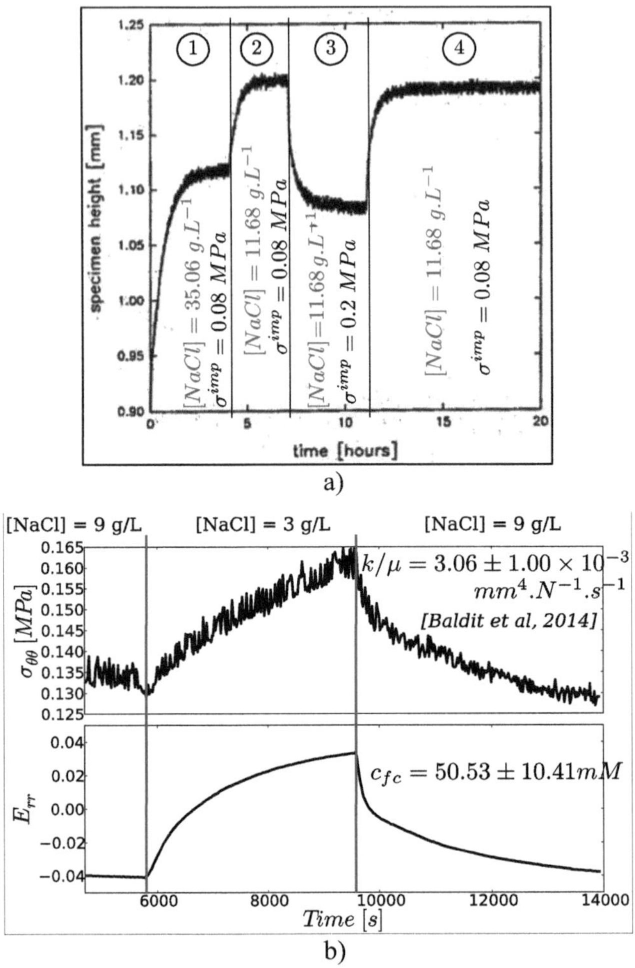

Figure 11.9. *chemo-sensitivity of annulus fibrosus tissue under a) confined compression test (adapted from [DRO 95]) and b) uni-axial tensile test [BAL 14b]. Salt concentration modifications induce a) height or b) stress reversible variations*

The data of these kinds of tests are very delicate to interpret without a numerical simulation due to the complexity of the couplings. Therefore, the next section is dedicated to introducing the modeling and simulation required.

11.5. Intervertebral disk's modeling and simulation

11.5.1. *Mechanical modeling*

Many models are used to characterize IVD tissues and they are very similar to the one presented in Chapter 13 devoted to ligaments. This section will only present one model, developed in a hyper-elastic framework, to illustrate the discussion regarding the phenomena presented in the previous section. This description allows us to take into account the large strains occurring within the tissue defining a strain energy density W as a function of the right Cauchy-Green stretch tensor $\boldsymbol{C}$ and related to the second Piola Kirchhoff stress tensor $\boldsymbol{S}$ by:

$$\boldsymbol{S} = 2\frac{dW(\boldsymbol{C})}{d\boldsymbol{C}} \qquad [11.1]$$

Results being often presented with Cauchy stress tensor, the link between the tensors is made by:

$$\boldsymbol{\sigma} = \frac{1}{J}\boldsymbol{F}.\boldsymbol{S}.\boldsymbol{F}^T \qquad [11.2]$$

where $\boldsymbol{F}$ is the transformation gradient and J the third invariant or Jacobian.

The strain energy function often used for IVD has been defined by [GAS 06] due to its combination of a Neo-Hookean model implemented by an exponential description of two fiber families oriented differently. Initially designed for arterial walls [HOL 00], it fits the annulus fibrosus structure with oriented fibers organized in concentric lamellae with an expression like:

$$W = \frac{G}{2}\left(\overline{I}_1 - 3\right) + \frac{1}{2K_2}\sum_{i\in[1,2]} K_{1i}\left\{e^{K_2\left[k_{disp}\overline{I}_1 + \left(1-3k_{disp}\right)\overline{I_{4i}} - 1\right]^2} - 1\right\} + \frac{\kappa}{2}\left(J-1\right)^2 \qquad [11.3]$$

where the first term $\frac{G}{2}\left(\overline{I}_1 - 3\right)$ represents an isotropic matrix like the nucleus pulposus or the embedding matrix of the annulus fibrosus oriented fiber structure depending on the shear modulus G and the first modified invariant $\overline{I}_1$. The second term $\frac{1}{2K_2}\sum_{i\in[1,2]} K_{1i}\left\{e^{K_2\left[k_{disp}\overline{I_1}+(1-3k_{disp})\overline{I_{4i}}-1\right]^2}-1\right\}$ translates the oriented fiber structure with two i families as a function of three fiber parameters K_{1i}, K_2 and k_{disp} plus the modified invariant $\overline{I_{4i}}$ related to the dilatation in fiber directions. Eventually, the last term $\frac{\kappa}{2}(J-1)^2$ deals with the compressibility of the material function of the Bulk modulus κ and the third invariant or Jacobian J leading with modified invariants to consider a quasi-incompressible material.

On the other hand, considering only the first and last terms is equivalent to a Neo-Hookean model that perfectly suits an isotropic gel like nucleus pulposus.

11.5.2. *Poro-mechanical modeling*

To take into account the bi-phasic constitution of the IVD tissues, a poro-mechanical coupling is required. Considering the two phases are incompressible the model can be simplified [BAC 98] to a compressibility only related to a fluid flow in a porous matrix. Then the Darcy law can describe the fluid phase behavior:

$$\boldsymbol{q}^{f/s} = -\frac{k}{\mu}\boldsymbol{\nabla p} \qquad [11.4]$$

where $\boldsymbol{q}^{f/s}$ represents the fluid velocity with respect to the solid phase, k is the intrinsic permeability, μ the fluid dynamic viscosity and p the fluid pressure. The hydraulic permeability $\bar{k} = k/\mu$ can be defined as strain dependent to take into account the influence of the material transformation on the fluid flow within itself [ALM 97, ARG 96].

This allows introducing the definition of the Cauchy stress tensor $\underline{\underline{\sigma}}$ as:

$$\boldsymbol{\sigma} = \boldsymbol{\sigma}^s - p\boldsymbol{I} \qquad [11.5]$$

where $\boldsymbol{\sigma}^s$ is the Cauchy stress tensor related to the solid phase with a pure mechanical behavior law and $\boldsymbol{I}$ the identity matrix [ARG 96, MAL 11].

11.5.3. *Poro-chemo-mechanical modeling*

With respect to the experimental results presented before, only a model translating the tissue sensitivity to negatively charged molecules of the environment will be presented. To maintain the electro-neutrality within the tissue, fluid flows are generated due to an osmotic hydrostatic pressure defined by [COU 04, DON 11, LAN 87]:

$$\Delta\pi = -2RT\left[\sqrt{c_{ext}^2 + \left(c_{fc}/2\right)^2} - c_{ext}\right] \tag{11.6}$$

where R is the Boltzmann constant, T the absolute temperature, c_{ext} the external solution (or environment) ion concentration and c_{fc} the fixed charge concentration in the solid phase.

With the same additive approach as eq [11.5], the osmotic pressure is added to the Cauchy stress tensor as:

$$\boldsymbol{\sigma} = \boldsymbol{\sigma}^s - p\boldsymbol{I} + \Delta\pi\boldsymbol{I} \tag{11.7}$$

More complex coupled models can be used, taking into account chemical elements transfers within the fluid phase [COU 04, HUY 04, LEM 07]. Those models, described as multi-phasic, require more parameters to be identified and reveal a lack of experimental data at this scale.

11.6. Conclusion and perspectives

These experimental and modeling developments allow the simulation of full IVD behavior taking into account more or less observed phenomena. Research has been done by [MAL 11] dealing with poro-chemo-mechanical modeling and solved by finite element analysis. Gathering data from the literature, they built the following table of parameter values:

	φ_0	k_0 [$mm^4\,N^{-1}\,s^{-1}$]	M	L	G [MPa]	K [MPa]	$\Delta\pi$ [MPa]	ρ_{cell} [10^6cells mm^{-3}]
Healthy IVD properties (▲H = 13.7 mm)								
Outer AF	0.73*	0.0002◇	1.18◇	-	0.28**	0.37**	-	0.063#
Inner AF	0.78*	0.0002◇	1.18◇	-	0.28**	0.37**	-	0.048#
NP	0.83*	0.0009‡	8.5++	-	0.12**	0.16**	0.15♠	0.032#
Degenerated IVD properties (▲H = 12.3 mm)								
Outer AF	0.57*	0.0002◇	1.18◇	-	0.41**	0.55**	-	0.048#
Inner AF	0.6*	0.0002◇	1.18◇	-	0.41**	0.55**	-	0.032#
NP	0.71*	0.0009‡	8.5++	-	0.19**	0.25**	0.05♠	0.02#
Bone and endplates properties								
CEP	0.8+	0.0025+	4.63+	0.08+	7.14‡	33.3‡	-	-
BEP	0.05++	26800$	-	-	3846++	8333++	-	-
Cortical	0.05++	5$	-	-	3846++	8333++	-	-
Trabecular	0.8++	26800$	-	-	42++	56++	-	-

Table 11.1. *Poro-chemo-mechanical parameters values from [MAL 11]*

The simulation allowed a modeling of the distribution of oxygen and lactate within the IVD while under physiological cyclic loadings as presented in Figure 11.10.

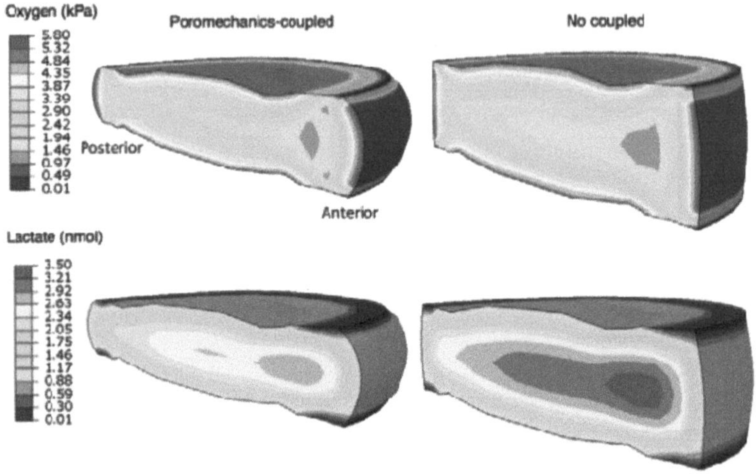

Figure 11.10. *Distribution of the independent oxygen and lactate concentrations [MAL 11]. For a color version of this figure, see www.iste.co.uk/ganghoffer/biomechanics.zip*

The results obtained have a major importance for the design of predictive tools for clinical practitioners who require accurate simulations pre- and post-treatment to ensure the quality of their therapeutic acts.

Although, as presented within the experimental and modeling sections, the parameters values that can be found in the literature are strongly dependent on the sample, the experimental protocol and the parameter identification that has ben performed. Therefore, gathering data from literature allows the validation of models and treatment processes but cannot be generalized to all tissues.

For instance, the prediction of tissue evolution is related to its cells' reactions to the environment. It means knowing how the macroscopic solicitations and chemical concentrations would lead to knowing how cell will proliferate and differentiate, in other words, a full mastering of whole scales in between the IVD disc scale and a single cell scale. Not considering the difficulty of getting consistent experimental data, it has already been shown that the homogeneity assumption realized by the model presented above can have a strong influence on local sample strain (Figure 11.11) leading to a complete different stress and fluid flow distribution around the cell within the tissue.

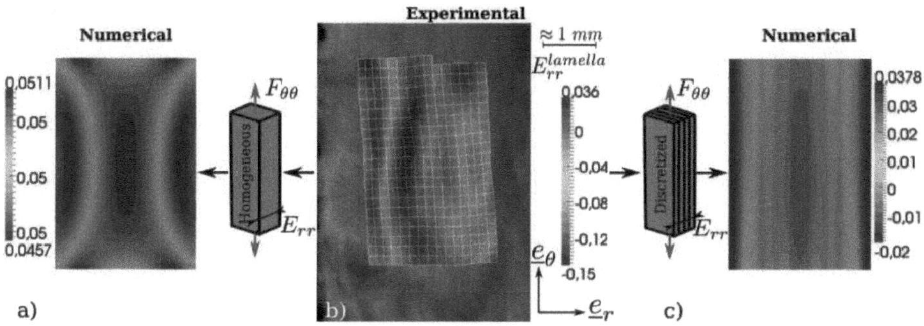

Figure 11.11. *Radial strain fields obtained for tensile relaxation test steady state using: a) a homogenous distribution of fibres, b) Digital Image Corralation on experimental loading pictures and c) a fiber distribution following the alternative changing from one lamellae to another (from [BAL 14b]). For a color version of this figure, see www.iste.co.uk/ganghoffer/biomechanics.zip*

This result doesn't mean that full simulations are wrong but highlights the sensitivity of the multi-scale regression that can lead to discrepancies in the results. Of course, the degree of accuracy has to deal with the computation cost but multi-scale sensitivity analysis has to be done [BAL 14c]. This rigorous analysis ensures that the model or simulation assumptions used will not mislead the results and the following discussions.

12

Effective Mechanical Response of Biological Membranes

The effective elastic response of biological membranes viewed as repetitive beam networks is evaluated based on the asymptotic homogenization method. A systematic methodology is set up to predict the overall mechanical properties of biological membranes in the nonlinear regime, reflecting the impact of the geometrical and mechanical network micro-parameters on the overall response of the effective substitution continuum. A classification of biomembranes networks is made based on nodal connectivity, so we analyze in this chapter 3, 4 and 6-connectivity networks, which are deemed to be representative of most biological encountered networks. The individual network filaments are modeled as undulated beams prone to entropic elasticity, and the persistence length is one important parameter entering the tensile modulus. An effective micropolar substitution continuum of the biological network is evaluated, with a kinematics including nodal displacements and rotations, reflecting the discrete network deformation modes. The ratio of the characteristic bending lengths of the effective micropolar continuum to the unit cell size is a non-dimensional parameter used to quantify the importance of micropolar effects. The peptidoglycan network may present re-entrant hexagonal configuration caused by thermal or pressure fluctuations, for which micropolar effects become important. The predictive nature of the employed homogenization technique allows identifying a strain energy density of a hyperelastic model.

12.1. Introduction

The membrane of biological cells is made of an assembly of filaments in the form of a network within the cell membrane. Such 2D biological networks may be

Chapter written by Khaled EL NADY, Jean-François GANGHOFFER and Ibrahim GODA.

wrapped around a cell wall or attached to its nuclear membrane. The soft character of the filaments which are the structural elements of biological cells account for their large deformability and easy motion, in contrast to most of the engineered man made thin structural sheets.

The mechanics of biological membranes depends on the architecture of the filaments network via the network connectivity, and on the entropic elasticity of the inherent filaments. Predictive nanomechanical models aim at understanding the impact of the network topology and properties on their behavior at the continuum level accounting for their large deformation behavior and evolving anisotropy, as the experimental determination of the mechanical properties of biological membranes is delicate. A review of the various techniques developed in the last two decades to measure the mechanical properties of cells (and especially the red blood cell, RBC in short, which has a complex structure), vesicles and biomembranes has been presented in [KIM12]. RBC's can sustain large deformations in the reversible domain to achieve shape changes while circulating through the microvasculature; it also determines oxygen transport via blood circulation, itself conditioned by the blood viscosity. This points towards the role of alteration of the deformability of RBC's on the development of pathological conditions. The understanding of the correlations between modifications of RBC properties and the development of diseases has raised the development of various types of measurement techniques over the past period, a short account of them being presented next.

The elastic properties of the RBC membrane are determined by its elastic moduli, namely the area expansion modulus, the shear modulus, and the bending modulus, associated with the corresponding fundamental deformation modes, area expansion, shear, and bending respectively. Both moduli can be measured by the microaspiration technique, which involves the application of a negative pressure onto the RBC membrane using a micropipette [EVA 75, SHI 90, HOC 00]; their respective value was obtained as $K = 450nM/m$ [EVA 97], $\mu = 9 \pm 1.7 \mu N/m$ [EVA 84], and $B = 43.5\ \mathrm{k}_B T$ [EVA 83]. The very small quadratic moment (it scales as the fourth power of the beam radius) entails that the threads are much weaker in bending than in tension; this has the consequence that membrane flexion is dominant over other modes of deformation.

Mechanical properties of RBCs can be measured by optical tweezers exerting an optical force onto two silica beads attached to opposite sides of the membrane [HEN 99], or by stretching RBCs using two diverging laser beams [GUC 01]. The shear modulus of the RBC may be accessed from the recorded change of its diameter, also relying on membrane models. The same technique also allows quantifying membrane fluctuations due to thermal agitation when the cell is subjected to a deformation [YOO 11].

By applying torques via oscillating magnetic fields on microbeads attached to the surface of the cell membrane, magnetic twisting cystometry is able to measure the stiffness and loss moduli of the membrane at different frequencies. Measurements lead to bending and shear moduli in the range $0.2-0.8 pN.\mu\text{m}$ and $6-12\mu N/m$ [PUI 07] respectively, while the torsional stiffness modulus is of the order of $10^{-3} Pa/nm$. The elastic and effective viscoelastic properties of the RBC can also be measured using quantitative phase imaging techniques [POP 06, WAN 11a].

The available mechanical models of biomembranes can be classified according to the scale, micro or nanostructural models at the finer scale versus continuum approaches at a more macroscopic level [LIM 06]. The microscopic-based models are by nature discrete and find their origin in [BOE 98, DIS 98] devoted to the red blood cell. They represent the membrane as a discrete spectrin network, employing nonlinear mechanical models for the filaments' elasticity, combined with an energetic description of the coarsed-grained network free energy incorporating in-plane and bending deformation models, under constant area and volume, as documented in [DAO 06]. Despite the easier development of continuum approaches, the identification of continuum models of biomembranes remains challenging, thus it is desirable to develop micromechanically-inspired approaches to link the existing scales and build constitutive laws with a physical basis originating from the nanostructural network parameters and accounting for the relative importance of entropic versus energetic effects at the molecular level of collagen fibrils and their cross-links [VAI 13, MAR 13]. Bottom-up approaches developed following this strategy are able to assess the effect of histological features such as inter-molecular interactions on the effective mechanical response of soft biological tissues up to damage and rupture [MAR 14, MAC 12].

In this chapter, we evaluated the homogenized nonlinear mechanical response of biomembranes considered as planar repetitive networks of soft filaments at the intermediate scale of a periodic unit cell. The asymptotic discrete homogenization method exposed in chapter 4 is exploited [ELN 15, ELN 16], considering the discrete architecture of biological networks, here idealized in their regular nearly periodic organization: filaments are assembled at regular nodes representing the cross links between them. The nonlinear effective response is essentially due to the modifications of network configurations (modifications of filaments orientations and lengths) under the effect of the local strain field, thus geometrically nonlinear effects are put at the forefront, considering that the individual filament response remains linear. In addition to the tensile and bending stiffness of the filaments which account for enthalpic effects, the persistence length enters as an additional parameter – a measure of the correlation length of the filaments waviness – to quantify entropic like effects that compete with energetic effects.

The assumption of the biological network represented as a quasi repetitive beam network is justified at low temperature for which thermal fluctuations may be discarded. The beam length in the network model can be identified with the average end-to-end distance of a given chain, a quantity that fluctuates with temperature and function of the persistence length [BOA 03]. Such fluctuations are indirectly incorporated into the present model by modifying the beam length and the persistence length. The individual network filaments are modeled as beam elements endowed with a tensile and bending stiffness in the linear regime, a situation representing the linear approximation of nonlinear models like the worm like chain model [UNT 13, HOL 14].

12.2. Mechanical response of biological networks from micromechanical models

Results from previous section allow the consideration of biological membranes as quasi repetitive periodical networks of threads representing the molecular chains within a unit cell, with a topology specific to each type of biological membranes.

Regular biological networks can be classified based on their nodal connectivity, defined as the number of filaments emanating from any connecting node within the network: one encounters essentially three fold, four fold, and six fold connectivity networks as the building blocks of biological membranes [BOA 03]; they are modeled by the square, hexagonal and triangular networks respectively, as illustrated in Figure 12.1.

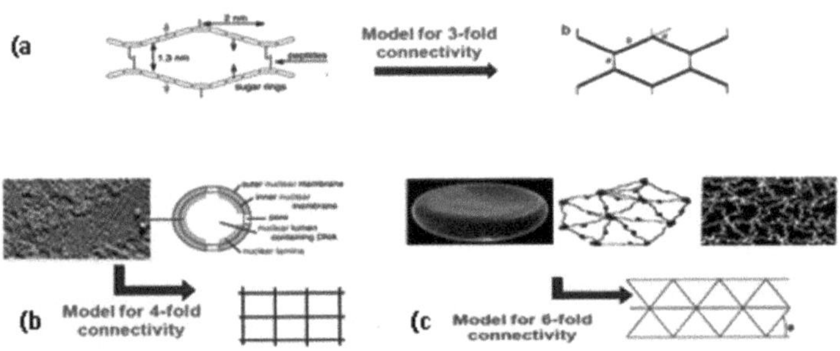

Figure 12.1. *a) Peptidoglycan (3-connectivity) network [KOC 92], b) Nuclear lamina (4-connectivity) [BOA 03] and c) Erythrocyte network (6-connectivity) [LIU 87). For a color version of this figure, see www.iste.co.uk/ganghoffer/biomechanics.zip*

The network geometrical and mechanical parameters are first exposed for the three considered types of biomembranes.

12.2.1. *Mechanical properties of the filaments of biological networks*

12.2.1.1. *Peptidoglycan network (3-fold connectivity)*

The peptidoglycan network is made of the assembly of two non-equivalent chains running in two perpendicular directions, as shown on the face view of a section (Figure 12.2); such an anisotropic molecular arrangement follows design principles correlated to the existing stress anisotropy, with the stiffer chains (sugar) sustaining the maximal stress along the girth of the bacterial wall. It is important to have a quantitative knowledge of the mechanical properties of the peptidoglycan network in order to understand bacterial growth and form.

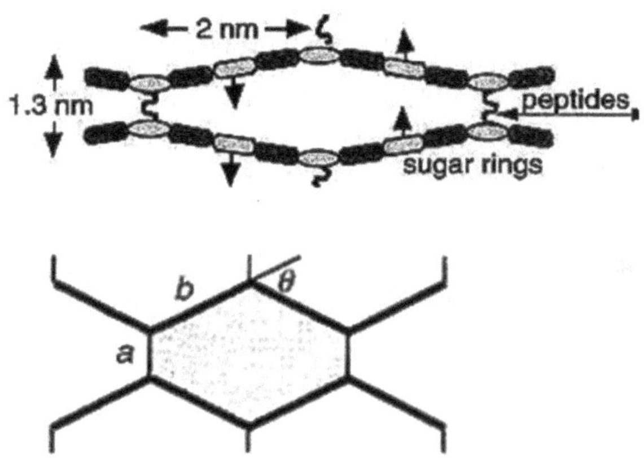

Figure 12.2. *Peptidoglycan network a) Molecular constitution (peptides and sugar rings) [KOC 92], b) Face view of the section of the peptidoglycan network and geometrical model*

The peptide and glycan chains are described as beams having a regular circular cross-section with radii 0.5 nm and 1 nm respectively, based on [BOA 03]; the geometry of the glycan network is parameterized by the angle θ (Figure 12.2b), which varies in the range [5-20 degrees], depending on the dimensions of the molecular chains of the unit cell shown in Figure 12.2. The average Young's modulus is estimated at low temperatures ($T \approx 273K$) as $4.799 \cdot 10^7 J/m^3$ for glycan and $1.53 \cdot 10^7 J/m^3$ for peptides. Young's moduli depend on the persistence length according to the theoretical relation [BOA 03]

$$\xi_p = \frac{YI}{K_B T}$$

In which Y, I, T, k_B are successively the tensile modulus, the quadratic moment (depending on the beam cross section), the absolute temperature and Boltzmann constant. As to the properties of individual chains, the peptide has an end-to-end length within the network $r_{ee} = 1.2nm$ lower than its contour length $L_c = 4.2nm$, so that it has the classical properties of entropic springs when restricted to the linear elastic regime. The peptide string has a persistence length computed from the relation $\langle r_{ee}{}^2 \rangle = 2\xi_{p,peptide} L_c$, leading to $\xi_{p,peptide} = 0.2nm$. In comparison, the glycan chains are much stiffer $\xi_{p,glycan} \approx 10nm$ [STO 90]. The assumption of ideal chains and filaments has been made, meaning that the length of any chain coincides with the end-to-end displacement, the scalar quantity r_{ee}.

12.2.1.2. *Nuclear lamina (4-fold connectivity)*

The nuclear lamina network is a four-fold connectivity network, with a square repetitive unit cell defining the biological network [AEB 86, MCK 86], pictured in Figure 12.3. It includes a nucleus bounded by two membranes, and a lamina in the interior of the nucleus, adjacent to the inner nuclear membranes. The cell cytoskeleton is made of three interconnected filament systems: actin-containing microfilaments (MFs), microtubules (MTs), and intermediate filaments (IFs), which are responsible for the mechanical stability of cells and tissues.

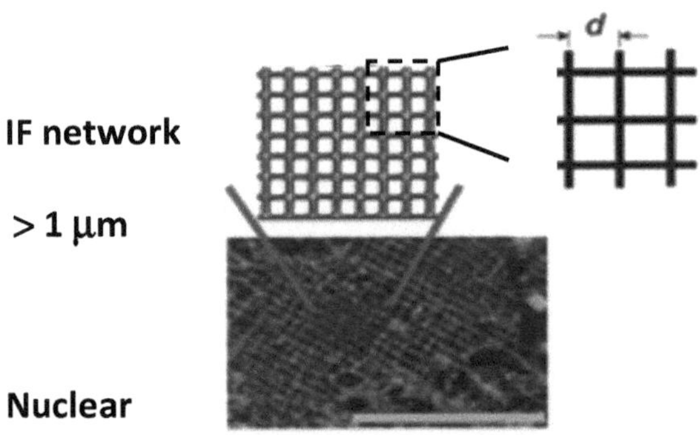

Figure 12.3. *Nuclear lamina and its square lattice of intermediate filaments having four-fold symmetry [QIN 11]. For a color version of this figure, see www.iste.co.uk/ganghoffer/biomechanics.zip*

Individual filaments are flexible, extensible but tough, all these features largely responsible of the observed mechanical behavior of biomembranes, especially their large deformability and shape change capability.

The nuclear lamina is a protein meshwork located at the nuclear membrane possessing a structural role, but also protecting the genetic material and regulating gene expressions. It includes two sets of nearly orthogonal intermediate filaments with a lattice constant d = 50 nm and a diameter of 10 nm [BOA 03], as shown in Figure 12.3. The persistence length of the protein filaments has been measured in the range 1-3 μm [HOH 99]. The persistence length and the Young's modulus are related by the theoretical expression $\xi_p = \frac{YI}{K_B T}$. Intermediate filaments have a persistence length of the order $\sim 1 \mu m$, corresponding to a Young's modulus $Y = 1 \cdot 10^9 \, Pa$.

	Diameter (nm)	Persistence length (μm)	Bending stiffness (Nm^2)	Young's modulus E (Pa)
Actin filaments	9-10	15	7×10^{-26}	$1.3\text{-}2.5\times10^{9}$
Microtubules	25	6000	2.6×10^{-23}	1.9×10^{9}
Intermediate filaments	10	1-3	$4\text{-}12\times10^{-27}$	$1\text{-}3\times10^{9}$

Table 12.1. *Main constituents of the cytoskeleton and their mechanical properties [KAM 10]*

12.2.1.3. *Erythrocyte network (six-fold connectivity)*

The erythrocyte network has a dominant six-fold connectivity, although 3% and 8% of the junctions exhibit five or seven-folds respectively [LIU 87, MOH 94]; it also has simultaneously a low density and a high flexibility. The topology of the in-plane network is modeled as illustrated in Figure 12.4, with configurations parameterized by the angular variable θ.

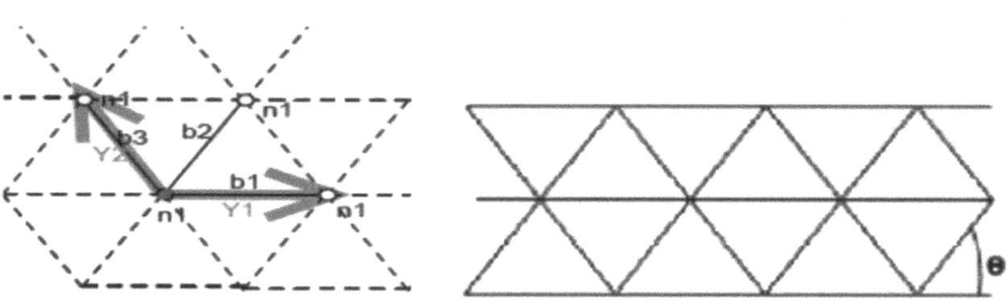

Figure 12.4. *a) Triangular network of the erythrocyte cytoskeleton, b) its idealized configuration parameterized by the angle θ. For a color version of this figure, see www.iste.co.uk/ganghoffer/biomechanics.zip*

At a low temperature close to T = 273K and and for an initially unstressed state, the spring constant for any chain within the network has an average constant value given in [BOE 98] as

$$\beta k_{eff} a^2 = 0.23 \qquad [12.1]$$

with β the inverse of the unit of thermal energy $(1/K_B T)$, controlled by the Boltzmann constant and a the bead diameter, taking the value $6.4nm$ [BOE 98]. From [12.1], the Young's modulus for any chain within the network is evaluated based on the expression

$$Y = \frac{0.23}{\beta a^2} \frac{l^b}{s} \qquad [12.2]$$

in which s denotes the area of the filament cross-section.

As to the geometry of the filaments modeled as beams, we consider a rectangular beam section with width $t = 15nm$ [BOA 94] and average end-to-end length $l^b = 200nm$ [LIU 87]. Filaments have a constant length in the present model and their angular orientation θ varies around the reference value $\theta = 60°$ (Figure 12.4(b)), thus the effective mechanical properties shall be determined over a continuous range of variations of this angle in order to characterize fluctuations of the membrane mechanical properties.

The computation of the effective nonlinear response of biomembranes by the discrete homogenization method relies on the evaluation of the ground properties in the linear regime as a first step, as explained in more detail in Chapter 4 of this book and in [ELN 16].

12.2.2. *First step: computation of the initial linear response*

We shall next evaluate the effective mechanical response of the three types of biological networks subjected to different loadings; the lattice geometrical input parameters need to be given as an input, which deserves the next section.

12.2.2.1. Ground state mechanical properties of the peptidoglycan network

The presence of thermal or pressure fluctuations lead to networks possessing different geometrical configurations; for instance, the classical (in its unperturbed state) hexagonal lattice may develop a re-entrant configuration [BOA 03] characterized by a negative angle θ, these changes possibly occurring in a reversible manner.

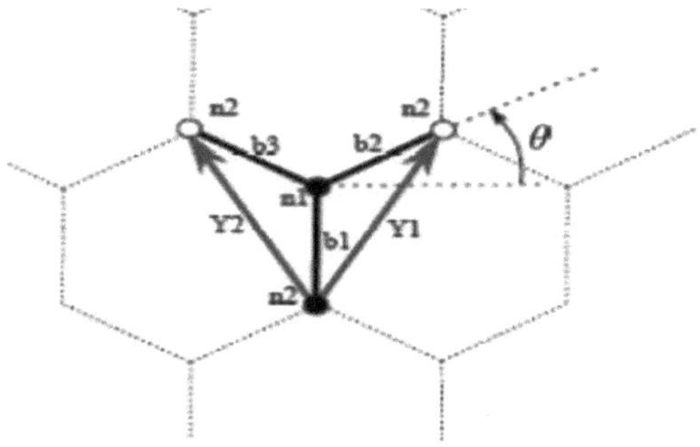

Figure 12.5. *Hexagonal lattice unit cell model parameterized by the angular variable θ. For a color version of this figure, see www.iste.co.uk/ganghoffer/biomechanics.zip*

The entire network is generated from the periodic repetition of the unit cell by translation into the tangent plane, thanks to two periodicity vectors $\mathbf{Y}_1, \mathbf{Y}_2$. The small strains effective moduli are identified from the effective compliance matrix, and are depending upon the geometrical and mechanical network microparameters. A micropolar effective continuum has been chosen in order to reflect the large filament flexibility, expecting to lead to a continuum microrotation and microcurvatures as additional degrees of freedom (in addition to the displacement that would suffice for a mere Cauchy continuum). The effective planar properties are characterized by five mechanical parameters: the two tensile moduli E_1^*, E_2^*, one shear modulus G^*, one Poisson ratio ν_{12} or ν_{21} as the structure proves to be orthotropic for all values of θ and the effective stiffness matrix is symmetrical, and an additional micropolar modulus γ. This set of continuum properties depends upon the filament tensile modulus, the slenderness ratio η and the angle θ; we adopt a slenderness ratio of the beam $\eta = 1/10$, based on images of filaments. The values of the effective properties in the ground state are very close to measurements and to values calculated from other approaches in the literature. The shear modulus is obtained as $3.6 \cdot 10^6 J/m^3$ in [BOA 03], close to the homogenized value $G^* = 2-3 \cdot 10^6 J/m^3$; the effective tensile moduli in the two principal directions, evaluated as $E_1^* = 1-3.2 \cdot 10^7 J/m^3$ and $E_2^* = 3-5 \cdot 10^6 J/m^3$, are close to the measured value $E^* = 2-3 \cdot 10^7 J/m^3$ [BOA 03].

12.2.2.2. Ground state mechanical properties of the nuclear lamina network

The nuclear lamina network with connectivity 4 is modeled as the periodic repetition of square unit cells, each of them characterized by two beams and two periodicity vectors $\mathbf{Y}_1, \mathbf{Y}_2$ as shown in Figure 12.6.

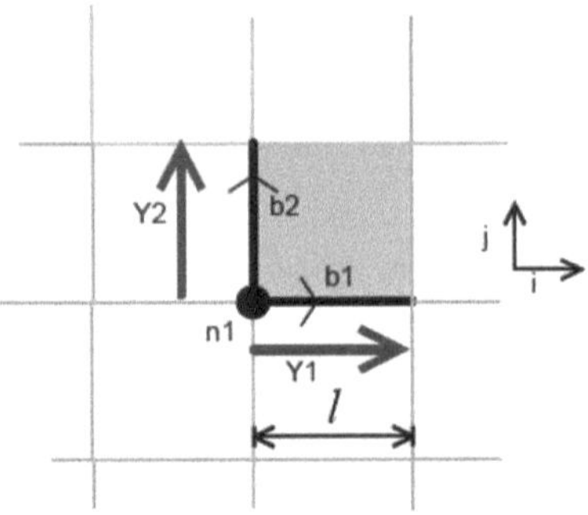

Figure 12.6. *Square unit cell of the nuclear lamina network. For a color version of this figure, see www.iste.co.uk/ganghoffer/biomechanics.zip*

The elastic moduli in the two principal directions are identical at low strains due to the square symmetry and they show identical evolutions with ongoing strains as expected (these evolutions are not shown here).

The Poisson ratio of the nuclear lamina membrane takes the very low value 0.000146, due to the fact that the square lattice is a relatively stiff structure thus prohibiting contraction changes. Moreover, the lamina meshwork is nearly isotropic, with two equal effective tensile moduli in the two principal directions, $E_1^* = E_2^* = 20 \cdot 10^7 J/m^3$; these homogenized values are close to the range of measurements, viz. $E^* \sim 10 - 34 \cdot 10^7 J/m^3$ [KAU 11]. The shear modulus computed by homogenization is of the order of $4 \cdot 10^6 J/m^3$.

12.2.2.3. *Ground state mechanical properties of the erythrocyte network*

Relying on the same methodology, the erythrocyte network with six-fold connectivity is modeled as a set of triangular repeated (periodic) unit cells, with the reference unit cell consisting of three beams and two periodicity vectors $\mathbf{Y}_1, \mathbf{Y}_2$. We select to be specific a configuration with the angular parameter value $\theta = 60^\circ$, (Figure 12.4a), which corresponds to an equilateral triangle ($\theta = 60^\circ$), the elastic moduli and Poisson's ratio in the two principal directions are identical, due to the fact that the equivalent continuum material is isotropic at this specific angle value. The elastic modulus in the transverse direction E_2^* becomes weak when the triangle is flattened in the vertical direction (the angle θ is decreased), whereas the modulus E_1^* increases, as all threads tend to align in the direction. The Poisson ratio of the RBC membrane is close to 0.33 for the reference angle $\theta = 60^\circ$. The effective shear modulus is close to the homogenized value $G^* = 9 \cdot 4710^{-6} J/m^3$, which agrees with [BOA 94, DAO 06]. Moreover, the effective tensile modulus

$E_1^* = E_2^* = 2.5 \cdot 10^{-5} J / m^3$ is rather close to the measurements performed in [DAO 06], who obtain a uniaxial tension Young's modulus $E \approx 2.21 \cdot 10^5 J / m^3$.

12.2.3. *Second step: computation of the nonlinear elastic response*

The nonlinear tensile and shear responses of the three types of biomembranes (relying on the previous classification based on nodal connectivity) are computed based on the incremental scheme described in details in Chapter 4.

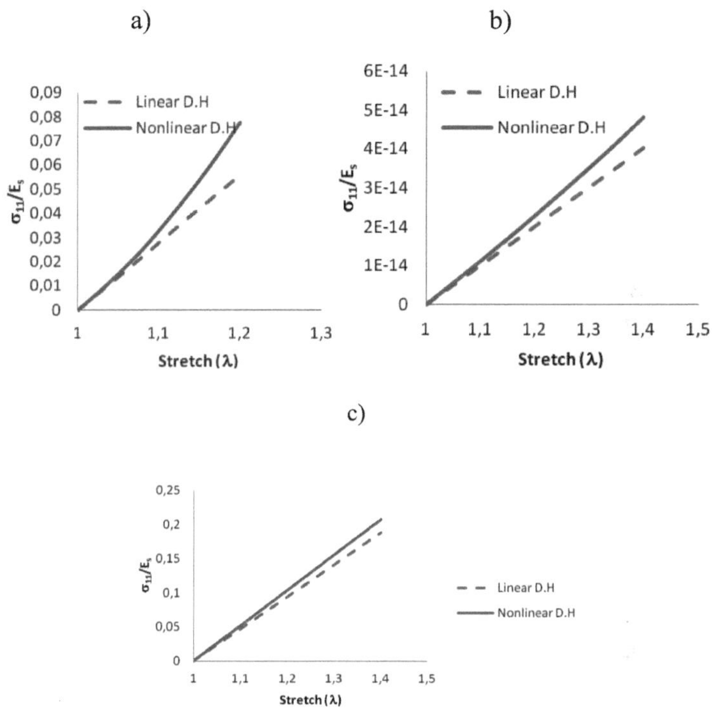

Figure 12.7. *Nonlinear tensile response of the three types of biomembranes: a) 3-connectivity b) 6-connectivity c) 4-connectivity (right). For a color version of this figure, see www.iste.co.uk/ganghoffer/biomechanics.zip*

The fictitious extrapolated linear response is plotted in (Figure 12.7) by extending the small strain response (relying on the initial tangent modulus), in order to quantitatively assess how the variation of network configuration impacts the effective mechanical response. The applied stretch is controlled in the direction of traction, but the large deformation aspects have the far reaching implication of evaluating the transverse stretch from the updated Poisson's ratio associated with each new geometrical configuration of the network.

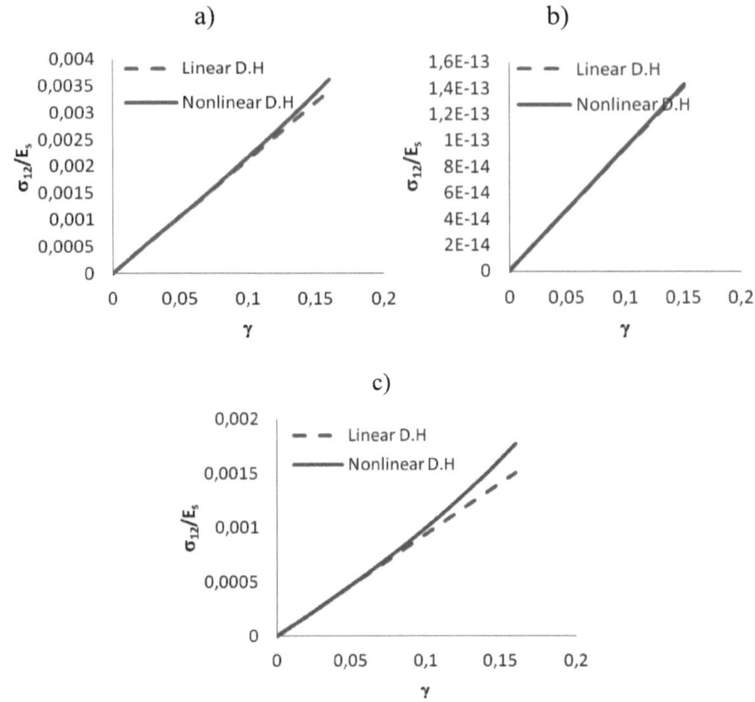

Figure 12.8. *Nonlinear shear response of the three types of biomembranes: a) 3-connectivity, b) 6-connectivity, c) 4-connectivity (right). For a color version of this figure, see www.iste.co.uk/ganghoffer/biomechanics.zip*

The nonlinear effect due to the change of network configuration is nearly 10% for the 3-connectivity and 6-connectivity networks, and is much more pronounced for the 4-connectivity network: as an explanation of this behaviour, the square lattice has no beam to resist the imposed shear strain, which makes it much softer (Figure 12.8).

12.2.4. *Contraction and planar area response*

The surface element (a vector) in the reference configuration, vector $\mathbf{N}dS$ is related to its Eulerian counterpart, vector $\mathbf{n}ds$ (Nanson formula) by

$$\frac{da}{dA} = \frac{\|\mathbf{n}ds\|}{\|\mathbf{N}dS\|} = \frac{\|J\mathbf{F}^{-T}.\mathbf{N}dS\|}{\|\mathbf{N}dS\|} = J\|\mathbf{F}^{-T}.\mathbf{N}\|$$

with $\mathbf{N}$ and $\mathbf{n}$ respectively the unit exterior normal to Lagrangian and Eulerian infinitesimal surface elements of magnitude dS and ds respectively, and $J = \det \mathbf{F}$ the local area change (a non-negative quantity). Under a uniaxial loading test, the area change is evaluated based on previous relation as

$$\frac{da}{dA} = \det(\mathbf{F}) = \lambda_1 \lambda_2 (\lambda_1) \rightarrow \frac{da - dA}{dA} = J - 1$$

with λ_1, λ_2 the two principal stretches. This entails the following expression of the Poisson's contraction coefficient and the area change:

$$\nu_{12} = -E_2 / E_1 \rightarrow \frac{da}{dA}(\lambda_1) = (\nu_{12}(\lambda_1) + 1)\lambda_1 - \nu_{12}(\lambda_1)\lambda_1^2$$

with E_1, E_2 the strains in the two loading directions, referring to a Cartesian basis. This last formula shows that the area changes are strongly controlled by the variation of Poisson's ratio, itself a nonlinear function of the applied stretch in general.

The contraction for the erythrocyte network (Figure 12.09 left), is close to 40% initially and it increases slowly at first, so that there are very little area variations; a stronger increase of the contraction occurs for strains higher than 25%, that entails a decrease of network area. The contraction remains very small for the 4-connectivity network (Figure 12.09 right), since Poisson's ratio is close to zero, thereby evidencing the strong impact of the network topology on its contraction.

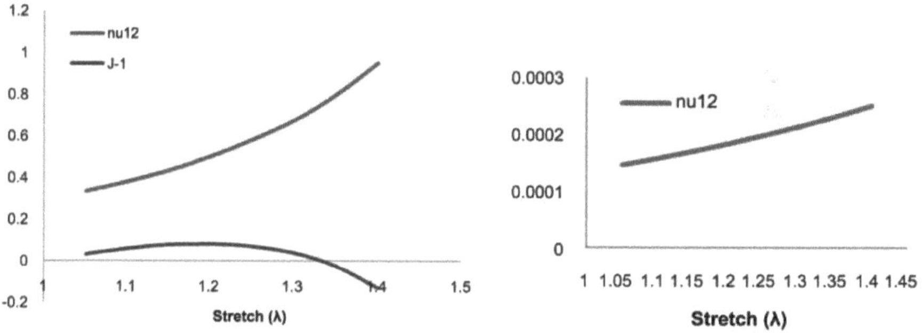

Figure 12.9. *Evolution of Poisson's ratio versus the applied uniaxial stretch for the erythrocyte network and area change for the erythrocyte (left view) and lamina (right view) networks. For a color version of this figure, see www.iste.co.uk/ganghoffer/biomechanics.zip*

The 2D in-plane area compressibility modulus is computed from an equibiaxial test; this positive scalar quantity is elaborated as the relative variation of area

$$K_A := \frac{Tr(\boldsymbol{\sigma})}{Tr(\boldsymbol{\varepsilon})}$$

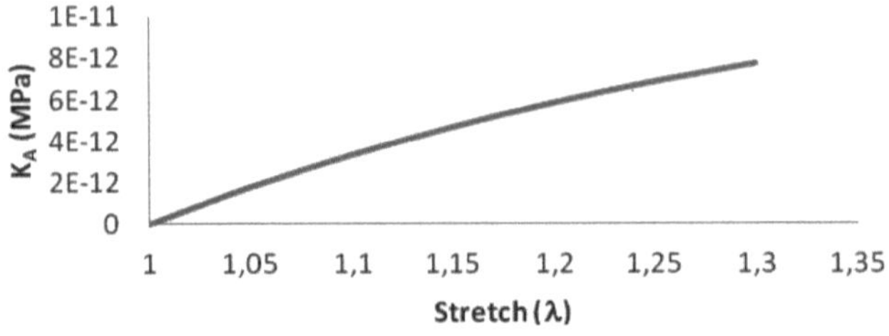

Figure 12.10. *Two-dimensional area compressibility modulus for the erythrocyte network*

The modulus increases versus the imposed (equibiaxial) stretch as shown on Figure 12.10, thereby showing the increased resistance of the network to area changes.

12.2.5. *Nonlinear out-of-plane bending of biomembranes*

Due to mechanical and thermal (entropic) effect, biomembranes undergo local out-of-plane fluctuations, which the present homogenization method aims to quantify at the level of the effective continuum. The bending strain is elaborated based on the geometrical picture shown in Figure 12.11, as

$$\varepsilon_b = \frac{L - L_o}{L_o} = -\frac{y}{\rho}$$

denoting by y the distance measured from the neutral axis, ρ the local curvature radius, and $L_0, L = (\rho - y)\theta$ successively the initial and actual lengths measured along the neutral axis (θ is the rotation angle of the bent configuration), as shown in Figure 12.11. We incorporate a local crimp in the initial membrane geometry to mimic the thermal fluctuations of height causes by Brownian motion [BOA 12].

The internal moment $m_{xz} = k_{33}^{\mu} \frac{\partial \phi_z}{\partial x}$ is recorded versus the curvature $\kappa_{zx} := \frac{\partial \phi_z}{\partial x}$ imposed as a control over the unit cell (Figure 12.12), to highlight the nonlinear bending response. The bending behavior shows a softer response in comparison to the (extrapolated) linear response: an explanation for this softening behavior lies in the fact that the outer face of the bent sheet is subjected to tensile stresses which act to reduce the initial crimp; this crimp is a local curvature enhancing the local bending rigidity, an effect similar to the resistance of shells to bending due to the initial curvature.

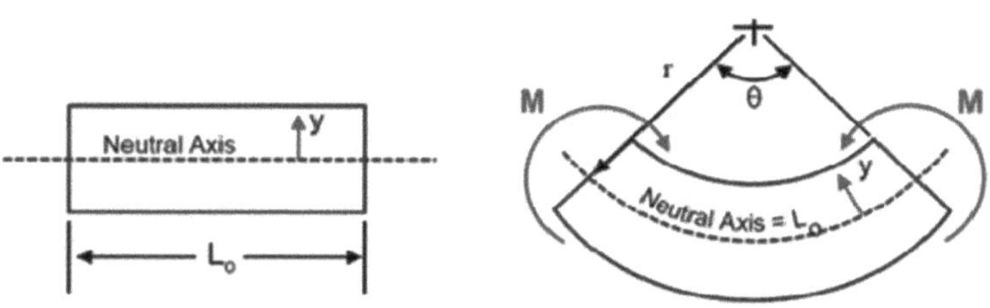

Figure 12.11. *Initial and bent configurations of the continuum biommebrane. For a color version of this figure, see www.iste.co.uk/ganghoffer/biomechanics.zip*

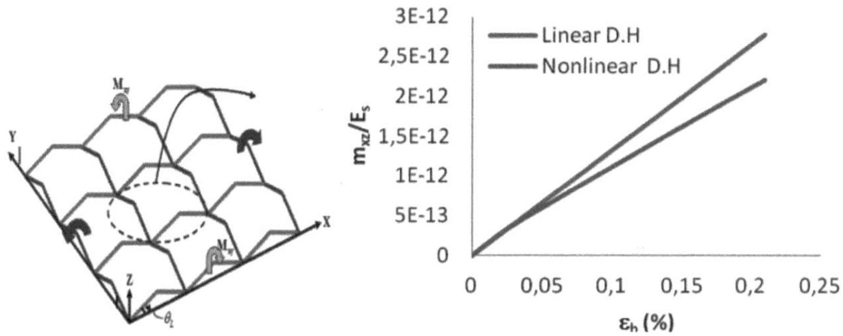

Figure 12.12. *Topology of the 4-connectiviy network (left) and its nonlinear bending reponse (right). For a color version of this figure, see www.iste.co.uk/ganghoffer/biomechanics.zip*

12.2.6. *Identification of a hyperelastic model for biological networks*

The predictive nature of the employed discrete homogenization method allows the identification of a hyperelastic model for biological networks. Note that the nonlinear response of such soft networks may be alternatively captured by hypoelastic models, which however have the drawback of exhibiting a fictitious hysteresis, thus hyperelastic models are preferred instead. Such a micromechanical approach based on virtual tests to identify a constitutive model for biomembranes is novel and contrasts with the widely employed classical phenomenological approach in soft material mechanic and soft tissues biomechanics, which postulates a strain energy potential including anisotropic features.

The selected strain energy density is chosen in a form expressed versus the principal stretches λ_1, λ_2 exerted in the tensile directions and the shear angle θ as:

$$\tilde{W} = \sum_{\alpha=0}^{m} \frac{1}{\alpha+2} a_\alpha \left(\lambda_1^{\alpha+2} - 1\right) + \sum_{\beta=0}^{n} \frac{1}{\beta+2} b_\beta \left(\lambda_2^{\beta+2} - 1\right) +$$

$$+ \sum_{\gamma=0}^{p} \frac{1}{\gamma+1} c_\gamma \cos^{\gamma+1}\theta + k_{r1}\lambda_1\lambda_2 + k_{r2}\left(\lambda_1\lambda_2\right)^2$$

with $a_\alpha, b_\beta, c_\gamma, k_{r1}, k_{r2}$ the set of material parameters to be identified. Both coefficients k_{r1}, k_{r2} are required to introduce a dependency of the two stress components S_{11}, S_{22} versus both stretches simultaneously. Previous density then entails the expression of the second Piola-Kirchoff stress formally as

$$\mathbf{S} = 2\frac{\partial \tilde{W}}{\partial \mathbf{C}} = 2\left[\frac{\partial \tilde{W}}{\partial \lambda_1}\frac{\partial \lambda_1}{\partial \mathbf{C}} + \frac{\partial \tilde{W}}{\partial \lambda_2}\frac{\partial \lambda_2}{\partial \mathbf{C}} + \frac{\partial \tilde{W}}{\partial \cos\theta}\frac{\partial \cos\theta}{\partial \mathbf{C}}\right]$$

Material parameters $a_\alpha, b_\beta, c_\gamma, k_{r1}, k_{r2}$ are identified from a combination of tensile, equibiaxial and simple shear numerical tests done over the unit cell of the biological networks; such a combination is needed to warrant that a sufficiently rich database of numerical tests is available so that the identified hyperelastic model has the capability of predicting the mechanical response of the biomembranes under an arbitrary combination of loadings.

The identification of the material parameters proceeds from the minimization of the following functional with respect to the set of parameters $a_\alpha, b_\beta, c_\gamma, k_{r1}, k_{r2}$

$$\underset{a_\alpha, b_\beta, c_\gamma, k_{r1}, k_{r2}}{\text{Min}} \; S^e\left(a_\alpha, b_\beta, c_\gamma, k_{r1}, k_{r2}\right) := \left\{\left\|\mathbf{S}_{11}^{DH} - \mathbf{S}_{11}\right\|^2 + \left\|\mathbf{S}_{22}^{DH} - \mathbf{S}_{22}\right\|^2 + \left\|\mathbf{S}_{12}^{DH} - \mathbf{S}_{12}\right\|^2\right\}^{1/2}$$

The function $S^e\left(a_\alpha, b_\beta, c_\gamma, k_{r1}, k_{r2}\right)$ of the parameters to be identified is constructed as the quadratic measure of the error between the stress components evaluated numerically by homogenization and their analytical counterpart given by the hyperelastic potential.

As an application, the strain energy density is identified for the peptidoglycan network, considering uniaxial tension in x and y directions, equibiaxial tension and simple shear as simulation loadings. The stress-strain responses obtained from the nonlinear homogenization are presented in Figure 12.13 as well as their 3rd order polynomial adjustments; we have used the function *Lsqcurvefit* of the Optimization Toolbox of MATLAB to identify the optimal parameters of the strain energy density; they are recorded in Table 12.2.

a_1	a_2	a_3	b_1	b_2	b_3	k_{r1}	k_{r2}	c_1	c_2	c_3
5.193	-30.7	20.57	-8.22	0.546	2.658	6.31	-0.67	0.786	-0.023	2.118

Table 12.2. *Optimal material parameters of the hyperelastic density function (all coefficients are in MPa)*

A few words about the limitations of the proposed models of biomembranes are in order. The developed micromechanical modeling of biological membranes essentially deals with the mesoscopic aspects and it largely ignores molecular biological mechanisms which also impact their response. More precisely, the developed modeling framework has the following limitations: the initial entropic undulations of the individual filaments have been accounted for by the persistence length, but the resulting perturbed geometrical shape needs to be described explicitly to provide a more realistic picture. The homogenization of the collagen fibril mechanics accounting for molecular cross-links deformation mechanisms [MAC 12] needs further developments. In the same spirit, geometrical nonlinearities also occur at the level of individual filaments and not only at the level of the homogenized continuum [MAR 12]. Although describing the membrane as a single layer is a common viewpoint of the literature (molecular based models treat the spectrin network and lipid membrane globally, [DAO 06]), it is is a strong approximation as to the in-plane shear behavior, whereas it has less influence on the in-plane stretch; biological membranes indeed consist of two leaflets with a relative sliding mobility.

The periodicity assumption made in this work is certainly valid locally, large configuration changes occur at larger scales, requiring the extension of micromechanical schemes based on perfect regularity and periodicity to account for the true randomness of biological networks.

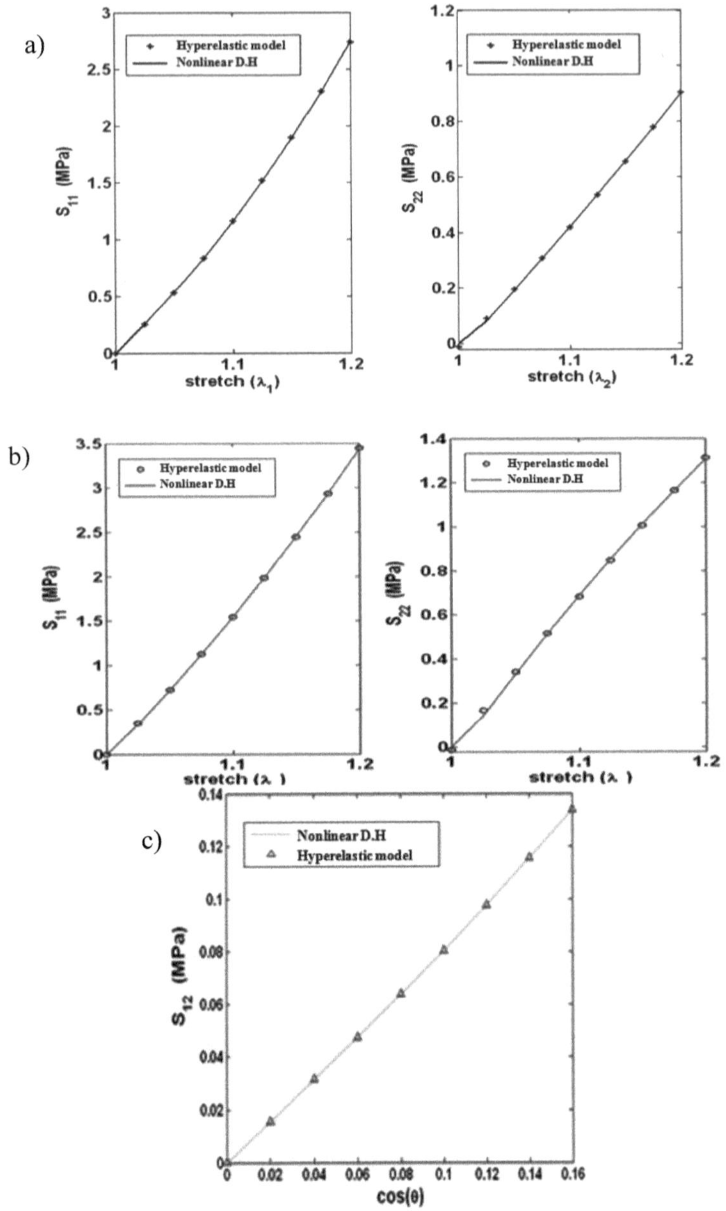

Figure 12.13. *Second Piola-Kirchhoff stress components for a) uniaxial tension in x-direction (left) and y-direction (right) b) equibiaxial tension c) simple shear responses from nonlinear discrete homogenization and from the identified hyperelastic model. For a color version of this figure, see www.iste.co.uk/ganghoffer/biomechanics.zip*

The derived nonlinear constitutive laws account for the large configuration changes of the biomembranes network; simulations results have shown especially that contraction effects are pronounced at large strains, a feature observed in the fetal membrane [BUE 13]. The DH method shall be extended to the consideration of a viscous behavior of the filaments, by extending the writing of the virtual work principle which is at the base of the method.

Micromechanical analyses of biological networks based on refined constitutive laws form the basis for FE-based simulations of biological cells and biomembranes submitted to loadings reflecting the mechanical effect of their biological environment.

13

Micromechanics of Ligaments and Tendons

13.1. Introduction

Tendons aim at linking muscles and skeleton, and have the twofold function of stabilizing joints and transmit muscular forces. They also decrease the energy needed for locomotion since they act as energy stores [BEN 97]. Ligaments link one bone to another, and restrict their relative movements during motion. Both tendinous and ligamentous materials are very resistant connective tissues based on a complex hierarchy of collagen fibres and fibrils. From the 1960s, the mechanical behaviour of these soft tissues has motivated numerous studies, concerning particularly:

1) the characterization of their mechanical behaviour, their evolution and the design of ad-hoc experimental methods,

2) the modelling of their specific mechanical behaviour,

3) the definition of relations between their structural properties and their macroscopic behaviour. These different issues are not fully solved, and numerous studies are published each year in order to better understand and diagnose tendon-related or ligament-related pathologies, often due to over-loading.

In the present chapter, a brief description of the structure and main properties of ligaments and tendons is firstly proposed, including their physiological function and environment. A non-exhaustive summary of the main *in vivo* and *in vitro* characterization methods used to determine their mechanical response is then exposed. A state of art of modeling approaches applied to tendon and ligament is then presented, including both phenomenological approaches and microstructural

---

Chapter written by Cédric LAURENT.

models. At last, a brief focus on ligament and tendon tissue engineering is proposed, in order to link the micromechanical models proposed in the literature with future applications concerning alternative reconstruction methods for these tissues. A conclusion concerning the barriers that are still to be broken down regarding tendon and ligament characterization or modeling will be the proposed.

13.2. Structure and properties of ligaments and tendons

13.2.1. *Structure and composition*

Tendons and ligaments are soft and dense collagenous tissues, whose structures are multiscale and highly organized. The shape and length of both tendons and ligaments largely vary from one tissue to another [BEN 97]. Nevertheless, their structure is similar and based on a highly organized and hierarchical assembly of collagen fibrils made of tropocollagen molecules, grouped into fibers that are collected into different levels of bundles and fascicles, both wrapped up in thin surfaces of connective tissue (Figure 13.1).

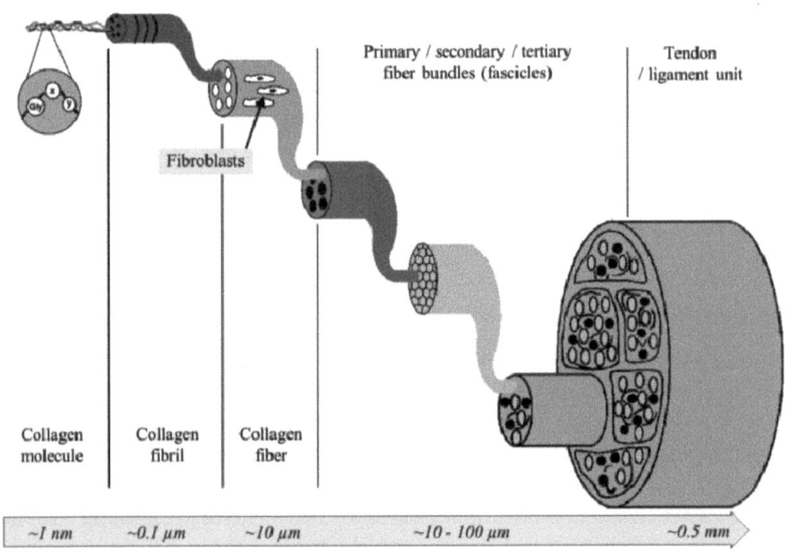

Figure 13.1. *Illustration of the multiscale hierarchical structure of ligament and tendons. Adapted from [WAN 06a]. For a color version of this figure, see www.iste.co.uk/ganghoffer/biomechanics.zip*

Bundles and fascicles have a helical arrangement within most tendons and ligaments [REE 10, SCR 04a, SHE 15b, THO 13] and collagen fibers exhibit a well-known crimp pattern [CAC 07, FRA 07, GRY 09, MIL 12]. The alignment of both the spiral-shaped bundles and the crimped fibers in the direction of loading (Figure 13.2) contributes to the flexibility of the overall tissue, together with the sliding which may occur between these different levels of organization. Consequently, this multiscale organization makes the overall tissue response difficult to interpret, since the role of each scale and their interactions are crucial and hardly separable.

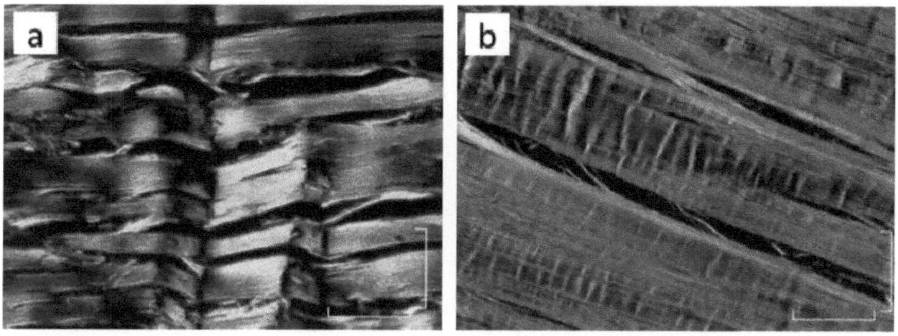

Figure 13.2. *a) Polarized light micrograph of relaxed Achilles tendon b) Polarized light micrograph of stretched Achilles tendon. Scale bar = 100 μm. Adapted from [FRA 07]. For a color version of this figure, see www.iste.co.uk/ganghoffer/biomechanics.zip*

Tendons and ligaments consist in a compact assembly of water, collagen fibres and different proteins contained in the Extra-Cellular Matrix (ECM). The interested reader is referred to specific reviews for a detailed description of the role of these components in the physiological function of the tendon [BEN 08, KER 07, KJA 04, WAN 06a]. Most of the similar reviews concerning ligament focus on the case of the frequently injured Anterior Cruciate Ligament (ACL) [BIC 10, DUT 05, KAR 03, KAR 04, VUN 04, WOO 06]. Briefly, while the proportions of their different components are subject to large variability, ligaments and tendons are biological composites essentially made of a fluid phase (65% to 75% of water by weight [BEN 97]). Collagen is the main component of the solid phase of tendons and ligaments (70 to 80% of the dry mass), and is mainly of its tensile properties. Collagen I is the most abundant, while other types of collagens may be present depending on the tissue. Proteoglycans exhibit a crucial role in binding water within the tendon and ligament, and participate in both its compressive properties and the lubrication allowing relative gliding between fibers. Glycosaminoglycans (GAG) filaments link the collagen fibrils together, as seen in Figure 13.3. Other components in lower proportions are present in the tendon: elastin, glycoproteins, a few other inorganic components (minerals) and cells (essentially fibroblasts).

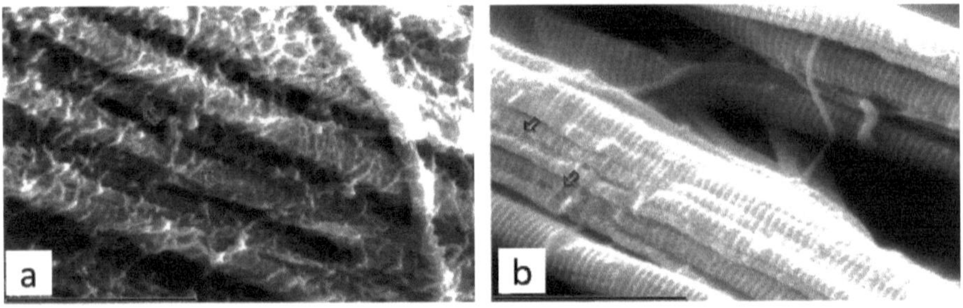

Figure 13.3. *Extra-cellular matrix or the rat tail tendon before a) and after (b) the digestion of glycosaminoglycans (GAG) filaments by chondroitinase. (a) GAG filaments are indicated by arrows. b) Without GAG filaments, the 67 nm period of single collagen fibrils is clearly visible at this resolution. Images obtained by high-resolution scanning electron microscopy, scale bar =1 µm. Adapted from [RAS 02]*

Tendons and ligaments are mostly highly innerved, and thus play a substantial role in proprioception. Noteworthy, a neurotendinous proprioceptive sensory receptor named "Golgi tendon organ" has been particularly shown to provide with a feedback on the forces present within tendons and thus to adapt the intensity of these forces [HOU 67]. On the contrary, tendons and ligamentous are poorly vascularized, limiting their potential to self-repair and heal in the case of injury. Tendons and ligaments are linked to bone through a fibrous or fibrocartilaginous connective tissue named *enthesis* [BEN 97]. This highly progressive interface constitutes a good example of a biological interface between two tissues whose properties are highly different, thus considerably decreasing stress concentration at this interface [BEN 02, BEN 06, THO 03].

13.2.2. *Adaptative behavior and healing*

Ligaments and tendons are subject to remodeling since their weight, size, turnover rate and mechanical properties are affected both by additional exercise or prolonged immobilization [BEN 97, WAN 06a]. Repeated or excessive loading of ligaments and tendons (and enthesis) may lead to injuries or ruptures. Low repetitive strains may also cause tendon inflammation and tendon degeneration [WAN 06a]. The healing process following tissue injury may be decomposed into three overlapping phases, spread over several months [WOO 99, WOO 06]. During a first inflammatory phase, a hematoma is formed and the inflammatory cells migrate and provide the fibroblasts with growth factors. It is followed by a second reparative phase during cells proliferate and produce of an extra-cellular collagenous matrix.

Finally, the collagen network is structured and matured during a last remodeling phase.

13.2.3. *Physiological loads and mechanical properties*

The estimation of both the physiological loadings applied to tendons and ligaments during daily activities and their native mechanical response to these loads constitutes the focus of many studies and requires the development of custom characterization methods described in the next section. Physiological loads applied to tendon may be very high for collagen-based connective tissues: for instance, these loads can reach 5000 N in the patellar tendon during isometric knee extension [HAN 06, OBR 10] and 2000 N in the Achilles tendon during isometric plantar flexion [KON 11]. Tendons and ligaments are obviously subject to fatigue loading, since they are cyclically loaded during normal locomotion.

Numerous *in vitro* studies have assessed the typical tensile response of different tendons and ligaments. This response is highly non-linear due to the arrangement of the tissue hierarchical structure, and more specifically due to the alignment of crimp fibres in the direction of loading [MIL 12, STO 85]. As a consequence, a typical tensile curve of a ligament or a tendon exhibits an initial "toe region" for strains up to 1–2%, depending on the subject, the type of tissue, and the sample site within the tissue. The tensile response then exhibits a linear part (where a Young modulus may be evaluated) up to the appearance of the first microscopic tearing of fibres, followed with the final macroscopic failure of the tissue (Figure 13.4). The physiological strains typically reach a few percent, and the strain at failure may exceed 15%. Depending on the tissue, the subject and the experimental protocol, typical values of the Young modulus range from some hundreds of MPa [JOH 94, WOO 06, WRE 01] to more than 1000 MPa [MAG 00]. While elastic properties of tendons and ligaments are mainly due to their collagenous composition, they also depend on the presence of glycosaminoglycan (GAG) interconnecting neighboring collagen fibers [RIG 09]. Due to the structural organization described previously, ligaments and tendons are highly anisotropic: for instance, the Young modulus of the supraspinatus tendon may vary by a factor 1000 depending on the loading direction and the sample site [LAK 10]. Due to its obvious privileged direction, ligaments or tendons have been often considered as transversely isotropic [GRY 09, MAC 11] or orthotropic [HIR 00].

Tendons and ligaments, like most of biological tissues, also exhibit viscoelastic behavior involving creep [HAW 09, PEA 07], dependency to strain-rate [PEA 07, PIO 00, WRE 01] and stress relaxation [JOH 94]. These viscoelastic phenomena may be explained both by fiber sliding [GUP 10] and by the presence of fluid movement [SWE 14].

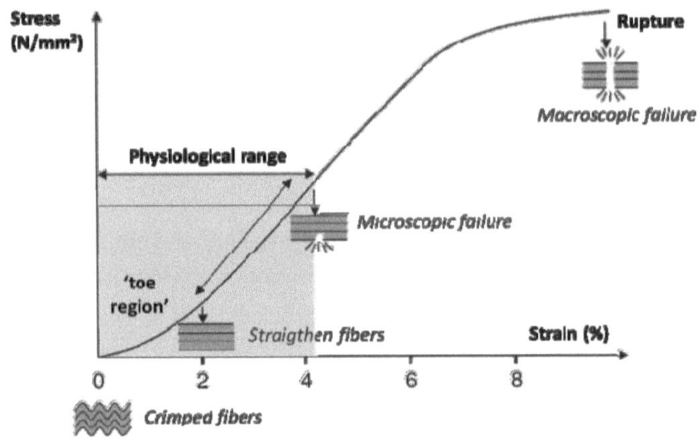

Figure 13.4. *Typical stress-strain curve for ligaments and tendons. Adapted from [WAN 06a]*

13.3. Characterization methods for tendons and ligaments

In the following sections, a brief review of experimental methods is provided concerning both *in vitro* and *in vivo* characterization. *In vitro* characterization allows to perform tests in perfectly controlled testing conditions, to precisely measure various mechanical parameters, and tests up to failure. On the opposite, *in vivo* characterization allows quantifying the physiological *in situ* loadings applied to the tissue, and to evaluate its properties in a native environment.

13.3.1. *In vitro characterization*

When realizing *in vitro* mechanical tests on biological soft tissues, a particular attention has to be drawn on the harvesting, the conservation and the preparation of the samples prior to the test. Samples are generally frozen until the test is performed, and this should be done right after the harvesting to avoid the tissue from drying. The hydration of the sample during the test is generally ensured by immersing [JOH 94, WRE 03] of spraying [DEF 04] the sample, and a physiological saline bath is generally used. For tendons and ligaments, the tissue may be characterized together with its bony insertions (and with its musculotendinous junction for tendon), which facilitates the gripping of the sample and allows to characterize the properties of these regions. In this case, the insertions are generally clamped using classical grips [KEN 76, NOY 74] or by embedding the bones within poly(methyl methacrylate) (PMMA) resin [JON 95]. Alternatively, the tissue may be tested isolated from its bony insertions or musculotendinous junction. In this case, the gripping of samples constitutes a real issue when testing soft tissues, since the slipping should be prevented without prescribing compression levels that may deteriorate the tissue. A common solution consists in embedding tissue extremities within PMMA [JOH 94], but the exothermic reaction associated with PMMA polymerization may damage the tissue. A reported alternative is to grip one or two extremities using insulated liquid nitrogen-cooled freeze clamps [WIE 04, WRE 01, WRE 03]. In this case, an associated problem may be to determine the actual temperature within the tissue during the mechanical test. Other gripping techniques such as sinusoidal clamps [DEF 04] or cyanoacrylate adhesive strips [RIG 09] have also been used.

Prior to the test, a pretension is generally applied in order to straighten the sample. No consensus has been established for this value, which can vary from 1 N [JOH 94] to 50 N [VER 11], with obvious consequences on the obtained toe region of the tensile response. Generally, a preconditioning is realized to minimize viscoelastic effects (e.g. 10 cycles at 2% of strain at 0.5 Hz [JOH 94, WRE 01]). Tests have been performed with various tensile speeds (e.g. from 20 mm/min [VER 11] to 10 mm/s [WRE 01]), with no clear consensus despite of its consequences on the obtained response. Because of these discrepancies in the adopted protocols, comparisons between studies are often delicate to formulate. Similarly, different methods are used to measure the sample cross-section (crucial for the calculation of stress). This may be realized using laser micrometer system [JOH 94], ultrasound [WRE 01], pre-built alginate molds [GOO 05] or rapid linear laser scanners [VER 10] allowing a real-time measurement, and therefore the assessment of effective Poisson's ratio. During these tests, a particular attention must be drawn concerning the definition of mechanical quantities, since large strains are often obtained. For instance, the engineering strain and stress should not be confounded with their large strain material forms (Green-Lagrange strain and the second Piola-Kirchoff stress), or with true strain and true stress. Similarly, the definition of a Young modulus may be particularly delicate since the response is highly non-linear. A *tangent modulus* should rather be defined, and its definition and calculation method should be clearly mentioned.

Measuring precisely the strain (local or global) also constitutes a true challenge when testing ligaments or tendons. One of the most basic method consists in using dye marks or lines onto the sample (for instance using Verhoeff's elastin stain [JOH 94, KWA 93] or small metallic beads glued on the tissue surface [WRE 01], associated with video analyzing. Radiopaque beads may also be used in association with fluoroscopic images [BEY 12, DEF 06]. These methods allow for calculating the macroscopic strain applied to the tissue, but are restricted to surface measurement. Alternatively, needles pinned across the tissue may be used [VER 11] together with a video tracking to get this macroscopic strain. However, it has been clearly shown that strains may be highly heterogeneous within tissues such as tendon, the strain in the central part being generally much lower that the strain in the insertions (with a factor 7 on the calculation of elastic modulus [DEF 06]). Considering a single macroscopic strain may therefore constitute a considerable source of error when characterizing the material properties of soft ligaments or tendons. Other methods based on image correlation from direct images [RIG 09], or ultrasound elastography [CHE 12b, CHE 14], allow us to build full strain maps. This second ultrasound-based method is also compatible with *in vivo* measurements. However, this is a 2D measurement which assumes that the deformation stays in a plane perpendicular to the acquisition, thus excluding out-of-plane motion. In order to characterize more precisely the local distribution of strains, various protocols have been reported: they may consist in using the cell nuclei used as local discrete markers and confocal microscopy in 2D [SCR 04b] or 3D [KHO 13] to build strain maps within the tissues. Using confocal microscopy to draw a photo bleached grid may also allow tracking local motions within tendon [CHE 07b]. These methods are performed under microscope, and are therefore not adapted to large samples. However, they have allowed us to emphasize the fact that local strain of collagen fibers are much lower that the global macroscopic strain, indicating that the deformation may be due attributed to sliding between fibers.

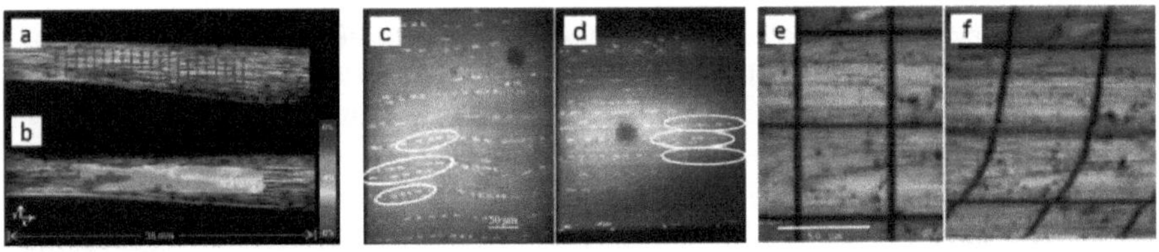

Figure 13.5. *Methods for measuring local strains in tendons or ligaments. a) and b) Image correlation from ultrasound elastography, before and after applying a 4% strain respectively. Adapted from [CHE 12b]. c) and d) Tracking of cell nuclei from confocal imageing, before and after a 6% strain respectively. Adapted from [SCR 04b] e) and f) Use of a photo bleached grid to compute strains before and after a 6% strain respectively. Adapted from [CHE 07b]. For a color version of this figure, see www.iste.co.uk/ganghoffer/biomechanics.zip*

13.3.2. *In vivo characterization*

One of the most basic methods for *in vivo* characterization of tendon or ligament properties is the direct implantation of gauges and sensors within the tissue. The main implantable gauges have already been listed and compared [BEY 12, FEL 4,

RAV 04]. Such an implantation is obviously invasive and requires volunteers; moreover the question of the impact of the surgery on the tissue properties comes up. To minimize the impact of the sensors on the tissue properties, it should be sufficiently deformable. Various sensors have therefore been proposed (e.g. liquid metal strain gauges, buckle transducers, fiber optic transducers, etc.). These different designs require the insertion and the prior calibration of the sensor. When possible, the sensor should be calibrated *in situ*: it is possible in the case of animal studies, by harvesting the tissue after the *in vivo* measurements and loading it in a controlled environment. These methods have permitted to provide with data concerning the physiological loadings associated with common rehabilitation exercises [BEY 98]. Such measurements may be coupled with measurements of muscular activity using electromyography (EMG), measurements of ground reaction forces using force platforms, of motion tracking using additional markers.

Ultrasound measurements may constitute a good alternative to the above techniques for measuring *in vivo* strains within ligaments or tendons. Indeed, some anatomical marks (bony insertions or musculotendinous junction) are visible using echography, and the global macroscopic strain can thus be computed from the motion of these marks during various physiological movements [KOR 10, OBR 10]. These methods are non-invasive, cheap and compatible with dynamic acquisitions. However, the principal associated limitation lies in the fact that it is a 2D measurement within a plane perpendicular to the acquisition device. It is therefore limited to specific tissues and to the measurements of planar strains. Algorithms based on image correlation have been developed and may increase the potential of these methods [KOR 10]. Moreover, if these measurements are associated to the quantification of *in situ* forces through the measurements joint torques, it is possible to calculate values of the *in vivo* stiffness of the tissues and therefore their elastic modulus [HAN 06, HAW 09, KON 11, MAG 02, OBR 10]. These methods are however limited by the errors associated to the evaluation of forces, tissue length and tissue cross-section. Concerning non-invasive measurement of forces within tendons, it has also been observed [CRE 09, POU 05] that the velocity of an ultrasonic wave propagating along the main axis of a tendon increases (in a non-linear way) with the force applied to this tendon, enabling a measure force after a calibration [VER 12].

Other *in vivo* indirect measurement methods have also been reported: it includes the insertion of implanted markers (tantalum beads [SCH 07] or radiopaque suture wire [BAR 11]) with a radiographic tracking. In the case of small animal studies, it is also possible to directly load a tissue using an external device and to visualize the strain after an incision [FUN 10]. This is however very invasive, and only adapted to a few accessible tissues. Lastly, the principle of ultrasound elastography (consisting in comparing echographic images before and after manual compressions) allows us to build stiffness maps within soft tissues using image correlation [KLA 10]. It is adapted to tendon, but do not provide with absolute values of material properties. Alternatively, shear wave elastography (Figure 13.6) may allow us to build maps of shear modulus within a tissue, and is a promising method in detecting tendon-related pathologies affected its properties [CHE 13].

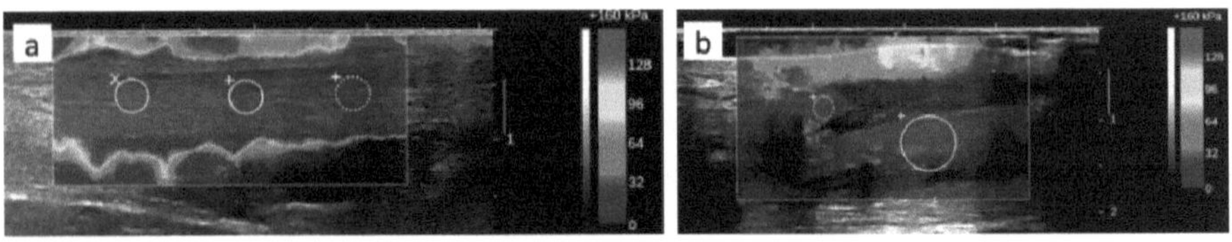

Figure 13.6. *Example of shear wave elastography applied to a healthy a) and freshly ruptured b) Achilles tendon. Adapted from [CHE 13]. For a color version of this figure, see www.iste.co.uk/ganghoffer/biomechanics.zip*

13.4. Modeling approaches for tendons and ligaments

13.4.1. *Development of mechanical models: rationale and challenges*

Modeling the mechanical behaviour of ligaments and tendons is crucial for various reasons: understanding the mechanisms of injury and healing, detecting and monitor pathologies, design new substitutes for repair, or modeling full joints for applications such as prosthesis design, transport, or ergonomics. Moreover, from the point of view of fundamental research, such a modeling constitutes a real scientific challenge since the ligamentous or tendinous tissues are highly non-linear, anisotropic, multiphasic, multiscale, in the framework of finite strains. Two types of modeling approaches are reported in the following sections: firstly, a focus is made on some well-known phenomenological approaches, tha consist in building a constitutive law reproducing experimental data within any tspecific link between the material parameters and the tissue microstructure detailed above. Secondly, some microstructure-based models (or *micromechanical* models) are detailed, aiming at considering the different microscopic components of the tissue so as to build macroscopic constitutive laws. If the second type of approaches theoretically allows us to directly measure the different parameters of the model, the phenomenological models are generally more efficient from a numerical point of view, and are therefore more suited to be integrated within simulations of full joints or limbs.

13.4.2. *Reminder and notations*

The basis of the constitutive laws detailed below are given in Chapter 3 of this book. Let us recall that for a hyperelastic material, the stress-strain relationship is provided starting from a strain energy function *W*:

$$\mathbf{S} = \frac{\partial W(\mathbf{E})}{\partial \mathbf{E}} = 2\frac{\partial W(\mathbf{C})}{\partial \mathbf{C}} \qquad [13.1]$$

where $\mathbf{S}$ stands for the PK2 stress tensor, $\mathbf{E}$ is the Green-Lagrange strain tensor, and $\mathbf{C}$ is the right Cauchy-Green stretch tensor. The strain energy density is a non-negative scalar-valued function of a tensorial strain measure. It vanishes during a rigid body motion if the media is initially in a stress-free configuration, leading to the so-called normalization condition:

$$W(\mathbf{I}) = 0 \qquad [13.2]$$

For an isotropic material, the strain-energy function is generally written in terms of the principal invariants of $\mathbf{C}$:

$$W = W\left(I_1(\mathbf{C}), I_2(\mathbf{C}), I_3(\mathbf{C})\right) \qquad [13.3]$$

which leads to the following constitutive equation:

$$\mathbf{S} = 2\left(\left(\frac{\partial W}{\partial I_1(\mathbf{C})} + I_1(\mathbf{C})\frac{\partial W}{\partial I_2(\mathbf{C})}\right)\mathbf{I} - \mathbf{C}\frac{\partial W}{\partial I_2(\mathbf{C})} + I_3(\mathbf{C})\frac{\partial W}{\partial I_3(\mathbf{C})}\mathbf{C}^{-1}\right) \qquad [13.4]$$

Alternatively, the strain-energy function may be written in terms on the principal stretches $\lambda_I, \lambda_{II}, \lambda_{III}$, leading to the following constitutive equation:

$$S_A = \frac{1}{\lambda_A}\frac{\partial W}{\partial \lambda_A} \qquad A \in [I, II, III] \qquad [13.5]$$

Under the assumption of incompressibility, the Jacobian of the transformation is unity ($J = \det \mathbf{F}$) and thus the strain-energy potential takes the form :

$$W = W - p(J-1) \qquad [13.6]$$

where p is a Lagrange multiplier enforcing the incompressibility condition, to be identified from equilibrium equations and boundary conditions.

If the material is not assumed to be isotropic, pseudo-invariants are generally introduced based on the structural tensor, accounting from the directions of fibers. In this case, the stress derives from a strain-energy function which depends on the n strain invariants, and equation [13.4] is thus extended to account for these new terms (see for instance [HOL 00a, HOL 00b]).

If the assumption of incompressibility is not formulated, it is common to split the deformation into volumetric and isochoric contributions, introducing a modified right Cauchy-Green stretch tensor associated with volume-preserving transformation:

$$\overline{\mathbf{C}} = J^{-2/3}\mathbf{C} \qquad [13.7]$$

In this case, the strain-energy function is generally decoupled into a volumetric density $W_{vol}(J)$ describing the volume changes, and an isochoric density $W_{iso}(\overline{\mathbf{C}})$ modeling shape changes:

$$W(\mathbf{C}) = W_{vol}(J) + W_{iso}(\overline{\mathbf{C}}) \qquad [13.8]$$

13.4.3. *Phenomenological approaches*

Even if initially developed for skin, it appears useful to mention one of the first hyperelastic model developed for soft tissues by Veronda and Westmann [VER 70] since it is still largely used in the field of numerical biomechanics. It is expressed in terms of principal stretches, assuming that the squares of transverse principal stretches λ_{II}^2 and λ_{III}^2 are identical and equal to $\lambda_{II}\lambda_{III}$. In this case, one may write the relation [13.5] on the form:

$$\frac{S_I}{2\left(1-\dfrac{\lambda_{II}\lambda_{III}}{\lambda_I^2}\right)} = \left(\frac{\partial W}{\partial I_1} + \lambda_{II}\lambda_{III}\frac{\partial W}{\partial I_2}\right) \qquad [13.9]$$

Based on experimental data, an exponential form for the strain-energy function has been proposed:

$$W = c_1\left(e^{\beta(I_1-3)} - 1\right) + c_2\left(I_2 - 3\right) + g\left(I_3\right) \qquad [13.10]$$

where c_1, β, c_2 are material parameters to be identified and $g(I_3)$ is deduced from the normalization condition [13.2]. This model has permitted to reproduce skin behaviour in a satisfying way. However, it considers the tissue to be isotropic, which may therefore restrict its use to ligament and tendon.

Other exponential forms for the strain-energy function have been proposed, and extended to the case of anisotropic soft tissues. In particular, Holzapfel developed an orthotropic model with two families of fibers initially to fit the behavior of arteries [HOL 07, HOL 00b, HOL 10]. As reminded above (equation [13.8]), the strain-energy function may be split into isochoric and volumetric contribution. In the model of Holzapfel, the isochoric contribution is written in terms of $\overline{\mathbf{C}}$ and the initial directions $\mathbf{a}_{01}, \mathbf{a}_{02}$ of two populations of fibers, and split into an isotropic part (due to the matrix) and an anisotropic part (due to the fibers):

$$W_{isochoric}\left(\overline{\mathbf{C}}, \mathbf{a}_{01}, \mathbf{a}_{02}\right) = W_{isotropic}\left(\overline{\mathbf{C}}\right) + W_{anisotropic}\left(\overline{\mathbf{C}}, \mathbf{a}_{01}, \mathbf{a}_{02}\right) \quad [13.11]$$

In the case of ligament or tendon, a single population of fibers may be considered. One may therefore introduce the following pseudo-invariant:

$$\overline{I_4}\left(\overline{\mathbf{C}}, \mathbf{a}_{01}\right) = \overline{\mathbf{C}} : \mathbf{A}_1 \quad \text{with} \quad \begin{cases} \mathbf{A}_i = \mathbf{a}_{0i} \otimes \mathbf{a}_{0i} \\ \left\| \mathbf{a}_{0i} \right\| = 1 \end{cases} \quad [13.12]$$

The isotropic contribution of the matrix may be modeled using a standard neo-Hookean model formulated in terms of $\overline{\mathbf{C}}$:

$$W_{isotropic}\left(\overline{\mathbf{C}}\right) = \frac{c}{2}\left(\overline{I_1}\left(\overline{\mathbf{C}}\right) - 3\right) \quad \text{with} \quad \overline{I_1}\left(\overline{\mathbf{C}}\right) = Tr\overline{\mathbf{C}} \quad [13.13]$$

The anisotropic contribution is written as previously on the form of an exponential expression:

$$W_{anisotropic}\left(\overline{I_4}\right) = \frac{k_1}{2k_1}\left(e^{k_2\left(\overline{I_4}-1\right)^2} - 1\right) \quad [13.14]$$

In equations [13.13] and [13.14], the material parameters c, k_1, k_2 have to be identified from experimental data. The final model is largely spread and often referred to as the HGO (Holzapfel–Gasser–Ogden) model.

Starting from the same assumption of two families of fibres, it is also possible to propose a decomposition of the strain-energy function W into three contributions respectively due to the matrix and the two families of fibers [HIR 00]. These different contributions may be weighted by the volume ratios of each component:

$$W(\mathbf{C}) = v_m W_m(\mathbf{C}) + v_a W_{fa}(\lambda_a) + v_b W_{fb}(\lambda_b) \quad \text{with} \quad v_m + v_a + v_b = 1 \quad [13.15]$$

where W_m, W_{fa}, W_{fb} are the stored energies in the matrix and the two families of fibers, v_m, v_a, v_b are the volume ratios of each component, and λ_a, λ_b are the stretches of the two families of fibers, respectively.

In this model, the matrix has been considered incompressible and follows a Mooney-Rivlin model. The stress-strain relationship of single collagen fibers is assumed to be bi-linear, characterized from a strain threshold ε_s .

$$\sigma = \begin{cases} k_S \varepsilon & 0 \leq \varepsilon \leq \varepsilon_s \\ K_e(\varepsilon - \varepsilon_s) + k_S \varepsilon & \varepsilon_s \leq \varepsilon \end{cases} \quad [13.16]$$

Fibres have been separated into three groups of different lengths, straightened at different strain levels, in order to smooth the macroscopic stress-strain relationship.

These two orthotropic models [HIR 00, HOL 10] have been implemented numerically within Finite Element (FE) codes, and have permitted to reproduce the non-linear response of soft tissues like ligament or tendons. However, since they are purely phenomenological, the issue of the physical meaning and the identification of the material parameters is a crucial issue.

Noteworthy, these phenomenological approaches may be also extended to the modeling of viscoelasticity within biological tissues in the same framework of continuum thermodynamics reminded in previous chapters. Pioletti [PIO 98] has proposed such a model, assuming the existence of a viscous potential $W^v(\mathbf{C})$ accounting for energy dissipation in addition to the previous strain-energy function $W(\mathbf{C}) = W^e(\mathbf{C})$. In this case, the second law of thermodynamics may be written:

$$\mathbf{S} - 2\frac{\partial W^e}{\partial \mathbf{C}} = 2\frac{\partial W^v}{\partial \dot{\mathbf{C}}} \quad \text{with the condition} \quad \frac{\partial W^v}{\partial \dot{\mathbf{C}}} : \dot{\mathbf{C}} \geq 0 \quad [13.17]$$

In this case, additional pseudo-invariants involving both $\mathbf{C}$ and $\dot{\mathbf{C}}$ have been proposed and derived in order to obtain the final three-dimensional constitutive law. In this specific model of Pioletti, the elastic part of this model may be written using one of the previously detailed model, such as the Veronda-Westmann model, while the viscous part $W^v(\mathbf{C})$ may be identified from experimental data:

$$W^v = \frac{\eta}{4} J_2\left(\dot{\mathbf{C}}\right)\left[I_1(\mathbf{C}) - 3\right] \quad \text{with} \quad J_2\left(\dot{\mathbf{C}}\right) = Tr\,\dot{\mathbf{C}}^2 \qquad [13.18]$$

where η is a viscous-like material parameter. This model succeeded in reproducing the strain rate dependency of different ligaments and tendons [PIO 98, PIO 00].

One should also note that some viscoelastic properties of tendons or ligaments may be reproduced by phenomenological models by considering the biphasic nature of such tissues [SWE 14]. This model has been originally developed in order to reproduce experimental observations concerning the very large Poisson's ratios of these tissues. After a numerical model of this couple solid-fluid model, stress relaxation is observed due to the fluid flow within the structure. In this transversely isotropic model, the contribution of ECM is modeled using a Neo-Hookean model and the contribution of fibers follows the model of Holzapfel described above. The volumetric part W_{vol} in equation [13.8] accounting for the compressibility of the media has been formulated in this model :

$$W_{vol} = \frac{\kappa}{2}\left(\ln\left(\frac{I_5 - I_1 I_4 + I_2}{\lambda^{4(m-\nu_0)} e^{-4m(\lambda-1)}}\right)\right)^2 \qquad [13.19]$$

assuming that the Poisson's ratio is a linear function of the stretch λ in the longitudinal direction.

13.4.4. *Microstructural models*

Microstructural models, or micromechanical models, aim to build macroscopic constitutive laws using material parameters directly measurable from experimental protocols (even if the data are not necessarily already available). These models have been the subject of numerous publications, and the present chapter does not pretend to be exhaustive and only presents some of the models considered to be relevant for ligaments and tendons.

A first model [CAC 07] was formulated to explicitly reproduce the progressive recruitment of collagen fibers observed for soft tissues. In this model, a strain-energy function W is once again decomposed into a contribution of the matrix considered as a Neo-Hookean solid, and a sum of different potentials due to each of the fibers. These fibers are progressively recruited following a probability density function $\mathcal{P}\left(\overline{\ell}\right)$ expressed from a statistical description of fibers geometry, where

$\bar{\ell}$ stands for a fibre-specific parameter accounting for fibre crimp. $\bar{\ell}$ is the threshold at which fibres have been considered mechanically active: therefore, the stress in a fibre is considered nil under this stretch (the fibre is not straight) and over an excessive stretch at which the fibre fails. Finally, the strain-energy function is written on a form:

$$W(\mathbf{C}) = \frac{c}{2}(I_1(\mathbf{C}) - 3) + \int_{\bar{\ell}=1}^{\lambda} W_f\left(\lambda, \bar{\ell}\right)\mathcal{P}\left(\bar{\ell}\right)d\bar{\ell} \quad \text{with} \quad W_f = \frac{k}{2}\left(\lambda - \bar{\ell}\right)^2 \qquad [13.20]$$

where k is a parameter corresponding to fibre stiffness, and W_f stands for the contribution of fibers. From this strain-energy function, a anisotropic constitutive law has been formulated. Various stress-strain relationships have been obtained depending on the statistical description of fibers geometry and the values of the strain at failure of collagen fibers.

Other models [GRY 09, REE 10, SHE 15b] explicitly consider the crimp pattern of collagen fibres as a constitutive element of tendons and ligaments, and explicitly represent the trajectories of fibres using geometrical measurable parameters. For instance, fibres may be modelled as helical beams [GRY 09], giving rise to stress-strain relationships for single fibres that can be integrated to obtain the full (anisotropic) contribution of fibres within the strain-energy function. In other cases [REE 10], micromechanical FE models have been combined with a homogenization procedure to numerically evaluate the 3D behaviour of an assembly of fibres (Figure 13.7). While this model permits to reproduce both the non-linear stress-strain response of ligaments and tendons and their large Poisson's ratio, it numerical cost is far much higher than the previous phenomenological models.

Another type of microstructural models considers [MAC 10, MAC 11] the real multiscale nature of tendons and ligaments within different levels of homogenization to go from the nanoscale (molecule) to the mesoscale (crimp fibres) and the macroscale (tissue). Collagen molecules have been modelled within a thermodynamical framework, considering two types of regimes: an entropic regime for small extension values (introducing a strain threshold) is firstly considered, followed by a transition from entropic to energetic elasticity. Fibres are then modelled as beams, whose properties are derived from the nanoscale using periodic homogenization. The tissue is then considered as transversely isotropic, whose properties are issued from matrix properties or fibres properties depending on the straightening of fibres. This complex multiscale model has permitted to satisfyingly reproduce the response of various ligaments and tendons, but it is once again very costly from a numerical point of view and involves numerous material parameters to be identified at different scales.

Consequently, it appears complicated to find a trade-off between the numerical efficiency of phenomenological approaches (requiring material parameters to be identified from experimental data) and the better physical meaning of microstructure-based models that are often restricted by their complexity and the number of parameters involved. However, it appears worthy to emphasize recent work that has been published by Shearer [SHE 15a, SHE 15b] and seems to be both numerically efficient and based on observable physical parameters. Within this model, the tendons or ligaments are described through their fascicles, considered firstly as an arrangement of crimp collagen fibres characterized by a crimp angle, distributed from 0 in the centre of the fascicle to θ_0 at its periphery [SHE 15a]. This model has been then extended in order to superpose this crimp pattern to a helical trajectory (Figure 13.7), similar to the one mentioned previously, and characterized by a helix angle α [GRY 09]. The material parameters needed in this model are restricted to six theoretically observable parameters, issued from dedicated experimental and imaging protocols. These parameters correspond to: the Young's modulus of collagen fibres E, the matrix shear modulus μ, the outer crimp angle θ_0, the fibres helix angle α, the collagen volume fraction φ, and a unit vector $\mathbf{M}$ pointing in fascicle direction in the initial configuration. At the basis of this model, it is considered that fibres contribute to fascicle stiffness only when straightened, introducing a critical fibre strain ε_c, distributed within the fascicle from a probability density function $p(\varepsilon_c)$. Considering the matrix as a Neo-Hookean material, and integrating an analytical formulation of the stress-strain relationship for tendon fascicle, a strain-energy function has been finally obtained and only involves the six observable parameters mentioned above. Interestingly, it has been shown that from fitting only the geometrical parameters α and θ_0, the non-linear mechanical response of various ligaments and tendons can be precisely reproduced. This model suggests that the consideration of the variability of crimped-helical collagen fibres may be the most significant aspect to be considered when modeling various ligaments and tendons.

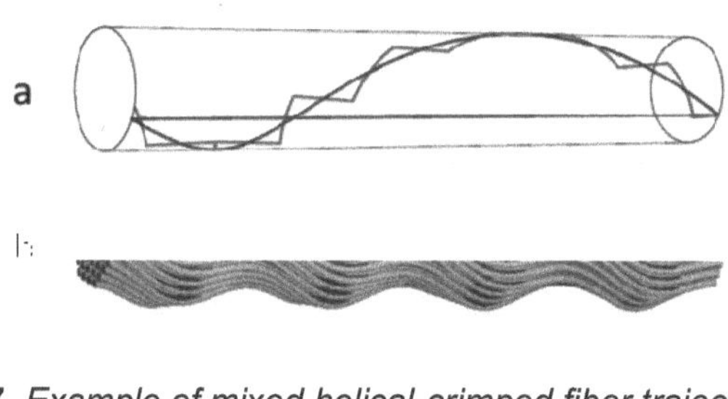

Figure 13.7. *Example of mixed helical-crimped fiber trajectories used in order to formulate a) an equivalent continuum b) or a micro-FE model. Adapted respectively from [SHE 15b] and [REE 10]. For a color version of this figure, see www.iste.co.uk/ganghoffer/biomechanics.zip*

13.5. Focus on ligament and tendon tissue engineering

13.5.1. *Principle and interest*

From the early 1990s, tissue engineering has been the main topic of thousands of publications at the border between cell biology, biomaterials, biomechanics and process engineering. Its principle consists in repairing injured tissues from the activity of cells seeded into a biodegradable temporary scaffold, subject to a suited biochemical and biophysical environment (growth factors, nutrients, physical stimuli, etc.). This principle constitutes a potential alternative to current surgical strategies for tendon and ligament injuries [BUT 08, VUN 04]. However, due to the multidisciplinary nature and the complexity of this scientific challenge, a few applications have actually reached the clinics [HOL 09], and many issues are still to be resolved especially in the case of soft tissues like ligament and tendon. Particularly, the choice of a suited scaffold constitutes a serious challenge in these special cases. Indeed, a suited scaffold should in the first hand replace the native tissue during the post-implantation period, and therefore offer similar mechanical properties. On the other hand, it should equally be biocompatible, biodegradable and promote tissue growth, which involves a suited porous morphology to facilitate the cell seeding and proliferation, as well as the flow of nutrients and degradation products [AHN 10, LEO 08, MEL 10, VUN 04]. Many scaffolds have been proposed in the literature for ligament or tendon, generally based on textile structures due to the high properties in tension that need to be mimicked. The present section restricts to the studies performed in the author's team [LAU 11, LAU 12, LAU 13, LAU 14a, LAU 14b]. However, the interested reader could refer to other studies and reviews in order to draw an overview on the currently proposed solutions [ALT 02, BUT 08, CAL 11, CHE 10, CHE 08b, COO 05, FRE 09, FRE 11, KUO 10, LAD 11, LEO 14, LI 14, LIU 08, LU 05, MAJ 05, NAU 15, NIN 12, NIR 07, PET 06, SAH 07, SAH 10, SHE 07, VIE 09].

13.5.2. *Mechanobiology*

Cells do respond to their mechanical environment, which affects their proliferation, differentiation, orientation, morphology, migration and ECM production. For more details concerning these observations and the associated mechanisms, interested reader should refer to numerous detailed reviews about mechanobiology and mechanotransduction [CHE 08a, HUM 14, ING 06, JAN 09, PED 05, WAN 06]. Numerous models have been proposed in order to understand and predict these cellular responses, among which one of the most famous and commonly used has been developed by Prendergast [LAC 02, PRE 10, PRE 97]. This predictive model focuses on cell differentiation from two types of mechanical stimuli: the fluid flow and the tissue shear strain. In the case of ligament or tendon tissue engineering, cells have to be differentiated into fibroblasts and an abundant ECM should be secreted. This is generally facilitated through the use of dynamic bioreactors, which permits to apply both fluid flow and shear strain to the seeded cells [BUT 09, JEO 08, KAH 08, LAU 14a]. It appears crucial for the design of new scaffolds and bioreactors to quantify the micromechanical environment to which the cells will be exposed. Numerical simulations using the Finite Element (FE) method allow performing such quantifications, on condition that the scaffold geometry is well known. Therefore, new numerical methods are regularly developed and increasingly applied in order to help to design scaffolds and to analyze biological observations. These computer-based approaches have sometimes been gathered on the term "computer-aided tissue engineering" [LAU 13].

13.5.3. *Defining a scaffold*

As detailed above, micromechanical approaches based on a crimped-helical arrangement of fibers allows us to reproduce the non-linear response of ligaments and tendons depending on geometrical parameters [SHE 15b]. Consequently, one can reasonably assume that a scaffold design incorporating this crimped-helical trajectory of fibers is biomimetic and should be suitable for regenerating ligaments and tendons. The author's team has developed a scaffold for ligament (and tendon) repair based on the following key requirements:

– the scaffold should be 3D, porous, with a variable pore size distribution (depending on the cell type);

– the scaffold should exhibit a high stiffness in the longitudinal direction, with adjustable properties close to the native tissue, and should reproduce the initial "toe region" in order to permit a sufficient joint laxity;

– the scaffold material should be biocompatible and slowly biodegradable;

– the scaffold geometry should be compatible with numerical simulations, in order to predict and optimize its morphological and mechanical properties.

Based on these requirements, a multi-layer braided scaffold based on poly(lactide-co- ε-caprolactone) (PLCL) fibers has been developed, as well as the associated custom process [LAU 11]. It has been shown that this scaffold exhibits a suited morphology, adjustable depending on process parameters, and with a pore size gradient from the core to the periphery. It is known that such a gradient facilitates the flow of nutrients within the scaffold [AHN 10].

13.5.4. *Mechanical and biological characterization*

As mentioned in the previous section, the proposed scaffold design exhibits a predictable geometry and is adapted to different type of simulations. Noteworthy, a dedicated finite element analyses based on a method developed for fiber-based materials [DUR 12] has been applied and validated in the case of the braided scaffold [LAU 12, LAU 14b]. It permits to predict and optimize its macroscopic properties and to finely describe the micromechanical environment that will be transferred to the cells during the culture in bioreactor and/or during and intra-articular implantation (Figure 13.8). This approach has enabled to propose a suited structure for the case of anterior cruciate ligament. Due to the versatility of the structure, it is perfectly adapted to different ligamentous or tendinous tissues, by varying the process parameters (fiber dimeter, number of layers, braiding angle, etc.).

Concerning the biological characterization, static cultures have been firstly performed using ovine stem cells during one month. It has permitted to emphasize that the scaffold is compatible with cell adhesion, proliferation, and with ECM creation (Figure 13.8). Secondly, preliminary studies within a dynamic bioreactor have permitted to validate the protocol [LAU 14a], and further dynamic cultures are currently performed. These data will permit to quantify both the decrease in properties due to scaffold degradation, and the increase in properties due to ECM creation and organization. Ideally, these two phenomena should compensate each other, which constitutes one of the principal issues concerning tissue engineering. Mechanical models detailed in the previous sections will permit to finely describe and model the evolution of scaffold response. At last, *in vivo* subcutaneous implantations have been performed on rats (unpublished at the date of redaction of the present chapter) and have concluded to a high suitability of the scaffold concerning colonization and vascularization.

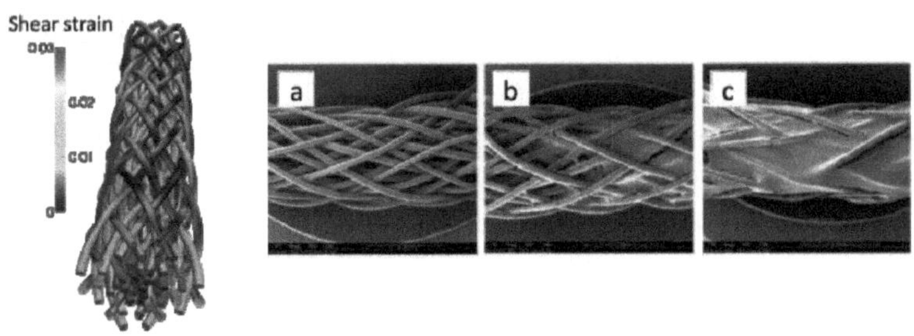

Figure 13.8. *(left) Example of finite element simulation of the braided scaffold for ligament or tendon tissue engineering, giving access to local microenvironment of cells [LAU 14b]. (right) Scanning Electron Microscopy results of ovine stem cells seeded on the braided scaffold after (a) 3 days (b) 14 days (c) 28 days [LAU 14a]. For a color version of this figure, see www.iste.co.uk/ganghoffer/biomechanics.zip*

13.6. Conclusion and perspectives

Tendons and ligaments are based on a highly hierarchical and structured assembly of fibres and fascicles, whose mechanical behaviour has been widely studied at every scale. Among these studies, a first group with a direct clinical application aims to help the detection of pathologies, understanding the different phases of healing, or the effect of drugs or rehabilitation protocols. They require the development of characterization methods, which progressively move from invasive *in vitro* protocols to *in vivo* methods in order to consider the tissue within its physiological environment. Recent advances in medical imaging and image processing allow to progressively overtake the reported limitations of *in vivo* approaches. A second group of studies focuses on the fundamental understanding and modeling of the mechanical response of these tissues. Phenomenological models are getting progressively enriched and numerically efficient, allowing ligaments and tendons to be integrated in large scale numerical models of joints or body. Micromechanical models based on the tissue composition progressively permits to identify the crucial and interdependent role of each scale in the response of ligaments and tendons. Due to the complexity of the issue, these models are obviously far from integrating every aspects of ligament and tendon mechanical behaviour (anisotropy, fluid flow, viscoelasticity, damage, rupture, etc.). Challenges are still to be taken up concerning the development of powerful, simple and complete models that could integrate these aspects. Ideally, micromechanical models tend to involve mechanical or geometrical parameters that could all be measurable using recent advances in imaging or testing techniques, avoiding the use of on-predictive fitting or inverse methods. Noteworthy, in order to include the large intra- and inter-variability inherent to biological tissues, there is still a need to incorporate statistics within these models. Among the potential applications of such models, tissue engineering constitutes a promising perspective. Indeed, the understanding of the role of each scale of the tissue microstructure as well as the effect of geometrical arrangement of fibres could permit in the future to imagine and process new biomimetic scaffolds, that could fulfil the numerous requirements associated with tissue engineering of soft tissues. Once again, numerous challenges are still to be risen up in this field, concerning the mechanisms of tissue growth and the effects of biochemical and biophysical environments. However, recent published studies tend to prove that in the next decades tissue engineering could provide clinicians with a new alternative to current surgical methods in order to treat the different pathologies associated with tendon and ligament.

Bibliography

[ABR 85] ABRAHAM R., MARSDEN J., *Foundations of Mechanics*, 2nd edition, Addison-Wesley, 1985.

[ACA 95] ACAROGLU E.R., IATRIDIS J.C., SETTON L.A. *et al.*, "Degeneration and aging affect the tensile behavior of human Lumbar Anulus Fibrosus", *Spine*, vol. 20, pp. 2690–2701, 1995.

[ADA 98] ADACHI T., TOMITA Y., TANAKA M., "Computational simulation of deformation behavior of 2D-lattice continuum", *International Journal of Mechanical Sciences*, vol. 40, pp. 857–866, 1998.

[ADA 10] ADACHI T., KAMEO Y., HOJO M., "Trabecular bone remodeling simulation considering osteocytic response to fluid-induced shear stress", *Philos. Trans. R. Soc. A Math. Phys. Eng. Sci.*, vol. 368, pp. 2669–2682, 2010.

[AEB 86] AEBI U., COHN J., BUHLE L. *et al.*, "The nuclear lamina is a meshwork of intermediate-type filaments", *Nature*, vol. 323, pp. 560–564, 1986.

[AHN 10] AHN G. *et al.*, "Effect of pore architecture on oxygen diffusion in 3D scaffolds for tissue engineering", *Journal of Biomechanical Engineering*, vol. 132, no. 10, p. 104506, 2010.

[AIM 08] AIMENE Y., HAGEGE B., SIDOROFF F., *et al.*, "Hyperelastic approach for composite reinforcing forming simulations", *International Journal of Material Forming*, vol. 1, pp. 811–814, 2008.

[ALK 08] ALKHADER M., VURAL M., "Mechanical Response of cellular solids: role of cellular topology and microstructural irregularity", *International Journal of Engineering Science*, vol. 46, no. 10, pp. 1035–1051, 2008.

[ALK 09] ALKHADER M., VURAL M., "An energy-based anisotropic yield criterion for cellular solids and validation by biaxial FE simulations", *Journal of the Mechanics and Physics of Solids*, vol. 57, pp. 871–890, 2009.

[ALM 97] ALMEIDA E.S., SPILKER R.L., "Mixed and penalty finite element models for the nonlinear behavior of biphasic soft tissues in finite deformation: Part i–alternate formulations", *Computer Methods in Biomechanics and Biomedical Engineering*, vol. 1, no. 1, pp. 25–46, 1997.

[ALT 02] ALTMAN G.H. *et al.*, "Silk matrix for tissue engineered anterior cruciate ligaments", *Biomaterials*, vol. 23, no. 20, pp. 4131–4141, 2002.

[AMB 02] AMBROSI D., MOLLICA F., "On the mechanics of a growing tumor", *International Journal of Engineering Science*, vol. 40, pp. 1297–1316, 2002.

[AMB 07a] AMBROSI D., GUILLOU A., "Growth and dissipation in biological tissues", *Continuum Mech. Thermodyn.*, vol. 19, pp. 245–251, 2007.

[AMB 07b] AMBROSI D., GUANA F., "Stress modulated growth", *Mathematics and Mechanics of Solids*, vol. 12, no. 3, pp. 319–342, 2007.

[AMB 09] AMBARD D., CHERBLANC F., "Mechanical behavior of annulus fibrosus: a microstructural model of fibers reorientation", *Ann Biomed Eng*, vol. 37, no. 11, pp. 2256–2265, 2009.

[ARG 96] ARGOUBI M., SHIRAZI-ADL A., "Poroelastic creep response analysis of a lumbar motion segment in compression", *Journal of Biomechanics*, vol. 29, no. 10, pp. 1331–1339, 1996.

[ASH 70] ASHKIN A., "Acceleration and trapping of particles by radiation pressure", *Physical Review Letters*, vol. 24, no. 4, pp. 156–159, 1970.

[ASH 86] ASHMAN R.B., TURNER C.H., COWIN S.C., "Ultrasonic technique for the measurement of the structural elastic modulus of cancellous bone", *Transactions of the Orthopedic Research Society*, vol. 43, 1986.

[ASH 89] ASHMAN R.B., RHO J.Y., TURNER C.H., "Anatomical variation of orthotropic elastic moduli of the proximal tibia", *Journal of Biomechanics*, vol. 22, pp. 895–900, 1989.

[ASS 11a] ASSIDI M., BEN BOUBAKER B., GANGHOFFER J.F., "Equivalent properties of monolayer fabric from mesoscopic modelling strategies", *International Journal of Solids and Structures*, vol. 48, pp. 2920–2930, 2011.

[ASS 11b] ASSIDI M., DOS REIS F., GANGHOFFER J.F., "Equivalent mechanical properties of biological membranes from lattice homogenization", *Journal of the Mechanical Behavior of Biomedical Materials*, vol. 4, no. 8, pp. 1833–1845, 2011.

[ATE 07] ATESHIAN G.A., "On the theory of reactive mixtures for modeling biological growth", *Biomechan. Model. Mechanobiol.*, vol. 6, pp. 423–445, 2007.

[AZN 05] AZNAR J.M., RUEBERG T., DOBLARE M., "A bone remodeling model coupling microdamage growth and repair by 3DBMU-activity", *Biomech. Model Mechanobiol.*, vol. 4, pp. 147–167, 2005.

[BAC 98] BACHRACH N.M., MOW V.C., GUILAK F., "Incompressibility of the solid matrix of articular cartilage under high hydrostatic pressures", *Journal of Biomechanics*, vol. 31, pp. 445–451, 1998.

[BAD 00] BADICHE X., FOREST S., GUIBERT T. *et al.*, "Mechanical properties and non-homogeneous deformation of open-cell nickel foams: application of the mechanics of cellular solids and of porous materials", *Materials Science and Engineering A*, vol. 289, pp. 276–288, 2000.

[BAG 00] BAGGE M., "A model of bone adaptation as an optimization process", *Journal of Biomechanics,* vol. 33, pp. 1349–1357, 2000.

[BAK 89] BAKHVALOV N.S., PANASENKO G.P., *Homogenization Averaging Processes in Periodic Media,* Kluwer Academic Publishing, Dordrecht, 1989.

[BAK 98] BAKER W.E., TOGAMI T.C., WEYDERT J.C., "Static and dynamic properties of high density metal honeycombs", *International Journal of Impact Engineering*, vol. 21, pp. 149–163, 1998.

[BAL 13] BALDIT A., Analysis of hydro-chemo-mechanical interactions in biological tissues: application to the cell nutrition of intervertebral disc, Doctoral thesis, University of Montpellier, Montpellier, 2013.

[BAL 14a] BALDIT A., AMBARD D., CHERBLANC F. *et al.*, "Experimental analysis of the transverse behaviour of annulus fibrosus tissue mechanical", *Biomechanics and Modeling in Mechanobiology,* vol. 13, no. 3, pp. 643-652, 2014.

[BAL 14b] BALDIT A., AMBARD D., CHERBLANC F. *et al.*, "Annulus Fibrosus microstructure: an explanation to local heterogeneities", *Proceedings of SB2014*, Valenciennes, France, 2014.

[BAL 14c] BALDIT A., CAMPOS A., BRUNELLI M. *et al.*, "Multi-scale modeling in tissue engineering: a Virtual Physiological approach", *Proceedings of VPH14,* Trondheim, Norway, 2014.

[BAN 02] BANSE X., SIMS T.J., BAILEY A.J., "Mechanical properties of adult vertebral cancellous bone: correlation with collagen intermolecular cross-links", *Journal of Bone and Mineral Research*, vol. 17, no. 9, pp. 1621–1628, 2002.

[BAN 05] BANERJEE S., BHASKAR A., "Free vibration of cellular structures using continuum modes", *Journal of Sound and Vibration*, vol. 287, nos 1–2, pp. 77–100, 2005.

[BAN 09] BANERJEE S., BHASKAR A., "The applicability of the effective medium theory to the dynamics of cellular beams", *International Journal of Mechanical Sciences*, vol. 51, pp. 598–608, 2009.

[BAR 11] BARING T.K.A. *et al.*, "Rotator cuff repair failure in vivo: a radiostereometric measurement study", *Journal of Shoulder and Elbow Surgery*, vol. 20, no. 8, pp. 1194–1199, 2011.

[BAS 00] BASAR Y, WEICHERT D., *Nonlinear Continuum Mechanics of Solids. Fundamental Mathematical and Physical Concepts*, Springer, 2000.

[BAS 03] BASEAU E., "Finite deformation of elastic–plastic filamentary networks", *International Journal of Nonlinear Mechanics*, vol. 38, pp. 1473–1479, 2003.

[BAS 04] BASS E.C., ASHFORD F.A., SEGAL M.R. *et al.*, "Biaxial testing of human annulus fibrosus and its implications for a constitutive formulation", *Ann Biomed Eng*, vol. 32, no. 9, pp. 1231–1242, 2004.

[BAY 04a] BAYRAKTAR H.H., GUPTA A., KWON R.Y. *et al.*, "The modified super-ellipsoid yield criterion for human trabecular bone", *Journal of Biomechanical Engineering*, vol. 126, no. 6, pp. 677–684, 2004.

[BAY 04b] BAYRAKTAR H.H., KEAVENY T.M., "A computational investigation of the nonlinear behavior of human trabecular bone", *Transactions of the 12th annual Pre-ORS Symposium on Computational Methods in Orthopaedic Biomechanics*, p. 2, 2004.

[BEA 90] BEAUPRE G.S., ORR T.E., CARTER D.R., "An approach for time-dependent bone modeling and remodeling-application: a preliminary remodeling simulation", *Journal of Orthopaedic Research*, vol. 8, no. 5, pp. 662–670, 1990.

[BEN 78] BENSOUSSAN A., LIONS J.L., PAPANICOLAOU G., *Asymptotic Analysis for Periodic Structures*, North Holland, Amsterdam, 1978.

[BEN 97] BENJAMIN M., RALPHS J.R., "Tendons and ligaments – an overview", *Histol Hostopathol*, vol. 12, pp. 1135–1144, 1997.

[BEN 02] BENJAMIN M. *et al.*, "The skeletal attachment of tendons—tendon 'entheses'", *Comparative Biochemistry and Physiology Part A: Molecular & Integrative Physiology*, vol. 133, no. 4, pp. 931–945, 2002.

[BEN 06] BENJAMIN M. *et al.*, Where tendons and ligaments meet bone: attachment sites ('entheses') in relation to exercise and/or mechanical load", *Journal of Anatomy*, vol. 208, no. 4, pp. 471–490, 2006.

[BEN 07a] BEN BOUBAKER B., HAUSSY B., GANGHOFFER J.F., "Consideration of the yarn-yarn interactions in meso/macro discrete model of the fabric. Part I: Single yarn behavior", *Mechanics Research Communications*, vol. 34, no. 4, pp. 359–370, 2007.

[BEN 07b] BEN BOUBAKER B., HAUSSY B., GANGHOFFER J.F., "Consideration of the yarn-yarn interactions in meso/macro discrete model of the fabric, Part II: Woven fabric under uniaxial and biaxial extension", *Mechanics Research Communications*, vol. 34, no. 4, pp. 371–378, 2007.

[BEN 07c] BEN BOUBAKER B., HAUSSY B., GANGHOFFER J.F., "Discrete woven structure model: yarn-on-yarn friction", *Comptes Rendus Mécanique*, vol. 335, pp. 150–158, 2007.

[BEN 07d] BEN BOUBAKER B., HAUSSY B., GANGHOFFER J.F., "Discrete models of woven structures. Macroscopic approach", *Composites B*, vol. 38, no. 4, 498–505, 2007.

[BEN 08] BENJAMIN M., KAISER E., MILZ S., "Structure-function relationships in tendons: a review", *Journal of Anatomy*, vol. 212, no. 3, pp. 211–228, 2008.

[BES 94] BEST B.A., GUILAK F., SETTON L.A. *et al.*, "Compressive mechanical properties of the human anulus fibrosus and their relationship to biochemical composition", *Spine (Phila Pa 1976)*, vol. 19, no. 2, pp. 212–221, 1994.

[BEY 98] BEYNNON B.D., FLEMING B.C., "Anterior cruciate ligament strain in-vivo: A review of previous work", *Journal of Biomechanics*, vol. 31, no. 6, pp. 519–525, 1998.

[BEY 12] BEY M.J., DERWIN K.A. "Measurement of in vivo tendon function", *Journal of Shoulder and Elbow Surgery*, vol. 21, no. 2, pp. 149–157, 2012.

[BIB 01] BIBBY S.R.S., JONES D.A., LEE R.B. *et al.*, "Biochimie, biologie et physiologie du disque intervertebral", *Rev Rhum*, vol. 68, pp. 903–907, 2001.

[BIC 10] BICER E.K. *et al.*, "Current knowledge in the anatomy of the human anterior cruciate ligament", *Knee Surgery, Sports Traumatology, Arthroscopy*, vol. 18, no. 8, p. 1075–1084, 2010.

[BIG 12] BIGONI D., *Nonlinear Solid Mechanics, Bifurcation Theory and Material Instability*, Cambridge University Press, 2012.

[BIR 13] BIRMINGHAM E., GROGAN J.A., NIEBUR G.L. *et al.*, "Computational modelling of the mechanics of trabecular bone and marrow using fluid structure interaction techniques", *Annals of Biomedical Engineering*, vol. 41, no. 4, pp. 814–826, 2013.

[BOA 93] BOAL D.H., "Negative poisson ratio for two-dimensional networks under tension", *Phys. Rev. E*, vol. 48, no. 6, pp. 4274–4283, 1993.

[BOA 94] BOAL D.H., "Computer simulation of a model network for the erythrocyte cytoskeleton", *Biophysical Journal,* vol. 67, no. 2, pp. 521–529, 1994.

[BOA 03] BOAL D.H., *Mechanics of the Cell*, Cambridge University Press, 2003.

[BOA 12] BOAL D.H., *Mechanics of the Cell*, 2nd edition, Cambridge University Press, 2012.

[BOE 98] BOEY S.K., BOAL D.H., DISCHER D.E., "Simulations of the erythrocyte cytoskeleton at large deformation", *Biophysical Journal*, vol. 75, pp. 1573–1583, 1998.

[BOI 97] BOISSE P., BORR M., BUET K. *et al.*, "Finite element simulations of textile composite forming including the biaxial fabric behavior", *Composites B*, vol. 28, no. 4, pp. 453–464, 1997.

[BOI 01] BOISSE P., GASSER A., HIVET G., "Analyses of fabric tensile behaviour: determination of the biaxial tension-strain surfaces and their use in forming simulations", *Composites A*, vol. 32, no. 10, pp. 1395–1414, 2001.

[BOI 05] BOISSE P., ZOUARI B., GASSER A., "A mesoscopic approach for the simulation of woven fibre composite forming", *Composites Science and Technology*, vol. 65, nos 3–4, pp. 429–436, 2005.

[BOI 06] BOISSE P., ZOUARI B., DANIEL J., "Importance of in-plane shear rigidity in finite element analyses of woven fabric composite performing", *Composites A*, vol. 37, no. 12, pp. 2201–2212, 2006.

[BON 11] BONFOH N., NOVINYO E., LIPINSKI P., "Modeling of bone adaptive behavior based on cells activities", *Biomechanics and Modeling in Microbiology*, vol. 5, pp. 789–798, 2011.

[BOU 00] BOUYGE F., A multiscale model of cancellous bone, Master's thesis, *Georgia Institute Of Technology*, 2000.

[BOU 01] BOUYGE F., JASIUK I., OSTOJA-STARZEWSKI M., "A micromechanically based couple-stress model of an elastic two-phase composite", *International Journal of Solids Structures*, vol. 38, pp. 1721–1735, 2001.

[BOU 02] BOUYGE F., JASIUK I., BOCCAR S. *et al.*, "A micromechanically based couple-stress model of an elastic orthotropic two-phase composite", *European Journal of Mechanics – A/Solids*, vol. 21, pp. 465–481, 2002.

[BOW 98] BOWMAN S.M. *et al.*, "Creep contributes to the fatigue behavior of bovine trabecular bone", *Journal of Biomechanical* Engineering, vol. 120, no. 5, pp. 647–654, 1998.

[BOY 93] BOYDE A., ELLIOT J.C., JONES S.J., "Stereology and histogram analysis of backscattered electron images: age changes in bone", *Bone*, vol. 14, no. 3, pp. 205–210, 1993.

[BRE 06] BREIVIK H., COLLETT B., VENTAFRIDDA V. *et al.*, "Survey of chronic pain in Europe: Prevalence, impact on daily life, and treatment", *European Journal of Pain*, vol. 10, no. 4, pp. 287–333, 2006.

[BRO 74] BROEK D., *Elementary Fracture Mechanics*, Noordhoff, 1974.

[BRU 04] BRUEHLMANN S.B., HULME P.A., DUNCAN N.A., "In situ intercellular mechanics of the bovine outer annulus fibrosus subjected to biaxial strains", *Journal of Biomechanics*, vol. 37, no. 2, pp. 223–231, 2004.

[BUE 13] BUERZLE W., MAZZA E., "On the deformation behavior of human amnion", *Journal of Biomechanics*, vol. 46, pp. 1777–1783, 2013.

[BUR 75] BURSTEIN A.H., ZIKA J.M., HEIPLE K.G. *et al.*, "Contribution of collagen and mineral to the elastic–plastic properties of bone", *Journal of Bone and Joint Surgery*, vol. 57-A, no. 7, pp. 956–961, 1975.

[BUT 08] BUTLER D.L. *et al.*, "Functional tissue engineering for tendon repair: A multidisciplinary strategy using mesenchymal stem cells, bioscaffolds, and mechanical stimulation", *Journal of Orthopaedic Research*, vol. 26, no. 1, pp. 1–9, 2008.

[BUT 09] BUTLER D.L. *et al.*, "Using functional tissue engineering and bioreactors to mechanically stimulate tissue-engineered constructs", *Tissue Engineering. Part A*, vol. 15, no. 4, pp. 741–749, 2009.

[CAC 07] CACHO F. *et al.*, "A constitutive model for fibrous tissues considering collagen fiber crimp", *International Journal of Non-Linear Mechanics*, Special Issue in Honour of Dr Ronald S. Rivlin, vol. 42, no. 2, pp. 391–402, 2007.

[CAI 06] CAILLERIE D., MOURAD A., RAOULT A., "Discrete homogenization in graphene sheet modeling", *Journal of Elasticity*, vol. 84, pp. 33–68, 2006.

[CAL 85] CALLEN H.B., *Thermodynamics and an Introduction to Thermostatics*, John Wiley & Sons, New York, 1985.

[CAL 11] CALIARI S.R., RAMIREZ M.A., HARLEY B.A.C., "The development of collagen-GAG scaffold-membrane composites for tendon tissue engineering", *Biomaterials*, vol. 32, no. 34, pp. 8990–8998, 2011.

[CAM 07] CAMPANA S., DE GUISE J.A., RILLARDON L. *et al.*, "Lumbar intervertebral disc mobility: effect of disc degeneration and of geometry", *Eur J Orthop Surg Traumatol*, vol. 17, pp. 533–541, 2007.

[CAN 95] VAN RIETBERGEN B., WEINANS H., HUISKES R. *et al.*, "A new method to determine trabecular bone elastic properties and loading using micromechanical finite-element models", *Journal of Biomechanics*, vol. 28, pp. 69–81, 1995.

[CAR 76] CARTER D.R., HAYES W.C., "Bone compressive strength: the influence of density and strain rate", *Science*, vol. 194, pp. 1174–1 176, 1976.

[CAR 87] CARTER D.R., FYHRIE D.P., WHALEN R.T., "Trabecular bone density and loading history: regulation of connective tissue biology by mechanical energy", *Journal of Biomechanics*, vol. 20, no. 8, pp. 785–794, 1987.

[CAR 96] CARTER D.R., VAN DER MEULEN M.C.H., BEAUPRÉ G.S., "Mechanical factors in bone growth and development", *Bone*, vol. 18, pp. S5–S10, 1996.

[CAR 01] CARTER D.R., BEAUPRE G.S., *Skeleton Function and Form (Mechanobiology of Skeleton Development, Aging, and Regeneration*, Cambridge University Press, 2001.

[CEN 99] CENDRE E., MITTON D., ROUX J.P. *et al.*, "High-Resolution Computed Tomography for Architectural Characterization of Human Lumbar Cancellous Bone: Relationships with Histomorphometry and Biomechanics", *Osteoporosis International*, vol. 10, no. 5, pp. 353–360, 1999.

[CHA 10] CHARALAMBAKIS N., "Homogenization techniques and micromechanics. A survey and perspectives", *Appl. Mech. Rev*, vol. 63, no. 3, pp. 1–10, 2010.

[CHA 11] CHARMETTANT A. *et al.*, "Hyperelastic modelling for mesoscopic analyses of composite reinforcements", *Composite Science and Technology*, vol. 71, pp. 1623–1631, 2011.

[CHE 99] CHEN T.J., LU C., FLECK N.A., "Effect of imperfections of the yielding of two dimensional foams", *Journal Of The Mechanics and Physics of Solids*, vol. 47, no.11, pp. 2235–2272, 1999.

[CHE 02] CHEN C., FLECK N.A., "Size effects in the constrained deformation of metallic foam", *Journal of The Mechanics and Physics of Solids*, vol. 50, no. 5, pp. 955–977, 2002.

[CHE 07a] CHEN G., PETTET G., PEARCY M. *et al.*, "Comparison of two numerical approaches for bone remodeling", *Medical Engineering Physics*, vol. 29, pp. 134–139, 2007.

[CHE 07b] CHENG V.W.T., SCREEN H.R.C., "The micro-structural strain response of tendon", *Journal of Materials Science*, vol. 42, no. 21, pp. 8957–8965, 2007.

[CHE 08a] CHEN C.S., "Mechanotransduction – a field pulling together?", *Journal of Cell Science*, vol. 121, no. 20, pp. 3285–3292, 2008.

[CHE 08b] CHEN X. *et al.*, "Ligament regeneration using a knitted silk scaffold combined with collagen matrix", *Biomaterials*, vol. 29, no. 27, pp. 3683–3692, 2008.

[CHE 10] CHEN J.L. *et al.*, "Efficacy of hESC-MSCs in knitted silk-collagen scaffold for tendon tissue engineering and their roles", *Biomaterials*, vol. 31, no. 36, pp. 9438–9451, 2010.

[CHE 12a] CHEN X.-M. *et al.*, "Shear wave elastographic characterization of normal and torn achilles tendons: A pilot study", *Journal of Ultrasound in Medicine*, vol. 32, no. 3, pp. 449–455, 2013.

[CHE 12b] CHERNAK L.A., THELEN D.G., "Tendon motion and strain patterns evaluated with two-dimensional ultrasound elastography", *Journal of Biomechanics*, vol. 45, no. 15, pp. 2618–2623, 2012.

[CHE 14] CHERNAK SLANE L., THELEN D.G., "The use of 2D ultrasound elastography for measuring tendon motion and strain", *Journal of Biomechanics*, vol. 47, no. 3, pp. 750–754, 2014.

[CHO 92] CHOI K., GOLDSTEIN S.A., "A comparison of the fatigue behavior of human trabecular and cortical bone tissue", *Journal of Biomechanics*, vol. 25, pp. 1371–1381, 1992.

[CHR 00] CHRISTENSEN R., "Mechanics of cellular and other low-density materials", *International Journal of Solids and Structures*, vol. 37, pp. 93–104, 2000.

[CHR 01] CHRISTIANSEN P., "The skeleton in primary hyperparathyroidism: a review focusing on bone remodeling, structure, mass, and fracture", *APMIS Suppl*, vol. 102, pp. 1–52, 2001.

[CHR 06] CHRISTIAN G.T., OGDEN R.W., HOLZAPFEL G.A., "Hyperelastic modelling of arterial layers with distributed collagen fibre orientations", *Journal of the Royal Society Interface*, vol. 3, pp. 15–35, February 2006.

[COO 05] COOPER J.A. *et al.*, "Fiber-based tissue-engineered scaffold for ligament replacement: design considerations and in vitro evaluation", *Biomaterials*, vol. 26, no. 13, pp. 1523–1532, 2005.

[COR 11] CORTES D., ELLIOTT D., "Extra-fibrillar matrix mechanics of annulus fibrosus in tension and compression", *Biomechanics and Modeling in Mechanobiology*, vol. 11, pp. 781–790, 2011.

[COS 07] COSTI J.J., STOKES I.A., GARDNER-MORSE M. *et al.*, "Direct measurement of intervertebral disc maximum shear strain in six degrees of freedom: motions that place disc tissue at risk of injury", *Journal of Biomechanics*, vol. 40, no. 11, pp. 2457–2466, 2007.

[COS 09] COSSERAT E., COSSERAT F., *Théorie des Corps Déformables*, A. Hermann et Fils, Paris, 1909.

[COU 80] COUCHMAN P.R., LINFORD R.G., "Aspects of solid surface thermodynamics: relationships of the Shuttleworth type", *Journal of Electroanalytical Chemistry*, vol. 115, pp. 143–148, 1980.

[COU 04] COUSSY O., *Poromechanics*, John Wiley & Sons, 2004.

[COW 70] COWIN S.C., "An incorrect inequality in micropolar elasticity theory", *Zeitschrift für angewandte Mathematik und Physik ZAMP*, vol. 21, no. 3, pp. 494–497, 1970.

[COW 76] COWIN S.C., HEGEDUS D.H., "Bone remodeling I: theory of adaptive elasticity", *Journal of Elasticity*, vol. 6, no. 3, pp. 313–325, 1976.

[COW 79] COWIN S.C., VAN BUSKIRK W.C., "Surface bone remodeling induced by a medullary pin", *Journal of Biomechanics*, vol. 12, no. 4, pp. 269–276, 1979.

[COW 81] COWIN S.C., FIROOZBAKHSH K., "Bone remodeling of diaphysial surfaces under constant load: theoretical predictions", *Journal of Biomechanics*, vol. 14, no. 7, pp. 471–484, 1981.

[COW 00] COWIN S.C., HUMPHREY J.D., Cardiovascular Soft Tissue Mechanics, Springer, 2000.

[COW 07] COWIN S.C., DOTY S.B., *Tissue Mechanics*, Springer, New York, 2007.

[CRE 09] CREVIER-DENOIX N. *et al.*, "Reproducibility of a non-invasive ultrasonic technique of tendon force measurement, determined in vitro in equine superficial digital flexor tendons", *Journal of Biomechanics*, vol. 42, no. 13, pp. 2210–2213, 2009.

[CRO 05] CROOKSTON J.J., LONG A.C., JONES I.A., "A summary review of mechanical properties prediction methods for textile reinforced polymer composites", *Proc. IMechE, Part L: J. Materials: Design and applications*, vol. 219, pp. 91–109, 2005.

[DAO 06] DAO M., LI J., SURESH S., "Molecularly based analysis of deformation of spectrin network and human erythrocyte", *Material Sciences and Engineering C*, vol. 26, pp. 1232–1244, 2006.

[DEF 04] DEFRATE L.E. *et al.*, "The effect of length on the structural properties of an Achilles tendon graft as used in posterior cruciate ligament reconstruction", *The American Journal of Sports Medicine*, vol. 32, no. 4, pp. 993–997, 2004.

[DEF 06] DEFRATE L.E. *et al.*, "The measurement of the variation in the surface strains of Achilles tendon grafts using imaging techniques", *Journal of Biomechanics*, vol. 39, no. 3, pp. 399–405, 2006.

[DEM 07] DEMIRAY S., BECKER W., HOHE J., "Numerical determination of initial and subsequent yield surfaces of open-celled model foams", *International Journal of Solids and Structures*, vol. 44, pp. 2093–2108, 2007.

[DER 97] DER BURG M.W., SHULMEISTER V., GEISSEN E. *et al.*, "On the linear elastic properties of regular and random open-cell foam models", *Journal of Cellular Plastics*, vol. 33, pp. 31–54, 1997.

[DES 01a] DESHPANDE V.S., ASHBY M.F., FLECK N.A., "Foam topology bending versus stretching dominated architectures, *Acta materialia*, vol. 49, pp. 1035–1040, 2001.

[DES 01b] DESHPANDE V., FLECK N., ASHBY M., "Effective properties of the octet-truss lattice material", *Journal of the Mechanics and Physics of Solids*, vol. 49, pp. 1747–1769, 2001.

[DEV 89] DEVRIES F., DUMONTET H., DUVAUT G. *et al.*, "Homogenization and damage for composite structures", *International Journal for Numerical Methods in Engineering*, vol. 27, pp. 285–298, 1989.

[DIA 07] DIAMANT I., SHAHAR R., MASHARAWI Y. *et al.*, "A method for patient-specific evaluation of vertebral cancellous bone strength in vitro validation", *Clin. Biomech*, vol. 22, pp. 282–291, 2007.

[DIC 05] DI CARLO A. "Surface and bulk growth unified", in STEINMANN P., MAUGIN G.A. (eds), *Mechanics of Material Forces, Advances in Mechanics and Mathematics*, Springer, vol. 11, 2005.

[DIE 03] DIEBELS S., STEEB H., "Stress and couple stress in foams", *Computational Materials Science*, vol. 28, no. 3–4, pp. 714–722, 2003.

[DIS 98] DISCHER D.E., BOAL D.H., BOEY S.K., "Simulations of the erythrocyte cytoskeleton at large deformation", *II. Biophysical Journal*, vol. 75, pp. 1584–1597, 1998.

[DOB 02] DOBLARÉ M., GARCIA J.M., "Anisotropic bone remodelling model based on a continuum damage-repair theory", *Journal of Biomechanics*, vol. 35, no. 1, pp. 1–17, 2002.

[DON 11] DONNAN F.G., HARRIS A.B., "The osmotic pressure and conductivity of aqueous solutions of congo-red, and reversible membrane equilibria", *J. Chem. Soc., Trans.*, vol. 99, pp. 1554–1577, 1911.

[DOS 10a] DOS REIS F., GANGHOFFER J.F., "Discrete homogenization of architectured materials: implementation of the method in a simulation tool for the systematic prediction of their effective elastic properties", *Technische Mechanik*, vol. 30, pp. 1–3, pp. 85–109, 2010.

[DOS 10b] DOS REIS F., Homogéniésation automatique de milieux discrets périodiques, Applications aux mousses polymères et aux milieux auxétique, PhD thesis, National Polytechnic Institute of Lorraine, 2010.

[DOS 12a] DOS REIS F., GANGHOFFER J.F., "Construction of micropolar continua from the asymptotic homogenization of beam lattices", *Computers and Structures*, vol. 112–113, pp. 354–363, 2012.

[DOS 12b] DOS REIS F., GANGHOFFER J.F., "Equivalent mechanical properties of auxetic lattices from discrete homogenization", *Computational Materials Science*, vol. 51, no. 1, pp. 314–321, 2012.

[DOY 06] DOYOYO M., HU J.W., "Plastic failure analysis of an auxetic foam or inverted strut lattice under longitudinal and shear loads", *Journal of the Mechanics and Physics of Solids*, vol. 54, pp. 1479–1492, 2006.

[DRO 95] DROST M.R., WILLEMS P., SNIJDERS H. *et al.*, "Confined compression of canine annulus fibrosus under chemical and mechanical loading", *Journal of Biomechanical Engineering*, 1995.

[DUA 06b] DUAN Y., KEEFE M., BOGETTI T.A. *et al.*, "Finite element modeling of transverse impact on a ballistic fabric", *International Journal of Mechanical Sciences*, vol. 48, no. 1, pp. 33–43, 2006.

[DUR 02] DURVILLE D., "Modélisation par éléments finis des propriétés mécaniques de structures textiles de la fibre au tissue", *European Journal of Computational Mechanics*, vol. 11, pp. 463–477, 2002.

[DUR 12] DURVILLE D., "Contact-friction modeling within elastic beam assemblies: an application to knot tightening", *Computational Mechanics*, vol. 49, no. 6, pp. 687–707, 2012.

[DUT 05] DUTHON V.B. *et al.*, "Anatomy of the anterior cruciate ligament", *Knee Surgery, Sports Traumatology, Arthroscopy*, vol. 14, no. 3, pp. 204–213, 2005.

[ELN 15] EL NADY K., GANGHOFFER J.F., "Computation of the effective mechanical response of biological networks accounting for large configuration changes", *Journal of the Mechanical Behavior of Biomedical Materials*, vol. 58, pp. 28–44, 2015.

[ELN 16] EL NADY K., GODA I., GANGHOFFER J.F., Computation of the effective nonlinear mechanical response of lattice materials considering geometrical nonlinearities, *Comp. Mech.*, vol. 58, pp. 1–23, 2016.

[ELN 17] EL NADY K., DOS REIS F., GANGHOFFER J.F., "Computation of the homogenized nonlinear elastic response of 2D and 3D auxetic structures based on micropolar continuum models", *Composite Structures*, 2017.

[EPS 00] EPSTEIN M., MAUGIN G.A., "Thermomechanics of volumetric growth in uniform bodies", *International Journal of Plasticity*, vol. 16, pp. 951–978, 2000.

[EPS 10] EPSTEIN M., "Kinetics of boundary growth", *Mechanics Research Communications*, vol. 37, no. 5, pp. 453–457, 2010.

[ERI 66] ERINGEN A.C., "Linear theory of micropolar elasticity", *Journal of Mathematics and Mechanics*, vol. 15, pp. 909–923, 1966.

[ERI 68] ERINGEN A.C., "Theory of Micropolar Elasticity", in LIEBOWITZ H. (ed.), *Fracture*, vol. 2, Academic Press, New York, 1968.

[ERI 76] ERINGEN A.C., *Continuum Physics – Non-local Field Theories*, Academic Press, New York, 1976.

[ERI 99] ERINGEN A.C., *Microcontinuum Field Theories: I Foundations and Solids*, Springer Verlag, New York, 1999.

[ERI 01] ERINGEN A.C., *Microcontinuum Field Theories II*, Springer, New York, 2001.

[ERI 06] ESWARAN S.K., GUPTA A., ADAMS M.F. *et al.*, "Cortical and trabecular load sharing in the human vertebral body", *Journal of Bone and Mineral Research*, vol. 21, pp. 307–314, 2006.

[ESH 51] ESHELBY J.D., "The force on an elastic singularity", *Philosophical Transactions of the Royal Society A*, vol. 244, no. 877, pp. 87–112, 1951.

[ESH 70] ESHELBY J.D., "Energy relations and the energy–momentum tensor in continuum mechanics", in KANNINEN M.F., ADLER W.F., ROSENFIELD A.R. *et al.* (eds), *Inelastic Behaviour of Solids*, McGraw-Hill, New York, 1970.

[EVA 75] EVANS E.A., LA CELLE P.L., "Intrinsic material properties of the erythrocyte membrane indicated by mechanical analysis of deformation", *Blood*, vol. 45, no. 1, pp. 29–43, 1975.

[EVA 83] EVANS E., "Bending elastic modulus of red blood cell membrane derived from buckling instability in micropipet aspiration tests", *Biophysical Journal*, vol. 43, no. 1, pp. 27–30, 1983.

[EVA 84] EVANS E., MOHANDAS N., "Static and dynamic rigidities of normal and sickle erythrocytes, Major influence of cell hemoglobin concentration", *Journal of Clinical Investigation*, vol. 73, no. 2, p. 477, 1984.

[EVA 97] EVANS E., WAUGH R., "Osmotic correction to elastic area compressibility measurements on red cell membrane", *Biophysical Journal*, vol. 20, no. 3, pp. 307–313, 1977.

[FAN 05] FANG Z., Image-guided modeling, fabrication and micromechanical analysis of bone and heterogeneous structure, PhD thesis, Drexel University, 2005.

[FAN 08] FAN H., JIN F., FANG D., "Nonlinear mechanical properties of lattice truss materials", *Materials and Design*, vol. 30, no. 3, pp. 511–517, 2008.

[FAT 02] FATEMI J., VAN KEULEN F., ONCK P.R., "Generalized Continuum Theories, Application to Stress Analysis in Bone", *Meccanica*, vol. 37, pp. 385–396, 2002.

[FAT 03] FATEMI J., ONCK P.R., POORT G. *et al.*, "Cosserat moduli of anisotropic cancellous bone a micromechanical analysis", *Journal de Physique IV France*, vol. 105, pp. 273–280, 2003.

[FER 99] FERNANDES P., RODRIGUES H., JACOBS C., "A model of bone adaptation using a global optimisation criterion based on the trajectorial theory of wolff", *Computer Methods in Biomechanics and Biomedical Engineering*, vol. 2, no. 2, pp. 125–138, 1999.

[FER 02] FERNANDES P.R., FOLGADOA J., JACOBS C. *et al.*, "A contact model with ingrowth control for bone remodelling around cementless stems", *Journal of Biomechanics*, vol. 35, pp. 167–176, 2002.

[FER 10] FERNANDEZ J., GARCIA-AZNAR J., MARTINEZ R., VIANO J., "Numerical analysis of a strain-adaptive bone remodelling problem", *Computer Methods in Applied Mechanics and Engineering*, vol. 199, nos. 23–24, pp. 1549–1557, 2010.

[FEY 00] FEYEL F., CHABOCHE J.-L., "FE2 multiscale approach for modelling the elastoviscoplastic behaviour of long fibre SiC/Ti composite materials", *Computer Methods in Applied Mechanics and Engineering*, vol. 183, no. 3, pp. 309–330, 2000.

[FIE 12] FIELD C., LI Q., LI W. *et al.*, "A comparative mechanical and bone remodelling study of all-ceramic posterior inlay and onlay fixed partial dentures", *Journal of Dentistry*, vol. 40, no. 1, pp. 48–56, 2012.

[FIS 99] FISH J., YU Q., SHEK K., "Computational damage mechanics for composite materials based on mathematical homogenization", *International Journal for Numerical Methods in Engineering*, vol. 45, no. 11, pp. 1657–1679, 1999.

[FLE 04] FLEMING B.C., BEYNNON B.D., "In vivo measurement of ligament/tendon strains and forces: A review", *Annals of Biomedical Engineering*, vol. 32, no. 3, pp. 318–328, 2004.

[FLO 06] FLORENCE C., SAB K., "A rigorous homogenization method for the determination of the overall ultimate strength of periodic discrete media and an application to general hexagonal lattices of beams", *European Journal of Mechanics A/Solids*, vol. 25, pp. 72–97, 2006.

[FOR 96a] FORD C.M., KEAVENY T.M., "The dependence of shear failure properties of trabecular bone on apparent density and trabecular orientation", *Journal of Biomechanics*, vol. 29, no. 10, pp. 1309–1317, 1996.

[FOR 96b] FORD C.M., KEAVENY T.M., HAYES W.C., "The effect of impact direction on the structural capacity of the proximal femur during falls", *Journal of Bone and Mineral Research*, vol. 11, no. 3, pp. 377–383, 1996.

[FOR 98a] FORD C.M., GIBSON L.J., "Uniaxial strength asymmetry in cellular materials: an analytical model", *International Journal of Mechanical Sciences*, vol. 40, no. 6, pp. 521–531, 1998.

[FOR 98b] FOREST S., SAB K., "Cosserat overall modeling of heterogeneous materials", *Mechanics Research Communications*, vol. 25, no. 4, pp. 449–454, 1998.

[FOR 99] FOREST S., DENDIEVEL R., CANOVA G.R., "Estimating the overall properties of heterogeneous Cosserat materials", *Modelling and Simulation in Materials Science and Engineering*, vol. 7, pp. 829–840, 1999.

[FOR 06] FOREST S., SIEVERT R., "Nonlinear microstrain theories", *International Journal of Solids and Structures*, vol. 43, no. 24, pp. 7224–7245, 2006.

[FRA 07] FRANCHI M. *et al.*, "Crimp morphology in relaxed and stretched rat Achilles tendon", *Journal of Anatomy*, vol. 210, no. 1, pp. 1–7, 2007.

[FRE 09] FREEMAN J.W. "Tissue engineered devices for ligament repair, replacement and regeneration", *African Journal of Biotechnology*, vol. 8, no. 25, 2009.

[FRE 11] FREEMAN J.W. *et al.*, "Evaluation of a hydrogel–fiber composite for ACL tissue engineering", *Journal of Biomechanics*, vol. 44, no. 4, pp. 694–699, 2011.

[FRI 97] FRIJNS A.J.H., HUYGHE J.M., JANSSEN J.D., "A validation of the quadriphasic mixture theory for intervertebral disc tissue", *International Journal of Engineering Science*, vol. 35, no. 15, pp. 1419–1429, 1997.

[FRI 98] FRIDEZ P., RAKOTOMANANA L., TERRIER A. *et al.*, "Three dimensional model of bone external adaptation", *Computer Methods in Biomechanics and Biomedical Engineering*, vol. 2, pp. 189–196, 1998.

[FRO 01] FROST H.M., "The Utah paradigm of skeletal physiology: an overview of its insights for bone, cartilage and collagenous tissue organs", *J.Bon Miner. Metab. 1*, vol. 4, pp. 179–184, 2001.

[FUN 81] FUNG Y.C., *Biomechanics, Mechanical Properties of Living Tissues*, Springer-Verlag, 1981.

[FUN 93] FUNG Y.C., *Biomechanics Mechanical Properties of Living Tissues*, Springer, 1993.

[FUN 10] FUNG D.T. *et al.*, "Early response to tendon fatigue damage accumulation in a novel in vivo model", *Journal of Biomechanics*, vol. 43, no. 2, pp. 274–279, 2010.

[GAB 16] GABRIELE S., RIZZI N., VARANO V.A., "1D nonlinear TWB model accounting for in plane cross-section deformation", *International Journal of Solids and Structures*, vol. 94–95, pp. 170–178, 2016.

[GAD 00] GADELETA S.J., BOSKEY A.L., PASCHALIS E. *et al.*, "A physical, chemical, and mechanical study of lumbar vertebrae from normal, ovariectomized, and nandrolone decanoate-treated cynomolgus monkeys (Macaca fascicularis)", *Bone*, vol. 27, no. 4, pp. 541–550, 2000.

[GAN 05] GANGHOFFER J.F., HAUSSY B., "Mechanical modeling of growth considering domain variation. Part I: Constitutive framework", *International Journal of Solids and Structures*, vol. 42, no. 15, pp. 4311–4337, 2005.

[GAN 10a] GANGHOFFER J.F., "Mechanical modeling of growth considering domain variation. Part II: Volumetric and surface growth involving Eshelby tensors", *J. Mech. Phys. Solids*, vol. 58, no. 9, pp. 1434–1459, 2010.

[GAN 10b] GANGHOFFER J.F., "On Eshelby tensors in the context of open systems application to volumetric growth", *International Journal of Engineering Science*, vol. 48, no. 12, pp. 2081–2098, 2010.

[GAN 11] GANGHOFFER J.F., "Mechanics and thermodynamics of surface growth viewed as moving discontinuities", *Mechanics Research Communications*, vol. 38, pp. 372–377, 2011.

[GAN 12] GANGHOFFER J.F., "A contribution to the mechanics and thermodynamics of surface growth. Application to bone external remodeling", *International Journal of Engineering Science*, vol. 50, pp. 166–191, 2012.

[GAN 13] GANGHOFFER J.F., "A kinematically and thermodynamically consistent volumetric growth model based on the stress-free configuration", *International Journal of Solids and Structures*, vol. 50, pp. 3446–3459, 2013.

[GAN 14] GANGHOFFER J.F., SOKOLOWSKI J., "A micromechanical approach to volumetric and surface growth in the framework of shape optimization", *International Journal of Engineering Science*, vol. 74, pp. 207–226, 2014.

[GAN 16a] GANGHOFFER J.F., RAHOUADJ R., BOISSE J. *et al.*, "Phase field approaches of bone remodeling based on TIP", *J. Non-Equilib. Thermodyn.*, vol. 41, no. 1, pp. 49–75, 2016.

[GAN 16b] GANGHOFFER J.F., "Micromechanical analysis of volumetric growth in the context of open systems thermodynamics and configurational mechanics', *Application to tumor growth. J. Cont. Mech. Thermodyn.*, 2016.

[GAN 16c] GANGHOFFER J.F., GODA I., MAURICE G., "Combined bone internal and external remodeling based on Eshelby stress", *International Journal of Solids and Structures*, vol. 94–95, pp. 138–157, 2016.

[GAR 03] GARIKIPATI K., NARAYANAN H., ARRUDA E.M., "Material forces in the context of biotissue remodeling In: Proc. Of the EUROMECH Colloquium 445", *Mechanics of material forces*, Kaiserslautern, Germany, 21–24 May 2003.

[GAR 14] GARIJO N., FERNÀNDEZ J.R., PÉREZ M.A. *et al.*, "Numerical stability and convergence analysis of bone remodeling model", *Computer Methods in Applied Mechanics and Engineering*, vol. 271, pp. 253–268, 2014.

[GAS 06] GASSER T.C., OGDEN R.W., HOLZAPFEL G.A., "Hyperelastic modeling of arterial layers with distributed collagen fibre orientations", *Journal of the Royal Society Interface*, vol. 22, nos. 3–6, pp. 15–35, 2006.

[GEF 08] GEFEN A., PORTNOY S., DIAMANT S., "Inhomogeneity of tissue-level strain distributions in individual trabeculae: Mathematical model studies of normal and osteoporosis cases", *Medical Engineering & Physics*, vol. 30, pp. 624–630, 2008, available at: http://www.journals. elsevier.com/medical-engineering-and-physics/.

[GER 04] GERE J.M., *Mechanics of Materials*, 6th ed. Bill Stenquist, 2004.

[GER 83] GERMAIN P., *Cours de Mécanique*, Ecole Polytechnique, édition 1983.

[GIB 82a] GIBSON LJ., ASHBY MF., "The mechanics of 3-dimensional cellular materials", *Proc. R. Soc. Lond. Ser. A: Math. Phys. Eng. Sc*i. vol. 382, no. 1782, p. 43, 1982.

[GIB 82b] GIBSON L.J., ASHBY M.F., SCHAJER G.S., "The mechanics of two-dimensional cellular materials", *Proc. R. Soc. Lond. Ser. A: Math. Phys. Eng. Sci.*, vol. 382, no. 1782, pp. 25–42, 1982.

[GIB 85] GIBSON LJ., "The mechanical behaviour of cancellous bone", *Journal of Biomechanics*, vol. 18, no. 5, pp. 317–328, 1985.

[GIB 97] GIBSON L.J., ASHBY M.F., *Cellular Solids: Structures and Properties*, Cambridge University Press, Cambridge, UK, 1997.

[GIB 05] GIBSON L.J., "Biomechanics of cellular solids", *Journal of Biomechanics*, vol. 38, pp. 377– 399, 2005.

[GIO 15] GIORGIO I., ANDREAUS U., LEKSZYCKI T. *et al.*, "The Influence of Different Geometries of Matrix/Scaffold on the Remodeling Process of a Bone and Bio-Resorbable Material Mixture with Voids", *Mathematics and Mechanics of Solids*, vol. 22, no. 5, pp. 969–987, 2015.

[GIO 16] GIORGIO I., ANDREAUS U., SCERRATO D. *et al.*, "A visco-poroelastic model of functional adaptation in bones reconstructed with bio-resorbable materials", *Biomechanics and Modeling in Mechanobiology*, vol. 15, no. 5, pp. 1325–1343, 2016.

[GIT 10] GITMAN I.M., ASKES H., KUHL K. *et al.*, "Stress concentrations in fractured compact bone simulated with a special class of anisotropic gradient elasticity", *International Journal of Solids and Structures*, vol. 47, pp. 1099–1107, 2010.

[GOD 12a] GODA I, ASSIDI M., GANGHOFFER J.F., "Micropolar constitutive model of cancellous bone from discrete homogenization", *Computer Methods in Biomechanics and Biomedical Engineering*, vol. 16, pp. 87–108, 2012.

[GOD 12b] GODA I., ASSIDI M., GANGHOFFER J.F., "Cosserat 3D anisotropic models of trabecular bone from the homogenization of the trabecular structure", *Biomechanics and Modeling in Mechanobiology*, vol. 15, Supplement 1, 2012.

[GOD 12c] GODA I., ASSIDI M., BELOUETTAR S. *et al.*, "A micropolar anisotropic constitutive model of cancellous bone from discrete homogenization", *J. Mech. Behav. Biomed. Mater.*, vol., 16, pp. 87–108, 2012.

[GOD 13a] GODA I., RAHOUADJ R., GANGHOFFER J.F., "Size dependent static and dynamic behavior of trabecular bone based on micromechanical models of the trabecular architecture", *International Journal of Engineering Science*, vol. 72, pp. 53–77, 2013.

[GOD 13b] GODA I., ASSIDI M., GANGHOFFER J.F., "Equivalent mechanical properties of textile monolayers from discrete asymptotic homogenization", *Journal of the Mechanics and Physics of solids*, vol. 61, no. 12, pp. 2537–2565, 2013.

[GOD 14] GODA I., ASSIDI M., GANGHOFFER J.F., "A 3D elastic micropolar model for vertebral trabecular bone from lattice homogenization of the trabecular structure", *J. Mech. Behavior of Biomed. Mat.*, vol. 13, no. 1, pp.53–83, 2014.

[GOD 15a] GODA I., GANGHOFFER J.F., "3D plastic collapse and brittle fracture surface models of trabecular bone from asymptotic homogenization method", *Int. J. Engng Sci.*, vol. 87, pp. 58–82, 2015.

[GOD 15b] GODA I., GANGHOFFER J.F., "Identification of couple-stress moduli of vertebral trabecular bone based on the 3D internal architectures", *J. Mech. Behav. of Biomed. Mat.*, vol. 51, pp. 99–118, 2015.

[GOD 15c] GODA I., Micromechanical models of network materials presenting internal length scales applications to trabecular bone under stable and evolutive conditions. PhD thesis, University of Lorraine, 2015.

[GOD 16a] GODA I., GANGHOFFER J.F., MAURICE G., "Combined bone internal and external remodeling based on Eshelby stress", *International Journal of Solids and Structures*, vol. 94–95, pp. 138–157, 2016.

[GOD 16b] GODA I., RAHOUADJ R., GANGHOFFER J.F. *et al.*, "3D couple-stress moduli of porous polymeric biomaterials using μCT image stack and FE characterization", *International Journal of Engineering Science*, vol. 100, pp. 25–44, 2016.

[GOD 16c] GODA I., GANGHOFFER J.F., "Construction of the effective plastic yield surfaces of vertebral trabecular bone under twisting and bending moments stresses using a 3D microstructural model", *ZAMM - Journal of Applied Mathematics and Mechanics*, 2016.

[GOD 16d] GODA I., GANGHOFFER J.F., CZARNECKI S. *et al.*, "Optimal internal architectures of femoral bone based on relaxation by homogenization and isotropic material design", *Mechanics Research Communications*, vol. 76, pp. 64–71, 2016.

[GON 08] GONELLA S., RUZZENE M., "Homogenization and equivalent in-plane properties of two-dimensional periodic lattices", *International Journal of Solids and Structures*, vol. 45, pp. 2897–2915, 2008.

[GOO 69] GOODRICH F.C., In Surface Science and Colloid Science, vol. 1, Edited by MATIJEVIC E., p. 1, 1969.

[GOO 75] GOODSTEIN J.D., *States of Matter*, Dover Phoenix Editions, 1975.

[GOO 05] GOODSHIP A.E., BIRCH H.L. "Cross sectional area measurement of tendon and ligament in vitro: a simple, rapid, non-destructive technique", *Journal of Biomechanics*, vol. 38, no. 3, pp. 605–608, 2005.

[GRE 12] GREGORY D.E., CALLAGHAN J.P., "An examination of the mechanical properties of the annulus fibrosus : The effect of vibration on the intra-lamellar matrix strength", *Medical Engineering & Physics*, vol. 34, no. 4, pp. 472–477, 2012.

[GRE 15] GRECO L, CUOMO M., "Consistent tangent operator for an exact Kirchhoff rod model", *Continuum Mechanics and Thermodynamics*, vol., 27, no. 4, pp. 861–877, 2015.

[GRY 09] GRYTZ R., MESCHKE G., "Constitutive modeling of crimped collagen fibrils in soft tissues", *Journal of the Mechanical Behavior of Biomedical Materials*, vol. 2, no. 5, pp. 522–533, 2009.

[GU 99] GU W.Y., MAO X.G., FOSTER R.J. *et al.*, "The anisotropic hydraulic permeability of human lumbar anulus fibrosus: Influence of age, degeneration, direction, and water content", *Spine*, vol. 24, no. 23, pp. 24–49, 1999.

[GUC 01] GUCK J., ANANTHAKRISHNAN R., "The optical stretcher: a novel laser tool to micromanipulate cells", *Biophysical Journal*, vol. 84, no. 2, pp. 767–784, 2001.

[GUE 90] GUEDES J.M., KIKUCHI N., "Preprocessing and post processing for materials based on the homogenization method with adaptive finite element methods", *Computer Methods in Applied Mechanics and Engineering*, vol. 83, pp. 143–198, 1990.

[GUK 01] GUCK J., ANANTHAKRISHNAN R., "The optical stretcher: a novel laser tool to micromanipulate cells", *Biophysical Journal*, vol. 81, no. 2, pp. 767–784, 2001.

[GUO 94] GUO X.-D.E., MCMAHON T.A., KEAVENY T.M. *et al.*, "Finite element modelling of damage accumulation in trabecular bone under cyclic loading", *Journal of Biomechanics*, vol. 27, pp. 145–155, 1994.

[GUP 10] GUPTA H.S. *et al.*, "In situ multi-level analysis of viscoelastic deformation mechanisms in tendon collagen", *Journal of Structural Biology*, vol. 169, no. 2, pp. 183–191, 2010.

[GUR 00] GURTIN M.E., *In Configurational forces as Basic Concepts of Continuum Physics*, Springer, New York, 2000.

[GUR 90] GURTIN M.E., STRUTHERS A., "Multiphase thermomechanics with interfacial structure", *Part 3. Arch. Rat. Mech. Anal*, vol. 112, p. 97–160, 1990.

[GUR 95] GURTIN M.E., "On the nature of configurational forces", *Arch. Rat. Mech. Anal*, vol. 131, pp. 67–100, 1995.

[HAD 11] HADJESFANDIARI A.R, DARGUSH G.F., "Couple stress theory for solids", *International Journal of Solids and Structures*, vol. 48, no. 18, pp. 2496–2510, 2011.

[HAG 14] HAGHPANAH B., PAPADOPOULOS J., VAZIRI A., "Plastic collapse of lattice structures under a general stress state", *Mechanics of Materials*, vol. 68, pp. 267–274, 2014.

[HAM 09] HAMILA N., BOISSE P., SABOURIN F. *et al.*, "A semi-discrete shell finite element for textile composite reinforcement forming simulation", *International Journal for Numerical Methods in Engineering*, vol. 79, pp. 1443–1466, 2009.

[HAM 11] HAMBLI R., "Apparent damage accumulation in cancellous bone using neural networks", *Journal of the Mechanical Behavior of Biomedical Materials*, vol. 4, no. 6, pp. 868–878, 2011.

[HAM 14] HAMBLI R., "Connecting mechanics and bone cell activities in the bone remodeling process: an integrated finite element modeling", *Front. Bioeng. Biotechnol*, pp. 2–6, 2014.

[HAN 97] HANSEN J.C., SKALAK R.S., CHEN HOGER A., "Influence of network topology on the elasticity of the red blood cell membrane skeleton", *Biophysical J.*, vol. 72, pp. 2369–2381, 1997.

HAN 06] HANSEN P. *et al.*, "Mechanical properties of the human patellar tendon, in vivo", *Clinical Biomechanics*, vol. 21, no. 1, pp. 54–58, 2006.

[HAN 12] HAN W., NERURKAR N., SMITH L., JACOBS N. *et al.*, "Multi-scale structural and tensile mechanical response of annulus fibrosus to osmotic loading", *Annals of Biomedical Engineering*, vol. 40, pp. 1610–1621, 2012.

[HAR 08] HARRISON N.M., MCDONNELL P.F., O'MAHONEY D.C. *et al.*, "Heterogeneous linear elastic trabecular bone modelling using micro-CT attenuation data and experimentally measured heterogeneous tissue properties", *Journal of Biomechanics*, vol. 41, pp. 2589–2596, 2008.

[HAR 10] HARTMANN D., "A multiscale model for red blood cell mechanics", *Biomechanics and Modelling in Mechanobiology*, vol . 9, no.1, pp. 1–17, 2010.

[HAV 10] HAVILL L.M., ALLEN M.R., BREDBENNER T.L. *et al.*, "Heritability of lumbar trabecular bone mechanical properties in baboons", *Bone*, vol. 46, pp. 835–840, 2010.

[HAW 09] HAWKINS D. *et al.*, "Dynamic creep and pre-conditioning of the Achilles tendon in-vivo", *Journal of Biomechanics*, vol. 42, no. 16, pp. 2813–2817, 2009.

[HAY 79] HAYES W.C., BLACK D., "Post-yield energy absorption characteristics of trabecular bone", in VAN BUSKIRK W.C. (ed.), *Proc. ASME Biomechanics Symposium*, New York, vol. 177, 1979.

[HAZ 01] HAZELWOOD S.J., MARTIN R.B., RASHID M.M. *et al.*, "A mechanistic model for internal bone remodeling exhibits different dynamic responses in disuse and overload", *Journal of Biomechanics*, vol. 34, pp. 299–308, 2001.

[HEA 69] HEARLE J.W.S., GROSBERG P., BAKER S., *Structural Mechanics of Fibers, Yarns, & Fabrics 1*, John Wiley & Sons, New York, USA, 1969.

[HEA 01] HEARLE J.W.S., POTLURI P., THAMMANDRA V.S., "Modelling fabric mechanics", *Journal of Textile Institute*, vol. 92, no. 1–3, pp. 53–69, 2001.

[HEC 05] HECHT F., PIRONNEAU O., LE HYARIC A., OHTSUKA K., FreeFem++ Manual, downloadable at http://www.freefem.org, 2005.

[HEL 04] HELLMICH C., ULM F.J., DORMIEUX L., "Can the diverse elastic properties of trabecular and cortical bone be attributed to only a few tissue-independent phase properties and their interactions? Arguments from a multiscale approach", *Biomechanics and Modeling in Mechanobiology*, vol. 2, pp. 219–238, 2004.

[HEN 99] HENON S., LENORMAND G., "A new determination of the shear modulus of the human erythrocyte membrane using optical tweezers", *Biophys J*, vol . 67, no. 2, pp. 1145–1151, 1999.

[HER 00] HERNANDEZ C.J., BEAUPRE G.S., CARTER D.R., "A model of mechanobiologic and metabolic infuences on bone adaptation", *Journal of Rehabilitation Research and Development*, vol. 37, no. 2, pp. 235–244, 2000.

[HIM 05] HIMPEL G., KUHL E., MENZEL A. *et al.*, "Computational modelling of isotropic multiplicative growth", *CMES, Tech Science Press*, vol. 8, no. 2, pp. 119–134, 2005.

[HIR 00] HIROKAWA S.; TSURUNO R., "Three-dimensional deformation and stress distribution in an analytical/computational model of the anterior cruciate ligament", *Journal of Biomechanics*, vol. 33, no. 9, pp. 1069–1077, 2000.

[HOC 00] HOCHMUTH R.M., "Micropipette aspiration of living cells", *J Biomech*, vol. 33, no.1, pp. 15–22, 2000.

[HOH 99] HOHENADL M., STORZ T., KIRPAL H. *et al.*, "Desmin filaments studied by quasi-elastic light scattering", *J. Biophys*, vol. 77, pp. 2199–2209, 1999.

[HOH 03] HOHE J., BECKER W., "Effective mechanical behavior of hyperelastic honeycombs and two dimensional model foams at finite strain", *International Journal of Mechanical Sciences*, vol. 45, pp. 891–913, 2003.

[HOL 81] HOLM S., MAROUDAS A., URBAN J.P. *et al.*, "Nutrition of the intervertebral disc: solute transport and metabolism", *Connect Tissue Res*, vol. 8, no. 2, pp. 101–119, 1981.

[HOL 92] HOLLISTER S.J., KIKUCHI N., "A comparison of homogenization and standard mechanics analysis for periodic porous composites", *Computational Mechanics*, vol. 10, pp. 73–95, 1992.

[HOL 94] HOLLISTER S.J., BRENNAN J.M., KIKUCHI N., "A homogenization sampling procedure for calculating trabecular bone effective stiffness and tissue level stress", *Journal of Biomechanics*, vol. 27, pp. 433–444, 1994.

[HOL 00] HOLZAPFEL G.A., GASSER T.C., OGDEN R.W. "A New Constitutive Framework for Arterial Wall Mechanics and a Comparative Study of Material Models", *Journal of Elasticity and the Physical Science of Solids*, vol. 61, nos. 1–3, pp. 1–48, 2000.

[HOL 01] HOLZAPFEL G., *Nonlinear Solid Mechanicsm, A continuum approach for engineering*, Wiley, 2001.

[HOL 02] HOLZAPFEL G.A., GASSER T.C., STADLER M., "A structural model for the viscoelastic behavior of arterial walls: continuum formulation and finite element analysis", *Eur. J. Mech. A/solids*, vol. 21, pp. 441–463, 2002.

[HOL 05] HOLZAPFEL G.A., SCHULZE-BAUER C.A.J., FEIGL G. *et al.*, "Single lamellar mechanics of the human lumbar anulus fibrosus", *Biomechanics and Modeling in Mechanobiology*, vol. 3, pp. 125–140, 2005.

[HOL 06] HOLZAPFEL G.A., OGDEN R.W. (eds), *Mechanics of Biological tissues*, Springer-Verlag, Heidelberg, 2006.

[HOL 07] HOLZAPFEL G.A., GASSER T.C., "Computational stress-deformation analysis of arterial walls including high-pressure response", *International Journal of Cardiology*, vol. 116, no. 1, pp. 78–85, 2007.

[HOL 09] HOLLISTER S.J. "Scaffold Design and Manufacturing: From Concept to Clinic", *Advanced Materials*, vol. 21, nos. 32–33, pp. 3330–3342, 2009.

[HOL 10] HOLZAPFEL G.A.; OGDEN R.W,. "Constitutive modelling of arteries", *Proceedings of the Royal Society of London A: Mathematical, Physical and Engineering Sciences*, vol. 466, no. 2118, pp. 1551–1597, 2010.

[HOL 14] HOLZAPFEL G.A., UNTERBERGER M.J., OGDEN R.W., "An affine continuum mechanical model for cross-linked f-actin networks with compliant linker proteins", *J. Mech. Behav. Biomed. Mater*, vol. 38, pp. 78–90, 2014.

[HOU 67] HOUK J., SIMON W. "Responses of Golgi tendon organs to forces applied to muscle tendon", *Journal of Neurophysiology*, vol. 30, no. 6, pp. 1466–1481, 1967.

[HU 02] HU J.H., DING M., SØBALLE K., "Effects of short-term alendronate treatment on the three-dimensional microstructural, physical, and mechanical properties of dog trabecular bone", *Bone*, vol. 31, no. 5, pp. 591–597, 2002.

[HUI 00] HUISKES R., RUIMERMAN R., VAN LENTHE G.H., JANSSEN J.D., "Effects of mechanical forces on maintenance and adaptation of form in trabecular bone", *Nature*, vol. 405, pp. 704–706, 2000.

[HUI 87] HUISKES R., WEINANS H., GROOTENBOER H.J. *et al.*, "Adaptive bone-remodeling theory applied to prosthetic-design analysis", *Journal of Biomechanics*, vol. 20, pp. 1135–1150, 1987.

[HUM 02] HUMPHREY J.D., RAJAGOPAL K.R., "A constrained mixture model for growth and remodeling of soft tissues", *Math. Meth. Mod. App. Sci*, vol. 12 no. 3, pp. 407–430, 2002.

[HUM 14] HUMPHREY J.D., DUFRESNE E., SCHWARTZ M., "Mechanotransduction and extracellular matrix homeostasis", *Nature Reviews, Molecular Cell Biology*, vol. 15, no. 12, pp. 802–812, 2014.

[HUT 06] HUTCHINSON R., FLECK N., "The structural performance of the periodic truss", *J. Mech.Phys.Solids*, vol. 54, pp. 756–82, 2006.

[HUY 04] HUYGHE J.M.R.J., DROST M., "Uniaxial tensile testing of canine annulus fibrosus tissue under changing salt concentrations", *Biorheology*, vol. 41, pp. 255–261, 2004.

[HUY 10] HUYGHE J.M.R.J., "Biaxial testing of canine annulus fibrosus tissue under changing salt concentration", *Ann. Braz. Ac. Sci*, vol. 82, no. 1, pp. 1–7, 2010.

[IAT 98] IATRIDIS J.C., SETTON L.A., FOSTER R.J. *et al.*, "Degeneration affects the anisotropic and nonlinear behaviors of human anulus fibrosus in compression", *J Biomech*, vol. 31, no. 6, pp. 535–544, 1998.

[ING 06] INGBER D.E., "Cellular mechanotransduction: putting all the pieces together again", *The FASEB Journal*, vol. 20, no. 7, pp. 811–827, 2006.

[INO 81] INOUE H., "Three-dimensional architecture of lumbar intervertebral discs", *Spine* (Phila Pa 1976), vol. 6, no. 2, pp. 139–146, 1981.

[ITS 13] ITSKOV M., *Tensor algebra and tensor analysis for engineers, With applications to continuum mechanics*, 3rd edition, Springer, 2013.

[IVA 04] IVANOV I., TABIEI A., "Loosely woven fabric model with viscoelastic crimped fibres for ballistic impact simulations", *International Journal for Numerical Methods in Engineering*, vol. 61, no. 10, pp. 1565–1583, 2004.

[JAN 03] JANUS-MICHALSKA M., PĘCHERSKI RP., "Macroscopic properties of open-cell foams based on micromechanical modeling", *Technische Mechanik, Band*, vol. 23, Heft, nos. 2–4, pp. 221–231, 2003.

[JAN 05] JANUS-MICHALSKA J., "Effective models describing elastic behavior of cellular materials", *Archives of Metallurgy and Materials*, vol. 50, pp. 595–608, 2005.

[JAN 08] JANG I.G., KIM I.Y., "Computational study of Wolff's law with trabecular architecture in the human proximal femur using topology optimization", *Journal of Biomechanics*, vol. 41, pp. 2353–2361, 2008.

[JAN 09] JANMEY P.A. *et al.*, "The hard life of soft cells", *Cell Motility and The Cytoskeleton*, vol. 66, no. 8, pp. 597–605, 2009.

[JAN 10] JANG I.G., KIM I.Y., "Application of design space optimization to bone remodeling simulation of trabecular architecture in human proximal femur for higher computational efficiency", *Finite Elem. Anal. Des*, vol. 4, pp. 311–319, 2010.

[JAN 11] JANUS-MICHALSKA J., "Hyperelastic behavior of cellular structures based on micromechanical modeling at small strain", *Arch. Mech.*, vol. 63, no. 1, pp. 3–23, 2011.

[JEO 08a] JEONG J., ABID-RAMEZANI H., AL-MUKHTAR, M., "Numerical simulation of elastic linear micropolar media based on the pore space length scale assumption", *Strength of materials*, vol. 40, pp. 425–438, 2008.

[JEO 08b] JEONG S.I., LEE Y.M., SHIN H., "Tissue engineering using a cyclic strain bioreactor and gelatin/PLCL scaffolds", *Macromolecular Research*, vol. 16, no. 6, pp. 567–569, 2008.

[JOG 00] JOG C.S., *Continuum mechanics, Foundations and Applications of Mechanics*, 2nd Edition, Alpha Science, vol. I, 2000.

[JOH 94] JOHNSON G.A. *et al.*, "Tensile and viscoelastic properties of human patellar tendon", *Journal of Orthopaedic Research*, vol. 12, no. 6, pp. 796–803, 1994.

[JON 95] JONES R. *et al.*, "Mechanical properties of the human anterior cruciate ligament", *Clinical Biomechanics*, vol. 10, no. 7, pp. 339–344, 1995.

[JOR 02] JORGENSEN C.S., KUNDU T., "Measurement of material elastic constants of trabecular bone: a micromechanical analytic study using a 1 GHz acoustic microscope", *Journal of Orthopaedic Research*, vol. 20, pp. 151–158, 2002.

[JU 91] JU J., Damage mechanics of composite materials: constitutive modeling and computational algorithms. Technical report, DTIC document, 1991.

[KAC 11] KACZMARCZYK L., PEARCE C.J., "Efficient numerical analysis of bone remodeling", *Journal of the Mechanical Behavior of Biomedical Materials*, vol. 4, pp. 858–867, 2011.

[KAH 08] KAHN C.J. *et al.* "A novel bioreactor for ligament tissue engineering", *Bio-Medical Materials and Engineering*, vol. 18, no. 4–5, pp. 283–287, 2008.

[KAM 10] KAMM R., LAMMERDING J., MOFRAD M., *Cellular nanomechanics*, Springer, New York, 2010.

[KAN 09] KANOUTÉ P., BOSO D., CHABOCHE J. *et al.*, "Multiscale methods for composites: a review", *Archives of Computational Methods in Engineering*, vol. 16, no. 1, pp. 31–75, 2009.

[KAP 85] KAPLAN S., HAYES W., STONE J. *et al.*, "Tensile strength of bovine trabecular bone", *Journal of Biomechanics*, vol. 18, pp. 723–727, 1985.

[KAR 03] KARMANI S., EMBER T. "The anterior cruciate ligament—1", *Current Orthopaedics*, vol. 17, no. 5, pp. 369–377, 2003.

[KAR 04] KARMANI S., EMBER T., "The anterior cruciate ligament—II", *Current Orthopaedics*, vol. 18, no. 1, pp. 49–57, 2004.

[KAR 10] KARIMI K., KHOEI A.R., "On the analysis of simple shear problem using the micropolar hypoelasticity Cosserat theory", *European Journal of Mechanics - A/Solids*, vol. 29, no. 4, pp. 664–674, 2010.

[KAS 07] KASRA M., GRYNPAS M.D., "On shear properties of trabecular bone under torsional loading: Effects of bone marrow and strain rate", *Journal of Biomechanics*, vol. 40, pp. 2898–2903, 2007.

[KAT 99] KATO S., YOSHIRO T., MINAMI H., "Formulation of constitutive equations for fabric membranes based on the concept of fabric lattice model", *Engineering Structures*, vol. 21, pp. 691–708, 1999.

[KAU 11] KAUFMANN A., HEINEMANN F., RADMACHER. *et al.*, Amphibian oocyte nuclei expressing lamin a with the progeria mutation e145k exhibit an increased elastic modulus, Nucleus, vol. 2, no. 4, pp. 310–319, 2011.

[KEA 94] KEAVEN T.M., WACHTEL E.F., FORD C.M. *et al.*, "Differences between the tensile and compressive strengths of bovine tibial trabecular bone depend on modulus", *Journal*, 1994.

[KEA 94] KEAVENY T.M., WACHTEL E.F., FORD C.M. *et al.*, "Differences between the tensile and compressive strengths of bovine tibial trabecular bone depend on modulus", *Journal of Biomechanics*, vol. 27, pp. 1137–1146, 1994.

[KEA 97] KEAVENY T.M., PINILLA T.P., CRAWFORD R.P. *et al.*, "Systematic and random errors in compression testing of trabecular bone", *Journal of Orthopaedic Research*, vol. 15, no. 1, pp. 101–110, 1997.

[KEN 76] KENNEDY J.C. *et al.*, "Tension studies of human knee ligaments. Yield point, ultimate failure, and disruption of the cruciate and tibial collateral ligaments", *The Journal of Bone and Joint Surgery*, American Volume, vol. 58, no. 3, pp. 350–355, 1976.

[KER 07] KER R.F., "Mechanics of tendon, from an engineering perspective", *International Journal of Fatigue*, vol. 29, no. 6, pp. 1001–1009, 2007.

[KHO 13] KHODABAKHSHI G. *et al.*, "Measuring three-dimensional strain distribution in tendon", *Journal of Microscopy*, vol. 249, no. 3, pp. 195–205, 2013.

[KIM 02] KIM H.S., AL-HASSANI, S.T.S., "A morphological model of vertebral trabecular bone", *Journal of Biomechanics*, vol. 35, pp. 1101–1114, 2002.

[KIM 12a] KIM Y., HIGGINS J.M., Anisotropic light scattering of individual sickle red blood cells, *Journal of Biomedical Optics*, vol. 17, no. 4, pp. 0405011–0405013, 2012.

[KIM 12b] KIM Y., KIM K., PARK Y., *Measurement Techniques for Red Blood Cell Deformability: Recent Advances*, In Tech, 2012.

[KIN 05] KING M.J., JEARANAISILAWONG P., SOCRATE S., "A continuum constitutive model for the mechanical behavior of woven fabrics", *International Journal of Solids and Structures*, vol. 42, pp. 3867–3896, 2005.

[KJA 04] KJÆR M., "Role of Extracellular Matrix in Adaptation of Tendon and Skeletal Muscle to Mechanical Loading", *Physiological Reviews*, vol. 84, no. 2, pp. 649–698, 2004.

[KLA 10] KLAUSER A.S., FASCHINGBAUER R., JASCHKE W.R., "Is sonoelastography of value in assessing tendons?", *Seminars in Musculoskeletal Radiology*, vol. 14, no. 3, pp. 323–333, 2010.

[KOC 92] KOCH A.L., WOESTE S., "Elasticity of the sacculus of escherichia coli", *Journal of Bacteriology*, vol. 174, pp. 327–341, 1992.

[KOI 64] KOITER W.T., "Couple stress in the theory of elasticity", *Proceeding of the Koninklijke Nederland Akademie van Wettenschappen*, vol. B-67, pp. 17–44, 1964.

[KOM 03] KOMAROVA S.V., SMITH R.J., DIXON S.J. *et al.*, "Mathematical model predicts a critical role for osteoclasts autocrine regulation in the control of bone modeling", *Bone*, vol. 33, pp. 206–215, 2003.

[KON 11] KONGSGAARD M. *et al.*, "Mechanical properties of the human Achilles tendon, in vivo", *Clinical Biomechanics*, vol. 26, no. 7, pp. 772–777, 2011.

[KOP 02] KOPPERDAHL D.L., MORGAN E.F., KEAVENY T.M., "Quantitative computed tomography estimates of the mechanical properties of human vertebral trabecular bone", *Journal of Orthopaedic Research*, vol. 20, pp. 801–805, 2002.

[KOP 98] KOPPERDAHL D.L., KEAVENY T.M., "Yield strain behavior of trabecular bone", *Journal of Biomechanics*, vol. 31, pp. 601–608, 1998.

[KOR 10] KORSTANJE J.-W.H. *et al.*, "Development and validation of ultrasound speckle tracking to quantify tendon displacement", *Journal of Biomechanics*, vol. 43, no. 7, pp. 1373–1379, 2010.

[KOW 10] KOWALCZYK P., "Simulation of orthotropic microstructure remodelling of cancellous bone", *Journal of Biomechanics*, vol. 43, pp. 563–569, 2010.

[KRA 91] KRAYNIK A.M., REINELT D.A., PRINCEN H.M., "The nonlinear elastic behaviour of polydisperse hexagonal foams and concentrated emulsion", *Journal of Rheology*, vol. 35, no. 6, pp. 1235–1253, 1991.

[KRA 09] KRAEMER J., *Intervertebral Disk Diseases: Causes, Diagnosis, Treatment and Prophylaxis*, 3rd edition, Thieme, 2009.

[KUH 03] KUHL E., MENZEL A., STEINMANN P., "Computational modeling of growth", *Computational Mechanics*, vol. 32, pp. 71–88, 2003.

[KUM 04] KUMAR R.S., MCDOWELL D.L., "Generalized continuum modelling of 2-d periodic cellular solids", *International Journal of Solids and Structures*, vol . 41, pp. 7399–7422, 2004.

[KUO 10] KUO C.K., MARTURANO J.E., TUAN R.S., "Novel strategies in tendon and ligament tissue engineering: Advanced biomaterials and regeneration motifs", *SMARTT: Sports Medicine, Arthroscopy, Rehabilitation, Therapy & Technology*, vol. 2, pp. 20–33, 2010.

[KWA 93] KWAN M.K., LIN T.H.-C., WOO S.L.-Y., "On the viscoelastic properties of the anteromedial bundle of the anterior cruciate ligament", *Journal of Biomechanics*, vol. 26, nos. 4–5, pp. 447–452, 1993.

[LAC 02] LACROIX D., PRENDERGAST P.J., "A mechano-regulation model for tissue differentiation during fracture healing: analysis of gap size and loading", *Journal of Biomechanics*, vol. 35, pp. 1163–1171, 2002.

[LAD 11] LADD M.R. *et al.*, "Co-electrospun dual scaffolding system with potential for muscle–tendon junction tissue engineering", *Biomaterials*, vol. 32, no. 6, pp. 1549–1559, 2011.

[LAI 93] LAI W.M., RUBIN D., KREMPL E., *Introduction to continuum mechanics*, Pergamon Press, New York, 1993.

[LAK 86] LAKES R., "Experimental microelasticity of two porous solids", *International Journal of Solids and Structures*, vol. 22, pp. 55-63, 1986.

[LAK 90] LAKES R., NAKAMURA S., BEHIRI J. *et al.*, "Fracture mechanics of bone with short cracks", *Journal of Biomechanics*, vol. 23, pp. 967–975, 1990.

[LAK 93] LAKES R., "Experimental methods for study of Cosserat elastic solids and other generalized elastic continua", in MUHLHAUS H.-B. (ed.), *Continuum Models for Materials with Microstructure*, Wiley, New York, 1995.

[LAK 01] LAKES R., Materials with structural hierarchy, *Nature*, vol. 361, pp. 511– 515, 1993.

[LAK 10] LAKE S.P. *et al.*, "Tensile properties and fiber alignment of human supraspinatus tendon in the transverse direction demonstrate inhomogeneity, nonlinearity, and regional isotropy", *Journal of Biomechanics*, vol. 43, no. 4, pp. 727–732, 2010.

[LAN 80] LANGER J.S., "Instabilities and pattern formation in crystal growth", *Rev. Mod. Phys.*, vol. 52, pp. 1–28, 1980.

[LAN 87] LANIR Y., Biorheology and fluid flux in swelling tissues. I. bicomponent theory for small deformations, including concentration effects, *Biorheology*, vol. 24, pp. 173–187, 1987.

[LAU 11] LAURENT C.P. *et al.*, "Morphological characterization of a novel scaffold for anterior cruciate ligament tissue engineering", *Journal of biomechanical engineering*, vol. 133, no. 6, p. 065001, 2011.

[LAU 12] LAURENT C.P. *et al.*, "A multilayer braided scaffold for Anterior Cruciate Ligament: Mechanical modeling at the fiber scale", *Journal of the Mechanical Behavior of Biomedical Materials*, vol. 12, pp. 184–196, 2012.

[LAU 13] LAURENT C.P. *et al.*, "Computer-Aided Tissue Engineering: application to the case of Anterior Cruciate Ligament Repair", in LECCA P. (ed.), *Biomechanics of cells and tissues*, Springer ed. vol. 9, pp. 1–44. 2013.

[LAU 14a] LAURENT C.P. *et al.*, "Towards a Tissue-Engineered Ligament: Design and Preliminary Evaluation of a Dedicated Multi-Chamber Tension-Torsion Bioreactor", *Processes*, vol. 2, no. 1, pp. 167–179, 2014.

[LAU 14b] LAURENT C.P. *et al.*, "Mechanical behaviour of a fibrous scaffold for ligament tissue engineering: Finite elements analysis vs. X-ray tomography imaging", *Journal of the Mechanical Behavior of Biomedical Materials*, vol. 40, pp. 222–233, 2014.

[LEE 99] LEE T.C., NOELKE L., MCMAHON G.T. *et al.*, Functional adaptation in bone, in PEDERSEN P., BENDSOE M.P. (eds), *Synthesis in Bio Solid Mechanics* , Kluwer Academic Publishers, Dordrecht, 1999.

[LEE 07] LEE S.H., GRIER D.G., "Holographic microscopy of holographically trapped threedimensional structures", Opt. Express, vol. 15, no. 4, pp. 1505–1512, 2007.

[LEM 78] LEMAITRE J., CHABOCHE J.L., "Aspect Phenomenologique de la Rupture par Endommagement", *Journal de Mécanique Appliquée*, vol. 2, pp. 317–365, 1978.

[LEM 84] LEMAITRE J., "How to Use Damage Mechanics", *Nuclear Engineering and Design*, vol. 80, pp. 233–245, 1984.

[LEM 07] LEMAIRE T., MOYNE C., STEMMELEN D., "Modeling of electro-osmosis in clayey materials including pH effects", *Physics and Chemistry of the Earth*, vol. 32, pp. 441–452, 2007.

[LEN 01] LENORMAND G., HÉNON S., RICHERT A. *et al.*, "Direct measurement of the area expansion and shear moduli of the human red blood cell membrane skeleton", *Biophysical Journal*, vol . 81, pp. 43–56, 2001.

[LEO 08] LEONG K.F. *et al.*, "Engineering functionally graded tissue engineering scaffolds", *Journal of the Mechanical Behavior of Biomedical Materials*, vol. 1, no. 2, pp. 140–152, 2008.

[LEO 14] LEONG N.L., PETRIGLIANO F.A., MCALLISTER D.R., "Current tissue engineering strategies in anterior cruciate ligament reconstruction", *Journal of Biomedical Materials Research Part A*, vol. 102, no. 5, pp. 1614–1624, 2014.

[LEV 98] LEVENSTON M.E., CARTER D.R., "An energy dissipation- based model for damage stimulated bone adaptation", *Journal of Biomech*, vol. 31, pp. 579–586, 1998.

[LI 14] LI X. *et al.* "A novel silk–TCP–PEEK construct for anterior cruciate ligament reconstruction: an off-the shelf alternative to a bone–tendon–bone autograft", *Biofabrication*, vol. 6, no. 1, p. 015010, 2014.

[LIM 06] LIM C.T., ZHOU E.H., QUEK S.T., "Mechanical models for living cells—a review", *Journal of Biomechanics*, vol . 39, no.2, pp. 195–216, 2006.

[LIU 87] LIU S.-C., DERICK L., PALEK J., "Visualization of the hex- agonal lattice in the erythrocyte membrane skeleton", *J.Cell Biol.*, vol. 104, pp. 527–536, 1987.

[LIU 08] LIU Y., RAMANATH H.S., WANG D.-A. "Tendon tissue engineering using scaffold enhancing strategies", *Trends in Biotechnology*, vol. 26, no. 4, pp. 201–209, 2008.

[LIU 09] LIU X.S., ZHANG X.H., GUO X.E., "Contributions of trabecular rods of various orientations in determining the elastic properties of human vertebral trabecular bone", *Bone*, vol. 45, pp. 158–163, 2009.

[LOM 07] LOMOV S., IVANOV D., VERPOEST I. *et al.*, "Meso-FE modelling of textile composites: Road map, data flow and algorithms", *Composites Science and Technology*, vol. 67, pp. 1870–1891, 2007.

[LOR 06] LORENZETTI S.R., New method to determine the Young's modulus of single trabeculae, PhD Thesis, Rutgers University, *Swiss Federal Institute of Technology*, 2006.

[LOT 91] LOTZ J.C., CHEAL E.J., HAYES W.C., "Fracture prediction for the proximal femur using finite element models: Part I-Linear analysis", *Journal of Biomechanical Engineering*, vol. 113, pp. 353–360, 1991.

[LOU 16] LOUNA Z., GODA I., GANGHOFFER J.F. *et al.*, Formulation of an effective growth response of trabecular bone based on micromechanical analyses at the trabecular level. *Archive of Applied Mechanics.*, 2016.

[LU 05] LU H.H. *et al.*, "Anterior cruciate ligament regeneration using braided biodegradable scaffolds: in vitro optimization studies", *Biomaterials*, vol. 26, no. 23, pp. 4805–4816, 2005.

[LUB 02] LUBARDA V.A., HOGER A., "On the mechanics of solids with a growing mass", *Int. J. Solids Struct.*, vol. 39, pp. 4627–4664, 2002.

[MA 08] MA H.M., GAO X.-L., REDDY J.N., " A microstructure-dependent Timoshenko beam model based on a modified couple stress theory", *Journal of the Mechanics and Physics of Solids*, vol. 56, pp. 3379–3391, 2008.

[MAC 92] MCNAMARA B.P., PRENDERGAST P.J., TAYLOR D., "Prediction of bone adaptation in the ulnar-osteotomized sheep's forelimb using an anatomical finite element model", *Journal of Biomedical Engineering*, vol. 14, pp. 209–216, 1992.

[MAC 07] MCNAMARA L.M., PRENDERGAST P.J., "Bone remodelling algorithms incorporating both strain and microdamage stimuli", *Journal of Biomechanics*, vol. 40, pp. 1381–1391, 2007.

[MAC 10] MACERI F., MARINO M., VAIRO G., "A unified multiscale mechanical model for soft collagenous tissues with regular fiber arrangement", *Journal of Biomechanics*, vol. 43, no. 2, pp. 355–363, 2010.

[MAC 11] MACERI F., MARINO M., VAIRO G., "An insight on multiscale tendon modeling in muscle–tendon integrated behavior", *Biomechanics and Modeling in Mechanobiology*, vol. 11, nos. 3–4, pp. 505–517, 2011.

[MAC 12] MACERI F., MARINO M., VAIRO G., "Integrated mechanical models for collagenous biostructures at different length scales", *Mathematical Modeling*, vol. 7, 2012.

[MAC 13] MACERI F., MARINO M., VAIRO G., "Age-dependent arterial mechanics via a multiscale elastic approach", *Int. J. Comput. Methods Eng. Sci. Mech.*, 2013.

[MAG 00] MAGANARIS C.N., PAUL J.P. "Hysteresis measurements in intact human tendon", *Journal of Biomechanics*, vol. 33, no. 12, pp. 1723–1727, 2000.

[MAG 02] MAGANARIS C.N., PAUL J.P. "Tensile properties of the in vivo human gastrocnemius tendon", *Journal of Biomechanics*, vol. 35, no. 12, pp. 1639–1646, 2002.

[MAG 09] MAGNIER C., BOIRON O., WENDLING-MANSUY S. *et al.*, "Nutrient distribution and metabolism in the intervertebral disc in the unloaded state: a parametric study", *J Biomech*, vol. 42, no. 2, pp. 100–108, 2009.

[MAJ 04] MAJID T., MANZARI, "Application of micropolar plasticity to post failure analysis in geomechanics", *International Journal for Numerical and Analytical Methods in Geomechanics*, vol. 28, no. 10, pp. 1011–1032, 2004.

[MAJ 05] MAJIMA T. *et al.* "Alginate and chitosan polyion complex hybrid fibers for scaffolds in ligament and tendon tissue engineering", *Journal of Orthopaedic Science*, vol. 10, no. 3, pp. 302–307, 2005.

[MAL 11] MALANDRINO A., NOAILLY J., LACROIX D., "The effect of sustained compression on oxygen metabolic transport in the intervertebral disc decreases with degenerative changes", *PLoS Comput. Biol.*, vol. 7, no. 8, p. e1002112, 2011.

[MAL 12] MALANDRINO A., NOAILLY J., LACROIX D., "Regional annulus fibre orientations used as a tool for the calibration of lumbar intervertebral disc finite element models", *Computer Methods in Biomechanics and Biomedical Engineering*, pp. 1–6, 2012.

[MAR 75] MAROUDAS A., STOCKWELL R.A., NACHEMSON A. *et al.*, "Factors involved in the nutrition of the human lumbar intervertebral disc: cellularity and diffusion of glucose in vitro", *J Anat.*, vol. 120, no. Pt 1, pp. 113–130, 1975.

[MAR 92] MARTIN R.B., "A theory of fatigue damage accumula- tion and repair in cortical bone", *Journal of Orthopaedic Research*, vol. 10, pp. 818–825, 1992.

[MAR 95] MARTIN R.B., "Mathematical model for repair of fatigue damage and stress fracture in osteonal bone", *Journal of Orthopaedic Research*, vol. 13, no. 3, pp. 309–316, 1995.

[MAR 12] MARINO M., VAIRO G., "Equivalent stiffness and compliance of curvilinear elastic fibers", in MACERI F., FR´EMOND M. (eds.), *Mechanics, Models* and *Methods In Civil Engineering, Lecture Notes In Applied & Computational Mechanics*, vol. 61, pp. 309–332. Springer Verlag, Berlin, 2012.

[MAR 13] MARINO M., MACERI F., VAIRO G., "Age-dependent arterial mechanics via a multiscale elastic approach", *Int. J. Comput. Methods Eng. Sci. Mech.*, 2013.

[MAR 14] MARINO M., VAIRO G., "Influence of inter-molecular interactions on the elasto-damage mechanics of collagen fibrils: a bottom-up approach towards macroscopic tissue modeling", *JMPS*, vol. 73, pp. 38–54, 2014.

[MAS 96] MASTERS I.G., EVANS K.E., "Models for the elastic deformation of honeycombs", *Composite Structures*, vol. 35, pp. 403–422, 1996.

[MAT 95] MATTHEW J., SILVA WILSON C., HAYES LORNA J. *et al.*, "The effect of non-periodic microstructure on the elastic properties of two dimensional cellular solids, international", *Journal of Mechanical Sciences*, vol. 37, no. 11, pp. 1161–1177, November 1995.

[MAT 97] MATTHEW J. SILVA LORNA J., GIBSON., "The effects of non-periodic microstructure and defects on the compressive strength of two-dimensional cellular solid, international", *Journal of Mechanical Sciences*, vol. 39, no. 5, pp. 549–563, May 1997.

[MAT 03] MATSUURA Y., OHARU S., TAKATA T. *et al.*, "Mathematical approaches to bone reformation phenomena and numerical simulations", *J. Comput. Appl. Math.*, vol. 158, pp. 107–119, 2003.

[MAU 93] MAUGIN G.A., *Material Inhomogeneities in Elasticit*, CRC Press, London, 1993.

[MAU 95] MAUGIN G.A., TRIMARCO C., "The Dynamics of Configurational Forces at Phase-Transition Fronts", *Meccanica*, vol. 30, pp. 605–619, 1995.

[MAU 06a] MAUGIN G.A., "On canonical equations of continuum thermomechanics", *Mech. Res. Comm.*, vol. 33, pp. 705–710, 2006.

[MAU 06b] MAUGIN G.A., "On the thermomechanics of continuous media with diffusion and /or weak nonlocality", *Arch. Appl. Mech.*, vol. 75, pp. 723–738, 2006.

[MAU 10] MAUGIN, G.A., "From the propagation of phase-transition fronts to the evolution of the growth plate in long bones", *Proc. Estonian Academy of Sci.*, vol. 59, no. 2, pp. 72–78, 2010.

[MAZ 13] MAZZA E., Introduction to nonlienar continuum mechanics, Lecture Notes, Academic year 2012–2013.

[MCD 10] MCDONALD K., LITTLE J., PEARCY M. *et al.*, "Development of a multi-scale finite element model of the osteoporotic lumbar vertebral body for the investigation of apparent level vertebra mechanics and micro-level trabecular mechanics", *Medical Engineering & Physics*, vol. 32, pp. 653–661, 2010.

[MCK 86] MCKEON F.D., KIRSCHNER M.W., CAPUT D., "Homologies in both primary and secondary structure between nuclear envelope and intermediate filament proteins", *Nature*, pp. 463–468, 1986.

[MCN 92] MCNAMARA B.P., PRENDERGAST P.J., TAYLOR D., "Prediction of bone adaptation in the ulnar-osteotomized sheep's forelimb using an anatomical finite element model", *Journal of Biomedical Engineering*, vol. 14, pp. 209–216, 1992.

[MCN 07] MCNAMARA L.M., PRENDERGAST P.J., "Bone remodelling algorithms incorporating both strain and microdamage stimuli", *Journal of Biomechanics*, vol. 40, pp. 1381–1391, 2007.

[MEL 10] MELCHELS F.P.W. *et al.*, "Effects of the architecture of tissue engineering scaffolds on cell seeding and culturing", *Acta Biomaterialia*, vol. 6, no. 11, pp. 4208–4217, 2010.

[MER 07] MERODIO J., GOICOLEA J., "On thermodynamically consistent constitutive equations for fiber-reinforced nonlinearly viscoelastic solids with application to biomechanics", *Mech. Res. Commun.*, vol. 34, no. 7, pp. 561–571, 2007.

[MIC 10] MICHALEK A.J., BUCKLEY M.R., BONASSAR L.J. *et al.*, "The effects of needle puncture injury on microscale shear strain in the intervertebral disc annulus fibrosus", *The Spine Journal*, vol. 10, no. 12, pp. 1098–1105, 2010.

[MIL 05] MILLER Z., FUCHS M.B., "Effect of trabecular curvature on the stiffness of trabecular bone", *Journal of Biomechanics*, vol. 38, pp. 1855–1864, 2005.

[MIL 07] MILLER L.M., LITTLE W., SCHIRMER A. *et al.*, "Accretion of bone quantity and quality in the developing mouse skeleton", *Journal of Bone and Mineral Research*, pp. 1037–1045, 2007.

[MIL 12] MILLER K.S. *et al.*, "Characterizing local collagen fiber re-alignment and crimp behavior throughout mechanical testing in a mature mouse supraspinatus tendon model", *Journal of Biomechanics*, vol. 45, no. 12, pp. 2061–2065, 2012.

[MIN 62] MINDLIN R.D., TIERSTEN H.F., "Effects of couple stresses in linear elasticity", *Archive for Rational Mechanics and Analysis*, vol. 11, pp. 415–448, 1962.

[MIN 63] MINDLIN R.D., "Influence of couple-stresses on stress concentrations", *Experimental Mechanics*, vol. 3, pp. 1–7, 1963.

[MIN 64] MINDLIN R.D., "Micro-structure in linear elasticity", *Archive for Rational Mechanics and Analysis*, vol. 16, pp. 51–78, 1964.

[MIN 95] MINDLIN R.D., "Second gradient of strain and surface-tension in linear elasticity", *Int. J. Solids Struct.*, vol. 1, pp. 417–438, 1995.

[MOE 14] MOES N., "Mécanique des milieux continus", Cours E11 de l'Ecole Centrale de Nantes, 2013.

[MOH 94] MOHANDAS N., EVAN E., "Mechanical properties of the red cell membrane in relation to molecular structure and genetic defects", *Annual Review of Biopysics and Biomolecular Structure*, vol. 23, pp. 787–818, 1994.

[MOH 05] MOHR D., "Mechanism-based multi-surface plasticity model for ideal truss lattice materials", *International Journal of Solids and Structures*, vol. 42, pp. 3235–3260, 2005.

[MOR 93] MORI S., BURR D.B., "Increased intracortical remodelling following fatigue damage", *Bone*, vol. 14, pp. 103–109, 1993.

[MOR 01] MORGAN E.F., KEAVENY T.M., "Dependence of yield strain of human trabecular bone on anatomic site", *Journal of Biomechanics*, vol. 34, pp. 569–577, 2001.

[MOR 03] MORGAN E.F., BAYRAKTAR H.H., KEAVENY T.M., "Trabecular bone modulus-density relationships depend on anatomic site", *Journal of Biomechanics*, vol. 36, pp. 897–904, 2003.

[MOS 88] MOSEKILDE L.I., "Age-related changes in vertebral trabecular bone architecture – assessed by a new method", *Bone*, vol. 9, pp. 247–250, 1988.

[MOS 89] MOSEKILDE L.I., "Sex differences in age-related loss of vertebral trabecular bone mass and structure – biomechanical consequences", *Bone*, vol. 10, pp. 425–432, 1989.

[MOU 03a] MOURAD A., Description topologique de l'architecture fibreuse et modélisation mécanique du myocarde, Ph.D. thesis, I.N.P.L. Grenoble, 2003.

[MOU 03b] MOURAD A., CAILLERIE D.A., RAOULT A., "A nonlinearly elastic homogenized constitutive law for the myocardium", *Computational Fluid and Solid Mechanics*, vol. 2, pp. 1779–1781, 2003.

[MUH 96] MUHLHAUS H.B., OKA F., "Dispersion and wave propagation in discrete and continuous models for granular materials", *International Journal of Solids and Structures*, vol. 33, pp. 2841–2858, 1996.

[MUL 08] MULDER L., KOOLSTRA J.H., DEN TOONDER J.M.J. *et al.*, "Relationship between tissue stiffness and degree of mineralization of developing trabecular bone", *Journal of Biomedical Materials Research*, vol. A84, no. 2, pp. 508–515, 2008.

[MUN 66] MUNSTER A., *Thermodynamique des processus irréversibles*, Presses Universitaires de France, 1966.

[NAC 11] NACEUR H., Non-linéarités géométriques des structures, grandes déformations, Cours ENSIAME, 2011.

[NAM 07] MC NAMARA L.M., PRENDERGAST J.P., "Bone remodeling algorithms incorporating both strain and microdamage stimuli", *J. Biomech.*, vol. 40, 2007, pp. 1381–1391.

[NAR 09] NARAYANAN H., ARRUDA E.M., GROSH K., *et al.*, "The micromechanics of fluid-solid interactions during growth in porous soft biological tissue", *Biomech Model Mechanobiol*, vol. 8, pp. 167–181, 2009.

[NAU 15] NAU T., TEUSCHL A., "Regeneration of the anterior cruciate ligament: Current strategies in tissue engineering", *World Journal of Orthopedics*, vol. 6, no. 1, pp. 127–136, 2015.

[NAW 13] NAWATHE S., JUILLARD F., KEAVENY T.M., "Theoretical bounds for the influence of tissue-level ductility on the apparent-level strength of human trabecular bone", *Journal of Biomechanics*, vol. 46, no. 7, pp. 1293–1299, 2013.

[NEG 07] NEGUS C., IMPELLUSO T., "Continuum remodeling revisited", *Biomechanics and Modeling in Mechanobiology*, vol. 6, no. 4, pp. 211–226, 2007.

[NG 98] NG S., TSE P., LAU K., "Numerical and experimental determination of the in-plane elastic properties of 2/2 Twill weave fabric composites", *Composites*, vol. B 29, pp. 735–744, 1998.

[NIE 99] NIEBUR G.L.,YUEN J.C., HSIA A.C. *et al.*, "Convergence behavior of high-resolution finite element models of trabecular bone", *Journal of Biomechanical Engineering*, vol. 121, no. 6, pp. 629–635, 1999.

[NIE 00] NIEBUR G.L., FELDSTEIN M.J., YUEN J.C. *et al.*, "High-resolution finite element models with tissue strength asymmetry accurately predict failure of trabecular bone", *Journal of Biomechanics*, vol. 33, pp. 1575–1583, 2000.

[NIN 12] NING LJ. *et al.*, "Preparation and characterization of decellularized tendon slices for tendon tissue engineering", *Journal of biomedical materials research Part A*, vol. 100, no. 6, pp. 1448–1456, 2012.

[NIR 07] NIRMALANANDHAN V.S. *et al.*, "Effect of length of the engineered tendon construct on its structure–function relationships in culture", *Journal of Biomechanics*, vol. 40, no. 11, pp. 2523–2529, 2007.

[NOA 05] NOAILLY J., LACROIX D., PLANELL J.A., "Finite element study of a novel intervertebral disc substitute", *Spine*, vol, 30, no. 20, 2005.

[NOE 18] NOETHER E., "Invariante Variations probleme", *Nachrichten von der Gesellschaft der Wissenschaften zu Göttingen, Mathematisch-Physikalische Klasse*, pp. 235–257, 1918.

[NOG 05] NOGUCHI H., GOMPPER G., "Shape transitions of fluid vesicles and red blood cells in capillary flows", PROC NATL ACAD SCI USA 102:14, vol. 164, pp. 159–14, 2005.

[NOL 72] NOLL W., "A new mathematical theory of simple materials", *Archive for Rational Mathematics and Analysis*, vol. 48, pp. 1–50, 1972.

[NOY 74] NOYES F.R., DELUCAS J.L., TORVIK P.J., "Biomechanics of anterior cruciate ligament failure: an analysis of strain-rate sensitivity and mechanisms of failure in primates", *Journal of Bone and Joint Surgery*, vol. 56, no. 2, 1974.

[NUN 79] NUNZIATO J.W., COWIN S.C., "A nonlinear theory of elastic materials with voids", *Archive for Rational Mechanics and Analysis*, vol. 72, no. 2, pp. 175–201, 1979.

[OBR 10] O'BRIEN T.D. *et al.*, "Mechanical properties of the patellar tendon in adults and children", *Journal of Biomechanics*, vol. 43, no. 6, pp. 1190–1195, 2010.

[OCO 12] O'CONNELL G., SEN S., ELLIOTT D., "Human annulus fibrosus material properties from biaxial testing and constitutive modeling are altered with degeneration", *Biomechanics and Modeling in Mechanobiology*, vol. 11, pp. 493–503, 2012.

[OGD 84] OGDEN R.W., *Non-linear elastic deformations*, Dover publications, 1984.

[OHA 17] OHAYON Y.P.J., *Hyperelastic Constitutive Laws for Finite Element Modeling, Biomechanics of Living Organs*, Academic Press, 2017.

[OLI 09] OLIVEIRA J.A., PINHO-DA-CRUZ J., TEIXEIRA-DIA F., "Asymptotic homogenisation in linear elasticity", Part II: Finite element procedures and multiscale applications, *Computational Materials Science*, vol. 45, pp. 1081–1096, 2009.

[ONC 01] ONCK P.R., ANDREWS E., GIBSON L.J., "Size effects in ductile cellular solids, Part I: modeling", *International Journal of Mechanical Sciences*, vol. 43, pp. 681–699, 2001.

[OST 99] OSTOJA-STARZEWSKI M., BOCCARA S., JASIUK I., "Couple-stress moduli and characteristic length of composite materials", *Mechanics Research Communications*, vol. 26, no. 4, pp. 387–397, 1999.

[OVE 97] OVERAKER D.W., Micromechanical modeling of open cell structures with application in finite element analysis of vertebral body mechanics, PhD Thesis, Rutgers University, 1997.

[PAA 94] PAARFIN M.A., "Osteonal and hemi-osteonal remodeling: The spatial and temporal framework for signal traffic in adult human bone", *Journal of Cellular Biochemistry*, vol. 55, no. 3, pp. 273–286, 1994.

[PAN 83] PANASENKO G.P., "Averaging of processes in frame constructions with random properties", *Zh. Vychisl. Mat. Fiz.*, vol. 23, pp. 1098–1109, 1983.

[PAR 87] PARK H.C., LAKES R.S., "Torsion of a micropolar elastic prism of square cross section", *International Journal of Solids and Structures*, vol. 23, pp. 485–503, 1987.

[PAR 10a] PARK Y.K., BEST C.A., "Measurement of red blood cell mechanics during morphological changes", *Proceedings of The National Academy of Sciences*, vol. 107, no. 15, p. 6731, 2010.

[PAR 10b] PARK T., HUANG W.S., HU J., "Plastic continuum models for truss lattice materials with cubic symmetry" *Journal of Mechanical Science and Technology*, vol. 24, no. 3, pp. 657–669, 2010.

[PAR 10c] PARSONS E.M., WEERASOORIYA T., SARVA S. *et al.*, "Impact of woven fabric: Experiments and mesostructure-based continuum-level simulations", *Journal of the Mechanics and Physics of Solids*, vol. 58, pp. 1995–2021, 2010.

[PAY 17] PAYAN Y., OHAYON J., "Biomechanics of living oegans", *Hyperelastic Constitutive Laws for Finite Element Modeling*, Academic press, 2017.

[PEA 07] PEARSON S.J., BURGESS K., ONAMBELE G.N.L., "Creep and the in vivo assessment of human patellar tendon mechanical properties", *Clinical Biomechanics*, vol. 22, no. 6, pp. 712–717, 2007.

[PED 05] PEDERSEN J.A., SWARTZ M.A., "Mechanobiology in the Third Dimension", *Annals of Biomedical Engineering*, vol. 33, no. 11, pp. 1469–1490, 2005.

[PET 06] PETRIGLIANO F.A., MCALLISTER D.R., WU B.M., "Tissue Engineering for Anterior Cruciate Ligament Reconstruction: A Review of Current Strategies", *Arthroscopy: The Journal of Arthroscopic & Related Surgery*, vol. 22, no. 4, pp. 441–451, 2006.

[PIE 09] PIETRASZKIEWICZ W., EREMEYEV VA., "On natural strain measures of the non-linear micropolar continuum", *International Journal of Solids and Structures*, vol. 46 no. 3, pp. 774–787, 2009.

[PIN 09] PINDERA M.J., KHATAM H., DRAGO A.S. *et al.*, "Micromechanics of spatially uniform heterogeneous media: A critical review and emerging approaches", *Compos. Struct.*, vol. 40, pp. 349–378, 2009.

[PIO 00] PIOLETTI D.P., RAKOTOMANANA L.R., "On the independence of time and strain effects in the stress relaxation of ligaments and tendons", *Journal of Biomechanics*, vol. 33, no. 12, pp. 1729–1732, 2000.

[PIO 98] PIOLETTI D.P. *et al.*, "Viscoelastic constitutive law in large deformations: application to human knee ligaments and tendons", *Journal of Biomechanics*, vol. 31, no. 8, pp. 753–757, 1998.

[PIV 08] PIVONKA P., ZIMAK J., SMITH D. *et al.*, "Model structure and control of bone remodeling : a theoretical study", *Bone*, vol. 43, pp. 249–263, 2008.

[POP 06] POPESCU G., IKEDA T., "Diffraction phase microscopy for quantifying cell structure and dynamics", *Optics Letters*, vol. 31, no. 6, pp. 775–777, 2006.

[POU 05] POURCELOT P. *et al.*, "A non-invasive method of tendon force measurement", *Journal of Biomechanics*, vol. 38, no. 10, pp. 2124–2129, 2005.

[PRA 98a] PRADEL F., Homogénéisation des milieux discrets périodiques orientés, Une application aux mousses, PhD Thesis, Ecole Nationale des Ponts et Chaussées, Champs-sur-Marne, 1998A.

[PRA 98b] PRADEL F., SAB K., "Cosserat modelling of elastic periodic lattice structures", *Comptes Rendus de l'Académie des Sciences - Series IIB - Mechanics-Physics-Astronomy*, vol. 326, pp. 699–704, 1998 B.

[PRE 94] PRENDERGAST P.J., TAYLOR D., "Prediction of bone adaptation using damage accumulation", *Journal of Biomechanics*, vol. 27, pp. 1067–1076, 1994.

[PRE 95] PRENDERGAST P.J., HUISKES R., "Mathematical modeling of microdamage in bone remodeling and adaptation", in ODGAARD A., WEINANS H. (eds.), *Bone Structure and Remodeling, Recent Advances in Human Biology 2*, World Scientific, Singapore, 1995.

[PRE 97] PRENDERGAST P.J., HUISKES R., SØBALLE K., "Biophysical stimuli on cells during tissue differentiation at implant interfaces", *Journal of Biomechanics*, vol. 30, no. 6, pp. 539–548, 1997.

[PRE 10] PRENDERGAST P.J., CHECA S., LACROIX D., "Computational Models of Tissue Differentiation", in DE S., GUILAK F., MOFRAD M., (eds), *Computational Modeling in Biomechanics*, Springer, Netherlands, 2010.

[PUI 07] PUIG-DE-MORALES-MARINKOVIC M., TURNER K.T., BUTLER J.P. *et al.*, "Viscoelasticity of the human red blood cell. American", *Journal of Physiology-Cell Physiology*, vol. 293, pp. C597–C605, 2007.

[QIN 98] QIN Y.X., RUBIN C.T., MCLEOD K.J., "Nonlinear dependence of loading intensity and cycle number in the maintenance of bone mass and morphology", *Journal of Orthopaedic Research*, vol. 16, pp. 482–489, 1998.

[QIN 10] QIN Z., BUEHLER M. J., KREPLAK L., "A multiscale approach to understand the mechanobiology of intermediate filaments", *J. Biomech.*, vol. 43, pp. 15–22. 2010.

RAH 12] RAHOUADJ R., Cours de Mécanique des milieux continus, ENSEM, 2012.

[RAJ 95] RAJAGOPAL K.R., "Multiple natural configurations in continuum mechanics", *Rep. Inst. Comput. Appl. Mech.*, vol. 6, 1995.

[RAJ 98] RAJAGOPAL K.R., SRINIVASA A.R., "Mechanics of the inelastic behavior of materials", *Part I: Theoretical underpinnings. Int. J. Plast.*, vol. 14, pp. 945–967 1998.

[RAM 99] RAMTANI S., ZIDI M., "Damaged-bone remodeling theory: thermodynamical approach", *Mechanics Research Communications*, vol. 26, no. 6, pp. 701–708, 1999.

[RAM 01] RAMTANI S., ZIDI M., "A theoretical model of the effect of continuum damage on a bone adaptation model", *Journal of Biomechanics*, vol. 34, pp. 471–479, 2001.

[RAM 02] RAMTANI S., ZIDI M., "Damaged-bone adaptation under steady homogeneous stress", *Journal of Biomechanical Engineering*, vol. 124, no. 3, pp. 322–327, 2002.

[RAN 00] RANNOU F., POIRAUDEAU S., CORVOL M. *et al.*, "Biochimie et biologie du disque intervertébral", *Rev Rhum [Éd Fr]*, vol. 67, no. l4, pp. 214–216, 2000.

[RAN 04] RANNOU F., MAYOUX-BENHAMOU M.A., POIRAUDEAU S. *et al.*, "Anatomy, biology, physiology, and biomechanics of intervertebral disk and other anatomical structures of the lumbar spine, *EMC-Rhumathologie Orthopdie*, vol. 1, pp. 487–507, 2004.

[RAO 07] RAO S.S., *Vibration of Continuous Systems*, John Wiley & Sons, Inc., Hoboken, New Jersey, 2007.

[RAO 08] RAOULT A., CAILLERIE D., MOURAD A., "Elastic lattices: equilibrium, invariant laws and homogenization", *Ann. Univ. Ferrara*, vol. 54, pp. 297–318, 2008.

[RAS 02] RASPANTI M., CONGIU T., GUIZZARDI S., "Structural aspects of the extracellular matrix of the tendon : An Atomic Force and Scanning Electron Microscopy Study", *Archives of Histology and Cytology*, vol. 65, no. 1, pp. 37–43, 2002.

[RAT 03] RATTENSPERGER H., EBERHARDSTEINER J., MANG H.A., "Numerical investigation of high pressure hydraulic hoses with steel wire braid", *Proceedings of the IUTAM Symposium on Computational Mechanics of Solid Materials at Large Strains*, pp. 407–416, 2003.

[RAV 04] RAVARY B. *et al.*, "Strain and force transducers used in human and veterinary tendon and ligament biomechanical studies", *Clinical Biomechanics*, vol. 19, no. 5, pp. 433–447, 2004.

[REA 92] REALFF M.L., Mechanical properties of fabrics woven from yarns produced by different spinning technologies, PhD Thesis", Massachusetts Institute of Technology, 1992.

[REA 97] REALFF M.L., BOYCE M.C., BACKER S., "A micromechanical model of the tensile behavior of woven fabric", *Textile Research Journal*, vol. 67, pp. 445–459, 1997.

[REE 10] REESE S.P., MAAS S.A., WEISS J.A., "Micromechanical models of helical superstructures in ligament and tendon fibers predict large Poisson's ratios", *Journal of Biomechanics*, vol. 43, no. 7, pp. 1394–1400, 2010.

[RHO 95] RHO J.Y., HO BA THO M.C., ASHMAN R.B., "Relations of mechanical properties to density and CT numbers in human bone", *Medical Engineering & Physics*, vol. 17, pp. 347–355, 1995.

[RIC 09] RITCHIE R.O., BUEHLER M.J., HANSMA P., "Plasticity and toughness in bone", *Physics Today*, vol. 62, no. 6, pp. 41–47, 2009.

[RIG 09] RIGOZZI S., MÜLLER R., SNEDEKER J.G., "Local strain measurement reveals a varied regional dependence of tensile tendon mechanics on glycosaminoglycan content", *Journal of Biomechanics*, vol. 42, no. 10, pp. 1547–1552, 2009.

[ROD 94] RODRIGUEZ E.K., HOGER A., MCCULLOGH A.D., "Stress-dependent finite growth in soft elastic tissues", *Journal of Biomechanics*, vol. 27, no. 4, pp. 455–467, 1994.

[ROS 97] ROSCHGER P., FRATZL P., KLAUSHOFER K. *et al.*, "Mineralization of cancellous bone after alendronate and sodium fluoride treatment: a quantitative backscattered electron imaging study on minipig ribs", *Bone*, vol. 20, no. 5, pp. 393–397, 1997.

[ROV 07] ROVATI M., VEBER D., "Optimal topologies for micropolar solids", *Structural and Multidisciplinary Optimization*, vol. 33, pp. 47–59, 2007.

[SAB 96] SAB K., "Déformations microscopiques et macroscopiques dans un assemblage dense de particules rigides", *C.R. Acad. Sci. Paris, Ser. IIb,* vol. 322, pp. 715–721, 1996.

[SAB 09] SAB K., PRADEL F., "Homogenisation of periodic Cosserat media", *International Journal of Computer Applications in Technology*, vol. 34, pp. 60–71, 2009.

[SAG 03] SAGAR T.V, POTLURI P., HEARLE J.W.S., "Mesoscale modelling of interlaced fibre assemblies using energy method", *Computational Materials Science*, vol. 28, no. 1, pp. 49–62, 2003.

[SAH 07] SAHOO S., CHO-HONG J.G., SIEW-LOK T., "Development of hybrid polymer scaffolds for potential applications in ligament and tendon tissue engineering", *Biomedical Materials*, vol. 2, no. 3, p. 169, 2007.

[SAH 10] SAHOO S., TOH S.L., GOH J.C.H., "A bFGF-releasing silk/PLGA-based biohybrid scaffold for ligament/tendon tissue engineering using mesenchymal progenitor cells", *Biomaterials*, vol. 31, no. 11, pp. 2990–2998, 2010.

[SAN 09] SANZ-HERRERA J., GARCIA-AZNAR J., DOBLARÉ M., "On scaffold designing for bone regeneration: a computational multiscale approach", *Acta Biomaterialia*, vol. 5, no. 1, pp. 219–229, 2009.

[SAN 80] SANCHEZ-PALENCIA E., *Non-Homogeneous Media and Vibration Theory*, Springer, Berlin, 1980.

[SCH 07] SCHEPULL T. *et al.*, "Mechanical properties during healing of Achilles tendon ruptures to predict final outcome: A pilot Roentgen stereophotogrammetric analysis in 10 patients", *BMC Musculoskeletal Disorders*, vol. 8, p. 116, 2007.

[SCH 10] SCHMIDT H., SHIRAZI-ADL A., GALBUSERA F. *et al.*, "Response analysis of the lumbar spine during regular daily activities - a finite element analysis", *Journal of Biomechanics*, vol. 43, no. 10, pp. 1849–1856, 2010.

[SCR 04a] SCREEN H.R.C. *et al.*, "An investigation into the effects of the hierarchical structure of tendon fascicles on micromechanical properties", *Proceedings of the Institution of Mechanical Engineers*, Part H, *Journal of Engineering in Medicine*, vol. 218, no. 2, pp. 109–119, 2004.

[SCR 04b] SCREEN H.R.C. *et al.*, "Local Strain Measurement within Tendon", *Strain*, vol. 40, no. 4, pp. 157–163, 1 November 2004.

[SEE 06] SEEMAN E., DELMAS P.D., "Bone quality-the material and structural basis of bone strength and fragility", *N. Engl. J. Med.*, vol. 25, no. 354(21), pp. 2250–2261, May 2006.

[SHA 07] SHAHKARAMI A., VAZIRI R., "A continuum shell finite element model for impact simulation of woven fabrics", *International Journal of Impact Engineering*, vol. 34, pp. 104–119, 2007.

[SHE 07] SHEARN J. *et al.*, "Mechanical stimulation of tendon tissue engineered constructs: effects on construct stiffness, repair biomechanics, and their correlation", *Journal of Biomechanical Engineering*, vol. 129, no. 6, pp. 848–854, 2007.

[SHE 15a] SHEARER T., "A new strain energy function for the hyperelastic modelling of ligaments and tendons based on fascicle microstructure", *Journal of Biomechanics*, vol. 48, no. 2, pp. 290–297, 2015.

[SHE 15b] SHEARER T., "A new strain energy function for modelling ligaments and tendons whose fascicles have a helical arrangement of fibrils", *Journal of Biomechanics*, vol. 48, no. 12, pp. 3017–3025, 2015.

[SHI 90] SHIGA T., MAEDA N., "Erythrocyte rheology", *Critical Reviews In Oncology/Hematology*, vol. 10, no. 1, pp. 9–48, 1990.

[SHI 10] SHIRAZI-ADL A., TAHERI M., URBAN J.P.G., "Analysis of cell viability in intervertebral disc: Effect of endplate permeability on cell population", *J Biomech*, vol. 43, no. 7, pp. 1330–1336, 2010.

[SHI 11] SHI X., LIU X.S., WANG X. *et al.*, "Type and orientation of yielded trabeculae during overloading of trabecular bone along orthogonal directions", *Journal of Biomechanics*, vol. 43, no. 13, pp. 2460–2466, 2011.

[SHM 07] SHMOYLOVA E., POTAPENKO S., ROTHENBURG L., "Stress distribution around a crack in plane micropolar elasticity", *Journal of Elasticity*, vol. 86, pp. 19–39, 2007.

[SHO 99a] SHOCKEY D.A., ELRICH D.C. *et al.*, Improved Barriers to Turbine Engine Fragments: Interim Report I. DOT/FAA AR-99/8,I, 1999.

[SHO 99b] SHOCKEY D.A., ELRICH D.C., SIMONS J.W., Improved Barriers to Turbine Engine Fragments: Interim Report II. DOT/FAA AR-99/8,II, 1999.

[SHO 01] SHOCKEY D.A., ELRICH D.C., SIMONS J.W., Improved Barriers to Turbine Engine Fragments: Interim Report III. DOT/FAA AR-99/8,III, 2001.

[SIL 95] SILVA M.J., HAYES W.C., GIBSON L.J., "The effects of non-periodic microstructure on the elastic properties of two-dimensional cellular solids", *International Journal of Mechanical Sciences*, vol. 37, pp. 1161–1177, 1995.

[SIL 97] SILVA M.J., GIBSON L.J., "Modeling the Mechanical Behavior of Vertebral Trabecular Bone: Effects of Age-Related Changes in Microstructure", *Bone*, vol. 21, pp. 191–199, 1997.

[SKA 82] SKALAK R., DASGUPTA G., MOSS M., "Analytical description of growth", *J. Theor. Biol.*, vol. 94, pp. 555–577, 1982.

[SKA 97] SKALAK R., FARROW D.A., HOGER A., "Kinematics of surface growth", *J. Math. Biol.*, vol. 35, pp. 869–907, 1997.

[SOM 01] SOMMERFELDT D.W., RUBIN C.T., "Biology of bone and how it orchestrates the form and function of the skeleton", *European Spine Journal*, vol. 10, pp. S86–S95, 2001.

[SPE 71] SPENCER A., "Part III, theory of invariants", *Continuum Physics*, vol. 1, pp. 239–353, 1971.

[SPE 84] SPENCER A.J.M., RIVLIN R.S., "Isotropic integrity bases for vectors and second-order tensors", *Part I, Archive Rational Mechanics and Analysis*, pp. 1–32, 1984.

[STE 10] STEFANOU I., SULEM J., "Continuum modeling of discontinuous rock structures, Homogenization of discrete particle assemblies", *European Journal of Environmental and Civil Engineering*, vol. 14, nos. 8–9, pp. 1199–1217, 2010.

[STE 94] STEINMANN P., "A micropolar theory of finite deformation and finite rotation multiplicative elastoplasticity", *Int J. Solids Structures*, vol. 31, no. 8, pp. 1063–1084, 1994.

[STO 85] STOUFFER D.C., BUTLER D.L., HOSNY D., "The relationship between crimp pattern and mechanical response of human patellar tendon-bone units", *Journal of Biomechanical Engineering*, vol. 107, no. 2, pp. 158–165, 1985.

[STO 90] STOKKE T., BRANT D.A., "The reliability of wormlike polysaccharide chain dimension estimated from electron micrographs", *Biopolymers*, vol. 30, pp. 1161–1181, 1990.

[STR 93] STRONGE W.J., YU T.X., *Dynamic Models for Structural Plasticity*, Springer, London, 1993.

[SUI 01] SUIKER A.S.J., DE BORST R., CHANG C.S., "Micro-mechanical modelling of granular material", *Part 1: Derivation of a secondgradient micro-polar constitutive theory*, *Acta Mechanca*, vol. 149, pp. 161–180, 2001.

[SUL 08] SULLIVAN R.M., GHOSN L.J., LERCH B.A., "A general tetrakaidecahedron model for open-celled foams", *International Journal of Solids and Structures?*, vol. 45, pp. 1754–1765, 2008.

[SWE 14] SWEDBERG A.M. *et al.* "Continuum description of the Poisson's ratio of ligament and tendon under finite deformation", *Journal of Biomechanics*, vol. 47, no. 12, pp. 3201–3209, 2014.

[TAB 95] TABER L., "Biomechanics of growth, Remodeling and morphogenesis", *Appl. Mech. Rev.*, vol. 48, pp. 487–545.

[TAB 01a] TABER L., HUMPHREY J.D., "Stress-modulated growth, Remodeling and morphogenesis", *Appl. Mech. Rev.*, vol. 48, pp. 487–545, 2001.

[TAB 01b] TABER L.A., "Nonlinear thoery of elastixity: applications in biomechanics", *World Scientific*, New York, 2004.

[TAB 02] TABIEI A., YI W., "Comparative study of predictive methods for woven fabric composite elastic properties", *Composite Structures*, vol. 58, pp. 149–164, 2002.

[TAL 09] TALEBI H., WONG S.V., HAMOUDA A.M.S., "Finite element evaluation of projectile nose angle effects in ballistic perforation of high strength fabric", *Composite Structures*, vol. 87, no. 4, pp. 314–320, 2009.

[TAN 97] TAN P., TONG L., STEVEN GP., "Modeling for predicting the mechanical properties of textile composites", *A Review. Composite Part A- Applied Sience and Manufacturing*, vol. 28, no. 11, pp. 903–922, 1997.

[TAN 99] TANAKA M., ADACHI T., "Lattice continuum model for bone remodeling considering microstructural optimality of trabecular architecture", in PEDERSEN P., BENDSOE M.P. (eds), *IUTAM Symposium on Synthesis in Bio Solid Mechanics*, Kluwer Academic Publishers, The Netherlands, pp. 43–54, 1999.

[TAN 03] TANOV R.R., BRUEGGERT M., "Finite element modeling of non-orthogonal loosely woven fabrics in advanced occupant restraint systems", *Finite Elements in Analysis and Design*, vol. 39, pp. 357–367, 2003.

[TAY 02] TAYLOR M., COTTON J., ZIOUPOS P., "Finite Element Simulation of the Fatigue Behaviour of Cancellous Bone", *Meccanica*, vol. 37, pp. 419–429, 2002.

[TAY 07] TAYLOR D., HAZENBERG J.G., LEE T.C., "Living with cracks: damage and repair in human bone", *Nature Materials*, vol. 6, no. 4, pp. 263–268, 2007.

[TEK 08] TEKOGLU C., ONCK P.R., "Size effects in two-dimensional Voronoi foams: a comparison between generalized continua and discrete models", *Journal of the Mechanics and Physics of Solids*, vol. 56, pp. 3541–3564, 2008.

[TEZ 05] TEZUKA K., WADA Y., TAKAHASHI A. *et al.*, "Computer-simulated bone architecture in a simple bone remodeling model based on a reaction-diffusion system", *J. Bone Miner. Metab.*, vol. 23, pp. 1–7, 2005.

[THO 92] THOMPSON D.W., *On Growth and Form*, 2nd ed., Dover publications, 1992.

[THO 03] THOMOPOULOS S. *et al.*, "Variation of biomechanical, structural, and compositional properties along the tendon to bone insertion site", *Journal of Orthopaedic Research*, vol. 21, no. 3, pp. 413–419, 2003.

[THO 13] THORPE C.T. *et al.*, "Helical sub-structures in energy-storing tendons provide a possible mechanism for efficient energy storage and return", *Acta Biomaterialia*, vol. 9, no. 8, pp. 7948–7956, 2013.

[TOL 87] TOLENADO A., MURAKAMI H., "A high-order mixture model for periodic particulate composites", *International Journal of Solids and Structures*, vol. 23, pp. 989–1002, 1987.

[TOL 98] TOLLENAERE H., CAILLERIE D., "Continuous modeling of lattice structures by homogenization", *Advances in Engineering Software*, vol. 29, pp. 699–705, 1998.

[TOV 06] TOVAR A., PATEL N.M., NIEBUR G.L. *et al.*, "Topology optimization using a hybrid cellular automat on method with local control rules", *J. Mech. Des.*, vol. 128, p. 1205, 2006.

[TOW 75] TOWNSEND P.R., ROSE R.M., RADIN E.L., "Buckling studies of single human trabeculae", *Journal of Biomechanics*, vol. 8, pp. 199–201, 1975.

[TRO 99] TROVALUSCI P., MASIANI R., "Material symmetries of micropolar continua equivalent to lattices", *Int J Solids Struct*, vol. 36, pp. 2091–108, 1999.

[TRU 65] TRUESDELL C., NOLL W., "The nonlinear field theory of mechanics", *Handbuch der Physik Band III/3*, Springer-Verlag, 1965.

[TSU 09] TSUBOTA K., SUZUKI Y., YAMADA T. *et al.*, "Computer simulation of trabecular remodeling in human proximal femur using large-scale voxel FE models: approach to understanding Wolff's law", *J. Biomec.*, vol. 42, pp. 1088–1094, 2009.

[UNT 13] UNTERBERGER M.J., SCHMOLLER K.M., BAUSCH A.R., "A new approach to model cross-linked actin networks: multi-scale continuum formulation and computational analysis", *Journal of The Mechanical Behavior of Biomedical Materials*, vol. 22, pp. 95–114, 2013.

[USH 13] USHIJIMA K., CANTWELL W., CHEN D., "Prediction of the mechanical properties of micro-lattices subjected to multi-axial loading", *International Journal of Mechanical Science*, vol. 68, pp. 47–55, 2013.

[VAI 13] VAIRO G., "Modeling and simulation in tissue biomechanics: modern tools to face an ancient challenge", *Journal of Biomedical Science and Engineering*, vol. 6, pp. 1–5, 2013.

[VAN 93] VAN DER MEULEN M.C.H., BEAUPRÉ G.S., CARTER D.R., "Mechanobiologic influences in long bone crosssectional growth", *Bone*, vol. 14, pp. 635–642, 1993.

[VAN 01] VAN DER LINDENA J.C., BIRKENHAGER-FRENKELB D.H. *et al.*, "Trabecular bone's mechanical properties are affected by its non-uniform mineral distribution", *Journal of Biomechanics*, vol. 34, pp. 1573–1580, 2001.

[VAN 04] VAN EIJDEN T.M.G.J., VAN RUIJVEN L.J., GIESEN E.B.W., "Bone tissue stiffness in the mandibular condyle is dependent on the direction and density of the cancellous structure", *Calcified Tissue International*, vol. 75, pp. 502–508, 2004.

[VER 70] VERONDA D.R., WESTMANN R.A., "Mechanical characterization of skin–finite deformations", *Journal of Biomechanics*, vol. 3, no. 1, pp. 111–124, 1970.

[VER 10] VERGARI C. *et al.*, "A linear laser scanner to measure cross-sectional shape and area of biological specimens during mechanical testing", *Journal of Biomechanical Engineering*, vol. 132, no. 10, p. 105001, 2010.

[VER 11] VERGARI C. *et al.*, "True stress and Poisson's ratio of tendons during loading", *Journal of Biomechanics*, vol. 44, no. 4, pp. 719–724, 2011.

[VER 12] VERGARI C. *et al.*, "Axial speed of sound is related to tendon's nonlinear elasticity", *Journal of Biomechanics*, vol. 45, no. 2, pp. 263–268, 2012.

[VIE 09] VIEIRA A.C., GUEDES R.M., MARQUES A.T., "Development of ligament tissue biodegradable devices: A review", *Journal of Biomechanics*, vol. 42, no. 15, pp. 2421–2430, 2009.

[VIG 14] VIGLIOTTI A., DEHSPANDE V.S., PASINI D., "Nonlinear constitutive models for lattice materials", *J. Mech. Phys. Solids*, vol. 64, pp. 44–60, 2014.

[VUN 04] VUNJAK-NOVAKOVIC G. *et al.*, "Tissue engineering of ligaments", *Annual Review of Biomedical Engineering*, vol. 6, pp. 131–156, 2004.

[WAG 04] WAGNER D.R., LOTZ J.C., "Theoretical model and experimental results for the nonlinear elastic behavior of human annulus fibrosus", *Journal Orthopaedic Research*, vol. 22, pp. 901–909, 2004.

[WAG 11] WAGNER D.W., LINDSEY D.P., BEAUPRE G.S., "Deriving tissue density and elastic modulus from micro-CT bone scans, *Bone*, vol. 49, no. 5, pp. 931–938, 2011.

[WAN 00] WANG Y., CUITINO AM., "Three-dimensional nonlinear open cell foams with large deformations", *Journals of the Mechanics and Physics of Solids*, vol. 48, pp. 961–988, 2000.

[WAN 01] WANG X.L., STRONGE W.J., "Micropolar theory for a periodic force on the edge of elastic honeycomb", *International Journal of Engineering Sciences*, vol. 39, no. 7, pp. 821–50, 2001.

[WAN 04] WANG AJ., MCDOWELL DL., "In plane stiffness and yield strength of periodic metal honeycombs", *J.Eng.Mater.Technol.Trans.ASME*, vol. 126, no. 2, pp. 137–156, 2004.

[WAN 05] WANG X., GUYETTE J., LIU X., *et al.*, "Axial-shear interaction effects on microdamage in bovine tibial trabecular bone", *European Journal of Morphology*, vol. 42, nos. 1–2, pp. 61–70, 2005.

[WAN 05] WANG A.J., MCDOWELL D., "Yield surfaces of various periodic metal honeycombs at intermediate relative density", *International Journal of Plasticity*, vol. 21, pp. 285–320, 2005.

[WAN 06a] WANG J.H.C., "Mechanobiology of tendon", *Journal of Biomechanics*, vol. 39, no. 9, pp. 1563–1582, 2006.

[WAN 06b] WANG J.H.C., THAMPATTY B.P., "An introductory review of cell mechanobiology", *Biomechanics and Modeling in Mechanobiology*, vol. 5, no. 1, pp. 1–16, 2006.

[WAN 11a] WANG R., DING H., "Effective 3D viscoelasticity of red blood cells measured by diffraction phase microscopy", *Biomedical Optics Express*, vol. 2, no. 3, pp. 485–490, 2011.

[WAN 11b] WANG C., ZHANG C., HAN J., *et al.*, "Simulated evolution of the vertebral body based on basic multicellular unit activities", *Journal of Bone and Mineral Metabolism*, vol. 29, no. 4, pp. 466–476, 2011.

[WAN 12] WANG C., HAN J., LI Q. *et al.*, "Simulation of bone remodeling in orthodontic treatment", *Computer Methods in Biomechanics and Biomedical Engineering*, vol. 17, no. 9, pp. 042–50, 2012.

[WAR 87] WARREN W.E., KRAYNIK A.M., "Foam mechanics: the linear elastic response of two dimensional spatially periodic cellular materials", *Mechanics of Materials*, vol. 6, pp. 27–37, 1987.

[WAR 89] WARREN W.E., KRAYNIK AM., STONE CM., "A constitutive model for two-dimensional nonlinear elastic foams", *Journal of the Mechanics and Physics of Solids and Structures*, vol. 37, pp. 717–733, 1989.

[WAR 91] WARREN WE., KRAYNIK AM., "The nonlinear elastic behaviour of open-cell foams", *Transactions of ASME*, vol. 58, pp. 375–381, 1991.

[WAR 02] WARREN WE., BYSKOV E., "Three-fold symmetry restrictions on two-dimensional micropolar materials", *European Journal of Mechanics A/Solids*, vol. 21, pp. 779–792, 2002.

[WAT 04] WATZKY A., NAILI S., "Orthotropic bone remodeling: case of plane stresses", *Mechanics Research Communications*, vol. 31, pp. 617–625, 2004.

[WEI 88] WEISS L., "Cell and Tissue Biology", *A Textbook of Histology,* Urban and Schwarzenberg, Baltimore, 1988.

[WEI 92] WEINANS H., HUISKES R., GROOTENBOER H.J., "The behavior of adaptive bone-remodeling simulation-models", *Journal of Biomechanics*, vol. 25, pp. 1425–1441, 1992.

[WIE 04] WIELOCH P. *et al.*, "A cryo-jaw designed for in vitro tensile testing of the healing Achilles tendons in rats", *Journal of Biomechanics*, vol. 37, no. 11, pp. 1719–1722, 2004.

[WOL 92] WOLFF J., *Das Gesetz Transformation der Knochen*, Hirschwald Verlag, Berlin, 1892.

[WOL 10] WOLFRAMA U., WILKE H.J., ZYSSET P.K., "Valid μ finite element models of vertebral trabecular bone can be obtained using tissue properties measured with nanoindentation under wet conditions", *Journal of Biomechanics*, vol. 43, pp. 1731–1737, 2010.

[WOO 99] WOO S. L. *et al.*, "Tissue engineering of ligament and tendon healing", *Clinical Orthopaedics and Related Research*, no. 367 Suppl, pp. S312-S323, 1999.

[WOO 06] WOO S.L.-Y. *et al.*, "Biomechanics of knee ligaments: injury, healing, and repair", *Journal of Biomechanics*, vol. 39, no. 1, pp. 1–20, 2006.

[WOO 10] WOODS B.I., SOWA G., VO N. *et al.*, "A change in strategy: the use of regenerative medicine and tissue engineering to augment the course of intervertebral disc degeneration", *Operative Techniques in Orthopaedics*, vol. 20, no. 2, 2010.

[WRE 01] WREN T.A.L. *et al.* "Mechanical properties of the human Achilles tendon", *Clinical Biomechanics*, vol. 16, no. 3, p. 245–251, 2001.

[WRE 03] WREN T.A.L. *et al.*, "Effects of Creep and Cyclic Loading on the Mechanical Properties and Failure of Human Achilles Tendons", *Annals of Biomedical Engineering*, vol. 31, no. 6, pp. 710–717, 2003.

[WU 01] WU C.H., "The role of Eshelby stress in composition-generated and stress-assisted diffusion", *J. Mech. Phys. Solids*, vol. 49, no. 8, pp. 1771–1794, 2001.

[WU 02] WU C.H., "Chemical potential and energy momentum tensor in single phase mixtures", *Mechanics Research Communications*, vol. 29, pp. 493–499, 2002.

[XUE 03] XUE P., PENG X., CAO J., "A non-orthogonal constitutive model for characterizing woven composites", *Composites* A., vol. 34, pp. 183–193, 2003.

[XUE 05] XUE P., CAO J., CHEN J., "Integrated micro/macro-mechanical model of woven fabric composites under large deformation", *Composite Structures*, vol, 70, pp. 69–80, 2005.

[YAN 81] YANG J.F.C., LAKES R.S., "Transient study of couple stress effects in compact bone: Torsion" *Journal of Biomechanical Engineering*, vol. 103, pp. 275–279, 1981.

[YAN 02] YANG F., CHONG A.C.M., LAM D.C.C. *et al.*, "Couple stress based strain gradient theory for elasticity", *International Journal of Solids and Structures*, vol. 39, no. 10, pp. 2731–2743, 2002.

[YEH 99] YEH O.C., KEAVENY T.M., "Biomechanical effects of intra-specimen variations in trabecular architecture a three-dimensional finite element study", *Bone*, vol. 25, 223–228, 1999.

[YOO 06] YOO A., JASIUK I., "Couple-stress moduli of a trabecular bone idealized as a 3D periodic cellular network", *Journal of Biomechanics*, vol. 39, pp. 2241–2252, 2006.

[YOO 11] YOON Y.Z., KOTAR J., "The Red blood cell dynamics: from spontaneous fluctuations to non-linear response", *Soft Matter*, vol. 7, no. 5, pp. 2024–2051, 2011.

[YOO 12] YOON Y.Z., KOTAR J., "Red blood cell dynamics: from spontaneous fluctuations to non-linear response", *Soft Matter*, vol. 7, no. 5, pp. 2042–2051, 2011.

[ZHA 08] ZHANG G.M., BATRA R.C., ZHENG J., "Effect of frame size, frame type and clamping pressure on the ballistic performance of soft body armor", *Composites*, vol. B 39, no. 3, pp. 476–489, 2008.

[ZHA 10] ZHANG H., WU J., FU. Z., "Extended multiscale finite element method for elasto-plastic analysis of 2d periodic lattice truss materials", *Computational Mechanics*, vol. 45, pp. 623–635, 2010

[ZHU 97] ZHU HX., KNOTT JF., MILLS NJ., "Analysis of the elastic properties of open-cell foams with tetrakaidecahedral cells", *J. Mech. Phys. Solids*, vol. 45, no. 3, p. 319, 1997.

[ZHU 00] ZHU C., BAO G., WANG N., “Cell Mechanics: Mechanical Response, Cell Adhesion, and Molecular Deformation”, *Annual Review Of Biomedical Engineering*, vol. 2, pp. 189–226, 2000.

[ZHU 01] ZHU H.X., HOBDELL J.R., WINDLE A.H., “Effects of cell irregularity on the elastic properties of 2D Voronoi honeycombs”, *Journal of the Mechanics and Physics of Solids*, vol. 49, pp. 857–870, 2001.

[ZHU 10] ZHU H.X., “Size-dependent elastic properties of micro-and nanohoneycombs”, *Journal of the Mechanics and Physics of Solids*, vol. 58, pp. 679–696, 2010.

[ZUL 04] ZULLIGER M.A., RACHEV A., STERGIOPULOS N., “A constitutive formulation of arterial mechanics including vascular smooth muscle tone”, *Am. J. Physiol*, *Heart CIRC, Physiol*, vol. 287, pp. H1335–H1343, 2004.

List of Authors

Adrien BALDIT
LEM3
University of Lorraine
Nancy
France

Khaled El NADY
Mechanical Department
Akhbar El Yom Academy
6th of October
Egypt

Jean-François GANGHOFFER
LEM3
University of Lorraine
Nancy
France

Ibrahim GODA
LPMT
University of Upper Alsace
Mulhouse
France

Cédric LAURENT
LEM3
University of Lorraine
Nancy
France

Rachid RAHOUADJ
LEM3
University of Lorraine
Nancy
France

Index

A, B

C

D, E

F, H, I

L

M, N

O, P

R, S

T, V, Z

FSC
www.fsc.org
MIX
Papier | Fördert gute Waldnutzung
FSC® C083411

Kolibri 360 GmbH
Ferdinand-Jühlke-Straße 7
99095 Erfurt, Deutschland
produktsicherheit@kolibri360.de